NEUROBIOLOGY of DEPRESSION

FRONTIERS IN NEUROSCIENCE

Series Editors
Sidney A. Simon, Ph.D.
Miguel A.L. Nicolelis, M.D., Ph.D.

Published Titles

Apoptosis in Neurobiology
Yusuf A. Hannun, M.D., Professor of Biomedical Research and Chairman, Department
of Biochemistry and Molecular Biology, Medical University of South Carolina, Charleston,
South Carolina
Rose-Mary Boustany, M.D., tenured Associate Professor of Pediatrics and Neurobiology, Duke
University Medical Center, Durham, North Carolina

Neural Prostheses for Restoration of Sensory and Motor Function
John K. Chapin, Ph.D., Professor of Physiology and Pharmacology, State University
of New York Health Science Center, Brooklyn, New York
Karen A. Moxon, Ph.D., Assistant Professor, School of Biomedical Engineering, Science,
and Health Systems, Drexel University, Philadelphia, Pennsylvania

Computational Neuroscience: Realistic Modeling for Experimentalists
Eric DeSchutter, M.D., Ph.D., Professor, Department of Medicine, University of Antwerp,
Antwerp, Belgium

Methods in Pain Research
Lawrence Kruger, Ph.D., Professor of Neurobiology (Emeritus), UCLA School of Medicine and
Brain Research Institute, Los Angeles, California

Motor Neurobiology of the Spinal Cord
Timothy C. Cope, Ph.D., Professor of Physiology, Wright State University, Dayton, Ohio

Nicotinic Receptors in the Nervous System
Edward D. Levin, Ph.D., Associate Professor, Department of Psychiatry and Pharmacology and
Molecular Cancer Biology and Department of Psychiatry and Behavioral Sciences,
Duke University School of Medicine, Durham, North Carolina

Methods in Genomic Neuroscience
Helmin R. Chin, Ph.D., Genetics Research Branch, NIMH, NIH, Bethesda, Maryland
Steven O. Moldin, Ph.D., University of Southern California, Washington, D.C.

Methods in Chemosensory Research
Sidney A. Simon, Ph.D., Professor of Neurobiology, Biomedical Engineering,
and Anesthesiology, Duke University, Durham, North Carolina
Miguel A.L. Nicolelis, M.D., Ph.D., Professor of Neurobiology and Biomedical Engineering,
Duke University, Durham, North Carolina

The Somatosensory System: Deciphering the Brain's Own Body Image
Randall J. Nelson, Ph.D., Professor of Anatomy and Neurobiology,
University of Tennessee Health Sciences Center, Memphis, Tennessee

The Superior Colliculus: New Approaches for Studying Sensorimotor Integration
William C. Hall, Ph.D., Department of Neuroscience, Duke University, Durham, North Carolina
Adonis Moschovakis, Ph.D., Department of Basic Sciences, University of Crete, Heraklion, Greece

New Concepts in Cerebral Ischemia
Rick C. S. Lin, Ph.D., Professor of Anatomy, University of Mississippi Medical Center,
    Jackson, Mississippi

DNA Arrays: Technologies and Experimental Strategies
Elena Grigorenko, Ph.D., Technology Development Group, Millennium Pharmaceuticals,
    Cambridge, Massachusetts

Methods for Alcohol-Related Neuroscience Research
Yuan Liu, Ph.D., National Institute of Neurological Disorders and Stroke,
    National Institutes of Health, Bethesda, Maryland
David M. Lovinger, Ph.D., Laboratory of Integrative Neuroscience, NIAAA,
    Nashville, Tennessee

Primate Audition: Behavior and Neurobiology
Asif A. Ghazanfar, Ph.D., Princeton University, Princeton, New Jersey

Methods in Drug Abuse Research: Cellular and Circuit Level Analyses
Barry D. Waterhouse, Ph.D., MCP-Hahnemann University, Philadelphia, Pennsylvania

Functional and Neural Mechanisms of Interval Timing
Warren H. Meck, Ph.D., Professor of Psychology, Duke University, Durham, North Carolina

Biomedical Imaging in Experimental Neuroscience
Nick Van Bruggen, Ph.D., Department of Neuroscience Genentech, Inc.
Timothy P.L. Roberts, Ph.D., Associate Professor, University of Toronto, Canada

The Primate Visual System
John H. Kaas, Department of Psychology, Vanderbilt University, Nashville, Tennessee
Christine Collins, Department of Psychology, Vanderbilt University, Nashville, Tennessee

Neurosteroid Effects in the Central Nervous System
Sheryl S. Smith, Ph.D., Department of Physiology, SUNY Health Science Center,
    Brooklyn, New York

Modern Neurosurgery: Clinical Translation of Neuroscience Advances
Dennis A. Turner, Department of Surgery, Division of Neurosurgery,
    Duke University Medical Center, Durham, North Carolina

Sleep: Circuits and Functions
Pierre-Hervé Luppi, Université Claude Bernard, Lyon, France

Methods in Insect Sensory Neuroscience
Thomas A. Christensen, Arizona Research Laboratories, Division of Neurobiology,
    University of Arizona, Tuscon, Arizona

Motor Cortex in Voluntary Movements
Alexa Riehle, INCM-CNRS, Marseille, France
Eilon Vaadia, The Hebrew University, Jerusalem, Israel

Neural Plasticity in Adult Somatic Sensory-Motor Systems
Ford F. Ebner, Vanderbilt University, Nashville, Tennessee

Advances in Vagal Afferent Neurobiology
Bradley J. Undem, Johns Hopkins Asthma Center, Baltimore, Maryland
Daniel Weinreich, University of Maryland, Baltimore, Maryland

The Dynamic Synapse: Molecular Methods in Ionotropic Receptor Biology
Josef T. Kittler, University College, London, England
Stephen J. Moss, University College, London, England

Animal Models of Cognitive Impairment
Edward D. Levin, Duke University Medical Center, Durham, North Carolina
Jerry J. Buccafusco, Medical College of Georgia, Augusta, Georgia

The Role of the Nucleus of the Solitary Tract in Gustatory Processing
Robert M. Bradley, University of Michigan, Ann Arbor, Michigan

Brain Aging: Models, Methods, and Mechanisms
David R. Riddle, Wake Forest University, Winston-Salem, North Carolina

Neural Plasticity and Memory: From Genes to Brain Imaging
Frederico Bermudez-Rattoni, National University of Mexico, Mexico City, Mexico

Serotonin Receptors in Neurobiology
Amitabha Chattopadhyay, Center for Cellular and Molecular Biology, Hyderabad, India

TRP Ion Channel Function in Sensory Transduction and Cellular Signaling Cascades
Wolfgang B. Liedtke, M.D., Ph.D., Duke University Medical Center, Durham, North Carolina
Stefan Heller, Ph.D., Stanford University School of Medicine, Stanford, California

Methods for Neural Ensemble Recordings, Second Edition
Miguel A.L. Nicolelis, M.D., Ph.D., Professor of Neurobiology and Biomedical Engineering,
    Duke University Medical Center, Durham, North Carolina

Biology of the NMDA Receptor
Antonius M. VanDongen, Duke University Medical Center, Durham, North Carolina

Methods of Behavioral Analysis in Neuroscience
Jerry J. Buccafusco, Ph.D., Alzheimer's Research Center, Professor of Pharmacology and Toxicology,
    Professor of Psychiatry and Health Behavior, Medical College of Georgia,
    Augusta, Georgia

***In Vivo* Optical Imaging of Brain Function, Second Edition**
Ron Frostig, Ph.D., Professor, Department of Neurobiology, University of California,
    Irvine, California

Fat Detection: Taste, Texture, and Post Ingestive Effects
Jean-Pierre Montmayeur, Ph.D., Centre National de la Recherche Scientifique, Dijon, France
Johannes le Coutre, Ph.D., Nestlé Research Center, Lausanne, Switzerland

The Neurobiology of Olfaction
Anna Menini, Ph.D., Neurobiology Sector International School for Advanced Studies, (S.I.S.S.A.),
    Trieste, Italy

Neuroproteomics
Oscar Alzate, Ph.D., Department of Cell and Developmental Biology,
    University of North Carolina, Chapel Hill, North Carolina

Translational Pain Research: From Mouse to Man
Lawrence Kruger, Ph.D., Department of Neurobiology, UCLA School of Medicine, Los Angeles,
    California
Alan R. Light, Ph.D., Department of Anesthesiology, University of Utah, Salt Lake City, Utah

Advances in the Neuroscience of Addiction
Cynthia M. Kuhn, Duke University Medical Center, Durham, North Carolina
George F. Koob, The Scripps Research Institute, La Jolla, California

NEUROBIOLOGY of DEPRESSION

Edited by

Francisco López-Muñoz

University of Alcalá
Spain

Cecilio Álamo

University of Alcalá
Spain

CRC Press
Taylor & Francis Group
Boca Raton London New York

CRC Press is an imprint of the
Taylor & Francis Group, an **informa** business

CRC Press
Taylor & Francis Group
6000 Broken Sound Parkway NW, Suite 300
Boca Raton, FL 33487-2742

© 2012 by Taylor & Francis Group, LLC
CRC Press is an imprint of Taylor & Francis Group, an Informa business

No claim to original U.S. Government works

Printed in the United States of America on acid-free paper
Version Date: 20110722

International Standard Book Number: 978-1-4398-3849-5 (Hardback)

Library of Congress Cataloging-in-Publication Data

Neurobiology of depression / editors, Francisco López-Muñoz and Cecilio Álamo.
     p. ; cm. -- (Frontiers in neuroscience)
   "A CRC title."
   Includes bibliographical references and index.
   ISBN 978-1-4398-3849-5 (hardcover : alk. paper)
   1. Depression, Mental--Pathophysiology. 2. Neurobiology. I. López-Muñoz, Francisco. II. Álamo González, Cecilio. III. Series: Frontiers in neuroscience (Boca Raton, Fla.)
   [DNLM: 1. Depressive Disorder--physiopathology. WM 171.5]

RC537.N465 2012
616.85'27--dc23
                                                                                   2011022666

Visit the Taylor & Francis Web site at
http://www.taylorandfrancis.com

and the CRC Press Web site at
http://www.crcpress.com

Contents

Foreword

Herman M. van Praag

BIOLOGICAL RESEARCH IN PSYCHIATRY; DIAGNOSIS AS THE RATE-LIMITING FACTOR

1 A PARADIGM SHIFT IN BODY/SOUL RESEARCH

In the early days of biological psychiatric research, iproniazid has been a prime mover. This drug was a close relative of isoniazid, a compound discovered in the early 1950s as a tuberculostatic. Iproniazid was also tested in severely ill tubercular patients. Although it appeared to have only weak effects on the tubercular process, it produced a remarkable symptomatic improvement. Appetite and libido increased, patients became more active, and mood improved dramatically. The drug thus seemed to possess antidepressant properties and was introduced as such in the late 1950s.

Iproniazid became the most remarkable psychotropic drug of its generation. Why? Because it was the only one of which, at the time of its introduction, a modicum was known about its actions in the brain. It had been demonstrated to be an inhibitor of monoamine oxidase (MAO), an enzyme involved in the degradation of catecholamines and serotonin (5-hydroxytryptamine). Serotonin had just been discovered as a compound occurring in blood platelets, the yellow cells of the gut, and in the brain, where it seemed to serve as a neurotransmitter. Being a MAO inhibitor, iproniazid supposedly increased the availability of monoamines in the brain. These observations raised three intriguing questions:

(1) Does iproniazid behave as a MAO inhibitor in humans too?
(2) Do MAO inhibition and antidepressant effect hang together?
(3) Is central monoamine functioning disturbed in depression, particularly in those patients who show a favorable response to a MAO inhibitor?

The introduction of iproniazid was a truly revolutionary event. Up until then, the body–soul relationship was an exclusively philosophical issue. Now, for the first time ever, this most fundamental enigma of the human existence had opened up to an experimental approach. This prospect I felt to be truly exciting. From the very first day of my residency in psychiatry, April 1, 1958, I fell upon the questions mentioned above. They would intrigue me for the rest of my life (see van Praag et al. 2004).

And it was not only me. For many years, monoamines played a major role in biological depression research. All our antidepressants so far are based on hypotheses assuming depression or certain features of depression to be related to disturbances in the functioning of monoaminergic neuronal systems. Monoamine research is still flourishing as far as depression is concerned. However, in recent years, many other inroads into the biology of depression opened up and are being explored. This book, edited by Professors Francisco López-Muñoz and Cecilio Álamo, attests to that. So far, this research has not resulted in new therapeutic agents, but I am optimistic that eventually it will.

2 ROADBLOCKS

However, we have to deal with a serious roadblock on the way ahead, that is, diagnosis. Biological approaches into the brain became ever more sophisticated and refined. The other side of the coin—definition, classification, and measurement of abnormal behavior, in this case, depressive behavior—lagged behind. There, stagnation is rife. For almost one and a half centuries now, we have been the prisoners of a strictly categorical system, introduced by Kraepelin (Hoff 1995). This system impedes progress in clarifying the biological determinants of abnormal behavior, as well as the development of new, innovative drugs.

Kraepelin inaugurated a new diagnostic paradigm in the realm of mental disorders, that of the nosological entity. Mental pathology, Kraepelin proposed, could be divided in discrete disorders, with one distinguishable from the other. Each has its own symptomatology and course; each has its own specific pathophysiology. Nosology, according to Kraepelin, should be the principal guideline in exploring the biological roots of mental pathology.

And so it came to be. The influence of his ideas has been enormous until this very day. Most biological psychiatric research is predicated upon the nosological presupposition and aspires to elucidate the biological underpinnings of discrete disease "packages" as presently defined by the Diagnostic and Statistical Manual of Mental Disorders (DSM) system, a classification system that heavily leans on Kraepelinean ideas. This approach has not been productive. After half a century of research, we cannot boast of much progress. Yes, knowledge of the functioning of the brain advanced tremendously; insight in the pathophysiology of mental disorders, however, hardly did. The reason for that, I think, is the diagnostic system in use. I will clarify my point of view, restricting myself to the construct of depression, as defined by the DSM system. The same reasoning holds, however, for most diagnoses mentioned in the DSM. The following sections are in part adopted from another paper (van Praag 2008).

The starting point of my reasoning is that refined diagnosing is the very bedrock of biological psychiatry; it is the precise definition of the psychopathological construct the pathophysiology of which one wants to elucidate. Hence, two questions have to be addressed. First, does the construct major depression, as proposed by the DSM, meet that clarity criterion? Second, is nosology the proper foundation for biological psychiatric research?

3 CLARITY OF THE MAJOR DEPRESSION CONSTRUCT

Symptomatologically, mood lowering is the anchor symptom of major depression, but for the rest, the symptoms vary considerably. In other words, the term *major depression* covers a variety of syndromes. Moreover, these syndromes appear more often than not in conjunction with other mental disorders, particularly with anxiety and personality disorders. Major depression without further specification is a diagnosis as general and thus as vague as, for instance, that of anemia. It gives some information about someone's mental condition, but far too little.

Course and prognosis of this depression type are unpredictable. It may start rather suddenly, sometimes even overnight, or gradually, in the course of weeks or months. It may last for a few days up to many months or even years, recurring frequently, with large time intervals or (more seldom) not at all. The symptoms of depression may be hardly noticeable to the outer world, may cripple the patient both socially and professionally, or may even reach a psychotic degree. Recovery may be full or partial—in which the patient remains hampered in his daily activities—or the disorder may take a chronic course.

The premorbid personality structure and previous living circumstances may be clearly disturbed, showing, for instance, considerable weaknesses in interpersonal skills and adaptability to changing living conditions, or may be quite undisturbed, leaving relatives and friends of the patient wondering how on earth this person could become depressed.

The depression may follow psychotraumatic life events or may occur "out of the blue," without any discernable provocation.

A variety of neurobiological disturbances have been reported to occur in depression, most prominently in the monoaminergic systems, the stress hormone system, and in the production of some growth factors, such as brain-derived neurotrophic factor. For instance, the concentration of 5-hydroxyindoleacetic acid in cerebrospinal fluid was found to be diminished, the density of serotonin 5-HT$_{1A}$ receptors in the brain to be decreased, and the synthesis rate of serotonin to be attenuated—all signs of disturbed serotonergic functioning in the brain (see van Praag et al. 2004). These disturbances, however, are found in some depressive patients, not in all. Moreover, they are not specific for depression, found as they are in other diagnostic categories as well.

Response to treatment, biological or psychological in nature, is hard to predict. Patients may become symptom-free, show residual symptoms, or hardly respond if they do at all.

One can thus hardly avoid the conclusion that the precision and clarity of the diagnosis major depression leave much to be desired and that the predictive validity of that construct is close to nil. Studying the pathophysiology of a construct of such heterogeneity is bound to fail, and surely, it has failed. We are, as said, not much closer to its biological roots than we were several decades ago. That outcome was predictable. What would one expect from a research program into the pathophysiology of, say, cardiac disorders? I presume little if anything. The diagnosis major depression I consider of comparable exactitude. The chance that utterly heterogeneous psychopathological constructs will be produced by well-defined brain disturbances is negligibly small.

These and similar observations convinced me that biological psychiatric research has been and still is proceeding toward a dead end, a street called nosological alley. I feel, and have felt for most of my professional life, that the diagnostic process in psychiatry should change directions, particularly if the goal is to explore the biological underpinnings of mental pathology. Two strategies I proposed should direct that effort. I have named them *verticalization* and *functionalization* (van Praag et al. 1987, 2004; van Praag 1997, 2000).

4   FUNCTIONALIZATION OF PSYCHIATRIC DIAGNOSIS

The concept of functionalization implies that diagnosing in psychiatry should proceed stepwise.

First, the diagnostic grouping to which the disorder belongs should be determined; that is, a categorical diagnosis should be made. For instance, the mental state in question is considered to belong to the basin of depressive disorders. This first diagnostic step provides no more than a global diagnostic indication. It is no more informative than the statement that a given person complaining of pain in the chest is possibly suffering from a cardiac disorder. Diagnostic basins are by definition heterogeneous.

Next, the syndrome is defined. This diagnostic information is also far from precise. Syndromes often appear in incomplete form, and many patients suffer simultaneously from more than one complete or incomplete syndrome.

Hence, a third diagnostic step seems to me crucial. It is one I have called functionalization of diagnosis. Functionalization means defining first of all the psychopathological symptoms constituting the syndrome and next—most importantly—examining and if possible measuring the psychic dysfunctions underlying the psychopathological symptoms. Psychopathological symptoms and psychic dysfunctions are not synonyms. The psychopathological symptom is the way the psychic dysfunction is experienced by the patient and observed by the investigator.

The last step I consider to be quintessential. If no methods are available to measure the assumed psychic dysfunctions, they should be developed. Functionalization of psychiatric diagnosing, thus, presupposes close collaboration between psychiatrists and experimental clinical psychologists.

Here are a few examples. In the case of dementia symptoms, the underlying cognitive disturbances should be tracked and measured. In the case of hallucinations, the same applies to the underlying perceptual disturbances. In the case of anhedonia, the defect in linking a particular perception with the corresponding emotion should be searched for.

Psychic dysfunctions underlying psychopathological symptoms should be, I propose, the focus of biological psychiatric research. It seems much more likely that brain dysfunctions correspond with disturbances in psychological regulatory systems than with largely man-designed categorical entities, or with symptom complexes rather arbitrarily designated as a syndrome or a dimension.

The search for biological determinants of psychic dysfunctions has indeed been proven to be much more fruitful than the search for the biological cause of a particular nosological entity, such as depression or schizophrenia (see Section 7).

5 ADVANTAGES OF FUNCTIONALIZATION OF DIAGNOSIS

Functionalization will advance scientification of psychiatric diagnosis. It will make this process more precise, more scientific, and more attuned to goal-directed biological studies and focused therapeutic interventions.

It is more precise and more scientific, because psychic dysfunctions are much more measurable than disease categories and syndromes, often even quantitatively.

Second, this approach provides the diagnostician with a detailed chart of those psychic domains that function abnormally and those functioning within normal limits. Ultimately, this approach will lead to what I have called a psychiatric physiology, a detailed chart of brain dysfunctions underlying abnormally functioning psychic systems.

Treatment, too, could benefit from this approach. Drug treatment as well as psychotherapy is presently pretty much unfocused. We prescribe drugs because someone is psychotic, depressed, anxious, or otherwise out of balance. Any further specification is generally lacking or deemed to be unnecessary. This is not the way to further psychopharmacological research, nor the way to increase the chance of finding new, innovative, and psychopathologically more specific psychotropic drugs.

The same reasoning holds for psychological treatment. We may recommend psychotherapy. For what exactly is seldom clear. What will be its focus? What do we hope to achieve? It is rarely defined in any detail. This holds in particular for psychodynamically oriented psychotherapies. Cognitive–behavioral therapists do somewhat better in this respect. Functionalization of diagnosis would make systematic detailing of therapeutic goals feasible. In fact, it would be its logical consequence.

To avoid misunderstanding, I am not pleading to abandon the categorical system (now). One should not throw away an existing system before a new and better one is available. What I suggest is to refine and enrich the categorical system by functionalizing it.

6 VERTICALIZATION OF PSYCHIATRIC DIAGNOSIS

Verticalization is the term I have used for the process of prioritizing the symptoms of a given disorder (van Praag 1997). Prioritizing means that their relationship to the underlying biological substratum is hypothesized (to begin with) and, if possible, established. Are they a direct consequence of the underlying substratum, or associated in a derivate manner? The former symptoms are called *primary*, the latter *secondary*. It seems possible that the secondary symptoms are a consequence of pathological processes elicited by the primary symptoms. If so, one can say that the primary symptoms possess "pacemaker qualities."

In medicine, verticalization of diagnosis is a common procedure. For example, a patient feels miserable and tired, is coughing, has a fever, expectorates, and has difficulties breathing. The doctor weighs the symptoms and considers the symptoms fever, coughing, and impaired breathing as primary, that is, diagnostically decisive. His tentative diagnosis is pneumonia.

In psychiatry, verticalization is not the common procedure. On the contrary, psychiatrists diagnose in a horizontal manner. The symptoms observed in a given psychopathological condition are counted and placed in a horizontal plane as if they all had the same diagnostic weight. This, however, is quite improbable. It is much more likely that some are the direct consequences of the underlying cerebral dysfunctions whereas others are of a subsidiary nature. The former (primary) phenomena are diagnostically the most important. They should be the focus of biological studies and should be the main target of psychopharmacological interventions.

Acknowledging that as of yet, we know little about biological determinants of abnormal behavior, what means do we presently have to verticalize the psychopathology constituting a psychiatric disorder, that is, to determine its relationship toward the underlying pathophysiology?

The first step is sequential analysis, that is, determination of the sequence of appearance of the psychopathological phenomena, preferably in a prospective manner. The hypothesis that the front-runners carry a primary character seems to make sense.

Next, the question is studied whether neurobiological disturbances that might have been revealed in a given case—such as dysfunctions in monoaminergic or other transmitter circuits or in stress-regulating systems such as the hypothalamic–pituitary–adrenal (HPA) axis—are associated with any of the phenomena hypothesized to be primary.

If indeed such associations are demonstrable, pharmacological interventions will be applied or developed aimed at normalizing the neurobiological abnormalities. If the associated psychopathological (primary) phenomena are responsive, the hypothesis about their primary nature is strengthened. If those psychopathological phenomena would have served as pacemakers of the psychiatric disorder, one would expect the remaining (secondary) symptoms to ameliorate subsequently.

It goes without saying that verticalization is not a diagnostic strategy ready for practical use. It is a research domain the development of which will, again, require long-term collaboration of psychiatrists and experimental (clinical) psychologists. It seems to be, however, a quintessential approach in elucidating the relationship between abnormal behavior and abnormal brain functions.

7   Verticalization and Functionalization in Practice

Functionalization and verticalization of psychiatric diagnosis are apt to increase the yield of biological studies. Our own research is a case in point. From the 1950s onward, we have studied the biology of depression, focusing mainly on monoaminergic systems (see van Praag et al. 2004).

First, our focus has been a nosological construct, that is, melancholia. It turned out to be the wrong focus. The construct seemed to be utterly heterogeneous; that is, few patients showed "pure" melancholia. The data we produced were confusing. We found in this group, among other things, disturbances in the serotonin system. On the average, the depressed group differed in this respect from the control group. However, such aberrations occurred in an unpredictable way, demonstrable as they were in some patients but not in others.

Next, we focused on a particular syndrome, the so-called syndrome of vital depression. This focus proved to be unsatisfactory as well, and for much the same reasons. Pure vital depression appeared to be rare. Most patients showed only parts of that syndrome, often in conjunction with (parts of) other syndromes. Again, we observed serotonin disturbances, but, as before, these were seen only in some patients, not in others.

Those were the reasons that prompted the development of the functionalization concept. We decided to dissect the syndrome in its component parts, that is, the psychopathological symptoms; the next step was to study the underlying psychic dysfunctions and then to search for relationships between psychic and biological dysfunctions.

This turned out to be a productive approach. The serotonin disturbances seemed indeed to correlate with psychic dysfunctions, that is, with disturbances in anxiety and aggression regulation. Those relationships, moreover, appeared not to be limited to depression but were also demonstrable in other diagnostic categories. They turned out to be functionally specific, that is, to be linked to disturbances in psychic regulatory systems, that is, anxiety and aggression regulation, not categorically or syndromally specific. They exist independently of nosological or syndromal diagnosis.

The functional approach thus led to specific brain–behavior relationships. The nosological and syndromal approach did not.

In applying the verticalization principle, we were able to tentatively delineate a new depression type called anxiety/aggression-driven depression (van Praag 2001). Those patients present the following characteristics:

(1) Increased anxiety and aggression are heralding the depressive state. Supposedly, they possess "pacemaker qualities" (see Section 6).
(2) Trait-related serotonin disturbances are demonstrable, probably related to labile anxiety and aggression regulation.
(3) Character neurotic traits are prevalent. Probably, they are responsible for the enhanced vulnerability for the destabilizing effects of (certain?) psychotraumatic events. Elevated stress levels might activate the corticotropin-releasing hormone/HPA axis, aggravating the serotonin disturbances to a level leading to rising levels of anxiety and aggression.
(4) Patients responsive to antidepressants will show first a fall in anxiety and aggression level and then mood elevation and a decrease in other depressive symptoms.

Via functionalization and verticalization of diagnosis, meaningful relationships were found between abnormal brain functions and abnormal psychic functions.

8 DISEASES OR REACTION FORMS?

Nosology has dominated psychiatry ever since its foundation as a scientific discipline by Kraepelin. Psychiatric disorders are regarded as true diseases, that is, as discrete entities each with their own causation, symptomatology, course, and outcome, hence identifiable. Biological research, as said, aims at uncovering "markers," and ultimately causes, of these entities.

Mental disorders, however, can be conceived of in a different way, that is, as reaction forms to noxious stimuli, with considerable interindividual variability and little consistency, rather than as discrete and separable entities. Meyer (1951) was the first to propose this diagnostic model. The noxious stimulus might be of a biological or psychological nature, could come from within or without, and might be genetically transmitted or acquired during life. That an individual cannot cope with them, physically and/or psychologically, is common to the various stimuli. According to this model, the co-occurrence of various discrete mental disorders is mainly appearance. In fact, we deal with ever-changing composites of psychopathological features.

In this model, the symptomatological variability of psychiatric conditions within and between individuals can be understood in the following manner: Noxious stimuli, that is, stimuli that an individual is unable to assimilate, will perturb a variety of neuronal circuits and hence a variety of psychic regulatory systems. The extent to which the various neuronal circuits will be involved varies individually, and consequently, psychiatric conditions will lack symptomatic consistency and predictability. For instance, mood lowering is blended with fluctuating measures of anxiety, anger, obsessional thoughts, addictive behavior, cognitive impairment, and psychotic features. Between subjects and, over time, within subjects, the appearance of psychiatric syndromes will thus be as variable as the shape of clouds in the sky. One recognizes the cloud; its shape, however, is variable and unpredictable.

The measure of neuronal disruption that a noxious stimulus will induce is, as noted, variable because it is contingent on a number of factors. Most important are the intrinsic qualities of the stimulus and the resilience of the brain. Preexistent neuronal defects—inborn or acquired—may cause certain brain circuits to function marginally. An equilibrium maintained only under normal conditions could fail if demands gain strength. Increased vulnerability can also be conceptualized on a psychological level in that imperfections in personality make-up cause stimuli that the average person can cope with to be psychologically disruptive.

The reaction form model of psychiatric disorders, if valid, would have profound consequences for biological psychiatry. The search for markers and eventually causes of discrete mental disorders would be largely futile. The farthest one could go is to group the multitude of reaction patterns in a limited number of diagnostic "basins," such as the group of the psychotic, the dementing, and the affective reaction forms, each of which, however, would show considerable heterogeneity. As much as it is futile to search for the antecedents and characteristics of, for instance, the group of abdominal disorders, it would equally lack wisdom to hope for the discovery of, for instance, the pathogenesis of affective reaction forms.

Within the scope of this model, the focus of biological psychiatric research has to shift from the alleged mental "disorders" to disordered psychological domains. Schizophrenia, panic disorder, or major depression as such will not be studied, but rather disturbances in perception, information processing, mood regulation, anxiety regulation, and impulse control, to mention only a few. A biology of psychic dysfunctions, as they occur in mental disorders, would thus be the ultimate goal of biological psychiatric research.

After World War II, the reaction form model was abandoned for no good reason, inasmuch as it has not been disproven. Adopting the three-tier diagnostic approach I proposed would provide the opportunity to explore the merits of both diagnostic standpoints for biological psychiatry.

9   TOWARD A NEW APPROACH IN BIOLOGICAL PSYCHIATRIC AND PSYCHOPHARMACOLOGICAL RESEARCH

General application of the principles of functionalization and verticalization of psychiatric diagnosing would have profound consequences for biological psychiatric and psychopharmacological research.

It would lead first of all to a change in focus. To unravel the biology of dysfunctioning psychic domains would be its ultimate goal, rather than elucidation of the cause of categorically defined entities such as major depression. The latter goal is considered to be too heterogeneous to make a uniform pathophysiology a likely supposition.

Moreover, drug research would also shift from categorical entities to dysfunctioning psychic domains, such as disturbed regulation of anxiety, aggression, mood and motoricity, and disturbances in perception, in hedonic functions, in information processing, and others. Psychopharmacological treatment would be geared toward dysfunctioning psychic domains—particularly those of a primary nature—rather than toward disease entities as such. This approach I have called *functional psychopharmacology* (van Praag et al. 1987).

Functional psychopharmacology would be more often than not polypharmaceutical in nature. In this case, however, polypharmacy would be scientifically based, not haphazardly applied.

Consequently systematic development of a functionalized and verticalized system of diagnosing in psychiatry would bring an era of psychiatric thinking gradually to a close. An era in which a static disease concept prevailed and mental disorders were purely categorically understood, while ushering in a new era of dynamic and functionally oriented psychiatry.

Conceptually as well as practically, such a change in orientation could be conceived as progress.

REFERENCES

Hoff, P. 1995. Kraepelin—Clinical section. In *A History of Clinical Psychiatry. The Origin and History of Psychiatric Disorders*, eds. G. E. Berrios, and R. Porter, 261–79. London: Athlone.

Meyer, A. 1951. *The Collected Papers of Adolph Meyer*. Baltimore: Johns Hopkins Press.

van Praag, H. M. 1997. Over the mainstream: Diagnostic requirements for biological psychiatric research. *Psychiatr Res* 72:201–12.

van Praag, H. M. 2000. Nosologomania: A disorder of psychiatry. *World J Biol Psychiatry* 1:151–8.

van Praag, H. M. 2001. Anxiety/aggression-driven depression. A paradigm of functionalization and verticalization of psychiatric diagnosing. *Prog Neuropsychopharm Biol Psychiatry* 12:28–39.

van Praag, H. M. 2008. Kraepelin, biological psychiatry, and beyond. *Eur Arch Psychiatry Clin Neurosci* 258 (Suppl 2):29–32.

van Praag, H. M., R. S. Kahn, G. M. Asnias, et al. 1987. Denosologization of biological psychiatry or the specificity of 5-HT disturbances in psychiatric disorders. *J Affect Disord* 13:1–8.

van Praag, H. M., R. De Kloet, and J. Van Os. 2004. *Stress, the Brain, and Depression*. Cambridge: Cambridge University Press.

Introduction

Francisco López-Muñoz and Cecilio Álamo

Depression is one of the leading public health problems in the world today and one of the most prevalent and serious psychiatric disorders, with significant socioeconomic and quality of life implications. According to the World Health Organization (2010), some 121 million people are currently suffering from depression, with an annual prevalence of 5.8% for men and 9.5% for women. Depression is the leading cause of disability, as measured by years lived with disability (YLDs), and was the fourth greatest contributor to the global burden of disease in 2000. By the year 2020, depression is projected to reach second place in the ranking of DALYs calculated for all ages and both sexes. Today, depression is already the second cause of DALYs in the age category of 15–44 years for the two sexes combined (Martín-Águeda et al. 2006).

Depression is a very multifaceted illness, with diverse symptoms and signs that probably respond to a complex neurobiology, involving multiple factors—biochemical, genetic, and environmental. Although its pathogenesis remains elusive, several neurotransmitters and neuropeptides have been implicated in the biological origin of this illness. Currently, a priority goal for neurosciences is to discover the precise nature of the central nervous system (CNS) circuits responsible for the modifications of neuronal functioning that lead to depression. Advances in this field, thanks to the huge development that has undergone basic disciplines, have multiplied over recent decades. In this sense, animal models of depression are an essential study tool. Modeling depression in animals is a challenging endeavor considering the complex nature of the disease. Jan M. Deussing and Damian Refojo (Max Planck Institute of Psychiatry, Germany) introduce some basic concepts with respect to what an animal model of depression can account for and what it is able to contribute to. In their chapter, these authors analyze the state of the art in this area and discriminate between true animal models, behavioral tests, and antidepressant screening paradigms.

In humans, the neuropsychological model of depression (Heller and Nitschke 1997) emphasizes left frontal and right posterior hypoactivation as a key feature of cortical activity pattern in depression. This model is supported by neurophysiological studies (lesion, EEG, and MEG studies) (Hughes and John 1999) that show, in depressed individuals, deficits that may be considered a vulnerability marker to depression related to serotonergic and stress mechanisms. In addition, new investigation tools such as coherence, synchronization, or complexity estimates are used for the evaluation of depression. These topics are discussed in this work by Gabriel Rubio and collaborators (Complutense University and Polytechnic University of Madrid, Spain).

Advances in the field of technology as modern neuroimaging techniques will have a significant influence on the future directions with respect to research in depressive disorders. Over the past two decades, investigators have used neuroimaging techniques to examine the neural substrates of mood disorders. In their review, Ian H. Gotlib and collaborators (Stanford University, USA) present findings from this body of research, identifying the major brain regions or structures that have been implicated in depression (the amygdala, the hippocampus, the subgenual anterior cingulate cortex, the dorsolateral prefrontal cortex, and the ventral striatum) and discussing their relation, when possible, to specific symptoms of depression. Results from these studies, and from investigations of individuals at elevated risk for major depression, are important in helping us understand the temporal relation between depression and neural functional and/or structural anomalies.

Similarly, imaging genetics has evolved from single pioneering studies to a frequently employed strategy that has certainly vitalized psychiatric genetics and advanced our understanding of how genes shape behavior (Pezawas and Meyer-Lindenberg 2010). However, the intermediate phenotype concept has been a matter of debate based on evidence that some candidate traits are akin to clinical phenotypes regarding their genetic complexity. Lukas Pezawas and collaborators (Medical University of Vienna, Austria) address in their chapter the genetic regulation of emotion brain circuitries and, through the candidate gene studies, basically monoaminergic genes (*SLC6A4, MAOA, TPH2, HTR1A, COMT*) and *BDNF*, provide valuable insights into pathogenetic pathways that may shape the brain and render it susceptible to depression. These studies open new perspectives for translating knowledge from preclinical studies to people. Hence, a translational approach that incorporates genome-wide studies for the identification of yet unknown genes as well as in vitro animal and neuroimaging studies to further analyze the related pathways may dramatically expand our neurobiological understanding of depression.

Brain development alterations may also play a role in the pathogenesis of mood disorders. An alteration during development may precipitate in the cortical layering and/or wiring, which may lead to an overall alteration of the behavioral programming of that individual. Marten P. Smidt (University Medical Center Utrecht, The Netherlands) analyzes the complex nature of the developing monoaminergic brain systems and notes that, due to their large influence on the whole brain and the importance of the correct homeostatic set points, aberrations during development in these systems may quickly lead to disorders that may precipitate in mood problems and many other psychiatric diseases.

Precisely, the first biochemical etiopathogenic hypothesis on depressive disorders was based on deficits in the functionalism of the monoaminergic systems. At this point, pharmacology has played a major role. The introduction by serendipity of iproniazid (the first monoamine oxidase inhibitor) and imipramine (pioneer of tricyclic antidepressants) in the 1950s ("the psychopharmacological revolution decade") substantially modified the conceptualization of the therapeutic approach (López-Muñoz and Álamo 2009). But in addition, these agents constituted an indispensable research tool for neurobiology, permitting, among other things, the postulation of the first etiopathogenic hypothesis of depressive disorders: "the monoamine hypothesis of depression," as Francisco López-Muñoz and Cecilio Álamo (University of Alcalá, Spain) show in their chapter. The current status in relation to the role of serotonin and noradrenaline in the etiology and pathophysiology of depression is analyzed, respectively, by Anne M. Andrews and collaborators (Pennsylvania State University, UCLA, and University of Pittsburgh, USA) and José Javier Meana and collaborators (University of the Basque Country and University of the Balearic Islands, Spain).

Despite the important advance represented by the monoaminergic hypotheses in relation to the etiopathogeny of depression, a feeling soon emerged that these theories reflected only a small part, probably an initial part, of this problem. More recent theories defend a dysregulation model, which would involve not only a single neurotransmitter pathway but also the alteration of several, including the noradrenergic, the serotonergic, and even the dopaminergic pathways. In line with these postulates, some authors propose an etiological model in which depression would result from a dysfunction of different brain areas, notably the frontal cortex, the hippocampus, the amygdala, and the basal ganglia, which, in turn, would be modulated by different monoaminergic neurotransmission systems.

Currently, there are considerable scientific data, from both animal and human experimentation, supporting the participation of nonmonoaminergic mechanisms in the physiopathology of depression. In this sense, ample evidence now exists that glutamate homeostasis and neurotransmission are disrupted in depression in humans (Zárate et al. 2003). For example, clinical studies suggest that *N*-methyl-D-aspartate (NMDA) receptor antagonists may produce robust antidepressant effects within hours, which are probably due to changes in synaptic activity rather than morphological changes. Similarly, there is dysregulation of alpha-amino-3-hydroxy-5-methyl-4-isoxazolepropionic acid (AMPA) receptors in depression, and AMPA receptor agonists act as antidepressant agents, perhaps by inducing brain-derived neurotrophic factor (BDNF) expression. Metabotropic glutamate receptors which regulate glutamate neuronal transmission can inhibit neurogenesis, and antagonists to these receptors also exhibit antidepressant-like properties in behavioral assays by acting downstream of AMPA receptors. These and other aspects related to glutamatergic transmission are developed by Rodrigo Machado-Vieira and Carlos A. Zárate (National Institute of Mental Health, USA, and University of Sao Paulo, Brazil).

The endocannabinoid system is a neuromodulatory system that is intricately involved in regulation of excitatory, inhibitory, and monoaminergic networks in the brain. Transmission by endocannabinoids in the CNS is a topic that is attracting more and more interest from the therapeutic point of view. In this regard, the manipulation of CB_1 receptors, the principal target of endocannabinoids, presents potent effects on behavior induced by stress and anxiety in rodents (Hill et al. 2009). It is not surprising that some ligands of this receptor, or substances that increase its endogenous ligands, may have antidepressant properties. However, the results so far are inconsistent. Gabriella Gobbi and collaborators (McGill University and University of British Columbia, Canada), in their chapter, describe the experimental bases for a possible use of endocannabinoids as antidepressants, highlighting their capacity (as is the case of other antidepressants) for boosting noradrenergic and serotoninergic mechanisms, at the same time as favoring hippocampal neurogenesis, via pathways distinct from those used by conventional drugs. This body of literature argues that enhancement of the endocannabinoid system is a key component involved in the etiopathology of major depression and may represent a novel therapeutic target for the treatment of this debilitating mental illness (Rodríguez-Bambico et al. 2009).

Some authors have also involved the opioid system in depression. This view is based fundamentally on the classic observation that some opiate agents may have certain antidepressant power. But moreover, it is known that electroconvulsive therapy increases plasmatic levels of β-endorphin and that the administration of amitryptiline increases the concentration of leu-enkephalin in the hypothalamus and spinal cord (Hamon et al. 1987). Also, research has detected an increase in the density of opioid receptors of the μ type, in both the frontal cortex and the caudate nucleus of suicide victims (Gabilondo et al. 1995), a density which decreases where there is long-term administration of antidepressants. As pointed out by Juan Antonio Micó and collaborators (University of Cádiz and Cajal Institute—CSIC, Spain), there are considerable preclinical data that involve endogenous opioid peptides and their receptors in the pathogeny of depression, including development of opioid receptor knockout mice. Moreover, μ-opioid and δ-opioid activation and/or κ-opioid blockade produce antidepressant-like effects in a number of preclinical assays, suggesting that these may be other pharmacological targets for treating depression in humans (Berrocoso et al. 2009). These authors also highlight the existence of common steps in the opiate and NMDA transduction pathways, which also supports the hypothesis of the possible antidepressant effect of opioids.

The melatonergic system constitutes an essential part of neurobiological circuits responsible to adapt the biological rhythms to periodic changes of the environment. At present, it seems evident that alterations in endogenous rhythms and abnormal secretion of melatonin can manifest through clinical disorders in the fields of sleep, cardiovascular, endocrinometabolic, neurology, and psychiatry (Schulz and Steimer 2009), including affective disorders. Recent progress in the understanding of molecular and cellular mechanisms responsible for control of the biological clock open new pathways to research to clarify the basis of the relations between circadian rhythm disorders and depression. The relevant role of melatonergic system in depression has become apparent with the recent clinical introduction of agomelatine, the first antidepressant with proven clinical efficacy that approaches the treatment of depression from a pharmacological perspective other than that of the other drugs used up to now. The primary mechanism of this agent is its agonistic action on MT_1 and MT_2 melatonin receptors (Álamo et al. 2008). Cecilio Álamo and Francisco López-Muñoz (University of Alcalá, Spain) discuss the role of this system and chronobiological rhythms in depression.

According to the dysregulation model, numerous factors, notably stress, could influence the correct functioning of certain brain areas, in either a selective or a generalized way, which concurs with the classic postulates of a heterogeneous etiology of depressive disorders. Appended to this hypothesis should be the possible existence of marked alterations in the intraneuronal pathways of signal transduction, which would also situate the source of depressive disorders in the sphere of molecular dysfunctions. For this purpose, research should not be limited to the classical amin-ergic mechanisms (reuptake and metabolization) or the more modern ones (receptor mechanisms) but should rather explore other sources of knowledge constantly being provided by biochemistry, molecular biology, and genomics.

Depressive condition is often considered a stress-related disorder because some form of stressful life event frequently triggers depressive symptoms. The discovery, around 20 years ago, of the role of corticotropin-releasing factor (CRF) in the control of the reactions of the hypothalamic-pituitary-adrenal (HPA) axis and other CNS areas to stress has revolutionized our knowledge of the neuro-biology of depression and has opened up interesting therapeutic perspectives. CRF is considered one of the body's major regulators of the stress response, and the function of CRF systems appears to be altered during depression. Hyperactivity of the HPA axis has been described in depressive patients, together with reduced control of negative feedback of glucocorticoids (Galard et al. 2002). CRF appears to act not only as a releasing factor of adrenocorticotropic hormone but also as a neu-rotransmitter that mediates emotional, cognitive, and behavioral reactions. Despite the fact that the molecular bases of this functional disorder of the HPA axis have not been fully clarified, numer-ous clinical studies have suggested that the functional normalization of this axis may be a neces-sary step for stabilization of the remission of depressive symptoms (Valdez 2009). In the light of these antecedents, Glenn R. Valdez (Grand Valley State University, USA) explores the role of CRF-related ligands, CRF receptors, HPA axis activity, and glucocorticoid receptors in the regulation of the stress response and depression. Research suggests that CRF_1 receptor antagonism and/or CRF_2 receptor activation may be a potentially promising novel mechanism of action for the development of antidepressants. Additionally, glucocorticoid receptor antagonism may be useful in reducing the neurocognitive symptoms associated with depression. Understanding the interactions between these systems may provide further insight into the biological foundations of depression, as well as stress-related psychiatric conditions.

Basic research has been instrumental in understanding the underlying molecular mechanisms and signaling pathways by which cytokines may affect complex depression-like symptoms and behavior. The hypothesis of the participation of cytokines in depression emerges from the obser-vation that the treatment of diverse pathologies with some of them, such as interferons, interleu-kin-1, interleukin-2, and tumor necrosis factor-α, can produce a series of symptoms similar to those detected in depression (anhedonia, reduced social interaction, fatigue, etc.) (Raison et al. 2006). Moreover, certain cytokines can be regulated by stress, and there are reports of alterations of the immunological system in depressive disorders (e.g., high levels of interleukin-6). These and other data are discussed by Bernhard T. Baune (University of Adelaide, Australia). In this sense, it is possible that in the future, at least some types of depression could be treated with new drugs with anti-inflammatory and antidepressant properties.

Substance P, a neurokinin that acts on NK-1 receptors, has aroused enormous interest in psychia-try due to its identification in the circuits related to fear and anxiety, circumstances involving the release of neurokinin, together with its location alongside serotonin and noradrenaline (Krishnan 2002). On the other hand, the administration of agonists of substance P results in characteristic defensive reactions, such as escape attempts, behavioral reactions, cardiovascular activation, and increased discharge from the locus coeruleus, related to responses to stress. In contrast, antagonists of these receptors appear to exhibit experimental anxiolytic and antidepressant properties. Even so, and as discussed by Tih-Shih Lee and K. Ranga Krishnan (Duke University Medical School, USA, and Duke—NUS Graduate Medical School, Singapore), there are still antidepressant therapeutic options for antagonists of other receptors (NK-2), as well as for different conditions that involve stress (Louis et al. 2008).

Neuropeptide Y (NPY) has also been involved in the etiology of depression. NPY and NPY recep-tors (Y_1, Y_2, Y_4, and Y_5) are abundantly expressed in relevant emotional-processing brain regions. Interestingly, NPY levels are decreased after stress as well as in animal models of depression-related behaviors whereas long-term treatments with antidepressants increase NPY levels (Morales-Medina et al. 2010). Moreover, the exogenous administration of NPY has been shown to reverse behavioral deficits noted in various animal models of depression. Finally, numerous studies have suggested that NPY levels are decreased in depressed subjects. Julio César Morales-Medina and Rémi Quirion (McGill University, Canada) review in this book the contribution of NPY and its receptors in various animal models of depression-related behaviors as well as in depressed subjects.

Nitric oxide (NO) serves an important role in the nervous system, where it acts as a messenger molecule in a number of physiological processes, including processes being linked to the major psychiatric diseases (Knott and Bossy-Wetzel 2009). Aleksander A. Mathé and Gregers Wegener (Karolinska Institutet, Sweden, and Aarhus University Hospital, Denmark) review in their chapter general aspects of the NO system in depression, as well as focus on inhibitors of NO production as putative therapeutic agents toward depression.

The role of G proteins (GP) in the modulation of receptor signals has aroused considerable interest in the past few years. This is because these proteins constitute the initial postreceptor step in the most important intracellular transduction pathways. Antidepressants, in long-term administration, are capable of modifying the functioning of different elements of GPs, which suggests that, in depression, there may be some disorder in the superfamily of GPs coupled to receptors and that antidepressants would act by modifying —supposedly toward "normality"— this dysfunction of GPs (Avissar and Schreiber 2006). Some research lines have been centered in the search of antidepressants targeting G-protein-coupled receptor kinases and β-arrestins responsible for the internalization and desensitization of GPs coupled to receptors. Moreover, the role of β-arrestins has increased considerably in the wake of its proven involvement in the mitogen-activated protein kinase cascade and its capacity to interact with regulators of transcription factors (Golan et al. 2009). In this sense, a substantial body of evidence has accumulated indicating that β-arrestins play a major role in the pathophysiology of mood disorders as well as in the mechanism of action of antidepressants (Schreiber et al. 2009). Sofia Avissar and Gabriel Schreiber (Ben Gurion University, Ashkelon, Israel) discuss these topics in their chapter.

Researchers have for many years been exploring the possibility of imitating the cascade of intracellular effects produced after prolonged administration of antidepressants. In such conditions, we can observe an increase in levels of cyclic adenosine monophosphate (cAMP), which is reflected in an increase in the expression of cAMP response element–binding and BDNF (Gonul et al. 2005), elements that appear to play a crucial role in antidepressant efficacy. BDNF, a major neuronal growth factor in the brain, promotes neuronal survival and maturation, as well as plasticity during development and adulthood (Castren and Rantamaki 2010). Kazuko Sakata (University of Tennessee Health Science Center, USA) reviews the current knowledge of the roles of BDNF in depression, focusing especially on BDNF gene regulation and the underlying molecular and cellular mechanisms. She also stressed that individual genetic and epigenetic variations of the BDNF gene may explain many of the differences between individuals and might be useful as a predictor of the antidepressant treatment efficacy.

Virtually all phosphodiesterases (PDEs) are expressed in the brain, some at high levels, suggesting their importance in overall CNS function. Specifically, PDE2, PDE4, and PDE9 are widely and highly expressed in the brain, in particular the limbic system, including the olfactory cortex, hippocampus, and amygdala. This distribution pattern is consistent with the roles of these PDEs in depression (Zhang 2009). Moreover, levels of cAMP have been successfully increased through blocking of its metabolization by inhibition of PDEs. In fact, PDE4 inhibitors, as rolipram, shows antidepressant activity (Li et al. 2009). Han-Ting Zhang and collaborators (West Virginia University Health Sciences Center, USA) report in their chapter the substantial progress made in the identification of roles of PDEs in the etiology of depression and antidepressant activity.

The link between vascular risk and depression has a long history. In fact, cerebrovascular risk in the form of conditions such as hypertension, diabetes, hyperlipidemia, and heart disease plays a role in the onset, maintenance, and exacerbation of depression among older people. Some authors have proposed that these chronic conditions contribute to small-vessel disease in the frontal and subcortical regions of the brain, which play a significant role in the regulation of mood (Lyness et al. 1998). In their chapter, Benjamin T. Mast and collaborators (University of Louisville and VA Palo Alto Health Care System, USA) review the evidence for the vascular depression hypothesis, with particular attention to the neuropsychological consequences and correlates.

With respect to the pharmacological treatment of depression, and despite the fact that the latest antidepressants introduced into clinical practice maintain the same action mechanism, that is, the modulation of monoaminergic neurotransmission at a synaptic level, the future of antidepressant therapy appears to be trying to turn toward extraneuronal nonaminergic mechanisms or mechanisms that modulate the intraneuronal biochemical pathways (López-Muñoz and Alamo 2009). In this way, the role of neurotrophic factors, CRF and glucocorticoids, excitatory amino acids, and NMDA receptor are being studied, as well as other possible targets (opioid system, interleukins, P substance, nitric oxide, etc.). According to this hypothesis, selective serotonin reuptake inhibitors, tricyclic antidepressants, and noradrenaline reuptake inhibitors would have a common intracellular mechanism, remote from their point of action in the synaptic cleft, which would modulate, through modifications of gene expression, the synthesis of particular substances such as proenkephaline, neurotensin, BDNF, or enzymes such as tyrosine-hydroxylase, which ultimately will provoke changes in functional adaptation, trophic actions, or synaptic remodeling, likely to counter possible anomalies present in depression. All these aspects are described by Cecilio Álamo and Francisco López-Muñoz (University of Alcalá, Spain), who analyze the products that today, within the field of pharmacology of depression, are being developed in different phases of clinical research. All of this undoubtedly constitutes a huge advance toward the knowledge of what would be the beginning of psychopharmacology of the present millennium.

Knowledge of the pathophysiology of depression has evolved from Galen's speculations in *Antiquity* about an excess of black bile ("melancholia") to theories that incorporate gene–environment interactions, endocrine, immunological and metabolic mediators, and neuronal forms of plasticity (López-Muñoz et al. 2009a, 2009b). In this sense, it should be stressed that many different research groups have considered a whole range of other neurotransmission and neuromodulation systems, including glia cells, as possible etiological models and therapeutic targets for depression. All such contributions are of great importance because the problem facing us probably has no single key, and it is the research effort as a whole that will finally help unlock the "Pandora's box" constituted by disorders of the mind. In any case, to improve the still-low remission rates of this psychiatric disorder, it will be imperative to look beyond monoamine mechanisms.

REFERENCES

Álamo, C., F. López-Muñoz, and M. J. Armada. 2008. Agomelatina: Un nuevo enfoque farmacológico en el tratamiento de la depresión con traducción clínica. *Psiquiatr Biol* 15:125–39.

Avissar, S., and G. Schreiber. 2006. The involvement of G proteins and regulators of receptor-G protein coupling in the pathophysiology, diagnosis and treatment of mood disorders. *Clin Chim Acta* 366:37–47.

Berrocoso, E., P. Sánchez-Blanque, J. Garzón, and J. A. Micó. 2009. Opiates as antidepressants. *Curr Pharm Des* 15:1612–22.

Castren, E., and T. Rantamaki. 2010. The role of BDNF and its receptors in depression and antidepressant drug action: Reactivation of developmental plasticity. *Dev Neurobiol* 70:289–97.

Gabilondo, A. M., J. J. Meana, and J. A García-Sevilla. 1995. Increased density of μ-opioid receptors in the postmortem brain of suicide victims. *Brain Res* 682:245–50.

Galard, R., R. Catalan, J. M. Castellanos, and J. M. Gallart. 2002. Plasma corticotropin-releasing factor in depressed patients before and after the dexamethasone suppression test. *Biol Psychiatry* 51:463–8.

Golan, M., G. Schreiber, and S. Avissar. 2009. Antidepressants, β arrestins and GRKs: From regulation of signal desensitization to intracellular multifunctional adaptor functions. *Curr Pharm Des* 15:1699–708.

Gonul, A. S., F. Akdeniz, F. Taneli, O. Donat, C. Eker, and S. Vaship. 2005. Effect of treatment on serum brain-derived neurotrophic factor levels in depressed patients. *Eur Arch Psychiatry Clin Neurosci* 255:381–6.

Hamon, M., H. Gozlan, S. Bourgoin, et al. 1987. Opioid receptors and neuropeptides in the CNS in rats treated chronically with amoxapine or amitriptyline. *Neuropharmacology* 26:531–9.

Heller, W., and J. B. Nitschke. 1997. Regional brain activity in emotion: A framework for understanding cognition in depression. *Cogn Emotion* 11:637–61.

Hill, M. N., C. J. Hillard, F. R. Bambico, S. Patel, B. B. Gorzalka, and G. Gobbi. 2009. The therapeutic potential of the endocannabinoid system for the development of a novel class of antidepressants. *Trends Pharmacol Sci* 30:484–93.

Hughes, J. R., and E. R. John. 1999. Conventional and quantitative electroencephalography in psychiatry. *J Neuropsychiatry Clin Neurosci* 11:190–208.

Knott, A. B., and E. Bossy-Wetzel. 2009. Nitric oxide in health and disease of the nervous system. *Antioxid Redox Signal* 11:541–54.

Krishnan, K. R. K. 2002. Clinical experience with substance P receptor (NK_1) antagonists in depression. *J Clin Psychiatry* 63 (Suppl 11):25–9.

Li, Y. F., Y. Huang, S. L. Amsdell, L. Xiao, J. M. O'Donnell, and H. T. Zhang. 2009. Antidepressant- and anxiolytic-like effects of the phosphodiesterase-4 inhibitor rolipram on behavior depend on cyclic AMP response element binding protein-mediated neurogenesis in the hippocampus. *Neuropsychopharmacology* 34:2404–19.

López-Muñoz, F., and C. Álamo. 2009. Monoaminergic neurotransmission: The history of the discovery of antidepressants from 1950s until today. *Curr Pharm Des* 15:1563–86.

López-Muñoz, F., G. Rubio, J. D. Molina, P. García-García, and C. Álamo. 2009a. La melancolía como enfermedad del alma (I): De la Antigüedad Clásica al Renacimiento. *An Psiquiatr* 25:146–59.

López-Muñoz, F., G. Rubio, J. D. Molina, P. García-García, and C. Álamo. 2009b. La melancolía como enfermedad del alma (II): Del periodo moderno a la actualidad. *An Psiquiatr* 25:197–209.

Louis, C., J. Stemmelin, D. Boulay, O. Bergis, C. Cohen, and G. Griebel. 2008. Additional evidence for anxiolytic and antidepressant-like activities of saredutant (SR48968), an antagonist at the neurokinin-2 receptor in various rodent-model. *Pharmacol Biochem Behav* 89:36–45.

Lyness, J. M., E. D. Caine, C. Cox, D. A. King, Y. Conwell, and T. Olivares. 1998. Cerebrovascular risk factors and later-life major depression. Testing a small-vessel brain disease model. *Am J Geriatr Psychiatry* 6:5–13.

Martín-Águeda, B., F. López-Muñoz, G. Rubio, P. García-García, A. Silva, and C. Álamo. 2006. Current situation of depression healthcare in Spain: Results of a psychiatrists' survey. *Eur J Psychiat* 20:211–23.

Morales-Medina, J. C., Y. Dumont, and R. Quirion. 2010. A possible role of neuropeptide Y in depression and stress. *Brain Res* 1314:194–205.

Pezawas, L., and A. Meyer-Lindenberg. 2010. Imaging genetics: Progressing by leaps and bounds *Neuroimage* 53: 801–3.

Raison, C. L., L. Capuron, and A. H. Miller. 2006. Cytokines sing the blues: Inflammation and the pathogenesis of depression. *Trends Immunol* 27:24–31.

Rodríguez-Bambico, F., A. Duranti, A. Tontini, G. Tarzia, and G. Gobbi. 2009. Endocannabinoids in the treatment of mood disorders: Evidence from animal models. *Curr Pharm Des* 15:1623–46.

Schreiber, G., Golan, M., Avissar S. 2009. β-Arrestin signaling complex as a target for antidepressants and a depression marker. *Drug News Perspect* 22:467–80.

Schulz, P., and T. Steimer. 2009. Neurobiology of circadian systems. *CNS Drugs* 23:3–13.

Valdez, G. R. 2009. CRF receptors as a potential target in the development of novel pharmacotherapies for depression. *Curr Pharm Des* 15:1587–94.

World Health Organization. 2010. Depression. Available at URL: www.who.int/mediacentre.http://www.who .int/mental_health/management/depression/definition/en/.

Zárate, C. A., Jr., J. Du, J. Quiroz, et al. 2003. Regulation of cellular plasticity cascades in the pathophysiology and treatment of mood disorders: Role of the glutamatergic system. *Ann N Y Acad Sci* 1003:273–91.

Zhang, H. T. 2009. Cyclic AMP-specific phosphodiesterase-4 as a target for the development of antidepressant drugs. *Curr Pharm Des* 15:1687–98.

Contributors

Cecilio Álamo
Neuropsychopharmacology Unit
Department of Pharmacology
University of Alcalá
Alcalá de Henares, Madrid, Spain

Cristina Alba-Delgado
Department of Neuroscience, Pharmacology
    and Psychiatry
School of Medicine
University of Cádiz
Mental Health CIBER (CIBERSAM)
Cádiz, Spain

Stefanie C. Altieri
Huck Institutes of the Life Sciences
The Pennsylvania State University
University Park, Pennsylvania
and
Department of Psychiatry
Semel Institute for Neuroscience and Human
    Behavior
Hatos Center for Neuropharmacology
David Geffen School of Medicine
University of California
Los Angeles, California

Anne M. Andrews
Department of Psychiatry
Semel Institute for Neuroscience and Human
    Behavior
Hatos Center for Neuropharmacology
David Geffen School of Medicine
University of California
Los Angeles, California
and
Department of Veterinary and Biomedical
    Sciences
The Pennsylvania State University
University Park, Pennsylvania

Sofia Avissar
Department of Pharmacology
Ben Gurion University of the Negev
Beer Sheva, Israel

Bernhard T. Baune
Discipline of Psychiatry
School of Medicine
University of Adelaide
Adelaide, Australia

Esther Berrocoso
Department of Neuroscience, Pharmacology
    and Psychiatry
School of Medicine
University of Cádiz
Mental Health CIBER (CIBERSAM)
Cádiz, Spain

Luis F. Callado
Department of Pharmacology
University of the Basque Country
Mental Health CIBER (CIBERSAM)
Leioa, Spain

Jeremy S. Carmasin
Department of Psychology and Brain Sciences
University of Louisville
Louisville, Kentucky

Jan M. Deussing
Molecular Neurogenetics
Max Planck Institute of Psychiatry
Munich, Germany

Alberto Fernández
Department of Psychiatry
Faculty of Medicine
Complutense University
Madrid, Spain

Michael Freissmuth
Department of Pharmacology
Center of Biomolecular Medicine and
    Pharmacology
Medical University of Vienna
Vienna, Austria

Daniella J. Furman
Department of Psychology
Stanford University
Stanford, California

Jesús A. García-Sevilla
Laboratory of Neuropharmacology
University Institute of Research in Health
    Sciences (IUNICS)
University of the Balearic Islands
Thematic Network on Addictive Disorders
    (RETICS)
Palma de Mallorca, Spain

Javier Garzón
Neuropharmacology Group
Department of Molecular, Cellular and
    Developmental Neurobiology
Cajal Institute, CSIC
Mental Health CIBER (CIBERSAM)
Madrid, Spain

Gabriella Gobbi
Neurobiological Psychiatry Unit
Department of Psychiatry
McGill University
Montreal, Canada

Ian H. Gotlib
Department of Psychology
Stanford University
Stanford, California

J. Paul Hamilton
Department of Psychology
Stanford University
Stanford, California

Tina Hofmaier
Department of Psychiatry and Psychotherapy
Divison of Biological Psychiatry
Medical University of Vienna
Vienna, Austria

Tih-Shih Lee
Department of Psychiatry and Behavioral
    Sciences
Duke University Medical School
Durham, North Carolina
and
Duke-NUS Graduate Medical School
College Road, Singapore

Francisco López-Muñoz
Neuropsychopharmacology Unit
Department of Pharmacology
University of Alcalá
Alcalá de Henares, Madrid, Spain
and
Faculty of Health Sciences
Camilo José Cela University
Villanueva de la Cañada, Madrid, Spain

Rodrigo Machado-Vieira
Experimental Therapeutics and
    Pathophysiology Branch
Division of Intramural Research Program
National Institute of Mental Health
National Institutes of Health
Bethesda, Maryland
and
Institute of Psychiatry
University of Sao Paulo
Sao Paulo, Brazil

Benjamin T. Mast
Department of Psychology and Brain Sciences
University of Louisville
Louisville, Kentucky

Aleksander A. Mathé
Department of Clinical Neuroscience
Karolinska Institutet
Stockholm, Sweden

Ryan J. McLaughlin
Department of Psychology
University of British Columbia
Vancouver, Canada

J. Javier Meana
Department of Pharmacology
University of the Basque Country
Mental Health CIBER (CIBERSAM)
Leioa, Spain

Juan Antonio Micó
Department of Neuroscience, Pharmacology
    and Psychiatry
School of Medicine
University of Cádiz
Mental Health CIBER (CIBERSAM)
Cádiz, Spain

Julio César Morales-Medina
Department of Neurology and Neurosurgery
McGill University
Montreal, Canada
and
Douglas Mental Health University Institute
McGill University
Montreal, Canada

Stephan Moratti
Department of Basic Psychology, Faculty of
    Psychology
Complutense University
Madrid, Spain
and
Centre of Biomedical Technology
Polytechnic University of Madrid
Madrid, Spain

James M. O'Donnell
Department of Behavioral Medicine and
    Psychiatry
West Virginia University Health Sciences
    Center
Morgantown, West Virginia

Lukas Pezawas
Department of Psychiatry and Psychotherapy
Divison of Biological Psychiatry
Medical University of Vienna
Vienna, Austria

Rémi Quirion
Douglas Mental Health University Institute
McGill University
Montreal, Canada
and
Department of Psychiatry
McGill University
Montreal, Canada

Ulrich Rabl
Department of Psychiatry and Psychotherapy
Divison of Biological Psychiatry
Medical University of Vienna
Vienna, Austria

K. Ranga Krishnan
Department of Psychiatry and Behavioral
    Sciences
Duke University Medical School
Durham, North Carolina
and
Duke-NUS Graduate Medical School
College Road, Singapore

Damian Refojo
Molecular Neurobiology
Max Planck Institute of Psychiatry
Munich, Germany

Francis Rodriguez Bambico
Neurobiological Psychiatry Unit
Department of Psychiatry
McGill University
Montreal, Canada

Sarah V. Rowe
Department of Psychology and Brain Sciences
University of Louisville
Louisville, Kentucky

Gabriel Rubio
Department of Psychiatry
Faculty of Medicine
Complutense University
Mental Health CIBER (CIBERSAM)
Madrid, Spain

Kazuko Sakata
Department of Pharmacology and
    Psychiatry, College of Medicine
The University of Tennessee Health Science
    Center
Memphis, Tennessee

Pilar Sánchez-Blázquez
Department of Neuroscience, Pharmacology
    and Psychiatry
School of Medicine
University of Cádiz
Mental Health CIBER (CIBERSAM)
Cádiz, Spain

Christian Scharinger
Department of Psychiatry and Psychotherapy
Divison of Biological Psychiatry
Medical University of Vienna
Vienna, Austria

Gabriel Schreiber
Department of Psychiatry
Ben Gurion University of the Negev
Beer Sheva, Israel
and
Barzilai Medical Center
Ashkelon, Israel

Etienne Sibille
Department of Psychiatry
University of Pittsburgh
Pittsburgh, Pennsylvania
and
Center for Neuroscience
University of Pittsburgh
Pittsburgh, Pennsylvania

Yogesh S. Singh
Department of Psychiatry
Semel Institute for Neuroscience and Human
    Behavior
Hatos Center for Neuropharmacology
David Geffen School of Medicine
University of California
Los Angeles, California
and
Department of Chemistry
The Pennsylvania State University
University Park, Pennsylvania

Marten P. Smidt
Department of Neuroscience and
    Pharmacology
Rudolf Magnus Institut of Neuroscience
University Medical Center Utrecht
Utrecht, The Netherlands

Glenn R. Valdez
Department of Psychology
Grand Valley State University
Allendale, Michigan

Herman M. van Praag
University of Groningen
Utrecht, Maastricht, The Netherlands
and
The Albert Einstein College of Medicine
Bronx, New York

Gregers Wegener
Centre for Psychiatric Research
Aarhus University Hospital
Risskov, Denmark

Ying Xu
Department of Behavioral Medicine and
    Psychiatry
West Virginia University Health Sciences
    Center
Morgantown, West Virginia

Brian P. Yochim
Sierra Pacific Mental Illness Research,
    Education, and Clinical Center (MIRECC)
VA Palo Alto Health Care System
Palo Alto, California

Carlos A. Zarate Jr.
Experimental Therapeutics and
    Pathophysiology Branch
Division of Intramural Research Program
National Institute of Mental Health
National Institutes of Health
Bethesda, Maryland

Han-Ting Zhang
Department of Behavioral Medicine and
    Psychiatry
West Virginia University Health Sciences
    Center
Morgantown, West Virginia

1 Animal Models of Depression

Damian Refojo and Jan M. Deussing

CONTENTS

1.1 INTRODUCTION

Major depression, also known as unipolar depression or major depressive disorder, is a main contributor to the global burden of disease. With respect to disability-adjusted life years, depression presently ranks third in the World Health Organization Global Burden of Disease study (World Health Organization 2008) and is projected to be the second leading cause of disability worldwide by 2030 (Mathers and Loncar 2006). Lifetime prevalence of the disease is 10–25% for women and 5–12% for men. Characteristic disease symptoms are depressed mood, loss of interest or pleasure, feelings of guilt or worthlessness, disturbed sleep and/or appetite, low energy, poor concentration, and suicidal ideations, which often become chronic or recurrent, substantially interfering with the

1

individual's ability to handle everyday life. Severe depression is life threatening—that is, without treatment, 10–15% of the affected subjects commit suicide.

Epidemiological studies suggest that 40–70% of the risk for developing depression is genetically determined (Fava and Kendler 2000; Malhi et al. 2000; Lesch 2004). However, not a single valid vulnerability gene or highly penetrant genetic variant has been identified so far, neither in classical linkage studies nor in recent large-scale genome-wide association studies (Bosker et al. 2010; Muglia et al. 2010). This missing heritability is largely attributable, on one hand, to the assumption that the contribution of individual genes to the disease is likely to be only minor and, on the other hand, to the lack of objective diagnostic tools/biomarkers that would allow a more differentiated diagnosis beyond the currently used and broad criteria contained in the *Diagnostic and Statistical Manual of Mental Disorders, Fourth Edition* (*DSM-IV*). Depression is a multifactorial disorder, which involves a complex interplay of genetic predispositions and largely unknown environmental factors, varying from emotional stress and trauma to the up-to-now unknown and probably stochastic processes during brain development (Nestler et al. 2002; Caspi et al. 2003; Lesch 2004). These interactions of environment and genes leave epigenetic footprints in the genome that might alter gene expression patterns and thus neural function, which ultimately could contribute to the development and manifestation of disease (Renthal and Nestler 2009; Petronis 2010).

The serendipitous discovery of tricyclic antidepressants and monoamine oxidase inhibitors in the 1950s has revolutionized the medication of depression and paved the way for the monoamine hypothesis of depression. Unfortunately, this has promoted the situation that almost all the subsequently developed antidepressants, including the most currently prescribed medications, have been developed based on the structures of those earliest compounds (Berton and Nestler 2006). In consequence, modern antidepressants have not dramatically improved their efficacy: they still require several weeks to exert their effects, side effects are still a problem, and a substantial proportion of patients do not respond adequately. Along these lines, the mechanism of action of monamine-based drugs is far from fully understood. Although these medications rapidly increase the activity of the brain's serotonergic and noradrenergic system, the medication has to be given for several weeks for their antidepressant actions to become manifest. Increasing evidence suggests that other mechanisms such as neurogenesis and neurotrophic processes could be relevant to the antidepressant effects of monoamine-based drugs, leaving the question open whether these are direct or secondary

---

TABLE 1.1
Modeling Symptoms of Major Depression According to *DSM-IV* in Rodents

DSM-IV Symptom	Modeling in Rodents
1. Depressed or irritable mood	Cannot be modeled
2. Diminished interest in pleasurable activities (anhedonia)	Reduced preference for palatable reward (e.g., sucrose consumption) or intracranial self-stimulation
3. Large changes in appetite and weight	Abnormal loss of weight after exposure to chronic stressors
4. Insomnia or hypersomnia	Abnormal sleep architecture in the electroencephalogram
5. Psychomotor agitation or retardation	Alteration of home-cage activity, treadmill running, or nest building
6. Fatigue or loss of energy	Reduced home-cage activity, treadmill running, or nest building
7. Feelings of worthlessness or excessive or inappropriate guilt	Cannot be modeled
8. Indecisiveness or diminished ability to think or concentrate	Deficits in working and spatial memory and impaired sustained attention
9. Recurrent thoughts of death or suicide	Cannot be modeled

Note: Five of the above symptoms have to be present in the same week, and at least one of the symptoms has to be either depressed mood or loss of interest or pleasure.

---

effects (Nestler et al. 2002; Dranovsky and Hen 2006). Moreover, animal models and antidepressant screening tests based on this hypothesis are circular in nature, since it is highly likely that they are unable to identify novel agents with actions independent of monoamines. Finally, despite the advent and rapid progress in the development of noninvasive technologies (e.g., functional magnetic resonance imaging) in the past decades (Agarwal et al. 2010), another major obstacle with which research on depression has to cope are the ethical and practical limitations to access the affected organ in human subjects—the living brain.

Considering the nature of the disease, it is obvious that none of the presently available animal models are—and most likely never will be—able to recapitulate all aspects of depression (Table 1.1). Nevertheless, existing models and paradigms have proven extremely useful with respect not only to the identification and improvement of antidepressant substances but also to the validation of neurobiological concepts. The challenge for the future is to further improve these animal models, an undertaking that will significantly profit from the tremendous development and refinement of genetic tools, although the contribution of current efforts to identify biomarkers and genetic underpinnings of the disease is less clear.

1.2 FEASIBILITY, PURPOSES, AND REQUIREMENTS OF AN ANIMAL MODEL OF DEPRESSION

An animal model of depression can serve a variety of purposes. It can be used (1) as a tool to investigate aspects of the neurobiology and pathophysiology of depression and (2) as a model to identify diagnostic biomarkers. Additionally, since antidepressant drugs have little or no effects in healthy individuals, an animal model can be used (3) as an experimental model to study the mechanism of action of antidepressant drugs and (4) as a screening test to elucidate antidepressant activity.

The problem of all animal models and particularly of those for mental disorders that are in part defined through subjective experience is to determine clear criteria that allow the assessment of the validity of a given model. The minimum requirements for a valid depression model were initially defined by McKinney and Bunney (1969), proposing that an animal model of depression should (1) reflect the human disorder in its manifestations or symptomatology, (2) display behavioral changes that can be objectively monitored, (3) present with behavioral changes that can be reversed by the same treatment modalities that are effective in humans, and (4) be highly reproducible between investigators. Willner (1984) further refined this concept and defined the dimensions of *construct*, *face*, and *predictive* validity as key criteria to assess the validity of an animal model of depression. The *construct* (or etiologic) validity addresses the theoretical rationale of the model. This, however, has several inherent problems since its evaluation naturally relies on the present knowledge of the etiology and pathophysiology of depression, which is far from being understood. In this sense, it is rather impossible to account for the plethora of factors involved in the development of depression, such as psychological (e.g., stressful life events, adverse experiences, and personality traits) or biological factors (e.g., genetic influences, physical illnesses, and medications). The *face* validity refers to the phenomenological similarity between the anatomical, biochemical, neuropathological, or behavioral features modeled in the animal and the symptoms of the disease in human subjects. However, there are few, if any, neurobiological abnormalities that are known with certainty to be hallmarks or biomarkers of depression. Finally, the *predictive* (or pharmacological) validity addresses the performance in a test, which should be predictive for the performance in the condition that is modeled. In practice, the predictive validity of animal models of depression is determined largely by their response to antidepressant drugs (Willner and Mitchell 2002). A valid model should be sensitive and specific. This means that it should respond to effective antidepressants, but not to nonselective drugs, and the responses should occur within an appropriate dose range.

Besides these criteria that assess the validity of a model, the utility of an animal model critically depends on its reliability, which refers to the consistency and stability with which variables of interest are observed and reproduced (Geyer and Markou 1995). Different laboratories commonly employ their own idiosyncratic versions of behavioral test apparatus and protocols, and every laboratory environment has additional unique features. A systematic study of the impact of the laboratory environment on mouse behavior has clearly demonstrated that despite all efforts to equate laboratory environments, significant and, in some cases, large effects of the environment were found (Crabbe et al. 1999; Wahlsten et al. 2003, 2006). In particular, in the case of behaviors where genetic effects are rather minor, such as those related to depression, major influences of environmental conditions specific to individual laboratories have to be carefully taken into account. Therefore, cautious replication in independent groups of animals and more careful evaluation of behavioral domains (e.g., anxiety-like behavior) with multiple tests are required (Crabbe et al. 1999). In this respect, the Eumorphia Consortium has pinpointed a clear direction with the European Mouse Phenotyping Resource for Standardized Screens (EMPReSS) by applying strict standard operating procedures, which have been demonstrated to markedly improve replicability across laboratories (Brown et al. 2005).

1.3 ADOPTING THE ENDOPHENOTYPE CONCEPT TO ANIMAL MODELS OF DEPRESSION

The currently used classification schemes in psychiatry were derived from clinical observations primarily dedicated to clinical description and communication. The criteria used to classify a major depressive episode according to *DSM-IV* are extremely heterogeneous and sometimes even opposite in their expression (e.g., substantial weight gain or loss, insomnia or hypersomnia), suggesting that depression should rather be viewed as a syndrome and not a disease (Berton and Nestler 2006). Considering this syndrome-like character of disease, it is difficult to envision an animal model that completely recapitulates the symptoms of depression observed in human patients. Moreover, animals lack consciousness of self, self-reflection, and consideration of others, and hallmarks of the disorder, such as depressed mood, low self-esteem, or suicidality, are hardly accessible in non-humans (Table 1.1). Nonetheless, depression, as with other mental disorders, constitutes intermediate traits that can be reproduced independently and evaluated in animals, including behavioral, physiological, endocrinological, and neuroanatomical alterations (McKinney and Bunney 1969; Hasler et al. 2004).

The lack of biological groundwork for the classification of psychiatric disorders accounts in part for the lack of success of studies seeking for the neurobiological and genetic roots of psychiatric disorders. Thus, the search for behavioral phenotypes in the mouse has often been guided by behavioral abnormalities that have been observed in human psychiatric disorders. However, rather than trying to find a "depressed" mouse, it is now widely acknowledged that it might be a more practical goal to focus on specific phenotype components rather than on entire syndromal models (Leboyer et al. 1998; Tarantino and Bucan 2000). This notion started with the description of the concept of endophenotype by Gottesman and Shields (Shields and Gottesman 1972; Gottesman and Gould 2003; Gould and Gottesman 2006). The endophenotype concept originally emerged as a strategic tool in neuropsychiatric research, which tries to dissect complex phenotypes into components that may be more amenable to genetic studies than the fully expressed psychiatric manifestation of a disease. Thus, the term *endophenotype* was coined to describe an internal phenotype as a conceptual instrument to search for the genetic influence on particular traits contributing to a wider neuropsychiatric syndrome whose mechanistic basis remains otherwise obscure because of the multiplicity of mechanisms that concurrently take place in a spatially and temporally intermingled manner.

Criteria for an endophenotype, distinguishing it from markers or biomarkers, have been defined as follows: (1) associated with illness in the population; (2) heritable; (3) primarily state independent (manifests in an individual whether or not illness is active); (4) also cosegregates with the disease within families; (5) and found in nonaffected family members at a higher rate than in the general population (Gottesman and Gould 2003; Gould and Gottesman 2006). Endophenotypes represent simpler clues to genetic underpinnings than the disease syndrome itself, promoting the view that psychiatric diagnoses can be decomposed or deconstructed. This concept can directly be translated to animal models, suggesting that animal models based on endophenotypes representing evolutionary selectable and quantifiable traits may better lend themselves to the investigation of psychiatric phenomena than models based on face-valid diagnostic phenotypes.

Nevertheless, the term has further evolved over the years and nowadays is often used less strictly with respect to potential genetic traits of the disease but is rather described as a tool to assess a particular attribute of biochemical, endocrinological, neurophysiological, neuroanatomical, cognitive, or neuropsychological nature that resembles with high construct and face validity a precise depressive sign or symptom regardless of its genetic or environmental mechanistic base. Accordingly, other terms with synonymous meanings, such as *intermediate phenotype*, *subclinical trait*, and *vulnerability marker*, have been used interchangeably. However, these terms rather reflect associated findings but not necessarily genetic underpinnings. In this respect, one major caveat in the original definition of endophenotypes is that the term was developed to dissect phenotypes associated to disease-causing genetic mutations or at least to highly penetrating genetic variants. However, such disease-causing genes or genetic variants have not been established with certainty for depression, and most reported genetic associations rather represent common variants of small effects. Until such highly penetrant gene mutations or variants are discovered, discussions about whether a particular phenotype is called *endophenotype* (caused by a genetic change) or *intermediate phenotype* (without explicit causal link) are neither conceptually fundamental nor technically instrumental. Nevertheless, the softening of the concept for the development of animal models of psychiatric diseases relies on the view that psychiatric syndromes can be decomposed in single phenotypes that can be more reliably modeled in animals. In this sense, the endophenotype concept will be used along this chapter.

1.4 CONCEPTUALIZING DIFFERENCES IN BEHAVIORAL ANALYSES OF DEPRESSION: WHAT IS A MODEL, WHAT IS A TEST, AND WHAT IS AN ANTIDEPRESSANT SCREENING PARADIGM?

When discussing about animal models of depression, it is necessary to clearly discriminate, on one hand, between models in a narrower sense, that is, models possessing a predisposition to depression or developing phenotypic features of the disease reflecting a precipitated depression (Willner and Mitchell 2002) and, on the other hand, behavioral tests that are used to assess phenotypic alterations relevant to depression. In addition, many researchers interpret behavioral drug-screening assays as models of depression and thereby add another layer of complexity to the attempt to unambiguously classify different paradigms of models and tests.

1.4.1 DISCRIMINATING BETWEEN ANIMAL MODELS OF DEPRESSION AND BEHAVIORAL TESTS

In the past decades, many neuropsychiatric disease models and tests were created aiming at the analyses of different behavioral readouts in animals. From our perspective, major confusions and obscurities have evolved in the field that persist until today, for example, that animal models of neuropsychiatric diseases and different tests or assays developed to study behavioral traits have been merged and confusingly mixed under the same concept of "models of depression."

With all the drawbacks mentioned above, animal models of depression in the proper meaning of the word are expected to reflect the underlying disease etiology with sufficient construct and face validity and at least some characteristic phenotypes of the disease. Thus, an animal model can be based on an environmental challenge, on a genetic change, or on a combination of both. In contrast, different assays and tests were developed to study changes in different behavioral or biological endophenotypes. In the same manner that a depressed patient would not be confounded with a Hamilton depression scale or a functional magnetic resonance imaging (MRI) study, an animal model of depression should be clearly distinguished from the different readouts employed to analyze or characterize a certain model.

In contrast to the true animal models of depression that will be described in the next section, the battery of assays currently available to analyze different aspects of depression is not solely based on behavioral changes. In fact, neurovegetative, neuroendocrine, neuroanatomical, and probably metabolic changes with high face validity can additionally be assessed in animal models of depression. In the following sections, an overview of currently used tests will be given.

1.4.1.1 Anhedonia- and Reward-Based Tests

Of the putative endophenotypes of major depression, anhedonia—the loss of interest in pleasurable and rewarding activities—is one of the most valid phenotypes within the behavioral dimension (Hasler et al. 2004). Ventral tegmental area dopamine neurons projecting to the prefrontal cortex, basolateral amygdala, and nucleus accumbens are critical for a wide variety of pleasurable and rewarding experiences (Wise 2002). Indeed, rewarding effects of dextroamphetamine in depressive patients suggest the presence of a hypofunction of the dopaminergic systems associated to hedonic responses (Tremblay et al. 2002). Along these lines, deep brain stimulation of the nucleus accumbens has recently been proven to alleviate anhedonia in refractory depressive patients (Schlaepfer et al. 2008).

A lack or decrease of pleasurable responses is typically addressed with *sucrose consumption and preference tests*. One drawback of these tests is that an absolute independence of sensorial, appetitive, or metabolic influences cannot be completely ruled out, although they are seemingly highly consistent in stress-based animal models of depression (Willner 1997). An additional limitation of these tests is the difficulty to define whether a lower consumption or preference of the appetitive cue represents a decreased reward response or, on the contrary, an enhanced sensitivity to reward inputs that elevate the response threshold. A way to partially circumvent these caveats is to use operant behaviors such as *intracranial self-stimulation*, which demands physical effort to receive the stimuli encoding reinforcement value. It should be noted that the test is strongly influenced by motor skills; thus, the potential influence of motor activity or strength on the behavioral endpoint must be carefully controlled. Finally, although it is often argued that anhedonia-based symptoms are not exclusively present in depression, suggesting a low face validity, it has to be emphasized that these tests evaluate a core symptom of depression.

1.4.1.2 Anxiety-Based Tests

Whereas depression is, by definition, a pathological mental condition, anxiety is a normal state of cognitive and behavioral preparedness that an organism mobilizes in response to a future or distant potential threat (Leonardo and Hen 2008). In the context of this definition, it should be noted that there is a cautious distinction between anxiety and fear. Fear behaviors are situation-evoked responses, triggered by explicit, imminent threats and are usually short-lived, evoking intense escape from and avoidance of threats. Anxiety occurs in response to less explicit, more generalized threats and sustains preparedness by increasing long-lasting arousal and risk assessment responses (Blanchard et al. 2003). Whereas anxiety in many settings is instrumental and protective, excessive anxiety can trigger disabling responses that, in time, culminate into anxiety disorders (e.g., generalized anxiety disorder, social phobia, simple phobia, panic disorder, posttraumatic stress disorder, and obsessive compulsive disorder) or emerge as part of a depressive syndrome (Stein and Bienvenu 2009).

Most of the current tests of anxiety are ethologically based "approach–avoidance" tasks that were developed and validated according to their effectiveness using classical anxiolytics. Because of an evolutionary selective pressure on defensive behaviors, rodents have an innate aversion to exposed, well-lit spaces. But the same selective pressure operated also inversely, favoring exploratory behaviors. The convergence of these antagonistic tendencies in a specific conflicting scenario, for example, approaching versus avoiding a potentially dangerous area, is exploited in exploration-based tests to analyze anxiety-like behavior. Different versions of the aversive arena have been implemented: a highly illuminated compartment, in the *dark–light box test*; elevated open arms, in the *elevated plus-maze*; and a central light area in the *open field test* (Lister 1990; Belzung and Griebel 2001). Other less frequently used tests such as the *elevated zero maze*, the *modified hole-board*, the *mirrored chamber test*, and the *staircase test* are based on similar principles (Lister 1990; Belzung and Griebel 2001). A recently developed variant of these tests, the *novelty-induced hypophagia*, in which animals placed in a novel environment show a latency to approach food, provides a further step in terms of predictive validity since it is responsive to chronic, but not acute or subchronic, antidepressant treatments (Dulawa and Hen 2005). The popularity of these tests is also because the evaluation does not require training sessions or complex schedules.

Although the approach–avoidance conflict tests continue to be critical for assaying anxiety-related behavior in rodents, some caveats have to be taken into account to avoid misinterpretation of the test results. First, these tests do not dissociate in a satisfactory manner a decreased anxiety-related response from other phenotypes, such as increased motivation, novelty seeking, or impulsivity, as all of them will manifest as increased time spent in the aversive area. Second, tests based on exploration can be strongly influenced by disparities in basal locomotor activity and cognition. In addition, animals may use different exploratory strategies, which can easily be misinterpreted as changes in anxiety-like behavior.

Different strategies should be considered to minimize the influence of these confounding factors: (1) the parallel use of several principally different anxiety tests; (2) the utilization of internal controls, for instance, measurements of general locomotor activity as distance traveled in the dark–light box or open field tests or the number of entries into the closed arm in the elevated plus-maze; (3) the assessment not only of the main common parameters but also of other more subtle behavioral phenotypes, such as stretched attendance, rearings, period of immobility, number of defecation boli, and locomotor adaptation, in each test.

The limitations of the exploration-based tests highlight the need for new tests and encourage the reconsideration of traditional but presently less prevalent models as the punishment-based conflict paradigms such as *Geller–Seifter test* and *Vogel conflict test*. In the Geller–Seifter test, rats are first trained to press a lever for a food reward and are then subsequently punished by application of mild electric shocks (Hartmann and Geller 1978). The extensive training of the animals (taking months rather than weeks) restricts its usefulness; however, it should also be considered that although the training period is laborious, once trained, the rats can be tested repeatedly over many months. The Vogel conflict paradigm is a more practical test. In its original version, male rats were water-deprived for 48 h and, during a test session of 3 min, drinking was punished by a mild but aversive shock delivered via the spout of the bottle every 20th lick. Accordingly, benzodiazepines attenuate the shock-induced suppression of drinking (Vogel et al. 1971). Subsequently, procedural variants have been introduced, in which the severity of the paradigm has been limited, for example, by reducing deprivation time or by permitting access to water for restricted periods during 4 days before the test (Millan and Brocco 2003).

Finally, the *defensive burying test* is based on the observation that rodents exposed to an electrified probe will bury the probe with sawdust as part of an innate response to prevent further shocks (Treit et al. 1981). The interesting feature of this test is that it is sort of a crossroad between ethologically (exploratory) based and experience (punishment)-based approaches.

1.4.1.3 Cognition-Based Tests

There is no doubt about the relevance of cognitive impairments in depression mostly related to focused attention toward negative stimuli and impaired planning and executive functions. Nonetheless, they are of course not exclusively found in depressive patients and do not belong to the core symptoms of depression. This is the most plausible reason why cognitive tasks are rarely included in the battery of tests for analyzing depression. We propose that this notion be reconsidered in the field, since cognitive alterations represent a very relevant comorbidity in depressed patients, and very well validated tests are available.

Along these lines, general spatial learning and memory tests, such as the *Morris water* or *Y-maze test*, might inform about more general cognitive deficits of hippocampal origin. So far, the models used to study cognitive features influenced by emotion have been focused on Pavlovian conditioning such as *classical fear conditioning* (Maren 2001) and *fear potentiated startle response* (Davis 1990). Pavlovian fear conditioning has been used to study the neural substrates of learning and memory rather than emotion per se. Nevertheless, the brain areas involved in fear conditioning response (prefrontal cortex, hippocampus, and amygdala) are closely superimposed to those circuits implicated in mood and anxiety disorders. Fear conditioning might better model posttraumatic stress disorder, phobias, or other anxiety-related disorders, which are mechanistically closer to fear and extinction memory deficits. Albeit the incorporation of modified schemes such as the reexposition to incomplete or ambiguous representations of the conditioned stimuli (Crestani et al. 1999) might facilitate the evaluation of fear responses in a conflictive situation, what would link this type of test closer to endophenotypes of mood disorders?

Neuropsychological studies on depressed patients have revealed cognitive impairments associated with frontal lobe dysfunction (Murphy et al. 1999; Fossati et al. 1999), including deficits in cognitive flexibility, perseveration, and response inhibition (Murphy et al. 1999; Fossati et al. 1999; Austin et al. 2001). Neuroimaging studies have further revealed an association between cognitive dysfunction in depression and alterations in activity and volume of regions of the prefrontal cortex (Sheline 2003; Price and Drevets 2010). Depressed patients show a narrowing of attentional focus to depression-relevant thoughts and difficulties to shift their cognitive set from one affective dimension to another (Murphy et al. 1999; Austin et al. 2001). Recently, a test to assess shifts in attention between perceptual dimensions of complex stimuli, successfully applied in humans and nonhuman primates, has been adapted and validated in rats: the *attentional set-shifting test* (Birrell and Brown 2000; Bondi et al. 2008). In this test, rats are trained to dig in bowls for a food reward. The bowls are presented in pairs, only one of which is baited. The rat has to select the bowl in which to dig, using as basis the odor of the medium that fills the bowl or the texture that covers its surface. On testing day, rats perform a series of discrimination tasks, including reversals, an intradimensional shift, and an extradimensional shift. Finally, the number of trials required to reach the criterion of six consecutive correct responses in different stages is scored. Most notably, lesions of the medial prefrontal cortex selectively disrupted extradimensional set shifting (Birrell and Brown 2000), and rats subjected to chronic unpredictable stress exhibited impairments in the attentional set-shifting test that were prevented by concurrent treatment with either desipramine or escitalopram (Bondi et al. 2008). Collectively, these results show a consistent construct, face, and predictive validity. Although the attentional set-shifting test protocol is extremely labor intensive, it represents a very interesting means to evaluate prefrontal cortex function on a cognitive trait relevant for depression.

Because of the importance of cognitive symptoms in mood disorders, we anticipate that, in the years to come, more cognitive tests will be included in the schedules for testing emotional behavior. A more systematic utilization of these tests will certainly help define in which cases or conditions other endophenotypes of depression cosegregate or are associated with cognitive impairments.

1.4.1.4   Neurovegetative and Neuroendocrine Changes

One frequently observed alteration that can be observed with the onset of major depressive symptoms is a *variation in body weight* (Hasler et al. 2004). These changes seem to be biologically plausible given that different monoamines and peptides mechanistically linked to depression, such as norepinephrine, dopamine, corticotropin-releasing hormone (CRH), and melanin-concentrating hormone, play important roles in the regulation of appetite, energy balance, and body weight (Nestler et al. 2002). However, both substantial loss and gain of weight are *DSM-IV* criteria used in the diagnosis of depression. This ambiguity complicates the analyses of putative mechanisms or conditions that might influence body weight in depression. The situation is further exacerbated by the fact that most depression models are based on different regimens of chronic stress that inherently implies high peripheral levels of hormones such as glucocorticoids and adrenaline. The direct effects of these hormones on metabolism are difficult to dissociate from the putative stress-induced changes in the central nervous system (CNS), which might be ultimately responsible for affecting metabolism. These obstacles might plausibly explain why body weight changes are usually not included in the battery of tests evaluating depression-related endophenotypes. Nevertheless, as with other endophenotypes for affective disorders, this scenario is gradually changing. For instance, animals that were subjected to chronic social defeat stress and then fed a diet high in triglycerides and cholesterol consistently displayed a metabolic syndrome characterized by weight gain as well as by insulin and leptin resistance (Chuang et al. 2010).

Another test related to neurovegetative symptoms is the *stress-induced hyperthermia* test, which measures the increase in body temperature elicited by an aversive event (Olivier et al. 2003). Stress-induced hyperthermia can be antagonized by benzodiazepines and 5-HT1A receptor agonists, but not by selective 5-HT reuptake inhibitors, and it provides a simple readout of autonomic reactivity to acute mild stress.

More than 90% of depressed patients complain about *impairments of sleep quality* (Hasler et al. 2004), and polysomnographic recordings show disinhibited rapid eye movement (REM) sleep and a decreased slow-wave sleep (Benca et al. 1992; Thase et al. 1997). Accordingly, animals chronically overexpressing CRH in forebrain structures showed constantly increased REM sleep (Kimura et al. 2010), whereas antidepressant medications with a different mechanism of action, such as paroxetine and tianeptine, have shown a pronounced inhibition of REM sleep in human patients (Murck et al. 2003). Through simultaneous monitoring of electroencephalographic and electromyographic activity, classification of a particular vigilance state as awake, non-REM sleep, or REM sleep can be achieved. It should be noted that these procedures are very demanding and require considerable expertise, not only for the surgical implantation of the probes and the recordings, but also for the subsequent analysis of data. Nevertheless, these procedures provide a fairly unique evaluation of an endophenotype.

As stated above, the genetic blueprint (nature) and the biographical impact of the environmental challenges (nurture) interact in every individual to determine either the resistance (resilience) or the predisposition to develop affective disorders. The *hypothalamic–pituitary–adrenal (HPA) axis* is a key controller of endocrine and behavioral adaptation to stress and as such plays a pivotal role in this gene–environment dialogue (Holsboer 2000; de Kloet et al. 2005a). The hypothalamic CRH has a neuroendocrine function acting as a secretagogue stimulating the secretion of adrenocorticotropic hormone from the pituitary, which will induce the synthesis and release of glucocorticoids. Additionally, this neuropeptide is also widely expressed throughout the brain, where it integrates stress-adaptation engrams that comprise many different behavioral programs and, by these, controls sleep architecture, appetite, sexual drive, and learning and memory, as well as locomotor activity and anxiety.

Clinical and preclinical studies have demonstrated a hyperactive HPA axis in a significant proportion of depressive patients, which is most likely produced by molecular impairments in negative feedback mechanisms exerted by glucocorticoid receptors on so far poorly characterized limbic areas of the brain (Holsboer 2000). This phenomenon produces an increment in the CRH activity that, on one hand, renders a hyperdrive of the HPA axis and, on the other hand, an imbalance in many of the extrahypothalamic functions controlled by CRH in the CNS. Thus, many of the observations found in clinical and preclinical studies might in fact be used as neurological/neuroendocrine readouts of depression: increased levels of corticosteroids (cortisol in humans and corticosterone in rodents), high levels of CRH in cerebrospinal fluid of depressive patients, and decreased CRH binding in the prefrontal cortex in suicide victims. In addition, a blunted adrenocorticotropic hormone response to exogenously administered CRH has been demonstrated, suggesting desensitized CRH receptors secondary to hypothalamic CRH hypersecretion (Nemeroff et al. 1984; Holsboer 1999; Holsboer 2000; Nemeroff and Vale 2005; de Kloet et al. 2005a).

1.4.2 DISCRIMINATING ANTIDEPRESSANT SCREENING PARADIGMS FROM ANIMAL MODELS OF DEPRESSION OR BEHAVIORAL TESTS

Another issue of controversy when discussing animal models of depression is the consideration of tests originally designed to screen monoamine-based antidepressant drugs as depression models (Cryan et al. 2005b). This is, in particular, the case for the Porsolt forced swim test (FST) and the tail suspension test (Porsolt et al. 1977; Steru et al. 1985).

The FST takes advantage of the observation that rodents, after initial escape-oriented movements, rapidly adopt a characteristic immobile posture in an inescapable cylinder filled with water. In this paradigm, immobility is often interpreted as a passive stress–coping strategy or depression-like behavior (behavioral despair) (Cryan et al. 2005a). However, this test is conducted in normal animals, with a strong acute stress that triggers passive versus active responses and where acute administration of antidepressant drugs decreases the time the animal remains immobile. In addition, these tests are essentially acute stress–based paradigms, and in consequence, every candidate compound, environmental challenge, or targeted gene affecting sensorial input, information processing, and stress-coping strategies will severely influence the final outcome of the test. A prominent example of this situation is the role of CRH, indisputably one of the systems whose mechanistic role in depression is better documented. Several lines of evidence emerging from preclinical and clinical research indicate that a hyperdriven CRH system is present in depressed patients (Holsboer 1999). Indeed, CRH receptor type 1 antagonists have been proven to diminish depressive symptoms with similar efficacy as paroxetine (Holsboer and Ising 2008). In fact, the central application of CRH or CRH receptor type 1 and the overexpression of CRH in the CNS result in a decrease in floating time in the FST (Butler et al. 1990; van Gaalen et al. 2002; Tezval et al. 2004; Lu et al. 2008). Should this be interpreted as an antidepressant response although previous work has clearly demonstrated that high levels of CRH are rather pro-depressive? The most arguable interpretation is that the CRH/CRH receptor type 1 system, which is a master regulator of stress response, also accounts for adjusting the response triggered by the test. Finally, there is an additional hermeneutic problem contained in the interpretation of the immobility posture, which was originally coined by Porsolt as "behavioral despair" under the assumption that animals "gave up hope of escaping," but this might instead reflect a successful strategy that conserves energy and allows the animal to float for prolonged periods, thereby improving its chances for survival (De Pablo et al. 1989; West 1990). This line of argumentation claims that immobility is largely dependent on learning and memory and that it would represent "learning to be immobile" rather than "behavioral despair."

In conclusion, we believe that these tests are without doubt very useful tools to analyze strategies of stress-coping behavior and monoamine-based antidepressant compounds (Lucki 1997), but they are simply not models for depression.

1.5 ANIMAL MODELS OF DEPRESSION IN A NARROWER SENSE

In general, most animal models of depression are based either on affecting the environment that becomes stressful and challenging for the animal or on affecting the animal's perception and processing of environmental stimuli by manipulating sensory and integrative functions of the brain. Animal models of depression have been established in various species, including hamsters, voles, tree shrews, primates, mice, and rats, but the rat has clearly been favored as the species of choice until the advent and propagation of genetic engineering tools in the mouse.

1.5.1 PHARMACOLOGICAL MODELS

The first pharmacological models of antidepressant-like activity significantly contributed to the coining of the term "monoamine theory of depression," which assumes that an elevation of serotonin and norepinephrine levels will improve depression-related symptoms.

Reserpine, an antihypertensive and antipsychotic drug, is capable of nonselectively depleting brain monoamines and thereby inducing a syndrome of locomotor hypomotility and reducing body temperature in rodents. Of note is the observed dysphoric effect of reserpine in patients (Bein 1978), which suggests that the syndrome observed in animals may have some equivalence to the human condition (O'Neil and Moore 2003). The reserpine model is based on the capacity of anti-depressants to reverse the inhibitory effects of reserpine on motility in rats and mice (Nutt 2006). These types of models offer good predictive validity in terms of monoamine-based antidepressant activity; however, their ability to model core symptoms of depression is limited and most of the pharmacological approaches are not models of depression in the sense of the criteria discussed previously.

A pharmacological model that is sensitive to antidepressants and displays characteristic symptoms of depression is the psychostimulant withdrawal paradigm. In humans, withdrawal from chronic psychostimulants produces symptoms that have strong behavioral and physiological parallels to depression, including diminished interest or pleasure. Therefore, the examination of the behavioral effects of amphetamine withdrawal in rodents has the potential to provide insights into the underlying neurobiological mechanisms. Animal models based on the withdrawal of psychostimulants have been established successfully. After withdrawal from drugs such as amphetamine or cocaine, rodents display behavioral changes that are highly analogous to some aspects of depression in humans, such as reward deficits, including elevations in brain reward thresholds (Barr and Markou 2005).

1.5.2 LESION MODELS

Lesion models are based on the assumption that depression is caused by regulatory deficits in neuronal circuits. The olfactory bulbectomy in mice and rats results in a disruption of the limbic–hypothalamic axis, with the consequence of behavioral, neurochemical, neuroendocrine, and neuroimmune alterations, of which many resemble changes seen in depressed patients (Song and Leonard 2005). The behavioral syndrome connected to olfactory bulbectomy is largely thought to result from compensatory mechanisms of neuronal reorganization. The observed behavioral alterations are clearly not just a consequence of loss of smell since peripheral anosmia does not produce a comparable syndrome (Mar et al. 2000). Such reorganizations involve changes in synaptic strength and/or loss of spine density in various limbic brain regions, including the amygdala and the hippocampus. Olfactory bulbectomy has predominantly been applied in rats, but studies in mice are increasingly available. In rats, it has been shown that there are marked changes in all major neurotransmitter systems after bulbectomy. The most consistent behavioral change of bulbectomy is a hyperactive response in the open field paradigm, which is reversed by antidepressant treatment (Cryan et al. 1999). Interestingly, this model of depression shows a high face validity as it mimics the slow onset of antidepressant action reported in clinical studies (Willner and Mitchell 2002). Unlike stress-related models, the bulbectomized rat represents an agitated, hyposerotonergic depression-related phenotype rather than a retarded depression (Lumia et al. 1992).

1.5.3 STRESS MODELS

Exposure to stress or to traumatic life events has a strong impact on the manifestation of depression, suggesting an impairment of proper stress-coping strategies in depressed patients (Kessler 1997; de Kloet et al. 2005a). Therefore, depression is also regarded as a stress-related disorder, and accordingly, most of the animal models of depression are based on the exposure to various types of acute or chronic stressors. These paradigms are capable of producing phenotypic changes reminiscent of symptoms of depression, which are at least to some extent reversible by antidepressant treatment. However, it is clear that stress per se is not sufficient to cause depression, and it has to be kept in mind that most people do not become depressed after serious stressful experiences but rather develop a kind of depression after stressors that most people would perceive as rather mild. In fact, severe stressors typically do not induce depression but instead cause posttraumatic stress disorder (Nestler et al. 2002).

1.5.3.1 Early Life Stress Models

Early life experience, particularly stress, infection, and maternal care, significantly programs the brain to confer vulnerability or resilience (Viltart and Vanbesien-Mailliot 2007; Bale et al. 2010). For instance, traumatic life events in childhood have been shown to result in an increased sensitivity to the effects of stress later in life and alter the individual's vulnerability to stress-related psychiatric disorders such as depression (Graham et al. 1999; Heim and Nemeroff 2001). Models of early life stress involve prenatal stress, early postnatal handling, and maternal separation. All paradigms have been demonstrated to produce significant effects, including behavioral and neuroendocrine phenotypes, that last until adulthood. Many of the induced alterations are reversible by antidepressant treatment. The most widely used model is the maternal separation paradigm of early life deprivation, in which pups are separated from the dam for 1–24 h per day during the first 2 postnatal weeks. This paradigm results in increased anxiety- and depression-like behaviors as well as an increased HPA axis response that can still be observed in adulthood (de Kloet et al. 2005b). It is noteworthy that shorter periods of separation tend to produce opposite effects compared with maternal separation. This handling effect is most likely attributed to qualitative changes in maternal care after separation (Meaney 2001). The exact psychological nature of the effects of postnatal maternal separation is not fully understood; nevertheless, the paradigm has demonstrated its value for studying the neurobiological basis of the impact of early life stress on emotional behavior (Holmes et al. 2005). Another recently developed model of early life stress omits the separation from the mother and thereby avoids undesirable interference common to most deprivation models, such as metabolic disturbances, exhaustion, or hypothermia of the pups. This novel early life stress model, which results in acute but also long-lasting consequences, is evoked via fragmented maternal care, generated by reducing the amount of nesting and bedding material available to the dam (Rice et al. 2008).

1.5.3.2 Adolescent and Adult Stress Models

An animal model of depression, which more closely incorporates disease etiology and predisposition, is the learned-helplessness paradigm. This paradigm is based on the observation that animals develop deficits in escape, cognitive, and rewarded behaviors (sucrose preference) when they have been subjected to repeated unavoidable and uncontrollable shocks (Seligman et al. 1975). The validity of the paradigm for depression-like behavior has been demonstrated by the sensitivity of these behavioral deficits to antidepressants. Unfortunately, the paradigm does not parallel clinical settings with regard to the slow onset of antidepressant action, that is, antidepressants are usually effective already after short-term treatment. As for most of the available models of depression, particularly in mice, the effects are only short lived and the animals recover a few days after cessation of uncontrollable shocks. Additionally, for yet unknown reasons, only a few of the animals develop signs of helplessness symptoms reflecting potential epigenetic phenomena related to gene–environment interactions. Finally, it has to be noted that the learned-helplessness paradigm is not fully selective for antidepressants since anxiolytics also have been demonstrated to revert the behavioral phenotype (Cryan and Mombereau 2004).

The chronic mild stress paradigm has been established as a model that is sensitive to chronic antidepressant treatment and emphasizes the predominant role of stress in the etiological cause of depression (Willner 1997). It involves the exposure of animals to a series of mild and unpredictable stressors (e.g., isolation or crowded housing, deprivation of food or water, disruption of dark–light cycle, tiling of home cages, and dampened bedding) for at least 2 weeks. Chronic mild stress has been reported to result in long-lasting changes in behavioral, neurochemical, neuroimmune, and neuroendocrinological parameters resembling alterations observed in depressed patients. The chronic mild stress paradigm interferes with brain reward functions, including decreased sucrose preference and intracranial self-stimulation, reflecting anhedonia reversed by chronic, but not acute, antidepressant treatment (Monleon et al. 1995). Although chronic mild stress has been hampered by poor interlaboratory reliability (Schweizer et al. 2009), successful application for rat and mouse studies have been reported repeatedly.

The above procedures do not take into account the etiology of human stress-related disorders, as most of the stressful stimuli in humans that have been shown to increase the risk of psychiatric disorders are of social nature (Brown and Prudo 1981). Attributional and social cognitive processes are unique human factors; however, other social dimensions such as poor interindividual relations and reaction to social and environmental stress can also be examined in animals (O'Neil and Moore 2003). Social stress is a chronic and recurring factor in the lives of virtually all higher animal species. Humans experiencing social defeat show increased symptoms of depression, loneliness, anxiety, social withdrawal, and loss of self-esteem (Bjorkqvist 2001). Therefore, social conflict models in animals have gained increasing attention, as they might render it useful to study certain endophenotypes of depression (Huhman 2006). Both dyadic and group social stress paradigms have been established. In the dyadic resident–intruder paradigm, an animal is repeatedly exposed to an aggressive and dominant conspecific with or without physical contact (Berton et al. 2006; Tsankova et al. 2006). The group social paradigms rely either on social instability, by setting up social groups and later mixing them, or on social disruption, where a selected highly aggressive male is introduced into a stable social group. The observation that different windows of stress vulnerability are probably existing across lifespan is also of importance (Lupien et al. 2009). Experiencing stress in these time windows will have a significant impact on further development; thus, an important variable in stress-based animal models of depression is the age at which the animals are exposed to the stressor. Adolescence has been demonstrated as a period of particular vulnerability, probably due to the still ongoing hormonal and neurodevelopmental processes (Schmidt et al. 2007). Several of these paradigms lead to a complex syndrome of behavioral and physiological changes. Stressed animals are reported to exhibit symptoms of anhedonia as well as various cardiovascular, metabolic, and neuroendocrine alterations, which could be reversed by long-term antidepressant treatment in some studies (Blanchard et al. 2001; Schmidt et al. 2007).

1.5.4 Genetic Models

Genetic approaches offer both the possibility to identify determinants of depression and the possibility to systematically generate models that reflect endophenotypes of depression or that possess a predisposition to develop these endophenotypes. In general, geneticists discriminate between forward and reverse genetic approaches. Forward genetics allows the identification of relevant genes without any prior knowledge of genetic or mechanistic underpinnings of a phenotype of interest. Classically, forward genetics involves large-scale random mutagenesis screens such as *N*-ethyl-*N*-nitrosourea- or gene trap–based approaches, which have resulted in a great number of mutants displaying alterations in endophenotypes of depression or antidepressant-like behavior (Bucan and Abel 2002). In contrast, reverse genetics is a gene-driven approach involving genetic manipulations that in their extremes result in loss- or gain-of-function mutants. However, among the many mutant mouse lines generated by forward or reverse genetics, only very few, if any, can be considered a valid genetic depression model, but rather a model of predisposition to depression (Cryan and Mombereau 2004).

1.5.4.1 Selective Breeding

A commonly used strategy in forward genetics is selective breeding, which has been successfully applied in rats and mice. Selective breeding takes advantage of differences in the individual responsiveness to paradigms relevant to depression-related endophenotypes. The breeding program usually begins with the evaluation of the trait of interest in a genetically heterogeneous population of outbred mice or rats. Individuals with responses at either extreme of the response curve are then selectively bred together for their opposing trait phenotypes over multiple generations. Prominent examples are, for instance, the Rouen "depressed" mice, which have been bred for high and low immobility levels in the FST (Scott et al. 1996; El Yacoubi ct al. 2003), or rats that display a higher susceptibility to learned helplessness compared to nonhelplessness rats (Edwards et al. 1999). Another recent example is mice selectively bred for high versus low reactivity of the HPA axis (Touma et al. 2008). These mice nicely reflect the dysfunctions of the HPA axis, which has been shown to play an important role in the etiology and pathophysiology of depression. Moreover, lines with different HPA axis reactivity might serve as models for particular subtypes of depression.

1.5.4.2 Genetic Engineering

Since the early times of classical transgenesis and gene targeting, the toolbox has significantly been expanded and refined by introducing sophisticated conditional strategies (Figure 1.1). This progress allows an increasingly refined control of spatial and temporal gene expression. In particular, genetically modified mice have proven to be an invaluable tool in dissecting complex molecular interactions and in studying the specific function of a single gene in the context of a whole organism, bearing the potential to identify and characterize putative novel pharmacological targets. Genetically engineered mice have been successfully used to validate hypotheses illuminating the etiology of depression (Urani et al. 2005). Mouse mutants targeting different components of the monoaminergic system, including receptors, transporters, and key enzymes involved in the synthesis and degradation of monoamines, have underscored their commonly accepted role in the pathophysiology of depression (Gingrich and Hen 2001). The neurotrophin hypothesis of depression has also been addressed in the mouse by targeting neurotrophic factors such as brain-derived neurotrophic factor and respective receptors. These studies have indicated that brain-derived neurotrophic factor may be essential for mediating aspects of antidepressant efficacy (Duman and Monteggia 2006). The generation and analysis of numerous constitutive and conditional mouse mutants affecting different constituents of the HPA axis is another example where genetically modified mice have demonstrated their enormous potential. The genetic dissection of the organism's major stress-integrating system in the mouse has confirmed a major role for corticosteroid receptors and the CRH system in the pathogenesis of affective disorders, including depression (Gass et al. 2001; Deussing and Wurst 2005; Muller and Holsboer 2006). Moreover, these preclinical studies have significantly contributed to establish the glucocorticoid receptor and the CRH receptor type 1 as promising molecular targets for a novel class of antidepressants (Holsboer 1999; Grigoriadis 2005; Debattista and Belanoff 2006; Holsboer and Ising 2008).

Finally, human genetic studies, including candidate gene or genome-wide association studies, will identify an increasing number of heritable quantitative traits that are risk factors for complex diseases such as depression. Only genetic engineering in the mouse provides the opportunity to model and validate these susceptibility loci in vivo. With a few exceptions (Lucae et al. 2006), the vast majority of single nucleotide polymorphisms that have been associated with mood disorders are located within noncoding or intergenic regions. With the combination of genome-wide association studies with next-generation sequencing technologies, it can be anticipated that more single-nucleotide polymorphisms will be found, and among them are novel nonsynonymous coding single-nucleotide polymorphisms. The intramolecular consequence of the substitution of a single amino acid depends on the primary sequence and the three-dimensional structure of the protein. In this context, species specificity might play an important role because single-point mutations found in humans might only have a role within the human protein itself but not on the same protein of another species. Thus, the only way to address this problem is to study the influence of a given

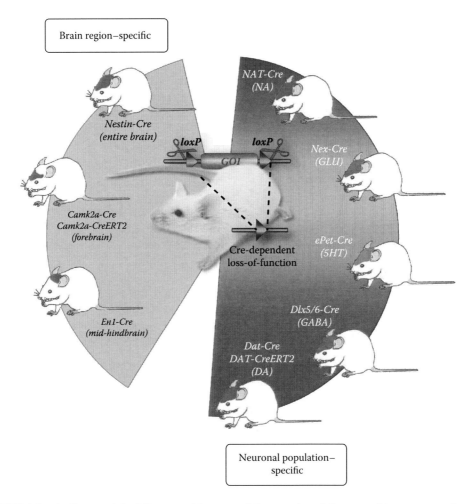

Brain region–specific

NAT-Cre
(NA)

Nestin-Cre
(entire brain)

loxP   loxP

GOI

Nex-Cre
(GLU)

Camk2a-Cre
Camk2a-CreERT2
(forebrain)

ePet-Cre
(5HT)

Cre-dependent
loss-of-function

En1-Cre
(mid-hindbrain)

Dlx5/6-Cre
(GABA)

Dat-Cre
DAT-CreERT2
(DA)

Neuronal population–
specific

FIGURE 1.1 A glimpse of the "Cre-recombinase zoo." Propagation of Cre-recombinase transgenic mouse lines allows addressing of gene function in a spatially restricted manner. *Nestin-Cre* mice (from Tronche, F., et al., *Nat Genet*, 23, 99–103, 1999, with permission) allow for gene targeting in the entire brain, whereas *Camk2a-Cre/CreERT2* (from Erdmann, G., et al., *BMC Neurosci*, 8, 63, 2007; Minichiello, L., et al., *Neuron*, 24, 401–14, 1999, with permission) and *En1-Cre* (from Kimmel, R.A., et al., *Genes Dev*, 14, 1377–89, 2000, with permission) mice confer specificity to forebrain or mid-hindbrain structures, respectively. Mouse lines expressing Cre-recombinase selectively in neurons of a specific neurotransmitter identity allow for gene targeting of specific populations of neurons. Moreover, increasing availability of mouse lines expressing tamoxifen-inducible Cre-recombinase variant CreERT2 offers additional temporal control and avoids obscurities due to developmental functions of targeted genes. *Dat-Cre/CreERT2* (Lemberger et al. 2007; Engblom et al. 2008); DA, dopamine; *Dlx5/6-Cre* (Monory et al. 2006); GABA, γ-aminobutyric acid; *Nex-Cre* (Goebbels et al. 2006); GLU, glutamate; *ePet-Cre* (Scott et al. 2005); 5HT, serotonin; *Nat-Cre* (http://www.gensat.org/cre.jsp); NA, noradrenaline; GOI, gene of interest.

single-nucleotide polymorphism on protein function by genetically engineering mice to entirely substitute the mouse protein by a human version. This strategy has been used in immunology and toxicology where humanized versions of specific antigens or cytochrome P450 enzymes were reintroduced via classical transgenic procedures into the genome of knockout mice lacking the murine version of the protein (Nishie et al. 2007; Cheung and Gonzalez 2008).

1.5.5 PREDICTIVE VALIDITY IN ANIMAL MODELS OF DEPRESSION: WHEN AND HOW ANTIDEPRESSANTS ARE ADMINISTERED MAKE A DIFFERENCE

The predictive or pharmacological validity implies that a model responds to antidepressant drugs in a way that resembles the effects of these treatments in humans. Thus, any model addressing potential mechanisms of antidepressant action should exhibit a similar time dependency as observed in humans, that is, a lag of several weeks for the onset of therapeutic effects. This goal is far from being modeled in animals, but another similar criterion has been adopted as a critical requirement in depression models: the behavioral alteration should be reverted by chronic, but not acute, administration of antidepressants. Therefore, the pharmacological validation of a model of depression should involve the comparison of acute and chronic treatments. This is a very sensitive issue mainly because, as discussed, antidepressants induce rapid effects in many behavioral tests such as the FST. In this regard, we want to emphasize that the acute effects of antidepressants should be evaluated in control groups at different time points, since pharmacokinetic and pharmacodynamic parameters are variable (e.g., when comparing different drugs, species, and strains), making it very difficult to predict a priori the precise time point when a single administration of the drug produces the maximal response. Accordingly, the selection of a single (and potentially wrong) time point for the behavioral evaluation can lead to the erroneous conclusion that a certain acute treatment does not produce any effect. For instance, in some chronic social defeat protocols, avoidance behavior was measured in defeated mice after acute (1 day after 27 days of saline) or chronic (1 day after 28 days) administration of imipramine or fluoxetine (Berton et al. 2006; Tsankova et al. 2006). Pharmacokinetic principles of accumulation and biodisponibility of drugs in the CNS suggest that the concentration of a bioactive drug cannot be the same 24 h after 28 daily injections of antidepressants compared with 24 h after a single injection, where perhaps even only traces of the drug are available in the brain. Thus, several other time points should be evaluated when looking for a predictive validation of a new test.

Another important consideration with respect to the temporal courses of antidepressant treatments is related to the onset and duration of the administration of the compound. In many models based on chronic stress paradigms, the consequences of the stress—the "inducer" of the depression-like behavior—is not lasting and dissipates soon after the stress ceases. This has forced researchers to coapply drugs (preventive component) and stress (inducer component) in parallel over the time. An obvious question arises at this point. Are drugs reverting misbehaviors already induced and established by stress, or are they just preventing the effects or perception of stress? Many times and because of the intrinsic features of the model, this question cannot be answered in a satisfactory manner. Nonetheless, if the coadministration of drug and "inducer" (e.g., chronic stress) cannot be avoided, one important way to minimize these obscuring factors would be to start the experiment with a rather long initial "drug-free" period of stress, which should be set up after the initiation of the pharmacological treatment. Precisely in this regard, a recent report took advantage of the persistence of the learned-helplessness model in rats to analyze the amelioration of behavioral deficits and synaptic changes induced by fluoxetine as well as its recurrence 3 months after fluoxetine withdrawal (Reines et al. 2008). Several features regarding face and predictive validity warrant emphasis in the context of the paradigm shown by these researchers. First, antidepressants are administered entirely after the application of stress and in already learned-helplessness animals. Second, fluoxetine is administered for 3 weeks and influences escape latencies only in learned-helplessness rats, but not in controls. Third, despair behavior reappears after pharmacological treatment is interrupted, which demonstrates that this might also be a suitable model to study relapse, a phenomenon totally underestimated at the current state of the art of modeling depression (Reines et al. 2008). All these features to a great extent mimic the clinically meaningful observations in humans and might inspire future studies attempting to develop better models of depression.

Similarly, Nestler and colleagues previously showed long-lasting effects of a 10-day chronic social defeat protocol on behavioral avoidance. In this case, chronic antidepressant treatment reverted the emotional effects of chronic stress when applied after (and not during) the defeat (Berton et al. 2006; Tsankova et al. 2006).

1.6 UPSURGE OF THE MOUSE

For decades, research in rodents has been critical for the advancement of our understanding of the neural systems mediating emotion, their dysfunctions, and how they can be therapeutically modulated. Historically, the rat was the species of choice in basic neuroscience research because rats perform well in many different cognitive and operant tasks. Furthermore, the body size and robustness of the rat versus the mouse, for instance, allows a better application of invasive techniques, such as electrodes and cannula implantation into the brain. Nonetheless, an explosion in the utilization of mice in neuropsychiatric research was and is driven by the development of a plethora of molecular and cellular tools, which allowed the development of targeted genetic manipulation strategies in mouse embryonic stem cells (Mak 2007). Although genetic engineering is now also possible in the rat and even in higher mammals (see below), it is only the mouse that has been amenable to these techniques for the past 20 years. The creation of genetic tools allowing the spatial and temporal control of gene expression (Lewandoski 2001), particularly the ingenious application of the *Cre/loxP* system in mouse genetics (Gu et al. 1993; Branda and Dymecki 2004), has certainly established mouse mutants as unique models in studying the influence of definite genes on brain physiology from neuronal function to complex behaviors (Figure 1.1). Furthermore, mice have practical and economic advantages: they are easy to breed, can be housed in large numbers, and have short generation times (3 months from being born to giving birth). Thus, the proliferation of mutant variants (transgenic, knockout, and knockin mice) with unique characteristics has stimulated researchers to adopt the use of mice in their batteries of behavioral tests.

1.6.1 AN OLD BUT STILL VALID STATEMENT: "A MOUSE IS NOT A SMALL RAT"

The maxim "a mouse is not a small rat" is as relevant in depression tests as in other behavioral paradigms. Rats and mice are ethologically different species with particular behavioral peculiarities. Historically, most of the currently used behavioral tests were initially developed, characterized, and validated in rats. Although there is an upsurge in the use of mice, which is basically driven by embryonic stem cells available as tools for genetic manipulation, necessary changes are required to translate tests from one species to the other, beyond simply reducing the arena size. For instance, many emotional and cognitive tests are based on swimming or navigation of rodents in inescapable or escapable tanks. But the inherent aversion to water ethologically present in mice is not present in rats, which might strongly influence the final interpretation of results. Along these lines, the response to chronic stress in mice and rats also strongly differs between these species. These differences might explain why several researchers have found difficulties in translating chronic stress models to mice, such as learned helplessness or chronic mild stress, which were well established and validated in rats (Willner 2005; Schweizer et al. 2009).

1.6.2 IMPORTANCE OF STRAIN BACKGROUND

Several studies have clearly described pronounced behavioral, neurovegetative, and neuroendocrine differences during anxiety- and depression-like behaviors among distinct mouse strains, including inbred (C57BL/6, A/J, BALB/c, DBA/2, C3H, or 129) and outbred strains (Swiss Webster and NMRI) (Anisman et al. 2001; Lucki et al. 2001; Ohl et al. 2001). These differences are particularly important when analyzing behavior in mutant mouse lines that most of the time carry different genetic backgrounds.

In these cases, two important cautions should be taken into account. First, the strain and therefore the genetic background onto which a mutation is backcrossed can affect the expression of behavioral readouts. For instance, DBA/2 mice show higher and C57BL/6 lower anxiety levels in multistrain comparisons (Anisman et al. 2001; Lucki et al. 2001; Ohl et al. 2003). Second, the genetic variability in nonbackcrossed lines might concomitantly increase the variation between behaviorally tested animals, thus further affecting statistical analysis. Backcrossing takes a long time, and often behavior in conditional mutant animals is studied in nonbackcrossed mice. Although new C57BL/6-derived embryonic stem cell lines recently have been generated and propagated (Pettitt et al. 2009), many floxed mice were formerly developed from 129-derived embryonic stem cells and germline transmitted to C57BL/6 backgrounds, rendering mixed 129/BL6 backgrounds. In addition, after crossing with the Cre lines, an additional contribution of that genetic background to the offspring is added. Since the contribution of every background is stochastically transmitted to the littermates, this will produce a high interindividual variation in the behavioral tests, which are variable per se (Crabbe et al. 1999; Wahlsten et al. 2003, 2006). To avoid these obstacles, backcrossing strategies should be pursued from early time points of the generation of a mutant mouse line.

1.6.3 Toward a Functional Annotation of the Mouse Genome

The sequencing of the human and mouse genomes has transformed the landscape of mammalian biology. The completion of these projects has determined a fairly accurate picture of the nature and content of protein-coding genes in these two mammalian genomes (Hayashizaki and Carninci 2006; Maeda et al. 2006). These analyses have reinforced the high levels of similarity between the two genomes; approximately 99% of the mouse genes have homologues in the human genome. Thus, different research institutions quickly identified the necessity of undertaking a comprehensive functional annotation of the mouse genome to understand how transformations in genetic networks lead to disease. The goal is the generation of a comprehensive database of molecular interventions in vivo via the generation of a collection of embryonic stem cell lines carrying site-directed genetic modifications. The means was the utilization of two independent techniques to genetically manipulate embryonic stem cells: gene trapping and gene targeting. In recent years, different scientific organizations have sponsored consortia to generate gene trapping–based embryonic stem cell lines for public use. The International Gene Trap Consortium launched in 2005 is composed of different consortia and laboratories, with the goal of integrating all publicly available gene trap information (Nord et al. 2006). At present, the International Gene Trap Consortium (http://www.genetrap.org/) offers about 430,000 publicly available gene trap cell lines from different resources covering about 12,000 genes, which are available on a noncollaborative basis for nominal handling fees.

The U.S.-based Knockout Mouse Project and Texas A&M Institute of Genomic Medicine, the European Conditional Mouse Mutagenesis Program, and the North American Conditional Mouse Mutagenesis Project in Canada joined together in the International Knockout Mouse Consortium, which aims at generating more than 40,000 targeted and gene-trapped embryonic stem cell lines (Austin et al. 2004; Wurst 2005; Friedel et al. 2007; Collins et al. 2007; Beckers et al. 2009). On the other hand, more than 900 new mouse mutant lines were derived from these embryonic stem cell resources and are currently available (http://www.knockoutmouse.org/). Thus, it is feasible to speculate that soon almost each mouse gene will be hit by at least one type of mutation. This fantastic resource is available to the whole scientific community.

1.7 FUTURE DIRECTIONS

Although we still lack a deeper understanding of the key molecules involved in the basic processes underlying depression, the currently increasing availability of genetic mouse models from large-scale mouse mutagenesis projects (http://www.knockoutmouse.org/) clearly demands that major efforts be directed toward the combination of genetic modifications and environmental challenges in the same subject, simulating the gene–environment interactions that more plausibly reflect the pathophysiological mechanisms of depression. Bridging the gap between the genetic composition and environment will allow researchers to address more specifically the question of vulnerability and resilience (Figure 1.2). Another risk factor often overlooked in animal models of depression is the gender difference. Women have a higher prevalence for stress-related disorders, including depression, than men. It is clear that social and cultural factors contribute to these differences; however, neurobiological differences in brain anatomy, chemistry, and function as well as in stress and drug response are involved in this gender difference (Goel and Bale 2009; Dalla et al. 2010). Depression is a chronic disorder with treatment options that are still far from optimal. Although up to 80% show partial responses, only about 50% of patients show full remission. In particular, patients who fail to reach full remission are threatened with relapse, a circumstance worth considering in future animal models of depression.

A frontier question in neuroscience in general and in behavioral neurobiology in particular is: Which are the neuronal subpopulations and specific circuits underlying mechanisms of complex brain diseases (Luo et al. 2008)? Without a doubt, the possibility of answering this type of question would present a giant step in our search for the mechanistic basis of depression. So far, the technical challenge has been to selectively activate or inactivate particular genes on specific subsets of neurons that are densely intermingled with millions of others. In recent years, hundreds of floxed alleles have been generated in the mouse through the efforts of independent laboratories or research consortia (see before). In parallel, a "zoo" of Cre recombinase driver lines has been created (e.g., NIH Neuroscience Blueprint Cre-driver network, http://www.credrivermice.org or Cre-X-Mice, http://nagy.mshri.on.ca/cre_new/index.php or) where the expression of the Cre recombinase is controlled by a specific promoter, which allows the deletion of a floxed allele with a unique spatial and temporal profile (Figure 1.1). As a proof of principle, this approach was recently used to genetically dissect the contribution of the cannabinoid receptor 1 on glutamatergic versus GABAergic neurons to epileptogenic circuits in the hippocampus (Monory et al. 2006).

In a similar direction, in the future optogenetics will offer a very interesting alternative to exogenously activate and inhibit specific neurons with an amazing temporal resolution. Although technically demanding, the combination of *genetics*, which allows the targeted expression of light-activated channels and enzymes in specific neuronal populations, and *optics*, which allows the manipulation

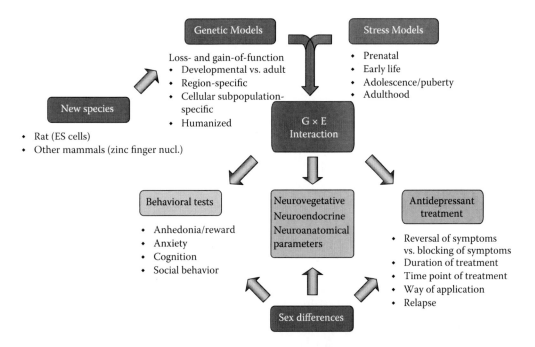

FIGURE 1.2 Modeling gene × environment interaction (G × E) by combining genetic and stress models.

of neural activity with millisecond precision, promises to provide amazing mechanistic information about circuits and behavior (Deisseroth et al. 2006; Miesenbock 2009). Another novel tool for regulating the activity of neuronal circuits is modified G-protein–coupled receptors known as *receptors activated solely by synthetic ligands* and *designer receptors exclusively activated by a designer drug* (Nichols and Roth 2009). These are G-protein–coupled receptors modified through either rational design or directed molecular evolution that functionally respond only to synthetic ligands. This technology, combined with transgenic approaches to selectively target expression, has the ability to regulate the activity of specific neurons or neuronal circuits through effector pathway modulation to study function and behavior (Nichols and Roth 2009).

Finally, two recent technological breakthroughs have the potential to revolutionize the development of genetic animal models of depression and to overcome the leading role of the mouse. On one hand is the establishment of embryonic stem cells of the rat and their successful application to produce the first knockout rats (Tong et al. 2010). This will sooner or later bring the rat back into the focus of depression research since it opens up the road to any desired genetic manipulations of the rat genome as already experienced in the mouse. On the other hand, zinc finger nucleases will not only enable the generation of knockout rats but also ultimately break the barrier to genetic engineering in virtually all mammals (Geurts et al. 2009; Jacob et al. 2010; van Boxtel and Cuppen 2010).

REFERENCES

Agarwal N., Port J. D., Bazzocchi M., and Renshaw P. F. 2010. Update on the use of MR for assessment and diagnosis of psychiatric diseases. *Radiology* 255:23–41.

Anisman H., Hayley S., Kelly O., Borowski T., and Merali Z. 2001. Psychogenic, neurogenic, and systemic stressor effects on plasma corticosterone and behavior: Mouse strain-dependent outcomes. *Behav Neurosci* 115:443–54.

Austin C. P., Battey J. F., Bradley A., et al. 2004. The knockout mouse project. *Nat Genet* 36:921–24.

Austin M. P., Mitchell P., and Goodwin G. M. 2001. Cognitive deficits in depression: Possible implications for functional neuropathology. *Br J Psychiatry* 178:200–6.

Bale T. L., Baram T. Z., Brown A. S., et al. 2010. Early life programming and neurodevelopmental disorders. *Biol Psychiatry* 68:314–9.

Barr A. M. and Markou A., 2005. Psychostimulant withdrawal as an inducing condition in animal models of depression. *Neurosci Biobehav Rev* 29:675–706.

Beckers J., Wurst W., and de Angelis M. H. 2009. Towards better mouse models: Enhanced genotypes, systemic phenotyping and envirotype modelling. *Nat Rev Genet* 10:371–80.

Bein H. J. 1978. Prejudices in pharmacology and pharmacotherapy: Reserpine as a model for experimental research in depression. *Pharmakopsychiatr Neuropsychopharmakol* 11:289–93.

Belzung C., and Griebel G. 2001. Measuring normal and pathological anxiety-like behaviour in mice: A review. *Behav Brain Res* 125:141–9.

Benca R. M., Obermeyer W. H., Thisted R. A., and Gillin J. C. 1992. Sleep and psychiatric disorders. A meta-analysis. *Arch Gen Psychiatry* 49:651–68.

Berton O., McClung C. A., DiLeone R. J., et al. 2006. Essential role of BDNF in the mesolimbic dopamine pathway in social defeat stress. *Science* 311:864–8.

Berton O., and Nestler E. J. 2006. New approaches to antidepressant drug discovery: Beyond monoamines. *Nat Rev Neurosci* 7:137–51.

Birrell J. M., and Brown V. J. 2000. Medial frontal cortex mediates perceptual attentional set shifting in the rat. *J Neurosci* 20:4320–4.

Bjorkqvist K. 2001. Social defeat as a stressor in humans. *Physiol Behav* 73:435–42.

Blanchard D. C., Griebel G., and Blanchard R. J. 2003. The mouse defense test battery: Pharmacological and behavioral assays for anxiety and panic. *Eur J Pharmacol* 463:97–116.

Blanchard R. J., McKittrick C. R., and Blanchard D. C. 2001. Animal models of social stress: Effects on behavior and brain neurochemical systems. *Physiol Behav* 73:261–71.

Bondi C. O., Rodriguez G., Gould G. G., Frazer A., and Morilak D. A. 2008. Chronic unpredictable stress induces a cognitive deficit and anxiety-like behavior in rats that is prevented by chronic antidepressant drug treatment. *Neuropsychopharmacology* 33:320–31.

Bosker F. J., Hartman C. A., Nolte I. M., et al. 2010. Poor replication of candidate genes for major depressive disorder using genome-wide association data. *Mol Psychiatry* doi:10.1038/mp.2010.38.

Branda C. S., and Dymecki S. M. 2004. Talking about a revolution: The impact of site-specific recombinases on genetic analyses in mice. *Dev Cell* 6:7–28.

Brown G. W., and Prudo R. 1981. Psychiatric disorder in a rural and an urban population: 1. Aetiology of depression. *Psychol Med* 11:581–99.

Brown S. D., Chambon P., and de Angelis M. H. 2005. EMPReSS: Standardized phenotype screens for functional annotation of the mouse genome. *Nat Genet* 37:1155.

Bucan M., and Abel T. 2002. The mouse: Genetics meets behaviour. *Nat Rev Genet* 3:114–23.

Butler P. D., Weiss J. M., Stout J. C. and Nemeroff C. B. 1990. Corticotropin-releasing factor produces fear-enhancing and behavioral activating effects following infusion into the locus coeruleus. *J Neurosci* 10:176–83.

Caspi A., Sugden K., Moffitt T. E., et al. 2003. Influence of life stress on depression: Moderation by a polymorphism in the 5-HTT gene. *Science* 301:386–9.

Cheung C., and Gonzalez F. J. 2008. Humanized mouse lines and their application for prediction of human drug metabolism and toxicological risk assessment. *J Pharmacol Exp Ther* 327:288–99.

Chuang J. C., Krishnan V., Yu H. G., et al. 2010. A beta3-adrenergic-leptin-melanocortin circuit regulates behavioral and metabolic changes induced by chronic stress. *Biol Psychiatry* 67:1075–82.

Collins F. S., Finnell R. H., Rossant J., and Wurst W. 2007. A new partner for the international knockout mouse consortium. *Cell* 129:235.

Crabbe J. C., Wahlsten D., and Dudek B. C. 1999. Genetics of mouse behavior: Interactions with laboratory environment. *Science* 284:1670–2.

Crestani F., Lorez M., Baer K., et al. 1999. Decreased GABAA-receptor clustering results in enhanced anxiety and a bias for threat cues. *Nat Neurosci* 2:833–9.

Cryan J. F., and Mombereau C. 2004. In search of a depressed mouse: Utility of models for studying depression-related behavior in genetically modified mice. *Mol Psychiatry* 9:326–57.

Cryan J. F., McGrath C., Leonard B. E., and Norman T. R. 1999. Onset of the effects of the 5-HT1A antagonist, WAY-100635, alone, and in combination with paroxetine, on olfactory bulbectomy and 8-OH-DPAT-induced changes in the rat. *Pharmacol Biochem Behav* 63:333–8.

Cryan J. F., Mombereau C., and Vassout A. 2005a. The tail suspension test as a model for assessing antidepressant activity: Review of pharmacological and genetic studies in mice. *Neurosci Biobehav Rev* 29:571–625.

Cryan J. F., Valentino R. J., and Lucki I. 2005b. Assessing substrates underlying the behavioral effects of antidepressants using the modified rat forced swimming test. *Neurosci Biobehav Rev* 29:547–69.

Dalla C., Pitychoutis P. M., Kokras N., and Papadopoulou-Daifoti Z. 2010. Sex differences in animal models of depression and antidepressant response. *Basic Clin Pharmacol Toxicol* 106:226–33.

Davis M. 1990. Animal models of anxiety based on classical conditioning: The conditioned emotional response (CER) and the fear-potentiated startle effect. *Pharmacol Ther* 47:147–65.

de Kloet E. R., Joels M., and Holsboer F. 2005a. Stress and the brain: From adaptation to disease. *Nat Rev Neurosci* 6:463–75.

de Kloet E. R., Sibug R. M., Helmerhorst F. M., and Schmidt M. V. 2005b. Stress, genes and the mechanism of programming the brain for later life. *Neurosci Biobehav Rev* 29:271–81.

De Pablo J. M., Parra A., Segovia S., and Guillamon A. 1989. Learned immobility explains the behavior of rats in the forced swimming test. *Physiol Behav* 46:229–37.

Debattista C., and Belanoff J. 2006. The use of mifepristone in the treatment of neuropsychiatric disorders. *Trends Endocrinol Metab* 17:117–21.

Deisseroth K., Feng G., Majewska A. K., et al. 2006. Next-generation optical technologies for illuminating genetically targeted brain circuits. *J Neurosci* 26:10380–6.

Deussing J. M., and Wurst W. 2005. Dissecting the genetic effect of the CRH system on anxiety and stress-related behaviour. *C R Biol* 328:199–212.

Dranovsky A., and Hen R. 2006. Hippocampal neurogenesis: Regulation by stress and antidepressants. *Biol Psychiatry* 59:1136–43.

Dulawa S. C., and Hen R. 2005. Recent advances in animal models of chronic antidepressant effects: The novelty-induced hypophagia test. *Neurosci Biobehav Rev* 29:771–83.

Duman R. S., and Monteggia L. M. 2006. A neurotrophic model for stress-related mood disorders. *Biol Psychiatry* 59:1116–27.

Edwards E., King J. A., and Fray J. C. 1999. Increased basal activity of the HPA axis and renin-angiotensin system in congenital learned helpless rats exposed to stress early in development. *Int J Dev Neurosci* 17:805–12.

El Yacoubi M., Bouali S., Popa D., et al. 2003. Behavioral, neurochemical, and electrophysiological characterization of a genetic mouse model of depression. *Proc Natl Acad Sci USA* 100:6227–32.

Engblom D., Bilbao A., Sanchis-Segura C., et al. 2008. Glutamate receptors on dopamine neurons control the persistence of cocaine seeking. *Neuron* 59:497–508.

Erdmann G., Schutz G. and Berger S. 2007. Inducible gene inactivation in neurons of the adult mouse forebrain. *BMC Neurosci* 8:63.

Fava M., and Kendler K. S. 2000. Major depressive disorder. *Neuron* 28:335–41.

Fossati P., Amar G., Raoux N., Ergis A. M., and Allilaire J. F. 1999. Executive functioning and verbal memory in young patients with unipolar depression and schizophrenia. *Psychiatry Res* 89:171–87.

Friedel R. H., Seisenberger C., Kaloff C., and Wurst W. 2007. EUCOMM—the European conditional mouse mutagenesis program. *Brief Funct Genomic Proteomic* 6:180–5.

Gass P., Reichardt H. M., Strekalova T., Henn F., and Tronche F. 2001. Mice with targeted mutations of glucocorticoid and mineralocorticoid receptors: Models for depression and anxiety? *Physiol Behav* 73:811–25.

Geurts A. M., Cost G. J., Freyvert Y., et al. 2009. Knockout rats via embryo microinjection of zinc-finger nucleases. *Science* 325:433.

Geyer M. A., and Markou, A. 1995. Animal models of psychiatric disorders. In *Psychopharmacology: The Fourth Generation of Progress*, eds F.E. Bloom and D.J. Kupfer, 777–98. New York: Raven Press.

Gingrich J. A., and Hen R. 2001. Dissecting the role of the serotonin system in neuropsychiatric disorders using knockout mice. *Psychopharmacology (Berl)* 155:1–10.

Goebbels S., Bormuth I., Bode U., et al. 2006. Genetic targeting of principal neurons in neocortex and hippocampus of NEX-Cre mice. *Genesis* 44:611–21.

Goel N., and Bale T. L. 2009. Examining the intersection of sex and stress in modelling neuropsychiatric disorders. *J Neuroendocrinol* 21:415–20.

Gottesman I. I., and Gould T. D. 2003. The endophenotype concept in psychiatry: Etymology and strategic intentions. *Am. J Psychiatry* 160:636–45.

Gould T. D., and Gottesman I. I. 2006. Psychiatric endophenotypes and the development of valid animal models. *Genes Brain Behav* 5:113–19.

Graham Y. P., Heim C., Goodman S. H., Miller A. H., and Nemeroff C. B. 1999. The effects of neonatal stress on brain development: Implications for psychopathology. *Dev Psychopathol* 11:545–65.

Grigoriadis D. E. 2005. The corticotropin-releasing factor receptor: A novel target for the treatment of depression and anxiety-related disorders. *Expert Opin Ther Targets* 9:651–84.

Gu H., Zou Y. R., and Rajewsky K. 1993. Independent control of immunoglobulin switch recombination at individual switch regions evidenced through Cre-loxP-mediated gene targeting. *Cell* 73:1155–64.

Hartmann R. J., and Geller I. 1978. Effects of brofoxine, a new anxiolytic, on experimentally induced conflict in rats. *Proc West Pharmacol Soc* 21:51–5.

Hasler G., Drevets W. C., Manji H. K., and Charney D. S. 2004. Discovering endophenotypes for major depression. *Neuropsychopharmacology* 29:1765–81.

Hayashizaki Y., and Carninci P. 2006. Genome Network and FANTOM3: Assessing the complexity of the transcriptome. *PLoS Genet* 2:e63.

Heim C., and Nemeroff C. B. 2001. The role of childhood trauma in the neurobiology of mood and anxiety disorders: Preclinical and clinical studies. *Biol Psychiatry* 49:1023–39.

Holmes A., le Guisquet A. M., Vogel E., et al. 2005. Early life genetic, epigenetic and environmental factors shaping emotionality in rodents. *Neurosci Biobehav Rev* 29:1335–46.

Holsboer F. 1999. The rationale for corticotropin-releasing hormone receptor (CRH-R) antagonists to treat depression and anxiety. *J Psychiatr Res* 33:181–214.

Holsboer F. 2000. The corticosteroid receptor hypothesis of depression. *Neuropsychopharmacology* 23:477–501.

Holsboer F., and Ising M. 2008. Central CRH system in depression and anxiety—evidence from clinical studies with CRH1 receptor antagonists. *Eur J Pharmacol* 583:350–7.

Huhman K. L. 2006. Social conflict models: Can they inform us about human psychopathology? *Horm Behav* 50:640–6.

Jacob H. J., Lazar J., Dwinell M. R., Moreno C., and Geurts A. M. 2010. Gene targeting in the rat: Advances and opportunities. *Trends Genet* doi:10.1016/j.tig.2010.08.006.

Kessler R. C. 1997. The effects of stressful life events on depression. *Annu Rev Psychol* 48:191–214.

Kimura M., Muller-Preuss P., Lu A., et al. 2010. Conditional corticotropin-releasing hormone overexpression in the mouse forebrain enhances rapid eye movement sleep. *Mol Psychiatry* 15:154–65.

Kimmel R. A., Turnbull D. H., Blanquet V., et al. 2000. Two lineage boundaries coordinate vertebrate apical ectodermal ridge formation. *Genes Dev* 14:1377–89.

Leboyer M., Bellivier F., Nosten-Bertrand M., et al. 1998. Psychiatric genetics: Search for phenotypes. *Trends Neurosci* 21:102–5.

Lemberger T., Parlato R., Dassesse D., et al. 2007. Expression of Cre recombinase in dopaminoceptive neurons. *BMC Neurosci* 8:4.

Leonardo E. D., and Hen R. 2008. Anxiety as a developmental disorder. *Neuropsychopharmacology* 33:134–40.

Lesch K. P. 2004. Gene–environment interaction and the genetics of depression. *J Psychiatry Neurosci* 29:174–84.

Lewandoski M. 2001. Conditional control of gene expression in the mouse. *Nat Rev Genet* 2:743–55.

Lister R. G. 1990. Ethologically-based animal models of anxiety disorders. *Pharmacol Ther* 46:321–40.

Lu A., Steiner M. A., Whittle N., et al. 2008. Conditional mouse mutants highlight mechanisms of corticotropin-releasing hormone effects on stress-coping behavior. *Mol Psychiatry* 13:1028–42.

Lucae S., Salyakina D., Barden N., et al. 2006. P2RX7, a gene coding for a purinergic ligand-gated ion channel, is associated with major depressive disorder. *Hum Mol Genet* 15:2438–45.

Lucki I. 1997. The forced swimming test as a model for core and component behavioral effects of antidepressant drugs. *Behav Pharmacol* 8:523–32.

Lucki I., Dalvi A., and Mayorga A. J. 2001. Sensitivity to the effects of pharmacologically selective antidepressants in different strains of mice. *Psychopharmacology (Berl)* 155:315–22.

Lumia A. R., Teicher M. H., Salchli F., Ayers E., and Possidente B. 1992. Olfactory bulbectomy as a model for agitated hyposerotonergic depression. *Brain Res* 587:181–5.

Luo L., Callaway E. M., and Svoboda K. 2008. Genetic dissection of neural circuits. *Neuron* 57:634–60.

Lupien S. J., McEwen B. S., Gunnar M. R., and Heim C. 2009. Effects of stress throughout the lifespan on the brain, behaviour and cognition. *Nat Rev Neurosci* 10:434–45.

Maeda N., Kasukawa T., Oyama R., et al. 2006. Transcript annotation in FANTOM3: Mouse gene catalog based on physical cDNAs. *PLoS Genet* 2:e62.

Mak T. W. 2007. Gene targeting in embryonic stem cells scores a knockout in Stockholm. *Cell* 131:1027–31.

Malhi G. S., Moore J., and McGuffin P. 2000. The genetics of major depressive disorder. *Curr Psychiatry Rep* 2:165–9.

Mar A., Spreekmeester E., and Rochford J. 2000. Antidepressants preferentially enhance habituation to novelty in the olfactory bulbectomized rat. *Psychopharmacology (Berl)* 150:52–60.

Maren S. 2001. Neurobiology of Pavlovian fear conditioning. *Annu Rev Neurosci* 24:897–931.

Mathers C. D., and Loncar D. 2006. Projections of global mortality and burden of disease from 2002 to 2030. *PLoS Med* 3:e442.

McKinney W. T., and Bunney W. E. 1969. Animal model of depression. I. Review of evidence: Implications for research. *Arch Gen Psychiatry* 21:240–8.

Meaney M. J. 2001. Maternal care, gene expression, and the transmission of individual differences in stress reactivity across generations. *Annu Rev Neurosci* 24:1161–92.

Miesenbock G. 2009. The optogenetic catechism. *Science* 326:395–9.

Millan M. J., and Brocco M. 2003. The Vogel conflict test: Procedural aspects, gamma-aminobutyric acid, glutamate and monoamines. *Eur J Pharmacol* 463:67–96.

Minichiello L., Korte M., Wolfer D., et al. 1999. Essential role for TrkB receptors in hippocampus-mediated learning. *Neuron* 24:401–14.

Monleon S., D'Aquila P., Parra A. et al. 1995. Attenuation of sucrose consumption in mice by chronic mild stress and its restoration by imipramine. *Psychopharmacology (Berl)* 117:453–7.

Monory K., Massa F., Egertova M., et al. 2006. The endocannabinoid system controls key epileptogenic circuits in the hippocampus. *Neuron* 51:455–66.

Muglia P., Tozzi F., Galwey N. W., et al. 2010. Genome-wide association study of recurrent major depressive disorder in two European case-control cohorts. *Mol Psychiatry* 15:589–601.

Muller M.B., and Holsboer F. 2006. Mice with mutations in the HPA-system as models for symptoms of depression. *Biol Psychiatry* 59:1104–15.

Murck H., Nickel T., Kunzel H., et al. 2003. State markers of depression in sleep EEG: Dependency on drug and gender in patients treated with tianeptine or paroxetine. *Neuropsychopharmacology* 28:348–58.

Murphy F. C., Sahakian B. J., Rubinsztein J. S., et al. 1999. Emotional bias and inhibitory control processes in mania and depression. *Psychol Med* 29:1307–21.

Nemeroff C. B., and Vale W. W. 2005. The neurobiology of depression: Inroads to treatment and new drug discovery. *J Clin Psychiatry* 66 (Suppl 7):5–13.

Nemeroff C. B., Widerlov E., Bissette G., et al. 1984. Elevated concentrations of CSF corticotropin-releasing factor-like immunoreactivity in depressed patients. *Science* 226:1342–4.

Nestler E. J., Barrot M., DiLeone R. J., et al. 2002. Neurobiology of depression. *Neuron* 34:13–25.

Nichols C. D., and Roth B. L. 2009. Engineered G-protein coupled receptors are powerful tools to investigate biological processes and behaviors. *Front Mol Neurosci* 2:16.

Nishie W., Sawamura D., Goto M., et al. 2007. Humanization of autoantigen. *Nat Med* 13:378–83.

Nord A. S., Chang P. J., Conklin B. R. et al. 2006. The International Gene Trap Consortium Website: A portal to all publicly available gene trap cell lines in mouse. *Nucleic Acids Res* 34:D642–48.

Nutt D. J. 2006. The role of dopamine and norepinephrine in depression and antidepressant treatment. *J Clin Psychiatry* 67 (Suppl 6):3–8.

O'Neil M. F., and Moore N. A. 2003. Animal models of depression: Are there any? *Hum Psychopharmacol* 18:239–54.

Ohl F., Roedel A., Binder E., and Holsboer F. 2003. Impact of high and low anxiety on cognitive performance in a modified hole board test in C57BL/6 and DBA/2 mice. *Eur J Neurosci* 17:128–36.

Ohl F., Sillaber I., Binder E., Keck M. E., and Holsboer F. 2001. Differential analysis of behavior and diazepam-induced alterations in C57BL/6N and BALB/c mice using the modified hole board test. *J Psychiatr Res* 35:147–54.

Olivier B., Zethof T., Pattij T., et al. 2003. Stress-induced hyperthermia and anxiety: Pharmacological validation. *Eur J Pharmacol* 463:117–32.

Petronis A. 2010. Epigenetics as a unifying principle in the aetiology of complex traits and diseases. *Nature* 465:721–7.

Pettitt S. J., Liang Q., Rairdan X. Y., et al. 2009. Agouti C57BL/6N embryonic stem cells for mouse genetic resources. *Nat Methods* 6:493–5.

Porsolt R. D., Le P. M., and Jalfre M. 1977. Depression: A new animal model sensitive to antidepressant treatments. *Nature* 266:730–2.

Price J. L., and Drevets W. C. 2010. Neurocircuitry of mood disorders. *Neuropsychopharmacology* 35:192–216.

Reines A., Cereseto M., Ferrero A., et al. 2008. Maintenance treatment with fluoxetine is necessary to sustain normal levels of synaptic markers in an experimental model of depression: Correlation with behavioral response. *Neuropsychopharmacology* 33:1896–908.

Renthal W., and Nestler E. J. 2009. Chromatin regulation in drug addiction and depression. *Dialogues Clin Neurosci* 11:257–68.

Rice C. J., Sandman C. A., Lenjavi M. R., and Baram T. Z. 2008. A novel mouse model for acute and long-lasting consequences of early life stress. *Endocrinology* 149:4892–900.

Schlaepfer T. E., Cohen M. X., Frick C., et al. 2008. Deep brain stimulation to reward circuitry alleviates anhedonia in refractory major depression. *Neuropsychopharmacology* 33:368–77.

Schmidt M. V., Sterlemann V., Ganea K., et al. 2007. Persistent neuroendocrine and behavioral effects of a novel, etiologically relevant mouse paradigm for chronic social stress during adolescence. *Psychoneuroendocrinology* 32:417–29.

Schweizer M. C., Henniger M. S., and Sillaber I. 2009. Chronic mild stress (CMS) in mice: Of anhedonia, 'anomalous anxiolysis' and activity. *PLoS One* 4:e4326.

Scott M. M., Wylie C. J., Lerch J. K., et al. 2005. A genetic approach to access serotonin neurons for in vivo and in vitro studies. *Proc Natl Acad Sci USA* 102:16472–7.

Scott P. A., Cierpial M. A., Kilts C. D., and Weiss J. M. 1996. Susceptibility and resistance of rats to stress-induced decreases in swim-test activity: A selective breeding study. *Brain Res* 725:217–30.

Seligman M. E., Rosellini R. A., and Kozak M. J. 1975. Learned helplessness in the rat: Time course, immunization, and reversibility. *J Comp Physiol Psychol* 88:542–7.

Sheline Y. I. 2003. Neuroimaging studies of mood disorder effects on the brain. *Biol Psychiatry* 54:338–52.

Shields J., and Gottesman I. I. 1972. Cross-national diagnosis of schizophrenia in twins. The heritability and specificity of schizophrenia. *Arch Gen Psychiatry* 27:725–30.

Song C., and Leonard B. E. 2005. The olfactory bulbectomised rat as a model of depression. *Neurosci Biobehav Rev* 29:627–47.

Stein M. B., and Bienvenu O. J. 2009. Diagnostic classification of anxiety disorders: *DSMV* and beyond. In *Neurobiology of Mental Illness*, eds. D.S. Charney and E.J. Nestler, 525–34. New York: Oxford University Press.

Steru L., Chermat R., Thierry B., and Simon P. 1985. The tail suspension test: A new method for screening antidepressants in mice. *Psychopharmacology (Berl)* 85:367–70.

Tarantino L. M. and Bucan M. 2000. Dissection of behavior and psychiatric disorders using the mouse as a model. *Hum Mol Genet* 9:953–65.

Tezval H., Jahn O., Todorovic C., et al. 2004. Cortagine, a specific agonist of corticotropin-releasing factor receptor subtype 1, is anxiogenic and antidepressive in the mouse model. *Proc Natl Acad Sci USA* 101:9468–73.

Thase M. E., Kupfer D. J., Fasiczka A. J., et al. 1997. Identifying an abnormal electroencephalographic sleep profile to characterize major depressive disorder. *Biol Psychiatry* 41:964–73.

Tong C., Li P., Wu N. L., Yan Y., and Ying Q. L. 2010. Production of p53 gene knockout rats by homologous recombination in embryonic stem cells. *Nature* 467:211–13.

Touma C., Bunck M., Glasl L., et al., 2008. Mice selected for high versus low stress reactivity: A new animal model for affective disorders. *Psychoneuroendocrinology* 33:839–62.

Treit D., Pinel J. P., and Fibiger H. C. 1981. Conditioned defensive burying: A new paradigm for the study of anxiolytic agents. *Pharmacol Biochem Behav* 15:619–26.

Tremblay L. K., Naranjo C. A., Cardenas L., Herrmann N., and Busto U. E. 2002. Probing brain reward system function in major depressive disorder: Altered response to dextroamphetamine. *Arch Gen Psychiatry* 59:409–16.

Tronche F., Kellendonk C., Kretz O., et al. 1999. Disruption of the glucocorticoid receptor gene in the nervous system results in reduced anxiety. *Nat Genet* 23:99–103.

Tsankova N. M., Berton O., Renthal W., et al. 2006. Sustained hippocampal chromatin regulation in a mouse model of depression and antidepressant action. *Nat Neurosci* 9:519–25.

Urani A., Chourbaji S., and Gass P. 2005. Mutant mouse models of depression: Candidate genes and current mouse lines. *Neurosci Biobehav Rev* 29:805–28.

van Boxtel R., and Cuppen E. 2010. Rat traps: Filling the toolbox for manipulating the rat genome. *Genome Biol* 11:217.

van Gaalen M. M., Stenzel-Poore M. P., Holsboer F., and Steckler T. 2002. Effects of transgenic overproduction of CRH on anxiety-like behaviour. *Eur J Neurosci* 15:2007–15.

Viltart O., and Vanbesien-Mailliot C. C. 2007. Impact of prenatal stress on neuroendocrine programming. *Sci World J* 7:1493–537.

Vogel J. R., Beer B., and Clody D. E. 1971. A simple and reliable conflict procedure for testing anti-anxiety agents. *Psychopharmacologia* 21:1–7.

Wahlsten D., Bachmanov A., Finn D. A., and Crabbe J. C. 2006. Stability of inbred mouse strain differences in behavior and brain size between laboratories and across decades. *Proc Natl Acad Sci USA* 103:16364–9.

Wahlsten D., Metten P., Phillips T. J., et al. 2003. Different data from different labs: Lessons from studies of gene–environment interaction. *J Neurobiol* 54:283–311.

West A. P. 1990. Neurobehavioral studies of forced swimming: The role of learning and memory in the forced swim test. *Prog. Neuropsychopharmacol. Biol Psychiatry* 14:863–77.

Willner P. 1984. The validity of animal models of depression. *Psychopharmacology (Berl)* 83:1–16.

Willner P. 1997. Validity, reliability and utility of the chronic mild stress model of depression: A 10-year review and evaluation. *Psychopharmacology (Berl)* 134:319–29.

Willner P. 2005. Chronic mild stress (CMS) revisited: Consistency and behavioural-neurobiological concordance in the effects of CMS. *Neuropsychobiology* 52:90–110.

Willner P., and Mitchell P. J. 2002. The validity of animal models of predisposition to depression. *Behav Pharmacol* 13:169–88.

Wise R. A. 2002. Brain reward circuitry: Insights from unsensed incentives. *Neuron* 36:229–40.

World Health Organization. 2008. *The global burden of disease: 2004 update*. WHO: Geneva.

Wurst W. 2005. Mouse geneticists need European strategy too. *Nature* 433:13.

2 Neurophysiological and Neuropsychological Models of Depression

Stephan Moratti, Alberto Fernández, and Gabriel Rubio

CONTENTS

2.1 INTRODUCTION

Depression is the most commonly occurring disorder of emotion regulation, affecting approximately 5% to 9% of women and 2% to 3% of men (American Psychiatric Association 1994). Depression is an affective disorder that manifests not only affective symptoms such as anhedonia, feelings of sadness, and low emotional arousal (Heller and Nitschke 1997; Loas et al. 1994) but also cognitive impairments in higher-order domains such as visual spatial attention, memory, and executive functions (Heller and Nitschke 1997).

In the past decade, there has been a growing interest in applying models of emotion and motivation emanating from basic research to the study of psychopathology. Several of the most influential models posit higher-order dimensions that reflect the operation of biologically based systems that organize and activate responses to rewarding, appetitive, or otherwise positive hedonic stimuli and to aversive, threatening, or otherwise negative hedonic stimuli (Tomarken et al., in press). Such models are particularly relevant to unipolar depression in the light of the proposal that this disorder is characterized by a combination of a hypoactivation in a biologically based approach system that mediates responses to appetitive stimuli and hyperactivation of a protective–defensive withdrawal system that mediates responses to aversive, threatening, or otherwise stressful stimuli (Fowles 1988).

Two of the most relevant models in this field have investigated the relationship between depression and positive and negative affects (Rottenberg et al. 2005) and also the association between emotional functioning and cortical brain asymmetries in subjects with affective disorders (Heller and Nitschke 1997).

2.1.1 DEPRESSION AND EMOTION-CONTEXT SENSITIVITY

On self-report measures that assess mood states and traits, unipolar depressed, compared with nondepressed, individuals consistently report (1) lower positive affect, lower behavioral activation, and increased anhedonia and (2) heightened negative affect, behavioral inhibition, and generalized distress (Kasch et al. 2002). On measures of reactivity to emotional stimuli, however, the story is less clear and often quite different. Depressed individuals consistently demonstrate deficits in the processing of and responses to positive hedonic stimuli on self-report (Sloan et al. 2001), behavioral (Henriques and Davidson 2000), and psychophysiological (Dichter et al. 2004) measures. However, the evidence concerning reactions to negative affective stimuli is more equivocal. Whereas a few studies have found exaggerated responses to aversive or other negative stimuli (Sigmon and Nelson-Gray 1992), others have not and/or have found inconsistent results across dependent measures (Yee and Miller 1988; Rottenberg et al. 2002). Furthermore, within-subject comparisons often indicate that depressed individuals are less responsive to variations in the affective valence of eliciting stimuli than are nondepressed individuals (Rottenberg et al. 2005). On the basis of these data, Rottenberg et al. (2005) and Rottenberg (2007) have argued that depression is associated with a broad emotion context insensitivity (ECI). They have argued that this view is consistent with evolutionary accounts of depression that conceptualize this syndrome as characterized by disengagement and a bias against action (Nesse 2000). Affective startle modulation during picture viewing is also well established. When nondepressed individuals view affective pictures and the latency between picture and startle probe onsets is relatively long (e.g., 3500 ms), response magnitude is modulated by the valence of the picture. Unpleasant pictures potentiate and pleasant pictures attenuate the magnitude of the startle blink relative to neutral pictures (Bradley et al. 1993). This linear pattern of startle modulation is thought to reflect the priming of neurobiologically based defensive and appetitive systems by unpleasant and pleasant foreground stimuli, respectively (Lang et al. 1998). However, when the latency between picture and startle probe onsets is relatively short (e.g., 300 ms), the magnitude of the eyeblink reflex is quadratic (i.e., both unpleasant and pleasant foreground stimuli attenuate blink magnitudes relative to neutral stimuli), possibly reflecting heightened allocation of attentional resources to the affective stimuli (Bradley et al. 1993; Dichter et al. 2004).

The ECI hypothesis can be contrasted with two other theories: the positive attenuation hypothesis and the negative potentiation hypothesis (Rottenberg et al. 2005). The positive attenuation hypothesis states that positive emotion is decreased in depression (i.e., anhedonia). The negative potentiation hypothesis states that depression is associated with high negative emotion. In their study of these three competing theories of depression, Rottenberg et al. (2005) found more support for the ECI hypothesis than for either the positive attenuation hypothesis or the negative potentiation hypothesis.

2.1.2 Cortical Brain Asymmetries and Depression

This model proposes that less relative left than right frontal brain activity is associated with reduced positive/increased negative mood in depression. Greater left frontal activity leads to positive valenced emotions, whereas greater right frontal activity is related to negative affect. Relative left frontal hypoactivation in depression has been related not only to low positive affect but also to cognitive deficits such as poor problem solving, memory problems, and tasks involving effortful processing (Heller and Nitschke 1997). Furthermore, right temporoparietal hypoactivation in depressive patients is associated with low emotional arousal and impairments of right posterior hemisphere functions such as emotional face processing (Deldin et al. 2000) and sustained attention even during remission (Weiland-Fiedler et al. 2004). However, right temporoparietal activity is supposed to covary with comorbid anxiety. High levels of comorbid anxiety lead to a right temporoparietal hyperactivation as emotional arousal increases. Therefore, this model suggests that comorbid anxiety has to be carefully controlled in studies investigating brain activity asymmetries in depression, as different levels of anxiety can result in opposing effects. However, over the years, the model proposed by Heller et al. has been refined. For example, anxious apprehension is now differentiated from anxious arousal, and the two have been associated with different brain asymmetry patterns. Interestingly, the Heller model adopts a dimensional approach of emotion that considers that all emotions can be described along a valence and an arousal dimension that are specifically related to frontal and posterior brain regions. This is in line with emotion research based on a motivated attention framework (Lang and Bradley 2010) that also considers emotions to be organized along valence and arousal dimensions. However, the motivated attention framework assumes that valence (pleasant vs. unpleasant) represents the activation of appetitive and defensive motivational circuits and that arousal is associated with the intensity of the activation of one of the motivational systems. This accords with the notion that frontal asymmetries in brain activity reflect approach and withdrawal motivational tendencies rather than direct emotional valence (Harmon-Jones et al. 2010). Hereby, left frontal cortex activity is associated with approach behavior that usually provokes positive affect, whereas right frontal activation reflects withdrawal behavior that is accompanied by negative emotional experience (Davidson 2003).

To clarify the most relevant findings published in the field of neurophysiological measures [such as startle, electroencephalography (EEG), and magnetoencephalography (MEG)] in depression, the following will be presented in this chapter: (1) major depression (MD) studies based on startle reflex; (2) studies focused on neurophysiological measures such as EEG and MEG; (3) an overview of new perspectives of EEG and MEG analysis, such as coherence, synchronization, and complexity in depression.

2.2 STARTLE REFLEX AND MD

The magnitude of the startle reflex may be an endophenotypic marker for vulnerability to recurrent depression. The startle reflex, which is widely used in the investigation of the physiology of affect and attention, consists of a cascade of physical reactions to an intense stimulus with sudden onset, for example, a loud noise. One of the earliest and most stable components is the eyeblink response. This is a burst of electromyographic activity in the orbicularis oculi muscle associated with the presentation of the startle stimulus (Andreassi 2000). Variation in the affective and attentional modulation of the eyeblink startle reflex is associated with a number of psychopathologies (Grillon and Baas 2003). In this paradigm, compared with neutral stimuli, the startle reflex is augmented during viewing of unpleasant pictures and inhibited when viewing pleasant pictures.

2.2.1 Affective Startle Modulation: Comparisons between Depressed and Nondepressed Individuals

Studies that have investigated affective modulation of the startle response in people with depression (Allen et al. 1999; Dichter et al. 2004; Forbes et al. 2005; Kaviani et al. 2004) have found conflicting results. The most common finding has been a lack of affective startle modulation among depressed patients (Allen et al. 1999; Dichter et al. 2004; Kaviani et al. 2004), a result that supports the ECI hypothesis. In contrast, although Forbes et al. (2005) found a significant linear trend in their depressed group's responses to different valences, similar to the controls, they found no potentiation of responses to unpleasant stimuli relative to neutral ones; this result only partially supports the ECI hypothesis. Finally, when the depressed group in the study of Allen et al. (1999) was divided by severity of depression, only those with scores of 30 or higher on the Beck Depression Inventory showed significantly greater startle responses to pleasant stimuli than to unpleasant stimuli, a result that supports the positive attenuation hypothesis. All of these studies included at least some patients medicated with antidepressants. In the studies by Allen et al. (1999) and Kaviani et al. (2004), most of the participants were medicated (14 of 14 in the study of Allen et al.; 18 of 22 in the study of Kaviani et al.). Forbes et al. (2005) had only 11 of 76 patients taking medication and reported that the results did not differ if those participants were removed. Dichter et al. (2004) examined the effects of bupropion on affective ratings and affective modulation of startle and found that this drug had no effect on the participants' responses, although their scores on measures of depression improved markedly. However, it has also been reported that among healthy volunteers, citalopram attenuated startle responses to negative stimuli (Harmer et al. 2004), whereas both citalopram and reboxetine have been found to reduce the recognition of fearful and angry facial expressions (Harmer et al. 2006). Accordingly, different psychotropic medications may have different effects on startle modulation.

In sum, in spite of these heterogeneous data, it seems that the affective startle-modulation paradigm may be a useful psychophysiological tool for studying depression.

2.2.2 Unmodulated Baseline Startle Reflex and Depression

Allen et al. (1999) reported that those with more severe depression had significantly lower baseline startle magnitude and also that those with the lowest levels of positive affect (anhedonia) displayed an inhibited baseline startle amplitude. In addition, Kaviani et al. (2004) reported that whereas only startle reactivity to pleasant stimuli is reduced at low levels of depression and anhedonia, all reactivity is reduced at higher levels. Furthermore, and consistent with the findings of Allen et al. (1999), it was also found that those with the highest levels of depression showed a lower startle amplitude. O'Brien-Simpson et al. (2009) conducted a 2-year follow-up study on depressed individuals and found that a relatively attenuated startle response at initial assessment was strongly predictive of both depressive symptomatology and possible relapse.

Taken together, these data show not only an association between depression and attenuated startle (Mneimne et al. 2008) but also that baseline startle is, however, highly heritable (Anokhin et al. 2003, 2007), suggesting that baseline startle may be a candidate endophenotypic marker of vulnerability to depression. Different mechanisms may account for the association between inhibited startle and depression. The serotonin transporter genotype has been involved in the dysfunction of both the startle reflex and depressive states (Brocke et al. 2006; Hariri et al. 2006).

The hypothalamic–pituitary–adrenocortical (HPA) system, which facilitates response to threat, is known to be dysregulated and hyperactive in depressive subjects, with increased central cortico-releasing hormone (CRH) production (De Winter et al. 2003; Krieg et al. 2001). Although increased levels of CRH are generally associated with an increase in anxiety behaviors and greater startle amplitude, Miller and Gronfier (2006) reported an inverse relationship between salivary cortisol and startle eyeblink magnitude in relation to diurnal variation in normal adults, and they suggested that this provides evidence for a possible link between startle reactivity and HPA axis activity. However, the relationship between cortisol levels and startle responsivity may not be linear. Buchanan et al. (2001) reported that whereas administration of low levels of exogenous cortisol (5 mg) increases startle, administration of higher levels (20 mg) attenuates startle.

In animal studies, Dirks et al. (2002) pointed out that chronic CRH hyperactivity in transgenic mice overexpressing CRH is correlated with lower startle reactivity. The Wistar–Kyoto rat, which is widely considered a model for depression (Solberg et al. 2003), provides ecological evidence for the connection between depressive behaviors, the HPA and the startle response, as it exhibits an exaggerated HPA, and attenuated baseline startle (Pardon et al. 2002). Furthermore, this hypercortisolemia, which might underlie inhibited startle magnitude, is reported as impacting on serotonergic function (Porter et al. 2004), and McAllister-Williams et al. (1998) reported that 5-HT$_{1A}$ receptor function was reduced by hypercortisolemia.

In addition, Bhagwagar et al. (2002) reported that although healthy participants upregulate 5-HT$_{1A}$ receptor function in response to acute administration of hydrocortisone, those with recurrent depression did not. Interestingly, the 5-HT$_{1A}$ receptor also has a role in the modulation of acoustic startle. Nanry and Tilson (1989) reported, using a rat model, that administration of a 5-HT$_{1A}$ receptor agonist increased the magnitude of startle reflex. It may be hypothesized from this that hypercortisolemia may reduce 5-HT$_{1A}$ receptor function and that this in turn is reflected in an attenuated startle response.

In other words, the lower startle reactivity displayed by depressive subjects may be considered a vulnerability marker to depression related to serotonergic and stress mechanisms.

2.3 STUDIES FOCUSED ON CENTRAL NEUROPHYSIOLOGICAL MEASURES SUCH AS EEG AND MEG

2.3.1 INTRODUCTION TO EEG AND MEG IN DEPRESSION

The acknowledgment that "mental illnesses" are, at least in part, caused by some kind of brain dysfunction yielded a radical change in the methods of basic research and clinical practice within the fields of psychiatry and psychopathology (Andreasen 1997). MD is a good example of this radical change. In the past three decades, considerable investigation efforts focused on brain imaging and neurophysiological/psychophysiological techniques, such as positron emission tomography, structural and functional magnetic resonance imaging, quantitative EEG (QEEG), event-related potentials (ERPs), and MEG. These brain imaging methods proved that MD has a definite neurobiological correlate. In fact, pharmacological interventions are usually the first choice for MD treatment, and such interventions are based on the assumption of an underlying neurobiological dysregulation. Nonetheless, attempts to use brain imaging techniques to assist the diagnosis, to select treatment alternatives, or to objectively evaluate clinical outcomes are scarce.

EEG (and its associated ERPs) and MEG (event-related fields) may be of particular importance since they offer a set of noninvasive tools to assess brain activity. The number of EEG/MEG-based studies on MD is difficult to estimate, but they undoubtedly represent a fundamental body of replicated evidence showing significant differences between patients with depression and healthy controls. In addition, EEG is an easily available and relatively inexpensive technique with widely used and replicated analysis algorithms. MEG is a more expensive and less broadly employed technique, but it presents some technical advantages when compared to EEG (such as no volume conduction effects and no reference problem) that make cortical source estimation easier. Consequently, MEG should be considered a suitable research and clinical tool in MD. Importantly, the suitability of EEG/ MEG-based techniques in MD investigation may have some physiological determinants. In a key article, Hughes and John (1999) reviewed the underlying physiology of EEG (and indirectly MEG) dynamics and its potential relationship with the pathophysiology of mental illnesses. According to their point of view, EEG rhythms derive from the synchronous activity of neural groups localized in some key brain structures such as the thalamus and the mesencephalic reticular formation, which project to the cortex. Those key structures are mediated by neurotransmitters and neuromodulators such as the γ-aminobutyric acid (GABA), serotonin, acetylcholine, norepinephrine, and dopamine, all of which are involved in neurobiological models of psychiatric diseases, including MD. As a consequence, any dysregulation in the neurochemical homeostasis underlying EEG generators should produce, in turn, a modification of the EEG spectrum. Since neurotransmitter perturbations are widely believed to contribute to the psychiatric pathophysiology, EEG/MEG-based techniques should be considered first-choice tools in psychiatry.

Despite this particular sensitivity, the acceptance of EEG (or QEEG) and MEG within the psychiatric community has been slow and scarce. Hughes and John proposed two major explanations for the situation: First, the abnormalities reported in early EEG studies of visual inspection have been considered too unspecific. Recent quantitative investigations showing more stable and robust differences have not been published in psychiatry journals but rather in neurology or neurophysiology publications. Consequently, results presented in those investigations might not have reached the right target audience. Second, position papers published by professional organizations over the past decades concluded that EEG is of limited and adjunctive clinical use in psychiatric practice.

Here we will present an overview of EEG and MEG studies on MD. We will focus not only on basic research but also on potential clinical applications.

2.3.2 FRONTAL BRAIN ASYMMETRY AND DEPRESSION

2.3.2.1 Classical Lesion Studies

It was observed in the 1930s that left versus right anterior cortex damage could produce different mood changes (Goldstein 1939). Later studies confirmed that left hemisphere lesions are more likely to produce depressive symptoms (Gainotti 1972; Robinson et al. 1984; Sackeim et al. 1982). Sackheim et al. (1982) observed by evaluating three retrospective studies that pathological laughing was associated with lesions of the right hemisphere, whereas pathological crying was related to left hemispheric damage. In the second study, right hemispherectomy produced euphoric mood changes. Finally, gelastic epilepsy patients who present ictal outbursts of laughing were characterized by left hemispheric epileptic foci. More specifically, Robinson et al. (1984) showed that severity of depression in poststroke patients was correlated with the proximity of the lesion to the left frontal pole. However, within the right hemisphere, they found an inverse relationship, with depression being more likely when the lesion location was more posterior (see below).

These findings were mirrored by observations during the Wada test, in which sodium amobarbital is injected into the left or right carotid artery to anesthetize the left or the right hemisphere to determine the lateralization of brain function before neurosurgery. For example, Lee et al. (1990) reported more frequent laughter-related mood changes during injection of sodium amobarbital in the right carotid artery (right cerebral hemisphere), whereas crying occurred more often after left-sided injections. In sum, suppression of left hemisphere functions by lesion or anesthesia produced depression-like symptoms. Right-sided lesions or anesthesia resulted in opposing euphoric behavioral reactions. These data were interpreted in the light of disinhibition. Thus, a lesion in the left hemisphere has been supposed to result in an overactive right hemisphere, producing negative affect, whereas an overactive left hemisphere released from contralateral inhibition has been believed to generate positive affect.

2.3.2.2 EEG Alpha Symmetry and Affective Style

In healthy humans, the mutual interplay between left and right frontal cortex activity and its relationship with emotion has been mainly studied by using EEG measures of brain activation. Thus, the power in the alpha band frequency has been considered as an inverse measure of tonal brain activation (Oakes et al. 2004). Higher alpha power at a specific electrode location has been interpreted as less tonal brain activity of the brain region beneath that electrode site (Coan and Allen 2004). Following the logic of the above-mentioned lesion studies, alpha band power asymmetries between homologous left and right electrode sites have served as indicators of relative left and right hemisphere activity. Frontal (at frontal electrode sites) alpha band asymmetry measures were reported to be reliable and stable electrophysiological measures (Towers and Allen 2009).

Usually, alpha band power is recorded during rest while subjects have their eyes open and closed for the same amount of time. Davidson et al. (1979) published the first paper linking positive and negative affects to frontal EEG alpha power asymmetry. Relative left frontal activity (less left frontal compared with right anterior alpha power) was associated with positive affect, and relative right anterior activity with negative emotion. If frontal alpha band asymmetry indexes relative left versus right frontal activity and this asymmetry is related to general mood, individual differences in resting tonal frontal brain asymmetries should account for the variance of affective styles between subjects. Tomarken et al. (1992a) assessed affective styles by measuring individual differences of general positive and negative affects using the Positive and Negative Affect Schedule (Watson et al. 1988). Participants who were characterized by increased relative left versus right anterior activation reported increased positive and decreased negative affect in comparison with subjects showing the opposite EEG asymmetry. Baseline anterior alpha power asymmetries seemed to index general affective styles that can vary across subjects. Affective styles should also be related to emotional reactivity to affective stimuli. As mentioned above, some authors have linked left frontal activity with approach motivational tendencies associated with positive affect, whereas right frontal activation reflects withdrawal tendencies accompanied by negative affective experience (Davidson 2003; Harmon-Jones et al. 2010). A general disposition to approach behavior and positive affect should result in greater experience of positive emotion during positive affective stimulation, whereas negative biased motivational tendencies should provoke more negative emotional experience to unpleasant stimuli. Indeed, a greater relative right frontal baseline activity predicted greater fear response to negative film clips (Tomarken et al. 1990). Wheeler et al. (1993) reported that subjects with greater left frontal baseline activity responded with more intense positive affect to positive film clips, whereas participants with greater right frontal activation reacted more negatively to unpleasant film material. A problem of these studies was that the samples consisted only of female subjects. However, these results were replicated in men later on (Jacobs and Snyder 1996). These studies accord with the neuropsychological model of emotion proposed by Heller and colleagues (Heller and Nitschke 1997; Heller et al. 1997a, 1998) implicating the left frontal cortex with positive valence and the right frontal cortex with negative affect.

Frontal baseline alpha asymmetry seems to be a stable individual trait measure with good to excellent reliability (Tomarken et al. 1992b; Towers and Allen 2009). Up to 60% of anterior alpha power asymmetry could be explained by individual differences on a latent trait variable (Hagemann et al. 2002). Because anterior alpha asymmetry measures trait-like emotional/motivational dispositions predicting emotive reactions to affective situations, it is of interest to know if this asymmetry is predominantly under genetic or environmental influence. Recent twin study data suggest that only a modest but significant amount of variance (11–28% in children and 27% in young adults) of frontal alpha power asymmetry is accounted for by genetic factors and that most of the variance is attributable to environmental influences (Anokhin et al. 2006; Gao et al. 2009). Taken together, the trait-like stability of frontal alpha asymmetry but its relatively low hereditability suggests that activity biases in the frontal cortex and its related affective styles are a product of environmental influences during development rather than a result of hard-wired neuronal circuits in motivational systems.

Given this trait-like behavior, the frontal alpha power asymmetry has been considered a good candidate to investigate affective disorders in the hope of finding a measure indexing risk of psychopathology. Hereby, the idea is that trait-like relative right frontal activity (and less left frontal activation) and its associated affective style predispose for affective disorders such as anxiety and depression (Davidson 1995). However, results of the hereditability studies (Anokhin et al. 2006; Gao et al. 2009) put into doubt that frontal alpha asymmetry could represent an endophenotype of depression. Nevertheless, the great environmental influence on alpha asymmetry offers the possibility to figure out environmental factors preventing the development of unfavorable frontal EEG asymmetries that could put an individual at risk for depression.

2.3.2.3 EEG Alpha Asymmetry and Depression

High levels of anhedonia, blunted affect (Loas et al. 1994), and low emotional arousal (Heller and Nitschke 1997) characterize depressed patients. Therefore, depression is considered as an affective disorder implicating disturbed processing of emotional information. For example, depressed patients show abnormal startle reflex modulation during the presentation of affective pictures (Forbes et al. 2005) and reduced facial expression (Gehricke and Shapiro 2000). From a more cognitive perspective, depressed people exhibit an attentional bias toward negative information and, as a consequence, fail to avoid negative valenced information (Gotlib and MacLeod 1997). Furthermore, they better recall negative events (Nitschke et al. 2004) and make more negative judgments about future and actual life events (Beck 1976). This resembles affective styles of subjects who react more intensely with negative emotions to unpleasant stimuli and show frontal EEG asymmetries reflecting greater relative right and less left frontal activation (Tomarken et al. 1990; Wheeler et al. 1993). It stands to reason that similar frontal EEG asymmetry patterns would be expected in depressed patients.

Numerous studies have shown that depressed patients exhibit less relative left and greater right anterior EEG activity (Gotlib et al. 1998; Henriques and Davidson 1990, 1991; Schaffer et al. 1983; Tomarken et al. 2004). Less left frontal activation has been interpreted as reflecting deficits in approach behavior associated with symptoms such as sadness and depression. Greater right anterior activation has been supposed to index withdrawal behavior related to fear, disgust, and anxiety (Harmon-Jones et al. 2010). Because comorbid anxiety in depression is frequent, it is likely that anxiety levels influence anterior alpha power asymmetry patterns (Heller 1998).

Remitted depressed patients showed the same frontal EEG asymmetry pattern as currently depressed subjects, suggesting that the observed anterior EEG asymmetry in depression represents a stable state-independent marker of depression (Henriques and Davidson 1990). Gotlib et al. (1998) also reported that currently and previously depressed subjects showed the same left frontal hypoactivity compared to healthy controls. This accords with a more recent study (Allen et al. 2004) that showed that frontal EEG asymmetry stability over time in depressed patients is comparable to that observed in nonclinical samples (Hagemann et al. 2002). EEG asymmetry scores remained stable over time, although patients improved clinically, indicating that frontal alpha asymmetry represents a state-independent measure of depression (Allen et al. 2004). If relative left frontal hypoactivity is a trait marker of depression, it should aid in identifying people at risk for depression. In a longitudinal study, Pössel et al. (2008) measured EEG alpha asymmetry and assessed depressiveness in 100 healthy adolescents at two time points that were 2 years apart. The research question was whether left frontal EEG hypoactivity as indexed by alpha power asymmetry measures at time point one can predict later increases in depressiveness. Indeed, left frontal and right parietal hypoactivity predicted depression, supporting the notion that alpha EEG asymmetries can serve as a risk marker of depression (Davidson 1995).

However, as stated above, interindividual differences in frontal alpha power asymmetries were explained mainly by environmental factors during development (Anokhin et al. 2006; Gao et al. 2009). Thus, genetic factors probably predispose for the development of depression-like frontal asymmetries in unfavorable conditions. This is in line with the observation that left frontal hypo-activation was associated with lifetime depression but not with parental depression status per se (Bruder et al. 2005). Stress could influence the development of frontal activity asymmetry. Frontal EEG asymmetry variance in adolescents was mainly explained by the lower socioeconomic status rather than by the history of depression of their mothers (Tomarken et al. 2004). Lower socioeconomic status is associated with elevated stress levels as indexed by increased postwaking cortisol levels (Li et al. 2007). Cortisol is a key hormone during stress responses. Elevated cortisol levels are involved in the development of depression (Goodyer et al. 2001). Kalin et al. (1998) demonstrated that young rhesus monkeys with extreme cortisol levels also showed extreme frontal asymmetric electric activity, with the right hemisphere being overactive compared to the left frontal cortex. In 6-month-old infants, extreme right versus low left EEG activity was associated with elevated cortisol levels and fear reactions (Buss et al. 2003). Right frontal and parietal hyperactivation has been related to anxiety (Davidson 1995; Heller and Nitschke 1997). Anxiety often precedes depression. Thus, elevated environmental stress factors such as low socioeconomic status probably favor the development of right frontal hyperactivity patterns associated with withdrawal tendencies and anxiety. Later, during chronic stress exposure, anxiety could turn into depression with a remaining EEG asymmetry of left frontal hypoactivation. Tops et al. (2005) demonstrated a direct relationship between acute cortisol administration and changes in frontal alpha power asymmetry. In comparison with a placebo condition, cortisol provoked increased right and reduced left frontal EEG activity as measured by inverse alpha power. In sum, stable trait-like frontal alpha band asymmetry patterns that index risk for depression could result from chronic stress exposure throughout life, especially during childhood and adolescence.

Although studies of alpha asymmetry hereditability point to environmental stress factors shaping the balance between left and right frontal cortex activity (Anokhin et al. 2006; Gao et al. 2009), polymorphic gene variations affecting neurotransmitter systems that play an important role in the processing of stress-related information could determine the individual's vulnerability for psychopathology. Serotonergic dysfunction is implicated in the onset, course, and recovery of depression (for a review, see Lesch 2001). If relative less left and increased right frontal brain activity puts individuals at risk for developing depression, there should be a close relationship between interindividual genetic polymorphisms in serotonin genes and frontal alpha power asymmetries. Recently, it has been shown that subjects with homozygous 5-HT$_{1A}$ (HTR1A) risk alleles had greater right than left frontal EEG activation, indicating that serotonin gene abnormalities may influence trait level brain activity (Bismark et al. 2010). Bismark et al. (2010) argued that this polymorphism leads to increased HTR1A receptor concentration, resulting in negative feedback inhibition of serotonergic activity. Thus, increased HTR1A receptor density leads to decreased serotonergic transmission (Albert and Lemonde 2004). Interestingly, Fink et al. (2009) demonstrated a greater density of the inhibitory HTR1A receptor in the right frontal cortex. In a recent study, Mekli et al. (2010) demonstrated that HTR1A receptor expression is associated with stress-related information processing probably predisposing to the development of stress-related psychopathology such as depression.

In sum, less relative left and greater right frontal EEG activity is considered a trait-like electrophysiological marker of motivational/emotional dispositions that represent a risk factor of depression. Less left frontal activity is associated with less approach-related behavior linked to less positive affect, whereas greater right frontal activity indexes greater withdrawal tendencies and negative affect. However, twin studies suggest that frontal EEG asymmetries linked to depression develop under environmental circumstances. Probably, an overexpression of HTR1A serotonin receptors in right frontal cortex creates a vulnerability to develop frontal cortex activity asymmetries associated with depression in response to stressful life events.

2.3.3 POSTERIOR BRAIN ASYMMETRY AND DEPRESSION

Heller's neuropsychological model of emotion (Heller and Nitschke 1997) not only emphasizes left and right frontal brain activity associated with valence (positive and negative mood, respectively) but also assumes that right posterior cortex functioning is related to emotional arousal. From that perspective, depression has been associated with low emotional arousal due to right temporoparietal cortex dysfunction (Heller and Nitschke 1997; Heller et al. 1997a; Moratti et al. 2008).

2.3.3.1 Classical Lesion Studies

Lesion studies investigating the association between lesion location and depression have found strong relationships between left frontal damage and the development of poststroke depression. For example, Robinson et al. (1984) observed the opposite pattern in the right hemisphere. The more posterior the damage was localized within the right hemisphere, the more likely it was for patients to develop depression. Interestingly, a follow-up of poststroke patients (Shimoda and Robinson 1999) revealed that the correlation between left frontal lesion site and depression appeared only during the hospitalization of the patients. At a 1- to 2-year follow-up of the same patients, depression was associated with right posterior damage, indicating that right posterior hemisphere lesion might be a more stable predictor of poststroke depression.

More specifically, right posterior lesions result in reduced skin conductance responses to emotional slides (Meadows and Kaplan 1994). Skin conductance responses to emotional stimuli index autonomic arousal and vary as a function of arousal rather than valence. Therefore, poststroke depression after right posterior damage is probably related to low emotional arousal and reflects reduced affective arousal in depression per se (Heller and Nitschke 1997).

2.3.3.2 Neuropsychological Findings

To evaluate right temporoparietal cortex functioning in depressed patients, early studies used the chimeric faces test (CFT), a free vision test to assess hemispatial biases in emotional face processing (Levy et al. 1983). Thereby, split faces are used. One-half of a face consists of a smiling expression and the other half of a neutral expression. The smiling part can be on either the left or right side. Then two identical but mirrored chimeric faces are mounted on the top and the bottom of a page, and the participant has to judge which of the two faces looks happier. Healthy subjects tend to choose the face with the smile on the left side (Levy et al. 1983). As visual information from the left hemispace is projected to the contralateral hemisphere, this bias is believed to reflect an advantage of the right temporoparietal cortex in processing emotional faces (Bruder et al. 2002).

In the late 1980s, Jaeger et al. (1987) first observed that depressed subjects showed a reduced right hemisphere bias in the CFT test, suggesting a right temporoparietal dysfunction. This was later confirmed by comparing unipolar depressed, bipolar disorder, and left and right hemisphere–damaged patients. In this study, only unipolar depressed and right hemisphere lesion patients showed a reduced right hemisphere perceptual bias in comparison with the other patient groups and healthy controls (Kucharska-Pietura and David 2003).

As stated above, comorbid anxiety has been hypothesized to result in opposite right versus left posterior hemisphere activity (Heller 1998). Depressed patients with low anxiety should be mainly characterized by low emotional arousal. As the right temporoparietal cortex has been implicated directly in emotional arousal (Heller et al. 1997a; Moratti et al. 2008), low-anxiety depressed people should show the above-mentioned reduced right hemisphere bias in the CFT test. However, depressed subjects with high anxiety associated with high anxious arousal should show the opposite pattern. Therefore, it is important to carefully control for comorbid anxiety as opposing effects could cancel out each other. Highly depressed students exhibited reduced right hemisphere advantages, whereas highly anxious subjects displayed increased right hemisphere advantages in the CFT task (Heller et al. 1995). In a more recent study, Keller et al. (2000) reported the same effects.

However, not only different levels of comorbid anxiety seem to affect right hemispheric functioning in depression. Bruder et al. (2002) applied the CFT test to a large sample of depressed patients. They divided the sample into those patients meeting the criteria for typical and those with atypical depression, following the Columbia criteria for atypical depression. Thereby, typical depression is considered as melancholia subsuming symptoms such as anhedonia and nonreactivity of mood and vegetative symptoms such as insomnia, anorexia, and psychomotor retardation. In contrast, atypical depression is defined as a nonmelancholic subtype of depression that differs in reactivity of mood with a preserved pleasure capacity and opposite vegetative symptoms such as hypersomnia and excessive eating. Comparing the hemispheric bias scores of typical and atypical patients obtained by applying the CFT test, typical depression was associated with the reduction of a right hemisphere advantage. Atypical patients even exhibited an increased right hemisphere bias compared with controls and the typical depression group. Interestingly, the same differences were observed after dividing typical and atypical patients into major depressive disorder (MDD) and dysthymia groups. Thus, with respect to right hemisphere function, typical and atypical depressed patients differ regardless of whether they meet the *Diagnostic and Statistical Manual of Mental Disorders, Fourth Edition*, criteria for MDD or dysthymia. In sum, right hemisphere dysfunction in depression seems to be associated with the melancholic subtype characterized by anhedonia and low emotional reactivity. This accords with the neuropsychological model of Heller and Nitschke (1997), which emphasizes that low emotional arousal in depression is related to right temporoparietal hypoactivation. However, this only seems to hold true for typical depression of the melancholic subtype (see also Bruder et al. 1989) using a dichotic listening task and low anxious patients.

2.3.3.3 Posterior EEG Asymmetries

Following the same line of reasoning as mentioned above, greater power in the alpha band at right versus left parietal EEG electrode sites is considered to reflect less relative right than left parietal cortex activity. Mirroring the neuropsychological findings of right hemisphere dysfunction, many studies have found that depressed adults and also adolescents are characterized by a parietal EEG asymmetry indicating right parietal EEG hypoactivity (Blackhart et al. 2006; Bruder et al. 1997; Henriques and Davidson 1990; Kentgen et al. 2000). Thereby, current and previously depressed patients showing reduced left frontal EEG hypoactivity also exhibited reduced right parietal EEG activity (Henriques and Davidson 1990).

However, comorbid anxiety in depression has been proven to have a great influence on right parietal brain activity. High anxious depressed patients can exhibit an opposite pattern of parietal EEG asymmetry than low anxious depressed subjects (Bruder et al. 1997) or at least attenuate the asymmetry pattern associated with less right posterior activity (Kentgen et al. 2000). It has been proposed that especially anxious arousal is indicated by increased right posterior brain activity, whereas anxious apprehension is indexed by greater left posterior activation (Heller et al. 1997b). Augmented right parietal cortex activity in patients with high levels of anxious arousal is in line with the notion that the right parietal cortex is part of a cortical attentional vigilance system (Fernandez-Duque and Posner 2001). Thus, high anxious patients probably have an overactive attentional vigilance system scanning the environment for potential danger. Posttraumatic stress disorder patients were characterized by increased right posterior EEG activity, probably representing such a hyperactive vigilance system (Metzger et al. 2004). In contrast, depressed patients with low anxious arousal are characterized by low emotional arousal and show right parietal EEG hypoactivity (Bruder et al. 1997). Children with low emotionality at risk for depression also show reduced right parietal hypoactivation (Shankman et al. 2005). The processing of high-arousing emotional stimuli (pleasant and unpleasant) depends on activity in cortical attention systems that probably exert top-down influences onto brain areas of basic visual processing (Vuilleumier and Driver 2007). The right parietal cortex is part of such a network active during the processing of high-arousing emotional stimuli (Moratti et al. 2004). It seems that cortical attention networks are not properly engaged by motivationally significant stimuli in depressed patients.

As mentioned above, frontal EEG asymmetry seems to develop more likely under environmental influences (Anokhin et al. 2006; Gao et al. 2009), whereby stress could result in the development of depression-like frontal alpha power asymmetries. In a study by Bruder et al. (2005), left frontal EEG hypoactivation was associated with lifetime MDD and not with parental MDD status, whereas reduced right parietal activity depended on MDD status of the participants' parents. This indicates that right parietal dysfunction in depression probably is under a heavier genetic influence than frontal EEG asymmetries. This is in line with the observation that grandchildren with a family history of MDD but no own history of depression already show posterior EEG asymmetries indicative of less relative right parietal activity (Bruder et al. 2007). However, no genetic hereditability studies of posterior EEG asymmetry have been conducted so far.

2.3.3.4 Event-Related Potentials and Fields

Neuropsychological findings and EEG alpha power asymmetries are rather indirect measures of brain activity. More critically, they do not directly measure activation during emotional arousal. Thus, a better test of right temporoparietal hypofunction and its association with low emotional arousal would consist of measuring brain activity in that brain region during high versus low emotional arousal in depressed and healthy subjects. From the motivated attention perspective of emotional picture processing, numerous studies using evoked potentials and hemodynamic brain responses investigated emotional arousal–related brain responses (a complete review of this literature is not the objective of this chapter so, for a review, see Lang and Bradley 2010). The Lang laboratory at the Center for Emotion and Attention Study developed the International Affective Picture System (IAPS) (Lang et al. 2005), with normative ratings for valence and arousal. The advantage of the vast literature about brain responses and affective arousal is that they all used comparable stimulus material (Lang and Bradley 2010).

Electrophysiological studies recording evoked potentials generated by affective pictures from the IAPS have reported arousal modulations of early and late visual ERP components, such as the N100 (between 100 and 150 ms after stimulus onset), the early negative potential (about 200 ms after stimulus onset), and the late positive potential (LPP, about 300 ms after stimulus onset). Whereas early effects were reported as less consistent in the literature, arousal modulations of the LPP were reliably observed (for a review, see Olofsson et al. 2008). The LPP is a P3-like positive slow wave that develops between 300 and 700 ms (or longer) after stimulus onset and shows its maximum at central/parietal electrode sites (Schupp et al. 2000). The LPP has been shown to be greater for high-arousing pleasant and unpleasant stimuli when compared with low-arousing neutral pictures (Cuthbert et al. 2000; De Cesarei and Codispoti 2006; Dolcos and Cabeza 2002; Hajcak et al. 2009; Hajcak and Olvet 2008; Huang and Luo 2006; Keil et al. 2002; Pastor et al. 2008; Sabatinelli et al. 2007; Schupp et al. 2000). Thus, the LPP represents an extremely replicable and stable electrophysiological measure that varies with emotional arousal. Therefore, the LPP would be ideally suited for clinical research.

Indeed, ERP studies of emotional perception in depression found impaired P3 modulations at right hemisphere electrode locations in depressed patients (Deldin et al. 2000; Kayser et al. 2000). However, the disadvantage of these studies was that they did not use IAPS pictures but utilized face stimuli or self-constructed unpleasant stimuli. Future research should appreciate the standardized emotional stimulus material available. The consistent ERP findings in healthy controls obtained using the IAPS could help elucidate differences in the processing of high-arousing emotional stimuli in depression. Given the above revised literature, reduced arousal modulation of the LPP across right parietal electrodes would be expected for depressed patients.

In clinical research, stable and reproducible experimental protocols not only are needed but should also be of short duration. In EEG and MEG research (MEG is a technique measuring the magnetic fields generated by the electric activity of synchronized neurons), the steady-state response represents a good candidate to fulfill those criteria. The steady-state response is an oscillatory evoked potential/field that develops by presenting the affective pictures in a luminance-modulated mode.

One could imagine the same picture presented for several seconds but flickering. The flickering occurs at a stable frequency of, for example, 10 Hz (other frequencies can be used as well), meaning that the same picture is presented in an on/off cycle of 50 ms, showing the picture for 50 ms followed by 50 ms of background color. This cycle is then repeated for several seconds. Presenting visual stimuli in this manner generates the so-called steady-state visual evoked potential (ssVEP in EEG) or field (ssVEF in MEG). It is an oscillatory brain response with the same frequency as the driving stimulus (Regan 1989). Thus, a 10-Hz flickering picture evokes an oscillatory brain wave at a frequency of 10 Hz.

The advantage of this technique is that within a short recording time, one can obtain very good EEG and MEG signals (Mast and Victor 1991). The flickering mode allows a densely repeated presentation of the stimuli during a short recording time. Imagine that a specific picture is depicted at a flickering mode of 10 Hz. Thus, the same picture is shown 10 times within 1 s (1 cycle is 100 ms). If the picture is presented for 6 s, the same picture is shown 60 times. This results in a very good signal-to-noise ratio with respect to EEG/MEG evoked potentials/fields. To obtain an evoked EEG/MEG response, the stimuli have to be repeated many times and the stimulus-locked signals have to be averaged. More repetitions lead to higher signal-to-noise ratios. Steady-state paradigms implicate many stimulus repetitions.

Another advantage of the steady-state paradigm is that attentional processes across several seconds of picture viewing can be observed. Numerous studies have found that attended visual stimuli that are presented in a steady-state fashion produce steady-state oscillations of greater amplitudes than nonattended stimuli (Morgan et al. 1996; Müller and Hillyard 2000; Müller and Hübner 2002). The motivated attention approach to emotion emphasizes that high-arousing emotional stimuli automatically attract attention (Keil et al. 2005). This leads to a top-down facilitation of the visual processing of high-arousing stimuli, involving visual cortex and higher-order cortical attention networks, such as the frontal and parietal cortices of the right hemisphere (Vuilleumier and Driver 2007). Keil et al. (2003) demonstrated that high-arousing pleasant and unpleasant pictures generated greater ssVEP responses than low-arousing neutral pictures. Using MEG, Moratti et al. (2004) showed that arousal-modulated ssVEF responses were generated in right occipital and frontoparietal brain regions.

Moratti et al. (2008) tested the hypothesis of temporoparietal dysfunction and low emotional arousal in depression more directly by using the above-mentioned steady-state paradigm. The rationale behind this study was that ssVEF responses of healthy subjects recorded by MEG were modulated by emotional arousal (Moratti et al. 2004). This arousal modulation was observed in cortical attention networks of the right hemisphere. As the right temporoparietal cortex is part of that network, depressed patients should show a reduced modulation of their ssVEF responses in the right temporoparietal cortex. Indeed, low anxious clinically depressed patients exhibited reduced emotional arousal modulation of their ssVEF response in the right temporoparietal cortex (Moratti et al. 2008). In contrast, healthy subjects showed the strongest emotional arousal modulation of ssVEF activity in that brain region. Although this report directly manipulated the arousal dimension of the stimulus material, future studies should incorporate measures of autonomic arousal, such as skin conductance (see Figure 2.1).

2.3.4 FREQUENCY BANDS OTHER THAN ALPHA

As described in the previous sections, most EEG studies on MD focused on alpha activity and particularly on the so-called alpha asymmetry. However, other frequency bands, especially in the low-frequency range, deserve an investigation as well. The presence of sporadic or generalized low-frequency activity, particularly in the delta range, is considered a major sign of brain disturbance (see, e.g., Gloor et al. 1977), and increased low-frequency activity is systematically observed in dementias, brain trauma, hypoxia, and so on. Interestingly, low-frequency investigations on MD tended to show reduced power or current density values. Early studies, such as that of Brenner et

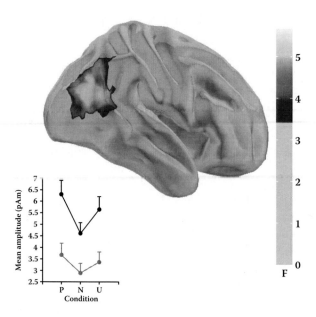

FIGURE 2.1 **(See color insert.)** A brain region of significant group (healthy controls and patients) by picture category (high arousing vs. low arousing) interaction, as indicated by a nonparametric cluster-based permutation statistics, is shown on a smoothed brain surface. Color bar indicates interaction F value. White panel depicts means and standard errors of brain activity within depicted brain region for each picture category and each group. Black lines represent healthy controls; red lines, depressed patients. P, pleasant pictures; N, neutral pictures; U, unpleasant pictures. Unpublished data.

al. (1986), showed that depressed patients differed from demented patients at the lower end of the spectrum, having significantly less delta and theta activities. The comparison of demented and aged depressed patients is important since both groups share some cognitive and emotional symptoms. Pozzi et al. (1995) performed QEEG evaluations in depressed Alzheimer disease patients, nondepressed Alzheimer's patients, nondemented depressed patients, and healthy aged controls. Similarly to the results of Brenner et al., they found reduced delta power in nondemented depressed patients as compared with both demented groups. Moreover, in the study of Pozzi et al., delta power values were also lower within the nondemented depressed group compared to healthy aged controls.

Conventional QEEG investigations have been criticized because of their technical dependence on a reference electrode. Results may vary depending on the reference's location, and different studies using different references are difficult to compare (Davidson 2004). Trying to overcome this technical problem, more recent investigations are employing new analysis techniques such as the low-resolution electromagnetic tomography (LORETA) (Pascual-Marqui 1999), which does not depend on a reference electrode. Mientus et al. (2002) found significantly lower source-current densities in unmedicated depressed patients in all frequency bands. Further delta and theta activity was characterized by reduced current density in anterior cingulate cortex. We will see that theta current density in anterior cingulate cortex has a critical role in the prediction of response to antidepressants. Lubar et al. (2003) obtained current density values by means of LORETA analysis in a group of depressed females and healthy controls. Overall, they found reduced current density values in the delta band within the depressed group. This effect was more evident in the right temporal lobe, and the authors interpreted the finding as produced by a lack of basic modulation properties in this region probably due to a lack of "small-scale local organization." LORETA solutions were also used by Flor-Henry et al. (2004). In their interesting study, they hypothesized that depression can be characterized by a prefrontal hypoactivation and, as usual, the hypoactivation is reflected by an alpha asymmetry. However, results did not support their hypothesis. The expected left anterior hypoactivation (increased resting alpha current density) was not seen in their results, but a reduced current density in the delta band was observed in the left hemisphere.

Decreased low-frequency activity also has been observed in MEG investigations (Fernandez et al. 2005; Wienbruch et al. 2003). Wienbruch et al. (2003) calculated dipole density values within the delta and theta range in a group of schizophrenic patients, depressed patients, and healthy controls. Whereas schizophrenic patients showed accentuated delta and theta activity in the temporal and parietal regions, patients with depression exhibited reduced frontal and prefrontal delta and theta dipole densities. As in previous investigations, such reduced low-frequency activity was associated with a dysfunctional pattern of local activation. Wienbruch et al. pointed out that previous EEG and MEG studies demonstrated increased delta power and dipole densities after effective electroconvulsive therapy (Sackeim et al. 1996; Sperling et al. 2000). Interestingly, the increase of delta dipole density was also observed after effective pharmacological treatment, but to a lesser and nonsignificant extent (see also Knott et al. 2002). This line of evidence suggests that delta and theta band activity seems to be reduced in MD. Effective treatment modifies this abnormal pattern, normalizing low-frequency values to those of controls.

High-frequency activity was also investigated in MD. For example, Knott et al. (2001) performed a comprehensive study trying to investigate not only the alpha asymmetry but also other parameters such as absolute and relative power in all bands, mean frequency values, and hemispheric coherence. Their results failed to show the expected alpha asymmetry. MD patients exhibited higher absolute and relative beta values, a higher mean frequency (tendency to values in the higher range of the EEG spectrum), and reduced interhemispheric coherence (coherence results will be discussed elsewhere). Pizzagalli et al. (2002) used LORETA solutions to investigate the influence of concomitant factors such as anxiety, melancholic features, and severity. They found no effects in the alpha band. Instead, depressed patients showed more beta3 (21.5–30 Hz) compared to controls on Brodmann areas 11 and in 9/10 of the right frontal lobe and less beta3 in the posterior cingulate gyrus and precuneus. Beck Depression Inventory values were positively correlated with frontal asymmetry in beta3. According to Pizzagalli et al.'s interpretation of their results, depressed patients were characterized by relative hyperactivity in right frontal regions and hypoactivity in the posterior cingulate and precuneus. Right frontal hyperactivity was most conspicuous in melancholic patients who were severely depressed. These results demonstrate that clinical characteristics (i.e., severity and melancholic features) are modulating the brain activity of MD patients. A study investigating high frequencies in MD (Strelets et al. 2007) reported that the power of gamma rhythm in frontal and temporal cortical areas was significantly greater in patients with depression than in healthy subjects. These results confirm the tendency to elevated power and current density values in the high-frequency bands in depression, whereas low-frequency power seems to be reduced.

2.3.5 EEG as a Predictor of Antidepressant Treatment Outcome in MD

MEG and specially EEG investigations demonstrated a high sensitivity in finding significant differences between MD patients and controls. Also, a strong correlation between mood variations and nonlinear estimates of EEG/MEG activity has been found. As briefly mentioned in the Introduction, such set of evidence was not accompanied by the acceptance of EEG or MEG as clinical tools by the psychiatry community. A good example of this apparent "contradiction" is the investigation of predictors of treatment outcome in MD. This is undoubtedly a critical field of research where more efforts are needed. According to some independent investigations (Trivedi et al. 2006), less than 30% of patients treated with first-choice antidepressants achieve total symptoms remission after 8 weeks of treatment, and only about 50–70% can be considered "responders" to treatment (usually defined as a 50% reduction of Hamilton-D scores). Unfortunately, nowadays, there is no way to predict which particular patient will benefit from a particular antidepressant. Current treatment of MD is characterized by a trial-and-error sequential strategy that results in response and remission delays (Leuchter et al. 2009a). A prolonged ineffective treatment can have a number of deleterious consequences. The most evident is that patients continue to suffer from the symptoms of depression, with well-known personal, familial, social, and economic consequences. Furthermore, this increases the risk of suicide, produces some deleterious effects on the central nervous system (CNS) (e.g., decrease of hippocampal volume), increases the risk for never receiving an adequate treatment, is associated with a poorer prognosis, and so on (see the review by Hunter et al. 2007). Several attempts have been made to find a good predictor of treatment outcome in MD, using genetic, brain imaging, or neurophysiological markers. Among these attempts, EEG-based studies have achieved highly valuable results. Here we will present a brief overview.

The utilization of EEG as a biomarker for the prediction of treatment outcome is based on early pharmaco-EEG studies, which demonstrated that the administration of antidepressants produces significant changes in brain activity within hours of dosing, both in patients and in healthy controls (Itil 1983). The administration of tricyclic and selective serotonin reuptake inhibitors (SSRIs) yields very similar effects, with an increase of delta, theta, and beta powers, and an important decrease of alpha power (Mucci et al. 2006). Some individual differences were found in response to antidepressants within healthy controls, and they were considered a sign of CNS traits of medication effects that, if correctly used, might have an important influence on the response prediction in MD patients.

As might be expected, early studies specifically focusing on the prediction of treatment outcome were devoted to the analysis of alpha power and asymmetry. For instance, in a follow-up study, Ulrich et al. (1984, 1986) found that responders to amitriptyline were characterized by a left lateralization of baseline alpha power, a decrease of absolute alpha power, and an increase in low-frequency bands after 4 weeks of treatment. Bruder et al. (2001) reported that lateralized alpha power was associated with response to fluoxetine. An increase in relative alpha power on the left hemisphere (indicating less relative left than right hemisphere activation) was associated with a 12-week response to SSRIs. In a more recent investigation, Bruder et al. (2008) reported that responders to fluoxetine had greater alpha power compared with nonresponders and healthy controls, with the largest differences at the occipital sites. There also were differences in alpha asymmetry. Responders showed greater alpha symmetry over the right than the left hemisphere, whereas nonresponders tended to show the opposite asymmetry. Baseline and posttreatment comparisons of theta activity also predicted response to antidepressant treatment. Responders to imipramine showed a significantly reduced theta activity at pretreatment EEGs and an increase of theta power after 2 weeks of treatment (Knott et al. 1996). It seems that a reduced baseline theta power is not specific for responses to tricyclic antidepressants. Knott et al. (2002) found that responders to paroxetine also exhibited lower baseline theta power as compared with nonresponders. In this case, reduced theta power was localized at frontal sites, whereas in the imipramine study, Knott reported an overall decrease of theta activity in responders. Below, we will show that the investigation of theta activity is currently one of the most promising lines of investigation within this field.

Evoked potentials as treatment predictors have been evaluated as well. Probably the most robust line of investigation is the loudness dependence of auditory evoked potentials (LDAEPs). This research line relies on the fact that LDAEPs may reflect activity in the brain's serotonergic system. Thus, the ratio of N1/P2 amplitude increases with increasing tone loudness during auditory stimulation, and LDAEP values are inversely correlated with central serotonergic activity (Hegerl and Juckel 1993; Hunter et al. 2007). Some EEG studies (Mulert et al. 2002, 2007) reported that patients displaying strong N1/P2 intensity dependence (i.e., greater ERP amplitude associated with LDAEPs and consequently lower serotonergic activity) respond better to serotonergic antidepressants than those who had lower LDAEPs (i.e., probably high or normal serotonergic activity).

An additional and promising research line is the examination of how the brain mediates emotional experience and how antidepressants modulate that experience. Using a steady-state paradigm (see above), the administration of citalopram, compared with placebo, attenuates the electrophysiological activation to unpleasant stimuli within the frontal and occipital cortices and potentiates the electrophysiological activation to pleasant stimuli within the parieto-occipital cortex (Kemp et al.

2002, 2004, 2008). According to Kemp et al. (2008), this effect is of particular importance because emotional response may be a key risk factor for the development of depression, and SSRIs are the most commonly prescribed antidepressants.

The above-cited studies represent a brief summary of different investigation threads. Notwithstanding, it is important to highlight for the reader that, in our point of view, three lines of research are currently offering results of potential clinical relevance because of their reported high sensitivity: (1) theta cordance studies, (2) antidepressant treatment response (ATR) index studies, and (3) studies on theta current density in rostral anterior cingulate cortex (rACC).

2.3.5.1 Theta Cordance as a Predictor of Treatment Outcome in MD

Andrew Leuchter's group coined the concept of *cordance* while trying to describe an EEG measure that combines information from relative and absolute power estimates. The cordance's key characteristic is its moderate correlation with metabolic and perfusion measures, a correlation that led authors to claim (in some of their papers) that cordance reflects the "cerebral energy utilization" (Leuchter et al. 1994, 1999). According to Leuchter, cordance extracts information with a higher physiological meaning as compared to traditional power spectrum calculation and therefore is more suitable for any kind of clinical research. The first cordance investigation in a depression sample was performed by Cook et al. (1998). The authors compared cordance measures for the traditional EEG bands in a group of late-life depressives and controls. They found an alteration of cordance values within the theta band in the patient group, interpreted as a reduction of energy utilization, which confirmed previous metabolic studies. Such alteration was not present in a subgroup of medicated patients, indicating that theta cordance was sensitive to drug effects. This early investigation also highlighted the role of the theta band. Other studies (see, e.g., Leuchter et al. 1997) explored a series of individual cases illustrating that a reduction of prefrontal theta cordance 48 h after beginning of medication predicts the clinical improvement in patients treated with SSRIs but also with serotonin–norepinephrine inhibitors. A prospective study revealed that a reduction of prefrontal theta cordance, 48 h after treatment onset, also predicted 2-months outcome in a group of seven MD patients (Cook and Leuchter 2001). In fact, changes in prefrontal theta cordance predicted treatment outcome with 100% sensitivity and 67% specificity, but obviously, the very small sample size must be considered. Marie-Mitchell et al. (2004) performed a more sophisticated study, combining not only EEG measures but also clinical data, to find predictors of "improved mood." Their results showed that a combination of variables such as pretreatment depressed mood, duration of depressive episode, verbalization of suicidal thoughts, and again the reduction of prefrontal theta cordance was capable of predicting improved mood in patients treated with fluoxetine or venlafaxine. Of note, this combination of variables predicted improved mood both for placebo and drug groups, since no significant differences in terms of treatment outcome were observed. When sources of theta cordance were analyzed in more detail, the authors found that two frontal areas (midline and right frontal) are responsible for the observed differences between responders and nonresponders (Cook et al. 2009). Thus, a reduction of theta cordance in midline and right frontal regions, 1 and 2 weeks after treatment onset, predicted response to fluoxetine in a placebo-controlled trial, with 90% sensitivity and 60% specificity. These promising cordance findings have been confirmed by an independent group (Bares et al. 2008), supporting their potential relevance.

2.3.5.2 ATR Index as a Predictor of Treatment Outcome in MD

The ATR index is a very recent tool for the prediction of treatment outcome, also proposed by the Leuchter group. In contrast to any of the measures presented in this chapter, the ATR index is integrated in a commercial tool (Aspect Medical Systems, Norwood, MA), although it was made available for a limited number of investigation groups to facilitate the replication of Leuchter et al.'s results. Obviously, this is not a problem or a limitation per se, but some particularities regarding the description of the ATR index calculation method and the utilization of an unconventional EEG montage are noteworthy. According to the description of Leuchter el al., the ATR index is a nonlinear "weighted" combination of three EEG features that previously have been associated with antidepressant outcome: relative combined theta and alpha power (3–12 Hz), alpha1 absolute power (8.5–12 Hz), and alpha2 absolute power (9–11.5 Hz). Relative combined theta and alpha power is calculated as the ratio of absolute combined theta and alpha power divided by the total power (2–20 Hz). The ATR is a weighted combination of the relative theta and alpha power at week 1 after treatment onset, and the difference between alpha1 power at baseline and alpha2 power at week 1, scaled to range from 0 (indicating low probability of response) to 100 (high probability of response). To obtain these values, power spectrum data are integrated in a relatively complex formula that includes some "constant" values (see Leuchter et al. 2009a, 2009b).

The ATR index has been included as one of the potential predictive variables in the Biomarkers for Rapid Identification of Treatment Effectiveness in Major Depression multicenter study. This is one of the strongest points of this measure. In the said study, patients were treated with escitalopram, and similarly to previously cited studies (Trivedi et al. 2006), the authors found 52.1% response and 38.4% remission rates. The ATR index predicted response and remission with 74% accuracy. Serum drug levels or genetic markers did not predict treatment outcome. Only Hamilton-D scores at day 7 after treatment onset showed some predictive capability. However, the ATR was the only predictor of remission (Leuchter et al. 2009c). In a parallel study, the ATR's capability for the prediction of treatment response was estimated in patients treated with escitalopram, bupropion, or a combination of both drugs. There were no significant differences between response and remission rates in the three medication groups. Using a threshold, the authors divided patients into those with high and low ATR values, and hypothesized that those with values above the threshold should be better responders to medication. As expected, results confirmed the hypothesis: 68% of patients with high ATR responded to medication (Leuchter et al. 2009c). This is a very attractive approach in the methodological perspective, but as previously noted, some aspects should be further clarified. Nevertheless, the ATR index has been used by an independent group (Iosifescu et al. 2009) that found very similar results. In the study of Iosifescu et al., the ATR index predicted treatment outcome to venlafaxine with 82% sensitivity and 54% specificity. Undoubtedly, research using the ATR index represents a very promising line of investigation.

2.3.5.3 Theta Current Density in Rostral Anterior Cingulate Cortex as a Predictor of Treatment Outcome in MD

As previously noted, theta band activity seems to play a key role in the field of the prediction of treatment outcome in MD. Patients with low theta power at baseline measures are better responders to antidepressants, and theta cordance predicts treatment outcome. Both lines of investigation are based (also cordance) on the calculation of absolute or relative powers in the theta band, but these studies usually do not estimate the sources of theta activity. Nowadays, methods of EEG source analysis such as LORETA are broadly used, and this approach is opening new windows to the understanding of response to antidepressants. In particular, a strong line of investigation has been devoted to theta band activity in the rACC. Pizzagalli et al. (2001) claimed that ACC has been the subject of increasing interest in neuroimaging research on depression, since several studies have shown low metabolism rates and blood flow in this area. The authors also pointed out that in contradiction with those reports showing low ACC activity in MD, treatment studies showed a relationship between higher metabolic rates in this region and response to treatment (Mayberg et al. 1997). This contradiction is probably attributable to the lack of knowledge on the function of the ACC in mood regulation.

Pizzagalli et al. performed one of the first studies that explicitly searched for a relationship between theta activity in the rACC and response to antidepressants. They calculated LORETA solutions in a group of 18 unmedicated MD patients treated with nortryptiline and 18 healthy controls. Of the 18 depressed patients, 16 were considered responders after 4 to 6 weeks of treatment. Those responders were characterized by higher current density values in the rACC at baseline. Further analyses indicated that the higher the theta activity in the rACC, the better the response to treatment, estimated as a reduction of the Beck Depression Inventory scores. Mulert et al. (2007) combined LDAEP measures and theta activity in the rACC to obtain better predictors of treatment outcome. In this case, responders to citalopram and reboxetine showed higher theta activity and LDAEP values, and the combination of both measures allowed a good discrimination between responders and nonresponders. The importance of rACC theta current density has been confirmed in a double-blinded placebo-controlled study. Korb et al. (2009) collected EEG data from 72 subjects with MD enrolled in three trials, receiving treatment with fluoxetine, venlafaxine, or placebo. As in previous reports, medication responders showed higher baseline current density in the rACC but also in the medial orbitofrontal cortex. Specifically, rACC theta activity predicted treatment response with 64% sensitivity and 67% specificity.

Compared to other lines of investigation with higher estimated sensitivity (see, e.g., cordance results), the role of theta activity in the rACC is supported by a robust neurobiological and neuropsychological basis. High theta activity in the rACC has been considered a sign of higher activation within this region, and such hyperactivation is a sign of an adaptive or compensatory reaction to the depressive situation that increases the likelihood of remission (Pizzagalli et al. 2001). Mayberg et al. (1997, 1999) integrated this perspective in a model of depression that considers a distinction between the roles of different cingulate regions. According to Mayberg et al., there is a "dorsal" compartment, which includes the dorsolateral prefrontal cortex, the dorsolateral ACC, the inferior parietal lobe, and the striatum, that modulates attentional and cognitive symptoms such as apathy and attentional and executive deficits. There is also a "ventral" compartment (including the hypothalamic–pituitary–adrenergical axis, insula, subgenual cingulate, and brainstem) involved in vegetative and somatic symptoms such as sleep disturbance and loss of appetite and libido. Mayberg's group poses that the rACC plays a key role in the response to treatment since it exerts a regulatory influence on integrating information of the dorsal and ventral compartments or, in other words, on the integration of salient affective and cognitive information. This is similar to the perspective of Devinsky et al. (1995), which emphasizes that rACC is a part of the "affect" subdivision of the ACC because of its connections with limbic and paralimbic structures, such as the amygdala, nucleus accumbens, and orbitofrontal cortex. Furthermore, the rACC is involved in evaluating the significance of environmental stimuli and in the prediction and evaluation of reward (Schultz 1998). According to Pizzagalli et al., both perspectives suggest that activation of the rACC may favor response to treatment by "fostering an individual's capacity to monitor present or future behavior with respect to reward and punishment."

2.4 OVERVIEW OF NEW PERSPECTIVES OF EEG AND MEG ANALYSIS, SUCH AS COHERENCE, SYNCHRONIZATION, AND COMPLEXITY IN DEPRESSION

2.4.1 CONNECTIVITY IN MD: COHERENCE AND SYNCHRONIZATION STUDIES

EEG and MEG connectivity studies are noninvasive tools for studying functional relationships between brain regions. Therefore, they are assumed to reflect functional interactions between neural networks represented in the cortex. Similar to other imaging techniques, connectivity investigations on MD are relatively recent and scarce. Coherence algorithms have been widely used in psychophysiology, and they usually compute the squared cross-correlation in the frequency domain between two EEG/MEG time series measured in two different scalp locations (Hogan et al. 2003). Roemer et al. (1992) obtained interhemispheric coherence values in elderly patients with depression and found that these patients had lower than normal anterior interhemispheric coherence in all frequency bands (i.e., delta, theta, and alpha beta). Reduced coherence values in MD are a common finding. Knott et al. (2001) also found reduced interhemispheric coherence values in MD patients for all frequency bands. Importantly, the combination of delta and beta interhemispheric coherence and alpha interhemispheric asymmetry yielded a 91.3% of overall patient classification capability in a discriminant analysis. As the authors have pointed out, the main problem with coherence estimates is their interpretation in terms of functional and/or anatomical connectivity. According to the results of Roemer et al. and Knott et al. (see also Ford et al. 1986; Lieber 1988), it might be posed that interhemispheric synaptic connection is reduced in MD and other affective disorders (Ford et al. 1986) or that an imbalance exists between the functional processes of both hemispheres. However, coherence results should be interpreted with caution.

Synchronization studies are even less frequent. Fingelkurts et al. (2007) conducted a study on EEG connectivity in unmedicated depressive patients. First, they criticized the current "state of the art" in the field of MD psychophysiology, mostly dominated by the alpha asymmetry paradigm (see the section on alpha asymmetry above). The authors claimed that modern models of mind disorders consider the disease as produced by a change in the autonomy and connectedness of different brain systems that sustain health. According to this, connectivity estimates seem to be more appropriate to detangle the pathophysiology of MD. Fingelkurts et al. (2003) used a synchronization measure called *structural synchrony index*, which estimates general nonlinear interdependences between dynamic systems. Contrary to coherence findings, the authors hypothesized that MD is characterized by an increase in functional connectivity. Their results confirmed the hypothesis that depressive patients showed a state of increased and "strengthened" functional connectivity of brain processes, especially for what they called *short-range distances* (i.e., closer electrode pairs). These results were difficult to explain considering previous coherence reports, but the authors suggested that an imbalance between serotonergic and GABAergic systems might imply an increase in functional interdependence in MD patients' brains.

Synchronization measures were also obtained during sleep in MD patients. It is well known that sleep disturbance is a major symptom of MD, and polysomnography recordings have revealed reduced total sleep time and sleep efficiency (Kemp et al. 2008). Therefore, sleep analysis should be a major issue in MD investigation. Leistedt et al. (2009) used synchronization likelihood, a measure of generalized synchronization (see Stam and van Dijk 2002), to assess connectivity patterns in depressed patients and healthy controls. Results indicated that (1) the acute state of depressive disease was characterized by a decrease in the global mean levels of the EEG sleep synchronization and (2) a major depressive episode displays a significant reorganization of the neural brain networks. These findings may be of particular relevance since slow-wave activity synchronization may play a role in the recently hypothesized importance of slow-wave activity during sleep for cognitive performance (Huber et al. 2004). Cognitive problems, including executive dysfunctions, attention alterations, and memory impairments, are strongly associated with a depressive episode.

2.4.2 Brain Complexity in MD

Complexity analysis is an emerging field of investigation within the theoretical background of nonlinear analysis methods. The fundamental assumption of nonlinear analysis is that EEG or MEG signals are generated by nonlinear deterministic processes with nonlinear coupling interactions between neuronal populations. Complexity analysis is a particular form of nonlinear brain activity analysis. Brain complexity, as derived from EEG–MEG signals, has received different interpretations that usually highlight the degree of randomness predictability and/or the number of independent oscillators underlying the observed signals (Goldberger et al. 1990; Lutzenberger 1995). In particular, the Lempel–Ziv complexity (LZC), one of the most frequently used complexity measures, represents an estimate of the number of different frequency components that actually compose the brain signals (Aboy et al. 2006). Overall, higher variability and number of independent oscillators yield higher complexity scores.

Early studies of complexity in MD (Nandrino et al. 1994) indicated a higher predictability of EEG signals in depressive patients and consequently a decrease in complexity scores. This lower complexity was interpreted as associated with a lower level of environmental interaction, but later investigations contradicted the results of Nandrino et al. For example, Thomasson et al. (2000) observed that averaged global entropy slightly decreased during treatment in patients with depression, suggesting an important correlation with mood modulation in all depressive patients. The main objective of Thomasson et al.

was the demonstration of a longitudinal covariance between brain dynamics complexity (represented by global entropy levels) and mood using different treatment approaches in patients with depression. This study demonstrated a clear association between clinical outcome and brain dynamics reorganization. Later, Thomasson et al. (2002) confirmed such an association. In their intriguing article, Thomasson's group presented a case study of a 48-h cyclic manic–depressive patient where they investigated the correlation between mood variations and EEG nonlinear characteristics, such as entropy values. The correlation between entropy values and clinical self-assessment scale's values was impressive ($\rho = 0.92$). But of even more interest was that the correlation was explained by higher entropy levels associated with depressive phases, whereas lower entropy levels were related to hypomanic phases.

According to these results, it may be hypothesized that depression is associated with higher complexity of electrocortical brain signals. Two recent investigations using LZC methodology supported this hypothesis. First, Li et al. (2008) measured LZC derived from EEG signals in patients with schizophrenia, patients with psychotic depression, and healthy controls. Both patient groups showed higher complexity values when compared with healthy controls. Recently, such a tendency for higher LZC in MD was further confirmed by Méndez et al. (2010). They performed baseline and follow-up evaluation of spontaneous MEG scans in MD patients. At baseline, patients with depression showed higher LZC values, as compared with controls, and failed to exhibit the "normal" tendency to increased complexity values as a function of age observed in controls and described in previous investigations (Anokhin et al. 1996; Fernandez et al. 2009). After 6 months of treatment with mirtazapine, LZC values were reduced in all patients, making them more similar to the controls' LZC scores. Moreover, MD patients recovered the "normal" tendency for increased complexity values as a function of age. Thomasson et al.'s and Méndez et al.'s articles show that depressive phases are associated with a higher complexity of EEG/MEG signals, and also that symptoms' remission produces a reduction of the initially elevated complexity levels.

How can we explain this abnormal increase in electrocortical brain signal complexity in MD? As noted above, complexity estimates assess the variability and number of generators of EEG/MEG signals. Some early electrophysiological studies in schizophrenia emphasized a so-called dysrhythmia (Itil 1977), defined as an abnormally elevated frequency variability and reduced amplitude of EEG traces. Such abnormally elevated frequency variability in schizophrenia correlates with higher complexity scores (see, e.g., Elbert et al. 1992). Is higher frequency variability a common finding in depression studies? Certainly, it is not, but the analysis of frequency variability was not a matter of interest in the field of depression, dominated by traditional power spectrum analyses. Recently, Fingelkurts et al. (2006) published a key investigation on this issue. They employed a novel method for EEG analysis called *probability-classification analysis of short-term EEG spectral patterns*, which is supposed to "extract" the composition of brain oscillations. Two major findings were reported: (1) MD patients' EEG traces were characterized by a reorganization of their oscillatory activity, which affected not only the frontal lobes but also, more importantly, the posterior cortex of the brain. (2) The reorganization of the oscillatory activity in MD was characterized by more segments of polyrhythmic/disorganized activity, as compared to control subjects. This study supports the notion of increased frequency variability in MD, which might explain the elevated complexity levels observed in EEG and MEG studies. Here, it is important to note that complexity estimates demonstrated a "particular" sensitivity to detect the dynamical changes observed in MD (see, e.g., the association between symptoms' remission and complexity variations), and such sensitivity might play an important role in the clinical application of EEG/MEG within this field.

2.5 SUMMARY AND CONCLUSIONS

Studies based on startle reflex paradigms agreed with Rottenberg's ECI model: depressed individuals consistently displayed deficits in the processing of and responses to positive hedonic stimuli on self-report behavioral and psychophysiological measures. These deficits may be considered a vulnerability marker to depression related to serotonergic and stress mechanisms.

The neuropsychological model of depression as put forward by Heller and Nitschke (1997) emphasizes left frontal and right posterior hypoactivation as a key feature of cortical activity pattern in depression. Lesion, EEG, and MEG studies have supported this model. However, comorbid anxiety and depression type exert an important influence on frontal and parietal activity asymmetry. Low anxious melancholic depression seems to result in the previously described cortical asymmetry pattern of left frontal and right parietal hypoactivation indexing negative affect and low emotional arousal, respectively. High anxious depressed patients and atypical depression–type patients may be characterized by increased right temporoparietal activity indexing augmented emotional arousal. Whereas depression-like frontal EEG alpha asymmetry may develop under chronic environmental stress exposure, posterior activity asymmetry could depend more strongly on genetic factors.

However, alpha asymmetry is not the only EEG sign observed in depression patients. Several studies demonstrated a reduced power in the delta and theta bands, accompanied by an increase in high-frequency activity (beta and gamma bands). Such increase was sometimes observed in right frontal areas and considered new evidence of right frontal hyperactivation in depression. EEG may also play a critical role in the prediction of treatment outcome. Since the administration of anti-depressants produces a significant modification of the EEG spectrum, several attempts have been made to use EEG estimates for the prediction of treatment outcome. Nowadays, three lines of investigation seem to offer very promising results with relatively high sensitivity and specificity: frontal theta cordance studies, antidepressant treatment response index, and theta current density on anterior cingulated cortex. As frontal alpha asymmetry measures have been shown to be stable trait-like depression indices, they are not a candidate for a treatment response prediction.

Finally, new investigation tools such as coherence, synchronization, and complexity estimates are used for the evaluation of depression. Results indicate reduced coherence and sleep synchronization but increased synchronization during an awake state. Effective therapies seem to reduce complexity values, making them closer to those values observed in healthy individuals.

REFERENCES

Aboy, M., Hornero, R., Abasolo, D., and Alvarez, D. 2006. Interpretation of the Lempel–Ziv complexity measure in the context of biomedical signal analysis. *IEEE Trans Biomed Eng* 53 2282–8.

Albert, P. R., and Lemonde, S. 2004. 5-HT1A receptors, gene repression, and depression: Guilt by association. *Neuroscientist* 10:575–93.

Allen, J. J., Urry, H. L., Hitt, S. K., and Coan, J. A. 2004. The stability of resting frontal electroencephalographic asymmetry in depression. *Psychophysiology* 41:269–80.

Allen, N. B., Trinder, J., and Brennan, C. 1999. Affective startle modulation in clinical depression: Preliminary findings. *Biol Psychiatry* 46:542–50.

American Psychiatric Association. 1994. *Diagnostic and Statistical Manual of Mental Disorders*, 4th ed. Washington, DC: APA.

Andreasen, N. C. 1997. Linking mind and brain in the study of mental illnesses: A project for a scientific psychopathology. *Science* 275:1586–93.

Andreassi, J. L. 2000. *Psychophysiology and Physiological Response*. Mahwah, NJ: Lawrence Erlbaum Associates.

Anokhin, A. P., Birbaumer, N., Lutzenberger, W., Nikolaev, A., and Vogel, F. 1996. Age increases brain complexity. *Electroencephalogr Clin Neurophysiol* 99:63–8.

Anokhin, A. P., Heath, A. C., Myers, E., Ralano, A., and Wood, S. 2003. Genetic influences on prepulse inhibition of startle reflex in humans. *Neurosci Lett* 353:45–8.

Anokhin, A. P., Heath, A. C., and Myers, E. 2006. Genetic and environmental influences on frontal EEG asymmetry: A twin study. *Biol Psychol* 71:289–95.

Anokhin, A. P., Golosheykin, S., and Heath, A. C. 2007. Genetic and environmental influences on emotion-modulated startle reflex: A twin study. *Psychophysiology* 44:106–12.

Bares, M., Brunovsky, M., Kopecek, M., et al. 2008. Early reduction in prefrontal theta QEEG cordance value predicts response to venlafaxine treatment in patients with resistant depressive disorder. *Eur Psychiatry* 23:350–5.

Beck, A. T. 1976. *Cognitive Therapy and the Emotional Disorders*. New York: International Universities Press.

Bhagwagar, Z., Hafizi, S., and Cowen, P. J. 2002. Cortisol modulation of 5-HT-mediated growth hormone release in recovered depressed patients. *J Affect Disord* 72:249–55.

Bismark, A. W., Moreno, F. A., Stewart, J. L., et al. 2010. Polymorphisms of the HTR1a allele are linked to frontal brain electrical asymmetry. *Biol Psychol* 83:153–8.

Blackhart, G. C., Minnix, J. A., and Kline, J. P. 2006. Can EEG asymmetry patterns predict future development of anxiety and depression? A preliminary study. *Biol Psychol* 72:46–50.

Bradley, M. M., Lang, P. J., and Cuthbert, B. N. 1993. Emotion, novelty, and the startle reflex: Habituation in humans. *Behav Neurosci* 107:970–80.

Brenner, R. P., Ulrich, R. F., Spiker, D. G., et al. 1986. Computerized EEG spectral analysis in elderly normal, demented and depressed subjects. *Electroencephalogr Clin Neurophysiol* 64:483–92.

Brocke, B., Armbruster, D., Muller, J., et al. 2006. Serotonin transporter gene variation impacts innate fear processing: Acoustic startle response and emotional startle. *Mol Psychiatry* 11:1106–12.

Bruder, G. E., Quitkin, F. M., Stewart, J. W., Martin, C., Voglmaier, M. M., and Harrison, W. M. 1989. Cerebral laterality and depression: Differences in perceptual asymmetry among diagnostic subtypes. *J Abnorm Psychol* 98:177–86.

Bruder, G. E., Fong, R., Tenke, C. E., et al. 1997. Regional brain asymmetries in major depression with or without an anxiety disorder: A quantitative electroencephalographic study. *Biol Psychiatry* 41:939–48.

Bruder, G. E., Stewart, J. W., Tenke, C. E., et al. 2001. Electroencephalographic and perceptual asymmetry differences between responders and nonresponders to an SSRI antidepressant. *Biol Psychiatry* 49:416–25.

Bruder, G. E., Stewart, J. W., McGrath, P. J., Ma, G. J., Wexler, B. E., and Quitkin, F. M. 2002. Atypical depression: Enhanced right hemispheric dominance for perceiving emotional chimeric faces. *J Abnorm Psychol* 111:446–54.

Bruder, G. E., Tenke, C. E., Warner, V., et al. 2005. Electroencephalographic measures of regional hemispheric activity in offspring at risk for depressive disorders. *Biol Psychiatry* 57:328–35.

Bruder, G. E., Tenke, C. E., Warner, V., and Weissman, M. M. 2007. Grandchildren at high and low risk for depression differ in EEG measures of regional brain asymmetry. *Biol Psychiatry* 62:1317–23.

Bruder, G. E., Sedoruk, J. P., Stewart, J. W., McGrath, P. J., Quitkin, F. M., and Tenke, C. E. 2008. Electroencephalographic alpha measures predict therapeutic response to a selective serotonin reuptake inhibitor antidepressant: Pre- and post-treatment findings. *Biol Psychiatry* 63:1171–7.

Buchanan, T. W., Brechtel, A., Sollers, J. J., and Lovallo, W. R. 2001. Exogenous cortisol exerts effects on the startle reflex independent of emotional modulation. *Pharmacol Biochem Behav* 68:203–10.

Buss, K. A., Schumacher, J. R., Dolski, I., Kalin, N. H., Goldsmith, H. H., and Davidson, R. J. 2003. Right frontal brain activity, cortisol, and withdrawal behavior in 6-month-old infants. *Behav Neurosci* 117:11–20.

Coan, J. A., and Allen, J. J. 2004. Frontal EEG asymmetry as a moderator and mediator of emotion. *Biol Psychol* 67:7–49.

Cook, I. A., Leuchter, A. F., Uijtdehaage, S. H., et al. 1998. Altered cerebral energy utilization in late life depression. *J Affect Disord* 49:89–99.

Cook, I. A., and Leuchter, A. F. 2001. Prefrontal changes and treatment response prediction in depression. *Semin Clin Neuropsychiatry* 6:113–20.

Cook, I. A., Hunter, A. M., Abrams, M., Siegman, B., and Leuchter, A. F. 2009. Midline and right frontal brain function as a physiologic biomarker of remission in major depression. *Psychiatry Res* 174:152–7.

Cuthbert, B. N., Schupp, H. T., Bradley, M. M., Birbaumer, N., and Lang, P. J. 2000. Brain potentials in affective picture processing: Covariation with autonomic arousal and affective report. *Biol Psychol* 52:95–111.

Davidson, R. J. 1995. Cerebral asymmetry, emotion, and affective style. In *Brain Asymmetry*, eds. R. J. Davidson and K. Hugdahl, 361–87. Cambridge: MIT.

Davidson, R. J. 2003. Affective neuroscience and psychophysiology: Toward a synthesis. *Psychophysiology* 40:655–65.

Davidson, R. J. 2004. What does the prefrontal cortex "do" in affect: Perspectives on frontal EEG asymmetry research. *Biol Psychol* 67:219–33.

Davidson, R. J., Schwartz, G. E., Saron, C., Bennet, J., and Goleman, D. J. 1979. Frontal versus parietal EEG asymmetry during positive and negative affect. *Psychophysiology* 16:202–3.

De Cesarei, A., and Codispoti, M. 2006. When does size not matter? Effects of stimulus size on affective modulation. *Psychophysiology* 43207–15.

De Winter, R. F., van Hemert, A. M., DeRijk, R. H., et al. 2003. Anxious-retarded depression: Relation with plasma vasopressin and cortisol. *Neuropsychopharmacology* 28:140–7.

Deldin, P. J., Keller, J., Gergen, J. A., and Miller, G. A. 2000. Right-posterior face processing anomaly in depression. *J Abnorm Psychol* 109:116–21.

Devinsky, O., Morrell, M. J., and Vogt, B. A. 1995. Contributions of anterior cingulate cortex to behaviour. *Brain* 118 (Pt 1):279–306.

Dichter, G. S., Tomarken, A. J., Shelton, R. C., and Sutton, S. K. 2004. Early- and late-onset startle modulation in unipolar depression. *Psychophysiology* 41:433–40.

Dirks, A., Groenink, L., Schipholt, M. I., et al. 2002. Reduced startle reactivity and plasticity in transgenic mice overexpressing corticotropin-releasing hormone. *Biol Psychiatry* 51:583–90.

Dolcos, F., and Cabeza, R. 2002. Event-related potentials of emotional memory: Encoding pleasant, unpleasant, and neutral pictures. *Cogn Affect Behav Neurosci* 2:252–63.

Elbert, T., Lutzenberger, W., Rockstroh, B., Berg, P., and Cohen, R. 1992. Physical aspects of the EEG in schizophrenics. *Biol Psychiatry* 32:595–606.

Fernandez-Duque, D., and Posner, M. I. 2001. Brain imaging of attentional networks in normal and pathological states. *J Clin Exp Neuropsychol* 23:74–93.

Fernandez, A., Rodriguez-Palancas, A., Lopez-Ibor, M., et al. 2005. Increased occipital delta dipole density in major depressive disorder determined by magnetoencephalography. *J Psychiatry Neurosci* 30:17–23.

Fernandez, A., Quintero, J., Hornero, R., et al. 2009. Complexity analysis of spontaneous brain activity in attention-deficit/hyperactivity disorder: Diagnostic implications. *Biol Psychiatry* 65:571–7.

Fingelkurts, A. A., Fingelkurts, A. A., Krause, C. M., Mottonen, R., and Sams, M. 2003. Cortical operational synchrony during audio-visual speech integration. *Brain Lang* 85:297–312.

Fingelkurts, A. A., Rytsala, H., Suominen, K., Isometsa, E., and Kahkonen, S. 2006. Composition of brain oscillations in ongoing EEG during major depression disorder. *Neurosci Res* 56:133–44.

Fingelkurts, A. A., Fingelkurts, A. A., Rytsala, H., Suominen, K., Isometsa, E., and Kahkonen, S. 2007. Impaired functional connectivity at EEG alpha and theta frequency bands in major depression. *Hum Brain Mapp* 28:247–61.

Fink, M., Wadsak, W., Savli, M., et al. 2009. Lateralization of the serotonin-1A receptor distribution in language areas revealed by PET. *NeuroImage* 45:598–605.

Flor-Henry, P., Lind, J. C., and Koles, Z. J. 2004. A source-imaging (low-resolution electromagnetic tomography) study of the EEGs from unmedicated males with depression. *Psychiatry Res* 130:191–207.

Forbes, E. E., Miller, A., Cohn, J. F., Fox, N. A., and Kovacs, M. 2005. Affect-modulated startle in adults with childhood-onset depression: Relations to bipolar course and number of lifetime depressive episodes. *Psychiatry Res* 134:11–25.

Ford, M. R., Goethe, J. W., and Dekker, D. K. 1986. EEG coherence and power in the discrimination of psychiatric disorders and medication effects. *Biol Psychiatry* 21:1175–88.

Fowles, D. C. 1988. Psychophysiology and psychopathology: A motivational approach. *Psychophysiology* 25:373–91.

Gainotti, G. 1972. Emotional behavior and hemispheric side of the lesion. *Cortex* 8:41–55.

Gao, Y., Tuvblad, C., Raine, A., Lozano, D. I., and Baker, L. A. 2009. Genetic and environmental influences on frontal EEG asymmetry and alpha power in 9–10-year-old twins. *Psychophysiology* 46:787–96.

Gehricke, J., and Shapiro, D. 2000. Reduced facial expression and social context in major depression: Discrepancies between facial muscle activity and self-reported emotion. *Psychiatry Res* 95:157–67.

Gloor, P., Ball, G., and Schaul, N. 1977. Brain lesions that produce delta waves in the EEG. *Neurology* 27:326–33.

Goldberger, A. L., Rigney, D. R., and West, B. J. 1990. Chaos and fractals in human physiology. *Sci Am* 262:42–9.

Goldstein, K. 1939. *The Organism: A Holistic Approach to Biology, Derived from Pathological Data in Man.* New York: American Book.

Goodyer, I. M., Park, R. J., Netherton, C. M., and Herbert, J. 2001. Possible role of cortisol and dehydroepiandrosterone in human development and psychopathology. *Br J Psychiatry* 179:243–9.

Gotlib, I. H., and MacLeod, C. 1997. Information processing in anxiety and depression: A cognitive developmental perspective. In *Attention, Development, and Psychopathology,* eds. J. Burack and J. Enns, 350–78. New York: Guilford Press.

Gotlib, I. H., Ranganath, C., and Rosenfeld, P. 1998. Frontal EEG alpha asymmetry, depression, and cognitive functioning. *Cogn Emotion* 12:449–78.

Grillon, C. and Baas, J. 2003. A review of the modulation of the startle reflex by affective states and its application in psychiatry. *Clin Neurophysiol* 114:1557–79.

Hagemann, D., Naumann, E., Thayer, J. F., and Bartussek, D. 2002. Does resting electroencephalograph asymmetry reflect a trait? An application of latent state–trait theory. *J Pers Soc Psychol* 82:619–41.

Hajcak, G., and Olvet, D. M. 2008. The persistence of attention to emotion: Brain potentials during and after picture presentation. *Emotion* 8:250–5.

Hajcak, G., Dunning, J. P., and Foti, D. 2009. Motivated and controlled attention to emotion: Time-course of the late positive potential. *Clin Neurophysiol* 120:505–10.

Hariri, A. R., Drabant, E. M., and Weinberger, D. R. 2006. Imaging genetics: Perspectives from studies of genetically driven variation in serotonin function and corticolimbic affective processing. *Biol Psychiatry* 59:888–97.

Harmer, C. J., Shelley, N. C., Cowen, P. J., and Goodwin, G. M. 2004. Increased positive versus negative affective perception and memory in healthy volunteers following selective serotonin and norepinephrine reuptake inhibition. *Am J Psychiatry* 161:1256–63.

Harmer, C. J., Mackay, C. E., Reid, C. B., Cowen, P. J., and Goodwin, G. M. 2006. Antidepressant drug treatment modifies the neural processing of nonconscious threat cues. *Biol Psychiatry* 59:816–20.

Harmon-Jones, E., Gable, P. A., and Peterson, C. K. 2010. The role of asymmetric frontal cortical activity in emotion-related phenomena: A review and update. *Biol Psychology* 84:451–62.

Hegerl, U., and Juckel, G. 1993. Intensity dependence of auditory evoked potentials as an indicator of central serotonergic neurotransmission: A new hypothesis. *Biol Psychiatry* 33:173–87.

Heller, W. 1998. The puzzle of regional brain activity in depression and anxiety: The importance of subtypes and comorbidity. *Cogn Emotion* 12:421–47.

Heller, W., and Nitschke, J. B. 1997. Regional brain activity in emotion: A framework for understanding cognition in depression. *Cogn Emotion* 11:637–61.

Heller, W., Etienne, M. A., and Miller, G. A. 1995. Patterns of perceptual asymmetry in depression and anxiety: Implications for neuropsychological models of emotion and psychopathology. *J Abnorm Psychol* 104:327–33.

Heller, W., Nitschke, J. B., and Lindsay, D. L. 1997a. Neuropsychological correlates of arousal in self-reported Emotion. *Cogn Emotion* 11:383–402.

Heller, W., Nitschke, J. B., Etienne, M. A., and Miller, G. A. 1997b. Patterns of regional brain activity differentiate types of anxiety. *J Abnorm Psychol* 106:376–85.

Heller, W., Nitschke, J. B., and Miller, G. A. 1998. Lateralization in emotion and emotional disorders. *Curr Direct Psychol Sci* 7:26–32.

Henriques, J. B., and Davidson, R. J. 1990. Regional brain electrical asymmetries discriminate between previously depressed and healthy control subjects. *J Abnorm Psychol* 99:22–31.

Henriques, J. B., and Davidson, R. J. 1991. Left frontal hypoactivation in depression. *J Abnorm Psychol* 100:535–45.

Henriques, J. B., and Davidson, R. J. 2000. Decreased responsiveness to reward in depression. *Cogn Emotion* 14:711–24.

Hogan, M. J., Swanwick, G. R., Kaiser, J., Rowan, M., and Lawlor, B. 2003. Memory-related EEG power and coherence reductions in mild Alzheimer's disease. *Int J Psychophysiol* 49:147–63.

Huang, Y. X., and Luo, Y. J. 2006. Temporal course of emotional negativity bias: An ERP study. *Neurosci Lett* 398:91–6.

Huber, R., Ghilardi, M. F., Massimini, M., and Tononi, G. 2004. Local sleep and learning. *Nature* 430:78–81.

Hughes, J. R., and John, E. R. 1999. Conventional and quantitative electroencephalography in psychiatry. *J Neuropsychiatry Clin Neurosci* 11:190–208.

Hunter, A. M., Cook, I. A., and Leuchter, A. F. 2007. The promise of the quantitative electroencephalogram as a predictor of antidepressant treatment outcomes in major depressive disorder. *Psychiatr Clin North Am* 30:105–24.

Iosifescu, D. V., Greenwald, S., Devlin, P., et al. 2009. Frontal EEG predictors of treatment outcome in major depressive disorder. *Eur Neuropsychopharmacol* 19:772–7.

Itil, T. M. 1977. Qualitative and quantitative EEG findings in schizophrenia. *Schizophr Bull* 3:61–79.

Itil, T. M. 1983. The discovery of antidepressant drugs by computer-analyzed human cerebral bio-electrical potentials (CEEG). *Prog Neurobiol* 20:185–249.

Jacobs, G. D., and Snyder, D. 1996. Frontal brain asymmetry predicts affective style in men. *Behav Neurosci* 110:3–6.

Jaeger, J., Borod, J. C., and Peselow, E. 1987. Depressed patients have atypical hemispace biases in the perception of emotional chimeric faces. *J Abnorm Psychol* 96:321–4.

Kalin, N. H., Larson, C., Shelton, S. E., and Davidson, R. J. 1998. Asymmetric frontal brain activity, cortisol, and behavior associated with fearful temperament in rhesus monkeys. *Behav Neurosci* 112:286–92.

Kasch, K. L., Rottenberg, J., Arnow, B. A., and Gotlib, I. H. 2002. Behavioral activation and inhibition systems and the severity and course of depression. *J Abnorm Psychol* 111:589–97.

Kaviani, H., Gray, J. A., Checkley, S. A., Raven, P. W., Wilson, G. D., and Kumari, V. 2004. Affective modulation of the startle response in depression: Influence of the severity of depression, anhedonia, and anxiety. *J Affect Disord* 83:21–31.

Kayser, J., Bruder, G. E., Tenke, C. E., Stewart, J. E., and Quitkin, F. M. 2000. Event-related potentials (ERPs) to hemifield presentations of emotional stimuli: Differences between depressed patients and healthy adults in P3 amplitude and asymmetry. *Int J Psychophysiol* 36:211–36.

Keil, A., Bradley, M. M., Hauk, O., Rockstroh, B., Elbert, T., and Lang, P. J. 2002. Large-scale neural correlates of affective picture processing. *Psychophysiology* 39:641–9.

Keil, A., Gruber, T., Müller, M. M., et al. 2003. Early modulation of visual perception by emotional arousal: Evidence from steady-state visual evoked brain potentials. *Cogn Affect Behav Neurosci* 3:195–206.

Keil, A., Moratti, S., Sabatinelli, D., Bradley, M. M., and Lang, P. J. 2005. Additive effects of emotional content and spatial selective attention on electrocortical facilitation. *Cereb Cortex* 15:1187–97.

Keller, J., Nitschke, J. B., Bhargava, T., et al. 2000. Neuropsychological differentiation of depression and anxiety. *J Abnorm Psychol* 109:3–10.

Kemp, A. H., Gray, M. A., Eide, P., Silberstein, R. B., and Nathan, P. J. 2002. Steady-state visually evoked potential topography during processing of emotional valence in healthy subjects. *NeuroImage* 17:1684–92.

Kemp, A. H., Gray, M. A., Silberstein, R. B., Armstrong, S. M., and Nathan, P. J. 2004. Augmentation of serotonin enhances pleasant and suppresses unpleasant cortical electrophysiological responses to visual emotional stimuli in humans. *NeuroImage* 22:1084–96.

Kemp, A. H., Gordon, E., Rush, A. J., and Williams, L. M. 2008. Improving the prediction of treatment response in depression: Integration of clinical, cognitive, psychophysiological, neuroimaging, and genetic measures. *CNS Spectr* 13:1066–88.

Kentgen, L. M., Tenke, C. E., Pine, D. S., Fong, R., Klein, R. G., and Bruder, G. E. 2000. Electroencephalographic asymmetries in adolescents with major depression: Influence of comorbidity with anxiety disorders. *J Abnorm Psychol* 109:797–802.

Knott, V., Mahoney, C., Kennedy, S., and Evans, K. 2001. EEG power, frequency, asymmetry and coherence in male depression. *Psychiatry Res* 106:123–40.

Knott, V., Mahoney, C., Kennedy, S., and Evans, K. 2002. EEG correlates of acute and chronic paroxetine treatment in depression. *J Affect Disord* 69:241–9.

Knott, V. J., Telner, J. I., Lapierre, Y. D., Browne, M., and Horn, E. R. 1996. Quantitative EEG in the prediction of antidepressant response to imipramine. *J Affect Disord* 39:175–84.

Korb, A. S., Hunter, A. M., Cook, I. A., and Leuchter, A. F. 2009. Rostral anterior cingulate cortex theta current density and response to antidepressants and placebo in major depression. *Clin Neurophysiol* 120:1313–9.

Krieg, J. C., Lauer, C. J., Schreiber, W., Modell, S., and Holsboer, F. 2001. Neuroendocrine, polysomnographic and psychometric observations in healthy subjects at high familial risk for affective disorders: The current state of the 'Munich vulnerability study'. *J Affect Disord* 62:33–7.

Kucharska-Pietura, K., and David, A. S. 2003. The perception of emotional chimeric faces in patients with depression, mania and unilateral brain damage. *Psychol Med* 33:739–45.

Lang, P. J., and Bradley, M. M. 2010. Emotion and the motivational brain. *Biol Psychol* 84:437–50.

Lang, P. J., Bradley, M. M., and Cuthbert, B. N. 1998. Emotion, motivation, and anxiety: Brain mechanisms and psychophysiology. *Biol Psychiatry* 44:1248–63.

Lang, P. J., Bradley, M. M., and Cuthbert, B. N. 2005. *International Affective Picture System (IAPS): Affective Ratings of Pictures and Instruction Manual. Technical Report A-6*. Gainesville: University of Florida.

Lee, G. P., Loring, D. W., Meador, K. J., and Brooks, B. B. 1990. Hemispheric specialization for emotional expression: A reexamination of results from intracarotid administration of sodium amobarbital. *Brain Cogn* 12:267–80.

Leistedt, S. J., Coumans, N., Dumont, M., Lanquart, J. P., Stam, C. J., and Linkowski, P. 2009. Altered sleep brain functional connectivity in acutely depressed patients. *Hum Brain Mapp* 30:2207–19.

Lesch, K. P. 2001. Serotonergic gene expression and depression: Implications for developing novel antidepressants. *J Affect Disord* 62:57–76.

Leuchter, A. F., Cook, I. A., Lufkin, R. B., et al. 1994. Cordance: A new method for assessment of cerebral perfusion and metabolism using quantitative electroencephalography. *NeuroImage* 1:208–19.

Leuchter, A. F., Cook, I. A., Uijtdehaage, S. H., et al. 1997. Brain structure and function and the outcomes of treatment for depression. *J Clin Psychiatry* 58 (Suppl 16):22–31.

Leuchter, A. F., Uijtdehaage, S. H., Cook, I. A., O'Hara, R., and Mandelkern, M. 1999. Relationship between brain electrical activity and cortical perfusion in normal subjects. *Psychiatry Res* 90:125–40.

Leuchter, A. F., Cook, I. A., Hunter, A. M., and Korb, A. S. 2009a. A new paradigm for the prediction of antidepressant treatment response. *Dialogues Clin Neurosci* 11:435–46.

Leuchter, A. F., Cook, I. A., Gilmer, W. S., et al. 2009b. Effectiveness of a quantitative electroencephalographic biomarker for predicting differential response or remission with escitalopram and bupropion in major depressive disorder. *Psychiatry Res* 169:132–8.

Leuchter, A. F., Cook, I. A., Marangell, L. B., et al. 2009c. Comparative effectiveness of biomarkers and clinical indicators for predicting outcomes of SSRI treatment in Major Depressive Disorder: Results of the BRITE-MD study. *Psychiatry Res* 169:124–31.

Levy, J., Heller, W., Banich, M. T., and Burton, L. A. 1983. Asymmetry of perception in free viewing of chimeric faces. *Brain Cogn* 2:404–19.

Li, L., Power, C., Kelly, S., Kirschbaum, C., and Hertzman, C. 2007. Life-time socio-economic position and cortisol patterns in mid-life. *Psychoneuroendocrinology* 32:824–33.

Li, Y., Tong, S., Liu, D., et al. 2008. Abnormal EEG complexity in patients with schizophrenia and depression. *Clin Neurophysiol* 119:1232–41.

Lieber, A. L. 1988. Diagnosis and subtyping of depressive disorders by quantitative electroencephalography: II. Interhemispheric measures are abnormal in major depressives and frequency analysis may discriminate certain subtypes. *Hillside J Clin Psychiatry* 10:84–97.

Loas, G., Salinas, E., Pierson, A., Guelfi, J. D., and Samuel-Lajeunesse, B. 1994. Anhedonia and blunted affect in major depressive disorder. *Compr Psychiatry* 35:366–72.

Lubar, J. F., Congedo, M., and Askew, J. H. 2003. Low-resolution electromagnetic tomography (LORETA) of cerebral activity in chronic depressive disorder. *Int J Psychophysiol* 49:175–85.

Lutzenberger, W., Preissl, H., and Pulvermuller, F. 1995. Fractal dimension of electroencephalographic time series and underlying brain processes. *Biol Cybern* 73:477–82.

Marie-Mitchell, A., Leuchter, A. F., Chou, C. P., James Gauderman, W., and Azen, S. P. 2004. Predictors of improved mood over time in clinical trials for major depression. *Psychiatry Res* 127:73–84.

Mast, J., and Victor, J. D. 1991. Fluctuations of steady-state VEPs: Interaction of driven evoked potentials and the EEG. *Electroencephalogr Clin Neurophysiol* 78:389–401.

Mayberg, H. S., Brannan, S. K., Mahurin, R. K., et al. 1997. Cingulate function in depression: A potential predictor of treatment response. *Neuroreport* 8:1057–61.

Mayberg, H. S., Liotti, M., Brannan, S. K., et al. 1999. Reciprocal limbic–cortical function and negative mood: Converging PET findings in depression and normal sadness. *Am J Psychiatry* 156:675–82.

McAllister-Williams, R. H., Ferrier, I. N., and Young, A. H. 1998. Mood and neuropsychological function in depression: The role of corticosteroids and serotonin. *Psychol Med* 28:573–84.

Meadows, M. E., and Kaplan, R. F. 1994. Dissociation of autonomic and subjective responses to emotional slides in right hemisphere damaged patients. *Neuropsychologia* 32:847–56.

Mekli, K., Payton, A., Miyajima, F., et al. 2010. The HTR1A and HTR1B receptor genes influence stress-related information processing. *Eur Neuropsychopharmacol* doi:10.1016/j.euroneuro.2010.06.013.

Méndez, M. A., Zuluaga, P., Rodríguez-Palancas, A., et al. 2010. Complexity analysis of spontaneous brain activity in major depression: An approach to understand the process of symptom remission. *Clin Neurophysiol* in press.

Metzger, L. J., Paige, S. R., Carson, M. A., et al. 2004. PTSD arousal and depression symptoms associated with increased right-sided parietal EEG asymmetry. *J Abnorm Psychol* 113:324–9.

Mientus, S., Gallinet, J., Wuebben, Y., et al. 2002. Cortical hypoactivation during resting EEG in schizophrenics but not in depressives and schizotypal subjects as revealed by low resolution electromagnetic tomography (LORETA). *Psychiatry Res Neuroimaging* 116:95–111.

Miller, M. W., and Gronfier, C. 2006. Diurnal variation of the startle reflex in relation to HPA-axis activity in humans. *Psychophysiology* 43:297–301.

Mneimne, M., McDermunt, W., and Powers, A.S. 2008. Affective ratings and startle modulation in people with nonclinical depression. *Emotion* 8:552–9.

Moratti, S., Keil, A., and Stolarova, M. 2004. Motivated attention in emotional picture processing is reflected by activity modulation in cortical attention networks. *NeuroImage* 21:954–64.

Moratti, S., Rubio, G., Campo, P., Keil, A., and Ortiz, T. 2008. Hypofunction of right temporoparietal cortex during emotional arousal in depression. *Arch Gen Psychiatry* 65:532–41.

Morgan, S. T., Hansen, J. C., and Hillyard, S. A. 1996. Selective attention to stimulus location modulates the steady-state visual evoked potential. *Proc Natl Acad Sci U S A* 93:4770–4.

Mucci, A., Volpe, U., Merlotti, E., Bucci, P., and Galderisi, S. 2006. Pharmaco-EEG in psychiatry. *Clin EEG Neurosci* 37:81–98.

Mulert, C., Juckel, G., Augustin, H., and Hegerl, U. 2002. Comparison between the analysis of the loudness dependency of the auditory N1/P2 component with LORETA and dipole source analysis in the prediction of treatment response to the selective serotonin reuptake inhibitor citalopram in major depression. *Clin Neurophysiol* 113:1566–72.

Mulert, C., Juckel, G., Brunnmeier, M., et al. 2007. Prediction of treatment response in major depression: Integration of concepts. *J Affect Disord* 98:215–25.

Müller, M. M., and Hillyard, S. 2000. Concurrent recording of steady-state and transient event-related potentials as indices of visual–spatial selective attention. *Clin Neurophysiol* 111:1544–52.

Müller, M. M., and Hübner, R. 2002. Can the spotlight of attention be shaped like a doughnut? Evidence from steady-state visual evoked potentials. *Psychol Sci* 13:119–24.

Nandrino, J. L., Pezard, L., Martinerie, J., et al. 1994. Decrease of complexity in EEG as a symptom of depression. *Neuroreport* 5:528–30.

Nanry, K. P., and Tilson, H. A. 1989. The role of 5HT1A receptors in the modulation of the acoustic startle reflex in rats. *Psychopharmacology (Berl)* 97:507–13.

Nesse, R. M. 2000. Is depression an adaptation? *Arch Gen Psychiatry* 57:14–20.

Nitschke, J. B., Heller, W., Etienne, M. A., and Miller, G. A. 2004. Prefrontal cortex activity differentiates processes affecting memory in depression. *Biol Psychol* 67:125–43.

Oakes, T. R., Pizzagalli, D. A., Hendrick, A. M., et al. 2004. Functional coupling of simultaneous electrical and metabolic activity in the human brain. *Hum Brain Mapp* 21:257–70.

O'Brien-Simpson, L., Di Parsia, P., Simmons, J.G., and Allen, N.B. 2009. Recurrence of major depressive disorder is predicted by inhibited startle magnitude while recovered. *J Affect Disord* 112:243–9.

Olofsson, J. K., Nordin, S., Sequeira, H., and Polich, J. 2008. Affective picture processing: An integrative review of ERP findings. *Biol Psychol* 77:247–65.

Pardon, M. C., Gould, G. G., Garcia, A., et al. 2002. Stress reactivity of the brain noradrenergic system in three rat strains differing in their neuroendocrine and behavioral responses to stress: Implications for susceptibility to stress-related neuropsychiatric disorders. *Neuroscience* 115:229–42.

Pascual-Marqui, R. D. 1999. Review of the methods for solving the EEG inverse problem. *Int J Bioelectromagnetism* 1:75–86.

Pastor, M. C., Bradley, M. M., Low, A., Versace, F., Molto, J., and Lang, P. J. 2008. Affective picture perception: Emotion, context, and the late positive potential. *Brain Res* 1189:145–51.

Pizzagalli, D., Pascual-Marqui, R. D., Nitschke, J. B., et al. 2001. Anterior cingulate activity as a predictor of degree of treatment response in major depression: Evidence from brain electrical tomography analysis. *Am J Psychiatry* 158:405–15.

Pizzagalli, D. A., Nitschke, J. B., Oakes, T. R., et al. 2002. Brain electrical tomography in depression: The importance of symptom severity, anxiety, and melancholic features. *Biol Psychiatry* 52:73–85.

Porter, R. J., Gallagher, P., Watson, S., and Young, A. H. 2004. Corticosteroid–serotonin interactions in depression: A review of the human evidence. *Psychopharmacology (Berl)* 173:1–17.

Pössel, P., Lo, H., Fritz, A., and Seemann, S. 2008. A longitudinal study of cortical EEG activity in adolescents. *Biol Psychol* 78:173–8.

Pozzi, D., Golimstock, A., Petracchi, M., Garcia, H., and Starkstein, S. 1995. Quantified electroencephalographic changes in depressed patients with and without dementia. *Biol Psychiatry* 38:677–83.

Regan, D. 1989. *Human Brain Electrophysiology: Evoked Potentials and Evoked Magnetic Fields in Science and Medicine*. New York, NY: Elsevier.

Robinson, R. G., Kubos, K. L., Starr, L. B., Rao, K., and Price, T. R. 1984. Mood disorders in stroke patients. Importance of location of lesion. *Brain* 107 (Pt. 1):81–93.

Roemer, R. A., Shagass, C., Dubin, W., Jaffe, R., and Siegal, L. 1992. Quantitative EEG in elderly depressives. *Brain Topogr* 4:285–90.

Rottenberg, J. 2005. Mood and emotion in major depression. *Curr Direct Psychol Sci* 14:167–70.

Rottenberg, J. 2007. Major depressive disorder: Emerging evidence for emotion context insensitivity. In *Emotion and Psychopathology: Bridging Affective and Clinical Science*, eds. J. Rottenberg and S. L. Johnson, 1. Washington, DC: American Psychological Association.

Rottenberg, J., Kasch, K. L., Gross, J. J., and Gotlib, I. H. 2002. Sadness and amusement reactivity differentially predict concurrent and prospective functioning in major depressive disorder. *Emotion* 2: 135–46.

Rottenberg, J., Gross, J. J., and Gotlib, I. H. 2005. Emotion context insensitivity in major depressive disorder. *J Abnorm Psychol* 114:627–39.

Sabatinelli, D., Lang, P. J., Keil, A., and Bradley, M. M. 2007. Emotional perception: Correlation of functional MRI and event-related potentials. *Cereb Cortex* 17:1085–91.

Sackeim, H. A., Greenberg, M. S., Weiman, A. L., Gur, R. C., Hungerbuhler, J. P., and Geschwind, N. 1982. Hemispheric asymmetry in the expression of positive and negative emotions. Neurologic evidence. *Arch Neurol* 39:210–8.

Sackeim, H. A., Luber, B., Katzman, G. P., et al. 1996. The effects of electroconvulsive therapy on quantitative electroencephalograms. Relationship to clinical outcome. *Arch Gen Psychiatry* 53:814–24.

Schaffer, C. E., Davidson, R. J., and Saron, C. 1983. Frontal and parietal electroencephalogram asymmetry in depressed and nondepressed subjects. *Biol Psychiatry* 18:753–62.

Schultz, W. 1998. Predictive reward signal of dopamine neurons. *J Neurophysiol* 80:1–27.

Schupp, H. T., Cuthbert, B. N., Bradley, M. M., Cacioppo, J. T., Ito, T., and Lang, P. J. 2000. Affective picture processing: The late positive potential is modulated by motivational relevance. *Psychophysiology* 37:257–61.

Shankman, S. A., Tenke, C. E., Bruder, G. E., Durbin, C. E., Hayden, E. P., and Klein, D. N. 2005. Low positive emotionality in young children: Association with EEG asymmetry. *Dev Psychopathol* 17:85–98.

Shimoda, K., and Robinson, R. G. 1999. The relationship between poststroke depression and lesion location in long-term follow-up. *Biol Psychiatry* 45:187–92.

Sigmon, S. T., and Nelson-Gray, R. O. 1992. Sensitivity to aversive events in depression: Antecedent, concomitant, or consequent. *J Psychopathol Behav Assess* 14:225–46.

Sloan, D. M., Strauss, M. E., and Wisner, K. L. 2001. Diminished response to pleasant stimuli by depressed women. *J Abnorm Psychol* 110:488–93.

Solberg, L. C., Ahmadiyeh, N., Baum, A. E., et al. 2003. Depressive-like behavior and stress reactivity are independent traits in a Wistar Kyoto × Fisher 344 cross. *Mol Psychiatry* 8:423–33.

Sperling, W., Martus, P., and Alschbach, M. 2000. Evaluation of neuronal effects of electroconvulsive therapy by magnetoencephalography (MEG). *Prog Neuropsychopharmacol Biol Psychiatry* 24:1339–54.

Stam, C. J., and van Dijk, B. W. 2002. Synchronization likelihood: An unbiased measure on generalized synchronization in multivariate datasets. *Physica D* 19:562–74.

Strelets, V. B., Garakh, Zh., V., and Novototskii-Vlasov, V. Y. 2007. Comparative study of the gamma rhythm in normal conditions, during examination stress, and in patients with first depressive episode. *Neurosci Behav Physiol* 37:387–94.

Thomasson, N., Pezard, L., Allilaire, J., Renault, B., and Jacques, M. 2000. Nonlinear EEG changes associated with clinical improvement in depressed patients. *Nonlinear Dynamics Psychol Life Sci* 4:203–18.

Thomasson, N., Pezard, L., Boyer, P., Renault, B., and Martinerie, J. 2002. Nonlinear EEG changes in a 48-hour cyclic manic-depressive patient. *Nonlinear Dynamics Psychol Life Sci* 6:259–67.

Tomarken, A. J., Davidson, R. J., and Henriques, J. B. 1990. Resting frontal brain asymmetry predicts affective responses to films. *J Pers Soc Psychol* 59:791–801.

Tomarken, A. J., Davidson, R. J., Wheeler, R. E., and Doss, R. C. 1992a. Individual differences in anterior brain asymmetry and fundamental dimensions of emotion. *J Pers Soc Psychol* 62:676–87.

Tomarken, A. J., Davidson, R. J., Wheeler, R. E., and Kinney, L. 1992b. Psychometric properties of resting anterior EEG asymmetry: Temporal stability and internal consistency. *Psychophysiology* 29:576–92.

Tomarken, A. J., Dichter, G. S., Garber, J., and Simien, C. 2004. Resting frontal brain activity: Linkages to maternal depression and socio-economic status among adolescents. *Biol Psychol* 67:77–102.

Tomarken, A. J., Shelton, R. C., and Holton, S. D. in press. Affective science as a framework for understanding the mechanisms and effects of antidepressant medications. In *Emotion and psychopathology: Bridging affective and clinical science*, eds. J. Rottenberg and S. L. Johnson. Washington DC: American Psychological Society.

Tops, M., Wijers, A. A., van Staveren, A. S., et al. 2005. Acute cortisol administration modulates EEG alpha asymmetry in volunteers: Relevance to depression. *Biol Psychol* 69:181–93.

Towers, D. N., and Allen, J. J. 2009. A better estimate of the internal consistency reliability of frontal EEG asymmetry scores. *Psychophysiology* 46:132–42.

Trivedi, M. H., Rush, A. J., Wisniewski, S. R., et al. 2006. Evaluation of outcomes with citalopram for depression using measurement-based care in STAR*D: Implications for clinical practice. *Am J Psychiatry* 163:28–40.

Ulrich, G., Renfordt, E., Zeller, G., and Frick, K. 1984. Interrelation between changes in the EEG and psychopathology under pharmacotherapy for endogenous depression. A contribution to the predictor question. *Pharmacopsychiatry* 17:178–83.

Ulrich, G., Renfordt, E., and Frick, K. 1986. The topographical distribution of alpha-activity in the resting EEG of endogenous-depressive in-patients with and without clinical response to pharmacotherapy. *Pharmacopsychiatria* 19:272–3.

Vuilleumier, P., and Driver, J. 2007. Modulation of visual processing by attention and emotion: Windows on causal interactions between human brain regions. *Philos Trans R Soc Lond B Biol Sci* 362:837–55.

Watson, D., Clark, L. A., and Tellegen, A. 1988. Development and validation of brief measures of positive and negative affect: The PANAS scales. *J Pers Soc Psychol* 54:1063–70.

Weiland-Fiedler, P., Erickson, K., Waldeck, T., et al. 2004. Evidence for continuing neuropsychological impairments in depression. *J Affect Disord* 82:253–8.

Wheeler, R. E., Davidson, R. J., and Tomarken, A. J. 1993. Frontal brain asymmetry and emotional reactivity: A biological substrate of affective style. *Psychophysiology* 30:82–9.

Wienbruch, C., Moratti, S., Elbert, T., et al. 2003. Source distribution of neuromagnetic slow wave activity in schizophrenic and depressive patients. *Clin Neurophysiol* 114:2052–60.

Yee, C. M., and Miller, G. A. 1988. Emotional information processing: Modulation of fear in normal and dysthymic subjects. *J Abnorm Psychol* 97:54–63.

3 Neural Foundations of Major Depression: Classical Approaches and New Frontiers

J. Paul Hamilton, Daniella J. Furman, and Ian H. Gotlib

CONTENTS

3.1 INTRODUCTION

Major depressive disorder (MDD) is among the most prevalent of all psychiatric disorders. Recent estimates indicate that almost 20% of the American population, or more than 30 million adults, will experience a clinically significant episode of depression during their lifetime (Kessler and Wang 2009). Moreover, depression is frequently comorbid with other mental and physical difficulties, including anxiety disorders, cardiac problems, and smoking (e.g., Freedland and Carney 2009). Depression also has significant economic and social costs. Kessler et al. (2006), for example, estimated that the annual salary-equivalent costs of depression-related lost productivity in the United States exceed $36 billion. Given the high prevalence, comorbidity, and costs of depression, it is not surprising that the World Health Organization Global Burden of Disease Study ranked this disorder as the single most burdensome disease worldwide (Murray and Lopez 1996). Finally, it is important to note that depression is a highly recurrent disorder. More than 75% of depressed patients have more than one depressive episode, often relapsing within 2 years of recovery from a depressive episode (Boland and Keller 2009). Indeed, between one-half and two-thirds of people who have ever been clinically depressed will be in an episode in any given year over the remainder of their lives (Kessler and Wang 2009).

Although MDD is primarily a disorder of emotion and its regulation, it is important to recognize that this disorder can be characterized by a full constellation of behavioral, emotional, and cognitive symptoms, including sad mood and/or a loss of interest or pleasure in almost all daily activities,

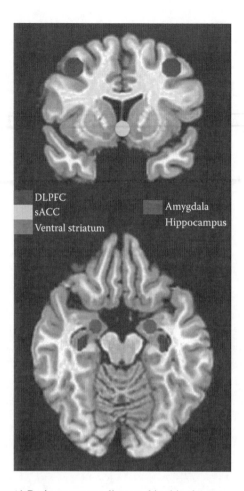

FIGURE 3.1 **(See color insert.)** Brain structures discussed in this chapter.

weight loss or gain, sleep disturbance, psychomotor agitation or retardation, fatigue, suicidal ide-ation, and concentration difficulties. In addition to the symptoms used to derive a formal diagnosis of MDD, investigators have also documented that depressed individuals exhibit enhanced process-ing of negative material and diminished processing of positive information, as well as blunted responsivity to rewarding stimuli (see Gotlib and Joormann 2010).

In attempting to gain a more comprehensive understanding of the development and maintenance of these symptoms, over the past two decades investigators have used neuroimaging techniques to examine the neural substrates of MDD. In this review, we present findings from this body of research, identifying the major brain regions or structures that have been implicated in depres-sion and discussing their relation, when possible, to specific *Diagnostic and Statistical Manual of Mental Disorders* symptoms of MDD. In this chapter, we focus on the structures that have received the most significant empirical attention: the amygdala, the hippocampus, the subgenual anterior cingulate cortex (sACC), the dorsolateral prefrontal cortex (DLPFC), and the ventral striatum (VS) (see Figure 3.1). For each of these structures, we present a brief overview of the general functions associated with the structure and then discuss findings of studies relating both volumetric and func-tional anomalies of the structure to MDD. In this context, we describe the types of tasks that have been used in the scanner to examine differences between depressed and nondepressed individuals in their patterns of neural activation in these structures.

For each structure, we also present the results of studies that have examined the relation between changes in functional or structural characteristics with recovery, or remission, of depression, either naturally or as a result of a specific intervention. Results from these studies, and from investigations of individuals at elevated risk for MDD, are important in helping us understand the temporal relation between depression and neural functional and/or structural anomalies. Following this presentation, we summarize the current state of our understanding of the neural aspects of depression based on our review of this literature. We then describe what we believe are three important (and necessary) directions for future research: conducting systems-level investigations of the neural foundations of MDD, examining the neural functioning of individuals at elevated risk for depression to gain a clearer understanding of the causal relation between neural anomalies and MDD, and assessing the effects of manipulating the level of activation in these neural structures on the course of depression. Before we begin our discussion, we should note that there are numerous studies that may be relevant to each of the points that we make; clearly, we cannot be exhaustive in our review of this literature. Instead, we cite representative studies and, where appropriate, existing reviews of relevant literatures.

3.2   BRAIN STRUCTURE AND FUNCTION: ASSOCIATIONS WITH DEPRESSION

3.2.1   Amygdala

The amygdala is a small, complex structure situated in the medial temporal lobe immediately adjacent to the anterior boundary of the hippocampus. Nuclei of the amygdala receive afferent projections from diverse regions of the brain, including the thalamus and hypothalamus; the cingulate, temporal, and insular cortices; and several midbrain structures. Efferent fibers project back to the thalamus, hypothalamus, and "limbic" cortical regions, in addition to brainstem nuclei and the striatum. Because of this pattern of connectivity, historically, the amygdala has been thought to integrate information from the senses and viscera, particularly in the service of detecting and mobilizing responses to signs of threat in the environment. Consistent with this formulation, stimulation of the amygdala in animals has been found to increase plasma corticosterone and autonomic signs of fear and anxiety (see Davis 1992). These findings are complemented by studies of amygdala lesions in humans, which have been found to result in decreased perception of emotionally significant stimuli, disrupted emotionality, and reduced fear learning (e.g., Anderson and Phelps 2001; Bechara et al. 1995).

In addition to a well-documented role in fear conditioning (see LeDoux 2003), there is considerable support for the involvement of the amygdala in the encoding of long-term emotional memories. In a seminal study with humans, Cahill et al. (1996) measured glucose metabolism using positron emission tomography (PET) as participants viewed emotionally arousing film clips and again as the same individuals viewed emotionally neutral clips. Several weeks later, participants were asked to recall as many of these film clips as possible. The number of recalled films was found to be positively correlated with the relative level of glucose metabolism in the right amygdala; importantly, this relation was obtained only for the emotional films. Investigators, using functional magnetic resonance imaging (fMRI), have extended this finding by relating moment-to-moment changes in amygdala activation to subsequent recall of emotional stimuli (Canli et al. 2000).

Neuroimaging studies have demonstrated reasonably consistent associations between exposure to emotionally salient material and amygdala activation. Meta-analyses of this work have shown that this association is present for both positively and negatively valenced material and is particularly robust when investigators use visual, gustatory, or olfactory stimuli (Costafreda et al. 2008). Interestingly, the amygdala also activates in response to neutral but *unpredictable* stimuli, such as the presentation of an irregular pattern of tones (Herry et al. 2007), leading researchers to posit that the amygdala functions in a much broader context than was originally believed. In fact, investigators have hypothesized that the amygdala operates at the level of a generalized self-relevance detection system (Sander et al. 2003), mediating vigilance, attentional resources, and behavioral responses in the face of a constantly changing stream of often ambiguous environmental cues.

Studies of differences in amygdala volume between samples of depressed and nondepressed individuals have yielded inconsistent results. Whereas some investigators have reported decreased amygdala volume in depression (e.g., Hastings et al. 2004), others have found increased amygdala volume in this disorder (e.g., Bremner et al. 2000). In attempting to account for these discrepant findings, we conducted a meta-analysis of studies examining the relation of amygdala volume and MDD, focusing on such characteristics as gender composition and medication status of the samples and chronicity of disorder. We found that whereas in unmedicated samples, depression is associated with decreased amygdala volume, in medicated samples depressed individuals are characterized by increased amygdala volume (Hamilton et al. 2008). The decrease in amygdala volume in unmedicated depression may be due to increased stress-induced glucocorticoid responding in MDD, which, itself, can lead to overstimulation and subsequent excitotoxic damage in glucocorticoid receptor–rich structures such as the amygdala (Sapolsky 1996). Similarly, the observed increase in amygdala volume in medicated depression may be due to the documented capacity of antidepressant medications to promote growth of new neurons in structures such as the amygdala, wherein neurogenesis is possible (Perera et al. 2007).

Investigations of amygdala activity in depression, both in response to affective stimuli and during wakeful resting state, have consistently shown aberrant amygdala functioning. Studies using techniques such as PET and single-photon computed emission tomography, which measure regional brain blood flow and/or metabolism (both of which are widely used estimates of regional brain activity), have documented increased tonic amygdala activity in MDD (e.g., Drevets et al. 1992) that has been found to normalize after various types of treatments, including antidepressant drugs (Drevets et al. 2002) and partial sleep deprivation (Clark et al. 2006). Consistent with this work, lower levels of amygdala activity in depressed individuals before treatment with transcranial magnetic stimulation (TMS), a procedure in which brief magnetic pulses are applied to and stimulate specific regions of the brain, predicted better therapeutic response (Nadeau et al. 2002).

Studies using fMRI have reported increased amygdala responsivity in MDD under a wide range of affectively negative conditions, including anticipating viewing aversive pictures (Abler et al. 2007), and anticipating and experiencing heat pain applied to the arm (Strigo et al. 2008). Similarly, Hamilton and Gotlib (2008) reported that individuals diagnosed with MDD were characterized by greater amygdala activation in response to viewing negative pictures that they recognized a week later than to negative pictures that they did not subsequently recognize. Moreover, in samples of depressed participants, amygdala hyperresponsivity has been found to correlate positively with both severity of depressive symptoms (Hamilton and Gotlib 2008) and level of ruminative responding (Siegle et al. 2002). Interestingly, unlike baseline amygdala activity in MDD, amygdala hyperreactivity has been found to persist following remission of depression (Hooley et al. 2009; Ramel et al. 2007). Given this discrepancy, it may be that whereas high tonic levels of amygdala activity characterize the depressed state, heightened amygdala reactivity is a stable "trait" that may play a role both in placing individuals at risk for the development of MDD and in increasing the likelihood of relapse among remitted depressed persons.

Although hyperreactivity of the amygdala has been found reliably in response to negative stimuli in depression, it is important to note that this pattern has also been observed as depressed individuals respond to various positive affective stimuli, such as positive, self-descriptive adjectives (Siegle et al. 2002) and happy faces (Sheline et al. 2001). Further complicating our understanding of amygdala functioning in the pathophysiology of MDD, researchers have noted that increased response in this structure to affectively valenced stimuli correlates positively with therapeutic response to cognitive–behavioral therapy (Siegle et al. 2006) and with symptom improvement at 8-month follow-up (Canli et al. 2005). More recent formulations of amygdala function as part of a personal saliency network (Seeley et al. 2007) may help reconcile these apparently contradictory findings. Thus, in depression, the amygdala may be tuned to respond to negative stimuli because of their congruence with depressed mood and with positive stimuli as a function of their representation of a desired mood state. Moreover, although the increased amygdala-driven impact of affective stimuli may worsen current mood, the salient distress caused by this negative mood may help to mobilize adaptive, motivational resources that predict subsequent improvement in depressive symptoms.

3.2.2 HIPPOCAMPUS

The hippocampus is a long, heterogeneous structure nested within the medial temporal lobe (MTL). The hippocampus plays an important role in episodic memory, that is, in the formation of new memories about experienced events (Preston and Wagner 2007). In this context, the hippocampus is involved in the detection of novel events, places, and stimuli (Kumaran and Maguire 2009). In fact, some researchers view the hippocampus as part of a larger MTL system that is responsible for general declarative memory (Eichenbaum 2000). More relevant to its role in depression, however, scientists increasingly have implicated the hippocampus in the inhibition of responses to negative emotional stimuli (Goldstein et al. 2007) and in the regulation of the stress response. Indeed, the hippocampus contains high levels of glucocorticoid receptors. An excess of glucocorticoids produced by the hypothalamic–pituitary–adrenal axis can be particularly deleterious to hippocampal neurons (Sapolsky 2000) and can reduce hippocampal neurogenesis (Gould and Tanapat 1999). Indeed, recent evidence indicates that people who have experienced significant traumatic stress are characterized by reduced hippocampal volume (Bremner et al. 2003b).

With respect to MDD, the results of early research examining hippocampal volume in this disorder were equivocal. More recent meta-analyses, however, have not only documented a reduction of hippocampal volume in MDD but have also demonstrated further that this decrease in hippocampal volume is correlated with the duration of depressive illness (Videbech and Ravnkilde 2004). At this point, however, the nature of the relation between smaller hippocampal volume and depressive chronicity is unclear. It may be, for example, that prolonged duration of depression leads progressively to atrophy of the hippocampus; alternatively, decreased hippocampal volume may contribute to a longer course of illness, or both depression and hippocampal volume reduction may be due to a third factor. In beginning to examine this question, Frodl et al. (2004) assessed hippocampal volume in depressed and never-disordered persons both at intake and at a 1-year follow-up assessment. Consistent with the formulation that decreased hippocampal volume predicts longer course of illness, they found no difference over the year in hippocampal volume change between MDD and control subjects; they did find, however, that depressed persons who did not recover between intake and follow-up had smaller hippocampal volumes at intake than depressed persons who were in remission at follow-up. Similarly, as we discuss in greater detail later in this chapter, in our laboratory we have scanned young girls who have a maternal history of recurrent depression but who have not themselves yet experienced a diagnosable episode of MDD. We recently reported that girls at high risk for depression had smaller hippocampal volume than their low-risk counterparts (Chen et al. 2010), suggesting that decreased hippocampal volume is present in high-risk individuals before the onset of a depressive episode.

Several investigators have now also used PET and similar procedures to examine baseline blood flow or metabolism in the hippocampus in samples of depressed individuals. In general, these researchers have reported greater activation in the hippocampus in depressed than in nondepressed individuals (e.g., Seminowicz et al. 2004). Moreover, greater severity of depression has been found to be associated with more activation in the hippocampus (Hornig et al. 1997). Although less consistent, studies of change in hippocampal activation in response to treatment of depression have found decreases in hippocampal activity after treatment (Aihara et al. 2007). Importantly, higher levels of hippocampal activation before treatment have been found to predict greater treatment efficacy (Ebert et al. 1994).

As we have noted above, the hippocampus is a functionally heterogeneous structure that has been implicated both in memory and in the regulation of the stress response. Not surprisingly, therefore, investigators have reported anomalous hippocampal activity in MDD in response to both cognitive and affective challenges. Compared with nondepressed controls, depressed individuals have been found reliably to exhibit lower levels of hippocampal activation during performance of hippocampus-dependent cognitive tasks, including declarative memory encoding of a paragraph (Bremner et al. 2004), explicit learning of cues predicting subsequent reward (Kumar et al. 2008), and navigation of a virtual water maze to find a hidden platform (Cornwell et al. 2010). Similarly, investigators have also found lower levels of hippocampal activation in response to affective challenges in depressed individuals, regardless of stimulus valence. For example, depressed individuals have been found to be characterized by attenuated hippocampal response during viewing of positive picture–caption pairs (Kumari et al. 2003) and of positive social stimuli (Fu et al. 2007). Similarly, investigators using pictures portraying negative scenes have found reduced hippocampal activation in depression (Lee et al. 2007).

3.2.3 Subgenual Anterior Cingulate Cortex

The sACC, including Brodmann's area (BA) 25 and the ventral portion of BA 24 immediately rostral to it, has extensive connections with the amygdala, periaqueductal gray, mediodorsal and anterior thalamic nuclei, nucleus accumbens (NAcc), and VS. Because of its unique interconnectivity with subcortical and limbic structures, this partition of the cingulate cortex is hypothesized to be involved in both affective processes and visceromotor control (Johansen-Berg et al. 2008; Vogt 2005).

Investigators examining neural aspects of emotional functioning have often associated the sACC with the induction of negative mood (e.g., Damasio et al. 2000). Liotti et al. (2000), for example, asked participants to generate short autobiographical scripts detailing a recent event in which they felt sad or anxious. During subsequent PET scanning, participants were shown their scripts, with the expectation that the participants' original mood states would be reconstituted. Liotti et al. reported increased regional blood flow within the sACC only as a function of participants' sadness, suggesting that the relation between sACC and negative mood is specific to this state. In fact, this specificity is partially supported by a meta-analysis of a wide range of emotion provocation studies, including those using visual, auditory, and memory recall methods to induce fear, anger, sadness, happiness, and disgust (Phan et al. 2002). The results of this meta-analysis indicate that the subgenual or subcallosal gyrus is activated significantly more frequently during sadness than during the experience of any other emotion. The association between sad mood and sACC appears to be particularly pronounced when autobiographical or self-relevant stimuli are used as part of the mood induction procedure. Interestingly, work in our laboratory has shown that the sACC is activated when participants try to recall positive autobiographical memories after, but not before, a sad mood induction (Cooney et al. 2007), suggesting a more complex role for the sACC in affective processing and regulation.

Although several investigators have reported decreased volume of the sACC in depressed individuals (e.g., Wagner et al. 2008), other researchers have failed to replicate this finding (e.g., Pizzagalli et al. 2004). More recent MRI studies have conducted finer-grained analyses of sACC volume, differentiating areas within this structure at the levels of both cytoarchitecture and white-matter projections (Johansen-Berg et al. 2008). Importantly, the results of these investigations may account for the discrepant findings described above. More specifically, these studies have documented marked reductions in the volume of the *posterior* (BA 25), but not of the *anterior* (BA 24), extents of the sACC in depressed individuals. These findings underscore the importance of differentiating the anterior and posterior extents of the sACC and may have significant implications for studies of functional activations in this structure.

Investigators examining regional blood flow and metabolism have also reported inconsistent findings concerning baseline sACC activity in MDD. Whereas some researchers have reported increased sACC activation in depression (Mayberg et al. 2005), others have documented decreased sACC activity in this disorder (Drevets et al. 1997). Drevets et al. (1997) posited that volumetric reductions in the sACC in MDD could account for observed decreases in activation in this structure. In fact, Drevets et al. argued that, on a per-unit-volume basis, sACC activity was actually *increased* in MDD. Although initially speculative, this formulation has now been supported by a clear majority of studies examining treatment response in MDD; these investigations have documented decreases in sACC activity after recovery from depression (e.g., Mayberg et al. 2005). Underscoring the functional significance of reduced sACC activation in depressed individuals after treatment, investigators have shown this reduction in sACC activation to be associated both with lower scores on the Hamilton Rating Scale for Depression (Clark et al. 2006) and with lower scores on factors assessing symptoms of anxiety and tension (Brody et al. 2001). Finally, and consistent with these findings, studies using symptom provocation paradigms, which reinstate depressive symptomatology through either neurochemical (e.g., serotonin depletion) or behavioral (e.g., sad mood induction) means, have found that as depressive symptoms increase, so does sACC activation (Hasler et al. 2008).

Complementing the research described above showing increased tonic activation of the sACC in depression, studies using fMRI to examine sACC function have found an increased phasic sACC response to affective stimuli in depressed individuals. This pattern has been reported both during passive viewing of happy and of sad faces (Gotlib et al. 2005) and during viewing of positive picture–caption pairs (Kumari et al. 2003). Importantly, this elevated sACC responding to affective stimuli has been shown not only to normalize after successful pharmacotherapy for MDD (Keedwell et al. 2009) but also to predict symptom change as a result of cognitive–behavioral therapy, that is, lower sACC activation in response to affective stimuli predicted better outcome in depressed patients (Siegle et al. 2006).

3.2.4 DORSOLATERAL PREFRONTAL CORTEX

The DLPFC, comprising BA 9 and BA 46, is highly interconnected with motor and premotor cortices, medial prefrontal cortex, and the basal ganglia. It is commonly associated with executive control processes. Studies examining the function of the DLPFC suggest a role for this cortical region in the representation of current task demands and in the attention, sensory processing, and behaviors required to match these demands, especially when the course of action is ambiguous or changing (see Miller and Cohen 2001). Consistent with this formulation, investigators have found that the DLPFC is involved in successful performance on the Stroop task. More specifically, this area is recruited when participants must suppress their prepotent word-reading responses in order to attend to the colors of the words; greater activation of the DLPFC is associated with lower Stroop interference (MacDonald et al. 2000). Successful performance on many cognitive tasks also involves the ability to maintain information in working memory. Interestingly, research in primates has demonstrated that neurons in the DLPFC become, and remain, activated during the delay period of delayed match-to-sample tasks (see Goldman-Rakic 1995), a finding corroborated by a meta-analysis of human neuroimaging studies (Wager and Smith 2003). The DLPFC, then, is critical in executive control of cognitive function.

Importantly, the DLPFC has also been implicated in the regulation of emotion. Investigators have found increased activation in the DLPFC when individuals attempt to reduce or modulate the negative impact of various types of stimuli (e.g., see Ochsner and Gross 2008), also reflected by increased coupling between DLPFC and amygdala during efforts to regulate emotion (Banks et al. 2007). Furthermore, and perhaps not surprising given cortical–limbic patterns of neural connectivity, Quirk et al. (2003) demonstrated that stimulation of regions medial to, and interconnected with, the DLPFC decreases the responsiveness of amygdala output neurons.

With respect to depression, studies of resting-state brain perfusion and metabolism have shown reliably lower levels of DLPFC activation in individuals who are diagnosed with MDD than in healthy controls (e.g., Biver et al. 1994). This diminished DLPFC activity appears to be specific to the state of depression; investigators have documented that DLPFC activation normalizes after both spontaneous recovery from depression (Bench et al. 1995) and successful pharmacotherapy for MDD (Kennedy et al. 2001). Moreover, decreases in baseline DLPFC activity have been reinstated in remitted depressed individuals who relapse in response to tryptophan (Bremner et al. 1997) and catecholamine (Bremner et al. 2003a) depletion procedures. Consistent with these findings, pretreatment levels of resting DLPFC activity have been found to predict better therapeutic outcome in MDD after TMS was applied over the left DLPFC (Baeken et al. 2009).

Studies examining correlations between resting DLPFC activity and clinical variables in MDD indicate that level of baseline functioning of DLPFC is related to cognitive aspects of depression, including negative cognitive biases and impaired regulation of affective processing. Using PET with depressed individuals, Dunn et al. (2002) found DLPFC activity to be inversely correlated with the "negative cognitions" cluster of the Beck Depression Inventory; moreover, improvement in the "cognitive disturbance" factor of the Hamilton Rating Scale for Depression was found to be associated with greater metabolic DLPFC activation after treatment of depression (Brody et al. 2001).

In addition to these investigations of baseline DLPFC activity, researchers have used fMRI to examine patterns of DLPFC activation in depressed individuals under different experimental conditions. The results of these studies point to decreased DLPFC response as depressed persons attempt to regulate their affect. For example, Fales et al. (2008) reported that depressed individuals exhibited reduced levels of DLPFC activation as they tried to ignore fear-related stimuli; importantly, this decrease normalized with pharmacotherapy (Fales et al. 2009). Similarly, Hooley et al. (2005) found an absence of DLPFC response to maternal criticism in depressed persons that, in the presence of an increase in amygdala activation, they interpreted as a failure of these individuals to regulate their affective response to their mothers' negative comments.

3.2.5 Ventral Striatum

The VS spans a region of the basal ganglia, including the NAcc, the ventromedial caudate, and the rostroventral putamen. The VS receives projections from multiple limbic and paralimbic regions, including the amygdala, hippocampus, orbitofrontal cortex, ventromedial prefrontal cortex, insula, and anterior cingulate cortex (ACC), as well as from the thalamus and dopaminergic midbrain. Like the dorsal striatum, the VS projects principally to the globus pallidus and substantia nigra, the main output nodes of the basal ganglia.

Cellular recordings conducted by Schultz (1998) and others identified large numbers of dopaminergic neurons in the monkey substantia nigra that released dopamine into the VS when the animal received an unexpected reward. Once the monkey had been repeatedly exposed to cue–reward pairings, however, the cells released dopamine only when the monkey saw cues predicting future rewards or resembling reward-predicting stimuli. This temporal transfer of activation, along with the depression of neuronal firing when an anticipated reward fails to occur ("error signal"), is posited to facilitate an important part of reward-based learning.

Neuroimaging studies have now extended these findings to humans (see Haber and Knutson 2010). Increased activity in the VS, and especially in the NAcc, has been found in response to the anticipation or to the receipt of rewarding stimuli ranging from pleasant tastes (O'Doherty et al. 2002) to monetary gains (Knutson et al. 2001) and social approval (Izuma et al. 2010). Indeed, a meta-analysis of PET and fMRI studies revealed that more than 60% of studies investigating responses to positive stimuli or happiness report activations within the basal ganglia (Phan et al. 2002). Moreover, the extent of regional cerebral blood flow (rCBF) in the VS has been found to be positively correlated with the level of pleasure experienced during exposure to chill-inducing music (Blood and Zatorre 2001), suggesting that the degree of VS activation is related to the intensity of an experienced reward. Similarly, Drevets et al. (2001) reported that the level of dopamine release in the VS is related to subjective levels of pleasure or euphoria. Finally, levels of VS activation have been found to be related to such individual differences as preference for immediate versus delayed reward (Hariri et al. 2006) and susceptibility to pathological gambling (Reuter et al. 2005).

Findings from other studies suggest that the VS plays a more general role in motivating behavior. For example, research studying rats has found that the NAcc mediates willingness to exert effort in pursuit of rewards; more specifically, interfering with dopamine neurotransmission in the NAcc reduces the likelihood that rats will work for a preferred food (Salamone et al. 2006). Furthermore, although reported less consistently than are its associations with reward valuation and prediction, the VS has been found to be activated during both receipt (Becerra et al. 2001) and anticipation (Jensen et al. 2003) of aversive stimulation, suggesting that this region responds to *salient* environmental events rather than exclusively to positive or reward-cuing events. Collectively, results from these lines of research suggest that a properly functioning VS responds to predictors of both positive and negative outcomes by activating the behavioral responses needed to approach desirable elements in the environment and avoid undesirable elements.

The lack of a clear anatomical definition of the VS precludes volumetric analysis of this region per se. A small number of volumetric analyses of the NAcc, however, have been conducted on depressed and never-disordered samples. Although Pizzagalli et al. (2009) reported in an MR-based volumetric study that there were no differences in accumbens volume between depressed and non-depressed participants, an arguably more sensitive study of postmortem brain tissue reported significant reduction in accumbens volume in MDD individuals (Baumann et al. 1999).

Recently, investigators have induced depressive relapse in formerly depressed individuals through catecholamine depletion, one of which is the neurotransmitter dopamine. Moreover, this depletion was found to lead to an increase in VS activity, which was likely due to a decrease in dopamine-mediated inhibition of the VS (Hasler et al. 2008). Similarly, depressive relapse after tryptophan depletion has been found to be associated with increased VS activity (Neumeister et al. 2004). Complementing these studies, Segawa et al. (2006) found a decrease in VS activity after successful treatment of depression using electroconvulsive therapy.

Finally, a number of investigators have reported reduced VS response to reward in depression, predominantly to rewarding outcomes. This reduced striatal response has been noted across positively valenced modalities, including monetary rewards (Knutson et al. 2008), praise (Steele et al. 2007), and amphetamine stimulation (Tremblay et al. 2005). Importantly, this reduction in VS responding to reward has also been found in remitted depressed individuals (McCabe et al. 2009), indicating that it may be a vulnerability factor for the development of MDD.

3.3 NEURAL ASPECTS OF DEPRESSION: CURRENT STATUS AND FUTURE DIRECTIONS

3.3.1 CURRENT STATUS

Overall, our review of the roles of the amygdala, hippocampus, sACC, VS, and DLPFC in depression provides a broad neural-level conceptualization of MDD in which limbic and perilimbic structures are overactive and dorsal cortical structures are underresponsive. Although this simplified neural characterization is currently the dominant neural formulation of MDD, it is far from comprehensive and leaves unaddressed several important issues. For example, there are not always clear links between anomalies in neural function or structure and specific *Diagnostic and Statistical Manual of Mental Disorders* symptoms of depression. Although findings from a small number of studies suggest associations between abnormalities in the VS and loss of pleasure or anhedonia (e.g., Pizzagalli et al. 2009) and anergia or fatigue (Salamone et al. 2006), between DLPFC dysfunction and psychomotor retardation (Videbech et al. 2002), and between amygdala reactivity and sad mood (Furman et al. 2010), we still have only a rudimentary and speculative conception of how the specific diagnostic symptoms of MDD map onto brain structures and activations. Moreover, we do not yet have a model, informed by neurophysiological data, for understanding the various subtypes of depression that may warrant different therapeutic approaches. It is likely that these gaps in the literature persist, at least in part, because of our limited characterization of the neural systems–level properties of MDD. Moreover, our knowledge of neural cause and effect in MDD, as well as of the neuroanatomical manifestations of risk for the development of this disorder, is also lacking. In the final section of this chapter, therefore, we discuss the importance of these issues in terms of future directions in the field and describe recent work that has attempted to address these deficits in understanding the neural foundations of MDD.

3.3.2 Future Directions

3.3.2.1 System-Level/Network Models

Much of the research examining the neural basis of MDD either has taken a whole-brain exploratory approach or has investigated the role of a single neural region in the pathophysiology of MDD. It is important to recognize that there is considerable agreement among clinical neuroscientists who study depression that MDD is a neural network-level disorder. Indeed, recent investigations have used methodological and analytic procedures and techniques that are capable of characterizing depression at a neural-network level.

Arguably, the most influential network-level model of depression is that formulated by Mayberg et al. (1999). This model posits that there is a reciprocal relation between cortical and limbic structures, and that in MDD, limbic activation is stronger than cortical activation, reflecting a decreased ability of depressed individuals to exert cognitive control over their negative affect. This model has been supported by the results of research indicating that both induced sadness and MDD are characterized by elevated levels of limbic and paralimbic activity and reduced dorsal cortical activity (Mayberg et al. 1999) and by findings that this pattern of activation normalizes after pharmacotherapy for MDD (Kennedy et al. 2001). More recent research using functional connectivity techniques has been able to offer a more detailed picture of corticolimbic relations in MDD. For example, researchers have demonstrated that under conditions of both affective stimulus processing and rest, individuals diagnosed with MDD are characterized by decreased functional coupling between the amygdala and the rostral ACC (Anand et al. 2005a)—a structure that, as we have noted above, is implicated in the regulation of affect—and between the amygdala and the DLPFC (Dannlowski et al. 2009). Importantly, this anomalous corticolimbic relation has been found to begin to normalize with pharmacotherapy for depression (Anand et al. 2005b).

Other network-level research has documented anomalous patterns of intralimbic system activity in depression. For example, in our laboratory, we have found increased functional connectivity between the amygdala and anterior hippocampus, and between the amygdala and VS, in depressed individuals during successful encoding of negative, but not of neutral or positive, stimuli (Hamilton and Gotlib 2008). Similarly, other investigators have found in depressed persons increased functional contributions of limbic structures to the default-mode network, which subserves self-reflective and prospective processing. Most notably, perhaps, Greicius et al. (2007) reported abnormally high levels of functional connectivity between the sACC and the default-mode network in patients diagnosed with MDD.

Further elucidating network-level neural models of MDD, investigators have recently begun to examine temporal patterns of neural activations in depressed participants. For example, using effective connectivity analyses capable of detecting temporal effects in functional neuroimaging data from depressed individuals, we found evidence of mutual excitation between sACC and other ventral prefrontal regions, and dampening of dorsal cortical activation by sACC and hippocampus (Hamilton et al. in press). From a different perspective, research examining neural changes resulting from deep-brain stimulation (DBS) in MDD designed to decrease sACC activity has shown subsequent increases in dorsal cortical activity and decreases in ventral prefrontal activity (Mayberg et al. 2005). Interestingly, similar changes have been noted after successful NAcc stimulation in MDD (Schlaepfer et al. 2008).

3.3.2.2 Investigations of Individuals at Risk for Depression

Although it is clear from our review that we have made substantial progress in specifying a neural-level model of MDD, this model is far from comprehensive, and several crucial issues remain unresolved. For example, although we now have a consistent picture of the neural anomalies in individuals with MDD with respect to both baseline and reactivity conditions, we know much less about abnormalities in neural functioning over the longer "arc" of depressive pathology, especially during the period of risk for the onset of a first depressive episode.

Although only a small number of studies have examined brain structure and volume as a function of level of risk for depression, they have yielded surprisingly consistent findings. In particular, investigators have implicated reduced hippocampal volume in risk for MDD. Three studies, two of which defined risk for MDD in terms of having a mother (Chen et al. 2010) or a twin (Baare et al. 2010) with depression and one of which defined risk for MDD as having high scores on a depression inventory (Dedovic et al. 2010), reported smaller hippocampal volume in high-risk than in low-risk individuals. Such studies do not address the question of whether environmental or genetic factors (or, of course, both) are driving the reduction in hippocampal volume in high-risk persons. In this context, de Geus et al. (2007) defined risk for depression as high self-reported levels of neuroticism and anxiety, and found that high-risk twins in monozygotic twin pairs who were discordant for risk for depression had smaller hippocampal volume than their low-risk siblings. Although these findings underscore the possibility that environmental factors can affect the volume of neural structures, it will be important for future work to assess whether putative environmental effects are due to direct effects of the environment on hormonal and neurotransmitter systems, or whether they may be mediated by changes in the expression of specific genes (Tsankova et al. 2010).

A small related body of literature documents aberrant neural responsivity to affective stimuli in individuals at risk for MDD. For example, two investigations have examined the neural functioning of young girls at elevated risk for MDD by virtue of having a depressed mother. Seeley et al. (2007) found that high-risk girls responded to loss of reward with increased activation of the dorsal ACC, a region implicated in coding for the personal saliency of stimuli. Interestingly, Gotlib et al. (2010) also found increased activation in dorsal ACC in high-risk girls, but as they were receiving punishment, suggesting that losing reward and being punished are functionally equivalent for girls at high familial risk for depression. Moreover, Gotlib et al. found that while they anticipated reward, young high-risk girls exhibited less activation than their low-risk counterparts in the putamen and left insula.

Other investigators have found that individuals at high risk for developing MDD, both offspring of parents with major depression (Monk et al. 2008) and persons who report high levels of neuroticism (Chan et al. 2009), show elevated amygdala activation in response to viewing angry faces. In the study described earlier, de Geus et al. (2007) found greater amygdala activation in response to viewing angry faces in the high-risk than in the low-risk twins of monozygotic twin pairs discordant for risk for MDD. Although limited, these findings suggest that anomalies in neural structure and function precede the onset of depression in individuals at elevated risk for developing MDD, and represent an important foundation on which to continue to explore this issue.

3.3.2.3 Manipulation of Neural Activation

Finally, based on the results of studies that have implicated activation in specific neural structures in the pathophysiology of MDD, investigators have begun to use new technologies to examine the clinical efficacy of modulating activity in these structures. In one such approach, DBS, localized neural activity is modulated in depressed individuals by implanting and activating electrical stimulation devices near critical neural regions. Investigators using this procedure have reported dramatic clinical effects in achieving remission of treatment refractory depression after both downregulation of the sACC (Mayberg et al. 2005) and upregulation of the NAcc (Schlaepfer et al. 2008).

Following in the spirit of this work, research groups have been developing techniques for helping individuals learn to manipulate regional brain activity *endogenously*. In these localized "neurofeedback" methods, individuals are presented, virtually in real time, indices of their neural activation in a targeted brain structure or region. These indices can be used as training signals to teach people to modulate localized brain activity. Indeed, real-time neurofeedback paradigms have been effective in teaching people to modulate activity in regions subserving sensory–motor function (DeCharms et al. 2004) and affect (Caria et al. 2007).

Given the clinical effectiveness of DBS in the reduction of sACC activity (Mayberg et al. 2005), we have examined in our laboratory the viability of using neurofeedback to train individuals to modulate activity in the sACC. In an initial proof-of-concept study, Hamilton et al. (2011) demonstrated that healthy participants who were shown a neurofeedback signal from the sACC could learn to reduce activity in this structure; importantly, participants who were shown a sham neurofeedback signal (signals from other participants) could not learn to modulate activity in the sACC. Moreover, individuals who were presented with real neurofeedback in this study showed functional decoupling of the sACC from the default-mode network. This latter finding is particularly encouraging in light of the findings of Greicius et al. (2007) that the sACC and default-mode network are *more* strongly coupled functionally in depressed than in nondepressed individuals. Finally, given findings of elevated levels of limbic activation in individuals at high risk for depression, we have been evaluating the effectiveness of real-time neurofeedback in young girls at familial risk for MDD. Preliminary results of this protocol indicate that successful neurofeedback reduces biological reactivity to external stressors, measured with psychophysiological indicators 1 week later, in these girls. Although it remains for future research to examine longer-term consequences of altering patterns of neural activation in depressed individuals and in people at elevated risk for this disorder, we believe that this approach to the study of neural function in depression nicely illustrates the promise that many of us believed would emerge when we began to examine neural aspects of MDD.

ACKNOWLEDGMENTS

This work was supported by NIMH grants MH59259 and MH74849 to Ian H. Gotlib.

REFERENCES

Abler, B., Erk, S., Herwig, U., and Walter, H. (2007). Anticipation of aversive stimuli activates extended amygdala in unipolar depression. *J Psychiatr Res* 41(6):511–22.

Aihara, M., Ida, I., Yuuki, N., et al. (2007). HPA axis dysfunction in unmedicated major depressive disorder and its normalization by pharmacotherapy correlates with alteration of neural activity in prefrontal cortex and limbic/paralimbic regions. *Psychiatr Res Neuroimag* 155(3):245–56.

Anand, A., Li, Y., Wang, Y., et al. (2005a). Activity and connectivity of brain mood regulating circuit in depression: A functional magnetic resonance study. *Biol Psychiatry* 57(10):1079–88.

Anand, A., Li, Y., Wang, Y., et al. (2005b). Antidepressant effect on connectivity of the mood-regulating circuit: An fMRI study. *Neuropsychopharmacology* 30(7):1334–44.

Anderson, A. K., and Phelps, E. A. (2001). Lesions of the human amygdala impair enhanced perception of emotionally salient events. *Nature* 411:305–9.

Baare, W. F. C., Vinberg, M., Knudsen, G. M., et al. (2010). Hippocampal volume changes in healthy subjects at risk of unipolar depression. *J Psychiatr Res* 44(10):655–62.

Baeken, C., De Raedt, R., Van Hove, C., Clerinx, P., De Mey, J., and Bossuyt, A. (2009). HF-rTMS treatment in medication-resistant melancholic depression: Results from (18)FDG-PET brain imaging. *CNS Spectr* 14(8):439–48.

Banks, S. J., Eddy, K. T., Angstadt, M., Nathan, P. J., and Phan, K. L. (2007). Amygdala-frontal connectivity during emotion regulation. *Soc Cogn Affect Neurosci* 2(4):303–12.

Baumann, B., Danos, P., Krell, D., et al. (1999). Reduced volume of limbic system-affiliated basal ganglia in mood disorders: Preliminary data from a postmortem study. *J Neuropsychiatry Clin Neurosci* 11(1):71–8.

Becerra, L., Breiter, H. C., Wise, R., Gonzalez, R. G., and Borsook, D. (2001). Reward circuitry activation by noxious thermal stimuli. *Neuron* 32(5):927–46.

Bechara, A., Tranel, D., Damasio, H., Adolphs, R., Rockland, C., and Damasio, A. R. (1995). Double dissociation of conditioning and declarative knowledge relative to the amygdala and hippocampus in humans. *Science* 269(5227):1115–8.

Bench, C. J., Frackowiak, R. S. J., and Dolan, R. J. (1995). Changes in regional cerebral blood-flow on recovery from depression. *Psychol Med* 25(2):247–61.

Biver, F., Goldman, S., Delvenne, V., et al. (1994). Frontal and parietal metabolic disturbances in unipolar depression. *Biol Psychiatry* 36(6):381–8.

Blood, A. J., and Zatorre, R. J. (2001). Intensely pleasurable responses to music correlate with activity in brain regions implicated in reward and emotion. *Proc Nat Acad Sci USA* 98(20):11818–23.

Boland, R. J., and Keller, M. B. (2009). Course and outcome of depression. In *Handbook of Depression*, ed. I. H. Gotlib and C. L. Hammen, 23–43. New York: Guilford.

Bremner, J. D., Innis, R. B., Salomon, R. M., et al. (1997). Positron emission tomography measurement of cerebral metabolic correlates of tryptophan depletion-induced depressive relapse. *Arch Gen Psychiatry* 54(4):364–74.

Bremner, J. D., Narayan, M., Anderson, E. R., Staib, L. H., Miller, H. L., and Charney, D. S. (2000). Hippocampal volume reduction in major depression. *Am J Psychiatry* 157(1):115–7.

Bremner, J. D., Vythilingam, M., Ng, C. K., et al. (2003a). Regional brain metabolic correlates of alpha-methylparatyrosine-induced depressive symptoms—Implications for the neural circuitry of depression. *J Am Med Assoc* 289(23):3125–34.

Bremner, J. D., Vythilingam, M., Vermetten, E., et al. (2003b). MRI and PET study of deficits in hippocampal structure and function in women with childhood sexual abuse and posttraumatic stress disorder. *Am J Psychiatry* 160:924–32.

Bremner, J. D., Vythilingam, M., Vermetten, E., Vaccarino, V., and Charney, D. S. (2004). Deficits in hippocampal and anterior cingulate functioning during verbal declarative memory encoding in midlife major depression. *Am J Psychiatry* 161(4):637–45.

Brody, A. L., Saxena, S., Mandelkern, M. A., Fairbanks, L. A., Ho, M. L., and Baxter, L. R. (2001). Brain metabolic changes associated with symptom factor improvement in major depressive disorder. *Biol Psychiatry* 50(3):171–8.

Cahill, L., Haier, R. J., Fallon, J., et al. (1996). Amygdala activity at encoding correlated with long-term, free recall of emotional information. *Proc Natl Acad Sci USA* 93(15):8016–21.

Canli, T., Zhao, Z., Brewer, J., Gabrieli, J. D. E., and Cahill, L. (2000). Event-related activation in the human amygdala associates with later memory for individual emotional experience. *J Neurosci* 20:RC99.

Canli, T., Cooney, R. E., Goldin, P., et al. (2005). Amygdala reactivity to emotional faces predicts improvement in major depression. *Neuroreport* 16(12):1267–70.

Caria, A., Veit, R., Sitaram, R., et al. (2007). Regulation of anterior insular cortex activity using real-time fMRI. *NeuroImage* 35(3):1238–46.

Chan, S. W. Y., Norbury, R., Goodwin, G. M., and Harmer, C. J. (2009). Risk for depression and neural responses to fearful facial expressions of emotion. *Br J Psychiatry* 194(2):139–45.

Chen, M. C., Hamilton, J. P., and Gotlib, I. H. (2010). Decreased hippocampal volume in healthy girls at risk of depression. *Arch Gen Psychiatry* 67(3):270–6.

Clark, C. P., Brown, G. G., Archibald, S. L., et al. (2006). Does amygdalar perfusion correlate with antidepressant response to partial sleep deprivation in major depression? *Psychiatry Res Neuroimag* 146(1):43–51.

Cooney, R. E., Joormann, J., Atlas, L. Y., Eugène, F., and Gotlib, I. H. (2007). Remembering the good times: Neural correlates of affect regulation. *Neuroreport* 18(17):1771–4.

Cornwell, B. R., Salvadore, G., Colon-Rosario, V., et al. (2010). Abnormal hippocampal functioning and impaired spatial navigation in depressed individuals: evidence from whole-head magnetoencephalography. *Am J Psychiatry* 167(7):836–44.

Costafreda, S. G., Brammer, M. J., David, A. S., and Fu, C. H. Y. (2008). Predictors of amygdala activation during the processing of emotional stimuli: A meta-analysis of 385 PET and fMRI studies. *Brain Res Rev* 58(1):57–70.

Damasio, A. R., Grabowski, T. J., Bechara, A., Damasio, H., Ponto, L. L. B., Parvizi, J., and Hichwa, R. D. (2000). Subcortical and cortical brain activity during the feeling of self-generated emotions. *Nat Neurosci* 3:1049–56.

Dannlowski, U., Ohrmann, P., Konrad, C., et al. (2009). Reduced amygdala-prefrontal coupling in major depression: Association with MAOA genotype and illness severity. *Int J Neuropsychopharmacology* 12(1):11–22.

Davis, M. (1992). The role of the amygdala in fear and anxiety. *Annu Rev Neurosci* 15:353–75.

de Geus, E. J. C., van't Ent, D., Wolfensberger, S. P. A., et al. (2007). Intrapair differences in hippocampal volume in monozygotic twins discordant for the risk for anxiety and depression. *Biol Psychiatry* 61(9):1062–71.

DeCharms, R. C., Christoff, K., Glover, G. H., Pauly, J. M., Whitfield, S., and Gabrieli, J. D. E. (2004). Learned regulation of spatially localized brain activation using real-time fMRI. *NeuroImage* 21(1):436–43.

Dedovic, K., Engert, V., Duchesne, A., Lue, S. D., Andrews, J., Efanov, S. I., Beaudry, T., and Pruessner, J. C. (2010). Cortisol awakening response and hippocampal volume: Vulnerability for major depressive disorder? *Biol Psychiatry* 68:847–53.

Drevets, W. C., Videen, T. O., Price, J. L., Preskorn, S. H., Carmichael, S. T., and Raichle, M. E. (1992). A functional anatomical study of unipolar depression. *J Neurosci* 12(9):3628–41.

Drevets, W. C., Price, J. L., Simpson, J. R., et al. (1997). Subgenual prefrontal cortex abnormalities in mood disorders. *Nature* 386(6627):824–7.

Drevets, W. C., Gautier, C., Price, J. C., et al. (2001). Amphetamine-induced dopamine release in human ventral striatum correlates with euphoria. *Biol Psychiatry* 49(2):81–96.

Drevets, W. C., Bogers, W., and Raichle, M. E. (2002). Functional anatomical correlates of antidepressant drug treatment assessed using PET measures of regional glucose metabolism. *Eur Neuropsychopharmacol* 12(6):527–44.

Dunn, R. T., Kimbrell, T. A., Ketter, T. A., et al. (2002). Principal components of the beck depression inventory and regional cerebral metabolism in unipolar and bipolar depression. *Biol Psychiatry* 51(5): 387–99.

Ebert, D., Feistel, H., Barocka, A., and Kaschka, W. (1994). Increased limbic blood flow and total sleep deprivation in major depression with melancholia. *Psychiatry Res Neuroimag* 55(2):101–9.

Eichenbaum, H. (2000). A cortical-hippocampal system for declarative memory. *Nat Rev Neurosci* 1(1):41–50.

Fales, C. L., Barch, D. M., Rundle, M. M., et al. (2008). Altered emotional interference processing in affective and cognitive-control brain circuitry in major depression. *Biol Psychiatry* 63(4):377–84.

Fales, C. L., Barch, D. M., Rundle, M. A., et al. (2009). Antidepressant treatment normalizes hypoactivity in dorsolateral prefrontal cortex during emotional interference processing in major depression. *J Affect Dis* 112(1–3):206–11.

Freedland, K. E., and Carney, R. M. (2009). Depression and medical illness. In *Handbook of Depression,* 2nd edn., ed. I. H. Gotlib and C. L. Hammen, 113–41. New York: Guilford.

Frodl, T., Meisenzahl, E. M., Zetzsche, T., et al. (2004). Hippocampal and amygdala changes in patients with major depressive disorder and healthy controls during a 1-year follow-up. *J Clin Psychiatry* 65(4):492–9.

Fu, C. H. Y., Williams, S. C. R., Brammer, M. J., et al. (2007). Neural responses to happy facial expressions in major depression following antidepressant treatment. *Am J Psychiatry* 164(4):599–607.

Furman, D. J., Hamilton, J. P., Joormann, J., and Gotlib, I. H. (2010). Altered timing of amygdala activation during sad mood elaboration as a function of 5-HTTLPR. *Soc Cog Affect Neurosci* doi: 10.1093/scan/nsq029.

Goldman-Rakic, P. S. (1995). Cellular basis of working memory. *Neuron* 14:477–85.

Goldstein, M., Brendel, G., Tuescher, O., et al. (2007). Neural substrates of the interaction of emotional stimulus processing and motor inhibitory control: An emotional linguistic *go/no-go* fMRI study. *NeuroImage* 36(3):1026–40.

Gotlib, I. H., and Joormann, J. (2010). Cognition and depression: Current status and future directions. *Ann Rev Clin Psychol* 6:285–312.

Gotlib, I. H., Sivers, H., Gabrieli, J. D. E., et al. (2005). Subgenual anterior cingulate activation to valenced emotional stimuli in major depression. *Neuroreport* 16(16):1731–4.

Gotlib, I. H., Hamilton, J. P., Cooney, R. E., Singh, M. K., Henry, M. L., and Joormann, J. (2010). Neural processing of reward and loss in girls at risk for major depression. *Arch Gen Psychiatry* 67(4):380–7.

Gould, E., and Tanapat, P. (1999). Stress and hippocampal neurogenesis. *Biol Psychiatry* 46(11):1472–9.

Greicius, M. D., Flores, B. H., Menon, V., et al. (2007). Resting-state functional connectivity in major depression: Abnormally increased contributions from subgenual cingulate cortex and thalamus. *Biol Psychiatry* 62(5):429–37.

Haber, S. N., and Knutson, B. (2010). The reward circuit: Linking primate anatomy and human imaging. *Neuropsychopharmacol Rev* 35:4–26.

Hamilton, J. P., and Gotlib, I. H. (2008). Neural substrates of increased memory sensitivity for negative stimuli in major depression. *Biol Psychiatry* 63(12):1155–62.

Hamilton, J. P., Siemer, M., and Gotlib, I. H. (2008). Amygdala volume in major depressive disorder: A meta-analysis of magnetic resonance imaging studies. *Mol Psychiatry* 13(11):993–1000.

Hamilton, J. P., Glover, G. H., Hsu, J. J., Johnson, R. F., and Gotlib, I. H. (2011). Modulation of subgenual anterior cingulate cortex activity with real-time neurofeedback. *Hum Brain Mapp* 32(1):22–31.

Hamilton, J. P., Chen, G., Thomason, M. E., Johnson, R. F., and Gotlib, I. H. (in press). Investigating neural primacy in Major Depressive Disorder: Multivariate granger causality analysis of resting state fMRI time-series data. *Mol Psychiatry*.

Hariri, A. R., Brown, S. M., Williams, D. E., Flory, J. D., de Wit, H., and Manuck, S. B. (2006). Preference for immediate over delayed rewards is associated with magnitude of ventral striatal activity. *J Neurosci* 26(51):13213–7.

Hasler, G., Fromm, S., Carlson, P. J., et al. (2008). Neural response to catecholamine depletion in unmedicated subjects with major depressive disorder in remission and healthy subjects. *Arch Gen Psychiatry* 65(5):521–31.

Hastings, R. S., Parsey, R. V., Oquendo, M. A., Arango, V., and Mann, J. J. (2004). Volumetric analysis of the prefrontal cortex, amygdala, and hippocampus in major depression. *Neuropsychopharmacology* 29(5):952–9.

Herry, C., Bach, D. R., Esposito, F., et al. (2007). Processing of temporal unpredictability in human and animal amygdala. *J Neurosci* 27(22):5958–66.

Hooley, J. M., Gruber, S. A., Scott, L. A., Hiller, J. B., and Yurgelun-Todd, D. A. (2005). Activation in dorsolateral prefrontal cortex in response to maternal criticism and praise in recovered depressed and healthy control participants. *Biol Psychiatry* 57(7):809–12.

Hooley, J. M., Gruber, S. A., Parker, H. A., Guillaumot, J., Rogowska, J., and Yurgelun-Todd, D. A. (2009). Cortico-limbic response to personally challenging emotional stimuli after complete recovery from depression. *Psychiatry Res Neuroimag* 171(2):106–19.

Hornig, M., Mozley, P. D., and Amsterdam, J. D. (1997). HMPAO SPECT brain imaging in treatment-resistant depression. *Prog Neuro-Psychopharmacol Biol Psychiatry* 21(7):1097–14.

Izuma, K., Saito, D. N., and Sadato, N. (2010). Processing of the incentive for social approval in the ventral striatum during charitable donation. *J Cog Neurosci* 22(4):621–31.

Jensen, J., McIntosh, A. R., Crawley, A. P., Mikulis, D. J., Remington, G., and Kapur, S. (2003). Direct activation of the ventral striatum in anticipation of aversive stimuli. *Neuron* 40(6):1251–7.

Johansen-Berg, H., Gutman, D. A., Behrens, T. E. J., et al. (2008). Anatomical connectivity of the subgenual cingulate region targeted with deep brain stimulation for treatment-resistant depression. *Cerebral Cortex* 18(6):1374–83.

Keedwell, P., Drapier, D., Surguladze, S., Giampietro, V., Brammer, M., and Phillips, M. (2009). Neural markers of symptomatic improvement during antidepressant therapy in severe depression: Subgenual cingulate and visual cortical responses to sad, but not happy, facial stimuli are correlated with changes in symptom score. *J Psychopharmacol* 23(7):775–88.

Kennedy, S. H., Evans, K. R., Kruger, S., et al. (2001). Changes in regional brain glucose metabolism measured with positron emission tomography after paroxetine treatment of major depression. *Am J Psychiatry* 158(6):899–905.

Kessler, R. C. and Wang, P. S. (2009). Epidemiology of depression. In *Handbook of Depression*, 2nd edn., ed. I. H. Gotlib and C. L. Hammen, 5–22. New York: Guilford.

Kessler, R. C., Akiskal, H. S., Ames, M., Birnbaum, H., Greenberg, P., Hirschfeld, R. M. A., Jin, R., Merikangas, K. R., Simon, G. E., and Wang, P. S. (2006). Prevalence and effects of mood disorders on work performance in a nationally representative sample of U.S. workers. *Am J Psychiatry* 163:1561–68

Knutson, B., Adams, C. M., Fong, G. W., and Hommer, D. (2001). Anticipation of increasing monetary reward selectively recruits nucleus accumbens. *J Neurosci* 21:RC159.

Knutson, B., Bhanji, J. P., Cooney, R. E., Atlas, L. Y., and Gotlib, I. H. (2008). Neural responses to monetary incentives in major depression. *Biol Psychiatry* 63(7):686–92.

Kumar, P., Waiter, G., Ahearn, T., Milders, M., Reid, I., and Steele, J. D. (2008). Abnormal temporal difference reward-learning signals in major depression. *Brain* 131(8):2084–93.

Kumaran, D., and Maguire, E. A. (2009). Novelty signals: A window into hippocampal information processing. *Trends Cog Sci* 13(2):47–54.

Kumari, V., Mitterschiffthaler, M. T., Teasdale, J. D., et al. (2003). Neural abnormalities during cognitive generation of affect in treatment-resistant depression. *Biol Psychiatry* 54(8):777–91.

LeDoux, J. (2003). The emotional brain, fear, and the amygdala. *Cell Mol Neurobiol* 23(4–5):727–38.

Lee, B. T., Seong Whi, C., Hyung Soo, K., et al. (2007). The neural substrates of affective processing toward positive and negative affective pictures in patients with major depressive disorder. *Prog Neuro-Psychopharmacol Biol Psychiatry* 31(7):1487–92.

Liotti, M., Mayberg, H. S., Brannan, S. K., McGinnis, S., Jerabek, P., and Fox, P. T. (2000). Differential limbic–cortical correlates of sadness and anxiety in healthy subjects: Implications for affective disorders. *Biol Psychiatry* 48(1):30–42.

MacDonald III, A. W., Cohen, J. D., Stenger, V. A., and Carter, C. S. (2000). Dissociating the role of the dorsolateral prefrontal and anterior cingulate cortex in cognitive control. *Science* 288(5472):1835–8.

Mayberg, H. S., Liotti, M., Brannan, S. K., et al. (1999). Reciprocal limbic-cortical function and negative mood: Converging PET findings in depression and normal sadness. *Am J Psychiatry* 156(5): 675–82.

Mayberg, H. S., Lozano, A. M., Voon, V., et al. (2005). Deep brain stimulation for treatment-resistant depression. *Neuron* 45(5):651–60.

McCabe, C., Cowen, P. J., and Harmer, C. J. (2009). Neural representation of reward in recovered depressed patients. *Psychopharmacology* 205(4):667–77.

Miller, E. K., and Cohen, J. D. (2001). An integrative theory of prefrontal cortex function. *Ann Rev Neurosci* 24:167–202.

Monk, C. S., Klein, R. G., Telzer, E. H., et al. (2008). Amygdala and nucleus accumbens activation to emotional facial expressions in children and adolescents at risk for major depression. *Am J Psychiatry* 165(1):90–8.

Murray, C., and Lopez, A. (1996). Evidence-based health policy—Lessons from the Global Burden of Disease Study. *Science* 274:740–3.

Nadeau, S. E., McCoy, K. J. M., Crucian, G. P., et al. (2002). Cerebral blood flow changes in depressed patients after treatment with repetitive transcranial magnetic stimulation: Evidence of individual variability. *Neuropsychiatr Neuropsychol Behav Neurol* 15(3):159–75.

Neumeister, A., Nugent, A. C., Waldeck, T., et al. (2004). Neural and behavioral responses to tryptophan depletion in unmedicated patients with remitted major depressive disorder and controls. *Arch Gen Psychiatry* 61(8):765–73.

O'Doherty, J. P., Deichmann, R., Critchley, H. D., and Dolan, R. J. (2002). Neural responses during anticipation of a primary taste reward. *Neuron* 33(5):815–26.

Ochsner, K. N., and Gross, J. J. (2008). Cognitive emotion regulation: Insights from social cognitive and affective neuroscience. *Curr Direct Psychol Sci* 17(2):153–8.

Perera, T. D., Coplan, J. D., Lisanby, S. H., et al. (2007). Antidepressant-induced neurogenesis in the hippocampus of adult nonhuman primates. *J Neurosci* 27(18):4894–901.

Phan, K. L., Wager, T., Taylor, S. F., and Liberzon, I. (2002). Functional neuroanatomy of emotion: A meta-analysis of emotion activation studies in PET and fMRI. *NeuroImage* 16(2):331–48.

Pizzagalli, D. A., Oakes, T. R., Fox, A. S., et al. (2004). Functional but not structural subgenual prefrontal cortex abnormalities in melancholia. *Mol Psychiatry* 9(4):393–405.

Pizzagalli, D. A., Holmes, A. J., Dillon, D. G., et al. (2009). Reduced caudate and nucleus accumbens response to rewards in unmedicated individuals with major depressive disorder. *Am J Psychiatry* 166(6):702–10.

Preston, A. R., and Wagner, A. D. (2007). The medial temporal lobe and memory. In *Neurobiology of Learning and Memory*, 2nd edn., ed. R. P. Kesner and J. L. Martinez, 305–37. New York: Elsevier, Inc.

Quirk, G. J., Likhtik, E., Pelletier, J. G., and Paré, D. (2003). Stimulation of medial prefrontal cortex decreases the responsiveness of central amygdala output neurons. *J Neurosci* 23:8800–7.

Ramel, W., Goldin, P. R., Eyler, L. T., Brown, G. G., Gotlib, I. H., and McQuaid, J. R. (2007). Amygdala reactivity and mood-congruent memory in individuals at risk for depressive relapse. *Biol Psychiatry* 61(2):231–9.

Reuter, J., Raedler, T., Rose, M., Hand, I., Gläscher, J., and Büchel, C. (2005). Pathological gambling is linked to reduced activation of the mesolimbic reward system. *Nat Neurosci* 8(2):147–8.

Salamone, J. D., Correa, M., Mingote, S. M., Weber, S. M., and Farrar, A. M. (2006). Nucleus accumbens dopamine and the forebrain circuitry involved in behavioral activation and effort-related decision making: Implications for understanding anergia and psychomotor slowing in depression. *Curr Psychiatr Rev* 2(2):267–80.

Sander, D., Grafman, J., and Zalla, T. (2003). The human amygdala: An evolved system for relevance detection. *Rev Neurosci* 14:303–16.

Sapolsky, R. M. (1996). Why stress is bad for your brain. *Science* 273(5276):749–50.

Sapolsky, R. M. (2000). The possibility of neurotoxicity in the hippocampus in major depression: A primer on neuron death. *Biol Psychiatry* 48(9):755–65.

Schlaepfer, T. E., Cohen, M. X., Frick, C., et al. (2008). Deep brain stimulation to reward circuitry alleviates anhedonia in refractory major depression. *Neuropsychopharmacology* 33(2):368–77.

Schultz, W. (1998). Predictive reward signal of dopamine neurons. *J Neurophysiol* 80(1):1–27.

Seeley, W. W., Menon, V., Schatzberg, A. F., et al. (2007). Dissociable intrinsic connectivity networks for salience processing and executive control. *J Neurosci* 27(9):2349–56.

Segawa, K., Azuma, H., Sato, K., et al. (2006). Regional cerebral blood flow changes in depression after electroconvulsive therapy. *Psychiatry Res Neuroimag* 147(2–3):135–43.

Seminowicz, D. A., Mayberg, H. S., McIntosh, A. R., et al. (2004). Limbic-frontal circuitry in major depression: A path modeling metanalysis. *NeuroImage* 22(1):409–18.

Sheline, Y. I., Barch, D. M., Donnelly, J. M., Ollinger, J. M., Snyder, A. Z., and Mintun, M. A. (2001). Increased amygdala response to masked emotional faces in depressed subjects resolves with antidepressant treatment: An fMRI study. *Biol Psychiatry* 50(9):651–8.

Siegle, G. J., Steinhauer, S. R., Thase, M. E., Stenger, V. A., and Carter, C. S. (2002). Can't shake that feeling: Assessment of sustained event-related fMRI amygdala activity in response to emotional information in depressed individuals. *Biol Psychiatry* 51(9):693–707.

Siegle, G. J., Carter, C. S., and Thase, M. E. (2006). Use of fMRI to predict recovery from unipolar depression with cognitive behavior therapy. *Am J Psychiatry* 163(4):735–8.

Steele, J. D., Kumar, P., and Ebmeier, K. P. (2007). Blunted response to feedback information in depressive illness. *Brain* 130:2367–74.

Strigo, I. A., Simmons, A. N., Matthews, S. C., Craig, A. D., and Paulus, M. P. (2008). Association of major depressive disorder with altered functional brain response during anticipation and processing of heat pain. *Arch Gen Psychiatry* 65(11):1275–84.

Tremblay, L. K., Naranjo, C. A., Graham, S. J., et al. (2005). Functional neuroanatomical substrates of altered reward processing in major depressive disorder revealed by a dopaminergic probe. *Arch Gen Psychiatry* 62(11):1228–36.

Tsankova, N., Renthal, W., Kumar, A., and Nestler, E. J. (2010). Epigenetic regulation in psychiatric disorders. *Focus* 8:435–48.

Videbech, P., and Ravnkilde, B. (2004). Hippocampal volume and depression: A meta-analysis of MRI studies. *Am J Psychiatry* 161(11):1957–66.

Videbech, P., Ravnkilde, B., Pedersen, T. H., et al. (2002). The Danish PET/depression project: Clinical symptoms and cerebral blood flow. A regions-of-interest analysis. *Acta Psychiatr Scand* 106(1):35–44.

Vogt, B. A. (2005). Pain and emotion interactions in subregions of the cingulate gyrus. *Nat Rev Neurosci* 6:533–44.

Wager, T. D., and Smith, E. E. (2003). Neuroimaging studies of working memory: A meta-analysis. *Cog Affect Behav Neurosci* 3:255–74.

Wagner, G., Koch, K., Schactitzabel, C., Reichenbach, J. R., Sauer, H., and Schlosser, R. G. M. (2008). Enhanced rostral anterior cingulate cortex activation during cognitive control is related to orbitofrontal volume reduction in unipolar depression. *J Psychiatry Neurosci* 33(3):199–208.

4 Genetic Regulation of Emotion Brain Circuitries

Ulrich Rabl, Christian Scharinger, Tina Hofmaier,
Michael Freissmuth, and Lukas Pezawas

CONTENTS

4.1 INTRODUCTION

The observation of familial aggregation of mood disorders has promoted the search for genetic factors related to the neurobiological underpinnings of major depressive disorder (MDD) (McGuffin and Katz 1989). Heritability is at least modest and has been estimated to range between 31% and 42% according to twin studies (Sullivan et al. 2000). Despite numerous reports of associations between candidate genes and MDD, results of clinical association studies are characterized by a high level of inconclusiveness (Levinson 2006; Burmeister et al. 2008), and several enthusiastically acclaimed findings (Lesch et al. 1996; Caspi et al. 2003) have recently been questioned (Risch et al. 2009; Munafo et al. 2009). Similarly, attempts to establish a clinical decision tree for therapeutic drug interventions or diagnostics based on genetics have shown to be only of modest clinical relevance (Schosser and Kasper 2009; Kato and Serretti 2008). Similar discouraging findings have been made in other common disorders ranging from schizophrenia to hypertension. This is in stark contrast to successful gene assignments in monogenetic diseases that are subject to Mendelian inheritance, for example, Huntington disease (Risch 2000). Several aspects are held responsible for such elusive results: as assumed for common disorders in general (Lohmueller et al. 2003), the heritability found in MDD likely originates from joint effects of multiple gene variants resulting in elusive effect sizes for single variants (Demirkan et al. 2010). Additional genetic mechanisms such as incomplete penetrance, interplay between genes, and interactions with nongenetic factors further complicate the neurobiology of mental illness (Burmeister et al. 2008).

Nevertheless, it is of interest to identify gene variants underlying the susceptibility to depression. Accordingly, the research efforts continue to focus on susceptibility and modifier genes because better diagnostic tools and personalized treatment opportunities may emerge with improved insights. Most importantly, these insights may further advance our understanding of pathophysiological mechanisms and hence facilitate the identification of new treatment targets. Strategies to overcome the aforementioned hurdles include increasing sample sizes, best practice guidelines for initial and replication studies (Chanock et al. 2007), and genome-wide-association approaches that are independent of a priori candidate gene hypotheses (McMahon 2010). These attempts aim at improving study designs. However, others have questioned the applicability of psychiatric phenotypes per se (Meyer-Lindenberg and Weinberger 2006).

4.2 BRIDGING THE GAP BETWEEN GENOTYPE AND PHENOTYPE

In general, common disorders are considered to be quantitative traits (Plomin et al. 2009), whereas mental disorders such as depression are usually treated as qualitative traits. A patient is diagnosed with MDD according to the onset of specific symptoms without consideration of the symptom severity. Moreover, patients may only share a minority of symptoms but are nevertheless lumped together into the diagnostic category "depression" (Miller 2010). Since different symptoms likely result from distinct biological pathways and neural systems, the symptom-based entity MDD presumably subsumes biologically heterogeneous conditions into one artificial qualitative trait.

Genes do not translate directly into psychiatric phenotypes but are assumed to exert their effects on complex paths that extend from a molecular level to altered function of neurons. In turn, these effects mediate neural activity on a brain systems level, which eventually modifies behavior, mood, or cognition (Pezawas and Meyer-Lindenberg 2010). Conversely, the closer one gets from the level of a given phenotype to the genotype level, the easier it will be to disentangle effects of single genes. The concept of *endophenotypes* is based on these assumptions: whereas the phenotype represents the clinical syndrome, an endophenotype is a measurable component of neurophysiological, biochemical, endocrinological, neuroanatomical, cognitive, or neuropsychological nature that accounts for a significant part of the phenotype (Gottesman and Gould 2003). Since a neurobiological endophenotype is less vague than a clinical phenotype and therefore tighter linked to the underlying genetics, it is probably more indicative of the genotype, resembling the simplicity of Mendelian genetics (Gottesman and Gould 2003). Accordingly, criteria for candidate biomarkers have been adapted from conventional genetics: Endophenotypes are defined as heritable and state-independent traits that are associated with illness in the population and cosegregate with illness within families. Moreover, an endophenotype found in affected family members is expected to be more common in nonaffected family members than in the general population (Gottesman and Gould 2003). Only a few, if any, proposed endophenotypes actually fulfill these strict criteria (Meyer-Lindenberg and Weinberger 2006) because most candidate traits represent broader subprocesses related to the disorder of interest instead of reflecting specific pathogenetic endpoints of single alleles (Gottesman and Gould 2003). Traits that are not "sufficiently" heritable to meet the endophenotype criteria are referred to as *intermediate phenotypes* (Meyer-Lindenberg and Weinberger 2006), although both terms are often used interchangeably (Gottesman and Gould 2003).

Intermediate phenotypes, although not as traceable as Mendelian traits, reduce genetic complexity. Moreover, they may break the surface of a clinical phenotype and inform about the underlying biology. Some proposed nonbiological intermediate phenotypes, for example, neuroticism, provide little—if any—advantages because they have been found to be nearly as complex as their clinical counterparts (Shifman et al. 2008). In contrast, biological intermediate phenotypes did yield more promising results (Pezawas and Meyer-Lindenberg 2010). Neuroimaging has been at the front line of these approaches because of the attractive possibility to study brain function and anatomy in vivo without causing any harm. Neuroimaging techniques have been applied to visualize intermediate phenotypes and to investigate linkage to psychiatric candidate genes. This burgeoning field, termed *imaging genetics*, has rapidly evolved (Hariri et al. 2006) and yielded encouraging results for the understanding of MDD and other mental disorders including schizophrenia, anxiety disorders, and attention-deficit hyperactivity disorder (Pezawas and Meyer-Lindenberg 2010). Approaches based on "imaging genetics" have been found to be at least one order of magnitude more sensitive and specific than classical association studies (Meyer-Lindenberg 2010).

4.3 TECHNIQUES USED IN IMAGING GENETICS

Although many neuroimaging methods have been used to study genetic effects, functional magnetic resonance imaging (fMRI) has emerged as the most popular technique. Briefly, fMRI takes advantage of the reduced magnetic susceptibility of oxygenated hemoglobin, which accumulates in active brain regions because of reactive blood flow in response to increased oxygen extraction by activated neurons. Accordingly, an increased blood oxygen level–dependent (BOLD) signal can be detected as a surrogate measure of neuronal mass activity (Logothetis 2008). Studies using fMRI typically expose subjects to neuropsychological tasks or stimuli that are designed to activate the brain regions of interest, or, without a specific task, measure default activity during rest (Poldrack et al. 2008). Neural activity can further be indirectly measured by positron emission tomography (PET), which uses radioactive tracers to assess regional cerebral blood flow or glucose metabolism. In addition, using tracers that specifically bind to target molecules, regional distributions of neurotransmitter receptors or transporters can be measured. The latter type of studies is generally carried out in smaller samples than fMRI studies because they are limited by their invasive nature (Bandettini 2009).

Another important field is computational neuroanatomy, which provides voxel-wise morphometric measures such as voxel-based morphometry (Ashburner and Friston 2000) or surface-based techniques (Tosun et al. 2004).

Other imaging techniques offer intriguing insights into the interplay of brain regions, stressing the importance of investigating circuitries instead of single structures. Structural and functional connectivities are such correlative measures indicating connection strengths in between distant regions of anatomical or functional network nodes (Meyer-Lindenberg 2009). Such putative circuits can further be investigated by diffusion tensor imaging (DTI), in which fractional anisotropy is used to study the orientation and integrity of white matter tracts (Sexton et al. 2009).

4.4 DEPRESSION-RELATED BRAIN CIRCUITS

Gene variants that confer risk for MDD presumably act on brain function and architecture, since psychiatric disorders are considered brain disorders (Meyer-Lindenberg 2010). In line with this model, a comprehensive network of brain regions has been found to be affected during MDD (Rigucci et al. 2010). Intermediate phenotypes located in these brain areas may partly precede diagnosis and therefore reflect risk factors, whereas others may only be accompanying factors without pathogenetic relevance. The ideal intermediate phenotype has initially been considered state-independent (Meyer-Lindenberg and Weinberger 2006). However, several intermediate phenotypes in MDD have been shown to change during the course of disease and treatment. The potential implications of these varying conditions remain to be evaluated (Rigucci et al. 2010). Nevertheless, several studies in clinically symptomatic patients highlighted brain areas that have been found to be under the influence of depression-related gene variants, encouraging an imaging genetics approach

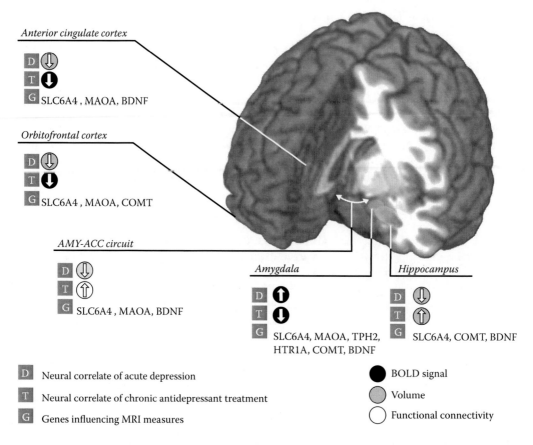

FIGURE 4.1 Neural correlates of acute MDD as well as chronic antidepressant treatment and genetic influence on MRI measures in brain regions of emotion processing in healthy subjects. Gene nomenclature corresponds to Online Mendelian Inheritance in Man. MRI, magnetic resonance imaging; BOLD, blood oxygen level dependent; AMY-ACC circuit, amygdala–anterior cingulate cortex circuit.

(Scharinger et al. 2010) (Figure 4.1). Unsurprisingly, most emphasis has been given to brain areas involved in emotion processing, notably the amygdala, anterior cingulate cortex (ACC), orbitofrontal cortex (OFC), and hippocampus (Price and Drevets 2010). In addition, other regions that are not directly linked to emotion have been subject to evaluation, for instance, the dorsolateral prefrontal cortex (DLPFC), which has been related to cognitive or executive dysfunction in depression (Rogers et al. 2004) or the reward system, and specifically the nucleus accumbens, which has been implicated in anhedonia, a prominent symptom of MDD (Pizzagalli et al. 2009). Recent studies exploring the contribution of brain networks in MDD further suggest an interplay between emotional, cognitive, and reward circuits, considering a broader network that may jointly contribute to the disorder (Hamilton et al. 2010; Sheline et al. 2010). Since most emphasis with regard to genetic factors has been given to regions of emotion processing, this chapter will mainly focus on the amygdala, ACC, OFC, and hippocampus.

4.4.1 ANTERIOR CINGULATE CORTEX

The ACC is considered a core region of depression (Drevets et al. 2008) and can be divided into a pregenual and a subgenual (sACC) portion according to cytoarchitectonic and pharmacological studies (Vogt 2009). Decreases in metabolism and gray matter volume of the sACC have been found in both unipolar and bipolar depressed samples (Drevets et al. 1997), although ACC volume reduction has predominantly been related to unipolar depression by meta-analytical data (Hajek et al. 2008). This finding is further supported by the observation of glial reduction in a corresponding region of postmortem brain samples from depressive patients (Ongur et al. 1998). Moreover, ACC volume reduction has been found to be related to subclinical depressive symptoms in adolescents with high familial risk for depression. This observation implicates the ACC as a possible target of genetic risk factors (Boes et al. 2008). Apart from its likely role in the pathogenesis of MDD, the ACC has been implicated in antidepressant treatment: higher ACC volume and metabolism have been found to be predictive for better treatment response to selective serotonin reuptake inhibitor (SSRI) treatment (Rabl et al. 2010). It is noteworthy that ACC activity or metabolism has been shown to be altered during acute and chronic antidepressant treatment (Rabl et al. 2010), as well as during deep brain stimulation (Mayberg et al. 2005). Since the ACC is part of an important neural circuitry comprising the amygdala, many studies have further focused on the role of this circuit in MDD (Scharinger et al. 2010), which we will discuss separately.

4.4.2 ORBITOFRONTAL CORTEX

Converging lines of evidence including neuropathological, lesion, and neuroimaging studies implicate the OFC in the pathophysiology of MDD (Drevets 2007). It shares connections with the amygdala, is involved in reward-guided decision making as well as error detection, and is therefore tightly related to the ACC (O'Doherty 2007; Drevets 2007). Neuroimaging has visualized reductions of gray matter volume and increased metabolism in the OFC of depressed subjects (Drevets 2007). In addition to the ACC, OFC activation or metabolism appears to decrease in response to successful antidepressant therapy (Rabl et al. 2010).

4.4.3 AMYGDALA

The amygdala is primarily recognized for its role in the formation of fear memories and response (LeDoux 2007). Thereby, the amygdala is thought to link negative emotions with other aspects of cognition such as learning and memory (Baxter and Murray 2002). Acutely depressed patients have been shown to exhibit increased amygdala activation in reaction to negatively valenced cues (Sheline et al. 2001; Siegle et al. 2002; Fu et al. 2004; Strigo et al. 2008) as well to subliminal aversive stimuli (Suslow et al. 2010). Notably, amygdala hyperactivity is also considered to be an intermediate phenotype of anxiety disorders (Domschke and Dannlowski 2010), reflecting the high level of comorbidity between both diagnoses (Zimmerman et al. 2002). As in the ACC, antidepressant treatment attenuates amygdala activity and metabolism (Rabl et al. 2010). Furthermore, volumetric alterations of the amygdala in MDD have been the focus of several studies but showed highly inconsistent results (Hamilton et al. 2008).

4.4.4 ACC–AMYGDALA CIRCUIT

A growing number of studies focus on brain circuits, encouraged by observations suggesting that features of connectivity may better relate to behavior than to regional parameters (Meyer-Lindenberg 2009). Much light has been shed on the interplay between the ACC and the amygdala. Decreases of coactivation (functional connectivity) between both structures in samples of acutely depressed patients suggest a role in the pathogenesis of MDD (Stein et al. 2007; Anand et al. 2005a; Chen et al. 2008). Similar observations have been made in healthy subjects showing a negative correlation between ACC–amygdala connectivity and harm avoidance or neuroticism, both trait markers related to depression (Pezawas et al. 2005; Cremers et al. 2010). Accordingly, it has been hypothesized that decreased coupling between both structures reflects diminished control of the ACC over amygdala response. This might at least partly explain findings of increased amygdala reactivity

(Pezawas et al. 2005; Cremers et al. 2010). DTI studies corroborate this conjecture by reporting that higher structural integrity of the ACC–amygdala circuitry predicts lower anxiety (Kim and Whalen 2009). Accordingly, drug treatment studies demonstrated that increased ACC–amygdala coupling is responsible for the alleviation of amygdala hyperactivity under chronic antidepressant medication (Anand et al. 2005b; Anand et al. 2007; Chen et al. 2008).

4.4.5 HIPPOCAMPUS

Together with the ACC, most evidence for volumetric characteristics in MDD has been accumulated for the hippocampus, which has consistently been found to exhibit volume reductions in MDD patients (Macqueen and Frodl 2010). Since hippocampal neurogenesis is vulnerable to glucocorticoids, hippocampal volume loss has primarily been viewed as stress induced (McEwen 2001). This is further underlined by a frequent observation of hippocampal volume reduction in posttraumatic stress disorder (Bremner et al. 2008). Importantly, animal studies clearly suggest that stress-inhibited neurogenesis recovers during antidepressant treatment (Santarelli et al. 2003).

It is a matter of debate whether hippocampal volume loss precedes the first major depressive episode or occurs during the course of the disease (Macqueen and Frodl 2010). Two recent studies in healthy adolescents at risk for depression emphasize the former hypothesis: adverse events in early life have been found to be related to decreased hippocampal volume, suggesting it as a mediator between adverse life events and risk for MDD (Rao et al. 2010). In contrast, others reported decreased volume in healthy participants at familial risk for MDD (Chen et al. 2010). The latter suggests the hippocampus as a target site of gene variants conveying risk for MDD, since there is a strong genetic component in determining hippocampal volume (Macqueen and Frodl 2010).

4.5 GENETIC REGULATION OF EMOTION BRAIN CIRCUITS

A number of imaging genetics studies have been carried out over the past couple of years (Scharinger et al. 2010). The strategy has been dominated by a candidate gene approach for historical reasons: it evolved from association studies with the intent to overcome the limitations inherent to clinical phenotypes (Meyer-Lindenberg 2010). Most studies applied reverse genetics: they investigated the effects of a given genetic variant with known molecular effects on a brain systems level and related those findings to behavior. This approach is the opposite of that used in classical genetics where gene discovery is driven by a given phenotype (Meyer-Lindenberg 2010). Gene variants were therefore selected by their candidate status, which was primarily based on association studies and on hypotheses gained from translational research (Pezawas and Meyer-Lindenberg 2010). This hypothesis-driven strategy links a candidate gene to a certain disorder; the inherent bias may give rise to controversies and leads to a circular argument because the genetic origin is usually not established beyond doubt (Pezawas and Meyer-Lindenberg 2010). Furthermore, the overwhelming number of studies has been carried out in healthy participants and therefore does not directly address clinical entities (Scharinger et al. 2010). Healthy subjects are believed to be less heterogeneous because mental health is probably a more biologically uniform phenotype than those states associated with psychiatric diagnoses. Consistently, studies following this approach are characterized by a high level of conclusiveness (Scharinger et al. 2010). At present, these studies have not yet revealed the whole story of genetics in depression and it is questionable if this goal can be achieved. However, they have yielded intriguing insights into the underlying biology. Accordingly, studying neural effects of specific functional variants can be seen in analogy to manipulations in animal studies that allow for understanding the downstream effects of specific pathways (Meyer-Lindenberg 2010). The two most prominent hypotheses assume that depression arises because of a change in monoamines (monoamine hypothesis) and/or in neurotrophins (neurotrophin hypothesis) (Castren 2005). Accordingly, the following section focuses on monoaminergic genes (*SLC6A4*, *MAOA*, *TPH2*, *HTR1A*, *COMT*) and *BDNF*.

4.5.1 Serotonin Transporter Gene

Alterations in the serotonergic system are widely recognized to be involved in the pathogenesis of MDD (Jans et al. 2007). The serotonergic system originates from the raphe nuclei of the brainstem and projects to all cortical areas and many subcortical regions. However, projection densities are highly variable depending on the location (Gaspar et al. 2003). In the adult brain, the serotonin transporter (5-HTT) is generally localized at the presynaptic terminals of serotonergic neurons. The 5-HTT terminates serotonergic transmission by eliminating serotonin (5-HT) from the synaptic cleft (Canli and Lesch 2007). Because it is the rate-limiting bottleneck in 5-HT clearance, 5-HTT is of high pharmacological relevance. In fact, blockage by SSRIs is established as the first-line treatment of MDD (Bauer et al. 2007). It is noteworthy that 5-HTT along with the 5-HT system is not equally distributed throughout the brain and occurs with highest cortical concentrations along the cingulate gyrus and specifically the sACC (Kranz et al. 2010; Varnas et al. 2004).

In the past decade, a polymorphic region (5-HTTLPR) within the promoter of the 5-HT transporter gene (*SLC6A4*) has attracted unique interest: it has become the most widely investigated genetic variant in psychiatry (Canli and Lesch 2007). 5-HTTLPR is a variable number of tandem repeats (VNTR) polymorphism, occurring as a short (S) and a long (L) allelic version (Heils et al. 1996). About 60% of Caucasians are carriers of the S allele (Lesch et al. 1996), but allele frequencies vary remarkably with ethnicity and likely interact with cultural styles (Chiao and Blizinsky 2010).

Studies in human cell lines showed the 5-HTTLPR genotype to be functionally relevant because it affects the transcriptional activity of the *SLC6A4* gene (Heils et al. 1996). The S allele results in a decrease in 5-HTT expression by about 30% to 40% (Lesch et al. 1996). Similarly, the 5-HTTLPR genotype is thought to affect 5-HTT expression in vivo and consequently to alter extracellular 5-HT concentrations (Canli and Lesch 2007). Since 5-HT levels cannot be measured directly in the living brain, evidence supporting this assumption comes from studies applying PET and single photon emission computed tomography using various radioligands (Willeit and Praschak-Rieder 2010). Accordingly, S allele carriers have been reported to exhibit decreases of 5-HTT binding potential in the midbrain as compared to individuals with L/L genotype (Reimold et al. 2007; Praschak-Rieder et al. 2007; Kalbitzer et al. 2009; Heinz et al. 2000). Other research groups, however, reported divergent results, which may indicate that the impact of the 5-HTTLPR genotype on 5-HTT expression in the human brain is more subtle than suggested by the initial in vitro experiments (Murthy et al. 2010; van Dyck et al. 2004; Willeit et al. 2001; Parsey et al. 2006a; Shioe et al. 2003).

Numerous association studies have been carried out in the past years to link the 5-HTTLPR genotype and susceptibility to MDD. However, attempts to replicate previous positive findings yielded little success (Lopez-Leon et al. 2008). Meta-analyses are similarly controversial: one recent study reported only a small association of the S allele and MDD (Clarke et al. 2010), but others failed to detect any effect (Risch et al. 2009; Lasky-Su et al. 2005). Comparatively, association studies on trait anxiety have initially linked neuroticism to S allele carrier status (Lesch et al. 1996), but subsequent studies failed or reported smaller effects than originally anticipated (Munafo et al. 2009).

In the light of this weak support for a direct pathogenetic role of the S allele in MDD or its vulnerability factors, a report on gene–environment interactions between 5-HTTLPR and environmental adversity has fueled new enthusiasm within the field (Caspi et al. 2003). However, replication studies failed to provide the desired consistent results, possibly due to several methodological issues (Caspi et al. 2010).

In contrast, imaging genetics studies could consistently relate S allele carrier status to amygdala hyperreactivity in healthy subjects and in patients (Hariri et al. 2005; Bertolino et al. 2005; Brown and Hariri 2006; Canli et al. 2005a; Dannlowski et al. 2008, 2009a; Friedel et al. 2009; Hariri et al. 2002; Heinz et al. 2005; Pezawas et al. 2005; Williams et al. 2009; Smolka et al. 2007; Gillihan et al. 2010; Dannlowski et al. 2007). According to meta-analytical data, 5-HTTLPR genotype has been estimated to account for up to 10% of variable amygdala response (Munafo et al. 2008). This finding is supported by animal studies that demonstrated BOLD signal increases after drug-induced

5-HT release (Preece et al. 2009). Furthermore, it has been shown that S allele carriers do not only exhibit increased amygdala activation but also show a faster response to negatively valenced stimuli (Furman et al. 2010). Given the high interconnectivity between amygdala and ACC, it is not surprising that the S allele is also expanding its effects toward this neural circuit, where the S allele has been identified to disinhibit feedback projections from the ACC to the amygdala (Pezawas et al. 2005).

Evidence for a direct link between trait anxiety and amygdala activity is lacking, but functional connectivity of the amygdala and the sACC has been found to account for up to 30% of trait anxiety (Pezawas et al. 2005). Hence, gene-induced reductions of this putative feed-forward loop reflect a biological correlate of neuroticism and lead to distant changes in neural activity within the amygdala. In addition to its role as a neurotransmitter, 5-HT is increasingly recognized as an important factor during brain development (Gaspar et al. 2003). It is noteworthy that during early developmental stages, 5-HTT is not only expressed on 5-HT neurons but also on glutaminergic neurons, where it serves as a "receptor" for the chemical gradient of 5-HT and thus shapes growth cone development and neuronal wiring: 5-HT is known to have an inhibitory influence on cell migration (Homberg et al. 2010). Given the strong structure–function relationship in the nervous system, it is obvious that this type of developmental impact on important neural circuitries has to be reflected in altered neural activation patterns in the adult brain (Sibille and Lewis 2006; Riccio et al. 2009; Ansorge et al. 2004). As noted before, 5-HTTLPR is characterized by variable 5-HTT expression; and since maximal cortical 5-HTT densities can be found along the cingulate cortex and specifically within the sACC (Kranz et al. 2010; Varnas et al. 2004), the S allele of 5-HTTLPR has been found to decrease sACC volume in healthy subjects (Pezawas et al. 2005; Canli et al. 2005a; Frodl et al. 2008). This intriguing finding has also been replicated in primates (Jedema et al. 2010) and likely reflects the altered integrity of white matter tracts between amygdala and sACC (Pacheco et al. 2009). These result in S allele–associated activation changes as detailed above. Apart from volumetric sACC reductions, the S allele has also been associated with morphometric alterations in other regions such as the amygdala (Pezawas et al. 2008; Pezawas et al. 2005; Frodl et al. 2008), OFC (Canli et al. 2005a), and hippocampus (Frodl et al. 2008).

4.5.2 MONOAMINE OXIDASE A GENE

The monoamine oxidase A (MAOA) metabolizes monoamines such as 5-HT, norepinephrine, and dopamine and is predominantly located in the outer mitochondrial membrane of monoaminergic neurons. The pharmacological blockage of MAOA by monoamine oxidase inhibitors increases the extracellular levels of those monoamines and has been found to have substantial antidepressant effects (Racagni and Popoli 2010). Furthermore, reports of relatively elevated levels of MAOA in acutely depressed patients (Meyer et al. 2006) implicated MAOA in the pathogenesis of MDD. This observation is also consistent with the monoamine deficiency hypothesis of depression (Belmaker and Agam 2008).

MAOA is encoded by the *MAOA* gene located on the X chromosome. A functional VNTR polymorphism in the upstream promoter region of the *MAOA* gene has been found to affect gene expression in vitro. Briefly, the transcriptional efficiency in individuals carrying 3.5 or 4 repeats has been found to be three- to fivefold higher than in those carrying 2, 3, or 5 repeats (Sabol et al. 1998). However, available evidence from PET studies suggests no differences in brain MAOA activity between the highly active MAOA-H allele and the less active MAOA-L allele (Fowler et al. 2007; Alia-Klein et al. 2008). In contrast, clinical association studies demonstrated a relative increase in the frequency of MAOA-H alleles in samples of acutely depressed patients (Yu et al. 2005; Schulze et al. 2000). However, most evidence today is available for an association with aggressive behavior (Meyer-Lindenberg et al. 2006b). Studies applying fMRI in samples of healthy volunteers related MAOA-L carrier status to activity increases in both amygdala (Lee and Ham 2008; Meyer-Lindenberg et al. 2006a) and hippocampus (Meyer-Lindenberg et al. 2006a; Lee and Ham

2008). Opposing effects have been found for the OFC (Passamonti et al. 2008; Meyer-Lindenberg et al. 2006a).

There are also studies that report a link between neuroanatomical alterations and the *MAOA* gene: OFC volume is higher (Meyer-Lindenberg et al. 2006a; Cerasa et al. 2008a) and ACC volume is lower (Meyer-Lindenberg et al. 2006a) in healthy MAOA-L carriers. On a brain systems level, the MAOA-L allele has been associated with increases of amygdala–ACC coupling in healthy subjects (Buckholtz et al. 2008) and in depressive patients (Dannlowski et al. 2009b).

4.5.3 Tryptophan Hydroxylase 2 Gene

In serotonergic neurons, tryptophan hydroxylase (TPH) is the rate-limiting enzyme for the biosynthesis of 5-HT (Walther et al. 2003). TPH occurs in two isoforms (TPH1 and TPH2), but only TPH2 is expressed within the brain (Zill et al. 2007; Patel et al. 2004). TPH2 has been associated with MDD based on postmortem studies (Bach-Mizrachi et al. 2006). Further evidence stems from genetic studies demonstrating the impact of variants in *TPH2* on TPH2 expression (Haghighi et al. 2008; Scheuch et al. 2007), 5-HT synthesis (Zhang et al. 2004), and risk for MDD (Zhang et al. 2005; Zhou et al. 2005). Accordingly, anxiety-related personality traits and emotional instability were linked to variation within *TPH2* (Gutknecht et al. 2007). On the level of neural activity, the T allele of G(–703)T TPH2 (rs4570625) was associated with relative increases in amygdala reactivity in healthy volunteers (Brown et al. 2005; Canli et al. 2005b). Finally, a recent study reported hippocampal volume reductions in healthy T allele carriers as compared to G allele homozygotes (Inoue et al. 2010).

4.5.4 Serotonin Receptor 1A Gene

Serotonin receptor 1A (5-HT1A) plays an important role in the regulation of 5-HT signaling (Blier 2010). Specifically, 5-HT1A autoreceptors have been implicated in the regulation of serotonergic neurotransmission throughout the prefrontal cortex (Blier 2010; Sharp et al. 2007) and also related to antidepressant response (Richardson-Jones et al. 2010). Moreover, a neuroimaging study demonstrated an inverse relationship between 5-HT1A autoreceptor density and amygdala activity in healthy volunteers (Fisher et al. 2006).

A functional SNP was identified within the promoter region of the 5-HT1A gene (*HTR1A*), C(–1019)G (rs6295). The G allele prevents binding of transcriptional repressors. This leads to enhanced 5-HT1A receptor expression (Lemonde et al. 2003). Imaging genetics studies are sparse, but preliminary evidence indicates that the G allele is related to increased 5-HT1A receptor binding (Parsey et al. 2006b) and decreased amygdala responsiveness (Fakra et al. 2009). On the phenotype level, the G allele is associated with higher levels of trait anxiety (Strobel et al. 2003) and increased risk for MDD and suicidality (Lemonde et al. 2003).

4.5.5 Catechol-O-Methyltransferase Gene

Catechol-*O*-methyltransferase (COMT) is a monoamine-degrading enzyme. It exists in two isoforms: a soluble version within the cytosol and a membrane-bound form, where the catalytic cleft faces the extracellular milieu (Jeffery and Roth 1984). The latter form is primarily found in neurons. COMT is specifically important for prefrontal dopaminergic neurotransmission because dopamine transporters are absent or of low abundance in the prefrontal cortex (Dickinson and Elvevag 2009). *COMT* has become a major candidate gene for psychiatric disorders with research mainly focusing on psychosis because of its implication in dopaminergic neurotransmission and because of the localization of the *COMT* gene in a chromosomal region conferring risk for psychiatric disorders (Craddock et al. 2006). The genetic regulation of COMT is complex (Nackley et al. 2006; Meyer-Lindenberg et al. 2006b). However, research efforts have focused on a nonsynonymous SNP in

COMT (Val[158]Met, rs4680), which results in lower enzymatic activity and thermostability (Chen et al. 2004). Both allelic variants of Val[158]Met COMT have been associated with several psychiatric disorders such as anxiety disorders, depression, alcohol dependence, and specifically schizophrenia (Craddock et al. 2006). In spite of discouraging results in clinical replication studies (Craddock et al. 2006), converging lines of evidence suggest an important role for cognition and emotion processing (Dickinson and Elvevag 2009). Cognitive performance crucially depends on available dopamine levels and follows an inverted U-shaped curve, with both high and low dopamine levels leading to suboptimal performance (Mattay et al. 2003). Val[158]Met has been found to modulate this relationship by altering the function of working memory areas of the DLPFC (Meyer-Lindenberg et al. 2006b; Mier et al. 2010). Moreover, the influence of Val[158]Met is not restricted to the cognitive domain but affects aspects of emotion regulation in a pleiotropic manner (Mier et al. 2010). Whereas impaired performance of the DLPFC has been linked to the Val allele, the same allele seems to be endowed with improved capacity for emotional regulation (Mier et al. 2010). This relationship seems to form a trade-off between cognitive and emotional capabilities that led to the "warrior vs. worrier" hypothesis, which suggests that both alleles may be advantageous in a specific evolutionary context (Stein et al. 2006). Whereas the Met allele may be associated with superior processing of memory or attention tasks (worrier strategy), the Val allele may be advantageous in the processing of aversive states (warrior strategy) (Stein et al. 2006). Notably, a comparable debate is currently taking place for 5-HTTLPR (Borg et al. 2009; Chiao and Blizinsky 2010). The "warrior vs. worrier" model provides a simplistic metaphor of the complex effects of Val[158]Met; this is substantiated by a recent study pointing to an advantage of the Val allele in rapid and implicit learning from rewards (Krugel et al. 2009). Nevertheless, the worrier/warrior dichotomy is corroborated by studies in transgenic mice (Papaleo et al. 2008) and imaging genetics studies in human subjects (Mier et al. 2010). The latter observed that the amygdala was more active in carriers of the Met allele (Smolka et al. 2005, 2007; Williams et al. 2010; Rasch et al. 2010). One study, however, did not detect any impact on amygdala activation; this is remarkable because the study was carried out in a large sample and was therefore not underpowered (Drabant et al. 2006). Other studies even reported an opposite relationship, particularly in women (Kempton et al. 2009) and in panic disorder patients (Domschke et al. 2008). These inconsistencies may be attributable to differences in the distribution of further allelic variations in *COMT* in the study samples (Rasch et al. 2010; Nackley et al. 2006). Furthermore, Met allele carriers presented increased activation of the ACC (Smolka et al. 2007; Williams et al. 2010) but decreased activation of the OFC (Bishop et al. 2006; Dreher et al. 2009). In contrast, both regions have been found to exhibit dysfunctional patterns of deactivation in Val homozygotes (Pomarol-Clotet et al. 2010). Moreover, greater functional coupling of the amygdala and the OFC has been linked to the Met allele, underscoring the tight relationship of both structures (Drabant et al. 2006; Rasch et al. 2010). The effects of Val[158]Met on brain function further extend to the hippocampus, where the Met allele consistently leads to decreased activation (Smolka et al. 2005, 2007; Drabant et al. 2006; Krach et al. 2010).

As highlighted by the deleterious effects of amphetamine abuse on brain structure, alterations in dopaminergic neurotransmission may not only affect brain function but also its anatomy (Berman et al. 2008). Accordingly, the impact of Val[158]Met on regional brain structure has been scrutinized. Pertinent studies have conclusively reported the Met allele to be related to larger volumes of the amygdala (Taylor et al. 2007; Cerasa et al. 2008b; Ehrlich et al. 2010), OFC (Cerasa et al. 2008b), and the hippocampus (Taylor et al. 2007; Cerasa et al. 2008b; Ehrlich et al. 2010; Honea et al. 2009). However, one study failed to detect the volume effects of Val[158]Met (Ohnishi et al. 2006). This is probably related to methodological issues as well as to ethnicity (Ohnishi et al. 2006).

4.5.6 BRAIN-DERIVED NEUROTROPHIC FACTOR GENE

Brain-derived neurotrophic factor (BDNF) plays a unique role in activity-dependent neuroplasticity (Castren and Rantamaki 2010). Its actions are mediated by two receptor types: the tropomyosin-related kinase B and p75 neurotrophin receptors (Lu et al. 2005). Notably, the precursor molecule of BDNF (pro-BDNF) binds to the p75 neurotrophin receptor, thereby triggering several signal cascades, which cause apoptosis, axonal degeneration, and reduced dendritic complexity. These are related to hippocampal long-term depression (Yang et al. 2009). In contrast, the mature form of BDNF predominantly activates the tropomyosin-related kinase B receptor, promoting neurotrophic and antiapoptotic effects such as neuronal survival, axonal growth, dendritic differentiation, and long-term potentiation (Lu et al. 2005). Over the past decade, a growing body of literature has implicated reductions in BDNF signaling in the pathogenesis of MDD (Castren 2005). Although the exact role of BDNF in depression is yet far from being understood (Groves 2007), decreases of BDNF expression are thought to lead to neuronal atrophy and cell loss in depression-related brain regions, particularly in the prefrontal cortex and the hippocampus (Duman and Monteggia 2006). Therefore, genetic variation within the *BDNF* gene has extensively been investigated and several polymorphic sites have been reported (Licinio et al. 2009). Among these, the focus has been on rs6265, a functional nonsynonymous SNP that leads to an amino acid substitution (Val[66]Met). The altered sequence impairs intracellular trafficking and activity-dependent secretion of BDNF (Egan et al. 2003). However, attempts at linking Val[66]Met directly to MDD have mainly failed (Verhagen et al. 2010; Chen et al. 2008; Lopez-Leon et al. 2008) with the exception of one meta-analysis, which took gender into account and reported a significant association of the Met allele with MDD in males (Verhagen et al. 2010).

In contrast to this weak support for a direct pathogenetic role of Val[66]Met in MDD, highly consistent findings exist for an impact on hippocampal volume where BDNF is known to occur in the highest concentration in the brain. A number of studies reported healthy and acutely depressed carriers of the Met allele to exhibit relatively smaller hippocampal volume (Frodl et al. 2007; Matsuo et al. 2009; Pezawas et al. 2004; Schofield et al. 2009; Montag et al. 2009; Chepenik et al. 2009; Bueller et al. 2006; Szeszko et al. 2005). However, a minority failed to detect volumetric effects (Jessen et al. 2009; Koolschijn et al. 2009). In addition to these effects on brain structure, functional studies with fMRI also demonstrated an impact of Val[66]Met BDNF on hippocampal activation (Egan et al. 2003; Hashimoto et al. 2008; Hariri et al. 2003; Schofield et al. 2009). Interestingly, imaging genetics studies further suggest that the effects of Val[66]Met BDNF on limbic regions are most pronounced in the hippocampus and may not be present in other depression-related regions (Scharinger et al. 2010). This conjecture is supported by the failure to reveal Val[66]Met BDNF-related alterations in amygdala volume (Frodl et al. 2007; Matsuo et al. 2009; Nemoto et al. 2006; Pezawas et al. 2004; Schofield et al. 2009) or in amygdala function (Egan et al. 2003; Schofield et al. 2009; Hashimoto et al. 2008; Hariri et al. 2003; Montag et al. 2008). Similarly, limited evidence is available for BDNF effects on ACC volume (Matsuo et al. 2009) and OFC activation (Gasic et al. 2009).

4.6 PLEIOTROPY, EPISTASIS, AND GENE-ENVIRONMENT INTERACTIONS

Intermediate phenotypes are considered less complex than clinical phenotypes. Nevertheless, they are still substantially more complex with respect to the underlying genetics than Mendelian traits (Meyer-Lindenberg 2010). This is not necessarily a disadvantage for the reverse genetic approach as long as effect sizes are large enough to disentangle the effects of single variants (Meyer-Lindenberg 2010). However, the inherent complexity may result in a large variation in the range of findings; in the absence of additional insights, these variable study outcomes are difficult to understand (Scharinger et al. 2010).

Several candidate loci have been found to exert pleiotropic effects on the level of intermediate phenotypes. For instance, Val[158]Met COMT does not only affect prefrontal regions linked to executive function (Mier et al. 2010) but also modulates an extensive network of limbic regions including the amygdala, OFC, and hippocampus (Scharinger et al. 2010). In turn, locus heterogeneity has frequently been observed: several candidate loci have been shown to impact on the same intermediate phenotypes, for example, Val[66]Met BDNF has been shown to be a regulator of hippocampal volume like Val[158]Met COMT, which in turn shares effects on amygdala activation with the 5-HT-TLPR of *SLC6A4* (Scharinger et al. 2010). Moreover, an abundant number of candidate genes has been shown to act on regions of emotion processing in single studies but yet require replication (Scharinger et al. 2010). Several of those genes such as *HTR2A*, *HTR3A*, *NPY*, *CREB1*, or *DISC1* are promising candidates that may play a similar role as *SLC6A4*, *BDNF*, or *COMT* (Scharinger et al. 2010). This stresses the notion that such variants may not only jointly act on the level of intermediate phenotypes but also interact with each other on the level of molecules, cells, or brain systems. Such gene–gene interactions can range from additive effects (Smolka et al. 2007) to complex interactions, termed *epistasis* (Pezawas et al. 2008). In a seminal study, it has been shown that the Met allele of Val[66]Met BDNF protects against the deleterious effects of the 5-HTTLPR S allele on the amygdala–ACC circuitry (Pezawas et al. 2008). These studies pinpoint the need for taking complex interactions into account and challenge the assumption of specific alleles as risk factors in depression. Further complexity is added by interactions with environmental factors (Caspi et al. 2010). Although gene–environment interactions have been a hot topic in psychiatric genetics in recent years, imaging genetics studies have been rather reluctant to adopt such models (Gatt et al. 2009; Canli et al. 2006). This reluctance is readily understood, if one takes into account the large sample size needed to study complex interactions. However, it is safe to predict that, soon, studies based on larger cohorts will address this fundamental issue because the nature/nurture dichotomy lies at the very heart of many theories of psychiatric diseases.

4.7 CONCLUSION AND FUTURE DIRECTIONS

Imaging genetics has evolved from single pioneering studies to a frequently employed strategy that has certainly vitalized psychiatric genetics and advanced our understanding of how genes shape behavior (Bigos and Weinberger 2010). However, the intermediate phenotype concept has been a matter of debate based on evidence that some candidate traits are akin to clinical phenotypes regarding their genetic complexity (Flint and Munafo 2007). Nevertheless, imaging genetics studies have proven the suitability of this approach by accomplishing robust effect sizes on a brain systems level (Mier et al. 2010; Munafo et al. 2008) for variants that are still questioned with respect to their clinical implication (Risch et al. 2009; Craddock et al. 2006). Candidate gene studies provided valuable insights into pathogenetic pathways that may shape the brain and render it susceptible to MDD. Furthermore, these studies open new perspectives for translating knowledge from preclinical studies to people (Soliman et al. 2010). However, regardless of whether clinical phenotypes or intermediate phenotypes were used, candidate gene studies did not yet fulfill expectations on genetically tailored diagnostics and treatment (Taylor et al. 2010; Risch et al. 2009). More agnostic approaches such as genome-wide association studies (GWAS) provide little support for traditional candidate genes (Bosker et al. 2011) but are still far from being transferrable from bench to bedside (Demirkan et al. 2010). Whereas the application of intermediate phenotypes is likely to enhance genome-wide approaches for gene discovery (Potkin et al. 2009), candidate variants derived from GWAS could be put under further scrutiny by imaging genetics studies as recently been shown for a newish risk variant for psychosis (Esslinger et al. 2009). Hence, a translational approach that incorporates genome-wide studies for the identification of yet unknown genes as well as in vitro, animal, and neuroimaging studies to further analyze the related pathways may dramatically expand our neurobiological understanding of depression.

REFERENCES

Alia-Klein, N., R. Z. Goldstein, A. Kriplani, et al. 2008. Brain monoamine oxidase A activity predicts trait aggression. *J Neurosci* 28 (19):5099–104.
Anand, A., Y. Li, Y. Wang, J. et al. 2005a. Activity and connectivity of brain mood regulating circuit in depression: A functional magnetic resonance study. *Biol Psychiatry* 57 (10):1079–88.

Anand, A., Y. Li, Y. Wang, J. et al. 2005b. Antidepressant effect on connectivity of the mood-regulating circuit: An FMRI study. *Neuropsychopharmacology* 30 (7):1334–44.

Anand, A., Y. Li, Y. Wang, K. Gardner, and M. J. Lowe. 2007. Reciprocal effects of antidepressant treatment on activity and connectivity of the mood regulating circuit: An FMRI study. *J Neuropsychiatry Clin Neurosci* 19 (3):274–82.

Ansorge, M. S., M. Zhou, A. Lira, R. Hen, and J. A. Gingrich. 2004. Early-life blockade of the 5-HT transporter alters emotional behavior in adult mice. *Science* 306 (5697):879–81.

Ashburner, J., and K. J. Friston. 2000. Voxel-based morphometry—the methods. *NeuroImage* 11:805–21.

Bach-Mizrachi, H., M. D. Underwood, S. A. Kassir, et al. 2006. Neuronal tryptophan hydroxylase mRNA expression in the human dorsal and median raphe nuclei: Major depression and suicide. *Neuropsycho-pharmacology* 31 (4):814–24.

Bandettini, P. A. 2009. What's new in neuroimaging methods? *Ann N Y Acad Sci* 1156:260–93.

Bauer, M., T. Bschor, A. Pfennig, et al. 2007. World Federation of Societies of Biological Psychiatry (WFSBP) guidelines for biological treatment of unipolar depressive disorders in primary care. *World J Biol Psychiatry* 8 (2):67–104.

Baxter, M. G., and E. A. Murray. 2002. The amygdala and reward. *Nat Rev Neurosci* 3 (7):563–73.

Belmaker, R. H., and G. Agam. 2008. Major depressive disorder. *N Engl J Med* 358 (1):55–68.

Berman, S., J. O'Neill, S. Fears, G. Bartzokis, and E. D. London. 2008. Abuse of amphetamines and structural abnormalities in the brain. *Ann N Y Acad Sci* 1141:195–220.

Bertolino, A., G. Arciero, V. Rubino, et al. 2005. Variation of human amygdala response during threatening stimuli as a function of 5′HTTLPR genotype and personality style. *Biol Psychiatry* 57 (12):1517–25.

Bigos, K. L., and D. R. Weinberger. 2010. Imaging genetics—days of future past. *NeuroImage* 53 (3):804–9.

Bishop, S. J., J. D. Cohen, J. Fossella, B. J. Casey, and M. J. Farah. 2006. COMT genotype influences prefrontal response to emotional distraction. *Cogn Affect Behav Neurosci* 6 (1):62–70.

Blier, P. 2010. Altered function of the serotonin 1A autoreceptor and the antidepressant response. *Neuron* 65 (1):1–2.

Boes, A. D., L. M. McCormick, W. H. Coryell, and P. Nopoulos. 2008. Rostral anterior cingulate cortex volume correlates with depressed mood in normal healthy children. *Biol Psychiatry* 63 (4):391–7.

Borg, J., S. Henningsson, T. Saijo, et al. 2009. Serotonin transporter genotype is associated with cognitive performance but not regional 5-HT1A receptor binding in humans. *Int J Neuropsychopharmacol* 12 (6):783–92.

Bosker, F. J., C. A. Hartman, I. M. Nolte, et al. 2011. Poor replication of candidate genes for major depressive disorder using genome-wide association data. *Mol Psychiatry* 16(5):516–32.

Bremner, J. D., B. Elzinga, C. Schmahl, and E. Vermetten. 2008. Structural and functional plasticity of the human brain in posttraumatic stress disorder. *Prog Brain Res* 167:171–86.

Brown, S. M., and A. R. Hariri. 2006. Neuroimaging studies of serotonin gene polymorphisms: Exploring the interplay of genes, brain, and behavior. *Cogn Affect Behav Neurosci* 6 (1):44–52.

Brown, S. M., E. Peet, S. B. Manuck, et al. 2005. A regulatory variant of the human tryptophan hydroxylase-2 gene biases amygdala reactivity. *Mol Psychiatry* 10 (9):884–8, 805.

Buckholtz, J. W., J. H. Callicott, B. Kolachana, et al. 2008. Genetic variation in MAOA modulates ventromedial prefrontal circuitry mediating individual differences in human personality. *Mol Psychiatry* 13 (3):313–24.

Bueller, J. A., M. Aftab, S. Sen, D. Gomez-Hassan, M. Burmeister, and J. K. Zubieta. 2006. BDNF Val66Met allele is associated with reduced hippocampal volume in healthy subjects. *Biol Psychiatry* 59 (9):812–5.

Burmeister, M., M. G. McInnis, and S. Zollner. 2008. Psychiatric genetics: Progress amid controversy. *Nat Rev Genet* 9 (7):527–40.

Canli, T., and K. P. Lesch. 2007. Long story short: The serotonin transporter in emotion regulation and social cognition. *Nat Neurosci* 10 (9):1103–9.

Canli, T., K. Omura, B. W. Haas, A. Fallgatter, R. T. Constable, and K. P. Lesch. 2005a. Beyond affect: A role for genetic variation of the serotonin transporter in neural activation during a cognitive attention task. *Proc Natl Acad Sci U S A* 102 (34):12224–9.

Canli, T., E. Congdon, L. Gutknecht, R. T. Constable, and K. P. Lesch. 2005b. Amygdala responsiveness is modulated by tryptophan hydroxylase-2 gene variation. *J Neural Transm* 112 (11):1479–85.

Canli, T., M. Qiu, K. Omura, E. et al. 2006. Neural correlates of epigenesis. *Proc Natl Acad Sci U S A* 103 (43):16033–8.

Caspi, A., K. Sugden, T. E. Moffitt, et al. 2003. Influence of life stress on depression: Moderation by a polymorphism in the 5-HTT gene. *Science* 301 (5631):386–9.

Caspi, A., A. R. Hariri, A. Holmes, R. Uher, and T. E. Moffitt. 2010. Genetic sensitivity to the environment: The case of the serotonin transporter gene and its implications for studying complex diseases and traits. *Am J Psychiatry* 167 (5):509–27.

Castren, E. 2005. Is mood chemistry? *Nat Rev Neurosci* 6 (3):241–6.

Castren, E., and T. Rantamaki. 2010. Role of brain-derived neurotrophic factor in the aetiology of depression: Implications for pharmacological treatment. *CNS Drugs* 24 (1):1–7.

Cerasa, A., M. C. Gioia, A. Labate, et al. 2008a. MAO A VNTR polymorphism and variation in human morphology: A VBM study. *Neuroreport* 19 (11):1107–10.

Cerasa, A., M. C. Gioia, A. Labate, M. Liguori, P. Lanza, and A. Quattrone. 2008b. Impact of catechol-*O*-methyltransferase Val(108/158) Met genotype on hippocampal and prefrontal gray matter volume. *Neuroreport* 19 (4):405–8.

Chanock, S. J., T. Manolio, M. Boehnke, et al. 2007. Replicating genotype–phenotype associations. *Nature* 447 (7145):655–60.

Chen, C. H., J. Suckling, C. Ooi, et al. 2008. Functional coupling of the amygdala in depressed patients treated with antidepressant medication. *Neuropsychopharmacology* 33 (8):1909–18.

Chen, J., B. K. Lipska, N. Halim, et al. 2004. Functional analysis of genetic variation in catechol-*O*-methyltransferase (COMT): Effects on mRNA, protein, and enzyme activity in postmortem human brain. *Am J Hum Genet* 75 (5):807–21.

Chen, L., D. A. Lawlor, S. J. Lewis, et al. 2008. Genetic association study of BDNF in depression: Finding from two cohort studies and a meta-analysis. *Am J Med Genet B Neuropsychiatr Genet* 147B (6): 814–21.

Chen, M. C., J. P. Hamilton, and I. H. Gotlib. 2010. Decreased hippocampal volume in healthy girls at risk of depression. *Arch Gen Psychiatry* 67 (3):270–6.

Chepenik, L. G., C. Fredericks, X. Papademetris, et al. 2009. Effects of the brain-derived neurotrophic growth factor val66met variation on hippocampus morphology in bipolar disorder. *Neuropsychopharmacology* 34 (4):944–51.

Chiao, J. Y., and K. D. Blizinsky. 2010. Culture–gene coevolution of individualism–collectivism and the serotonin transporter gene. *Proc Biol Sci* 277 (1681):529–37.

Clarke, H., J. Flint, A. S. Attwood, and M. R. Munafo. 2010. Association of the 5-HTTLPR genotype and unipolar depression: A meta-analysis. *Psychol Med* (Epub):1–12.

Craddock, N., M. J. Owen, and M. C. O'Donovan. 2006. The catechol-*O*-methyl transferase (COMT) gene as a candidate for psychiatric phenotypes: Evidence and lessons. *Mol Psychiatry* 11 (5):446–58.

Cremers, H. R., L. R. Demenescu, A. Aleman, et al. 2010. Neuroticism modulates amygdala–prefrontal connectivity in response to negative emotional facial expressions. *NeuroImage* 49 (1):963–70.

Dannlowski, U., P. Ohrmann, J. Bauer, et al. 2007. Serotonergic genes modulate amygdala activity in major depression. *Genes Brain Behav* 6 (7):672–6.

Dannlowski, U., P. Ohrmann, J. Bauer, et al. 2008. 5-HTTLPR biases amygdala activity in response to masked facial expressions in major depression. *Neuropsychopharmacology* 33 (2):418–24.

Dannlowski, U., C. Konrad, H. Kugel, et al. 2009a. Emotion specific modulation of automatic amygdala responses by 5-HTTLPR genotype. *NeuroImage* 53 (3):893–8.

Dannlowski, U., P. Ohrmann, C. Konrad, et al. 2009b. Reduced amygdala–prefrontal coupling in major depression: Association with MAOA genotype and illness severity. *Int J Neuropsychopharmacol* 12 (1): 11–22.

Demirkan, A., B. W. Penninx, K. Hek, et al. 2010. Genetic risk profiles for depression and anxiety in adult and elderly cohorts. *Mol Psychiatry* doi: 10.1038/mp.2010.65.

Dickinson, D., and B. Elvevag. 2009. Genes, cognition and brain through a COMT lens. *Neuroscience* 164 (1): 72–87.

Domschke, K., and U. Dannlowski. 2010. Imaging genetics of anxiety disorders. *NeuroImage* 53 (3):822–31.

Domschke, K., P. Ohrmann, M. Braun, et al. 2008. Influence of the catechol-*O*-methyltransferase val158met genotype on amygdala and prefrontal cortex emotional processing in panic disorder. *Psychiatry Res* 163 (1):13–20.

Drabant, E. M., A. R. Hariri, A. Meyer-Lindenberg, et al. 2006. Catechol *O*-methyltransferase val158met genotype and neural mechanisms related to affective arousal and regulation. *Arch Gen Psychiatry* 63 (12): 1396–406.

Dreher, J. C., P. Kohn, B. Kolachana, D. R. Weinberger, and K. F. Berman. 2009. Variation in dopamine genes influences responsivity of the human reward system. *Proc Natl Acad Sci U S A* 106 (2):617–22.

Drevets, W. C. 2007. Orbitofrontal cortex function and structure in depression. *Ann N Y Acad Sci* 1121: 499–527.

Drevets, W. C., J. L. Price, J. R. Simpson, et al. 1997. Subgenual prefrontal cortex abnormalities in mood disorders. *Nature* 386 (6627):824–7.

Drevets, W. C., J. Savitz, and M. Trimble. 2008. The subgenual anterior cingulate cortex in mood disorders. *CNS Spectr* 13 (8):663–81.

Duman, R. S., and L. M. Monteggia. 2006. A neurotrophic model for stress-related mood disorders. *Biol Psychiatry* 59 (12):1116–27.

Egan, M. F., M. Kojima, J. H. Callicott, et al. 2003. The BDNF val66met polymorphism affects activity-dependent secretion of BDNF and human memory and hippocampal function. *Cell* 112 (2):257–69.

Ehrlich, S., E. M. Morrow, J. L. Roffman, et al. 2010. The COMT Val108/158Met polymorphism and medial temporal lobe volumetry in patients with schizophrenia and healthy adults. *NeuroImage* 53 (3):992–1000.

Esslinger, C., H. Walter, P. Kirsch, et al. 2009. Neural mechanisms of a genome-wide supported psychosis variant. *Science* 324 (5927):605.

Fakra, E., L. W. Hyde, A. Gorka, et al. 2009. Effects of HTR1A C(–1019)G on amygdala reactivity and trait anxiety. *Arch Gen Psychiatry* 66 (1):33–40.

Fisher, P. M., C. C. Meltzer, S. K. Ziolko, et al. 2006. Capacity for 5-HT1A-mediated autoregulation predicts amygdala reactivity. *Nat Neurosci* 9 (11):1362–3.

Flint, J., and M. R. Munafo. 2007. The endophenotype concept in psychiatric genetics. *Psychol Med* 37 (2): 163–80.

Fowler, J. S., N. Alia-Klein, A. Kriplani, et al. 2007. Evidence that brain MAO A activity does not correspond to MAO A genotype in healthy male subjects. *Biol Psychiatry* 62 (4):355–8.

Friedel, E., F. Schlagenhauf, P. Sterzer, et al. 2009. 5-HTT genotype effect on prefrontal–amygdala coupling differs between major depression and controls. *Psychopharmacology (Berl)* 205 (2):261–71.

Frodl, T., C. Schule, G. Schmitt, et al. 2007. Association of the brain-derived neurotrophic factor Val66Met polymorphism with reduced hippocampal volumes in major depression. *Arch Gen Psychiatry* 64 (4): 410–6.

Frodl, T., N. Koutsouleris, R. Bottlender, et al. 2008. Reduced gray matter brain volumes are associated with variants of the serotonin transporter gene in major depression. *Mol Psychiatry* 13 (12):1093–101.

Fu, C. H., S. C. Williams, A. J. Cleare, et al. 2004. Attenuation of the neural response to sad faces in major depression by antidepressant treatment: A prospective, event-related functional magnetic resonance imaging study. *Arch Gen Psychiatry* 61 (9):877–89.

Furman, D. J., J. P. Hamilton, J. Joormann, and I. H. Gotlib. 2010. Altered timing of amygdala activation during sad mood elaboration as a function of 5-HTTLPR. *Soc Cogn Affect Neurosci* doi: 10.1093/scan/nsq029.

Gasic, G. P., J. W. Smoller, R. H. Perlis, et al. 2009. BDNF, relative preference, and reward circuitry responses to emotional communication. *Am J Med Genet B Neuropsychiatr Genet* doi: 10.1002/ajmg.b.30944.

Gaspar, P., O. Cases, and L. Maroteaux. 2003. The developmental role of serotonin: News from mouse molecular genetics. *Nat Rev Neurosci* 4 (12):1002–12.

Gatt, J. M., C. B. Nemeroff, C. Dobson-Stone, et al. 2009. Interactions between BDNF Val66Met polymorphism and early life stress predict brain and arousal pathways to syndromal depression and anxiety. *Mol Psychiatry* 14 (7):681–95.

Gillihan, S. J., H. Rao, J. Wang, et al. 2010. Serotonin transporter genotype modulates amygdala activity during mood regulation. *Soc Cogn Affect Neurosci* 5 (1):1–10.

Gottesman, II, and T. D. Gould. 2003. The endophenotype concept in psychiatry: Etymology and strategic intentions. *Am J Psychiatry* 160 (4):636–45.

Groves, J. O. 2007. Is it time to reassess the BDNF hypothesis of depression? *Mol Psychiatry* 12 (12): 1079–88.

Gutknecht, L., C. Jacob, A. Strobel, et al. 2007. Tryptophan hydroxylase-2 gene variation influences personality traits and disorders related to emotional dysregulation. *Int J Neuropsychopharmacol* 10 (3):309–20.

Haghighi, F., H. Bach-Mizrachi, Y. Y. Huang, et al. 2008. Genetic architecture of the human tryptophan hydroxylase 2 Gene: Existence of neural isoforms and relevance for major depression. *Mol Psychiatry* 13 (8):813–20.

Hajek, T., J. Kozeny, M. Kopecek, M. Alda, and C. Hoschl. 2008. Reduced subgenual cingulate volumes in mood disorders: A meta-analysis. *J Psychiatry Neurosci* 33 (2):91–9.

Hamilton, J. P., M. Siemer, and I. H. Gotlib. 2008. Amygdala volume in major depressive disorder: A meta-analysis of magnetic resonance imaging studies. *Mol Psychiatry* 13 (11):993–1000.

Hamilton, J. P., G. Chen, M. E. Thomason, M. E. Schwartz, and I. H. Gotlib. 2010. Investigating neural primacy in major depressive disorder: Multivariate Granger causality analysis of resting-state fMRI time-series data. *Mol Psychiatry* doi:10.1038/mp.2010.46.

Hariri, A. R., V. S. Mattay, A. Tessitore, B. et al. 2002. Serotonin transporter genetic variation and the response of the human amygdala. *Science* 297 (5580):400–3.

Hariri, A. R., T. E. Goldberg, V. S. Mattay, et al. 2003. Brain-derived neurotrophic factor val66met polymorphism affects human memory-related hippocampal activity and predicts memory performance. *J Neurosci* 23 (17):6690–4.

Hariri, A. R., E. M. Drabant, K. E. Munoz, et al. 2005. A susceptibility gene for affective disorders and the response of the human amygdala. *Arch Gen Psychiatry* 62 (2):146–52.

Hariri, A. R., E. M. Drabant, and D. R. Weinberger. 2006. Imaging genetics: Perspectives from studies of genetically driven variation in serotonin function and corticolimbic affective processing. *Biol Psychiatry* 59 (10):888–97.

Hashimoto, R., Y. Moriguchi, F. Yamashita, et al. 2008. Dose-dependent effect of the Val66Met polymorphism of the brain-derived neurotrophic factor gene on memory-related hippocampal activity. *Neurosci Res* 61 (4):360–7.

Heils, A., A. Teufel, S. Petri, G. Stober, P. Riederer, D. Bengel, and K. P. Lesch. 1996. Allelic variation of human serotonin transporter gene expression. *J Neurochem* 66 (6):2621–4.

Heinz, A., D. W. Jones, C. Mazzanti, et al. 2000. A relationship between serotonin transporter genotype and in vivo protein expression and alcohol neurotoxicity. *Biol Psychiatry* 47 (7):643–9.

Heinz, A., D. F. Braus, M. N. Smolka, et al. 2005. Amygdala–prefrontal coupling depends on a genetic variation of the serotonin transporter. *Nat Neurosci* 8 (1):20–1.

Homberg, J. R., D. Schubert, and P. Gaspar. 2010. New perspectives on the neurodevelopmental effects of SSRIs. *Trends Pharmacol Sci* 31 (2):60–5.

Honea, R., B. A. Verchinski, L. Pezawas, et al. 2009. Impact of interacting functional variants in COMT on regional gray matter volume in human brain. *NeuroImage* 45 (1):44–51.

Inoue, H., H. Yamasue, M. Tochigi, et al. 2010. Effect of tryptophan hydroxylase-2 gene variants on amygdalar and hippocampal volumes. *Brain Res* 1331:51–7.

Jans, L. A., W. J. Riedel, C. R. Markus, and A. Blokland. 2007. Serotonergic vulnerability and depression: Assumptions, experimental evidence and implications. *Mol Psychiatry* 12 (6):522–43.

Jedema, H. P., P. J. Gianaros, P. J. Greer, et al. 2010. Cognitive impact of genetic variation of the serotonin transporter in primates is associated with differences in brain morphology rather than serotonin neurotransmission. *Mol Psychiatry* 15 (5):512–22, 446.

Jeffery, D. R., and J. A. Roth. 1984. Characterization of membrane-bound and soluble catechol-*O*-methyltransferase from human frontal cortex. *J Neurochem* 42 (3):826–32.

Jessen, F., A. Schuhmacher, O. von Widdern, et al. 2009. No association of the Val66Met polymorphism of the brain-derived neurotrophic factor with hippocampal volume in major depression. *Psychiatr Genet* 19 (2):99–101.

Kalbitzer, J., V. G. Frokjaer, D. Erritzoe, et al. 2009. The personality trait openness is related to cerebral 5-HTT levels. *NeuroImage* 45 (2):280–5.

Kato, M., and A. Serretti. 2008. Review and meta-analysis of antidepressant pharmacogenetic findings in major depressive disorder. *Mol Psychiatry* 15 (5):473–500.

Kempton, M. J., M. Haldane, J. Jogia, et al. 2009. The effects of gender and COMT Val158Met polymorphism on fearful facial affect recognition: A fMRI study. *Int J Neuropsychopharmacol* 12 (3):371–81.

Kim, M. J., and P. J. Whalen. 2009. The structural integrity of an amygdala–prefrontal pathway predicts trait anxiety. *J Neurosci* 29 (37):11614–8.

Koolschijn, P. C., N. E. van Haren, S. C. Bakker, M. L. Hoogendoorn, H. E. Pol, and R. S. Kahn. 2009. Effects of brain-derived neurotrophic factor Val66Met polymorphism on hippocampal volume change in schizophrenia. *Hippocampus* 20 (9):1010–7.

Krach, S., A. Jansen, A. Krug, et al. 2010. COMT genotype and its role on hippocampal–prefrontal regions in declarative memory. *NeuroImage* 53 (3):978–84.

Kranz, G. S., S. Kasper, and R. Lanzenberger. 2010. Reward and the serotonergic system. *Neuroscience* 166 (4): 1023–35.

Krugel, L. K., G. Biele, P. N. Mohr, S. C. Li, and H. R. Heekeren. 2009. Genetic variation in dopaminergic neuromodulation influences the ability to rapidly and flexibly adapt decisions. *Proc Natl Acad Sci U S A* 106 (42):17951–6.

Lasky-Su, J. A., S. V. Faraone, S. J. Glatt, and M. T. Tsuang. 2005. Meta-analysis of the association between two polymorphisms in the serotonin transporter gene and affective disorders. *Am J Med Genet B Neuropsychiatr Genet* 133B (1):110–5.

LeDoux, J. 2007. The amygdala. *Curr Biol* 17 (20):R868–74.

Lee, B. T., and B. J. Ham. 2008. Monoamine oxidase A-uVNTR genotype affects limbic brain activity in response to affective facial stimuli. *Neuroreport* 19 (5):515–9.

Lemonde, S., G. Turecki, D. Bakish, et al. 2003. Impaired repression at a 5-hydroxytryptamine 1A receptor gene polymorphism associated with major depression and suicide. *J Neurosci* 23 (25):8788–99.

Lesch, K. P., D. Bengel, A. Heils, et al. 1996. Association of anxiety-related traits with a polymorphism in the serotonin transporter gene regulatory region. *Science* 274 (5292):1527–31.

Levinson, D. F. 2006. The genetics of depression: A review. *Biol Psychiatry* 60 (2):84–92.

Licinio, J., C. Dong, and M. L. Wong. 2009. Novel sequence variations in the brain-derived neurotrophic factor gene and association with major depression and antidepressant treatment response. *Arch Gen Psychiatry* 66 (5):488–97.

Logothetis, N. K. 2008. What we can do and what we cannot do with fMRI. *Nature* 453 (7197):869–78.

Lohmueller, K. E., C. L. Pearce, M. Pike, E. S. Lander, and J. N. Hirschhorn. 2003. Meta-analysis of genetic association studies supports a contribution of common variants to susceptibility to common disease. *Nat Genet* 33 (2):177–82.

Lopez-Leon, S., A. C. Janssens, A. M. Gonzalez-Zuloeta Ladd, et al. 2008. Meta-analyses of genetic studies on major depressive disorder. *Mol Psychiatry* 13 (8):772–85.

Lu, B., P. T. Pang, and N. H. Woo. 2005. The yin and yang of neurotrophin action. *Nat Rev Neurosci* 6 (8): 603–14.

Macqueen, G., and T. Frodl. 2011. The hippocampus in major depression: Evidence for the convergence of the bench and bedside in psychiatric research? *Mol Psychiatry* 16 (3):252–64.

Matsuo, K., C. Walss-Bass, F. G. Nery, et al. 2009. Neuronal correlates of brain-derived neurotrophic factor Val66Met polymorphism and morphometric abnormalities in bipolar disorder. *Neuropsychopharmacology* 34 (8):1904–13.

Mattay, V. S., T. E. Goldberg, F. Fera, et al. 2003. Catechol *O*-methyltransferase val158-met genotype and individual variation in the brain response to amphetamine. *Proc Natl Acad Sci U S A* 100 (10):6186–91.

Mayberg, H. S., A. M. Lozano, V. Voon, et al. 2005. Deep brain stimulation for treatment-resistant depression. *Neuron* 45 (5):651–60.

McEwen, B. S. 2001. Plasticity of the hippocampus: Adaptation to chronic stress and allostatic load. *Ann N Y Acad Sci* 933:265–77.

McGuffin, P., and R. Katz. 1989. The genetics of depression and manic–depressive disorder. *Br J Psychiatry* 155:294–304.

McMahon, F. J. 2010. Pioneering first steps and cautious conclusions. *Biol Psychiatry* 67 (2):99–100.

Meyer, J. H., N. Ginovart, A. Boovariwala, et al. 2006. Elevated monoamine oxidase A levels in the brain: An explanation for the monoamine imbalance of major depression. *Arch Gen Psychiatry* 63 (11): 1209–16.

Meyer-Lindenberg, A. 2009. Neural connectivity as an intermediate phenotype: Brain networks under genetic control. *Hum Brain Mapp* 30 (7):1938–46.

Meyer-Lindenberg, A. 2010. Intermediate or brainless phenotypes for psychiatric research? *Psychol Med* 40 (7):1057–62.

Meyer-Lindenberg, A., and D. R. Weinberger. 2006. Intermediate phenotypes and genetic mechanisms of psychiatric disorders. *Nat Rev Neurosci* 7 (10):818–27.

Meyer-Lindenberg, A., J. W. Buckholtz, B. Kolachana, et al. 2006a. Neural mechanisms of genetic risk for impulsivity and violence in humans. *Proc Natl Acad Sci U S A* 103 (16):6269–74.

Meyer-Lindenberg, A., T. Nichols, J. H. Callicott, et al. 2006b. Impact of complex genetic variation in COMT on human brain function. *Mol Psychiatry* 11 (9):867–77, 797.

Mier, D., P. Kirsch, and A. Meyer-Lindenberg. 2010. Neural substrates of pleiotropic action of genetic variation in COMT: A meta-analysis. *Mol Psychiatry* 15:918–27.

Miller, G. 2010. Psychiatry. Beyond DSM: Seeking a brain-based classification of mental illness. *Science* 327 (5972):1437.

Montag, C., M. Reuter, B. Newport, C. Elger, and B. Weber. 2008. The BDNF Val66Met polymorphism affects amygdala activity in response to emotional stimuli: Evidence from a genetic imaging study. *NeuroImage* 42 (4):1554–9.

Montag, C., B. Weber, K. Fliessbach, C. Elger, and M. Reuter. 2009. The BDNF Val66Met polymorphism impacts parahippocampal and amygdala volume in healthy humans: Incremental support for a genetic risk factor for depression. *Psychol Med*:1–9.

Munafo, M. R., S. M. Brown, and A. R. Hariri. 2008. Serotonin transporter (5-HTTLPR) genotype and amygdala activation: A meta-analysis. *Biol Psychiatry* 63 (9):852–7.

Munafo, M. R., N. B. Freimer, W. Ng, et al. 2009. 5-HTTLPR genotype and anxiety-related personality traits: A meta-analysis and new data. *Am J Med Genet B Neuropsychiatr Genet* 150B (2):271–81.

Murthy, N. V., S. Selvaraj, P. J. Cowen, et al. 2010. Serotonin transporter polymorphisms (SLC6A4 insertion/deletion and rs25531) do not affect the availability of 5-HTT to [11C] DASB binding in the living human brain. *NeuroImage* 52 (1):50–4.

Nackley, A. G., S. A. Shabalina, I. E. Tchivileva, et al. 2006. Human catechol-*O*-methyltransferase haplotypes modulate protein expression by altering mRNA secondary structure. *Science* 314 (5807):1930–3.

Nemoto, K., T. Ohnishi, T. Mori, et al. 2006. The Val66Met polymorphism of the brain-derived neurotrophic factor gene affects age-related brain morphology. *Neurosci Lett* 397 (1–2):25–9.

O'Doherty, J. P. 2007. Lights, camembert, action! The role of human orbitofrontal cortex in encoding stimuli, rewards, and choices. *Ann N Y Acad Sci* 1121:254–72.

Ohnishi, T., R. Hashimoto, T. Mori, et al. 2006. The association between the Val158Met polymorphism of the catechol-*O*-methyl transferase gene and morphological abnormalities of the brain in chronic schizophrenia. *Brain* 129 (Pt 2):399–410.

Ongur, D., W. C. Drevets, and J. L. Price. 1998. Glial reduction in the subgenual prefrontal cortex in mood disorders. *Proc Natl Acad Sci U S A* 95 (22):13290–5.

Pacheco, J., C. G. Beevers, C. Benavides, J. McGeary, E. Stice, and D. M. Schnyer. 2009. Frontal–limbic white matter pathway associations with the serotonin transporter gene promoter region (5-HTTLPR) polymorphism. *J Neurosci* 29 (19):6229–33.

Papaleo, F., J. N. Crawley, J. Song, et al. 2008. Genetic dissection of the role of catechol-*O*-methyltransferase in cognition and stress reactivity in mice. *J Neurosci* 28 (35):8709–23.

Parsey, R. V., R. S. Hastings, M. A. Oquendo, et al. 2006a. Effect of a triallelic functional polymorphism of the serotonin-transporter-linked promoter region on expression of serotonin transporter in the human brain. *Am J Psychiatry* 163 (1):48–51.

Parsey, R. V., M. A. Oquendo, R. T. Ogden, et al. 2006b. Altered serotonin 1A binding in major depression: A [carbonyl-C-11]WAY100635 positron emission tomography study. *Biol Psychiatry* 59 (2):106–13.

Passamonti, L., A. Cerasa, M. C. Gioia, et al. 2008. Genetically dependent modulation of serotonergic inactivation in the human prefrontal cortex. *NeuroImage* 40 (3):1264–73.

Patel, P. D., C. Pontrello, and S. Burke. 2004. Robust and tissue-specific expression of TPH2 versus TPH1 in rat raphe and pineal gland. *Biol Psychiatry* 55 (4):428–33.

Pezawas, L., and A. Meyer-Lindenberg. 2010. Imaging genetics: Progressing by leaps and bounds *NeuroImage* 53(3):801–3.

Pezawas, L., B. A. Verchinski, V. S. Mattay, et al. 2004. The brain-derived neurotrophic factor val66met polymorphism and variation in human cortical morphology. *J Neurosci* 24 (45):10099–102.

Pezawas, L., A. Meyer-Lindenberg, E. M. Drabant, et al. 2005. 5-HTTLPR polymorphism impacts human cingulate–amygdala interactions: A genetic susceptibility mechanism for depression. *Nat Neurosci* 8 (6):828–34.

Pezawas, L., A. Meyer-Lindenberg, A. L. Goldman, et al. 2008. Evidence of biologic epistasis between BDNF and SLC6A4 and implications for depression. *Mol Psychiatry* 13 (7):709–16.

Pizzagalli, D. A., A. J. Holmes, D. G. Dillon, et al. 2009. Reduced caudate and nucleus accumbens response to rewards in unmedicated individuals with major depressive disorder. *Am J Psychiatry* 166 (6):702–10.

Plomin, R., C. M. Haworth, and O. S. Davis. 2009. Common disorders are quantitative traits. *Nat Rev Genet* 10 (12):872–8.

Poldrack, R. A., P. C. Fletcher, R. N. Henson, K. J. Worsley, M. Brett, and T. E. Nichols. 2008. Guidelines for reporting an fMRI study. *NeuroImage* 40 (2):409–14.

Pomarol-Clotet, E., M. Fatjo-Vilas, P. J. McKenna, et al. 2010. COMT Val158Met polymorphism in relation to activation and de-activation in the prefrontal cortex: a study in patients with schizophrenia and healthy subjects. *NeuroImage* 53 (3):899–907.

Potkin, S. G., J. A. Turner, J. A. Fallon, et al. 2009. Gene discovery through imaging genetics: Identification of two novel genes associated with schizophrenia. *Mol Psychiatry* 14 (4):416–28.

Praschak-Rieder, N., J. Kennedy, A. A. Wilson, et al. 2007. Novel 5-HTTLPR allele associates with higher serotonin transporter binding in putamen: A [(11)C] DASB positron emission tomography study. *Biol Psychiatry* 62 (4):327–31.

Preece, M. A., M. J. Taylor, J. Raley, A. Blamire, T. Sharp, and N. R. Sibson. 2009. Evidence that increased 5-HT release evokes region-specific effects on blood-oxygenation level–dependent functional magnetic resonance imaging responses in the rat brain. *Neuroscience* 159 (2):751–9.

Price, J. L., and W. C. Drevets. 2010. Neurocircuitry of mood disorders. *Neuropsychopharmacology* 35 (1):192–216.

Rabl, U., C. Scharinger, M. Müller, and L. Pezawas. 2010. Imaging genetics: Implications for research on variable antidepressant drug response. *Exp Rev Clin Pharmacol* 3:471–89.

Racagni, G., and M. Popoli. 2010. The pharmacological properties of antidepressants. *Int Clin Psychopharmacol* 25 (3):117–31.

Rao, U., L. A. Chen, A. S. Bidesi, M. U. Shad, M. A. Thomas, and C. L. Hammen. 2010. Hippocampal changes associated with early-life adversity and vulnerability to depression. *Biol Psychiatry* 67 (4):357–64.

Rasch, B., K. Spalek, S. Buholzer, et al. 2010. Aversive stimuli lead to differential amygdala activation and connectivity patterns depending on catechol-*O*-methyltransferase Val158Met genotype. *NeuroImage* 52 (4): 1712–9.

Reimold, M., M. N. Smolka, G. Schumann, et al. 2007. Midbrain serotonin transporter binding potential measured with [^{11}C]DASB is affected by serotonin transporter genotype. *J Neural Transm* 114 (5):635–9.

Riccio, O., G. Potter, C. Walzer, et al. 2009. Excess of serotonin affects embryonic interneuron migration through activation of the serotonin receptor 6. *Mol Psychiatry* 14 (3):280–90.

Richardson-Jones, J. W., C. P. Craige, B. P. Guiard, et al. 2010. 5-HT1A autoreceptor levels determine vulnerability to stress and response to antidepressants. *Neuron* 65 (1):40–52.

Rigucci, S., G. Serafini, M. Pompili, G. D. Kotzalidis, and R. Tatarelli. 2010. Anatomical and functional correlates in major depressive disorder: The contribution of neuroimaging studies. *World J Biol Psychiatry* 11 (2 Pt 2):165–80.

Risch, N. J. 2000. Searching for genetic determinants in the new millennium. *Nature* 405 (6788):847–56.

Risch, N., R. Herrell, T. Lehner, et al. 2009. Interaction between the serotonin transporter gene (5-HTTLPR), stressful life events, and risk of depression: A meta-analysis. *JAMA* 301 (23):2462–71.

Rogers, M. A., K. Kasai, M. Koji, et al. 2004. Executive and prefrontal dysfunction in unipolar depression: A review of neuropsychological and imaging evidence. *Neurosci Res* 50 (1):1–11.

Sabol, S. Z., S. Hu, and D. Hamer. 1998. A functional polymorphism in the monoamine oxidase A gene promoter. *Hum Genet* 103 (3):273–9.

Santarelli, L., M. Saxe, C. Gross, et al. 2003. Requirement of hippocampal neurogenesis for the behavioral effects of antidepressants. *Science* 301 (5634):805–9.

Scharinger, C., U. Rabl, H. H. Sitte, and L. Pezawas. 2010. Imaging genetics of mood disorders. *NeuroImage* 53 (3):810–21.

Scheuch, K., M. Lautenschlager, M. Grohmann, et al. 2007. Characterization of a functional promoter polymorphism of the human tryptophan hydroxylase 2 gene in serotonergic raphe neurons. *Biol Psychiatry* 62 (11):1288–94.

Schofield, P. R., L. M. Williams, R. H. Paul, et al. 2009. Disturbances in selective information processing associated with the BDNF Val66Met polymorphism: Evidence from cognition, the P300 and fronto-hippocampal systems. *Biol Psychol* 80 (2):176–88.

Schosser, A., and S. Kasper. 2009. The role of pharmacogenetics in the treatment of depression and anxiety disorders. *Int Clin Psychopharmacol* 24 (6):277–88.

Schulze, T. G., D. J. Muller, H. Krauss, et al. 2000. Association between a functional polymorphism in the monoamine oxidase A gene promoter and major depressive disorder. *Am J Med Genet* 96 (6):801–3.

Sexton, C. E., C. E. Mackay, and K. P. Ebmeier. 2009. A systematic review of diffusion tensor imaging studies in affective disorders. *Biol Psychiatry* 66 (9):814–23.

Sharp, T., L. Boothman, J. Raley, and P. Queree. 2007. Important messages in the 'post': Recent discoveries in 5-HT neurone feedback control. *Trends Pharmacol Sci* 28 (12):629–36.

Sheline, Y. I., D. M. Barch, J. M. Donnelly, J. M. Ollinger, A. Z. Snyder, and M. A. Mintun. 2001. Increased amygdala response to masked emotional faces in depressed subjects resolves with antidepressant treatment: An fMRI study. *Biol Psychiatry* 50 (9):651–8.

Sheline, Y. I., J. L. Price, Z. Yan, and M. A. Mintun. 2010. Resting-state functional MRI in depression unmasks increased connectivity between networks via the dorsal nexus. *Proc Natl Acad Sci U S A* 107 (24): 11020–5.

Shifman, S., A. Bhomra, S. Smiley, et al. 2008. A whole genome association study of neuroticism using DNA pooling. *Mol Psychiatry* 13 (3):302–12.

Shioe, K., T. Ichimiya, T. Suhara, A. Takano, et al. 2003. No association between genotype of the promoter region of serotonin transporter gene and serotonin transporter binding in human brain measured by PET. *Synapse* 48 (4):184–8.

Sibille, E., and D. A. Lewis. 2006. SERT-ainly involved in depression, but when? *Am J Psychiatry* 163 (1):8–11.

Siegle, G. J., S. R. Steinhauer, M. E. Thase, V. A. Stenger, and C. S. Carter. 2002. Can't shake that feeling: Event-related fMRI assessment of sustained amygdala activity in response to emotional information in depressed individuals. *Biol Psychiatry* 51 (9):693–707.

Smolka, M. N., M. Buhler, G. Schumann, et al. 2007. Gene–gene effects on central processing of aversive stimuli. *Mol Psychiatry* 12 (3):307–17.

Smolka, M. N., G. Schumann, J. Wrase, et al. 2005. Catechol-*O*-methyltransferase val158met genotype affects processing of emotional stimuli in the amygdala and prefrontal cortex. *J Neurosci* 25 (4): 836–42.

Soliman, F., C. E. Glatt, K. G. Bath, et al. 2010. A genetic variant BDNF polymorphism alters extinction learning in both mouse and human. *Science* 327 (5967):863–6.

Stein, D. J., T. K. Newman, J. Savitz, and R. Ramesar. 2006. Warriors versus worriers: The role of COMT gene variants. *CNS Spectr* 11 (10):745–8.

Stein, J. L., L. M. Wiedholz, D. S. Bassett, et al. 2007. A validated network of effective amygdala connectivity. *NeuroImage* 36 (3):736–45.

Strigo, I. A., A. N. Simmons, S. C. Matthews, A. D. Craig, and M. P. Paulus. 2008. Association of major depressive disorder with altered functional brain response during anticipation and processing of heat pain. *Arch Gen Psychiatry* 65 (11):1275–84.

Strobel, A., L. Gutknecht, C. Rothe, et al. 2003. Allelic variation in 5-HT1A receptor expression is associated with anxiety- and depression-related personality traits. *J Neural Transm* 110 (12):1445–53.

Sullivan, P. F., M. C. Neale, and K. S. Kendler. 2000. Genetic epidemiology of major depression: Review and meta-analysis. *Am J Psychiatry* 157 (10):1552–62.

Suslow, T., C. Konrad, H. Kugel, et al. 2010. Automatic mood-congruent amygdala responses to masked facial expressions in major depression. *Biol Psychiatry* 67 (2):155–60.

Szeszko, P. R., R. Lipsky, C. Mentschel, et al. 2005. Brain-derived neurotrophic factor val66met polymorphism and volume of the hippocampal formation. *Mol Psychiatry* 10 (7):631–6.

Taylor, M. J., S. Sen, and Z. Bhagwagar. 2010. Antidepressant response and the serotonin transporter gene-linked polymorphic region. *Biol Psychiatry* 68:536–43.

Taylor, W. D., S. Zuchner, M. E. Payne, et al. 2007. The COMT Val158Met polymorphism and temporal lobe morphometry in healthy adults. *Psychiatry Res* 155 (2):173–7.

Tosun, D., M. E. Rettmann, X. Han, et al. 2004. Cortical surface segmentation and mapping. *NeuroImage* 23 (Suppl 1):S108–18.

van Dyck, C. H., R. T. Malison, J. K. Staley, et al. 2004. Central serotonin transporter availability measured with [^{123}I]beta-CIT SPECT in relation to serotonin transporter genotype. *Am J Psychiatry* 161 (3):525–31.

Varnas, K., C. Halldin, and H. Hall. 2004. Autoradiographic distribution of serotonin transporters and receptor subtypes in human brain. *Hum Brain Mapp* 22 (3):246–60.

Verhagen, M., A. van der Meij, P. A. van Deurzen, et al. 2010. Meta-analysis of the BDNF Val66Met polymorphism in major depressive disorder: Effects of gender and ethnicity. *Mol Psychiatry* 15:260–71.

Vogt, B. A. 2009. *Cingulate Neurobiology and Disease*. New York: Oxford University Press.

Walther, D. J., J. U. Peter, S. Bashammakh, et al. 2003. Synthesis of serotonin by a second tryptophan hydroxylase isoform. *Science* 299 (5603):76.

Willeit, M., and N. Praschak-Rieder. 2010. Imaging the effects of genetic polymorphisms on radioligand binding in the living human brain: A review on genetic neuroreceptor imaging of monoaminergic systems in psychiatry. *NeuroImage* 53 (3):878–92.

Willeit, M., J. Stastny, W. Pirker, et al. 2001. No evidence for in vivo regulation of midbrain serotonin transporter availability by serotonin transporter promoter gene polymorphism. *Biol Psychiatry* 50 (1):8–12.

Williams, L. M., J. M. Gatt, P. R. Schofield, G. Olivieri, A. Peduto, and E. Gordon. 2009. 'Negativity bias' in risk for depression and anxiety: Brain–body fear circuitry correlates, 5-HTT-LPR and early life stress. *NeuroImage* 47 (3):804–14.

Williams, L. M., J. M. Gatt, S. M. Grieve, et al. 2010. COMT Val(108/158)Met polymorphism effects on emotional brain function and negativity bias. *NeuroImage* 53 (3):918–25.

Yang, F., H. S. Je, Y. Ji, G. Nagappan, B. Hempstead, and B. Lu. 2009. Pro-BDNF-induced synaptic depression and retraction at developing neuromuscular synapses. *J Cell Biol* 185 (4):727–41.

Yu, Y. W., S. J. Tsai, C. J. Hong, T. J. Chen, M. C. Chen, and C. W. Yang. 2005. Association study of a monoamine oxidase A gene promoter polymorphism with major depressive disorder and antidepressant response. *Neuropsychopharmacology* 30 (9):1719–23.

Zhang, X., J. M. Beaulieu, T. D. Sotnikova, R. R. Gainetdinov, and M. G. Caron. 2004. Tryptophan hydroxylase-2 controls brain serotonin synthesis. *Science* 305 (5681):217.

Zhang, X., R. R. Gainetdinov, J. M. Beaulieu, et al. 2005. Loss-of-function mutation in tryptophan hydroxylase-2 identified in unipolar major depression. *Neuron* 45 (1):11–6.

Zhou, Z., A. Roy, R. Lipsky, et al. 2005. Haplotype-based linkage of tryptophan hydroxylase 2 to suicide attempt, major depression, and cerebrospinal fluid 5-hydroxyindoleacetic acid in 4 populations. *Arch Gen Psychiatry* 62 (10):1109–18.

Zill, P., A. Buttner, W. Eisenmenger, H. J. Moller, M. Ackenheil, and B. Bondy. 2007. Analysis of tryptophan hydroxylase I and II mRNA expression in the human brain: A post-mortem study. *J Psychiatr Res* 41 (1–2):168–73.

Zimmerman, M., I. Chelminski, and W. McDermut. 2002. Major depressive disorder and axis I diagnostic comorbidity. *J Clin Psychiatry* 63 (3):187–93.

5 Development of Brain Monoaminergic Systems

Marten P. Smidt

CONTENTS

5.1 MONOAMINERGIC SYSTEMS IN THE CENTRAL NERVOUS SYSTEM

The main monoaminergic systems in the central nervous system (CNS) are the following: histamine, dopamine, norepinephrine [noradrenaline (NA)], epinephrine (adrenaline), serotonin (5-HT), and melatonin. Relevant to the human pathological condition known as "major depression" are the dopaminergic, noradrenergic, and serotonergic systems (Figure 5.1). The development of these systems is in a region around the mid–hindbrain border and may rely on shared molecular mechanisms, although the final terminal differentiation is acquired through the recruitment of specific factors (Goridis and Rohrer 2002). Transmitter synthesis relies on the conversion of the amino acids tyrosine and tryptophan, which will lead to dopamine, noradrenaline, and 5-HT, respectively (Goridis and Rohrer 2002). Some of the enzymes that perform the conversion are required by all the pathways [like L-aromatic amino acid decarboxylase (Aadc)] and are therefore present in all different neuronal groups. The dopamine and noradrenaline syntheses share the first two steps of the pathway since noradrenaline is generated by dopamine-β-hydroxylase (Dbh). These two neuronal groups therefore share the enzymes tyrosine hydroxylase (Th) and Aadc, whereas the latter group additionally contains Dbh.

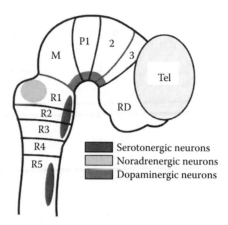

FIGURE 5.1 Representation of monoaminergic systems and their location in developing central nervous system (Smidt, M., and Burbach, J. P. H., *Nat Rev Neurosci*, 8 (1), 21–32, 2007; Goridis, C., and Rohrer, H., *Nat Rev Neurosci*, 3 (7), 531–41, 2002). Anatomical regions are numbered A4 to A6 for noradrenergic neurons (LC, locus coeruleus), A8 to A10 for mdDA neurons (VTA, ventral tegmental area; RR, retrorubral field; and SNc, substantia nigra compacta), and B1 to B4 rostral and B5 to B9 caudal for 5-HT neurons (raphe nucleus).

5.2 DEVELOPMENTAL PROCESSES THAT DETERMINE THE MID–HINDBRAIN REGION

Among the earliest events fundamental to monoaminergic neuronal development is the specification of the permissive region to allow these neurons to be generated (Figure 5.2a). Initial CNS division through the formation of molecular borders (Baek et al. 2006) is essential in this process. Central in defining the permissive region is the formation and positioning of the mid–hindbrain border, the isthmus. The formation and signaling from the isthmus (Fgf8/Fgf4) together with signaling from

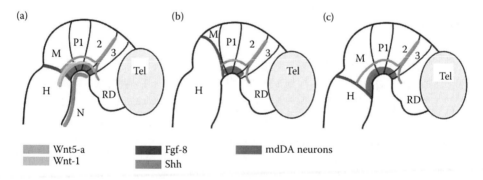

FIGURE 5.2 **(See color insert.)** Region within CNS where momoaminergic neurons emerge. Instructive signals such as Fgf-8 emerging from mid–hindbrain border (isthmus) and Shh expressed in ventricular zone determine site of mdDA generation. Wnt-1 and Wnt-5 expression is present in this region and is essential for formation of mesodiencephalon and is important in inducing early mdDA-specific gene expression as En1 (Danielian P. S., and McMahon, A. P., *Nature*, 383 (6598), 332–4, 1996; McMahon, A. P., et al., *Cell*, 69 (4), 581–95, 1992). (a) In wild-type animals, the region encompasses midbrain and prosomer 1–3. Important extrinsic and intrinsic signals are indicated in their respective positions in developing CNS. (b) In animals where isthmus has shifted to a rostral position (ectopic Gbx-2 expression in midbrain), midbrain section of mdDA neurons has been ablated. (c) Caudal shift of isthmus (overexpression of Otx2 in rostral hindbrain) induces expansion of mdDA neuronal field. N, notochord; H, hindbrain; M, midbrain; P1–3, prosomer 1–3; RD, rostral diencephalon; Tel, telencephalon.

the notochord (Shh) designate, at a specific signaling intersection point, the region where dopaminergic, noradrenergic, and serotonergic neurons are born (Hynes et al. 1995; Hynes and Rosenthal 1999; Hynes et al. 2000). Influencing the position of the isthmus, through manipulation of key transcription factors such as Otx2 and Gbx2 (Rhinn and Brand 2001; Simeone et al. 2002; Simeone 2002; Glavic et al. 2002), indirectly influences the emergence of dopaminergic neurons through the ablation (Figure 5.2b) (Simeone et al. 2002; Acampora et al. 1999, 2000, 2001, 2005) or expansion (Figure 5.2c) (Wassarman et al. 1997; Millet et al. 1999; Joyner et al. 2000) of the midbrain being the main site of dopaminergic neuronal generation. Through elegant ex vivo experiments in rats and chickens, investigators have shown that Tgf-β is essential for the early Shh signaling and subsequent induction of floor plate–derived neuronal systems (Farkas et al. 2003; Blum 1998). This signaling positions Tgf-β together with Fgf and Shh central in defining the molecular signaling to prepare the region. Wnt signaling (Wnt-1 and Wnt-5a) is essential for the establishment of the mid–hindbrain region and is involved in activating Engrailed (En) genes, which are essential in later stages of neuronal specification (Danielian and McMahon 1996; Castelo-Branco et al. 2003, 2004). Finally, retinoic acid (RA) is crucial for proper positioning of the mid–hindbrain border and is therefore indirectly essential for the proper organization of the mid- and hindbrain (Avantaggiato et al. 1996; Clotman et al. 1997; Holder and Hill 1991). RA signaling can occur through local synthesis (see also below) and perhaps also through an extraembryonal source (Otto et al. 2003). Dividing cells in the ventral mesodiencephalic ventricular zone specifically express retinoid aldehyde dehydrogenases (Wallén et al. 1999), which generates cell-specific synthesis of RA.

An essential step in providing cellular diversity is the subdivision of the developing CNS in longitudinal and transverse domains, which are specified through specific gene expression patterns. The longitudinal domains are designated floor plate, basal plate, alar plate, and roof plate. The transverse domains along the anterior/posterior axis lead to the following domains in a rostral to caudal order: telencephalon, rostral diencephalon, prosomer 1–3, midbrain, and hindbrain (for reviews on the prosomeric model, see Puelles and Rubenstein 1993, 2003; Rubenstein et al. 1994, 1998). Early molecular coding of the ventricular zone influences the fate of newborn neurons at specific dorsal/ventral positions along the anterior–posterior axis (Craven et al. 2004; Puelles et al. 2004; Vernay et al. 2005; Smits et al. 2006). The coding is generated by early players that are involved in early instructive signals during CNS patterning. Transcription factors expressed in the ventricular zone instruct newborn neurons for their early differentiation steps into the dopaminergic/noradrenergic and serotonergic phenotypes. In conclusion, early signaling from organizing centers generates a permissive region defined by specific gene expression in the di/mes/myelencephalic ventricular zone. This gene expression induces the switch from mitotic cells toward postmitotic young neurons that are destined to become fully differentiated neurons.

5.3 DEVELOPMENT OF SEROTONERGIC NEURONS

5.3.1 SPECIFICATION OF SEROTONERGIC NEURONS

Serotonergic neurons are generated in the CNS, born between E10.5 and E12.5 in the mouse (Pattyn et al. 2004), and make up the anatomical locations designated as B1 to B9 in the adult system (Goridis and Rohrer 2002). These neurons are identified by the enzymes that produce 5-HT through the hydroxylation (tryptophan hydroxylase, or Tph2) of tryptophan to 5-hydroxytryptophan and the subsequent decarboxylation (L-Aadc) toward serotonin (5-hydroxytryptamine, or 5-HT). The latter enzyme is shared with dopaminergic, noradrenergic, and adrenergic neurons, since it also catalyzes the decarboxylation of L-dopa to dopamine. Early development of the isthmic area creates a rostral ventral hindbrain region where 5-HT progenitors are induced close to the floor plate. These progenitors, aligned along a rostral caudal position in rhombomeres 1–5, migrate out to form two distinct populations: a rostral cell group (pontine group) very close to the caudal edge of the isthmus and a caudal cell group (medullary) in the myelencephalon, which are divided by a region that does not contain any 5-HT neurons (brachial motor area) (Pattyn et al. 2004). The rostral cell groups connect to several areas in the forebrain, and the caudal group projects mainly to the spinal cord area (for a detailed overview on neuroanatomy, see Hornung 2003).

The origin of the rostral cell group relies on floor plate signals such as Shh. The different response of cells to go into dopaminergic differentiation or serotonergic differentiation seems to rely on the specific Fgf signal. The combination of Shh and Fgf8 leads to the dopaminergic phenotype, and the combination of Shh with Fgf4 leads to the serotonergic phenotype. It has been suggested that the transcription factor Nkx2-2 may be a relay system of the early signaling events since in the Nkx2-2 knockout, few 5-HT neurons develop, although one dorsal, raphe nucleus, cluster of 5-HT neurons seems to be independent of Nkx2-2 activity (Pattyn et al. 2004). Analysis of Ascl1/Mash1 mutants has indicated that Ascl1 together with Nkx2-2 is essential in activating the 5-HT specification factors Lmx1b, Pet1, and Gata3 (Figure 5.3). Interestingly, recent data suggest that Insm1 might cooperatively act with Ascl1 in the activation of the just mentioned 5-HT specification factors and moreover is shown to be essential for the expression of Tph2, the essential enzyme in 5-HT production (Jacob et al. 2009).

The Ets domain factors Gata2 and Gata3 have essential function in the rostral and caudal 5-HT cell groups, respectively. In rostral 5-HT neurons, Gata2 is essential and sufficient to induce 5-HT neurons (Craven et al. 2004). Interestingly, Gata2 is unable to rescue the loss of Gata2 in these developing rostral 5-HT neurons. On the other hand, Gata3 is required for the proper specification of caudal 5-HT neurons (van Doorninck et al. 1999). Thus, these two distinct groups of 5-HT neurons rely on two different Gata factors, which suggests that these neurons rely on analogous but dissimilar gene activation programs to create the same transmitter phenotype.

In parallel to the activation of Gata2/3 is the activation of the Lim homeobox gene *Lmx1b*, which is coexpressed in all 5-HT neurons (Ding et al. 2003). It has been suggested that Lmx1b acts as an intermediate step in the terminal differentiation between Nkx2-2 activity and activation of Pet1. Interestingly, Lmx1b ablation studies have shown that all 5-HT neurons rely on this transcription factor for their normal developmental program. Through elegant studies of mouse mutants, it was shown that Lmx1b depends on the expression of Nkx2-2 in the rostral 5-HT neuronal group. In the Lmx1b knockout, the expression of Nkx2-2 is unaffected, which places Lmx1b downstream of Nkx2-2. Interestingly, Gata3 was downregulated in the caudal 5-HT neurons as a consequence of Nkx2-2 deletion, leading to an ablation of the 5-HT phenotype as discussed earlier for Gata3 mutants. These data suggest once more that the programming of the dorsal and caudal 5-HT neurons is distinct in nature. In Lmx1b mutants, the expression of Pet1, driving the 5-HT terminal differentiation markers, is lost, leading to the absence of any 5-HT phenotypic characteristics. In

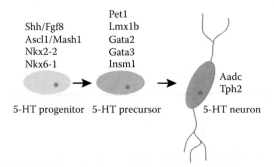

FIGURE 5.3 Schematic representation of 5-HT neuronal differentiation. 5-HT progenitors are induced by combined signals of Ascl1/Mash1, Nkx2.2, and Nkx6.1. As a result, 5-HT differentiation critical genes *Pet1*, *Lmx1b*, *Gata2*, *Gata3*, and *Insm1* are induced, which drive serotonergic phenotype marked by specific expression of Aadc and Tph2. (Godiris, C., and Rohrer, H., *Nat. Rev. Neurosci.*, 3 (7), 531–41, 2002; Pattyn, A., et al., *Nat. Neurosci.*, 7 (6), 589–95, 2004; Jacob, J., et al., *Development*, 136 (14), 2477–85, 2009; Liu, H., et al., *Dev. Biol.*, 286 (2), 521–36, 2005.)

Pet1 mutants, the expression of Lmx1b is unaffected, placing Lmx1b upstream of Pet1. Mouse that was conditionally ablated for Lmx1b, driven by Cre under the control of Pet1 (Zhao et al. 2006), in 5-HT neurons displayed a normal initial generation of 5-HT neurons. At E12.5, the amount of 5-HT neurons was markedly decreased; and by E14.5, almost all neurons are lost as measured by 5-HT markers such as Tph2 and Sert. The cells themselves were not ablated as was proven through a lacZ crossing into the 5-HT neurons. Thus, the 5-HT gene expression loss is a consequence of lost gene activation by Lmx1b and not due to cell loss itself. The data clearly suggest that in addition to the induction of Pet1, and subsequent activation of 5-HT specific markers, Lmx1b is essential for maintaining the expression of Pet1 and therefore the 5-HT phenotype.

The essential step in terminal differentiation, inducing the 5-HT phenotype, is established by the action of Pet1 (Hendricks et al. 1999, 2003). This Ets factor is expressed just before the cells are starting to display 5-HT production. In ablation studies, it was shown that Pet1 is essential, since in its absence most 5-HT precursors fail to develop and the cells lack the genes for synthesis, uptake, and storage of 5-HT. This failure of proper specification leads to increased anxiety and aggressive behavior in mice.

The above-described factors may act in the following linear cascade of events: Shh/Fgf4/Fgf8 → Nkx2-2/Nkx6.1/Ascl1/Insm1 → Lmx1b → Gata2/3/Insm1 → Pet1/Insm1 → 5-HT phenotype. It is intriguing to note that within the current knowledge of the development of the central 5-HT system, no description has been made about the involvement of orphan nuclear hormone receptors. The other two monoaminergic systems discussed below rely on activity of these types of transcription factors. The expression in the region has been described for Nurr77 (Nr4a1) (Xiao et al. 1996), but no functional studies have been described.

5.3.2 Serotonin as Regulator of Cortical Cytoarchitecture

Raphe nucleus 5-HT neurons project to almost all areas of the brain in an even manner. This predicts the involvement of 5-HT modulation on many CNS systems. Surprisingly, 5-HT knockout animals are able to develop and mature to adulthood (Daubert and Condron 2010), although it should be mentioned that a real null for 5-HT signaling is difficult to assess since even maternal 5-HT enters the embryo and can therefore supply just enough 5-HT for development to proceed. Overall, the relative importance of the 5-HT system is reflected by the redundancy of the 5-HT system. 5-HT signaling modulates many factors of cytoarchitecture of the brain (Daubert and Condron 2010), such as axonal outgrowth, cellular migration, branch morphology, and dendritic arborization. In addition to target systems, such modulation of morphology has also been reported in an autoregulatory fashion. Since 5-HT is targeting most of the CNS and is able to modulate so many modalities, it is feasible to assume that 5-HT has a role in early brain development, which may be visible in pathological conditions of hampered 5-HT signaling. One of the major pharmacological tools in the treatment of adult depression is selective serotonin reuptake inhibitors (SSRIs). These compounds block the reuptake transporter (Sert) and thereby induce higher levels of synaptic 5-HT. Interestingly, Sert ablation or blocking during early development leads to anxiety in animal models (Noorlander et al. 2008). Thus, the increased 5-HT tone during development results in increased pathological behavior, whereas the 5-HT increase in adult systems through treatment with SSRIs relieves the symptoms. The relationship between the cytoarchitectural changes driven by 5-HT signaling and the development of specific behaviors (or deficits) remains to be elucidated.

5.4 DEVELOPMENT OF NORADRENERGIC NEURONS

5.4.1 Specification of Noradrenergic Neurons

Early development of the isthmic area creates a rostral dorsal hindbrain niche where NA progenitors are induced through combined BMP and FGF8 signals. Impairment of BMP signaling by

BMP blockers as noggin suppresses the initiation of the NA phenotype as marked by the failure of Phox2A induction (Goridis and Rohrer 2002). In addition, the change in BMP signaling may also lead to ectopic induction of NA progenitors indicative that an exact BMP signaling strength is required for proper locus coeruleus (LC) development (Goridis and Rohrer 2002). Similar as in the early induction of 5-HT progenitors is the initial induction of Scl1/Mash1 essential in the formation of LC NA neurons. Through a series of elegant ablation experiments, the main transcriptional program required for the generation of LC NA neurons has been described, which follows a specific hierarchy (Figure 5.4).

Through the analysis of recently analyzed knockout mice, six critical genes have been identified in the differentiation of NA neurons: *Ascl1/Mash1, Ear2/Nr2f6, Phox2A, Phox2B, Tlx/Rnx*, and *Insm1* (Goridis and Rohrer 2002; Pattyn et al. 2004; Jacob et al. 2009). *Ascl1/Mash1* functions early in commitment to the NA phenotype. Targeted deletion of *Ascl1* in mice has indicated that the central adrenergic and noradrenergic nuclei are missing, whereas other brainstem nuclei are still present (Hirsch et al. 1998; Pattyn et al. 2004; Stanke et al. 2004). In addition, it was reported that the critical downstream transcription factor Phox2A is absent or mainly altered in expression in Mash1 mutants. This suggests that Mash is essential in the early phases of NA neuronal development and acts probably through activation of Phox2A. Genetic ablation of *Phox2A* resulted in the reported absence of the LC (Morin et al. 1997) as analyzed through in situ hybridization for Dbh and Th. In these animals, some other (nor)adrenergic nuclei are spared; and it has been suggested that *Phox2B* might play a part in the rescue of these nuclei through functional redundancy. Indeed, it has been reported that the functional loss of *Phox2B* abrogates all NA centers in the brain, including the LC (Pattyn et al. 2000). In these animals, the expression of Th and Dbh is lost and *Phox2B* is therefore sometimes referred to as being a master switch for the NA phenotype in central and peripheral NA systems. Interestingly, a recent paper described the function of the orphan nuclear hormone receptor, Ear2 (Nr2f6), being in between the transcriptional cascade of Ascl2/Mash1 and Phox2A/Phox2B (Warnecke et al. 2005). In Ear2 knockout mice, the amount of LC neurons is reduced by about threefold, as measured through the expression of *Phox2A* and *Phox2B* in the LC progenitor domain. Staining of Th and Dbh confirmed the 70% reduction in LC neurons. The expression of *Ascl1/Mash1* is unaffected; therefore, the authors conclude that Ear2 is in between Ascl1/Mash1 activation and activation of Phox2A. Independently from Phox2A and Phox2B, another transcription factor, the homeodomain gene *Rnx/Tlx3*, has been shown to be involved in NA neuronal

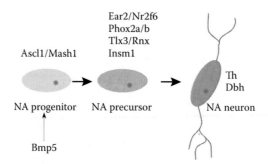

FIGURE 5.4 Schematic representation of locus coeruleus NA neuronal specification. Dorsal/rostral region of hindbrain is under inducing control of FGF8 and strict BMP tone. These signals induce NA progenitors that start to express Ascl1/Mash1. These NA progenitors will differentiate into Ear2/Nr2f6, Phox2, Tlx3, and Insm1 expression cells that will differentiate into NA neurons marked by specific expression of Th and Dbh to generate NA neurotransmitter phenotype. (Goridis, C., and Rohrer, H., *Nat Rev Neurosci*, 3 (7), 531–41, 2002; Pattyn, A., et al., *Nat Neurosci*, 7 (6), 589–95, 2004; Jacob, J., et al., *Development*, 136 (14), 2477–85, 2009; Warnecke, M., et al., *Genes Dev*, 19 (5), 614–25, 2005; Vogel-Hopker, A., and Rohrer, H., *Development*, 129 (4), 983–91, 2002.)

specification (Qian et al. 2001). In Rnx/Tlx3 knockout mice, the expression of Dbh is abolished in most NA centers. Interestingly, some subsets of LC NA neurons still arise and are spared by the Rnx deletion. Suggested to be final in the transcriptional cascade is the zinc-finger transcription factor Insm1. Ablation of Insm1 leads to a delay in Th expression in NA centers, although Dbh is unaffected (Jacob et al. 2009). This suggests that Insm1 is essential for fine-tuning of some aspects of the NA phenotype but is dispensable for Dbh expression.

In addition to the described transcription factor code for the generation of NA neurons of the LC, TrkB ligands (BDNF, NT4) are described to be involved in proper NA neuronal development (Holm et al. 2003). In TrkB null mice, a marked reduction in LC neurons has been described. The dependence of the NA neurons to the TrkB signal was confirmed by induction of the system through ex vivo application of TrkB ligands, which induced the survival of LC neurons. The data suggest that TrkB ligands may have a promoting effect or a prosurvival effect on the development of NA neurons located in the LC.

5.4.2 Migration and Origin of NA Neurons

The origin and migratory pattern of LC NA neurons are well described through an elegant mix of in situ hybridization experiments with the NA markers Dbh and Phox2A/B and fate mapping experiments by quail-chick homotopic grafts (Aroca et al. 2006). The data suggest that LC neurons first appear along the middle one-third of rhombomere 1 arranged in a linear stripe. From this position they tangentially migrate in the ventral direction to the lateral basal plate. The migratory pattern is restricted to rhombomere 1 and does not include translocation to other areas in the rostral/caudal region.

The molecular mechanisms involved in the migratory pattern of LC neurons are poorly understood; however, a recent paper described that the netrin receptor DCC is critically involved in the process (Shi et al. 2008). The DCC-ablated animals showed normal phenotypic characteristics, as described above, but were failing/delayed in the initiation of tangential migration. As a result, many NA neurons appeared scattered in the rostral pons and cerebellum.

Interestingly, a recent paper described the influence of the DAT/Sert/NET transporter blocker cocaine on the neurite outgrowth potential of NA neurons (Dey et al. 2006). The authors confirmed earlier in vivo data with in vitro experiments to analyze the specificity of the repressing effect of cocaine compared to dopaminergic neurons. The results suggest that the cocaine defect is specific for NA neurons and has an effect on numerous parameters of neurite outgrowth. The authors indicate that prenatal cocaine exposure might have significant effects on the wiring of the LC toward the target regions with the accompanying behavioral deficits. The importance of correct wiring has been emphasized by the fact that the NA innervation in the cortex is involved in regulating the amount of Cajal–Retzius cells, which might directly influence the cortical layering during development (Naqui et al. 1999).

5.5 DEVELOPMENT OF MESODIENCEPHALIC DOPAMINERGIC NEURONS

Mesodiencephalic dopaminergic (mdDA) neurons are involved in voluntary movement control and regulation of emotion-related behavior and are affected in many neurological and psychiatric disorders. MdDA neurons form a specific neural group that shares the neurotransmitter identity with several other functionally distinct dopaminergic cell groups in the CNS (Smidt and Burbach 2007). The important link between the mdDA neurons in human CNS disorders and behavioral dysfunction has led to the intense study of this neuronal group, first in pharmacology and recently in developmental neurobiology. The former has led to the generation of many pharmacological intervention strategies to alleviate disease symptoms, whereas the latter has provided general insight into the origin of the disease and the molecular mechanisms that define mdDA development, physiology, and its failure in human pathology.

5.5.1 Differentiation of mdDA Neurons

Early cloning experiments have identified several transcription factors that, based on their spatial/temporal expression pattern, were implicated in mdDA differentiation. Through in vivo ablation studies, several of those genes have been positioned in transcriptional cascades leading to specific phenotypic characteristics of mdDA neurons such as transmitter phenotype and specific neuronal vulnerability (Smidt and Burbach 2007). The developing CNS can be divided in specific molecular domains that contain their own molecular coding and specify the emergence of specific neuronal lineages, the prosomeric model (Puelles and Rubenstein 2003). Recently, it has been described that dopaminergic neurons receive region-specific signals subdividing mdDA neurons into specific subsets that are distinguishable through molecular and physiological parameters (Engele and Schilling 1996; Neuhoff et al. 2002; Smits et al. 2006). Within the adult mdDA system, these differences may add to the specification of specific connectivity that ultimately results in a multitude of functional units only sharing the use of the transmitter dopamine. In addition, the molecular subdivision may define the molecular basis of selective vulnerability as described for substantia nigra compacta (SNc) neurons in Parkinson disease.

5.5.1.1 Essential Factors for Generation of Differentiated mdDA Neurons

Knockout mice of Pitx3, Lmx1b, En 1/2, Ngn2, Nurr1, Phox2A/B, and Tgf-β have led to the generation of a gene-functional map during mdDA terminal differentiation and function (Smidt et al. 2000; Saucedo-Cardenas et al. 1998; Smidt et al. 2004a, 2004b; Asbreuk et al. 2002; Roussa et al. 2006; Smidt and Burbach 2007) (Figure 5.5). The neurotransmitter phenotype is, among others,

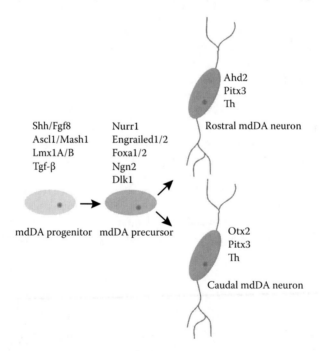

FIGURE 5.5 Schematic representation of mesodiencephalic dopamine neuron differentiation. At an intersecting point of Shh and Fgf8 signaling in caudal midbrain mdDA, progenitors are born that start to express Ascl1/Mash1, Lmx1a/b, and respond to Tgf-β signaling. These cells become postmitotic and start to express Nurr1, Engrailed1/2, Foxa1/2, Ngn2 (switched off by Nurr1 expression), and Dlk1. These cells then become terminally differentiated and can be distinguished as a rostral cell group expressing specific markers such as Ahd2 and a caudal group marked by Otx2 expression. (Smidt, M. P., and Burbach, J. P. H., *Nat Rev Neurosci*, 8 (1), 21–32, 2007; Ang, S.-L., *Adv Exp Med Biol*, 651, 58–65, 2009; Ferri, A. L. M., et al., *Development*, 134 (15), 2761–9, 2007; Di Salvio, M., et al., *Int J Dev Biol*, 54 (5), 939–45, 2010.)

determined by Nurrl, since Th, Vmat, DAT, and cRET are regulated through Nurrl(Smits et al. 2003; Wallén et al. 2001; Zetterström et al. 1997; Saucedo-Cardenas et al. 1998). In addition, the Nurrl–/–phenotype displays an mdDA maintenance defect, which cannot be attributed to the cRET ablation alone. MdDA neurons are born and developed in the absence of Nurrl. However, they fail to be maintained and have clear defects in transmitter synthesis and release. Interestingly, it has been suggested that mdDA neuronal induction by Nurrl is not sufficient to provide the fully differentiated phenotype (Kim et al. 2003), which is correct as later data on Pitx3 (see below) suggest that Nurrl activity is modulated through other transcription factors. Recent microarray analysis in the Nurrl mutant has identified at least three new target genes: *Ptpru*, *Dlk1*, and *Klhl1* (Jacobs et al. 2009a). Future research into the function of these genes might provide new clues to the exact terminal differentiation program of mdDA neurons. A small first result indicated already that Dlk1 might be involved in the repression of Dat from the most caudal parts of the mdDA system as described through Dlk1-ko analysis (Jacobs et al. 2009a).

Through elegant studies on the role of bHLH proteins (Kageyama et al. 2005), it has been shown that in the Nurrl-induced mdDA neuronal population, *Ngn2* suppresses Nurrl-induced gene transcription. On the other hand, *Mash1*, expressed in Nurrl-positive young mdDA neurons (Park et al. 2006), stimulates mdDA neuronal maturation unrelated to *Nurrl* gene activation or activation of terminal differentiation markers such as Pitx3 (Smidt et al. 1997). These data suggest a general role of Mash1 in activating mdDA neuronal maturation (Park et al. 2006) as described for other neuronal systems (Hatakeyama et al. 2001). Moreover, the data nicely complement the early functional role of Ngn2 (see below) and provide better understanding of the strict separation of Ngn2 and Nurrl expression as described (Andersson et al. 2006a). One of the early factors involved in mdDA neuronal generation is Lmx1b. Initial studies on the Lmx1b knockout showed that in the mdDA neuronal region, no proper molecularly coded mdDA neurons were born. Instead, Th-positive cells are present at E12.5 that do not express Pitx3; also, an ectopic cell population was described that expresses Pitx3 and does not coexpress Th (Smidt et al. 2000). Therefore, the *Lmx1b* gene is essential for the generation of properly differentiated mdDA neurons. An additional role for *Lmx1b* is the regional specification of the mid–hindbrain area probably through the activation of Wnt-1 (Adams et al. 2000).

A critical paper was published that describes the importance of *Lmx1a* in specifying the mdDA neuronal phenotype (Andersson et al. 2006b). *Lmx1a* is an early activator in mdDA differentiation and is crucial for activating Msx1. Both transcription factors then cooperate in activating the proneural gene *Ngn2*. This complete transcriptional complex is essential for sufficient induction of young mdDA neurons. Mutants of *Ngn2* itself display impairment and delay in mdDA development, but still part of the system is normally set up (Kele et al. 2006; Andersson et al. 2006a). *Ngn2* acts in activating Sox2-positive progenitors that will later develop into Nurrl-positive postmitotic neurons that will eventually differentiate into Th-positive mdDA neurons (Kele et al. 2006). Taken together, the pathway starting with Lmx1a through Msx1 and Ngn2 is essential for the normal development of mdDA neurons. Notably, the functional data on *Lmx1a* are derived from gene dosage experiments through in ovo electroporation in chicks. It remains to be determined whether the described molecular pathways are the same in mammals. However, initial ES-cell engineering performed by the same group suggests that this is the case (Andersson et al. 2006b).

Recent data on the *Wnt-1* gene have highlighted its function in specifying the mdDA neuronal phenotype by activating Otx2 and thereby suppressing expression of Nkx2.2. The latter factor is important in inducing the serotonergic neuronal lineage and acts as a suppressor of the dopaminergic lineage (Prakash et al. 2006). In mutant animals expressing one allele of Wnt-1 in the En1 expression domain, it was shown that the Otx2 border was shifted to a more caudal position, whereas Gbx2 was repressed in the same area. In the caudal position, ectopic mdDA neurons arise and these persist into the adult stage (Prakash et al. 2006). In addition, earlier data suggest that ectopic expression of Wnt-1 in the P1 region of the ventral CNS expands the SNc region of the mdDA neuronal population (Panhuysen et al. 2004; Smits et al. 2006), possibly by enhancing proliferation of mdDA precursors. However, the molecular coding of the floor plate and basal plate P1/P2 region might also be changed through the ectopic Wnt-1 expression.

The early mdDA expression of Hnf3-alpha/Foxa1 and beta/Foxa2 has been known for a long time, but the exact function of these factors in the development of the mdDA system became clear only recently (Ang 2009; Arenas 2008; Ferri et al. 2007). From knockout studies, it was concluded that Foxa1/2 is continuously required and is essential for the activation of Nurr1, En1, Th, and Aadc, although the latter two markers are described to be under the control of Nurr1 (Jacobs et al. 2009a). This activity pattern of Phoxa1/2 positions these factors in the early phase of progenitor development. Interestingly, it has been described that the in double-knockout animals (one conditional), a large amount of neurons were captured in an immature state, marked by expression of Nurr1 and absence of Th. The authors therefore concluded that Phoxa1/2 may also be essential in driving the terminal differentiation program.

The *En* genes were shown to be essential for the generation of mdDA neurons (Albéri et al. 2004; Simon et al. 2001, 2003, 2004; Thuret et al. 2004). The *En1* gene is widely expressed in mdDA neurons, whereas En2 is expressed in a subset and with different ontogeny. This specificity is highlighted by the difference in phenotype of En1+/–;En2–/– animals compared to that in En1–/–; En2+/–animals. The latter animals are far more severely affected. The initial defect in *En1/2* double knockouts is diminished generation of mdDA neurons, which later in development disappear completely (Simon et al. 2001). This suggests a dual role: one in generation and/or differentiation and one in maintenance. Therefore, the observed phenotype may rely on an indirect effect of the *En1/2* ablation since the organization of the midbrain is affected as well. Confirming the suggested dual role, it was shown that En has a direct effect on mdDA maintenance and cell death through apoptosis. Using elegant cell-mixing experiments, it was shown that the cell-autonomous action of En is essential to maintain mdDA neurons (Albéri et al. 2004). Taken together, the *En* genes seem to function at multiple levels during mdDA neuronal development—initially in the generation and differentiation of young mdDA neurons and later in maintenance of terminally differentiated mdDA neurons.

The *Pitx3* (*Ptx3*) gene was identified and shown to be expressed essentially in all mdDA neurons within the CNS (Smidt et al. 1997, 2000, 2004a, 2004b; Zhao et al. 2004). Data derived from the analysis of Aphakia (Ak) mice (Pitx3–/–) suggested that Pitx3 is specifically involved in the terminal differentiation and/or early maintenance of SNc neurons, since these neurons are absent in Ak animals (Smidt et al. 1997, 2003, 2004a, 2004b; Nunes et al. 2003; Hwang et al. 2003; Zhao et al. 2004). Since expression data suggested that *Pitx3* is expressed in all mdDA neurons (Smidt et al. 1997, 2000), the specific defect detected in the SNc could not easily be explained. Initially, it was described that *Pitx3* is expressed in only those mdDA neurons that are lost in the Aphakia mouse (van den Munckhof et al. 2003). However, data from two other groups showed that this is not the case (Smidt et al. 2004a, 2004b; Zhao et al. 2004). Therefore, it has been suggested that the specific vulnerability of the SNc in Ak mice as a consequence of Pitx3 ablation is derived from molecular differences between specific subsets of mdDA neurons within the SNc and ventral tegmental area (VTA) (Smidt et al. 2004a, 2004b; Smits et al. 2006). Interestingly, in a recent paper, the convergence of Nurr1 and Pitx3 activity has been described. The authors show that Nurr1 activity is potentiated by the interaction with Pitx3 (Jacobs et al. 2009b). In the mechanisms, Pitx3 acts as a "ligand" since the interaction to the complex, also containing PSF, removes the repressor Smrt from the protein complex. This leads to the release of Hdac proteins, and the *Nurr1* target gene is activated. The data suggest that Nurr1 acts as master switch and that other factors such as Pitx3 might modulate the activity in specific neurons.

Finally, another link toward subset specification has been provided through the description of caudal mdDA-specific expression of *Otx2* in terminally differentiated neurons (Di Salvio et al. 2010). This pattern suggests that specific roles are important for caudal VTA neurons relying on *Otx2* expression. The functional significance of the specific *Otx2* expression remains to be elucidated.

5.5.1.2 Retinaldehyde Dehydrogenase and RA Signaling

RA, the active derivative of vitamin A (retinol), is generated by three retinaldehyde dehydrogenases, Raldh1–3 (Smith et al. 2001), after initial retinol oxidation by alcohol dehydrogenases (Westerlund et al. 2005). The expression patterns of these enzymes are an indication of the main site of production of functional RA, as was shown through comparison of Raldh expression and RA-reporter mice (McCaffery and Dräger 1994a, 1994b; Smith et al. 2001; Westerlund et al. 2005). The expression of Raldh1 (Ahd2) in the midbrain ventricular zone (Wallén et al. 1999; Westerlund et al. 2005), in early developing and adult mdDA neurons (McCaffery and Dräger 1994a, 1994b; Smith et al. 2001; Galter et al. 2003), suggests that RA is locally synthesized during all steps in development and may have a function in mdDA generation and differentiation (Chung et al. 2005b). However, the actual RA level and which isomer is present in the midbrain region are unknown.

Very elegant experiments to unravel the function of RXR-Nurr1 heterodimers (Perlmann and Jansson 1995; Aarnisalo et al. 2002) as putative downstream effectors of RA (9-*cis*-RA) signaling were performed by using chimeric forms of Nurr1 coupled to Gal4. This experimental approach revealed the RXR-ligand-bound dependence of Nurr1 signaling in the ventral midbrain during mdDA development (Wallen-Mackenzie et al. 2003). After isolation of the most active fraction in a Nurr1-RXR reporter assay derived from E15.5 midbrain tissue, the RXR ligand was shown to be a dodecosahexanoic acid (de Urquiza et al. 2000; Wallen-Mackenzie et al. 2003) instead of 9-*cis*-RA. Signaling through the RXR-Nurr1 heterodimer was found to be involved in mdDA neuronal survival, which might be crucial in relation to the described survival defect in Nurr1−/− mice. Interestingly, activation of the RAR pathway (through RXR/RAR heterodimers and 9-*cis*-RA) negatively influences this function of RXR-Nurr1. The question remains whether direct signaling of all-*trans* RA through RAR plays a part in developing mdDA neurons. From in vitro studies, in a dopaminergic cell line, MN9D, it has become clear that RA can induce neuronal differentiation toward the dopaminergic phenotype (Wallén et al. 1999; Castro et al. 2001). Moreover, the described vulnerability of SNc neurons may be critically influenced through the appearance of active aldehydes. Highly expressed retinaldehyde dehydrogenases might therefore have an additional role in detoxifying these cells from such substances (Westerlund et al. 2005). A clear link toward a role for RA in the terminal differentiation was recently established, since it was shown that Pitx3 drives the expression of an aldehyde dehydrogenase (Ahd2) in the anterior mdDA neuronal field (Jacobs et al. 2007). Thus, the generation of RA from retinaldehyde is mdDA subset specific performed under the control of Pitx3. In Pitx3−/− animals, a part of the defect as described above could be counteracted through treatment of RA in a time window of E10.5 to E13.5. The rescue phenotype is stable until at least E18.5. This suggests that parts of the terminal differentiation are under the control of RA signaling in a subset of mdDA neurons. The elucidation of the exact mechanisms may shed new light on the issue of subset-specific gene activation and vulnerability as seen in Parkinson disease.

5.5.1.3 Tgf-α and Tgf-β

Several studies have suggested that Tgf signaling is essential for the proper development of mdDA neurons. An ablation study with Tgf-α has shown that a subset of mdDA neurons is absent in these animals (Blum 1998). Until now, however, this study did not have any follow-ups, nor is it known which subset is affected. It was described that Tgf-β2 and Tgf-β3 are expressed in the midbrain floor together with their cognate receptor TβR-II in the region where mdDA neurons arise (Farkas et al. 2003). Although no colocalization has been shown, an effect of Tgf-β on the emergence of Th-positive cells in E12 midbrain neuronal cultures was described. This effect has been confirmed in chicken embryos through antibody interference of Tgf signaling. However, the expression data were not conclusive about the cellular localization (Farkas et al. 2003). Later studies show essentially the same data but in addition try to position Tgf signaling in relation to other factors such as Shh and Fgf8, known for their essential role in induction of mdDA neurons. The data suggest that Tgf-β functions parallel to Shh and Fgf-8 in inducing Th-positive cells in culture (Roussa et al. 2004). Since no proliferation effects were seen in a response to Tgf-β, the effect seems to be restricted by inducing Th in putative mdDA precursors. Unfortunately, it is unknown whether these effects are direct or indirect; neither is it shown whether these Th-induced cells are proper mdDA neurons and which subtype they represent. In an additional study, Pitx3 and Nurr1, as mdDA specific markers, were analyzed in neurospheres and primary cultures derived from Tgf-β2 and Tgf-β3 knockout animals (Roussa et al. 2006). Careful analysis of the data showed that the neurospheres treated with Tgf-β, Shh, and Fgf-8 induced Th but failed to induce Pitx3 and Nurr1 to a similar extent. Moreover, in this experimental setup, there was a clear deviation between Th, Nurr1, and Pitx3 expression in these neurospheres and primary cultures of ventral midbrain tissue. In addition, the colocalization of these markers with Th was not demonstrated. Summarizing these data, the treatment of neurospheres derived from ventral midbrain tissue, with Tgf-β, influences the number of Nurr1/Th-positive neurons but leaves the number of Pitx3-positive neurons unaltered. This suggests that no proper mdDA neurons are induced in such cell systems, which indicates that Tgf-β is not sufficient to induce the full mdDA molecular repertoire. In the double knockout (Tgf-β2/3−/−), the amount of Th/Nurr1 cells was diminished (Roussa et al. 2006). Unfortunately, no extensive analysis of the entire mdDA systems was shown, which leaves the specific defect open for debate. Interestingly, there seems to be a dosage effect of Tgf-β signaling, when Th cell numbers are compared. This suggests that Tgf-β signaling, as suggested before, is important but not essential for mdDA neuronal development. Moreover, cross-talk between Tgf-β signaling and neurotrophic support has been reported. Tgf-β can support or induce Gdnf signaling through the recruitment of GfR-α1, being part of the Gdnf receptor complex, to the plasma membrane (Krieglstein et al. 1998; Peterziel et al. 2002). This indicates that parts of Tgf-β function in the generation and survival of mdDA neurons may rely on induction of neurotrophic support (see below). In conclusion, the present data suggest an important role for Tgf signaling. This needs to be elaborated to understand whether this is a direct effect on mdDA neurons and how the effect relates to the vulnerability in response to failing Tgf signaling between, for example, the SNc and VTA or other subsets within the mdDA neuronal field.

5.5.2 Migration of Young mdDA Neurons

The migration of young mdDA neurons to their specific ventral position and the subsequent formation of the anatomical specified VTA, retrorubral field, and SNc have been described. Two major theories are suggested based on the observed migration pattern: (1) only radial migration perpendicular to the ventricular zone (Kawano et al. 1995) and (2) a combination of radial and tangential migration (Hanaway et al. 1971). The necessary interaction of mdDA neurons with radial glia has been confirmed during E12 to E20 in the developing rat brain (Shults et al. 1990). Surprisingly, in most analyses, the anterior/posterior axis is neglected; therefore, the data described in the studies might reflect specific migration events at certain anterior/posterior position of migrating mdDA neuronal subsets. Detailed studies following labeled young mdDA neurons are needed to resolve all the specific aspects of mdDA neuronal migration and to link these processes to molecular determinants and specific function in the adult situation.

From studies on the axon guidance cue, Netrin-1 has been suggested to be important for defining mdDA neuronal organization (Flores et al. 2005). Netrin together with its receptor DCC forms a signaling complex that is present in mdDA neurons. The DCC knockout has a similar overall phenotype compared with the Netrin-1 knockout (Fazeli et al. 1997). However, the mdDA phenotype has not been described in DCC-ablated mice. Therefore, it is not proven that the DCC-Netrin-1 complex is essential for mdDA migration, but it is suggested that both proteins may have a role in determining the mdDA organization in terms of migration as well as axonal guidance (see below). Interestingly, Pax6 mutations display a dorsal expansion of mdDA neurons in prosomer 1 (Vitalis et al. 2000). It has been suggested that this correlates to disturbed ventral migration (Vitalis et al. 2000).

Two studies have linked Reelin to the organization and function of the mdDA system (Nishikawa et al. 2003; Ballmaier et al. 2002). The data are conflicting in such a way that the effects are seen in either the SNc or the VTA specifically. At this point, the exact role of Reelin in mdDA migration is therefore uncertain.

Proteocan/phosphacan 6B4 was shown to be expressed in mdDA neurons. It has been suggested that this protein together with the adhesion molecule L1 has a role in migratory pattern formation (Ohyama et al. 1998). This has been substantiated by the data derived from the L1 knockout (Demyanenko et al. 2001). Since these reports have not been followed by more thorough evaluations, the role of these proteins remains uncertain. Finally, the expression of neural cell adhesion molecule and polysialic acid was described in migrating mdDA neurons (Shults and Kimber 1992). The authors suggest that neural cell adhesion molecule and polysialic acid are involved in migration events. These observations have not been extended to provide more detail, and therefore the exact role has not been firmly established. Taken together, mdDA neuronal migration has been connected to a small number of molecules possibly involved in this process. Most of the data, however, are incomplete and require more in-depth analysis.

5.5.3 Axonal Outgrowth and Generation of Specific Connections

The mdDA neuronal field is suggested to comprise several molecular and functional units. A second level of complexity is introduced through the generation of a specific connectivity map to target areas in the rostral CNS. Although not much is known about subset-specific connectivity, recent molecular studies have identified important players in the establishment of mdDA connectivity.

5.5.3.1 MdDA Connectivity Patterns in CNS

Early anatomical studies have divided the medial forebrain bundle into three main pathways: the nigrostriatal pathway, the mesolimbic pathway, and the mesocortical pathway (Roffler-Tarlov and Graybiel 1984; Joel and Weiner 2000). This anatomical distinction is cofounded by functional studies, linking these pathways to motor control, motivation (addiction), and mood regulation, respectively. Most studies have described the relative simple connectivity of mdDA neurons to striatal tissue and have tried to understand the striatal patchy organization in terms of function (Lança et al. 1986; Holmes et al. 1995, 1997; Polli et al. 1991). The initiation of dopaminergic projections starts at E12.5 when fibers leave the mdDA field and dive down and sideways into the diencephalon (Van den Heuvel and Pasterkamp 2008). At E14.5, the first fibers reach the ventral striatum and a decision is made whether the fibers go through toward the prefrontal cortex or innervate the striatum. At E19, fibers are in the cortex and are still growing toward their final destinations. It has been suggested that the initial targeting during development is nonspecific, meaning that fibers originating from the SNc and VTA innervate overlapping areas in the striatum. The refinement occurs in later stages that form the nigrostriatal and mesolimbic pathways (Hu et al. 2004). The connections in the mesolimbic, mesostriatal, and mesocortical systems are well described. However, mdDA neurons project to many other areas in the brain (like the amygdala), and data on these projection pathways are lacking.

5.5.3.2 Guidance Factors Involved in Formation of Specific Connectivity

With the notion that mdDA subsets form specific connections, to establish separately controlled functional networks, the specificity and regulation of guidance of mdDA axons have become increasingly important. Studies focused on the identification of guidance factors and cues have produced the first data that describe the events that evolve during mdDA axonal pathfinding, synapse initiation, and stabilization (for a recent review, see Van den Heuvel and Pasterkamp 2008). The data include Netrin-1/DCC, Robo/Slit, and Semaphorin/Plexin/Neuropilin signaling (for reviews on axon guidance mechanisms, see Wen and Zheng 2006; Huber et al. 2003).

The localization of Netrin 1 and DCC in the mdDA system suggests a role for these factors in axon guidance (Osborne et al. 2005; Lin et al. 2005; Livesey and Hunt 1997). Although Netrin-1 and DCC null mutants have been described (Serafini et al. 1996; Fazeli et al. 1997), it is unknown whether these mice have mdDA axon guidance defects. Despite the observation that heterozygote Netrin-1 and DCC mice have no apparent neuroanatomical phenotype, functional changes such as a loss of amphetamine sensitization have been described (Flores et al. 2005). This clearly hints toward a function for DCC and Netrin-1 in the specification of mdDA neuronal connectivity.

The combined action of Slit (Holmes et al. 1998) and Robo receptors (Hivert et al. 2002)—Robo-1 (SNc and VTA) and Robo-2 (mainly SNc), Slit-1, Slit-2, and Slit-3 (Lin et al. 2005; Marillat et al. 2002; Bagri et al. 2002)—may initiate a strong repellent signal for mdDA axons (Lin et al. 2005). During development, Robo2 is expressed in the ventral part of the midbrain and ventral p1 and p2, and Slit-3 is expressed in the caudal midbrain (Marillat et al. 2002). Slit-3 is possibly involved in the reported repulsive action of the caudal midbrain that initiates the turning of growing fibers toward the rostral brain (Gates et al. 2004; Holmes et al. 1995). The dorsal midbrain (superior colliculus region) expresses Slit-1 at E15. This may cause the described repulsion of this tissue (Marillat et al. 2002) and may form the repulsive cue that triggers, after initial dorsal growth, the turning of mdDA axons toward the rostral/ventral axis. The combination of described repulsive cues in combination with attractive cues from main forebrain bundle tissue is an important signal that guides the early axons toward the ventral forebrain. This mechanism is very elegantly confirmed through a ventral midbrain transplant study in rat embryos in which a ventral midbrain piece was juxtaposed and transplanted in another embryo at the same position. As soon as the axons left the transplanted piece, they started to grow in the rostral position (Nakamura et al. 2000). These data clearly suggest that signals within the midbrain exist that guide mdDA axons toward the rostral/ventral axis. The diencephalon expresses Slit-1 and/or Slit-2 in specific regions as the ventral and dorsal thalamus (Marillat et al. 2002). These Slit signals may contribute to the ventral and lateral positioning of the main forebrain bundle. In Slit-1/2 double knockouts, the dopamine fibers leave the main forebrain bundle and split into two groups. One of these groups grows toward the hypothalamic area and approaches the midline (Van den Heuvel and Pasterkamp 2008).

Interestingly, in the striatal area, no Slit expression is detected before the adult phase, which indicates that the region is permissive during development to receive Robo carrying mdDA axon terminals. In the adult, Slit-2 is detected in scattered neurons, presumably cholinergic neurons; and at postnatal day 5, Slit-3 is widely expressed and remains in the adult stage. It has been suggested that this expression might influence mdDA nerve terminal plasticity (Marillat et al. 2002).

Additional guidance roles have been described for Shh in combination with the Shh receptor Smoothend (Smo) (Charron et al. 2003). This function of Shh signaling occurs in the spinal cord, but both factors are also present in the developing mdDA system and may have a similar function in guiding axons to the ventral CNS. Additional cues may be provided by the semaphorin/neuropilin/plexin system. Analysis of Nkx2.1 mutant animals showed that the medial forebrain bundle aberrantly crosses the midline (Kawano et al. 2003). This crossing was ascribed to a loss of repulsive cues such as Semaphorin 3A (Sema3A) (Pasterkamp and Kolodkin 2003) and Slit-2 in the hypothalamic area. Immunohistochemical staining for Neuropilin 1 and microarray studies on mdDA neurons confirmed that mdDA axons carry this semaphorin receptor component (Chung et al. 2005a). This would imply that plexinAs, besides the observed PlexinC1 (Chung et al. 2005a), should be present to form a functional signaling complex making these axons susceptible to Sema3A.

Initial mdDA pathfinding is disturbed in the Pax6 knockout. In these animals, axons derived from the mesencephalic/P1 region grow toward a dorsal position until they loop back to ventral in the diencephalic roof (Vitalis et al. 2000). Interestingly, mdDA neurons derived from P2/P3 still grow toward a ventral/rostral position. This indicates that the initial pathfinding instructions are different in the specific mdDA subsets. Pax6 might be an upstream activator or be essential for the generation of nonpermissive regions through the expression of Slit-1 and/or Slit-2 in the dorsal mesodiencephalic region.

The group of Ephs (receptors) and Ephrins (ligands) also has a role in mdDA neuronal connectivity. EphA5 is detected in mdDA neurons, and genetic alteration of this factor results in clear defects of the connectivity pattern (Van den Heuvel and Pasterkamp 2008). Moreover, mice that lack ephrina2 and/or ephrina5 have also been reported for their malformed connectivity in the striatum of mdDA axons. A recent study clearly indicated that semaphorin3F (Sema3F) and its receptor neuropilin2 (Npn2) play a crucial role in the formation of the meso-prefrontal pathway (Kolk et al. 2009). The function of these two proteins is described to act on two different regions. First, they are important for the fasciculation and channeling of the early outgrowing axons; and second, they are involved in the chemoattraction of fibers toward the medial prefrontal cortex. Interestingly, Sema3F acted more on rostral mdDA neurons compared with caudal mdDA neurons, suggesting that subset specification is reflected by specific guidance rules.

5.6 DEVELOPMENTAL DYSFUNCTION AND ACQUISITION OF MOOD DISORDERS

The data described above suggest that the monoaminergic systems involved in shaping the behavioral repertoire can be subjected to many disturbances if any of the molecular programs are hampered or influenced by external factors such as prenatal exposure to medication. The examples of prenatal exposure to SSRIs that are mentioned clearly suggest that developmental set points need to be generated in a normal fashion or the offspring will suffer from a "mood pathology." In addition to prenatal exposure to medication, the proper function of the transcription factors involved in building and wiring the monoaminergic systems is critical for the adult function of the brain. As has been shown for CNS serotonergic system, an alteration during development may precipitate in the cortical layering and/or wiring, which may lead to an overall alteration of the behavioral programming of that individual. Similarly, the exposure to drugs of abuse during development changes the correct homeostatic set points of the dopamine system, leading to a lifelong change in dopaminergic functioning. This may also be applicable to the intervention of dopamine signaling in the adolescent phase. In view of this, the treatment of young children with methylphenidate (an amphetamine derivative) should be carefully monitored and should not be used during pregnancy. In conclusion, because of the complex nature of the developing monoaminergic systems, their significant influence on the whole brain, and the importance of the correct homeostatic set points, aberrations during development in these systems may quickly lead to disorders that may precipitate in mood problems as described for depression, anxiety disorders, and many other psychiatric diseases.

REFERENCES

Aarnisalo, P., Kim, C.-H., Lee, J. W., et al. 2002. Defining requirements for heterodimerization between the retinoid X receptor and the orphan nuclear receptor Nurr1. *J Biol Chem* 277 (38):35118–23.

Acampora, D., Gulisano, M., and Simeone, A. 1999. Otx genes and the genetic control of brain morphogenesis. *Mol Cell Neurosci* 13 (1):1–8.

Acampora, D., Postiglione, M. P., Avantaggiato, V., et al. 2000. The role of Otx and Otp genes in brain development. *Int J Dev Biol* 44 (6):669–77.

Acampora, D., Gulisano, M., Broccoli, V., et al. 2001. Otx genes in brain morphogenesis. *Prog Neurobiol* 64 (1):69–95.

Acampora, D., Annino, A., Tuorto, F., et al. 2005. Otx genes in the evolution of the vertebrate brain. *Brain Res Bull* 66 (4–6):410–20.

Adams, K. A., Maida, J. M., Golden, J. A., et al. 2000. The transcription factor Lmx1b maintains Wnt1 expression within the isthmic organizer. *Development* 127 (9):1857–67.

Alberi, L., Sgado, P., and Simon, H. H. 2004. Engrailed genes are cell-autonomously required to prevent apoptosis in mesencephalic dopaminergic neurons. *Development* 131 (13):3229–36.

Andersson, E., Jensen, J. B., Parmar, et al. 2006a. Development of the mesencephalic dopaminergic neuron system is compromised in the absence of neurogenin 2. *Development* 133 (3):507–16.

Andersson, E., Tryggvason, U., Deng, Q., et al. 2006b. Identification of intrinsic determinants of midbrain dopamine neurons. *Cell* 124 (2):393–405.

Ang, S.-L. 2009. Foxa1 and Foxa2 transcription factors regulate differentiation of midbrain dopaminergic neurons. *Adv Exp Med Biol* 651:58–65.

Arenas, E. 2008. Foxa2: The rise and fall of dopamine neurons. *Cell Stem Cell* 2 (2):110–12.

Aroca, P., Lorente-Cánovas, B., Mateos, F. R., and Puelles, L. 2006. Locus coeruleus neurons originate in alar rhombomere 1 and migrate into the basal plate: Studies in chick and mouse embryos. *J Comp Neurol* 496 (6):802–18.

Asbreuk, C. H. J., Vogelaar, C. F., Hellemons, A., et al. 2002.CNS expression pattern of Lmx1b and coexpression with ptx genes suggest functional cooperativity in the development of forebrain motor control systems. *Mol Cell Neurosci* 21 (3):410–20.

Avantaggiato, V., Acampora, D., Tuorto, F., et al. 1996. Retinoic acid induces stage-specific repatterning of the rostral central nervous system. *Dev Biol* 175 (2):347–57.

Baek, J. H., Hatakeyama, J., Sakamoto, S., et al. 2006. Persistent and high levels of Hes1 expression regulate boundary formation in the developing central nervous system. *Development* 133 (13):2467–76.

Bagri, A., Marin, O., Plump, A. S., et al. 2002. Slit proteins prevent midline crossing and determine the dorsoventral position of major axonal pathways in the mammalian forebrain. *Neuron* 33 (2):233–48.

Ballmaier, M., Zoli, M., Leo, G., et al. 2002. Preferential alterations in the mesolimbic dopamine pathway of heterozygous reeler mice: An emerging animal-based model of schizophrenia. *Eur J Neurosci* 15 (7):1197–205.

Blum, M. 1998. A null mutation in TGF-alpha leads to a reduction in midbrain dopaminergic neurons in the substantia nigra. *Nat Neurosci* 1 (5):374–7.

Castelo-Branco, G., Wagner, J., Rodriguez, F. J., et al. 2003. Differential regulation of midbrain dopaminergic neuron development by Wnt-1, Wnt-3a, and Wnt-5a. *Proc Natl Acad Sci U S A* 100 (22):12747–52.

Castelo-Branco, G., Rawal, N., and Arenas, E. 2004. GSK-3beta inhibition/beta-catenin stabilization in ventral midbrain precursors increases differentiation into dopamine neurons. *J Cell Sci* 117 (Pt 24):5731–7.

Castro, D. S., Hermanson, E., Joseph, B., et al. 2001. Induction of cell cycle arrest and morphological differentiation by Nurr1 and retinoids in dopamine MN9D cells. *J Biol Chem* 276 (46):43277–84.

Charron, F., Stein, E., Jeong, J., et al. 2003. Themorphogen sonic hedgehog is an axonal chemoattractant that collaborates with netrin-1 in midline axon guidance. *Cell* 113 (1):11–23.

Chung, C. Y., Seo, H., Sonntag, K. C., et al. 2005a. Cell type-specific gene expression of midbrain dopaminergic neurons reveals molecules involved in their vulnerability and protection. *Hum Mol Genet* 14 (13):1709–25.

Chung, S., Hedlund, E., Hwang, M., et al. 2005b. The homeodomain transcription factor Pitx3 facilitates differentiation of mouse embryonic stem cells into AHD2-expressing dopaminergic neurons. *Mol Cell Neurosci* 28 (2):241–52.

Clotman, F., Maele-Fabry, G. Van, and Picard, J. J. 1997. Retinoic acid induces a tissue-specific deletion in the expression domain of Otx2. *Neurotoxicol Teratol* 19 (3):163–9.

Craven, S. E., Lim, K.-C., Ye, W., et al. 2004. Gata2 specifies serotonergic neurons downstream of sonic hedgehog. *Development* 131 (5):1165–73.

Danielian, P. S., and McMahon, A. P. 1996. Engrailed-1 as a target of the Wnt-1 signalling pathway in vertebrate midbrain development. *Nature* 383 (6598):332–4.

Daubert, E. A., and Condron, B. G. 2010. Serotonin: A regulator of neuronal morphology and circuitry. *Trends Neurosci* 33 (9):424–34.

de Urquiza, A. M., Liu, S., Sjoberg, M., et al. 2000. Docosahexaenoic acid, a ligand for the retinoid X receptor in mouse brain. *Science* 290 (5499):2140–4.

Demyanenko, G. P., Shibata, Y., and Maness, P. F. 2001. Altered distribution of dopaminergic neurons in the brain of L1 null mice. *Brain Res Dev Brain Res* 126 (1):21–30.

Dey, S., Mactutus, C. F., Booze, R. M., et al. 2006. Specificity of prenatal cocaine on inhibition of locus coeruleus neurite outgrowth. *Neuroscience* 139 (3):899–907.

Di Salvio, M., Di Giovannantonio, L. G., Omodei, D., et al. 2010. Otx2 expression is restricted to dopaminergic neurons of the ventral tegmental area in the adult brain. *Int J Dev Biol* 54 (5):939–45.

Ding, Y.-Q., Marklund, U., Yuan, W., et al. 2003. Lmx1b is essential for the development of serotonergic neurons. *Nat Neurosci* 6 (9):933–8.

Engele, J., and Schilling, K. 1996. Growth factor–induced c-*fos* expression defines distinct subsets of midbrain dopaminergic neurons. *Neuroscience* 73 (2):397–406.

Farkas, L. M., Dunker, N., Roussa, E., et al. 2003. Transforming growth factor-beta(s) are essential for the development of midbrain dopaminergic neurons in vitro and in vivo. *J Neurosci* 23 (12):5178–86.

Fazeli, A., Dickinson, S. L., Hermiston, M. L., et al. 1997. Phenotype of mice lacking functional deleted in colorectal cancer (DCC) gene. *Nature* 386 (6627):796–804.

Ferri, A. L. M., Lin, W., Mavromatakis, Y. E., et al. 2007. Foxa1 and Foxa2 regulate multiple phases of midbrain dopaminergic neuron development in a dosage-dependent manner. *Development* 134 (15): 2761–9.

Flores, C., Manitt, C., Rodaros, D., et al. 2005. Netrin receptor deficient mice exhibit functional reorganization of dopaminergic systems and do not sensitize to amphetamine. *Mol Psychiatry* 10 (6):606–12.

Galter, D., Buervenich, S., Carmine, A., et al. 2003. ALDH1 mRNA: Presence in human dopamine neurons and decreases in substantia nigra in Parkinson's disease and in the ventral tegmental area in schizophrenia. *Neurobiol Dis* 14 (3):637–47.

Gates, M. A., Coupe, V. M., Torres, E. M., et al. 2004. Spatially and temporally restricted chemoattractive and chemorepulsive cues direct the formation of the nigro-striatal circuit. *Eur J Neurosci* 19 (4):831–44.

Glavic, A., Gomez-Skarmeta, J. L., and Mayor, R. 2002. The homeoprotein Xiro1 is required for midbrain-hindbrain boundary formation. *Development* 129 (7):1609–21.

Goridis, C., and Rohrer, H. 2002. Specification of catecholaminergic and serotonergic neurons. *Nat Rev Neurosci* 3 (7):531–41.

Hanaway, J., McConnell, J. A., and Netsky, M. G. 1971. Histogenesis of the substantia nigra, ventral tegmental area of Tsai and interpeduncular nucleus: An autoradiographic study of the mesencephalon in the rat. *J Comp Neurol* 142 (1):59–73.

Hatakeyama, J., Tomita, K., Inoue, T., et al. 2001. Roles of homeobox and bHLH genes in specification of a retinal cell type. *Development* 128 (8):1313–22.

Hendricks, T., Francis, N., Fyodorov, D., et al. 1999. The ETS domain factor Pet-1 is an early and precise marker of central serotonin neurons and interacts with a conserved element in serotonergic genes. *J Neurosci* 19 (23):10348–56.

Hendricks, T. J., Fyodorov, D. V., Wegman, L. J., et al. 2003. Pet-1 ETS gene plays a critical role in 5-HT neuron development and is required for normal anxiety-like and aggressive behavior. *Neuron* 37 (2):233–47.

Hirsch, M. R., Tiveron, M. C., Guillemot, F., et al. 1998. Control of noradrenergic differentiation and Phox2a expression by MASH1 in the central and peripheral nervous system. *Development* 125 (4):599–608.

Hivert, B., Liu, Z., Chuang, C.-Y., et al. 2002. Robo1 and Robo2 are homophilic binding molecules that promote axonal growth. *Mol Cell Neurosci* 21 (4):534–45.

Holder, N, and Hill, J. 1991. Retinoic acid modifies development of the midbrain–hindbrain border and affects cranial ganglion formation in zebrafish embryos. *Development* 113 (4):1159–70.

Holm, P. C., Rodreaguez, F. J., Kresse, A., et al. 2003. Crucial role of TrkB ligands in the survival and phenotypic differentiation of developing locus coeruleus noradrenergic neurons. *Development* 130 (15): 3535–45.

Holmes, C., Jones, S. A., and Greenfield, S. A. 1995. The influence of target and non-target brain regions on the development of mid-brain dopaminergic neurons in organotypic slice culture. *Brain Res Dev Brain Res* 88 (2):212–9.

Holmes, C., Jones, S. A., Budd, T. C., et al. 1997. Non-cholinergic, trophic action of recombinant acetylcholinesterase on mid-brain dopaminergic neurons. *J Neurosci Res* 49 (2):207–18.

Holmes, G. P., Negus, K., Burridge, L., et al. 1998. Distinct but overlapping expression patterns of two vertebrate slit homologs implies functional roles in CNS development and organogenesis. *Mech Dev* 79 (1–2):57–72.

Hornung, J.-P. 2003. The human raphe nuclei and the serotonergic system. *J Chem Neuroanat* 26 (4):331–43.

Hu, Z., Cooper, M., Crockett, D. P., et al. 2004. Differentiation of the midbrain dopaminergic pathways during mouse development. *J Comp Neurol* 476 (3):301–11.

Huber, A. B., Kolodkin, A. L., Ginty, D. D., et al. 2003. Signaling at the growth cone: Ligand–receptor complexes and the control of axon growth and guidance. *Annu Rev Neurosci* 26:509–63.

Hwang, D.-Y., Ardayfio, P., Kang, U. J., et al. 2003. Selective loss of dopaminergic neurons in the substantia nigra of Pitx3-deficient Aphakia mice. *Brain Res Mol Brain Res* 114 (2):123–31.

Hynes, M., and Rosenthal, A. 1999. Specification of dopaminergic and serotonergic neurons in the vertebrate CNS. *Curr Opin Neurobiol* 9 (1):26–36.

Hynes, M., Porter, J. A., Chiang, C., et al. 1995. Induction of midbrain dopaminergic neurons by Sonic hedgehog. *Neuron* 15 (1):35–44.

Hynes, M., Ye, W., Wang, K., et al. 2000. The seven-transmembrane receptor smoothened cell-autonomously induces multiple ventral cell types. *Nat Neurosci* 3 (1):41–6.

Jacobs, F. M. J., Smits, S. M., Noorlander, C. W., et al. 2007. Retinoic acid counteracts developmental defects in the substantia nigra caused by Pitx3-deficiency. *Development* 134:2673–84.

Jacob, J., Storm, R., Castro, D. S., et al. 2009. Insm1 (IA-1) is an essential component of the regulatory network that specifies monoaminergic neuronal phenotypes in the vertebrate hindbrain. *Development* 136 (14):2477–85.

Jacobs, F. M. J., van der Linden, A. J. A., Wang, Y., et al. 2009a. Identification of Dlk1, Ptpru and Klhl1 as novel Nurr1 target genes in meso diencephalic dopamine neurons. *Development* 136 (14):2363–73.

Jacobs, F. M. J., van Erp, S., van der Linden, A. J. A., et al. 2009b. Pitx3 potentiates Nurr1 in dopamine neuron terminal differentiation through release of SMRT-mediated repression. *Development* 136 (4):531–40.

Joel, D., and Weiner, I. 2000. The connections of the dopaminergic system with the striatum in rats and primates: An analysis with respect to the functional and compartmental organization of the striatum. *Neuroscience* 96 (3):451–74.

Joyner, A. L., Liu, A., and Millet, S. 2000. Otx2, Gbx2 and Fgf8 interact to position and maintain a mid–hindbrain organizer. *Curr Opin Cell Biol* 12 (6):736–41.

Kageyama, R., Ohtsuka, T., Hatakeyama, J., et al. 2005. Roles of bHLH genes in neural stem cell differentiation. *Exp Cell Res* 306 (2):343–8.

Kawano, H., Ohyama, K., Kawamura, K., and Nagatsu, I. 1995. Migration of dopaminergic neurons in the embryonic mesencephalon of mice. *Brain Res Dev Brain Res* 86 (1–2):101–13.

Kawano, H., Horie, M., Honma, S., et al. 2003. Aberrant trajectory of ascending dopaminergic pathway in mice lacking Nkx2.1. *Exp Neurol* 182 (1):103–12.

Kele, J., Simplicio, N., Ferri, A. L. M., et al. 2006. Neurogenin 2 is required for the development of ventral midbrain dopaminergic neurons. *Development* 133 (3):495–505.

Kim, J.-Y., Koh, H. C., Lee, J.-Y., et al. 2003. Dopaminergic neuronal differentiation from rat embryonic neural precursors by Nurr1 overexpression. *J Neurochem* 85 (6):1443–54.

Kolk, S. M., Gunput, R.-A. F., Tran, T. S., et al. 2009. Semaphorin 3F is a bifunctional guidance cue for dopaminergic axons and controls their fasciculation, channeling, rostral growth, and intracortical targeting. *J Neurosci* 29 (40):12542–57.

Krieglstein, K., Henheik, P., Farkas, L., et al. 1998. Glial cell line–derived neurotrophic factor requires transforming growth factor-beta for exerting its full neurotrophic potential on peripheral and CNS neurons. *J Neurosci* 18 (23):9822–34.

Lança, A. J., Boyd, S., Kolb, B. E., et al. 1986. The development of a patchy organization of the rat striatum. *Brain Res* 392 (1–2):1–10.

Lin, L., Rao, Y., and Isacson, O. 2005. Netrin-1 and slit-2 regulate and direct neurite growth of ventral midbrain dopaminergic neurons. *Mol Cell Neurosci* 28 (3):547–55.

Liu, H., Margiotta, J. F., and Howard, M. J. 2005. BMP4 supports noradrenergic differentiation by a PKA-dependent mechanism. *Dev Biol* 286 (2):521–36.

Livesey, F. J., and Hunt, S. P. 1997. Netrin and netrin receptor expression in the embryonic mammalian nervous system suggests roles in retinal, striatal, nigral, and cerebellar development. *Mol Cell Neurosci* 8 (6):417–29.

Marillat, V., Cases, O., Nguyen-Ba-Charvet, K. T., et al. 2002. Spatiotemporal expression patterns of slit and robo genes in the rat brain. *J Comp Neurol* 442 (2):130–55.

McCaffery, P., and Dräger, U. C. 1994a. High levels of a retinoic acid-generating dehydrogenase in the meso-telencephalic dopamine system. *Proc Natl Acad Sci U S A* 91 (16):7772–6.

McCaffery, P., and Dräger, U. C. 1994b. Hot spots of retinoic acid synthesis in the developing spinal cord. *Proc Natl Acad Sci U S A* 91 (15):7194–7.

McMahon, A. P., Joyner, A. L., Bradley, A., et al. 1992. The midbrain hindbrain phenotype of Wnt1-/Wnt-1-mice results from stepwise deletion of engrailed-expressing cells by 9.5 days postcoitum. *Cell* 69 (4):581–95.

Millet, S., Campbell, K., Epstein, D. J., et al. 1999. A role for Gbx2 in repression of Otx2 and positioning the mid/hindbrain organizer. *Nature* 401 (6749):161–4.

Morin, X., Cremer, H., Hirsch, M. R., et al. 1997. Defects in sensory and autonomic ganglia and absence of locus coeruleus in mice deficient for the homeobox gene Phox2a. *Neuron* 18 (3):411–23.

Nakamura, S., Ito, Y., Shirasaki, R., et al. 2000. Local directional cues control growth polarity of dopaminergic axons along the rostro caudal axis. *J Neurosci* 20 (11):4112–9.

Naqui, S. Z., Harris, B. S., Thomaidou, D., et al. 1999. The noradrenergic system influences the fate of Cajal-Retzius cells in the developing cerebral cortex. *Brain Res Dev Brain Res* 113 (1–2):75–82.

Neuhoff, H., Neu, A., Liss, B., et al. 2002. I(h) channels contribute to the different functional properties of identified dopaminergic subpopulations in the midbrain. *J Neurosci* 22 (4):1290–302.

Nishikawa, S., Goto, S., Yamada, K., et al. 2003. Lack of Reelin causes mal positioning of nigral dopaminergic neurons: Evidence from comparison of normal and Reln(rl) mutant mice. *J Comp Neurol* 461 (2):166–73.

Noorlander, C. W., Ververs, F. F. T., Nikkels, P. G. J., et al. 2008. Modulation of serotonin transporter function during fetal development causes dilated heart cardiomyopathy and lifelong behavioral abnormalities. *PLoS One* 3 (7):e2782.

Nunes, I., Tovmasian, L. T., Silva, R. M., et al. 2003. Pitx3 is required for development of substantia nigra dopaminergic neurons. *Proc Natl Acad Sci U S A* 100 (7):4245–50.

Ohyama, K., Kawano, H., Asou, H., et al. 1998. Coordinate expression of L1 and 6B4 proteoglycan/phosphacan is correlated with the migration of mesencephalic dopaminergic neurons in mice. *Brain Res Dev Brain Res* 107 (2):219–26.

Osborne, P. B., Halliday, G. M., Cooper, H. M., et al. 2005. Localization of immunoreactivity for deleted in colorectal cancer (DCC), the receptor for the guidance factor netrin-1, in ventral tier dopamine projection pathways in adult rodents. *Neuroscience* 131 (3):671–81.

Otto, D. M. E., Henderson, C. J., Carrie, D., et al. 2003. Identification of novel roles of the cytochrome p450 system in early embryogenesis: Effects on vasculogenesis and retinoic acid homeostasis. *Mol Cell Biol* 23 (17):6103–16.

Panhuysen, M., Weisenhorn, D. M., Vogt, B., et al. 2004. Effects of Wnt1 signaling on proliferation in the developing mid-/hindbrain region. *Mol Cell Neurosci* 26 (1):101–11.

Park, C.-H., Kang, J. S., Kim, J.-S., et al. 2006. Differential actions of the proneural genes encoding Mash1 and neurogenins in Nurr1-induced dopamine neuron differentiation. *J Cell Sci* 119 (Pt 11):2310–20.

Pasterkamp, R. J., and Kolodkin, A. L. 2003. Semaphorin junction: Making tracks toward neural connectivity. *Curr Opin Neurobiol* 13 (1):79–89.

Pattyn, A., Goridis, C., and Brunet, J. F. 2000. Specification of the central noradrenergic phenotype by the homeobox gene Phox2b. *Mol Cell Neurosci* 15 (3):235–43.

Pattyn, A., Simplicio, N., van Doorninck, J. H., et al. 2004. Ascl1/Mash1 is required for the development of central serotonergic neurons. *Nat Neurosci* 7 (6):589–95.

Perlmann, T, and Jansson, L. 1995. A novel pathway for vitamin A signaling mediated by RXR heterodimerization with NGFI-B and NURR1. *Genes Dev* 9 (7):769–82.

Peterziel, H., Unsicker, K., and Krieglstein, K. 2002. TGFbeta induces GDNF responsiveness in neurons by recruitment of GFRalpha1 to the plasma membrane. *J Cell Biol* 159 (1):157–67.

Polli, J. W., Billingsley, M. L., and Kincaid, R. L. 1991. Expression of the calmodulin-dependent protein phosphatase, calcineurin, in rat brain: Developmental patterns and the role of nigrostriatal innervation. *Brain Res Dev Brain Res* 63 (1–2):105–19.

Prakash, N., Brodski, C., Naserke, T., et al. 2006. A Wnt1-regulated genetic network controls the identity and fate of midbrain-dopaminergic progenitors in vivo. *Development* 133 (1):89–98.

Puelles, E., Annino, A., Tuorto, F., et al. 2004. Otx2 regulates the extent, identity and fate of neuronal progenitor domains in the ventral midbrain. *Development* 131 (9):2037–48.

Puelles, L., and Rubenstein, J. L. 1993. Expression patterns of homeobox and other putative regulatory genes in the embryonic mouse forebrain suggest a neuromeric organization. *Trends Neurosci* 16 (11):472–9.

Puelles, L., and Rubenstein, J. L. R. 2003. Forebrain gene expression domains and the evolving prosomeric model. *Trends Neurosci* 26 (9):469–76.

Qian, Y., Fritzsch, B., Shirasawa, S., et al. 2001. Formation of brainstem (nor)adrenergic centers and first-order relay visceral sensory neurons is dependent on homeodomain protein Rnx/Tlx3. *Genes Dev* 15 (19):2533–45.

Rhinn, M., and Brand, M. 2001. The midbrain–hindbrain boundary organizer. *Curr Opin Neurobiol* 11 (1):34–42.

Roffler-Tarlov, S., and Graybiel, A. M. 1984. Weaver mutation has differential effects on the dopamine-containing innervation of the limbic and nonlimbic striatum. *Nature* 307 (5946):62–6.

Roussa, E., Farkas, L. M., and Krieglstein, K. 2004. TGF-beta promotes survival on mesencephalic dopaminergic neurons in cooperation with Shh and FGF-8. *Neurobiol Dis* 16 (2):300–10.

Roussa, E., Wiehle, M., Dunker, N., et al. 2006. TGF-beta is required for differentiation of mouse mesencephalic progenitors into dopaminergic neurons in vitro and in vivo. Ectopic induction in dorsal mesencephalon. *Stem Cells* 24 (9):2120–9.

Rubenstein, J. L., Martinez, S., Shimamura, K., et al. 1994. The embryonic vertebrate forebrain: The prosomeric model. *Science* 266 (5185):578–80.

Rubenstein, J. L., Shimamura, K., Martinez, S. et al. 1998. Regionalization of the prosencephalic neural plate. *Annu Rev Neurosci* 21:445–77.

Saucedo-Cardenas, O., Quintana-Hau, J. D., Le, W. D., et al. 1998. Nurr1 is essential for the induction of the dopaminergic phenotype and the survival of ventral mesencephalic late dopaminergic precursor neurons. *Proc Natl Acad Sci U S A* 95 (7):4013–8.

Serafini, T., Colamarino, S. A., Leonardo, E. D., et al. 1996. Netrin-1 is required for commissural axon guidance in the developing vertebrate nervous system. *Cell* 87 (6): 1001–14.

Shi, M., Guo, C., Dai, J.-X., et al. 2008. DCC is required for the tangential migration of noradrenergic neurons in locus coeruleus of mouse brain. *Mol Cell Neurosci* 39 (4): 529–38.

Shults, C. W., and Kimber, T. A. 1992. Mesencephalic dopaminergic cells exhibit increased density of neural cell adhesion molecule and polysialic acid during development. *Brain Res Dev Brain Res* 65 (2): 161–72.

Shults, C. W., Hashimoto, R., Brady, R. M., et al. 1990. Dopaminergic cells align along radial glia in the developing mesencephalon of the rat. *Neuroscience* 38 (2):427–36.

Simeone, A. 2002. Towards the comprehension of genetic mechanisms controlling brain morphogenesis. *Trends Neurosci* 25 (3):119–21.

Simeone, A., Puelles, E., and Acampora, D. 2002. The Otx family. *Curr Opin Genet Dev* 12 (4):409–15.

Simon, H. H., Saueressig, H., Wurst, W., et al. 2001. Fate of midbrain dopaminergic neurons controlled by the engrailed genes. *J Neurosci* 21 (9):3126–34.

Simon, H. H., Bhatt, L., Gherbassi, D., et al. 2003. Midbrain dopaminergic neurons: Determination of their developmental fate by transcription factors. *Ann N Y Acad Sci* 991 (Jun):36–47.

Simon, H. H., Thuret, S., and Alberi, L. 2004. Midbrain dopaminergic neurons: Control of their cell fate by the engrailed transcription factors. *Cell Tissue Res* 318 (1):53–61.

Smidt, M. P., and Burbach, J. P. H. 2007. How to make a mesodiencephalic dopaminergic neuron. *Nat Rev Neurosci* 8 (1):21–32.

Smidt, M. P., van Schaick, H. S., Lanctot, C., et al. 1997. A homeodomain gene Ptx3 has highly restricted brain expression in mesencephalic dopaminergic neurons. *Proc Natl Acad Sci U S A* 94 (24):13305–10.

Smidt, M. P., Asbreuk, C. H., Cox, J. J., et al. 2000. A second independent pathway for development of mesencephalic dopaminergic neurons requires Lmx1b. *Nat Neurosci* 3 (4):337–41.

Smidt, M. P., Smits, S. M., and Burbach, J. P. H. 2003. Molecular mechanisms underlying midbrain dopamine neuron development and function. *Eur J Pharmacol* 480 (1–3):75–88.

Smidt, M. P., Smits, S. M., Bouwmeester, H., et al. 2004a. Early developmental failure of substantia nigra dopamine neurons in mice lacking the homeodomain gene Pitx3. *Development* 131 (5):1145–55.

Smidt, M. P., Smits, S. M., and Burbach, J. P. H. 2004b. Homeobox gene Pitx3 and its role in the development of dopamine neurons of the substantia nigra. *Cell Tissue Res* 318 (1):35–43.

Smith, D., Wagner, E., Koul, O., et al. 2001. Retinoic acid synthesis for the developing telencephalon. *Cereb Cortex* 11 (10):894–905.

Smits, S. M., Ponnio, T., Conneely, O. M., et al. 2003. Involvement of Nurr1 in specifying the neurotransmitter identity of ventral midbrain dopaminergic neurons. *Eur J Neurosci* 18 (7):1731–8.

Smits, S. M., Burbach, J. P. H., and Smidt, M. P. 2006. Developmental origin and fate of mesodiencephalic dopamine neurons. *Prog Neurobiol* 78 (1):1–16.

Stanke, M., Stubbusch, J., and Rohrer, H. 2004. Interaction of Mash1 and Phox2b in sympathetic neuron development. *Mol Cell Neurosci* 25 (3):374–82.

Thuret, S., Bhatt, L., O'Leary, D. D. M., et al. 2004. Identification and developmental analysis of genes expressed by dopaminergic neurons of the substantia nigra pars compacta. *Mol Cell Neurosci* 25 (3):394–405.

Van den Heuvel, D. M. A., and Pasterkamp, R. J. 2008. Getting connected in the dopamine system. *Prog Neurobiol* 85 (1):75–93.

van den Munckhof, P., Luk, K. C., Ste-Marie, L., et al. 2003. Pitx3 is required for motor activity and for survival of a subset of midbrain dopaminergic neurons. *Development* 130 (11):2535–42.

van Doorninck, J. H., van Der Wees, J., Karis, A., et al. 1999. GATA-3 is involved in the development of serotonergic neurons in the caudal raphe nuclei. *J Neurosci* 19 (12):RC12.

Vernay, B., Koch, M., Vaccarino, F., et al. 2005. Otx2 regulates subtype specification and neurogenesis in the midbrain. *J Neurosci* 25 (19):4856–67.

Vitalis, T., Cases, O., Engelkamp, D., et al. 2000. Defect of tyrosine hydroxylase immunoreactive neurons in the brains of mice lacking the transcription factor Pax6. *J Neurosci* 20 (17):6501–16.

Vogel-Hopker, A., and Rohrer, H. 2002. The specification of noradrenergic locus coeruleus (LC) neurones depends on bone morphogenetic proteins (BMPs). *Development* 129 (4):983–91.

Wallen, A., Zetterström, R. H., Solomin, L., et al. 1999. Fate of mesencephalic AHD2-expressing dopamine progenitor cells in NURR1 mutant mice. *Exp Cell Res* 253 (2):737–46.

Wallen, A. A., Castro, D. S., Zetterstrom, R. H., et al. 2001. Orphan nuclear receptor Nurr1 is essential for Ret expression in midbrain dopamine neurons and in the brain stem. *Mol Cell Neurosci* 18 (6):649–63.

Wallen-Mackenzie, A., de Urquiza, A. M., Petersson, S., et al. 2003. Nurr1-RXR heterodimers mediate RXR ligand-induced signaling in neuronal cells. *Genes Dev* 17 (24):3036–47.

Warnecke, M., Oster, H., Revelli, J.-P., et al. 2005. Abnormal development of the locus coeruleus in Ear2(Nr2f6)-deficient mice impairs the functionality of the forebrain clock and affects nociception. *Genes Dev* 19 (5):614–25.

Wassarman, K. M., Lewandoski, M., Campbell, K., et al. 1997. Specification of the anterior hindbrain and establishment of a normal mid/hindbrain organizer is dependent on *Gbx2* gene function. *Development* 124 (15):2923–34.

Wen, Z., and Zheng, J. Q. 2006. Directional guidance of nerve growth cones. *Curr Opin Neurobiol* 16 (1): 52–8.

Westerlund, M., Galter, D., Carmine, A., et al. 2005. Tissue- and species-specific expression patterns of class I, III, and IV Adh and Aldh1 mRNAs in rodent embryos. *Cell Tissue Res* 322 (2):227–36.

Xiao, Q., Castillo, S. O., and Nikodem, V. M. 1996. Distribution of messenger RNAs for the orphan nuclear receptors Nurr1 and Nur77 (NGFI-B) in adult rat brain using in situ hybridization. *Neuroscience* 75 (1): 221–30.

Zetterstrom, R. H., Solomin, L., Jansson, L., et al. 1997. Dopamine neuron agenesis in Nurr1-deficient mice. *Science* 276 (5310):248–50.

Zhao, S., Maxwell, S., Jimenez-Beristain, A., et al. 2004. Generation of embryonic stem cells and transgenic mice expressing green fluorescence protein in midbrain dopaminergic neurons. *Eur J Neurosci* 19 (5): 1133–40.

Zhao, Z.-Q., Scott, M., Chiechio, S., et al. 2006. Lmx1b is required for maintenance of central serotonergic neurons and mice lacking central serotonergic system exhibit normal locomotor activity. *J Neurosci* 26 (49):12781–8.

6 Contribution of Pharmacology to Development of Monoaminergic Hypotheses of Depression

Francisco López-Muñoz and Cecilio Álamo

CONTENTS

6.1 INTRODUCTION

Until the mid-nineteenth century, etiological hypotheses on melancholy and depression postulated that these mental disorders were the result of humor dysregulation or mechanical alterations in the correct circulation of biological fluids, or were considered diseases of the soul (López-Muñoz et al. 2009a, 2009b). However, the great advances that took place during the second half of the nineteenth century in the field of neurobiology and in the knowledge of the structure of the nervous system, as well as the positivist mentality of psychiatry of this time, definitely put a stop to these postulates. These developments led to a process called *mental illness somatization*, in which mental disorders were regarded as products of an organic lesion (Alexander and Selesnick 1970; Huertas 1993). In this condition, the origin of affective disorders was associated, for example, with a local effect on the brain due to a spasm in the dura mater (*status striatus*), a state of exhaustion or decrease of brain energy, or a hormonal or metabolic alteration (Jackson 1990). But all these changed with the discovery, in the first half of the twentieth century, of the chemical messengers responsible for interneuronal communication (López-Muñoz and Alamo 2009a) and, especially, with the clinical introduction in the 1950s of the main groups of psychotropic drugs still used today, which resulted to a veritable revolution in the field of psychiatry.

The 1950s is considered the "golden decade" of psychopharmacology (López-Muñoz et al. 2000). It suffices to highlight the discovery of the antimanic action of lithium in 1949, the clinical introduction of chlorpromazine in 1952 and meprobamate in 1954, the publication of the antipsychotic properties of reserpine in 1954, the discovery of imipramine in 1955 and the psychiatric use of iproniazid in 1957, and the introduction, finally, of chlordiazepoxide in 1960. Thus, when we speak of a "revolution" in the area of psychiatry during the 1950s, our intention is to highlight the crucial importance of the introduction of truly effective therapeutic tools for treating different psychiatric disorders. And this is the case of iproniazid, the first monoamine oxidate inhibitor (MAOI), and, most importantly, of imipramine, a prototype of trycyclic antidepressant (TCA)—two drugs that marked a new era in the treatment of depression (López-Muñoz et al. 1998a; López-Muñoz and Alamo 2009b). Even so, the clinical introduction of these drugs had many critics, since psychoanalytic currents, doctrinally dominant in the field of psychiatry at the time, considered depression as a symptomatological manifestation of certain internal personality conflicts. According to this view, such conditions were even deemed to have positive characteristics, in that they were a form of externalizing a whole series of subconscious and traumatic internal conflicts, supposedly processed by the patients themselves. In this framework, the pharmacological treatment of depressive symptoms (as would occur some years later in the case of anxiety disorders) was viewed by a large part of the psychiatric community as a real error, since it would prevent patients from discovering the "true" roots of their internal conflicts. However, this clinical incorporation gave rise to a Copernican shift in the way the origin of mental illness was understood. Although they had only a very mild effect on people with normal mood state, these drugs brought about a marked reduction of symptoms in depressed patients, especially after a period of 2 or 3 weeks. Thus, support began to grow among the scientific community for the notion that these drugs might be correcting a kind of specific "chemical imbalance," the underlying cause of the illness. Furthermore, this concept would be highly convenient and liberating for society in general, and for the mentally ill and their families in particular, since it would relieve patients of a heavy moral burden and allow the symptoms of the illness to be removed through a series of administration of chemical substances (Kirkby 2007). This concept of "chemical imbalance" then revolutionized the view held by the most traditional sectors of society about mental illness, entrenched in pseudomedieval concepts such as states of possession, and even by psychiatrist themselves, some of whom continued to see psychiatric patients as deranged individuals with moral defects who needed moral therapy.

Thus, one of the main consequences of the clinical introduction of iproniazid, imipramine, and reserpine in the 1950s, together with the simultaneous advances in the knowledge of the neurochemical aspects of brain functioning—as in the case of neurotransmitters—was the possibility of postulating the first biological hypotheses on the genesis of depressive disorders and laying the bases of the so-called biological psychiatry (López-Muñoz et al. 2005a). Finally, the clinical introduction of selective serotonin reuptake inhibitors in the late 1980s definitively confirmed these monoaminergic hypotheses of depression.

6.2 FIRST ADVANCE: DISCOVERY OF NEUROTRANSMITTERS AND CHEMICAL HYPOTHESIS OF NEUROTRANSMISSION

The discovery of the chemical messengers responsible for interneuronal communication was one of the great scientific advances of the twentieth century. The hypothesis of chemical synaptic transmission was postulated in 1904 by Thomas Renton Elliott and his mentor, John Newport Langley (Department of Physiology at Cambridge University), thanks to their experiments, initially, on sympathetic stimulation and administration of suprarenal extracts and, later on, with adrenaline in the peripheral nervous system (Elliott 1904). Forty years later, in 1946, Ulf von Euler, from the Karolinska Institute in Stockholm, demonstrated that the substance responsible for chemical transmission at the postganglionic adrenergic synapses of the sympathetic system was noradrenaline and not adrenaline (Von Euler 1946). For his part, Langley, a year after Elliott's crucial contribution, anticipated the existence of the receptor, which he called the *receptive substance of target cells* (Langley 1905). This theory of neurotransmission was consolidated over the following decades, with the implication of acetylcholine as a potential neurotransmitter (Dale 1914; Loewi 1921).

The year 1933 marks another crucial contribution in this field by Edinburgh pharmacologist Alfred Joseph Clark: the quantitative analysis of the actions of drugs on biological systems. Clark applied the action mass law to the study of the relationship between drug and receptor, establishing the basic concepts of the theory of occupancy and describing the dose–response curves and their parameters (Clark 1933). These pharmacological parameters, in addition to the development of knowledge about the central nervous system (CNS), permitted the scientific community to create the first classifications of receptors. Thus, observing the pharmacological activity of a series of agonists on highly diverse tissues, Raymond P. Ahlquist proposed in 1948 in Augusta (Medical College of Georgia) the classification of adrenoceptors into two types, designated as α and β, based on the range of activity and potency of noradrenaline, adrenaline, and synthetic analogues in different tissues (Ahlquist 1948). The classification of many other subtypes of receptors, such as serotonin, was also based, subsequently, on these pharmacological criteria.

The second half of the 1950s saw the confirmation of the central neurotransmitter role of the principal biogenic amines (Valenstein 2005). In 1954, Marthe L. Vogt, from Edinburgh University, demonstrated the presence of noradrenaline in the brain not arising from the sympathetic innervation of the blood vessels (Stjärne 1999; Robinson 2001), although she herself did not see it as evidence for chemical transmission. Shortly afterward, the group led by John Carew Eccles at the Australian National University in Canberra (Eccles et al. 1956), using the novel techniques of microiontophoresis, identified acetylcholine as the transmitter present at the synapse between the axonic collaterals of the motor neurons of the spinal cord of an intact cat and the Renshaw cells (Bloom 1988). Likewise, other amines were identified in the CNS, such as serotonin (5-hydroxytryptamine),which was identified in 1953 by Betty M. Twarog's team (Twarog and Page 1953) and whose detailed anatomical location was communicated the following year by John H. Gaddum, from the University of Edinburgh (Amin et al. 1954), or dopamine, which is another catecholamine and was identified through the work begun in 1956 by the group of Arvid Carlsson (Figure 6.1), from the University of Lund (Carlsson et al. 1958). Crucial to such work was the development of the spectrophotofluorometric methodology by Robert L. Bowman and Sidney Udenfriend (Figure 6.1), from the National Institutes of Health (NIH) in Bethesda. This analytical technique substantially contributed to speeding up our understanding of how psychoactive drugs work and, in turn, of the physiopathology of mental disorders (Ban 2007). With this technique, it was possible to begin detecting changes in the levels of monoamines in the brain and of their metabolites through direct research (Bowman et al. 1955). Using this technique, Bernard B. Brodie and his colleagues at the Laboratory of Chemical Pharmacology of the National Heart Institute (a division of the NIH) (Figure 6.2) precisely confirmed that reserpine and other compounds produced a release of serotonin and noradrenaline in the brain (Brodie et al. 1957).

Another important technical advance in this field, resulting from research at the University of Lund and the University of Gothenburg, was the development of the fluorescence histochemical method by Bengt Falck and Nils-Ake Hillarp in the early 1960s. This method is for the visualization of catecholamine and indolamine neurons and their pathways in the brain (Falk et al. 1962). Thanks to this new technology, Carlsson provided, also in the early 1960s, indisputable evidence for the existence of neurotransmitters at a central level (Carlsson et al. 1964).

We cite another relevant development that took place in the early 1960s. Julius Axelrod (Figure 6.3), from the NIH in Bethesda, showed that the exogenous administration of noradrenaline labeled with tritium gave rise to an accumulation in tissues innervated by the sympathetic nervous system, an accumulation that did not occur in cases of prior denervation or after preadministration of drugs such as cocaine or imipramine (Axelrod et al. 1961). He described for the first time the

FIGURE 6.1   Arvid Carlsson (1923–) (left) and Sidney Udenfriend (1918–2001) (right) during the 1960s, standing near a scheme on action mechanism of newly discovered psychotropic drugs, such as reserpine. Carlsson, a professor at the University of Gothenburg since 1959, confirmed that dopamine was a brain neurotransmitter and not a precursor for noradrenaline, and he was one of the first authors to postulate the dopaminergic hypothesis of schizophrenia. Thanks to these merits, he won the Nobel Prize in Physiology or Medicine in 2000. Carlsson also discovered that trycyclic antidepressants blocked the reuptake neuronal of serotonin at a brain level, an event of great importance in postulation of the monoaminergic hypothesis of affective disorders. Sidney Udenfriend worked as a researcher at National Heart Institute (NIH, Bethesda) and since 1967 was the director of Roche Institute of Molecular Biology (Nutley). Udenfriend collaborated in development of spectrophotofluorometric methodology, a tool of vital importance in understanding the action mechanism of psychotropic drugs and physiopathology of mental disorders. (Reprinted with permission from Ban, T. H., Beckmann, H., and Ray, O. S., *CINP International Photo Archives in Neuropsychopharmacology CD-ROM*, MC&CD, Budaörs, 2000 photo 759.)

FIGURE 6.2   Bernard B. Brodie (1901–1989) (a) and Alfred Pletscher (1917–2006) (b). Research work of Brodie (director of Laboratory of Chemical Pharmacology at NIH, Bethesda) on serotonin and catecholamines in 1950s contributed highly valuable data for subsequent postulation of neurobiological hypotheses of depression. In 1967, Brodie was awarded the Lasker Prize for Basic Medical Research "for his extraordinary contribution to biochemical pharmacology." Work of Pletscher (director of Research at F. Hoffmann-LaRoche, S. A., in its headquarters in Basle, and professor of Pharmacology at the city's university) with serotonin–reserpine–iproniazid, initially in collaboration with Brodie, opened the way to an understanding of biochemical bases of affective disorders. (Reprinted with permission from López-Muñoz, F., and Alamo, C., *Curr Pharm Des*, 15, 1563–86, 2009. Copyright by Bentham Science Publishers.)

FIGURE 6.3   Julius Axelrod (1912–2004), director of the Pharmacology Section of the Laboratory of Clinical Science at the NIH and recipient of a Nobel Prize in Physiology and Medicine in 1970 for his contribution to the knowledge of mechanisms of neurotransmission. His research on the neurotransmitter in CNS, employing spectrophotofluorometric techniques, also developed at NIH, allowed understanding of phenomena of storage, release, and reuptake of noradrenaline, which was key in postulation of initial theories on etiology of depression and action mechanism of antidepressants. (Reprinted from National Library of Medicine, 2007. Profiles in Science. Julius Axelrod measuring chemicals. Available at: http://profiles.nlm.nih.gov/ps/retrieve/ResourceMetadata/HHAABC. This photo is in the public domain and may be used without permission.)

process of noradrenaline reuptake and the action mechanism of certain antidepressants. Finally, the use of brain neuroimaging techniques began in the 1980s, allowing the gathering of new data in neurobiological depression research, such as the density of monoaminergic receptors in different areas of the CNS or the metabolic rate of monoamines in brain in vivo.

6.3   SECOND ADVANCE: CONTRIBUTION OF PSYCHOTROPIC DRUGS TO NEUROBIOLOGICAL RESEARCH ON DEPRESSION

The clinical introduction of reserpine, iproniazid, and imipramine (Figure 6.4) helped in revolutionizing therapy in the field of psychiatry (López-Muñoz et al. 2007a, 2008; Fangmann et al. 2008). However, despite the importance of this contribution, it is not the only one of historical relevance to the progress of biological psychiatry. From a strictly scientific point of view, reserpine, TCAs, and MAOIs played a decisive role in the development of the first biological etiopathogenic theories of affective disorders in the 1960s (Coppen 1967; López-Muñoz et al. 2007b). Thus, it could finally be demonstrated that psychotropic agents had the capacity to modify altered mental states and that psychiatric disorders should no longer be filed under "alterations of the soul" (López-Muñoz et al. 2009b). Moreover, the development of these drugs made possible the introduction of new methods for assessing the antidepressant activity of different substances (Costa et al. 1960). This, in turn, permitted the enlargement of the arsenal of antidepressant therapy over the following decades, always in search—in line with the postulates set half a century earlier by Paul Ehrlich—of a kind of "magic bullet" that would eradicate the illness without harming the patient.

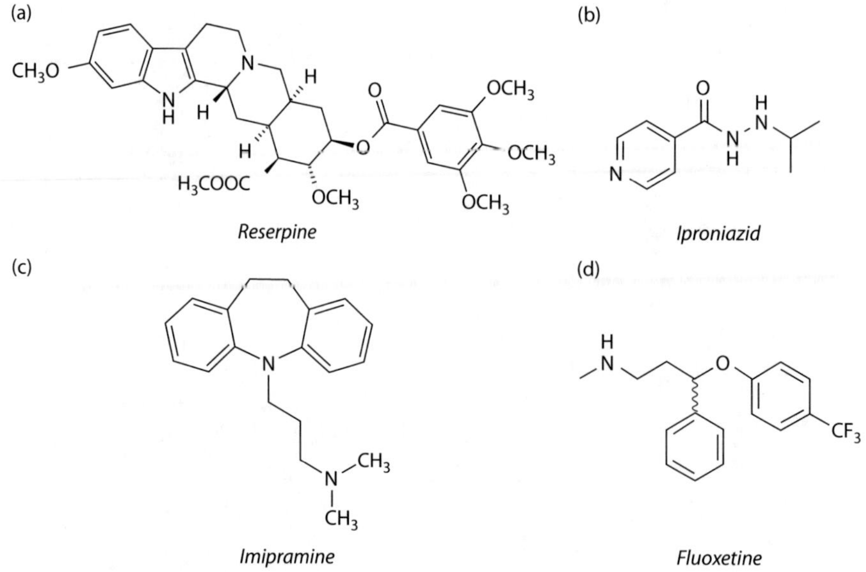

FIGURE 6.4 Chemical structure of main psychotropic drugs involved in development of monoaminergic theories of depression: reserpine (a), iproniazid (b), imipramine (c), and fluoxetine (d).

6.3.1 ROLE OF RESERPINE

One of the first antipsychotic agents was reserpine, an alkaloid isolated out of *Rauwolfia serpentina*. This plant, from the Apocynaceae family, comes from India, and its alleged medicinal properties were widely used by traditional Hindu medicine for many centuries (Bhatara et al. 2004). The first scientific report on the effects of *Rauwolfia* in humans was made in 1931 by Calcutta physicians Gananath Sen and Kartick Chandra Bose, who described its sedative properties and its ability to reduce blood pressure (Bhatara et al. 2004, 2007). In 1951, a group of researchers from the Swiss pharmaceutical company Ciba at Basel, composed of Emil Schlittler, Johannes Müller, and Hugo J. Bein, managed to isolate the alkaloid responsible for most of the sedative and hypotensive activities of the *Rauwolfia* root, called *reserpine* (Figure 6.4a) (Müller et al. 1952). Two years later, the company (Ciba Pharmaceutical Products) marketed reserpine under the trade name Serpasol®, which was chemically synthesized in 1956 by Professor Robert B. Woodward from Harvard University. Hugo Bein, in 1953, described the pharmacological properties of reserpine as an initial reversible hypnotic action similar to that reported with chlorpromazine, an absence of anticonvulsant action, and the preservation of pupillary reflexes and different painful reflexes, indicating an absence of analgesic action (Bein 1953).

The use of reserpine in the treatment of psychosis was pioneered by Nathan S. Kline (Figure 6.5a) of the Rockland State Hospital in New York (López-Muñoz et al. 2004). In 1954, he designed a double-blind, placebo-controlled clinical trial that enrolled 411 patients (94.4% schizophrenics) and evaluated the effectiveness of reserpine (Kline 1954). In the succeeding years, reserpine was widely used based on its two major pharmacological activities: antipsychotic and hypotensive. However, the introduction of new antihypertensive agents that were more effective orally, its association with some cases of death by thrombosis, its association with breast cancer (later refuted by controlled studies), and, most importantly, the description of depressive conditions induced by reserpine, with the consequent risk of suicide, significantly reduced its use. Finally, the clinical introduction of phenothiazines gradually overshadowed the therapeutic use of reserpine in psychiatry starting in 1957 (López-Muñoz et al. 2004). However, reserpine has a pivotal role in

FIGURE 6.5   (a) Nathan S. Kline (1916–1983) and (b) Roland Kuhn (1912–2005). Kline was one of the great pioneers of psychopharmacology. In 1952, he set up the research unit at Rockland Psychiatric Center (later called Rockland State Hospital) in Orangeburg (New York). Kline was responsible for clinical introduction of iproniazid and reserpine in psychiatry. He was twice awarded the prestigious Lasker Prize: in 1957 for his studies of the antipsychotic effect of reserpine and in 1964 for his contribution to development of MAOIs. Kuhn, medical director at the Psychiatric Clinic of Thurgau Canton in Münsterlingen (near Basle), was also one of the great pioneers of psychopharmacological research and the discoverer of antidepressant properties of imipramine. (Reprinted with permission from López-Muñoz, F., and Alamo, C., *Curr Pharm Des*, 15, 1563–86, 2009. Copyright by Bentham Science Publishers.)

the origin of the monoaminergic hypothesis of mental illnesses (López-Muñoz et al. 1998b), as it contributed indirectly to Arvid Carlsson's (now at the University of Gothenburg) (Figure 6.1) postulation in 1963 of his dopaminergic theory of schizophrenia and more directly to the development of the serotonergic hypothesis of depression, thanks in part to the work of Bernard Brodie (Figure 6.2a) in the NIH (Alamo et al. 2004b).

Brodie and his colleagues at the Laboratory of Chemical Pharmacology, Parkhurst A. Shore and Alfred Pletscher (the latter being Director of Research at Hoffmann-La Roche in Switzerland) (Figure 6.2b), using spectrophotofluorimetry, discovered that the administration of reserpine to rabbits brought about a depletion of brain levels of serotonin and, more importantly, that the sedative and other effects of this substance remitted as the serotonin returned to its normal levels (Brodie et al. 1955). These and other observations carried out with reserpine between 1955 and 1957, such as depletion of serotonin in the nervous system, platelets, and intestine (Pletscher et al. 1956; Brodie et al. 1957) or modifications in levels of noradrenaline in different tissues, in countries such as the United States, Britain, Switzerland, Spain, or Sweden, permitted considerable progress in the knowledge of the physiology of the autonomic and CNS (Bein 1984; Alamo et al. 2004b). Soon it was observed that reserpine also caused a depletion of noradrenaline levels in the brain (Holzbauer and Vogt 1956) and at the peripheral level (Carlsson and Hillarp 1956), which would explain the evident hypotensive effect of this drug. On the other hand, pretreatment with a precursor to serotonin, such as 5-hydroxytryptophan, prevented the serotonin depletion but did not affect the behavioral effects of reserpine, whereas pretreatment with a catecholaminergic precursor, such as L-dihydroxy-phenyl-alanine [levodopa (L-DOPA)], prevented the effects of reserpine but did not avoid the depletion of noradrenaline (Carlsson et al. 1957). These important roles were highlighted in 1959 by Francisco G. Valdecasas, from Barcelona University (Spain), and some of his colleagues, such as Eduardo Cuenca, who trained in Bethesda with Brodie, and José Antonio Salvá, thanks to their work during the second half of the 1950s. The authors concluded that reserpine produced a depletion of sympathetic amines in the tissues, which, like denervation, made the organs more sensitive to exogenous transmitting substances. This would explain the enhancement of the pressure effect of adrenaline and noradrenaline induced by the alkaloid. Additionally, this depletion of amines would explain the blocking effect of reserpine on the responses to the sympathetic centrifuge stimulation provoked by direct means (electrical stimulation of the cardioaccelerator nerve, sympathetic fibers that innervate the nictitant membrane) or by indirect means (inhalation of CO_2), which demonstrated that reserpine behaved as a sympatholytic (depleting amines) but not as an adrenolytic (blocking receptors). This sympatholytic effect would also explain the paradoxical actions of reserpine. When administered before amphetamines, similar to the spinal hyperreflex induced by some MAOIs, it was antagonized by the administration of phentolamine (adrenolytic agent). Finally, the period of initial intranquilization induced by reserpine was antagonized by mephenamine, a drug with parasympatholytic properties (Valdecasas et al. 1959). The group of paradoxical reactions observed with reserpine was called the *reserpine phenomenon*.

The reserpine phenomenon is a very important contribution from the group of Valdecasas and consists of reserpine's induction of a hypertensive or contractile response in the spinal cord of a cat or in the vas deferens of a rat when it is administered after amphetamine, cocaine, and other substances (sympathomimetics, MAOIs) (Valdecasas et al. 1958, 1959, 1965). This effect was initially attributed to a MAO inhibition because some of the substances that provoked it possessed this property, but it was later found to be independent of the said condition. From all of this experimental work, Valdecasas's group managed to create a pharmacological profile of reserpine, with a clear orientation toward the knowledge of its action mechanism and the role of monoamines, both at the central and peripheral levels, as well as of its safety as a psychotropic drug.

In this sense, the role of reserpine in basic research of affective disorders has been so important during various decades, between the 1960s and the 1980s, that one of the most used experimental models in the search for new antidepressant agents was the capacity of the candidate drug to antagonize the behavioral effects induced by reserpine (Costa et al. 1960). Of the many symptoms induced by reserpine in mouse, the most widely used psychopharmacological tests to find new antidepressant agents were hypothermia, ptosis, and akathisia (Bourin et al. 1983). The antagonism of reserpine-induced hypothermia has been an indicator of substances with β-adrenergic activity since this symptom is attributable to the lack of stimulus of β-adrenergic receptors due to decrease or absence of catecholamine (Slater et al. 1979). Reserpine-induced ptosis would be secondary, in rats, to diminution of central sympathetic discharges (Tedeschi et al. 1967) or to reduction, in mouse, of the brain serotonin levels (Bourin et al. 1983), by which the antagonism of this symptom would be an indicator of substances with α-adrenergic or serotonergic activity. Finally, reduction of motor activity by reserpine is attributable to a decline in brain dopamine rate (Hollinger 1969), and thus, akinesia antagonism is an indicator of drugs with dopaminergic activity.

However, as Bein (1978) notes, this important tool in pharmacological research played a prominent role in the consolidation of the black history of reserpine clinically, while establishing a close nexus between experimental reserpine and an archetype of depression-inducing drugs.

6.3.2 ROLE OF IPRONIAZID

The origin of the first specifically antidepressant family of drugs, the MAOIs, lies in the antitubercular hydrazide agents that had been used since the early 1950s (López-Muñoz et al. 1998a, 2007a). It was precisely in 1952 that studies began, at the Sea View Hospital on Staten Island (New York), on the clinical effects of iproniazid (Figure 6.4b), carried out by Irving J. Selikoff and Edward Robitzek, who observed that this drug possessed greater power to stimulate the CNS compared to that of isoniazid, an effect initially interpreted as a side effect (Selikoff et al. 1952). Finally, Nathan S. Kline (Figure 6.5a) and colleagues (Harry P. Loomer and John C. Saunders) from the Rockland State Hospital were the first psychiatrists to assess the efficacy of iproniazid, marketed as an antitubercular agent under the trade name Marsilid®, in nontuberculosis depressed patients (chronic psychotic depression). For their study, they recruited 17 highly inhibited subjects with severe schizophrenia and 7 with depression. Their results indicated that iproniazid had a stimulant effect on depressed patients and that 70% of the patients who received iproniazid had undergone substantial improvement (raised mood, weight gain, better interpersonal capacity, increased interest in their surroundings and themselves, etc.) (Loomer et al. 1958). In 1957, Kline published a report on the first neuropsychiatric experiences with iproniazid (Kline 1984), proposing the term *psychic energizer* to refer to the drug's action (Loomer et al. 1957).

Before the discovery of the antitubercular properties of hydrazines, Mary L. Hare (Mary Bernheim), a researcher at Cambridge University, described for the first time in 1928 how an enzyme (which she called tyramine oxidase) was capable of bringing about oxidative disamination of the biogenic amines (Hare 1928). This enzyme was identified in 1937 by the groups led by Herman Blaschko and Derek Richter, from the Physiology Department of Cambridge University (Blaschko et al. 1937), and by those of Caecilia E. Pugh and Juda H. Quastel, from the Biochemical Laboratory of Cardiff City Mental Hospital (Pugh and Quastel 1937), and was given the name *monoamine oxidase* (MAO). Blaschko's group showed that MAO, isolated from cells of the liver, kidney, and small intestine, was capable of metabolizing adrenaline through oxidation, a process that could be inhibited by ephedrine (Blaschko et al. 1937). These authors concluded that the MAOIs would act in a very similar manner to ephedrine, activating the adrenergic system. Quastel and Pugh also identified this enzyme in the brain, although its significance remained unclear (Pugh and Quastel 1937). It would later be definitively demonstrated that MAO constitutes a flavoprotein enzymatic system located at the level of the mitochondrial membrane and whose function is to produce oxidative disamination not only of the biogenic amines (serotonin and catecholamines) but also of other sympathomimetic amines (tyramine, benzylamine, β-phenylethylamine, etc.).

In 1952, the team led by Ernst Albert Zeller, from the Northwestern University Medical School (Chicago, Illinois), observed for the first time that iproniazid was capable of inhibiting MAO (Zeller et al. 1952). Subsequent studies found that serotonin in the brain was converted into hydroxyindoleacetic acid by MAO; indeed, for a long period this process was thought to be the only path for the metabolization of serotonin. In 1957, Sidney Udenfriend (Figure 6.1) and colleagues, from the NIH, observed that administration of iproniazid to experimental animals produced a rapid increase in brain levels of serotonin, similar to those resulting from the administration of 5-hydroxytryptophan, a substance capable of crossing the blood–brain barrier and producing serotonin through decarboxylation (Udenfriend et al. 1957). Simultaneously, Brodie's team, also from the NIH, confirmed the ability of reserpine to release serotonin in the brain and other tissues (Pletscher et al. 1956; Brodie et al. 1957).

All such studies opened up an interesting research line on brain functions, in the framework of which are the contributions of Charles Scott from Warner-Lambert Research Laboratories (Morris Plains) (López-Muñoz et al. 2007b). Scott believed that the tranquillizing effects observed in animals given reserpine were due to the release of serotonin caused by reserpine. With this hypothesis, and knowing the results of Zeller's work, Scott administered iproniazid with the aim of limiting the enzymatic destruction of serotonin. However, the pretreatment with iproniazid carried out by Scott before administering reserpine had the opposite result to that expected by the researcher: a stimulant effect, rather than the predicted tranquilizing effect (Chessin et al. 1956). Similar results were obtained by Brodie's team from the NIH. In 1956, Scott's group described this effect of experimental alertization with iproniazid, which he called *marsilization*, in reference to the trade name of this agent (Chessin et al. 1956). On the other hand, as in what had occurred with TCAs, the MAOIs antagonized, at an experimental level, the effects of reserpine and tetrabenazine and boosted the effects of amphetamine, hexobarbital, 5-hydroxytryptophan, and DOPA (López-Muñoz et al. 2007b).

6.3.3 Role of Imipramine and TCAs

The history of tricyclic and tetracyclic antidepressants began in the 1930s, with the Swiss chemical company J. R. Geigy's search for antihistamines that promised to be commercially successful as hypnotics or sedatives. Robert Domenjoz, who was the director of the Pharmacology section of J. R. Geigy and later became the director of the Pharmacology Institute at the University of Bonn, encouraged his team to look into the effects of phenothiazines, for which no important application had been found at that time, in the hope of their use as sedatives (Fangmann et al. 2008; López-Muñoz et al. 2008). Subsequently, in 1952, the news came that Pierre Deniker and Jean Delay had made an important discovery while testing a phenothiazine called *chlorpromazine* at the Saint-Anne University Hospital in Paris (López-Muñoz et al. 2002, 2005b). These findings spurred an intensification of the search for substances with similar properties. The result was that some long-forgotten antihistamines filed away by J. R. Geigy were dusted off in the hope that they might prove useful to psychiatry (Kuhn 1988; Healy 1997; Shorter 1997; Fangmann et al. 2008). One of these substances, known internally as G-22355 and which had been synthesized in 1948, was sent to Roland Kuhn (Figure 6.5b), from the Thurgausiche Heil- und Pflegeanstalt in Münsterlingen (close to Lake Constance), in early 1956 to be tested whether it could be used as an antipsychotic (Fangmann et al. 2008; López-Muñoz et al. 2008). The extensive clinical research that took place in 1956 at the Kantonsspital Münsterlingen, near Basel, soon made it clear that agent G-22355 lacked any appreciable neuroleptic effect. However, Kuhn observed that three patients diagnosed with depressive psychosis showed marked improvement in their general state within just a few weeks. Subsequently, another 37 depressive patients were given this drug, now formally called *imipramine* (Figure 6.4c), thus demonstrating its special efficacy in the treatment of depressive disorders (Kuhn 1957, 1958, 1988). The antidepressant effect of imipramine was, therefore, totally unexpected, and its discovery completely accidental. Imipramine was put onto the Swiss market by J. R. Geigy by the end of 1957, under the trade name Tofranil. The drug came onto the rest of the European market in the spring of 1958 (Healy 1997) and represented a giant step in the treatment of depression as the first example of a new family of drugs known as imipraminic or tricyclic antidepressants. Indeed, imipramine has maintained its status as one of the most effective antidepressants until today.

The efficacy of TCAs in the treatment of depression was indisputable by the end of the 1950s (López-Muñoz et al. 1998a). However, their action mechanism was not well understood. Although the initial studies carried out with imipramine showed that the drug had multiple pharmacological actions, it could not be determined which of them was responsible for the positive effect on mood alterations. In 1959, however, Robert Domenjoz and W. Theobald confirmed that imipramine produced an antagonism of the effects of reserpine (Domenjoz and Theobald 1959). Two years later, Brodie's team from the NIH demonstrated the physiopathological role of biogenic amines in depression from studies with laboratory animals, finding that imipramine inhibited the absorption of noradrenaline. Another NIH team, led by Julius Axelrod (Figure 6.3), demonstrated a reduction in the uptake of noradrenaline in the synaptic nerve endings during treatment with TCAs (Matussek 1988). In Spain, the group led by Valdecasas also studied this topic. A series of experiments with guanethidine, bretylium, amphetamine, and tyramine suggested an inhibiting effect of noradrenaline uptake by demethylimipramine (Cuenca et al. 1964), which has been one of the basic mechanisms used over the past 50 years for the study of new antidepressants. Subsequently, it was demonstrated that demethylimipramine was capable of antagonizing the response to tyramine at the vascular level and in the deferent behavior of the rat, indicating that TCAs also inhibited the reuptake of this indirect sympathicomimetic amine (López-Muñoz et al. 2007c).

6.4 POSTULATE OF MONOAMINERGIC THEORIES OF DEPRESSION

Thanks to the discovery and subsequent therapeutic use of reserpine, TCAs, and MAOs, monoaminergic theories of depression prospered in the 1960s (van Praag 2007), which postulated a functional deficiency of noradrenergic or serotonergic neurotransmission in certain brain areas as a primary cause of these pathologies (Coppen 1967), as will be discussed in this section. With respect to dopamine, during the past five decades, researchers could not provide significant evidence in favor of or against its involvement in the pathophysiology of depression, and antidepressants lack relevant effects on this catecholaminergic system (van Praag 2006). In any case, these findings equipped psychiatry, for the first time in its history, with a series of biological bases similar to those available in other areas of internal medicine. Moreover, this theoretical approach, eminently pharmacocentric, opened the door for the first time to an understanding of the neurobiological bases of depressive disorders.

6.4.1 Catecholaminergic Hypothesis of Depression

As previously noted, the presence of noradrenaline in the CNS was demonstrated in 1954 by Marthe Vogt (Stjärne 1999; Robinson 2001), although the definitive confirmation of the central neurotransmitter role of the principal biogenic amines took place during the second half of the 1950s (López-Muñoz and Alamo 2009a). In 1961, Julius Axelrod described the process of noradrenaline reuptake and the role played by some drugs such as imipramine or cocaine, which blocked this process (Axelrod et al. 1961). On these biochemical bases, the "catecholaminergic hypothesis" of depression was the first to be postulated, based on the observations of the effects of the antidepressants recently discovered: the inhibitory action of iproniazid on MAO (Zeller et al. 1952), the blocking of synaptic reuptake of noradrenaline by imipramine (Glowinski and Axelrod 1964), and the fact that reserpine, an alkaloid that produces an emptying of noradrenaline in nerve endings, caused depressive symptoms in a large percentage of patients when used as an antihypertensive (Goodwin and Bunney 1971).

Among these facts, it was observed how imipramine antagonized and reversed sedation, hypothermia, ptosis, and diarrhea induced by reserpine in rats, and how, in turn, the pharmacological actions of reserpine were not confined to the depletion of noradrenaline and serotonin but rather included parasympathicomimetic effects. The effect of imipramine on the disorders induced by reserpine did not provide the key to unlocking the mysteries of the way this antidepressant drug functioned (Ban 2007; López-Muñoz et al. 2007b). However, such studies did help produce a series of pharmacological models of the antagonism of reserpine, which, in turn, helped in the quest for new antidepressant agents with imipraminic effects (Ban 2001). One of those developed agents was desipramine, the demethylated metabolite of imipramine itself. This imipramine metabolite did indeed make a greater contribution to the formation of the hypothesis that the effect of imipraminic antidepressants was mediated by their action at the level of noradrenergic neurotransmission, thanks to the discovery that the reversal of the effects of reserpine by desipramine did not occur in those animals in which a selective depletion of catecholamines had been induced through administration of α-methylparatyrosine, a selective antagonist of tyrosine hydroxylase (Alamo et al. 2004b). But this model also increased the confusion about the action mechanism of antidepressants and the pathogenic mechanism of the illness since the notion that depression was the result of a chemical imbalance in which noradrenaline deficiency played a certain role was called into question after it was demonstrated that administration of α-methylparatyrosine did not produce relapses in patients treated successfully with imipramine or that it is tryptophan, precursor to serotonin, and not phenylalanine, precursor of noradrenaline, that triggers the antidepressant effect of MAOIs (Coppen et al. 1963).

The results of all this work finally served as the basis for the subsequent work of William Bunney Jr. and Joseph J. Schildkraut, who, in 1965, proposed the catecholamine-deficit hypothesis of the etiology of depression (López-Muñoz et al. 2007b). This hypothesis on the biological mechanism of depression, presented by Schildkraut (Figure 6.6), then from the Laboratory of Clinical Science (Section on Psychiatry) of the NIH (Bethesda), in a classic work published in 1965, suggested that this pathology was due to a drop in noradrenaline level in the intersynaptic cleft: "some depressions, if not all, are associated with an absolute or relative deficit of catecholamines, particularly noradrenaline, in important adrenergic receptors in the brain. Contrariwise, elation may be associated with an excess of such amines" (Schildkraut 1965). To date, Schildkraut's article is still the most cited in the entire history of the *American Journal of Psychiatry*. In support of this theory, Schildkraut and his colleagues also highlighted the fact that lithium salts, effective in the treatment of the manic phases of bipolar disorders, reduced cerebral levels of noradrenaline (Schildkraut et al. 1967), an

FIGURE 6.6 Joseph Jacob Schildkraut (1934–2006), professor of Psychiatry at Harvard University and director of Neuropsychopharmacology Laboratory at Massachusetts Mental Health Center (Boston). In his 1965 article, "The Catecholamine Hypothesis of Affective Disorders," he postulated the noradrenergic hypothesis of depression. (Reprinted with permission from López-Muñoz, F., and and Alamo, C., *Curr Pharm Des*, 15, 1563–86, 2009. Copyright by Bentham Science Publishers.)

effect opposite to that observed with TCAs. This so-called noradrenergic hypothesis of depression set off an avalanche of studies on the role of the noradrenergic system in the genesis of affective and other psychiatric disorders.

6.4.2 SEROTONERGIC HYPOTHESIS OF DEPRESSION

Since 1953, it had been known that 5-hydroxytryptamine (serotonin) was a brain neurotransmitter, thanks to the work of Betty Twarog, a researcher in Professor John Welsh's laboratory at Harvard University (Twarog 1988). At the same time, a "serotonergic hypothesis" of depression began to develop, and reserpine also contributed directly to this development, thanks, basically, to the work of Brodie, Shore, and Pletscher, from the Laboratory of Chemical Pharmacology of the NIH, who correlated in 1957 the onset of depressive symptoms with the depletion of serotonin deposits after the administration of reserpine (Brodie et al. 1957). Moreover, when the animals were given an MAOI (iproniazid) after the administration of reserpine, there was no modification of the sedative effect induced by this agent (Brodie et al. 1956), whereas if administration of iproniazid preceded that of reserpine, there was neither depletion of serotonin nor the effects of reserpine, although the animals displayed hyperactivity and signs of increased sympathetic nerve activity (Chessin et al. 1957). In the words of Pletscher (1998), "these experiments with the iproniazid–reserpine combination also supported the hypothesis that the action of the psychotropic drugs is mediated by biogenic amines, and that biogenic amines play a role in the functioning of the brain."

Thus, it could be confirmed that the administration of reserpine reduced the levels of serotonin (Besendorf and Pletscher 1956) and noradrenaline (Carlsson 1998) in the brain, whereas the administration of iproniazid increased these levels. This, combined with the clinical observations that iproniazid induced euphoric conditions in some tuberculosis patients (Crane 1957) and that reserpine induced depression in some hypertensive patients (Freis 1954), allowed the postulation of the initial hypothesis that inhibition of MAO and the resulting increase in serotonin and noradrenaline in the brain were responsible for the stimulant effect of iproniazid on mood (antidepressant effect) (Hollister 1998); these factors also lent support to the theory that alterations of mood were mediated by modifications of the levels of serotonin, noradrenaline, or both (Pletscher et al. 1956).

FIGURE 6.7   Serotonergic hypothesis of depression is based largely on the pioneering work of (a) Alec James Coppen (1923–), from Neuropsychiatric Research Unit of UK Medical Research Council, and (b) Herman M. van Praag (1929–), from Department of Biological Psychiatry of University of Groningen (The Netherlands). (Reprinted with permission from López-Muñoz, F., and and Alamo, C., *Curr Pharm Des*, 15, 1563–86, 2009. Copyright by Bentham Science Publishers.)

The efforts of Brodie's group, as noted above and in relation to reserpine and the tissular depletion of serotonin, including that in the brain (Pletscher et al. 1956), were complemented by those of other groups such as that led by Alec J. Coppen (Figure 6.7a) of the Neuropsychiatric Research Institute, which belonged to the Medical Research Council of London. Coppen's group demonstrated that the administration of tryptophan, a precursor of serotonin, to depressed animals boosted the therapeutic effects of MAOIs (Coppen et al. 1963). Another important group was that of Dutch psychiatrist Herman M. van Praag (Figure 6.7b), from the Department of Biological Psychiatry at Groningen University. van Praag, working initially with the biochemist Bart Leijnse, concluded that there were reasons to acknowledge a relationship between MAO inhibition and antidepressant action, and between serotonergic dysfunctions and the appearance of certain types of depression (van Praag and Leijnse 1964).

However, this serotonergic hypothesis was postulated without a clear demonstration of neurobiochemical correlates at a central level (Coppen 1967) but rather based on studies of variables related to peripheral serotonergic dysfunction, basically at a platelet level (van Praag 2007). The definitive extrapolation of these hypotheses to CNS functioning did not take place until the introduction of more modern techniques. Thus, in 1968, Arvid Carlsson and colleagues at the University of Gothenburg described for the first time how TCAs blocked the reuptake of serotonin at a brain level (Carlsson et al. 1968), permitting Izyaslav P. Lapin and Gregory F. Oxenkrug (Figure 6.8), from the Bekhterev Psychoneurological Research Institute of Leningrad, to postulate in 1970 the serotonergic theory of depression, as opposed to the catecholaminergic hypothesis, based on a deficit of serotonin at an intersynaptic level in certain brain regions (Lapin and Oxenkrug 1969). Finally, it would be confirmed that TCAs (and electroconvulsive therapy) improved the efficiency of serotonergic transmission, above all that which was mediated by 5-HT$_{1A}$ receptors, either through sensitization of postsynaptic receptors or through desensitization of presynaptic receptors, which usually reduces the release of serotonin in the synaptic cleft or inhibits the frequency of discharge of serotonergic neurons (Blier and De Montigny 1994). With all of these experimental observations, it could also be concluded that a drop in synaptic levels of serotonin in certain brain areas was one of the biochemical causes of depressive disorders.

6.4.2.1   Pharmacological Contribution of Fluoxetine and Selective Serotonin Reuptake Inhibitors to Serotonergic Hypothesis of Depression

At the end of the 1960s, the serotonergic hypothesis of depression began gaining momentum among researchers after studies had demonstrated the powerful inhibition of cerebral reuptake of serotonin caused by imipramine and other tertiary derivatives (Carlsson 1970)—an inhibition that was much more powerful in the case of clomipramine (a molecule with a chloride group added to the triple ring of imipramine). Clinical evidence also supported this serotonergic hypothesis, such as a

FIGURE 6.8 Izyaslav P. Lapin (1930–) (left), an unidentified researcher (middle; not identified in the *CINP International Photo Archives in Neuropsychopharmacology*) and Gregory F. Oxenkrug (1941–) (right) in the Laboratory of Psychopharmacology at Bekhterev Psychoneurological Research Institute (Leningrad) in 1965. In 1970, these researchers confirmed that an intensification of central serotonergic mechanisms could be associated to a thymoleptic effect and postulated the serotonergic hypothesis of depression, as opposed to noradrenergic hypothesis. Oxenkrug emigrated to the United States and joined the Department of Psychiatry at Tufts University (Boston). (Reprinted with permission from Ban, T. H., Beckmann, H., and Ray, O. S., *CINP International Photo Archives in Neuropsychopharmacology* CD-ROM, MC&CD, Budaörs, 2000, photo 203.)

decrease in levels of serotonin and its metabolite, 5-hydroxyindoleacetic acid, in the rhombencephalon of depressive patients who had committed suicide (Shaw et al. 1967), and reduced concentration of 5-hydroxyindoleacetic acid in the cerebrospinal fluid of depressive patients (Ashcroft et al. 1966). Moreover, treatments with precursors of serotonin, such as tryptophan or 5-hydroxytryptophan, showed antidepressant effects (Coppen 1967).

In 1971, Ray W. Fuller (Figure 6.9), a prestigious pharmacologist with experience in serotonin research, joined Lilly Research Laboratories (Indianapolis). In the same year, Solomon H. Snyder from Johns Hopkins University—one of the founding fathers of modern biological psychiatry—was honored by Lilly Research Laboratories and was invited to give a lecture. The theme he chose was neurotransmission, and his lecture highlighted the great utility for biological research of the so-called brain synaptosomes, a procedure that he himself developed to determine the high-affinity kinetics in the uptake of serotonin (Engleman et al. 2007). A "serotonin-depression study team" was then created, composed of Fuller, the biochemist David T. Wong (Figure 6.9), the organic chemist Bryan B. Molloy (Figure 6.9), and Robert Rathbun. This group, led by Wong, devoted its research efforts in the early 1970s to obtaining molecules capable of selectively inhibiting the reuptake of serotonin as potential antidepressant agents (López-Muñoz et al. 1998a; López-Muñoz and Alamo 2009b). Having observed that diphenhydramine and other antihistamines were capable of inhibiting the reuptake of monoamines (Carlsson and Lindqvist 1969) and of blocking, to the same extent as imipramine and amitryptyline, the ptosis induced by tetrabenazine in mice (Barnett et al. 1969)—a standard test of antidepressant activity—Molloy synthesized a series of phenoxyphenylpropylamines as analogues of diphenhydramine. One of these substances, LY-110140 (later known as fluoxetine chlorhydrate) (Figure 6.4d), which was tested on July 24, 1972, was the most powerful and selective inhibitor of serotonin uptake in the entire series (Wong et al. 1974), with a potency for inhibiting the uptake of serotonin six times greater than N-methyl-phenoxyphenylpropylamine, the father compound of the series, whereas its potency was 100 times smaller in the uptake of noradrenaline. Moreover, the affinity of fluoxetine for different neuroreceptors was seen to be very low

FIGURE 6.9 David T. Wong (1936–) (left), biochemist at Lilly Research Laboratories and leader of the research group that developed and introduced fluoxetine, seen here at his laboratory in Indianapolis. Bryan B. Molloy (1939–2004) (right), research chemist at Lilly Research Laboratories and responsible for synthesis of fluoxetine, among other phenoxyphenylpropylamine derivatives, in 1972. In the center is Ray W. Fuller (1935–1996). With discovery of fluoxetine, the serotonergic hypothesis of depression was definitively confirmed. (Reprinted with permission from López-Muñoz, F., and Alamo, C., *Curr Pharm Des*, 15, 1563–86, 2009. Copyright by Bentham Science Publishers.)

(Wong et al. 1983), which explained the scarcity of adverse effects associated with it, especially compared with the range of side effects typical of TCAs (constipation, urine retention, blurred vision, orthostatic hypotension, sedation, memory disorders, dizziness, etc.).

In 1980, Lilly decided to commit its efforts definitively to the new molecule, entrusting the clinical research task to John Feighner, who carried out his studies at his private psychiatric clinic in La Mesa (California). In 1983, the first positive results began to appear: fluoxetine was as effective an antidepressant as the classic TCAs and, moreover, showed far fewer adverse effects (López-Muñoz et al. 1998a; López-Muñoz and Alamo 2009b). Between 1984 and 1987, clinical trials with fluoxetine multiplied, and finally, in December 1987, the Food and Drug Administration definitively approved its clinical use, under the trade name Prozac®.

In vivo neurochemical studies also confirmed the specificity of fluoxetine in the inhibition of serotonin reuptake, thus reinforcing the serotonergic hypothesis of depression. Among the results of research in this context, it is also important to highlight the reduction of serotonin reuptake ex vivo by rat brain synaptosomes treated with fluoxetine (Wong et al. 1975) or the blocking of toxicity induced by neurotoxic agents, such as *p*-chloroamphetamine, at the level of the membrane transporter responsible for the reuptake (Fuller et al. 1975). Despite the lack of techniques for determining serotonin concentration in the synaptic cleft, various indirect tests permitted confirmation of an increase in extraneuronal concentrations of this neurotransmitter, such as a cytofluorimetric technique called *fading*, which reveals that fluoxetine increases the extracellular serotonin concentration in areas of the brain raphe in rats (Geyer et al. 1978), or in vivo voltimetric techniques, confirming a slight and prolonged increase of the signal in the corpus striatum of the rat as a result of fluoxetine, as well as its prevention of a rapid and intense increase in the signal after administration of *p*-chloroamphetamine, a competitor for the transporter of the reuptake pump of this amine (Marsden et al. 1979). Finally, through push–pull cannulation in the nucleus accumbens of the rat, a sevenfold increase in serotonin concentration was observed in the first hour after intraperitoneal injection of fluoxetine (Guam and McBride 1986).

However, despite the excellent results obtained at a biochemical level, fluoxetine was not effective in some models of tests carried out on experimental animals and used in preclinical screening of potential antidepressants, such as those related to avoidance of hypothermia induced by reserpine or apomorphine, whose result was negative with fluoxetine (Slater et al. 1979), or to the incapacity to reduce mobility in the forced swimming test in rats (Porsolt et al. 1979). In any case, all the above-mentioned data indicate that fluoxetine not only has constituted an extremely important therapeutic tool for treating different mental disorders but, as Fuller and Wong (1987) rightly claimed, also "has represented a valuable pharmacological instrument for the study of serotonergic transmission mechanisms and the physiological functions of serotonergic brain neurones."

6.5 NEW INTEGRATED APPROACHES IN ETIOPATHOGENIC THEORIES OF DEPRESSION

Despite the important advance represented by the monoaminergic hypotheses in relation to the etiopathogeny of depressive disorders and the action mechanism of antidepressants, a feeling soon emerged that these theories reflected only a small part, probably an initial part, of this action mechanism (López-Muñoz et al. 1998a). This realization led in the 1980s to the theory of receptor adaptation. According to this theory, the persistent activation of receptors, as a consequence of the increase in serotonin and noradrenaline in the synaptic cleft, led to the downregulation of 5-HT$_2$ and β-adrenergic receptors, a phenomenon that coincided with the onset of the therapeutic effect of the antidepressant (Sulser et al. 1978). However, the fact that this regulation phenomenon was not universal to antidepressants and that, furthermore, the blockers of these receptors lacked antidepressant effect—and could even induce depression in some people (Paykel et al. 1982)—called into question the possibility that this receptor-adaptation mechanism was not the sole factor responsible for the therapeutic effect of antidepressants.

More recent theories (Siever and Davis 1985) defended a "dysregulation" model that would involve not just a single neurotransmitter pathway but the alteration of several, including the noradrenergic, the serotonergic, and even the dopaminergic pathway. In line with these postulates, Pedro Delgado and Francisco Moreno, from the Department of Psychiatry at the University of Arizona Medical Center (Tucson), proposed an etiological model in which depression would result from a dysfunction of different brain areas, notably the frontal cortex, the hippocampus, the amygdala, and the basal ganglia, which, in turn, would be modulated by different monoaminergic neurotransmission systems (Delgado and Moreno 2000). According to this theory, numerous factors, notably stress, could influence the correct functioning of these areas, in either a selective or a generalized manner, which concurs with the classic postulates of a heterogeneous etiology of depressive disorders. Appended to this hypothesis should be the possible existence of marked alterations in the intraneuronal pathways of signal transduction, which would also situate the source of depressive disorders in the sphere of molecular dysfunctions (Lopez-Muñoz and Alamo 2009b).

6.5.1 NEW HYPOTHESES: FROM SYNAPSE TO INTRANEURONAL BIOCHEMICAL PATHWAYS

As pointed out earlier, in the 1960s and 1970s, most of the studies that set out to demonstrate the effects of drugs on the CNS focused on extracellular aspects of synaptic transmission (Figure 6.10), basically involving the interaction of the neurotransmitter with its receptor. This interaction was the consequence of acting on the inhibition of the systems of reuptake (TCAs and selective serotonin reuptake inhibitors) and metabolization (MAOIs), which increased the levels of monoamines in the synaptic cleft, thus facilitating their action on the receptor. However, these events occur rapidly and are detectable from the first administration of the antidepressant drugs, although the therapeutic effect does not begin to manifest until after a few weeks of treatment. These correlates suggested that the cited antidepressant effect occurs after a series of adaptations at a neuronal level, as a consequence of the chronic administration of these drugs (Alamo et al. 2007).

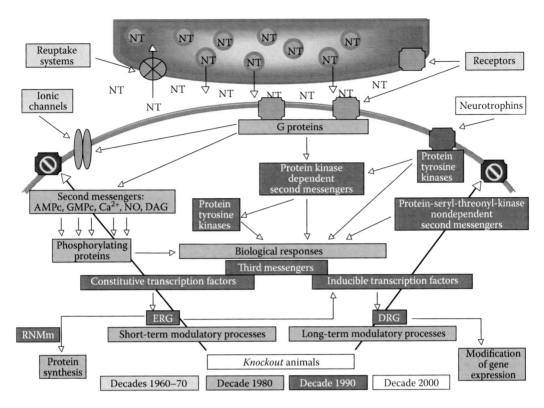

FIGURE 6.10   Advances in the understanding of neurobiology of neurotransmission systems and action mechanisms of psychotropic drugs, including intracellular systems of transduction of external signals. NT, neurotransmitter; NO, nitric oxide; DAG, diacylglycerol; ERG, early response genes; DRG, delayed response genes. (Reprinted with permission from López-Muñoz, F., and Alamo, C., *Curr Pharm Des*, 15, 1563–86, 2009. Copyright by Bentham Science Publishers.)

It became increasingly obvious that the union of the neurotransmitter and the receptor represented only the beginning of the effects of neurotransmitters on their neuronal targets. In fact, it became clear over the past two decades that these neurotransmitter substances regulated cell functioning (downregulation of receptors, protein synthesis, release of neurotransmitters, etc.) through the activation of biochemical pathways of intraneuronal messengers (G proteins, second messengers, such as AMPc or intracellular calcium and phosphorylating proteins), which eventually induced modifications in the gene expression of neurons, responsible for a wide range of biological responses (Figure 6.10) (Alamo et al. 2007; López-Muñoz and Alamo 2009b). Moreover, neurons possess a high quantity of tyrosine kinase proteins encrusted in their cell membrane, which act as receptors for neurotrophins and other growth factors (Figure 6.10), and this complicates even further the understanding of the diffusion of nerve information.

Thus, the increase of monoamines in the synaptic cleft triggers a series of intracellular neurochemical changes that, as recently postulated, have a decisive influence on the therapeutic effect of antidepressants. The intracellular effects of antidepressants have made it possible to propose the hypothesis that these drugs, after increasing the action of monoamines, the first messengers, on their corresponding receptors and regardless of the transduction pathway of the second messengers triggered (AMPc, diacylglycerol, etc.), would have a point of convergence in the protein kinases (PKA, PKC, PKCaM), considered as third messengers, which would control the genetic expression through the phosphorylation of transcription factors that could be considered as fourth messengers (Figure 6.10) (Alamo et al. 2004a, 2007). According to this hypothesis, antidepressant drugs would have a common intracellular mechanism widely removed from the point of action in the synaptic cleft, which would modulate through modifications of gene expression the synthesis of certain substances such as proenkephalin, neurotensine, brain-derived neurotrophic factor, or enzymes such as tyrosine hydroxylase, which, in sum, would provoke changes in functional adaptation, trophic actions, or synaptic remodeling, tending to offset the possible anomalies present in depression (Alamo et al. 2007; López-Muñoz and Alamo 2009b).

In any case, the scientific evidence accumulated over the past decade appears to confirm that there is no unitary biochemical mechanism to explain the origin of depression and the antidepressant effect of all the substances currently available. Thus, research is in progress on the roles of neurotrophic factors (neurotrophins and brain-derived neurotrophic factor), corticotrophin-releasing hormone and glucocorticoids, excitatory amino acids, and the N-methyl-D-aspartate receptor, as well as on other possible targets (opioid system, interleukins, substance P, somatostatin, neuropeptide Y, melatonin, nitric oxide, etc.). The role of all these targets will be analyzed in the different chapters of this book.

6.6 CONCLUSIONS

The clinical introduction of psychotropic drugs in the 1950s constitutes one of the great medical advances of the twentieth century, and the importance of the event has been compared with the discovery of antibiotics and vaccines. Although it is clear that in these early phases of psychopharmacology, serendipity played an important role in the discovery of most of the psychotropic drugs (Judd 1998; Ban 2007)—in addition, naturally, to large doses of wisdom and astute clinical observation (Baumeister et al. 2010)—what was truly important were the final results of these research processes.

Iproniazid, in spite of its short psychiatric life, and imipramine deserve a privileged place in the history of psychiatry, since not only did they open the door to the specific treatment of mood disorders—lending great prestige to the psychiatry of the time, improving everyday clinical practice, and increasing patients' quality of life—they also made possible, together with reserpine, the development of the first serious hypotheses on the biological nature of these affective disorders (López-Muñoz et al. 1998a, 2007b; Pletscher 1998, 2004; López-Muñoz and Alamo 2009b), the monoaminergic theories of depression, around which scientific discussion revolved in specialized journals in the 1960s and 1970s (Coppen 1967). Psychoactive drugs, in general, and reserpine, iproniazid, and imipramine, in the particular case that concerns us here, thus permitted the gradual definition of the neurochemical process underlying mental illness and the generation of a physiopathological theory on it, permitting the advent of the so-called biological psychiatry. We are talking, then, about an approach that could be considered "pharmacocentric" (Colodrón 1999), with far-reaching heuristic implications for psychiatry (Ban 2007). This situation emerged as unique in the history of medicine, since a large quantity of etiological hypotheses was based on the action of a series of drugs whose application to psychiatric pathologies resulted, in many cases, as today, from the intervention of chance. In any case, with iproniazid and imipramine, depression ceased to be an illness of the mind to become an illness of the brain (López-Muñoz et al. 2007b).

REFERENCES

Ahlquist, R. P. A. 1948. A study of the adrenotropic receptors. *Am J Physiol* 155:586–600.
Alamo, C., E. Cuenca, and F. López-Muñoz. 2004a. Avances en psicofarmacología y perspectivas de futuro. In *Nuevos Avances en Medicamentos,* ed. M. C. Avendaño and J. Tamargo, 351–429. Madrid: Real Academia Nacional de Farmacia.
Alamo, C., F. López-Muñoz, V. S. Bhatara, and E. Cuenca. 2004b. La contribución de la reserpina al desarrollo de la psiquiatría y su papel en la investigación neurobiológica. *Rev Psiquiatr Fac Med Barc* 31:130–9.
Alamo, C., F. López-Muñoz, and E. Cuenca. 2007. Nuevas perspectivas en psicofarmacología. In *Historia de la Psicofarmacología. Volume 3. La consolidación de la psicofarmacología como disciplina científica:*

Aspectos ético-legales y perspectivas de futuro, ed. F. López-Muñoz and C. Alamo, 1795–847. Madrid: Editorial Médica Panamericana.

Alexander, F., and S. Selesnick. 1970. *Historia de la Psiquiatría*. Barcelona: Ed. Expaxs.

Amin, A. H., T. B. B. Crawford, and J. H. Gaddum. 1954. The distribution of substance P and 5-hydroxytryptamine in the central nervous system of the dog. *J Physiol* 126:596–618.

Ashcroft, G. W., T. B. B. Crawford, D. Eccleston et al. 1966. 5-Hydroxy-indole compounds in the cerebrospinal fluid of patients with psychiatric or neurological disease. *Lancet* 2:1049–52.

Axelrod, J., L. G. Whitby, and G. Hertting. 1961. Effects of psychotropic drugs on the uptake of H^3-norepinephrine by tissues. *Science* 133:383–4.

Ban, T. A. 2001. Pharmacotherapy of depression: A historical analysis. *J Neural Transm* 108:707–16.

Ban, T. A. 2007. Psicofarmacología: El nacimiento de una nueva disciplina. In *Historia de la Psicofarmacología. Volume 2. La revolución de la psicofarmacología: Sobre el descubrimiento y desarrollo de los psicofármacos*, ed. F. López-Muñoz and C. Alamo, 577–97. Madrid: Editorial Médica Panamericana.

Ban, T. H., H. Beckmann, and O. S. Ray. 2000. *CINP International Photo Archives in Neuropsychopharmacology CD-ROM*, Budaörs: MC&CD.

Barnett, A., R. U. Taber, and F. E. Roth. 1969. Activity of antihistamines in laboratory antidepressant tests. *Int J Neuropharmacol* 8:73–9.

Baumeister, A. A., M. F. Hawkins, and F. López-Muñoz. 2010. Toward standardized usage of the word serendipity in the historiography of psychopharmacology. *J Hist Neurosci* 19:254–71.

Bein, H. J. 1953. Zur Pharmakologie des Reserpin, eines neuen Alkaloids aus *Rauwolfia serpentina* Benth. *Experientia* 9:107–10.

Bein, H. J. 1978. Prejudices in pharmacology and pharmacotherapy: Reserpine as a model for experimental research in depression. *Pharmakopsychiatr Neuropsychopharmakol* 11:289–93.

Bein, H. J. 1984. Biological research in the pharmaceutical industry with reserpine. In *Discoveries in Biological Psychiatry*, ed. F. J. Ayd and B. Blackwell, 142–54. Baltimore: Ayd Medical Communications.

Besendorf, H., and A. Pletscher. 1956. Beeinflussung zentraler Wirkungen von Reserpin und 5-Hydroxytryptamin durch Isonicotinsäurehydrazide. *Helv Physiol Acta* 14:383–90.

Bhatara, V. S., F. López-Muñoz, and C. Alamo. 2004. El papel de la medicina herbal ayurvédica en el descubrimiento de las propiedades neurolépticas de la reserpina: A propósito de la *Rauwolfia serpentina* y los orígenes de la era antipsicótica. *An Psiquiatr* 20:274–81.

Bhatara, V. S., F. López-Muñoz, and C. Alamo. 2007. Antipsicóticos (I). Los orígenes de la era antipsicótica: El descubrimiento de las propiedades neurolépticas de la reserpina. In *Historia de la Psicofarmacología. Volume 2. La revolución de la psicofarmacología: Sobre el descubrimiento y desarrollo de los psicofármacos*, ed. F. López-Muñoz and C. Alamo, 599–630. Madrid: Editorial Médica Panamericana.

Blaschko, H., D. Richter, and H. Scholassman. 1937. The inactivation of adrenaline. *J Physiol London*; 90:1–19.

Blier, P., and C. De Montigny. 1994. Current advances and trends in the treatment of depression. *Trends Pharm Sci* 15:220–6.

Bloom, F. E. 1988. Neurotransmitter: Past, present and future directions. *FASEB J* 2:22–41.

Bourin, M., M. Poncelet, R. Chermat, and P. Simon. 1983. The value of the reserpine test in psychopharmacology. *Arzneim Forsch Drug Res* 33:1173–6.

Bowman, R. L., P. A. Caulfield, and S. Udenfriend. 1955. Spectrophotofluorometric assay throughout the ultraviolet and visible range. *Science* 122:32–3.

Brodie, B. B., A. P. Pletscher, and P. A. Shore. 1955. Evidence that serotonin has a role in brain function. *Science* 122:968.

Brodie, B. B., A. P. Pletscher, and P. A. Shore. 1956. Possible role of serotonin in brain function and in reserpine action. *J Pharmacol Exp Ther* 116:9.

Brodie, B. B., J. S. Olin, R. Kuntzman, and P. A. Shore. 1957. Possible interrelationship between release of brain norepinephrine and serotonin by reserpine. *Science* 125:1293–4.

Carlsson, A. 1970. Structural specificity for inhibition for 14C-5-hydroxytryptamine uptake by cerebral slices. *J Pharm Pharmacol* 22:729–32.

Carlsson, A. 1998. Neuropharmacology. In *The rise of psychopharmacology and the story of CINP*, ed. T. A. Ban, D. Healy, and E. Shorter, 124–8. Budapest: Animula Publishing House.

Carlsson, A., and N. A. Hillarp. 1956. Release of adrenaline from the adrenal medulla of rabbits produced by reserpine. *Kungl Fysiogr Sällsk Lund Förhandl* 26:8.

Carlsson, A., and M. Lindqvist. 1969. Central and peripheral monoaminergic membrane pump blockade by some addictive analgesic and antihistamines. *J Pharm Pharmacol* 21:460–4.

Carlsson, A., M. Lindqvist, and T. Magnusson. 1957. 3,4-Dihydroxy phenylalanine and 5-hydroxy tryptophan as reserpine antagonists. *Nature* 180:1200.

Carlsson, A., M. Lindqvist, T. Magnusson, and B. Waldeck. 1958. On the presence of 3-hydroxytyramine in brain. *Science* 127:471.

Carlsson, A., B. Flack, K. Fuxe, and N. A. Hillarp. 1964. Cellular localization of monoamines in the spinal cord. *Acta Physiol Scand* 60:112–9.

Carlsson, A., K. Fuxe, and U. Ungerstedt. 1968. The effect of imipramine on central 5-hydroxytryptamine neurons. *J Pharm Pharmacol* 20:150–1.

Chessin, M., B. Dubnick, E. R. Kramer, and C. C. Scott. 1956. Modifications of pharmacology of reserpine and serotonin by iproniazid. *Fed Proc* 15:409.

Chessin, M., E. R. Kramer, and C. C. Scott. 1957. Modification of the pharmacology of reserpine and serotonin by iproniazid. *J Pharmacol Exp Ther* 119:453–60.

Clark, A. J. 1933. *The Mode of Action of Drugs on Cells.* Baltimore: Williams and Wilkins.

Colodrón, A. 1999. Psiquiatría Biológica: Historia y método. In *Fundamentos biológicos en psiquiatría,* ed. J. A. Cervilla and C. García-Ribera, 3–9. Barcelona: Masson S. A.

Coppen, A. 1967. The biochemistry of affective disorders. *Br J Psychiatry* 113:1237–64.

Coppen, A., D. M. Shaw, and J. P. Farrell. 1963. Potentiation of the antidepressive effect of a monoamine-oxidase inhibitor by tryptophan. *Lancet* 1:79–81.

Costa, E., S. Garattini, and S. Valzelli. 1960. Interactions between reserpine, chlorpromazine, and imipramine. *Experientia* 16:461–3.

Crane, G. 1957. Iproniazid (Marsilid) phosphate, a therapeutic agent for mental disorders and debilitating diseases. *Psychiatr Res Rep* 8:142–52.

Cuenca, E., J. A. Salvá, and F. G. Valdecasas. 1964. Some pharmacological effects of desmethylimipramine (DMI). *Int J Neuropharmacol* 3:167–71.

Dale, H. H. 1914. The action of certain esters and ethers of choline, and their relation to muscarine. *J Pharmacol* 6:147–90.

Delgado, P. L., and F. A. Moreno. 2000. Role of norepinephrine in depression. *J Clin Psychiatry* 61 (Suppl 1): 5–12.

Domenjoz, R., and W. Theobald. 1959. Zur Pharmakologie des Tofranil (*N*-3-dimethylaminopropyl-iminodibenzylhydrochlorid. *Arch Int Pharmakodyn Ther* 120:450–89.

Eccles, J. C., R. M. Eccles, and P. Fatt. 1956. Pharmacological investigations on a central synapse operated by acetylcholine. *J Physiol* 131:154–69.

Elliott, T. R. 1904. On the action of adrenaline. *J Physiol London* 31:xx–xxi.

Engleman, E. A., D. T. Wong, and F. P. Bymaster. 2007. Antidepresivos (III). El triunfo de la política de diseño racional de psicofármacos: Descubrimiento de la fluoxetina e introducción clínica de los ISRS. In *Historia de la Psicofarmacología. Volume 2. La revolución de la psicofarmacología: Sobre el descubrimiento y desarrollo de los psicofármacos,* ed. F. López-Muñoz and C. Alamo, 719–46. Madrid: Editorial Médica Panamericana.

Falk, B., N. A. Hillarp, G. Thieme, and A. Torp. 1962. Fluorescence of catecholamine and related compounds condensed with formaldehyde. *J Histochem Cytochem* 10:348–54.

Fangmann, P., H. J. Assion, G. Juckel, C. Alamo, and F. López-Muñoz. 2008. Half a century of antidepressant drugs. On the clinical introduction of monoamine oxidase inhibitors, tricyclics and tetracyclics: Part II. Tricyclics and tetracyclics. *J Clin Psychopharmacol* 28:1–4.

Freis, E. D. 1954. Mental depression in hypertensive patients treated for long periods with large doses of reserpine. *New Engl J Med* 251:1006–8.

Fuller, R. W., and D. T. Wong. 1987. Selective re-uptake blockers in vitro and in vivo. *J Clin Psychopharmacol* 7:365–435.

Fuller, R. W., K. W. Perry, and B. B. Molloy. 1975. Effect of 3-(*p*-trifluoromethylphenoxy)-*n*-methyl-3-phenylpropylamine on the depletion of brain serotonin by 4-chloroamphetamine. *J Pharmacol Exp Ther* 193:796–803.

Geyer, M. A., W. J. Dawsey, and A. J. Mandell. 1978. Fading: A new cytofluorimetric measure quantifying serotonin in the presence of catecholamines at the cellular level in brain. *J Pharmacol Exp Ther* 207:650–67.

Glowinski, J., and J. Axelrod. 1964. Inhibition of uptake of tritiated-noradrenaline in the intact rat brain by imipramine and structurally related compounds. *Nature* 204:1318–9.

Goodwin, P. K., and W. E. Bunney. 1971. Depression following reserpine: A re-evaluation. *Semin Psychiatr* 3:435–48.

Guam, X. M., and W. J. McBride. 1986. Selective action of fluoxetine on the extracellular poof of serotonin in the nucleus accumbens. *Soc Neurosci Abstr* 12:428.

Hare, M. L. C. 1928. Tyramine oxidase: A new enzyme system in the liver. *Biochem J* 22:968–79.

Healy, D. 1997. *The Antidepressant Era*. Cambridge, MA: Harvard University Press.

Hollinger, M. 1969. Effect of reserpine, alpha-methyl-*p*-tyrosine, *p*-chlorophenylalanine and pargyline on levorphanol-induced running activity in mice. *Arch Int Pharmacodyn Ther* 179:419–24.

Hollister, L. E. 1998. The beginning of psychopharmacology. A personal account. In *The Rise of Psychopharmacology and the Story of CINP*, ed. T. A. Ban, D. Healy, and E. Shorter, 41–4. Budapest: Animula Publishing House.

Holzbauer, M., and Vogt, M. 1956. Depression by reserpine of the noradrenergic concentration in the hypothalamus of the cat. *J Neurochem* 1:8–11.

Huertas, R. 1993. El saber psiquiátrico en la segunda mitad del siglo XIX: La somatización de la enfermedad mental. *Historia 16* 18 (211):66–73.

Jackson, S. W. 1990. *Historia de la Melancolía y de la Depresión. Desde los Tiempos Hipocráticos a la Época Moderna*. Madrid: Turner.

Judd, L. L. 1998. A decade of antidepressant development: The SSRIs and beyond. *J Affect Disord* 51: 211–3.

Kirkby, K. C. 2007. Consecuencias socio-sanitarias de la introducción clínica de los psicofármacos. In *Historia de la Psicofarmacología. Volume 3. La consolidación de la psicofarmacología como disciplina científica: Aspectos ético-legales y perspectivas de futuro*, ed. F. López-Muñoz and C. Alamo, 1435–50. Madrid: Editorial Médica Panamericana.

Kline, N. S. 1954. Use of *Rauwolfia serpentina* Benth. in neuropsychiatric conditions. *Ann N Y Acad Sci* 59:107–32.

Kline, N. S. 1984. Monoamine oxidase inhibitors: An unfinished picaresque tale. In *Discoveries in Biological Psychiatry*, ed. F. J. Ayd and B. Blackwell, 194–204. Baltimore: Ayd Medical Communications.

Kuhn, R. 1957. Über die Behandlung depressiver Zustände mit einem Iminodibenzylderivat (G 22355). *Schweiz Med Wchnschr* 87:1135–40.

Kuhn, R. 1958. The treatment of depressive states with G 22355 (imipramine hydrochloride). *Am J Psychiatr* 115:459–64.

Kuhn, R. 1988. Geschichte der medikamentösen Depressionsbehandlung. In *Pharmakopsychiatrie im Wandel der Zeit*, ed. O. K. Linde, 10–27. Klingenmünster: Tilia-Verlag.

Langley, J. N. 1905. On the reactions of cells and nerve-endings to certain poisons, chiefly as regards the reaction of striated muscle to nicotine and to curari. *J Physiol London* 33:374–413.

Lapin, J. P., and G. F. Oxenkrug. 1969. Intensification of the central serotonergic processes as a possible determinant of the thymoleptic effect. *Lancet* 1:132–6.

Loewi, O. 1921. Über humorale Übertragbarkeit der Herznervenwirkung. *Pflügers Arch* 189:239–42.

Loomer, H. P., I. C. Saunders, and N. S. Kline. 1957. Iproniazid, an amine oxidase inhibitor, as an example of a psychic energizer. *Congress Rec* 1:1382–90.

Loomer, H. P., I. C. Saunders, and N. S. Kline. 1958. A clinical and pharmacodynamic evaluation of iproniazid as a psychic energizer. *Psychiatr Res Rep Am Psychiatr Assoc* 8:129–41.

López-Muñoz, F., and C. Alamo. 2009a. Historical evolution of the neurotransmission concept. *J Neural Transm* 116:515–33.

López-Muñoz, F., and C. Alamo. 2009b. Monoaminergic neurotransmission: The history of the discovery of antidepressants from 1950s until today. *Curr Pharm Des* 15:1563–86.

López-Muñoz, F., C. Alamo, and E. Cuenca. 1998a. Fármacos antidepresivos. In *Historia de la Neuropsicofarmacología. Una Nueva Aportación a la Terapéutica Farmacológica de los Trastornos del Sistema Nervioso Central*, ed. F. López-Muñoz and C. Alamo, 269–303. Madrid: Ediciones Eurobook S. L.

López-Muñoz, F., C. Alamo, and E, Cuenca. 1998b. Fármacos antipsicóticos. In *Historia de la Neuropsicofarmacología. Una Nueva Aportación a la Terapéutica Farmacológica de los Trastornos del Sistema Nervioso Central*, ed. F. López-Muñoz and C. Alamo, 207–43. Madrid: Ediciones Eurobook S. L.

López-Muñoz, F., C. Alamo, and E. Cuenca. 2000. La "Década de Oro" de la Psicofarmacología (1950–1960): Trascendencia histórica de la introducción clínica de los psicofármacos clásicos. *Psiquiatria.COM* (electronic journal); 4 (3). http://www.psiquiatria.com/psiquiatria/revista/47/1800/?++interactivo.

López-Muñoz, F., C. Alamo, and E. Cuenca. 2002. Aspectos históricos del descubrimiento y de la introducción clínica de la clorpromazina: Medio siglo de psicofarmacología. *Frenia Rev Hist Psiquiatr* II (1): 77–107.

López-Muñoz, F., V. S. Bhatara, C. Alamo, and E. Cuenca. 2004. Aproximación histórica al descubrimiento de la reserpina y su introducción en la clínica psiquiátrica. *Actas Esp Psiquiatr* 32:387–95.

López-Muñoz, F., C. Alamo, and E. Cuenca. 2005a. Historia de la psicofarmacología. In *Tratado de Psiquiatría*, ed. J. Vallejo and C. Leal, 1709–36. Barcelona: Ars Medica.

López-Muñoz, F., C. Alamo, E. Cuenca, W. W. Shen, P. Clervoy, and G. Rubio. 2005b. History of the discovery and clinical introduction of chlorpromazine. *Ann Clin Psychiatr* 17:113–5.

López-Muñoz, F., C. Alamo, G. Juckel, and H. J. Assion. 2007a. Half a century of antidepressant drugs. On the clinical introduction of monoamine oxidase inhibitors, tricyclics and tetracyclics: Part I. Monoamine oxidase inhibitors. *J Clin Psychopharmacol* 27:555–9.

López-Muñoz, F., H. J. Assion, C. Alamo, P. García-García, and P. Fangmann. 2007b. Contribución de la iproniazida y la imipramina al desarrollo de la psiquiatría biológica: Primeras hipótesis etiopatogénicas de los trastornos afectivos. *Psiquiatr Biol* 14:217–29.

López-Muñoz, F., E. Cuenca, and C. Alamo. 2007c. Historia de la psicofarmacología preclínica en España. In *Historia de la Psicofarmacología. Volume 3. La Consolidación de la Psicofarmacología como Disciplina Científica: Aspectos Ético-Legales y Perspectivas de Futuro*, ed. F. López-Muñoz and C. Alamo, 1911–42. Madrid: Editorial Médica Panamericana.

López-Muñoz, F., H. J. Assion, C. Alamo, P. García-García, and P. Fangmann. 2008. La introducción clínica de la iproniazida y la imipramina: Medio siglo de terapéutica antidepresiva. *An Psiquiatr* 24:56–70.

López-Muñoz, F., G. Rubio, J. D. Molina, P. García-García, and C. Alamo. 2009a. La melancolía como enfermedad del alma (I): De la Antigüedad Clásica al Renacimiento. *An Psiquiatr* 25:146–59.

López-Muñoz, F., G. Rubio, J. D. Molina, P. García-García, and C. Alamo. 2009b. La melancolía como enfermedad del alma (II): Del periodo moderno a la actualidad. *An Psiquiatr* 25:197–209.

Marsden, C. A., J. Conti, E. Strope, G. Curzon, and R. N. Adams. 1979. Monitoring 5-hydroxytryptamine release in the brain of the freely moving unanaesthetized rat using in vivo voltammetry. *Brain Res* 171:86–99.

Matussek, N. 1988. Anfänge der biochemisch-psychiatrischen Depressionsforschung. In *Pharmakopsychiatrie im Wandel der Zeit*, ed. O. K. Linde, 190–5. Klingenmünster: Tilia-Verlag.

Müller, J. M., E. Schlittler, and H. J. Bein. 1952. Reserpin, der sedative Wirkstoff aus *Rauwolfia serpentina* Benth. *Experientia* 8:338–9.

National Library of Medicine. 2007. Profiles in Science. Julius Axelrod measuring chemicals. Available at http://profiles.nlm.nih.gov/ps/retrieve/ResourceMetadata/HHAABC

Paykel, E. S., R. Fleminger, and J. P. Watson. 1982. Psychiatric side effects of antihypertensive drugs other than reserpine. *J Clin Psychopharmacol* 2:14–39.

Pletscher, A. 1998. On the eve of the neurotransmitter era in biological psychiatry. In *The Rise of Psychopharmacology and the Story of CINP*, ed. T. A. Ban, D. Healy, and E. Shorter, 110–5. Budapest: Animula Publishing House.

Pletscher, A. 2004. Iproniazid: Prototype of antidepressant MAO-inhibitors. In *Reflections on Twentieth-Century Psychopharmacology*, ed. T. A. Ban, D. Healy, and E. Shorter, 174–7. Budapest: Animula Publishing House.

Pletscher, A., P. A. Shore, and B. B. Brodie. 1956. Serotonin release as a possible mechanism of reserpine action. *Science* 122:374–5.

Porsolt, R. D., A. Bertin, N. Blavet, M. Deniel, and M. Jalfre. 1979. Immobility induced by forced swimming in rats; effects of agents which modify central catecholamine and serotonin activity. *Eur J Pharmacol* 57:201–10.

Pugh, C. E., and J. H. Quastel. 1937. Oxidase of aliphatic amines by brain and other tissues. *Biochem J* 31:286–91.

Robinson, J. D. 2001. *Mechanisms of Synaptic Transmission. Bridging the Gaps (1890–1990)*. London: Oxford University Press.

Schildkraut, J. J. 1965. The catecholamine hypothesis of affective disorders: A review of supporting evidence. *Am J Psychiatry* 122:509–22.

Schildkraut, J. J., S. M. Schanberg, G. R. Breese, and I. J. Kopin. 1967. Norepinephrine metabolism and drugs used in the affective disorders: A possible mechanism of action. *Am J Psychiatry* 124:600–8.

Selikoff, I. J., E. H. Robitzek, and G. G. Ornstein. 1952. Treatment of pulmonary tuberculosis with hydrazine derivatives of isonicotinic acid. *JAMA* 150:973–80.

Shaw, D. M., F. E. Camps, and E. G. Eccleston. 1967. 5-Hydroxytryptamine in the hind-brain of depressive suicides. *Br J Psychiatry* 113:1407–11.

Shorter, E. 1997. *A history of Psychiatry. From the Era of the Asylum to the Age of Prozac*. New York: John Wiley & Sons, Inc.

Siever, L. J., and K. L. Davis. 1985. Overview: Toward dysregulation hypothesis of depression. *Am J Psychiatry* 142:1017–31.

Slater, I. H., R. C. Rathbun, and R. Kattau. 1979. Role of 5-hydroxytryptaminergic and adrenergic mechanism in antagonism of reserpine-induced hypothermia in mice. *J Pharm Pharmacol* 31:108–10.

Stjärne, L. 1999. Catecholaminergic neurotransmission: Flagship of all neurobiology. *Acta Physiol Scand* 166:251–9.

Sulser, F., J. Vetulani, and P. Mobley. 1978. Mode of action of antidepressant drugs. *Biochem Pharmacol* 27:257–61.

Tedeschi, D. H., P. J. Fowler, T. Fujita, and R. B. Miller. 1967. Mechanisms underlying reserpine-induced ptosis and blepharospasm: Evidence that reserpine decreases central sympathetic outflow in rats. *Life Sci* 6:515–23.

Twarog, B. M. 1988. Serotonin: History of a discovery. *Comp Biochem Physiol* 91C:21–4.

Twarog, B. M., and I. H. Page. 1953. Serotonin content of some mammalian tissues and urine and method for its determination. *Am J Physiol* 175:157–61.

Udenfriend, S., H. Weissbach, and D. F. Bogdanski. 1957. Effect of iproniazid on serotonin metabolism in vivo. *J Pharmacol Exp Ther* 120:255–60.

Valdecasas, F. G., E. Cuenca, and M. Morales. 1958. Acción de la reserpina sobre la presión arterial en gatos espinales tratados con inhibidores de la aminoxidasa. In *Libro de Actas de la V Reunión Nacional de la Sociedad Española de Ciencias Fisiológicas*, 305–9. Granada: S. E. C. F.

Valdecasas, F. G., J. A. Salvá, and E. Cuenca. 1959. Acción periférica y central de la reserpina. In *Libro de Actas de la V Reunión Nacional de la Sociedad Española de Ciencias Fisiológicas*, 333–4. Madrid: S. E. C. F.

Valdecasas, F. G., E. Cuenca, and L. Rodríguez. 1965. Fenómeno reserpínico en el conducto deferente de la rata. In *Libro de Actas de la IX Reunión Nacional de la Sociedad Española de Ciencias Fisiológicas*, 177–9. Pamplona: S. E. C. F.

Valenstein, E. S. 2005. *The War of the Soups and the Sparks. The Discovery of the Neurotransmitters and the Dispute over How Nerves Communicate*. New York: Columbia University Press.

van Praag, H. M. 2006. The neurotransmitter era in psychiatry. Monoamines and depression. A retrospective. In *The Neurotransmitter Era in Neuropsychopharmacology*, ed. T. A. Ban and R. Ucha-Udabe, 167–90. Buenos Aires: Editorial Polemos.

van Praag, H. M. 2007. Monoaminas y depresión: Una visión retrospectiva. In *Historia de la Psicofarmacología. Volume 1. De los orígenes a la medicina científica: Sobre los pilares biológicos del nacimiento de la psicofarmacología*, ed. F. López-Muñoz and C. Alamo, 517–44. Madrid: Editorial Médica Panamericana.

van Praag, H. M., and B. Leijnse. 1964. Die Bedeutung der Psychopharmakologie für die klinische Psychiatrie. Systematik als notwendinger Ausgangspunkt. *Nervenartzt* 34:530–7.

Von Euler, U. S. 1946. A specific sympathomimetic ergone in adrenergic nerve fibres (sympathin) and its relations to adrenaline and noradrenaline. *Acta Physiol Scand* 12:73–97.

Wong, D. T., J. S. Horng, F. P. Bymaster, K. L. Hauser, and B. B. Molloy. 1974. A selective inhibitor of serotonin uptake: Lilly 110140, 3-(*p*-trifluoromethylphenoxy)-*N*-methyl-3-phenylpropylamine. *Life Sci* 15:471–9.

Wong, D. T., J. S. Horng, and F. P. Bymaster. 1975. DL-*N*-methyl-3-(*o*-methoxyphenoxy)-3-phenylpropylamine hydrochloride, Lilly 94939, a potent inhibitor for uptake of norepinephrine into rat brain synaptosomes and heart. *Life Sci* 17:755–60.

Wong, D. T., F. P. Bymaster, L. R. Reid, and P. G. Threlkeld. 1983. Fluoxetine and two other serotonin uptake inhibitors without affinity for neuronal receptors. *Biochem Pharmacol* 32:1287–93.

Zeller, E. A., J. Barsky, J. R. Fouts, W. F. Kirchheimer, and L. S. Van Orden. 1952. Influence of isonicotinic acid hydrazide (INH) and 1-isonicotinic-2-isopropyl-hydrazide (IIH) on bacterial and mammalian enzymes. *Experientia* 8:349.

7 Serotonergic Pathways in Depression

Stefanie C. Altieri, Yogesh S. Singh,
Etienne Sibille, and Anne M. Andrews

CONTENTS

7.1 INTRODUCTION

The magnitude of evidence supporting a role for the serotonin neurotransmitter system in major depressive disorder (MDD) has been forthcoming over the past five decades and is indispensable to our understanding of this complex and heterogeneous illness. In this chapter, we will explore the historical perspectives that led to the monoamine hypothesis of depression and, specifically, the ideas regarding the role of serotonin in the etiology and pathophysiology of depression. We will present recent views on factors influencing susceptibility to depression, including the discovery of gene polymorphisms in humans and studies focused on the developmental sensitivity of the serotonin system, which provide approaches to unraveling the complexity of depression. Additionally, investigations on the mechanisms of action of different classes of antidepressant drugs have also contributed insight into the role of serotonin in depression. Finally, we will touch on peripheral biomarkers of the central serotonin system that might lead to personalized medicine approaches to improve therapeutic outcomes.

7.2 SEROTONIN: THE BASICS

7.2.1 Discovery and Biosynthetic Pathway

Serotonin (5-hydroxytryptamine, 5-HT) was first discovered in 1937 by Vialli and Erspamer and given the name *enteramine*. These investigators observed enteramine in the vesicles of enterochromaffin cells of the gut and determined that enteramine was associated with smooth muscle contraction (Vialli and Erspamer 1937, 1940). In an unrelated study, Rapport et al. (1948) isolated and crystallized a compound from beef serum with vasoconstricting properties, which they named *serotonin* (serum tonic). It was realized afterward that enteramine and serotonin were the same compound (Erspamer and Asero 1952, 1953). The discovery of serotonin in mammalian brain extracts led to its recognition as a chemical neurotransmitter (Twarog and Page 1953; Amin et al. 1954).

Serotonin is synthesized from tryptophan (L-tryptophan), an essential dietary amino acid (Figure 7.1). Tryptophan crosses the blood–brain barrier via the large neutral amino acid transporter (LNAAT) and is converted into 5-hydroxytryptophan (5-HTP) in serotonergic neurons, which selectively express the rate-limiting synthetic enzyme tryptophan hydroxylase (TPH) (Figure 7.1).

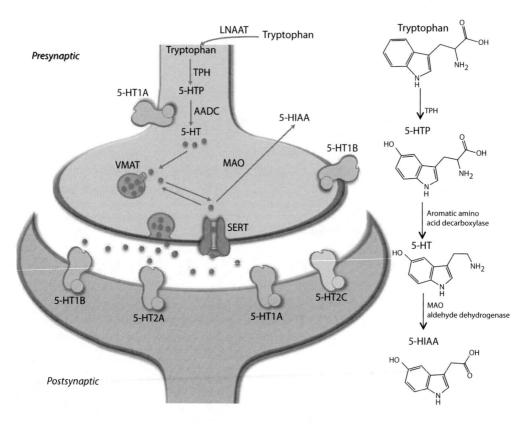

FIGURE 7.1 Key components of the serotonergic system. Serotonin is synthesized from the dietary amino acid L-tryptophan in a two-step process. Tryptophan is first transported into the brain via the large neutral amino acid transporter (LNAAT), where it is converted to 5-hydroxytryptophan (5-HTP) by the rate-limiting enzyme tryptophan hydroxylase (TPH). Next, 5-hydroxytryptophan is converted to 5-hydroxytryptamine (5-HT, serotonin) by amino acid decarboxylase (AADC). Newly synthesized serotonin is transported into synaptic vesicles within presynaptic neurons via vesicular monoamine transporters. Upon neuronal activation, serotonin is released from vesicles into the extracellular space where it interacts with a number of different serotonin receptors located both pre- and postsynaptically. The actions of serotonin are terminated by (1) transport back into presynaptic neurons by SERT where serotonin can be repackaged into vesicles or (2) breakdown into 5-hydroxyindoleacetic acid (5-HIAA) by MAO.

Tryptophan depletion or pharmacologic inhibition of TPH has been used to alter brain serotonin levels and to study the involvement of serotonin in depression. There are two gene isoforms that encode TPH termed *Tph1* and *Tph2*. The *Tph2* isoform is predominantly expressed in the central nervous system in humans and rodents (Gutknecht et al. 2009), whereas 5-HT is synthesized via *Tph1* mainly in the gastrointestinal system. The enzyme aromatic L-amino acid decarboxylase converts 5-hydroxytryptophan into serotonin to complete the two-step synthesis process (Figure 7.1). Only a small proportion of available plasma tryptophan is actually converted to serotonin; thus, the overall concentration of serotonin in the brain is low (Tagliamonte et al. 1973).

Serotonin is broken down by monoamine oxidase (MAO) into 5-hydroxyindoleacetic acid (5-HIAA) (Figure 7.1). Similar to TPH, MAO is encoded by two gene isoforms, *Mao-a* and *Mao-b*, which are coexpressed in serotonergic neurons during development (Vitalis et al. 2002). The MAO-B subtype predominates in serotonergic neurons later in life. Investigation of postmortem brain tissue, cerebrospinal fluid (CSF), platelets, and brain imaging studies has been used to detect variations in the key components of the serotonin system as they relate to depression. Evidence from these studies and their contributions to depression research are described below.

7.2.2 SEROTONIN RECEPTORS AND TRANSPORTERS

Once synthesized, serotonin is sequestered into presynaptic vesicles via vesicular monoamine transporters (VMAT1 and VMAT2), whereby its stimulated and/or constitutive release into the extracellular space produces extensive effects through interactions with a remarkably large class of receptors. The extracellular actions of serotonin are terminated primarily via uptake by the serotonin transporter (SERT; 5-HTT) (Figure 7.1). To date, there are 15 molecularly identified serotonin receptor subtypes categorized into seven major families termed 5-HT_1 (5-HT_{1A}, 5-HT_{1B}, 5-HT_{1D}, 5-HT_{1E}, and 5-HT_{1F}), 5-HT_2 (5-HT_{2A}, 5-HT_{2B}, 5-HT_{2C}), 5-HT_3 (5-HT_{3A}, 5-HT_{3B}), 5-HT_4, 5-HT_5 (5-HT_{5A}, 5-HT_{5B}), 5-HT_6, and 5-HT_7 (Hoyer et al. 2002). Serotonin receptors are known to have complexity at many different levels. For instance, 5-HT_{1B} and 5-HT_{1D} receptor subtypes, which have high sequence homology, were originally thought to occur exclusively in rats and mice or guinea pigs, pigs, dogs, cows, and primates, respectively. However, molecular cloning revealed that 5-HT_{1D} receptors in humans comprised two separate subtypes first named $5\text{-HT}_{1D\alpha}$ and $5\text{-HT}_{1D\beta}$ receptors. The discovery of the 5-HT1Dβ gene in humans prompted reclassification of 5-HT_{1B} receptors to include 5-HT_{1B} and $5\text{-HT}_{1D\beta}$ receptors, as these differ by only a single amino acid (Hannon and Hoyer 2008). The 5-HT_{1D} receptor subtype is now referred to as 5-HT_{1D} receptors across species. Another example of serotonin receptor heterogeneity involves the 5-HT_{2C} subtype, which was discovered to undergo RNA editing (adenosine → inosine) at multiple sites to produce several receptor proteins with different affinities for serotonin (Gurevich et al. 2002; Sanders-Bush et al. 2003). The presynaptic serotonin system is regulated by multiple autoreceptors including 5-HT_{1A}, 5-HT_{1B}, and 5-HT_{1D} receptors, which also function as heteroreceptors. The 5-HT_3 receptor family includes various combinations of 5-HT_{3A} and 5-HT_{3B} subtypes that form pentameric ligand-gated ion channels. All other subtypes of serotonin receptors function as G-protein-coupled receptors.

Serotonin interactions at 5-HT_1 and 5-HT_5 receptor families activate $G_\alpha i$ proteins, whereas 5-HT_4, 5-HT_6, and 5-HT_7 classes activate $G_\alpha s$ proteins to produce opposing effects to reduce or to increase, respectively, the activation of cyclic adenosine monophosphate (cAMP), which further influences downstream protein targets. The 5-HT_2 family of receptors is coupled to $G_\alpha q$ and regulates the activity of phospholipase C with the downstream consequences of protein kinase C phosphorylation and/or alteration of intracellular calcium levels. The large variety of receptor subtypes present in the serotonin system, in combination with their multiple G-protein-coupled pathways and overlapping regional and cellular expression patterns, highlights the complexity and the number of targets within the serotonin system that might be related to the etiology and/or pathophysiology of depression. Many of these serotonin receptors have not been fully characterized regarding localization and functional features. For receptors whose actions have been investigated in greater detail and for which functions have been identified, their involvement in depression and its treatment has been explored. The predominant focus here will be on members of the 5-HT$_1$ and 5-HT$_2$ serotonin receptor families (predominantly 5-HT$_{1A}$ and 5-HT$_{2A}$ subtypes) as these receptors have been extensively studied in the context of depression. Information on the mechanisms of additional serotonin receptors are given by Hensler (2006) and Hannon and Hoyer (2008).

The serotonin transporter is a central modulator of serotonergic neurotransmission. The SERT is a 12-transmembrane domain protein localized on soma and presynaptic terminals of serotonergic neurons, and also glia. SERT uses cotransport of Na$^+$ and Cl$^-$ and countertransport of K$^+$ to drive the clearance of serotonin from the extracellular space (Kanner and Schuldiner 1987). In addition to its role in modulating the pre- and postsynaptic effects of extracellular serotonin, SERT is also the principal target of widely prescribed classes of antidepressants, including the tricyclic antidepressants (TCAs), discovered in the 1960s, the selective serotonin reuptake inhibitors (SSRIs), and also the mixed serotonin and norepinephrine reuptake inhibitors (SNRIs) (White et al. 2005). Other important antidepressant targets within the serotonin system include the 5-HT$_1$ and 5-HT$_2$ classes of receptors, MAO-A and MAO-B, and TPH2; and these are also being investigated for their roles in the pathophysiology and etiology of MDD.

Serotonin is present in most brain regions in the central nervous system and in multiple peripheral organs and cell types. Serotonin and other constituents of the serotonin pathway appear early in embryonic development (during the first trimester in humans and higher-order primates and at day 10 of rodent gestation) and they continue to mature throughout early postnatal development (Levitt and Rakic 1982; Hendricks et al. 1999). In the brain, serotonergic neurons originate deep within the brain stem. Serotonin-containing neurons are anatomically grouped into caudal (B1–B5) and rostral (B6–B9) nuclei, with the former projecting to areas of the deep cerebellar

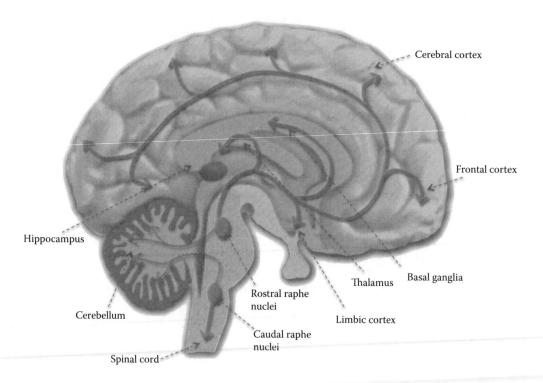

FIGURE 7.2 Serotonin projections in human brain. Anatomically, serotonin cell bodies are clustered in small groups called nuclei in the brain stem. From rostral raphe nuclei arise projections to many areas of cerebral cortex and midbrain, whereas projections from caudal raphe nuclei extend to the spinal cord and portions of cerebellum.

nuclei, cortex, and spinal cord, whereas the latter extend an axonal network throughout the fore-brain and cortices (Figure 7.2). Major targets of the rostral raphe nuclei include limbic regions (e.g., hippocampus and amygdala), hypothalamus, thalamus, and neocortex (Hornung 2003). A similar organization is observed in rodents and other mammals (Dahlstrom and Fuxe 1964). Of particular relevance to mechanisms of depression are projections to structural correlates of emotionality including the amygdala, prefrontal and cingulate cortices, hypothalamus, and thalamus (Figure 7.2). Recently, serotonin neurons have been classified based on genetic lineages. Specifically, serotonin neuronal progenitors can be subdivided into subsets defined by rhombomeres (r1, r2, r3, r5), which are discriminated by differing genetic cues (transcription factors) during development (Jensen et al. 2008).

The broad functions of serotonin in the central nervous system include its involvement in the modulation of aggression, sleep, appetite, mood, thermoregulation, and sexual function; however, serotonin levels in the brain account for only 10% of the total serotonin in the body (Berger et al. 2009). In the periphery, serotonin is predominantly found in enterochromaffin cells of the gut, where it functions in smooth muscle contraction and gut motility (Spiller 2008). Additionally, serotonin is taken up into platelets, where it is released to cause vasoconstriction and is involved in cardiovascular regulation (Ramage and Villalon 2008). Serotonin is also found in lymphocytes and monocytes and is associated with immune system function (Gordon and Barnes 2003). Peripheral tissues express SERT and multiple serotonin receptors, and thus possess many of the components of the central serotonin system (Berger et al. 2009). As will be discussed later, correlations between central and peripheral serotonin system components are important for advances in depression research, whereby peripheral cells might be used as diagnostic tools to study treatment responses in cases where direct brain measurements are not feasible.

7.3 THE MONOAMINE HYPOTHESIS OF DEPRESSION

7.3.1 DISCOVERY OF THE ROLE OF SEROTONIN

Early evidence for the involvement of monoamine neurotransmitters, including serotonin, in depression and other affective disorders arose serendipitously. Reserpine (*Rauwolfia serpentina*) was first used as an ayurvedic treatment for insanity, producing calming effects and lowering blood pressure (Sen and Bose 1931; Gupta et al. 1947; Vakil 1949). Wilkins and Judson (1953) confirmed the effects of reserpine to control hypertension. Later reports identified additional sedative properties of this drug, along with evidence reserpine might be useful for alleviating anxiety and symptoms of obsessive compulsive disorder (Kline 1954). By contrast, another study identified the depressogenic properties of reserpine. Here, five patients displayed depressive symptoms after receiving a chronic regimen of reserpine but had remission upon discontinuation of treatment (Freis 1954). Shortly thereafter, it was found that reserpine administration leads to a severe depletion of monoamine neurotransmitters in the mammalian brain (Holzbauer and Vogt 1956; Paasonen and Vogt 1956). Based on these findings, the monoamine hypothesis was formulated, which stated that decreases in monoamine neurotransmitters in the brain might be causative in depression.

Further support for the role of monoamines in depression occurred inadvertently via the discovery of two other drugs: iproniazid [a monoamine oxidase inhibitor (MAOI)] and imipramine (a monoamine uptake inhibitor). Clinical trials using iproniazid, a compound originally designed as an antibacterial agent, also reported antidepressant-like effects (Saunders et al. 1959). The mechanism of action of iproniazid was later shown to involve inhibition of MAO, which is responsible for the degradation of monoamine neurotransmitters (Davison 1957). These findings suggested that antidepressant effects could be achieved by increasing the concentrations of monoamines. Subsequently, the investigation of imipramine, a drug designed to mimic the effects of the antipsychotic agent chloropromazine, likewise revealed its antidepressant properties (Kuhn 1958). Axelrod and coworkers (1961) went on to show that imipramine increases monoamine concentrations by blocking the reuptake of serotonin and norepinephrine into presynaptic terminals. These early studies on iproniazid and imipramine together reinforced the monoamine hypothesis of depression.

Research carried out by Pollin and colleagues (1961) proposed that among the monoamine neurotransmitters (serotonin, dopamine, and norepinephrine), serotonin contributes foremost to the etiology of depression. While studying the effects of amino acids in schizophrenic patients, a paradigm was used whereby amino acid mixtures in the presence or absence of iproniazid were administered to potentiate levels of endogenous monoamines. These investigators observed that only tryptophan resulted in an elevation in mood in schizophrenic patients and these changes were altered after removal of tryptophan causing sudden increases in hostility and depression in some patients. Similar experiments in depressed patients receiving the MAO inhibitor tranylcypromine showed accelerated improvements when given in combination with tryptophan compared to tranylcypromine alone (Coppen et al. 1963). Together, these studies suggested that the key neurochemical disruption in depression and the therapeutic action of antidepressants are related to variations in serotonin levels.

7.3.2 Serotonin Metabolite Levels in Depression

Serotonin is broken down into 5-HIAA by the actions of MAO and aldehyde dehydrogenase (Figure 7.1). Thus, 5-HIAA has been used as an indirect marker for investigating alterations in serotonin neurotransmission. The concentration of 5-HIAA in CSF has been used primarily to make inferences about brain function because of its high concentration compared with that of serotonin, the accessibility of CSF, and the high degree of correlation between CSF and brain 5-HIAA levels in nonhuman primates (Eccleston et al. 1968) and humans (Stanley et al. 1985).

In initial studies examining 5-HIAA as an index of brain serotonin concentrations, decreases in CSF 5-HIAA levels were reported in psychiatric patients suffering from depressive psychosis compared to patients with neurological disorders without depression (Ashcroft and Sharman 1960). This finding prompted several subsequent studies aimed at measuring CSF 5-HIAA in patients diagnosed with psychiatric disorders to understand further the involvement of the serotonin system. However, the results of these studies were inconclusive, with findings ranging from significant decreases (Agren 1980a, 1980b) to a trend toward decreases (Papeschi and McClure 1971) and to no changes (Oreland et al. 1981) in 5-HIAA levels in depressed patients compared with healthy controls. Investigations in the relationship between CSF 5-HIAA levels and the severity of depression symptoms have also yielded conflicting results, with some studies observing significant associations (Peabody et al. 1987), whereas others did not detect significant correlations between CSF 5-HIAA levels and self-report scores on the Hamilton Depression Rating Scale (Papeschi and McClure 1971; Banki and Arató 1983).

Further studies have suggested a bimodal distribution of CSF 5-HIAA levels in depressed patients (Asberg et al. 1976a). In a longitudinal study, Träskman-Bendz and coworkers (1984) replicated this finding by observing a significant increase in 5-HIAA levels after recovery in a subset of depressed patients who exhibited lower CSF 5-HIAA concentrations before improvements. A subset of depressed patients with low 5-HIAA levels revealed a strong association between severity of depression, the amount of 5-HIAA in CSF (Asberg et al. 1976a), and increased vulnerability to suicide attempts (Asberg et al. 1976b). This was further supported by findings reporting a negative relationship between suicidal tendencies or seriousness of intent to harm oneself with CSF 5-HIAA levels in unipolar and bipolar depressed patients (Agren 1980a). Many studies in depressed individuals have reported low CSF 5-HIAA levels in patients who attempt suicide (Oreland et al. 1981; Edman et al. 1986; Jones et al. 1990; Nordström et al. 1994; Samuelsson et al. 2006) or have suicidal ideation (Lopez-Ibor et al. 1985; Mann et al. 1996) compared to depressed nonsuicidal patients or healthy controls. Furthermore, suicide attempters with no history of diagnosed depression also have been reported to have low CSF 5-HIAA levels compared with healthy controls (Träskman et al. 1981). However, some studies in patients with affective disorder have found no association between suicide attempt and low CSF 5-HIAA levels (Roy-Byrne et al. 1983; Secunda et al. 1986; Hou et al. 2006; Sullivan et al. 2006).

Collectively, the information obtained from these studies suggests that low CSF 5-HIAA levels might be a marker for decreased serotonin neurotransmission, and particularly as associated with suicide. However, how CSF 5-HIAA levels definitively pertain to MDD is as yet unclear. It appears as though low CSF 5-HIAA is linked to a higher probability for suicidal ideation or attempt, which is sometimes but not always comorbid with depression. The high variability in CSF levels in healthy subjects might contribute to the variability in the results of studies in this area (Kennedy et al. 2002). Furthermore, CSF levels can be affected by environmental factors such as daily diet or the time of day of sampling (Grimes et al. 2000; Kennedy et al. 2002).

7.3.3 Postmortem Studies of Brain Tissue

In comparison to CSF measurements, postmortem brain tissue provides direct access to central serotonin levels, in addition to other important serotonin system markers, including the expression of SERT and serotonin receptors. Early research on postmortem tissue serotonin and 5-HIAA levels in multiple brain regions of depressed and control subjects supported the monoamine hypothesis of depression. In depressed patients, significant reductions in serotonin were reported in the brain stem (Shaw et al. 1967; Pare et al. 1969; Lloyd et al. 1974; Birkmayer and Riederer 1975), nucleus accumbens, caudate, putamen, substantia nigra, and amygdala (Birkmayer and Riederer 1975). Moreover, this deficit in tissue serotonin content was not observed in brain tissue from remitted depressed cases, suggesting that treatment restored serotonin levels (Birkmayer and Riederer 1975). Similarly, decreases in tissue 5-HIAA levels were also observed in depressed patients (Pare et al. 1969; Birkmayer and Riederer 1975; Beskow et al. 1976). However, other reports of postmortem findings measuring neurochemicals in the brain failed to detect changes in serotonin (Bourne et al. 1968; Beskow et al. 1976; Gottfries et al. 1976) and 5-HIAA levels (Crow et al. 1984; Ferrier et al. 1986; McKeith et al. 1987) in depressed patients. The collection and processing of human postmortem tissues is lengthy (~6–12 h); and serotonin and other monoamines, which are highly susceptible to biological and chemical oxidation, might undergo substantial and variable degradation thereby contributing to these discrepancies (Roubein and Embree 1979; Siddiqui et al. 1990).

In addition to serotonin and 5-HIAA levels, SERT, which is the molecular target for a variety of antidepressants, has been investigated in postmortem human brain. Most of the studies have observed a significant decrease in SERT in depressed patients in many serotonergic-rich brain regions including the hippocampus (Perry et al. 1983), putamen (Lawrence et al. 1989), occipital cortex (Perry et al. 1983), frontal cortex (Leake et al. 1991), prefrontal cortex (Mann et al. 2000), and dorsal raphe nuclei (Arango et al. 2001). However, other studies have reported no significant alterations in SERT in the hippocampus, frontal cortex, midbrain (Little et al. 1997; Bligh-Glover et al. 2000), and prefrontal cortex (Thomas et al. 2006).

Decreases in 5-HT$_1$ receptor family expression were first observed in the frontal cortex of depressed patients (McKeith et al. 1987). In subsequent research focusing on 5-HT$_{1A}$ autoreceptor expression in the dorsal raphe nuclei, some studies report reductions in depressed patients (Drevets et al. 1999; Arango et al. 2001; Boldrini et al. 2008), whereas another report indicates an increase in 5-HT$_{1A}$ autoreceptor expression in this same brain region (Stockmeier et al. 1998). Discrepancies have been attributed to a reduced number of SERT-positive neurons in the dorsal raphe of depressed patients (Arango et al. 2001) and differences in brain tissue volume. In terms of postsynaptic 5-HT$_{1A}$ receptors in depressed patients, 5-HT$_{1A}$ receptor mRNA and protein expression are decreased in the hippocampus (Lowther et al. 1997; Lopez et al. 1998; Drevets et al. 1999). Variable results have been observed in the frontal cortex with one group reporting a trend toward an increase in postsynaptic 5-HT$_{1A}$ binding (Lowther et al. 1997), whereas others failed to observe differences in this brain region (Owen et al. 1983).

Alterations in the 5-HT$_2$ family of receptors in depressed patients were first reported by Stanley and coworkers (1983). They observed a significant increase in 5-HT$_2$ binding in the frontal cortex of depressed suicide victims. However, these results are in contrast to a similar study where no differences were found in 5-HT$_2$ binding in the frontal cortex (Owen et al. 1983). Moreover, in the case of suicide victims, reports were again divided in that some found a significant increase in 5-HT$_2$ receptor expression in the frontal cortex (Yates et al. 1990; Arango et al. 1992; Hrdina et al. 1993), whereas others failed to find similar changes (Cheetham et al. 1988). Investigations into other brain regions including the hippocampus and amygdala have also produced inconsistent findings. For example, in the amygdala, one study has reported an increase in 5 HT$_2$ receptor binding (Hrdina et al. 1993), whereas another showed no change (Cheetham et al. 1988) in depressed suicide victims. Cheetham and coworkers (1988) observed a trend toward increases in 5-HT$_2$ receptor binding in the frontal cortex and amygdala, which were accompanied by a decrease in 5-HT$_2$ binding in the hippocampus of violent suicide cases.

The bulk of postmortem studies have been carried out in suicide victims or elderly depressed patients. However, these groups of patients might not represent the general population of depressed subjects and instead are specific subsets with different serotonergic alterations compared to the majority of individuals with depression. Additionally, early studies did not take into account patient histories of antidepressant treatment. Only a few studies within the past decade have separated patients into antidepressant-exposed versus antidepressant-naive groups when considering changes in the expression of key components of the serotonin system, despite the observation that prior antidepressant exposure alters expression of these components (Messa et al. 2003; Parsey et al. 2006a; Hirvonen et al. 2008). Lastly, recent findings regarding associations between gene × environment interactions having an effect on the serotonin system warrant interpreting postmortem data with caution. Although it appears as though SERT expression in the dorsal raphe and serotonin and 5-HIAA levels in midbrain regions are reduced in depressed patients in a majority of studies, inconsistencies in these findings warrant further investigation while controlling for prior antidepressant history, substance abuse, suicidal ideation, and other factors.

7.3.4 Brain Imaging Studies

Recent advances in functional brain imaging methods including positron emission tomography (PET) and single photon emission computed tomography (SPECT), in addition to the design of radioactive ligands for use with these techniques, allow for noninvasive measurements of different aspects of the serotonin system to be carried out in vivo in humans. Currently, the targets that can be investigated using PET and SPECT include SERT, 5-HT$_1$ and 5-HT$_2$ receptor families, and serotonin synthesis (for review, see D'haenen et al. 1992; Zipursky et al. 2007).

In vivo visualization and measurement of SERT using PET and SPECT have revealed a significant decrease in SERT binding potential in the midbrain region in depressed patients (Malison et al. 1998; Cannon et al. 2006; Parsey et al. 2006a; Oquendo et al. 2007). However, information obtained from other brain regions has not resulted in straightforward findings. Some studies describe a decrease in SERT binding in the amygdala (Parsey et al. 2006a; Oquendo et al. 2007), hippocampus, thalamus, putamen, and anterior cingulate cortex (Oquendo et al. 2007) in depressed patients compared with healthy subjects. By contrast, in other studies, increases in SERT binding in the thalamus, dorsal cingulate cortex (Miller et al. 2009a), and medial frontal cortex have been reported in depressed patients (Cannon et al. 2006). Furthermore, subjects with a history of suicide attempt showed a pronounced reduction in SERT binding potential in the midbrain and an increase in SERT binding in the anterior cingulate cortex compared with both depressed patients without suicidal histories and healthy controls (Cannon et al. 2006). These results are in agreement with the hypothesis that individuals with suicidal tendencies might represent a specific subset of depressed patients characterized by serotonergic dysfunction. However, Parsey and coworkers (2006a) failed to find differences in SERT binding in any of the brain regions examined when comparing suicide attempters and nonattempters.

In an example of how environmental factors might influence serotonergic function, using PET imaging, Miller and colleagues (2009b) found that depressed patients with a history of childhood abuse displayed significant decreases in SERT binding in all brain regions, which were not present in depressed patients with no history of childhood abuse. Similar results were observed in rhesus monkeys, where decreased SERT binding in the brain stem, thalamus, and striatum was evident in peer-reared monkeys compared to mother-reared animals (Ichise et al. 2006). These findings suggest that environmental factors such as stress induced by abuse or maternal separation can modulate the serotonergic system and potentially manifest themselves as risk factors for developing depression.

Compared to investigations on the expression of SERT, results from studies of 5-HT_{1A} binding are more variable. Some groups have reported a global decrease in 5-HT_{1A} binding in antidepressant-naive depressed patients (Hirvonen et al. 2008), whereas others have observed a global increase in 5-HT_{1A} binding (Parsey et al. 2006c; Miller et al. 2009a), although in both cases, significant changes in 5-HT_{1A} binding potential were not pinpointed to specific brain regions. By contrast, 5-HT_{1A} binding has been reported to be reduced only in the dorsal raphe in elderly depressed patients compared to age-matched healthy control subjects (Meltzer et al. 2004). The relationship between 5-HT_{1A} binding potential and the severity of depression has also been taken into consideration whereby findings indicate a high correlation (Meltzer et al. 2004), although this has not been a consistent finding either (Hirvonen et al. 2008). Prior antidepressant treatment can also affect the results of imaging studies. Antidepressant-naive depressed patients show increased 5-HT_{1A} binding potential compared to healthy controls, but no changes were reported in depressed patients with a history of antidepressant exposure compared to healthy individuals (Parsey et al. 2006c; Miller et al. 2009a).

Initial PET imaging studies aimed at investigating 5-HT_2 receptors in depressed patients showed higher binding in the parietal cortex (D'haenen et al. 1992) and reduced binding in the right hemispherical region of the inferofrontal cortex (Biver et al. 1997). However, in another study of unmedicated depressed patients without a history of suicidal ideation, there were no changes in 5-HT_2 receptor binding (Meyer et al. 1999). When looking specifically at 5-HT_2 receptor subtypes, there are reports of decreased 5-HT_{2A} receptor binding in the hippocampus (Sheline et al. 2004), and frontal, occipital, and cingulate cortices (Messa et al. 2003) in antidepressant-naive depressed patients compared with controls, but similar findings were not observed in elderly depressed patients compared with age-matched healthy controls (Meltzer et al. 1999).

In vivo brain imaging techniques provide valuable information on changes in neuronal systems in relation to human disease and its treatment with the possibility for longitudinal studies. Nonetheless, the complexity of the serotonin system, most notably the heterogeneity of its receptor subtypes, necessitates the continued development of highly selective ligands. Additionally, single nucleotide polymorphisms (SNPs) and other common polymorphisms are rapidly being identified in serotonergic genes. This increasing genetic complexity will need to be considered in terms of its influence on the expression and function, including ligand binding, of in vivo imaging targets. A related issue is the fact that sample size is an increasingly important factor in the design and interpretation of depression-related studies, including brain imaging studies, and particularly in light of genetically heterogeneous human populations.

7.3.5 BLOOD PLATELET SEROTONERGIC MARKERS

Another common avenue to study the serotonin system in humans is through the use of blood cells, plasma, and serum. Platelets and lymphocytes express SERT, as well as a number of different types of serotonin receptors including 5-HT_{1A} receptors (Yang et al. 2006, 2007). SERT-mediated uptake of serotonin in platelets has been correlated with maximal uptake rates in brain in postmortem and SPECT imaging studies (Rausch et al. 2005; Uebelhack et al. 2006). In depressed patients, a number of studies have reported decreases in SERT expression in platelets using [^{3}H]imipramine binding (Briley et al. 1980; Paul et al. 1981; Ellis et al. 1990; Nemeroff et al. 1994); however, not all studies are in agreement and some have reported no changes in SERT expression in depressed individuals compared with control subjects (Gentsch et al. 1985; Lawrence et al. 1993). In terms of correlating the severity of depression with platelet SERT expression and function, consensus also has not been reached and results refuting (Kanof et al. 1987) or in support of this association are reported (Ellis et al. 1990; Fisar et al. 2008). Use of the SERT-selective ligand [^{3}H]paroxetine has also produced contrasting results (Kanof et al. 1987; D'haenen et al. 1988; Lawrence et al. 1993; Nemeroff et al. 1994). Additional factors hypothesized to affect the outcome of platelet studies include prior antidepressant exposure of patients, seasonal effects on SERT expression, genetic polymorphisms, and gender (Nemeroff et al. 1994; D'Hondt et al. 1996; Parsey et al. 2006c). A few studies have controlled for some of these factors, including seasonal effects (Kanof et al. 1987) and antidepressant exposure (Lawrence et al. 1993; Nemeroff et al. 1994), but none have controlled for all of these variables.

7.3.6 SEROTONIN DEPLETION STUDIES

Pharmacologic depletion of serotonin has been used as an approach to evaluate the hypothesis of the association between low serotonin levels and depressed mood. There are a number of steps in the serotonin synthesis pathway that can be manipulated; for example, decreasing the bioavailability of the serotonin precursor L-tryptophan or inhibiting the rate-limiting enzyme in serotonin synthesis. Factors to consider when using this type of paradigm include the relative specificity of drugs used to block these targets, the ethics of carrying out studies in humans (particularly patients), and interpreting behavioral changes. Despite these complexities, information has been obtained through pharmacologic studies that have contributed to our knowledge about the role of serotonin in mood disorders.

L-Tryptophan, the precursor for serotonin synthesis, is an essential dietary amino acid (Rose et al. 1954). Experimental protocols have been developed to reduce central levels of serotonin by decreasing plasma tryptophan levels, thereby depleting the amount of tryptophan available to cross the blood–brain barrier (Biggio et al. 1974). Brain tryptophan levels are dependent on plasma levels of other LNAAs as all compete to cross the blood–brain barrier using the same LNAA transporter (Hood et al. 2005). Reduced plasma tryptophan/LNAA ratios decrease the likelihood of tryptophan being transported into the brain. By orally administering an amino acid mixture to fasted subjects that contains large amounts of all LNAAs except tryptophan, it is possible to reduce plasma tryptophan levels, plasma tryptophan/LNAA ratios, brain serotonin synthesis as measured by PET, and CSF levels of 5-HIAA (Nishizawa et al. 1997; Carpenter et al. 1998; Klaassen et al. 1999; Moreno et al. 2000; Booij et al. 2003).

Recent reviews and meta-analyses highlight the reproducibility of results obtained using acute tryptophan depletion (ATD) methods to investigate the role of serotonin and other monoamines in depressive states (Booij et al. 2003; Fusar-Poli et al. 2006; Ruhé et al. 2007). The use of ATD does not appear to stimulate depressed mood in healthy individuals lacking an existing or previous diagnosis of depression nor does it increase the severity of depressed mood in patients clinically diagnosed with depression who are not currently receiving antidepressant treatment (Delgado et al. 1994; Neumeister et al. 1997; Murphy et al. 2002; Evers et al. 2005; Robinson and Sahakian 2009). However, ATD has been found to reinstate a transient depressed mood state in remitted patients with a prior diagnosis of MDD and who are actively receiving antidepressant therapy (Delgado et al. 1990; Moreno et al. 2006; Merens et al. 2008). Additional evidence suggests that female participants show greater responses to ATD compared to males as indicated by self-reports of affective measures. Furthermore, a family history, that is, first-degree relatives with MDD, serves as a risk factor for vulnerability to the depressogenic effects of ATD, irrespective of a clinical diagnosis of depression (Ellenbogen et al. 1996; Moreno et al. 2006). Moreover, in response to ATD, "*s*" allele carriers of the *5-HTTLPR SERT*-gene polymorphism (vide infra), particularly females, tend to be more vulnerable to relapse into a depressive state, and also to perform worse in motivational tasks and to have slower responses to antidepressants (Roiser et al. 2006; Walderhaug et al. 2007).

Animal models have also been used to investigate the effects of brain serotonin depletion on depression-related behavior. Degeneration of serotonergic neurons in the dorsal and medial raphe nuclei has been achieved using intracerebral administration of the neurotoxin 5,7-dihydroxytryptamine. The TPH inhibitor *p*-chlorophenylalanine also has been used to deplete serotonin levels, similar to the ATD paradigm in humans. These agents produce reductions in brain serotonin levels in rodents without affecting other monoamines (Lorens et al. 1976). However, decreased brain serotonin levels caused by neurotoxins or electrolytic lesion have not been associated with increases in behavioral despair, a depression-related endophenotype modeled in rats and mice by increases in immobility times in the forced swim test (Borsini 1995; Lieben et al. 2006). In contrast, neonatal rats administered 5,7-DHT displayed increases in behavioral despair in adulthood (Kostowski and Krzaścik 2003). The latter result supports the hypothesis that developmental disturbances in serotonin system function produce persistent changes in behavior in adulthood, and these ideas are developed below.

7.3.7 GENETICALLY ENGINEERED ANIMAL MODELS

In light of the difficulties associated with investigating the monoamine hypothesis of depression in humans, a number of animal models having targeted loss of expression of specific genes have been used and found to be particularly advantageous in understanding the role of serotonergic pathway components and their involvement in depression. Gardier et al. (2009) reviewed strategies to create knockout animal models for the study of neurochemical imbalances in the etiology of depression. Mice engineered to lack one or two functional copies of the *Sert* gene (SERT+/− and SERT−/−, respectively) (Bengel et al. 1998) possess a complex phenotype most often characterized by increased anxiety-like behavior that is dependent on background strain and environmental influences such as stress (Li et al. 1999; Tjurmina et al. 2002; Holmes et al. 2003; Joeyen-Waldorf et al. 2009). SERT-deficient mice also show increases in depression-like behaviors that are strain- and test-dependent and modest hypolocomotion (Holmes et al. 2002a; Kalueff et al. 2006). Constitutive reductions in SERT are associated with gene dose-dependent decreases in brain serotonin uptake rates (Montanez et al. 2003; Perez and Andrews 2005; Perez et al. 2006) and increases in extracellular serotonin levels (Fabre et al. 2000; Mathews et al. 2004) with the highest levels observed in SERT−/− mice. In addition, 5-HT$_{1A}$ and 5-HT$_{1B}$ receptor function is desensitized (Li et al. 1999; Fabre et al. 2000; Murphy and Lesch 2008). For recent review of the complex central and peripheral phenotypic changes associated with loss of SERT expression in mice, see Murphy et al. (2008). Collectively, findings in SERT-deficient mice have led to the hypothesis that increases in extracellular serotonin occurring throughout life, and particularly during early development, result in behavioral changes that are different from and sometimes opposite to those associated with antidepressant administration in adults (Murphy et al. 2004; Sibille and Lewis 2006; Murphy and Lesch 2008). Along the same lines, inactivating the genes encoding MAO, the enzyme that breaks down monoamine neurotransmitters including serotonin, might initially be predicted to result in an antidepressant effect. However, mice genetically lacking MAO-A are characterized by enhanced aggression (Cases et al. 1995), whereas MAO-B-deficient mice show exaggerated responses to stress (Grimsby et al. 1997).

Altering other serotonergic genes in mice to investigate their roles in depression has produced valuable insights. As discussed above, there are two isoforms of TPH produced from two separate genes, termed *TPH1* and *TPH2*, with the latter being the predominant brain isoform (Gutknecht et al. 2009). Mice lacking functional copies of TPH1 (TPH1−/− mice) show reduced serotonin levels restricted to the periphery (Côte et al. 2003; Savelieva et al. 2008) and deficits in the normal functioning of peripheral organs (Trowbridge et al. 2010). The discovery of TPH2 occurred more recently (Walther et al. 2003), and the initial report of behavioral changes in TPH2−/− mice, although somewhat inconclusive, suggests contrasting changes in depression-like behaviors across tests of behavioral despair that are more prominent in mice lacking both isoforms of TPH (Savelieva et al. 2008). In mice where the gene encoding the 5-HT$_{1A}$ receptor (*Htr1A*) has been deleted, the primary phenotype is one of enhanced anxiety and stress responsiveness, similar to that found in SERT–/– mice (Heisler et al. 1998; Parks et al. 1998; Ramboz et al. 1998). Interestingly, a strategy to restore 5-HT$_{1A}$ function in postsynaptic neurons in brain regions involved in emotional regulation (hippocampus, frontal cortex) reversed the increased anxiety-like phenotype associated with constitutive 5-HT$_{1A}$ receptor loss suggesting that postsynaptic 5-HT$_{1A}$ receptors modulate anxiety and that these effects occur during specific developmental time frames (Gross et al. 2002). Additional serotonin receptor subtypes have been investigated using knockout strategies producing a range of behavioral phenotypes, although uncertainty remains about their potential roles in depression (Gaspar et al. 2003).

7.3.8 GENETIC AND ENVIRONMENTAL INTERACTIONS

Genetic factors can increase predisposition to depression by as much as 40% to 50% (Fava and Kendler 2000). However, understanding the contribution of different genes is complicated by many factors including the high degree of heterogeneity associated with the pathophysiology of depression, the involvement of multiple neurotransmitter systems and brain regions, the growing multitude of genetic variants being identified and the small effects that most of these variants contribute, and a variety of possible environmental influences (Burmeister 1999; Caspi et al. 2010). Despite these limitations, genes that code for important regulators of the serotonin system have been investigated to attempt to identify the hereditary influences that might increase vulnerability to depression.

In depression, the most highly investigated gene variants occur in the SERT (*SLC6A4*) gene, and in particular, include the serotonin transporter–linked polymorphic region (*5-HTTLPR*) first reported by Heils and coworkers (1996). This polymorphism occurs only in humans and higher-order primates (Lesch et al. 1997). It is hypothesized that the short "*s*" allele is associated with reduced transcriptional efficiency and thus decreases in SERT expression that manifest as decreased serotonin transport (Lesch et al. 1996; Singh et al. 2010). Furthermore, there is a correlation between stress-associated depression and increased neuroticism in carriers of the "*s*" allele (Caspi et al. 2003; Uher and McGuffin 2008), although this has not always been consistently reported (Flory et al. 1999). Discrepancies might arise as a result of the genetic complexity of this polymorphic region, and additional functional allelic variants have been described (Hu et al. 2005; Wendland et al. 2008). For example, an SNP that results in a base-pair difference in the repeat elements, termed *LA* or *LG*, has been discovered near the *5-HTTLPR*. The *LG* variant is functionally similar to the low-expressing "*s*" allele and is likewise linked to stress-induced depression (Zalsman et al. 2006).

Brain imaging studies using PET or SPECT are associated with contrasting results regarding the association between SERT binding and *5-HTTLPR* genotype (Heinz et al. 2000; Willeit et al. 2001; van Dyck et al. 2004; Parsey et al. 2006b; Murthy et al. 2010). Brain imaging might be poor at detecting modest changes in SERT binding for a number of reasons including low sensitivity, small sample sizes, and the presence of other polymorphisms and SNPs within the *SERT* gene.

Ongoing studies provide further evidence for the complexities of the *SERT* gene and the genetics of depression. Recent genome-wide association studies from the Sequenced Treatment Alternatives to Relieve Depression (STAR*D) trial (described below) have sought to identify key genetic components involved in the etiology and treatment of depression, although definitive results have not been obtained because of the involvement of multiple genetic factors (alone or in combination) that may manifest in heterogeneous diseases such as depression (Garriock and Hamilton 2009; Laje et al. 2009).

7.3.9 SEROTONIN FUNCTION DURING DEVELOPMENT

In addition to its role as a neurotransmitter, serotonin is important in driving the development of neuronal circuitry and associated phenotypes. As discussed above, serotonin and markers for serotonergic neurons are expressed beginning in early embryonic development (Gaspar et al. 2003).

Additionally, some neurons express SERT only during brief periods of early postnatal life, although these particular neurons do not continue to express SERT or become serotonergic neurons in the adult brain (Lebrand et al. 1996; Verney et al. 2002). Transient expression of SERT in rodents is limited to sensory thalamic neurons (Lebrand et al. 1996), and neurons in the hippocampus (Lebrand et al. 1998a), anterior cingulate cortex (Lebrand et al. 1998a), and the superior olive (Cases et al. 1998). In humans, transient SERT is reported to occur in thalamocortical neurons, although it has been more difficult to ascertain additional subsets of neuronal populations expressing SERT during development in human tissues (Verney et al. 2002). The machinery in neurons with a transient partial serotonergic phenotype is limited to expression of SERT and VMAT (Lebrand et al. 1998b). Thus, although these neurons lack the capability to synthesize serotonin, serotonin uptake and release at these sites might have a morphogenic effect (Lebrand et al. 1996). Studies to identify a role for SERT during embryonic stages through the first few weeks of postnatal life have revealed that transient SERT expression is necessary for the development of the barrel field cortex in rodents (Vitalis et al. 1998). Other roles for developmentally sensitive SERT expression have yet to be fully identified, but there is evidence to suggest that disruptions in SERT during key windows of development might be important in mediating persistent adulthood phenotypes.

A critical window (the first few weeks of postnatal development in rodents) during which the serotonin system can be disrupted via the administration of antidepressants has been associated with changes in depression-related behaviors in adulthood. Clomipramine administered to rats during early life is correlated with reduced sexual activity (Neill et al. 1990; Vogel et al. 1990), increased rapid eye movement sleep (Mirmiran et al. 1981), and increased immobility in tests of behavioral despair (Velazquez-Moctezuma and Diaz Ruiz 1992). In mice, administration of the SSRIs fluoxetine or escitalopram during early postnatal development produces adult phenotypes characterized by reduced exploratory locomotor behavior (Ansorge et al. 2004; Ansorge et al. 2008; Popa et al. 2008), reduced sucrose consumption (a measure of anhedonia) (Popa et al. 2008), and increased immobility in behavioral despair tests (Popa et al. 2008). This early influence of altered serotonin reuptake on adult mood and anxiety-related behaviors in rodents might be used to further our understanding of how the serotonin system shapes affect-associated brain circuitry in development. Research in this area also has implications for pregnant women taking SSRIs and for establishing recommendations for treating childhood and adolescent mood and anxiety disorders with serotonergic drugs. For example, clinically relevant drug levels have been detected in the amniotic fluid and the umbilical cords of pregnant women treated with SSRIs (Hostetter et al. 2000; Hendrick et al. 2003b), and newborns display SSRI withdrawal effects and low birth weights (Hendrick et al. 2003a; Sanz et al. 2005). Some motor coordination effects have also been reported in young children (Casper et al. 2003), although it is not yet clear how human behavior might be influenced by exposure to prenatal antidepressants and also by untreated maternal depression. Maternal serotonin has recently been suggested as a necessary component for morphogenesis as embryos arising from TPH1-deficient mice show developmental deficiencies (Côté et al. 2007). The effects of low serotonin levels on pups from mothers lacking the *TPH2* gene are as yet unknown. The developmental influence of serotonin is a growing avenue of investigation in the field of depression research that has the possibility to elucidate susceptibility factors and developmental pathways that predispose individuals to MDD.

7.4 TREATMENT STRATEGIES FOR DEPRESSION

7.4.1 OVERVIEW OF ANTIDEPRESSANTS

Agents developed decades ago, including MAOIs (e.g., selegiline, moclobimide) and TCAs (e.g., imipramine, amitriptyline), remain effective for the treatment of MDD. However, newer classes of drugs, including those that have been specifically designed to target the serotonin system, are now the most widely prescribed antidepressants. Recently developed drug classes include SSRIs (e.g., fluoxetine, fluvoxamine, paroxetine, citalopram), mixed SNRIs (e.g., venlafaxine, duloxetine), norepinephrine reuptake inhibitors (e.g., reboxetine, mazindol), and drugs that have sites of action at both serotonin receptors and uptake transporters (e.g., trazodone, nefazodone) (Ghanbari et al. 2010). The ongoing search for novel classes of antidepressants is driven by efforts to decrease side effects and to increase therapeutic effectiveness, partly by reducing the delay in the onset of anti-depressant effects and also with the hope of treating higher percentages of patients, including those who are resistant to current therapeutic approaches. Compared to SSRIs and SNRIs, the use of MAOIs and TCAs is associated with more serious adverse side effects and potential toxicity caused by interactions with other transmitter systems, such as cholinergic and histaminergic receptors in the case of TCAs (Jefferson 1975; Remick et al. 1989) and hypertensive crisis associated with the effects of dietary biogenic amines in the case of MAOIs. Reduced side effect profiles of SSRIs and SNRIs result in better compliance, but not necessarily increased percentages of response and remis-sion compared to older antidepressants.

Most antidepressant drugs have a primary mechanism of action to inhibit the uptake of mono-amine neurotransmitters, primarily serotonin, but also norepinephrine and dopamine, or to prevent monoamine catabolism. Thus, all current antidepressants are hypothesized to produce elevated syn-aptic neurotransmitter levels, which are then available to elicit further homeostatic modifications in downstream targets (Langer and Schoemaker 1988). In contrast to this prevailing hypothesis of antidepressant mechanism of action, we found that half of the approximately 50 studies that have been carried out using in vivo microdialysis fail to report increases in extracellular serotonin levels in response to chronic antidepressant administration in mice and rats (Andrews 2009; Luellen et al. 2010). Electrophysiological studies have shown that acute SSRI treatment leads to activation of autoreceptors to reduce serotonergic firing rates, whereas chronic SSRI administration results in desensitization of autoreceptors, which allows normalization of the rates of serotonergic firing (Blier et al. 1986; Chaput et al. 1986a; Pineyro and Blier 1999). These data, along with many other studies, highlight the need to continue investigations into the long-term molecular mechanisms of antidepressant medications. Moreover, despite the reductions in side effects associated with newer classes of antidepressants, treatment with all existing antidepressants continues to be characterized by two major problems: (1) prolonged treatment from weeks to months is required to observe clini-cal effects; and (2) a large fraction of patients suffering from major depression, recently estimated to be as high as 50%, fail to respond or respond poorly to current antidepressant therapies (Rush 2007; Sinyor et al. 2010).

7.4.2 Mechanisms for Delayed Therapeutic Responses to Antidepressant Treatment

Although it is generally accepted that SSRIs and many TCAs have a primary mechanism of action associated with their ability to inhibit SERT, adaptive changes in brain function that develop with continued administration accounting for delayed efficacy are still not well understood, particularly those occurring beyond the serotonin system. However, many studies point to a variety of pre-synaptic and postsynaptic mechanisms that are evoked after prolonged antidepressant administra-tion. The combination of SSRIs and $5-HT_{1A}$ antagonists, such as WAY100635, reportedly hastens or potentiates changes in serotonin neurotransmission by inhibiting the negative feedback inher-ent in $5-HT_{1A}$ autoreceptor function (Hjorth et al. 2000), and the enhanced serotonergic tone that develops more quickly might be correlated with faster clinical benefits (Richardson-Jones et al. 2010). Alternatively, it might also be that $5-HT_{1A}$ receptor desensitization is one of the mechanisms responsible for producing therapeutic effects independent of extracellular serotonin levels (Popa et al. 2010). Research in rodents has established that chronic treatment with SSRIs desensitizes $5-HT_{1A}$ autoreceptor function (Chaput et al. 1986b; Kreiss and Lucki 1995; Pineyro and Blier 1999; Newman et al. 2004). Further investigation into compensatory changes has revealed that whereas fluoxetine-induced $5-HT_{1A}$ desensitization affects G-protein function (Hensler 2002; Pejchal et al. 2002), the same is not true for all SSRIs (Rossi et al. 2008).

Presynaptic 5-HT$_{1A}$ receptors have been studied with regard to antidepressant mechanisms owing to their purported role in regulating serotonin neurotransmission. However, postsynaptic 5-HT$_{1A}$ receptors have also been investigated, and changes in these receptors are better correlated with the mechanism of action of TCAs compared with SSRIs (Chaput et al. 1991; Rossi et al. 2006). Given that most of the serotonin receptors operate through G-protein-coupled mechanisms, there are many downstream molecular targets associated with postsynaptic receptors that might be modulated in response to chronic antidepressant administration. For example, the enhanced production of cAMP and the recruitment of the transcription factor cAMP response element-binding protein have been suggested to alter levels of brain-derived neurotrophic factor (BDNF) (Tanis et al. 2007). In human suicide victims diagnosed with depression, Dwivedi et al. (2003) reported a decrease in BDNF mRNA and protein levels in postmortem hippocampus and frontal cortex. Regarding peripheral BDNF levels, two recent meta-analyses conclude that serum BDNF protein levels in unmedicated patients with depression are subnormal, and recovery of serum BDNF occurs after 8 to 12 weeks of antidepressant therapy (Brunoni et al. 2008; Sen et al. 2008). These data suggest that increases in central and/or peripheral BDNF levels might be an important long-term therapeutic modification resulting from antidepressant treatment. Other postsynaptic alterations (e.g., changes in the expression, function, and/or activation of other serotonin receptor subtypes and respective signaling cascades) are also likely to play a role in delayed treatment efficacy.

7.4.3 NONRESPONDERS TO CURRENT ANTIDEPRESSANTS

In an effort to delineate the heterogeneity of antidepressant responsiveness among individuals suffering from depression, pharmacogenetic studies have been carried out seeking relationships between gene polymorphisms and variable responses to antidepressant treatment. Early studies showed promise in associating *5-HTTLPR* short "*s*" allele carriers with poor/slower antidepressant response. However, some recent studies have failed to replicate these associations (for review, see Table 1 of Kraft et al. 2007). Contradictory findings have been attributed to a number of factors, not the least of which is underpowered sample sizes. Nonetheless, mixed reports continue to emerge from the large STAR*D trial involving 2876 patients. For example, Kraft et al. genotyped STAR*D participants for transcriptionally active *5-HTTLPR* variants, as well as eight marker SNPs, but found no significant single locus or haplotype associations with response to citalopram (Kraft et al. 2007). Hu et al. (2007) examined associations between *5-HTTLPR* genotypes using a triallelic *5-HTTLPR* classification and similarly reported a lack of significant association with citalopram response. However, a weak association between *5-HTTLPR* genotype and remission was reported in white non-Hispanic subjects (Mrazek et al. 2009). Discrepancies arising from the STAR*D studies are hypothesized to involve different definitions of remission, considerations (or lack thereof) of ethnicity, and other subtle differences in study design and interpretation (Garriock and Hamilton 2009; Mrazek et al. 2009). Overall, findings from pharmacogenetic studies imply that common genetic polymorphisms in the *SERT* gene appear only modestly associated, at best, with SSRI responses in patients with depression.

Multiple factors, many of which remain unidentified, influence SERT expression (Lipsky et al. 2009). In human lymphoblasts, SERT mRNA levels vary by 5- to 10-fold even when controlling for triallelic genotype with only 8% of the variance arising from measurement considerations (Hu et al. 2006). In the postmortem brain, individuals segregated by *5-HTTLPR* genotype showed 10-fold differences in SERT mRNA and SERT binding levels in midbrain regions (Little et al. 1998) but not in the hippocampus (Naylor et al. 1998). Thus, genome-wide polymorphisms, in addition to those identified in the *SERT* gene, likely contribute to individual variations in SERT expression. Furthermore, nongenetic influences, for example, epigenetic factors (Philibert et al. 2008), posttranslational modifications, and plasma membrane localization, are expected to contribute to additional variability in SERT expression and function (Blakely et al. 2005; Jayanthi and Ramamoorthy 2005; Rudnick 2006; Iceta et al. 2008). Thus, identifying and accounting for all factors in terms of their combined potential to influence SERT and relationships to antidepressant response will require a substantially greater effort.

Nonetheless, variability in the expression and, ultimately, the function of SERT, which is the proximal molecular target for SSRIs, is expected to impact treatment effectiveness. Reduced SERT function during treatment might prevent adequate increases in extracellular serotonin levels from developing in response to SSRIs. Moreover, decreased SERT function beginning early in life has the potential to alter the development of pre- and postsynaptic targets (Hariri et al. 2002; Pezawas et al. 2005; Sibille and Lewis 2006; Murphy and Lesch 2008) such that the associated circuitry is less responsive to increases in extracellular serotonin caused by SSRIs in adulthood. Preclinical studies in mice and rats with intermediate and complete constitutive reductions in SERT expression support these hypotheses (Holmes et al. 2002b; Kim et al. 2005; Ansorge et al. 2008; Popa et al. 2008).

Additional genetic variations that might contribute to differential responses to antidepressants include those associated with the gene encoding TPH2. An SNP has been identified in rodents whereby the amino acid proline is substituted with arginine (Pro447Arg; C1437G). This *Tph2* SNP is associated with a reduction in enzyme activity and central serotonin content (Zhang et al. 2004; Sakowski et al. 2006). Differential responses to the SSRI citalopram in mice were accounted for by allelic variations across mouse strains at this locus in the *Tph2* gene (Cervo et al. 2005). However, evidence linking a similar *Tph2* SNP to major depression and antidepressant responses in humans (Zhang et al. 2005) has been hindered by the rarity of this SNP and inconsistent identification across sample populations (Bicalho et al. 2006; Delorme et al. 2006; Garriock and Hamilton 2009; Zhang et al. 2010).

7.5 FUTURE PROSPECTS

From the discovery of serotonin in 1937 until today, numerous lines of investigation have revolved around identifying the peripheral and central nervous system functions of this evolutionarily ancient signaling molecule, but notably as they pertain to the origins and treatment of human MDD. Taken on the whole, the studies discussed here, which necessarily represent only a subset of the work carried out in this area of research, strongly suggest that serotonin plays an important role in the etiology and pathophysiology of depression, despite the inability to pinpoint precise serotonergic pathways. Many studies in depressed patients describe alterations in one or more aspects of the serotonin system, with SERT playing a central role. However, the high degree of variability in the results of studies attempting to correlate various serotonergic markers with depression likely reflects the heterogeneous nature of the illness and of human populations. Indeed, as depression is currently diagnosed based on sets of unrelated symptoms, which do not necessarily correspond with underlying biological mechanisms, it remains likely that human studies will continue to be confounded by biological heterogeneity.

The SSRIs and other newer antidepressants continue to represent drugs of choice for the treatment of depression because they improve symptoms for many people while having low side effect profiles and a large safety margin. However, current antidepressant therapies, regardless of drug class, are also associated with a high rate of inadequate response and medication intolerance. For many patients, it takes months of treatment and multiple trials of different drugs and/or combinations of drugs before a satisfactory outcome is attained. Even the STAR*D trial, the largest of its kind to date, which sought to optimize treatment strategies, was unable to produce high response rates. Thus, improving therapeutic approaches remains a high priority in depression research. Elucidating the downstream mechanisms associated with chronic administration of current antidepressants will lead to a better understanding of the multiple etiopathological origins of depression and to the identification of new targets for the development of future antidepressant therapies having improved efficacies and side effect profiles.

Although a great deal of knowledge has been gained in the past decade related to the neurobiology of depression, our understanding of underlying causes, pathophysiology, and current antidepressant mechanisms is still lacking. The search for potential biomarkers to guide treatment selection is a promising new avenue of research that will help to improve the clinical efficacy of current and future drugs used to treat depression. Investigations focused on biomarkers of brain structure and function have yielded fruitful results in terms of making associations with features of depression (e.g., reduced gray matter in the hippocampus and anterior cingulate cortex) and for tracking early treatment responsiveness (Leuchter et al. 2009), but they are not predictive such that they might guide individualized therapeutic approaches (McKinnon et al. 2009; Leuchter et al. 2010). Furthermore, gene expression studies have identified candidate genes expressed in the amygdala that might be used as biomarkers for depression (Sibille et al. 2009). Alternately, peripheral blood cells such as lymphocytes and platelets that express serotonergic components whose expression and function have been suggested to be correlated with serotonergic function in the brain have the potential to allow for genetic screening (Blakely 2001) and for investigation of the effects of drugs ex vivo.

ACKNOWLEDGMENTS

The authors acknowledge Dr. Andrew Leuchter (UCLA) for helpful discussions on antidepressant therapy and Dr. Beatriz Campo Fernandez (UCLA) for helpful advice regarding conceptual aspects of this work. The authors also acknowledge Scully7491 (http://scully7491.deviantart.com/) for freely providing the Photoshop brush that was used to design Figure 7.2. Support from NARSAD and the National Institute of Mental Health (MH086108 and MH064756) is gratefully acknowledged. The content is solely the responsibility of the authors and does not necessarily represent the official views of the National Institute of Mental Health or the National Institutes of Health.

REFERENCES

Agren, H. 1980a. Symptom patterns in unipolar and bipolar depression correlating with monoamine metabolites in the cerebrospinal fluid: II. Suicide. *Psychiatry Res* 3:225–36.

Agren, H. 1980b. Symptom patterns in unipolar and bipolar depression correlating with monoamine metabolites in the cerebrospinal fluid: I. General patterns. *Psychiatry Res* 3:211–23.

Amin, A. H., Crawford, T. B., and Gaddum, J. H. 1954. The distribution of substance P and 5-hydroxytryptamine in the central nervous system of the dog. *J Physiol (Lond)* 126:596–618.

Andrews, A. M. 2009. Does chronic antidepressant treatment increase extracellular serotonin? *Front Neurosci* 3:246–7.

Ansorge, M. S., Zhou, M., Lira, A., Hen, R., and Gingrich, J. A. 2004. Early-life blockade of the 5-HT transporter alters emotional behavior in adult mice. *Science* 306:879–81.

Ansorge, M. S., Morelli, E., and Gingrich, J. A. 2008. Inhibition of serotonin but not norepinephrine transport during development produces delayed, persistent perturbations of emotional behaviors in mice. *J Neurosci* 28:199–207.

Arango, V., Underwood, M. D., and Mann, J. J. 1992. Alterations in monoamine receptors in the brain of suicide victims. *J Clin Psychopharmacol* 12:8S–12S.

Arango, V., Underwood, M. D., Boldrini, M., et al. 2001. Serotonin 1A receptors, serotonin transporter binding and serotonin transporter mRNA expression in the brainstem of depressed suicide victims. *Neuropsychopharmacology* 25:892–903.

Asberg, M., Thorén, P., Träskman, L., Bertilsson, L., and Ringberger, V. 1976a. "Serotonin depression"—a biochemical subgroup within the affective disorders? *Science* 191:478–80.

Asberg, M., Träskman, L., and Thorén, P. 1976b. 5-HIAA in the cerebrospinal fluid. A biochemical suicide predictor? *Arch Gen Psychiatry* 33:1193–7.

Ashcroft, G. W., and Sharman, D. F. 1960. 5-Hydroxyindoles in human cerebrospinal fluids. *Nature* 186: 1050–1.

Axelrod, J., Whitby, L. G., and Hertting, G. 1961. Effect of psychotropic drugs on the uptake of H3-norepinephrine by tissues. *Science* 133:383–4.

Banki, C. M., and Arató, M. 1983. Amine metabolites and neuroendocrine responses related to depression and suicide. *J Affect Disord* 5:223–32.

Bengel, D., Murphy, D. L., Andrews, A. M., et al. 1998. Altered brain serotonin homeostasis and locomotor insensitivity to 3,4-methylenedioxymethamphetamine ("Ecstasy") in serotonin transporter–deficient mice. *Mol Pharmacol* 53:649–55.

Berger, M., Gray, J. A., and Roth, B. L. 2009. The expanded biology of serotonin. *Annu Rev Med* 60:355–66.

Beskow, J., Gottfries, C. G., Roos, B. E., and Winblad, B. 1976. Determination of monoamine and monoamine metabolites in the human brain: Post mortem studies in a group of suicides and in a control group. *Acta Psychiatr Scand* 53:7–20.

Bicalho, M. A., Pimenta, G. J., Neves, F. S., et al. 2006. Genotyping of the G1463A (Arg441His) TPH2 polymorphism in a geriatric population of patients with major depression. *Mol Psychiatry* 11:799–800.

Biggio, G., Fadda, F., Fanni, P., Tagliamonte, A., and Gessa, G. L. 1974. Rapid depletion of serum tryptophan, brain tryptophan, serotonin and 5-hydroxyindoleacetic acid by a tryptophan-free diet. *Life Sci* 14:1321–9.

Birkmayer, W., and Riederer, P. 1975. Biochemical post-mortem findings in depressed patients. *J Neural Transm* 37:95–109.

Biver, F., Wikler, D., Lotstra, F., Damhaut, P., Goldman, S., and Mendlewicz, J. 1997. Serotonin 5-HT2 receptor imaging in major depression: Focal changes in orbito-insular cortex. *Br J Psychiatry* 171:444–8.

Blakely, R. D. 2001. Physiological genomics of antidepressant targets: Keeping the periphery in mind. *J Neurosci* 21:8319–23.

Blakely, R. D., Defelice, L. J., and Galli, A. 2005. Biogenic amine neurotransmitter transporters: Just when you thought you knew them. *Physiology (Bethesda)* 20:225–31.

Blier, P., De Montigny, C., and Azzaro, A. J. 1986. Modification of serotonergic and noradrenergic neurotransmissions by repeated administration of monoamine oxidase inhibitors: Electrophysiological studies in the rat central nervous system. *J Pharmacol Exp Ther* 237:987–94.

Bligh-Glover, W., Kolli, T. N., Shapiro-Kulnane, L., et al. 2000. The serotonin transporter in the midbrain of suicide victims with major depression. *Biol Psychiatry* 47:1015–24.

Boldrini, M., Underwood, M. D., Mann, J. J., and Arango, V. 2008. Serotonin-1A autoreceptor binding in the dorsal raphe nucleus of depressed suicides. *J Psychiatr Res* 42:433–42.

Booij, L., Van der Does, A. J. W., and Riedel, W. J. 2003. Monoamine depletion in psychiatric and healthy populations: Review. *Mol Psychiatry* 8:951–73.

Borsini, F. 1995. Role of the serotonergic system in the forced swimming test. *Neurosci Biobehav Rev* 19:377–95.

Bourne, H. R., Bunney, W. E., Colburn, R. W., et al. 1968. Noradrenaline, 5-hydroxytryptamine, and 5-hydroxyindoleacetic acid in hindbrains of suicidal patients. *Lancet* 292:805–8.

Briley, M. S., Langer, S. Z., Raisman, R., Sechter, D., and Zarifian, E. 1980. Tritiated imipramine binding sites are decreased in platelets of untreated depressed patients. *Science* 209:303–5.

Brunoni, A. R., Lopes, M., and Fregni, F. 2008. A systematic review and meta-analysis of clinical studies on major depression and BDNF levels: Implications for the role of neuroplasticity in depression. *Int J Neuropsychopharmacol* 11:1169–80.

Burmeister, M. 1999. Basic concepts in the study of diseases with complex genetics. *Biol Psychiatry* 45:522–32.

Cannon, D. M., Ichise, M., Fromm, S. J., et al. 2006. Serotonin transporter binding in bipolar disorder assessed using [^{11}C]DASB and positron emission tomography. *Biol Psychiatry* 60:207–17.

Carpenter, L. L., Anderson, G. M., Pelton, G. H., et al. 1998. Tryptophan depletion during continuous CSF sampling in healthy human subjects. *Neuropsychopharmacology* 19:26–35.

Cases, O., Seif, I., Grimsby, J., et al. 1995. Aggressive behavior and altered amounts of brain serotonin and norepinephrine in mice lacking MAOA. *Science* 268:1763–6.

Cases, O., Lebrand, C., Giros, B., et al. 1998. Plasma membrane transporters of serotonin, dopamine, and norepinephrine mediate serotonin accumulation in atypical locations in the developing brain of monoamine oxidase A knock-outs. *J Neurosci* 18:6914–27.

Casper, R. C., Fleisher, B. E., Lee-Ancajas, et al. 2003. Follow-up of children of depressed mothers exposed or not exposed to antidepressant drugs during pregnancy. *J Pediatr* 142:402–8.

Caspi, A., Sugden, K., Moffitt, T. E., et al. 2003. Influence of life stress on depression: Moderation by a polymorphism in the 5-HTT gene. *Science* 301(5631):386–9.

Caspi, A., Hariri, A. R., Holmes, A., Uher, R., and Moffitt, T. E. 2010. Genetic sensitivity to the environment: The case of the serotonin transporter gene and its implications for studying complex diseases and traits. *Am J Psychiatry* 167:509–27.

Cervo, L., Canetta, A., Calcagno, E., et al. 2005. Genotype-dependent activity of tryptophan hydroxylase-2 determines the response to citalopram in a mouse model of depression. *J Neurosci* 25:8165–72.

Chaput, Y., Blier, P., and de Montigny, C. 1986a. In vivo electrophysiological evidence for the regulatory role of autoreceptors on serotonergic terminals. *J Neurosci* 6:2796–801.

Chaput, Y., de Montigny, C., and Blier, P. 1986b. Effects of a selective 5-HT reuptake blocker, citalopram, on the sensitivity of 5-HT autoreceptors: Electrophysiological studies in the rat brain. *Naunyn Schmiedebergs Arch Pharmacol* 333:342–8.

Chaput, Y., de Montigny, C., and Blier, P. 1991. Presynaptic and postsynaptic modifications of the serotonin system by long-term administration of antidepressant treatments. An in vivo electrophysiologic study in the rat. *Neuropsychopharmacology* 5:219–29.

Cheetham, S. C., Crompton, M. R., Katona, C. L., and Horton, R. W. 1988. Brain 5-HT2 receptor binding sites in depressed suicide victims. *Brain Res* 443:272–80.

Coppen, A., Shaw, D. M., and Farrell, J. P. 1963. Potentiation of the antidepressive effect of a monoamine-oxidase inhibitor by tryptophan. *Lancet* 281:79–81.

Côte, F., Thevenot, E., Fligny, C., et al. 2003. Disruption of the nonneuronal tph1 gene demonstrates the importance of peripheral serotonin in cardiac function. *Proc Natl Acad Sci USA* 100:13525–30.

Côté, F., Fligny, C., Bayard, E., et al. 2007. Maternal serotonin is crucial for murine embryonic development. *Proc Natl Acad Sci USA* 104:329–34.

Crow, T. J., Cross, A. J., Cooper, S. J., et al. 1984. Neurotransmitter receptors and monoamine metabolites in the brains of patients with Alzheimer-type dementia and depression, and suicides. *Neuropharmacology* 23:1561–9.

D'haenen, H., De Waele, M., and Leysen, J. E. 1988. Platelet 3H-paroxetine binding in depressed patients. *Psychiatry Res* 26:11–7.

D'haenen, H., Bossuyt, A., Mertens, J., Bossuyt-Piron, C., Gijsemans, M., and Kaufman, L. 1992. SPECT imaging of serotonin2 receptors in depression. *Psychiatry Res* 45:227–37.

D'Hondt, P., Maes, M., Leysen, J. E., et al. 1996. Seasonal variation in platelet [³H]paroxetine binding in healthy volunteers. Relationship to climatic variables. *Neuropsychopharmacology* 15:187–98.

Dahlstrom, A., and Fuxe, K. 1964. Localization of monoamines in the lower brain stem. *Experientia* 20:398–9.

Davison, A. N. 1957. The mechanism of the irreversible inhibition of rat-liver monoamine oxidase by iproniazid (Marsilid). *Biochem J* 67:316–22.

Delgado, P. L., Charney, D. S., Price, L. H., Aghajanian, G. K., Landis, H., and Heninger, G. R. 1990. Serotonin function and the mechanism of antidepressant action. Reversal of antidepressant-induced remission by rapid depletion of plasma tryptophan. *Arch Gen Psychiatry* 47:411–8.

Delgado, P. L., Price, L. H., Miller, H. L., et al. 1994. Serotonin and the neurobiology of depression. Effects of tryptophan depletion in drug-free depressed patients. *Arch Gen Psychiatry* 51:865–74.

Delorme, R., Durand, C. M., Betancur, C., et al. 2006. No human tryptophan hydroxylase-2 gene R441H mutation in a large cohort of psychiatric patients and control subjects. *Biol Psychiatry* 60:202–3.

Drevets, W. C., Frank, E., Price, J. C., et al. 1999. PET imaging of serotonin 1A receptor binding in depression. *Biol Psychiatry* 46:1375–87.

Dwivedi, Y., Rizavi, H. S., Conley, R. R., Roberts, R. C., Tamminga, C. A., and Pandey, G. N. 2003. Altered gene expression of brain-derived neurotrophic factor and receptor tyrosine kinase B in postmortem brain of suicide subjects. *Arch Gen Psychiatry* 60:804–15.

Eccleston, D., Ashcroft, G. W., Moir, A. T., Parker-Rhodes, A., Lutz, W., and O'Mahoney, D. P. 1968. A comparison of 5-hydroxyindoles in various regions of dog brain and cerebrospinal fluid. *J Neurochem* 15:947–57.

Edman, G., Asberg, M., Levander, S., and Schalling, D. 1986. Skin conductance habituation and cerebrospinal fluid 5-hydroxyindoleacetic acid in suicidal patients. *Arch Gen Psychiatry* 43:586–92.

Ellenbogen, M. A., Young, S. N., Dean, P., Palmour, R. M., and Benkelfat, C. 1996. Mood response to acute tryptophan depletion in healthy volunteers: Sex differences and temporal stability. *Neuropsychopharmacology* 15:465–74.

Ellis, P. M., McIntosh, C. J., Beeston, R., Salmond, C. E., Cooke, R. R., and Mellsop, G. 1990. Platelet tritiated imipramine binding in psychiatric patients: Relationship to symptoms and severity of depression. *Acta Psychiatr Scand* 82:275–82.

Erspamer, V., and Asero, B. 1952. Identification of enteramine, the specific hormone of the enterochromaffin cell system, as 5-hydroxytryptamine. *Nature* 169:800–1.

Erspamer, V., and Asero, B. 1953. Isolation of enteramine from extracts of posterior salivary glands of *Octopus vulgaris* and of *Discoglossus pictus* skin. *J Biol Chem* 200:311–8.

Evers, E. A. T., Cools, R., Clark, L., et al. 2005. Serotonergic modulation of prefrontal cortex during negative feedback in probabilistic reversal learning. *Neuropsychopharmacology* 30:1138–47.

Fabre, V., Beaufour, C., Evrard, A., et al. 2000. Altered expression and functions of serotonin 5-HT1A and 5-HT1B receptors in knock-out mice lacking the 5-HT transporter. *Eur J Neurosci* 12:2299–310.

Fava, M., and Kendler, K. S. 2000. Major depressive disorder. *Neuron* 28:335–41.

Ferrier, I. N., McKeith, I. G., Cross, A. J., Perry, E. K., Candy, J. M., Perry, R. H. 1986. Postmortem neurochemical studies in depression. *Ann N Y Acad Sci* 487:128–42.

Fisar, Z., Kalisová, L., Paclt, I., Anders, M., and Vevera, J. 2008. Platelet serotonin uptake in drug-naïve depressive patients before and after treatment with citalopram. *Psychiatry Res* 161:185–94.

Flory, J. D., Manuck, S. B., Ferrell, R. E., Dent, K. M., Peters, D. G., and Muldoon, M. F. 1999. Neuroticism is not associated with the serotonin transporter (*5-HTTLPR*) polymorphism. *Mol Psychiatry* 4:93–6.

Freis, E. D. 1954. Mental depression in hypertensive patients treated for long periods with large doses of reserpine. *N Engl J Med* 251:1006–8.

Fusar-Poli, P., Allen, P., McGuire, P., Placentino, A., Cortesi, M., and Perez, J. 2006. Neuroimaging and electrophysiological studies of the effects of acute tryptophan depletion: A systematic review of the literature. *Psychopharmacology (Berl)* 188:131–43.

Gardier, A. M. et al. 2009. Interest of using genetically manipulated mice as models of depression to evaluate antidepressant drugs activity: A review. *Fundam Clin Pharmacol* 23(1):23–42.

Garriock, H. A., and Hamilton, S. P. 2009. Genetic studies of drug response and side effects in the STAR*D study, part 1. *J Clin Psychiatry* 70:1186–7.

Gaspar, P., Cases, O., and Maroteaux, L. 2003. The developmental role of serotonin: News from mouse molecular genetics. *Nat Rev Neurosci* 4:1002–12.

Gentsch, C., Lichtsteiner, M., Gastpar, M., Gastpar, G., and Feer, H. 1985. 3H-Imipramine binding sites in platelets of hospitalized psychiatric patients. *Psychiatry Res* 14:177–87.

Ghanbari, R., El Mansari, M., and Blier, P. 2010. Sustained administration of trazodone enhances serotonergic neurotransmission: In vivo electrophysiological study in the rat brain. *J Pharmacol Exp Ther* 335:197–206.

Gordon, J., and Barnes, N. M. 2003. Lymphocytes transport serotonin and dopamine: Agony or ecstasy? *Trends Immunol* 24:438–43.

Gottfries, C. G., von Knorring, L., and Perris, C. 1976. Neurophysiological measures related to levels of 5-hydroxyindoleacetic acid, homovanillic acid and tryptophan in cerebrospinal fluid of psychiatric patients. *Neuropsychobiology* 2:1–8.

Grimes, M. A., Cameron, J. L., and Fernstrom, J. D. 2000. Cerebrospinal fluid concentrations of tryptophan and 5-hydroxyindoleacetic acid in *Macaca mulatta*: Diurnal variations and response to chronic changes in dietary protein intake. *Neurochem Res* 25:413–22.

Grimsby, J., Toth, M., Chen, K., et al. 1997. Increased stress response and beta-phenylethylamine in MAOB-deficient mice. *Nat Genet* 17:206–10.

Gross, C., Zhuang, X., Stark, K., et al. 2002. Serotonin1A receptor acts during development to establish normal anxiety-like behaviour in the adult. *Nature* 416:396–400.

Gupta, J. C., Ghosh, S., Dutta, A. T., and Kahali, B. S. 1947. A note on the hypnotic principle of *Rauwolfia serpentina*. *J Am Pharm Assoc Am Pharm Assoc* 36:416.

Gurevich, I., Englander, M. T., Adlersberg, M., Siegal, N. B., and Schmauss, C. 2002. Modulation of serotonin 2C receptor editing by sustained changes in serotonergic neurotransmission. *J Neurosci* 22:10529–32.

Gutknecht, L., Kriegebaum, C., Waider, J., Schmitt, A., and Lesch, K.-P. 2009. Spatio-temporal expression of tryptophan hydroxylase isoforms in murine and human brain: Convergent data from Tph2 knockout mice. *Eur Neuropsychopharmacol* 19:266–82.

Hannon, J., and Hoyer, D. 2008. Molecular biology of 5-HT receptors. *Behav Brain Res* 195:198–213.

Hariri, A. R., Mattay, V. S., Tessitore, A., et al. 2002. Serotonin transporter genetic variation and the response of the human amygdala. *Science* 297:400–3.

Heils, A., Teufel, A., Petri, S., et al. 1996. Allelic variation of human serotonin transporter gene expression. *J Neurochem* 66:2621–4.

Heinz, A., Jones, D. W., Mazzanti, C., et al. 2000. A relationship between serotonin transporter genotype and in vivo protein expression and alcohol neurotoxicity. *Biol Psychiatry* 47:643–9.

Heisler, L. K., Chu, H. M., Brennan, et al. 1998. Elevated anxiety and antidepressant-like responses in serotonin 5-HT1A receptor mutant mice. *Proc Natl Acad Sci USA* 95:15049–54.

Hendrick, V., Smith, L. M., Suri, R., Hwang, S., Haynes, D., and Altshuler, L. 2003a. Birth outcomes after prenatal exposure to antidepressant medication. *Am J Obstet Gynecol* 188:812–5.

Hendrick, V., Stowe, Z. N., Altshuler, L. L., Hwang, S., Lee, E., and Haynes, D. 2003b. Placental passage of antidepressant medications. *Am J Psychiatry* 160:993–6.

Hendricks, T., Francis, N., Fyodorov, D., and Deneris, E. S. 1999. The ETS domain factor Pet-1 is an early and precise marker of central serotonin neurons and interacts with a conserved element in serotonergic genes. *J Neurosci* 19:10348–56.

Hensler, J. G. 2002. Differential regulation of 5-HT1A receptor–G protein interactions in brain following chronic antidepressant administration. *Neuropsychopharmacology* 26:565–73.

Hensler, J. G. 2006. Serotonin. In *Basic Neurochemistry: Molecular, Cellular, and Medical Aspects*, 7th edn, ed. Siegel, G. J., Albers, R. W., Brady, S. T., and Price, D. L., 227–48. Burlington: Elsevier Academic Press.

Hirvonen, J., Karlsson, H., Kajander, J., et al. 2008. Decreased brain serotonin 5-HT1A receptor availability in medication-naive patients with major depressive disorder: An in-vivo imaging study using PET and [carbonyl-11C]WAY-100635. *Int J Neuropsychopharmacol* 11:465–76.

Hjorth, S., Bengtsson, H. J., Kullberg, A., Carlzon, D., Peilot, H., and Auerbach, S. B. 2000. Serotonin autoreceptor function and antidepressant drug action. *J Psychopharmacol (Oxford)* 14:177–85.

Holmes, A., Yang, R. J., and Crawley, J. N. 2002a. Evaluation of an anxiety-related phenotype in galanin overexpressing transgenic mice. *J Mol Neurosci* 18:151–65.

Holmes, A., Yang, R. J., Murphy, D. L., and Crawley, J. N. 2002b. Evaluation of antidepressant-related behavioral responses in mice lacking the serotonin transporter. *Neuropsychopharmacology* 27: 914–23.

Holmes, A., Lit, Q., Murphy, D. L., Gold, E., and Crawley, J. N. 2003. Abnormal anxiety-related behavior in serotonin transporter null mutant mice: The influence of genetic background. *Genes Brain Behav* 2:365–80.

Holzbauer, M., and Vogt, M. 1956. Depression by reserpine of the noradrenaline concentration in the hypothalamus of the cat. *J Neurochem* 1:8–11.

Hood, S. D., Bell, C. J., and Nutt, D. J. 2005. Acute tryptophan depletion. Part I: Rationale and methodology. *Austral N Z J Psychiatry* 39:558–64.

Hornung, J. P. 2003. The human raphe nuclei and the serotonergic system. *J Chem Neuroanat* 26:331–43.

Hostetter, A., Ritchie, J. C., and Stowe, Z. N. 2000. Amniotic fluid and umbilical cord blood concentrations of antidepressants in three women. *Biol Psychiatry* 48:1032–4.

Hou, C., Jia, F., Liu, Y., and Li, L. 2006. CSF serotonin, 5-hydroxyindolacetic acid and neuropeptide Y levels in severe major depressive disorder. *Brain Res* 1095:154–8.

Hoyer, D., Hannon, J. P., and Martin, G. R. 2002. Molecular, pharmacological and functional diversity of 5-HT receptors. *Pharmacol Biochem Behav* 71:533–54.

Hrdina, P. D., Demeter, E., Vu, T. B., Sótónyi, P., and Palkovits, M. 1993. 5-HT uptake sites and 5-HT2 receptors in brain of antidepressant-free suicide victims/depressives: Increase in 5-HT2 sites in cortex and amygdala. *Brain Res* 614:37–44.

Hu, X., Oroszi, G., Chun, J., Smith, T. L., Goldman, D., and Schuckit, M. A. 2005. An expanded evaluation of the relationship of four alleles to the level of response to alcohol and the alcoholism risk. *Alcohol Clin Exp Res* 29:8–16.

Hu, X.-Z., Lipsky, R. H., Zhu, G., et al. 2006. Serotonin transporter promoter gain-of-function genotypes are linked to obsessive-compulsive disorder. *Am J Hum Genet* 78:815–26.

Hu, X. Z., Rush, A. J., Charney, D., et al. 2007. Association between a functional serotonin transporter promoter polymorphism and citalopram treatment in adult outpatients with major depression. *Arch Gen Psychiatry* 64:783–92.

Iceta, R., Aramayona, J. J., Mesonero, J. E., and Alcalde, A. I. 2008. Regulation of the human serotonin transporter mediated by long-term action of serotonin in Caco-2 cells. *Acta Physiol (Oxf)* 193:57–65.

Ichise, M., Vines, D. C., Gura, T., et al. 2006. Effects of early life stress on [11C]DASB positron emission tomography imaging of serotonin transporters in adolescent peer- and mother-reared rhesus monkeys. *J Neurosci* 26:4638–43.

Jayanthi, L. D., and Ramamoorthy, S. 2005. Regulation of monoamine transporters: Influence of psychostimulants and therapeutic antidepressants. *AAPS J* 7:E728–38.

Jefferson, J. W. 1975. A review of the cardiovascular effects and toxicity of tricyclic antidepressants. *Psychosom Med* 37:160–79.

Jensen, P., Farago, A. F., Awatramani, R. B., Scott, M. M., Deneris, E. S., and Dymecki, S. M. 2008. Redefining the serotonergic system by genetic lineage. *Nat Neurosci* 11:417–9.

Joeyen-Waldorf, J., Edgar, N., and Sibille, E. 2009. The roles of sex and serotonin transporter levels in age- and stress-related emotionality in mice. *Brain Res* 1286:84–93.

Jones, J. S., Stanley, B., Mann, J. J., et al. 1990. CSF 5-HIAA and HVA concentrations in elderly depressed patients who attempted suicide. *Am J Psychiatry* 147:1225–7.

Kalueff, A. V., Gallagher, P. S., and Murphy, D. L. 2006. Are serotonin transporter knockout mice 'depressed'?: Hypoactivity but no anhedonia. *Neuroreport* 17:1347–51.

Kanner, B. I., and Schuldiner, S. 1987. Mechanism of transport and storage of neurotransmitters. *CRC Crit Rev Biochem* 22:1–38.

Kanof, P. D., Coccaro, E. F., Johns, C. A., Siever, L. J., and Davis, K. L. 1987. Platelet [3H]imipramine binding in psychiatric disorders. *Biol Psychiatry* 22:278–86.

Kennedy, J. S., Gwirtsman, H. E., Schmidt, D. E., et al. 2002. Serial cerebrospinal fluid tryptophan and 5-hydroxy indoleacetic acid concentrations in healthy human subjects. *Life Sci* 71:1703–15.

Kim, D. K., Tolliver, T. J., Huang, S.-J., et al. 2005. Altered serotonin synthesis, turnover and dynamic regulation in multiple brain regions of mice lacking the serotonin transporter. *Neuropharmacology* 49:798–810.

Klaassen, T., Riedel, W. J., van Someren, A., Deutz, N. E., Honig, A., and van Praag, H. M. 1999. Mood effects of 24-hour tryptophan depletion in healthy first-degree relatives of patients with affective disorders. *Biol Psychiatry* 46:489–97.

Kline, N. S. 1954. Use of *Rauwolfia serpentina* Benth. in neuropsychiatric conditions. *Ann N Y Acad Sci* 59:107–32.

Kostowski, W., and Krzaścik, P. 2003. Neonatal 5-hydroxytryptamine depletion induces depressive-like behavior in adult rats. *Pol J Pharmacol* 55:957–63.

Kraft, J. B., Peters, E. J., Slager, S. L., et al. 2007. Analysis of association between the serotonin transporter and antidepressant response in a large clinical sample. *Biol Psychiatry* 61:734–42.

Kreiss, D. S., and Lucki, I. 1995. Effects of acute and repeated administration of antidepressant drugs on extracellular levels of 5-hydroxytryptamine measured in vivo. *J Pharmacol Exp Ther* 274:866–76.

Kuhn, R. 1958. The treatment of depressive states with G 22355 (imipramine hydrochloride). *Am J Psychiatry* 115:459–64.

Laje, G., Perlis, R. H., Rush, A. J., and McMahon, F. J. 2009. Pharmacogenetics studies in STAR*D: Strengths, limitations, and results. *Psychiatr Serv* 60:1446–57.

Langer, S. Z., and Schoemaker, H. 1988. Effects of antidepressants on monoamine transporters. *Prog Neuropsychopharmacol Biol Psychiatry* 12:193–216.

Lawrence, K. M., De Paermentier, F., Cheetham, S. C., Crompton, M. R., Katona, C. L., and Horton, R. W. 1989. Brain 5-HT uptake sites, labelled with [^{3}H]-paroxetine, in post-mortem samples from depressed suicide victims. *Br J Pharmacol* 98 (Suppl):812P.

Lawrence, K. M., Falkowski, J., Jacobson, R. R., and Horton, R. W. 1993. Platelet 5-HT uptake sites in depression: Three concurrent measures using [^{3}H]imipramine and [^{3}H]paroxetine. *Psychopharmacology (Berl)* 110:235–9.

Leake, A., Fairbairn, A. F., McKeith, I. G., and Ferrier, I. N. 1991. Studies on the serotonin uptake binding site in major depressive disorder and control post-mortem brain: Neurochemical and clinical correlates. *Psychiatry Res* 39:155–65.

Lebrand, C., Cases, O., Adelbrecht, C., et al. 1996. Transient uptake and storage of serotonin in developing thalamic neurons. *Neuron* 17:823–35.

Lebrand, C., Cases, O., Wehrlé, R., Blakely, R. D., Edwards, R. H., and Gaspar, P. 1998a. Transient developmental expression of monoamine transporters in the rodent forebrain. *J Comp Neurol* 401:506–24.

Lebrand, C., Cases, O., Wehrlé, R., Blakely, R. D., Edwards, R. H., and Gaspar, P. 1998b. Transient developmental expression of monoamine transporters in the rodent forebrain. *J Comp Neurol* 401:506–24.

Lesch, K. P., Bengel, D., Heils, A., et al. 1996. Association of anxiety-related traits with a polymorphism in the serotonin transporter gene regulatory region. *Science* 274:1527–31.

Lesch, K. P., Meyer, J., Glatz, K., et al. 1997. The 5-HT transporter gene-linked polymorphic region (*5-HT-TLPR*) in evolutionary perspective: Alternative biallelic variation in rhesus monkeys. Rapid communication. *J Neural Transm* 104:1259–66.

Leuchter, A. F., Cook, I. A., Hunter, A. M., and Korb, A. S. 2009. A new paradigm for the prediction of antidepressant treatment response. *Dialogues Clin Neurosci* 11:435–46.

Leuchter, A. F., Cook, I. A., Hamilton, S. P., et al. 2010. Biomarkers to predict antidepressant response. *Curr Psychiatry Rep* 12:553–62.

Levitt, P., and Rakic, P. 1982. The time of genesis, embryonic origin and differentiation of the brain stem monoamine neurons in the rhesus monkey. *Brain Res* 256:35–57.

Li, Q., Wichems, C., Heils, A., Van De Kar, L. D., Lesch, K. P., Murphy, D. L. 1999. Reduction of 5-hydroxytryptamine (5-HT)(1A)-mediated temperature and neuroendocrine responses and 5-HT(1A) binding sites in 5-HT transporter knockout mice. *J Pharmacol Exp Ther* 291:999–1007.

Lieben, C. K. J., Steinbusch, H. W. M., and Blokland, A. 2006. 5,7-DHT lesion of the dorsal raphe nuclei impairs object recognition but not affective behavior and corticosterone response to stressor in the rat. *Behav Brain Res* 168:197–207.

Lipsky, R. H., Hu, X. Z., and Goldman, D. 2009. Additional functional variation at the SLC6A4 gene. *Am J Med Genet B Neuropsychiatr Genet* 150B:153.

Little, K. Y., McLauglin, D. P., Ranc, J., et al. 1997. Serotonin transporter binding sites and mRNA levels in depressed persons committing suicide. *Biol Psychiatry* 41:1156–64.

Little, K. Y., McLaughlin, D. P., Zhang, L., et al. 1998. Cocaine, ethanol, and genotype effects on human midbrain serotonin transporter binding sites and mRNA levels. *Am J Psychiatry* 155:207–13.

Lloyd, K. G., Farley, I. J., Deck, J. H., and Hornykiewicz, O. 1974. Serotonin and 5-hydroxyindoleacetic acid in discrete areas of the brainstem of suicide victims and control patients. *Adv Biochem Psychopharmacol* 11:387–97.

Lopez, J. F., Chalmers, D. T., Little, K. Y., and Watson, S. J. 1998. A.E. Bennett Research Award. Regulation of serotonin1A, glucocorticoid, and mineralocorticoid receptor in rat and human hippocampus: Implications for the neurobiology of depression. *Biol Psychiatry* 43:547–73.

Lopez-Ibor, J. J., Saiz-Ruiz, J., and Pérez de los Cobos, J. C. 1985. Biological correlations of suicide and aggressivity in major depressions (with melancholia): 5-Hydroxyindoleacetic acid and cortisol in cerebral spinal fluid, dexamethasone suppression test and therapeutic response to 5-hydroxytryptophan. *Neuropsychobiology* 14:67–74.

Lorens, S. A., Guldberg, H. C., Hole, K., Köhler, C., and Srebro, B. 1976. Activity, avoidance learning and regional 5-hydroxytryptamine following intra-brain stem 5,7-dihydroxytryptamine and electrolytic midbrain raphe lesions in the rat. *Brain Res* 108:97–113.

Lowther, S., De Paermentier, F., Cheetham, S. C., Crompton, M. R., Katona, C. L., and Horton, R. W. 1997. 5-HT1A receptor binding sites in post-mortem brain samples from depressed suicides and controls. *J Affect Disord* 42:199–207.

Luellen, B. A., Gilman, T. L., and Andrews, A. M. 2010. Presynaptic adaptive responses to constitutive versus adult pharmacologic inhibition of serotonin uptake. In *Experimental Models in Serotonin Transporter Research,* ed. Kalueff, A. V., and LaPorte, J. L., 1–42. Cambridge: Cambridge University Press.

Malison, R. T., Price, L. H., Berman, R., et al. 1998. Reduced brain serotonin transporter availability in major depression as measured by [^{123}I]-2 beta-carbomethoxy-3 beta-(4-iodophenyl)tropane and single photon emission computed tomography. *Biol Psychiatry* 44:1090–8.

Mann, J. J., Malone, K. M., Psych, M. R., et al. 1996. Attempted suicide characteristics and cerebrospinal fluid amine metabolites in depressed inpatients. *Neuropsychopharmacology* 15:576–86.

Mann, J. J., Huang, Y. Y., Underwood, M. D., et al. 2000. A serotonin transporter gene promoter polymorphism (*5-HTTLPR*) and prefrontal cortical binding in major depression and suicide. *Arch Gen Psychiatry* 57:729–38.

Mathews, T. A., Fedele, D. E., Coppelli, F. M., Avila, A. M., Murphy, D. L., and Andrews, A. M. 2004. Gene dose-dependent alterations in extraneuronal serotonin but not dopamine in mice with reduced serotonin transporter expression. *J Neurosci Methods* 140:169–81.

McKeith, I. G., Marshall, E. F., Ferrier, I. N., et al. 1987. 5-HT receptor binding in post-mortem brain from patients with affective disorder. *J Affect Disord* 13:67–74.

McKinnon, M. C., Yucel, K., Nazarov, A., and MacQueen, G. M. 2009. A meta-analysis examining clinical predictors of hippocampal volume in patients with major depressive disorder. *J Psychiatry Neurosci* 34:41–54.

Meltzer, C. C., Price, J. C., Mathis, C. A., et al. 1999. PET imaging of serotonin type 2A receptors in late-life neuropsychiatric disorders. *Am J Psychiatry* 156:1871–8.

Meltzer, C. C., Price, J. C., Mathis, C. A., et al. 2004. Serotonin 1A receptor binding and treatment response in late-life depression. *Neuropsychopharmacology* 29:2258–65.

Merens, W., Booij, L., Haffmans, P. J., and van der Does, A. 2008. The effects of experimentally lowered serotonin function on emotional information processing and memory in remitted depressed patients. *J Psychopharmacol (Oxford)* 22:653–62.

Messa, C., Colombo, C., Moresco, R. M., et al. 2003. 5-HT(2A) receptor binding is reduced in drug-naive and unchanged in SSRI-responder depressed patients compared to healthy controls: A PET study. *Psychopharmacology (Berl)* 167:72–8.

Meyer, J. H., Kapur, S., Houle, S., et al. 1999. Prefrontal cortex 5-HT2 receptors in depression: An [18F]setoperone PET imaging study. *Am J Psychiatry* 156:1029–34.

Miller, J. M., Brennan, K. G., Ogden, T. R., et al. 2009a. Elevated serotonin 1A binding in remitted major depressive disorder: Evidence for a trait biological abnormality. *Neuropsychopharmacology* 34:2275–84.

Miller, J. M., Kinnally, E. L., Ogden, R. T., Oquendo, M. A., Mann, J. J., and Parsey, R. V. 2009b. Reported childhood abuse is associated with low serotonin transporter binding in vivo in major depressive disorder. *Synapse* 63:565–73.

Mirmiran, M., van de Poll, N. E., Corner, M. A., van Oyen, H. G., and Bour, H. L. 1981. Suppression of active sleep by chronic treatment with chlorimipramine during early postnatal development: Effects upon adult sleep and behavior in the rat. *Brain Res* 204:129–46.

Montanez, S., Owens, W. A., Gould, G. G., Murphy, D. L., and Daws, L. C. 2003. Exaggerated effect of fluvoxamine in heterozygote serotonin transporter knockout mice. *J Neurochem* 86:210–9.

Moreno, F. A., McGavin, C., Malan, T. P., et al. 2000. Tryptophan depletion selectively reduces CSF 5-HT metabolites in healthy young men: Results from single lumbar puncture sampling technique. *Int J Neuropsychopharmacol* 3:277–83.

Moreno, F. A., McGahuey, C. A., Freeman, M. P., and Delgado, P. L. 2006. Sex differences in depressive response during monoamine depletions in remitted depressive subjects. *J Clin Psychiatry* 67:1618–23.

Mrazek, D. A., Rush, A. J., Biernacka, J. M., et al. 2009. SLC6A4 variation and citalopram response. *Am J Med Genet B Neuropsychiatr Genet* 150B:341–51.

Murphy, D. L., and Lesch, K.-P. 2008. Targeting the murine serotonin transporter: Insights into human neurobiology. *Nat Rev Neurosci* 9:85–96.

Murphy, D. L., Lerner, A., Rudnick, G., Lesch, K.-P. 2004. Serotonin transporter: Gene, genetic disorders, and pharmacogenetics. *Mol Interv* 4:109–23.

Murphy, D. L., Fox, M. A., Timpano, K. R., et al. 2008. How the serotonin story is being rewritten by new gene-based discoveries principally related to SLC6A4, the serotonin transporter gene, which functions to influence all cellular serotonin systems. *Neuropharmacology* 55:932–60.

Murphy, F. C., Smith, K. A., Cowen, P. J., Robbins, T. W., and Sahakian, B. J. 2002. The effects of tryptophan depletion on cognitive and affective processing in healthy volunteers. *Psychopharmacology (Berl)* 163:42–53.

Murthy, N. V., Selvaraj, S., Cowen, P. J., et al. 2010. Serotonin transporter polymorphisms (SLC6A4 insertion/deletion and rs25531) do not affect the availability of 5-HTT to [^{11}C] DASB binding in the living human brain. *NeuroImage* 52:50–4.

Naylor, L., Dean, B., Pereira, A., Mackinnon, A., Kouzmenko, A., and Copolov, D. 1998. No association between the serotonin transporter–linked promoter region polymorphism and either schizophrenia or density of the serotonin transporter in human hippocampus. *Mol Med* 4:671–4.

Neill, D., Vogel, G., Hagler, M., Kors, D., and Hennessey, A. 1990. Diminished sexual activity in a new animal model of endogenous depression. *Neurosci Biobehav Rev* 14:73–6.

Nemeroff, C. B., Knight, D. L., Franks, J., Craighead, W. E., and Krishnan, K. R. 1994. Further studies on platelet serotonin transporter binding in depression. *Am J Psychiatry* 151:1623–5.

Neumeister, A., Praschak-Rieder, N., Hesselmann, B., et al. 1997. Rapid tryptophan depletion in drug-free depressed patients with seasonal affective disorder. *Am J Psychiatry* 154:1153–5.

Newman, M. E., Shalom, G., Ran, A., Gur, E., and Van de Kar, L. D. 2004. Chronic fluoxetine-induced desensitization of 5-HT1A and 5-HT1B autoreceptors: Regional differences and effects of WAY-100635. *Eur J Pharmacol* 486:25–30.

Nishizawa, S., Benkelfat, C., Young, S. N., et al. 1997. Differences between males and females in rates of serotonin synthesis in human brain. *Proc Natl Acad Sci USA* 94:5308–13.

Nordström, P., Samuelsson, M., Asberg, M., et al. 1994. CSF 5-HIAA predicts suicide risk after attempted suicide. *Suicide Life Threat Behav* 24:1–9.

Oquendo, M. A., Hastings, R. S., Huang, Y.-Y., et al. 2007. Brain serotonin transporter binding in depressed patients with bipolar disorder using positron emission tomography. *Arch Gen Psychiatry* 64:201–8.

Oreland, L., Wiberg, A., Asberg, M., et al. 1981. Platelet MAO activity and monoamine metabolites in cerebrospinal fluid in depressed and suicidal patients and in healthy controls. *Psychiatry Res* 4:21–9.

Owen, F., Cross, A. J., Crow, T. J., et al. 1983. Brain 5-HT-2 receptors and suicide. *Lancet* 2:1256.

Paasonen, M. K., and Vogt, M. 1956. The effect of drugs on the amounts of substance P and 5-hydroxytryptamine in mammalian brain. *J Physiol (Lond)* 131:617–26.

Papeschi, R., and McClure, D. J. 1971. Homovanillic and 5-hydroxyindoleacetic acid in cerebrospinal fluid of depressed patients. *Arch Gen Psychiatry* 25:354–8.

Pare, C. M., Yeung, D. P., Price, K., and Stacey, R. S. 1969. 5-Hydroxytryptamine, noradrenaline, and dopamine in brainstem, hypothalamus, and caudate nucleus of controls and of patients committing suicide by coalgas poisoning. *Lancet* 294:133–5.

Parks, C. L., Robinson, P. S., Sibille, E., Shenk, T., and Toth, M. 1998. Increased anxiety of mice lacking the serotonin 1A receptor. *Proc Natl Acad Sci USA* 95:10734–9.

Parsey, R. V., Hastings, R. S., Oquendo, M. A., et al. 2006a. Lower serotonin transporter binding potential in the human brain during major depressive episodes. *Am J Psychiatry* 163:52–8.

Parsey, R. V., Hastings, R. S., Oquendo, M. A., et al. 2006b. Effect of a triallelic functional polymorphism of the serotonin-transporter-linked promoter region on expression of serotonin transporter in the human brain. *Am J Psychiatry* 163:48–51.

Parsey, R. V., Oquendo, M. A., Ogden, R. T., et al. 2006c. Altered serotonin 1A binding in major depression: A [carbonyl-C-11]WAY100635 positron emission tomography study. *Biol Psychiatry* 59:106–13.

Paul, S. M., Rehavi, M., Skolnick, P., Ballenger, J. C., and Goodwin, F. K. 1981. Depressed patients have decreased binding of tritiated imipramine to platelet serotonin "transporter". *Arch Gen Psychiatry* 38:1315–7.

Peabody, C. A., Faull, K. F., King, R. J., Whiteford, H. A., Barchas, J. D., and Berger, P. A. 1987. CSF amine metabolites and depression. *Psychiatry Res* 21:1–7.

Pejchal, T., Foley, M. A., Kosofsky, B. E., and Waeber, C. 2002. Chronic fluoxetine treatment selectively uncouples raphe 5-HT(1A) receptors as measured by [(35)S]-GTP gamma S autoradiography. *Br J Pharmacol* 135:1115–22.

Perez, X. A., and Andrews, A. M. 2005. Chronoamperometry to determine differential reductions in uptake in brain synaptosomes from serotonin transporter knockout mice. *Anal Chem* 77:818–26.

Perez, X. A., Bianco, L. E., and Andrews, A. M. 2006. Filtration disrupts synaptosomes during radiochemical analysis of serotonin uptake: Comparison with chronoamperometry in SERT knockout mice. *J Neurosci Methods* 154:245–55.

Perry, E. K., Marshall, E. F., Blessed, G., Tomlinson, B. E., and Perry, R. H. 1983. Decreased imipramine binding in the brains of patients with depressive illness. *Br J Psychiatry* 142:188–92.

Pezawas, L., Meyer-Lindenberg, A., Drabant, E. M., et al. 2005. *5-HTTLPR* polymorphism impacts human cingulate–amygdala interactions: A genetic susceptibility mechanism for depression. *Nat Neurosci* 8:828–34.

Philibert, R. A., Sandhu, H., Hollenbeck, N., Gunter, T., Adams, W., and Madan, A. 2008. The relationship of 5HTT (SLC6A4) methylation and genotype on mRNA expression and liability to major depression and alcohol dependence in subjects from the Iowa Adoption Studies. *Am J Med Genet B Neuropsychiatr Genet* 147B:543–9.

Pineyro, G., and Blier, P. 1999. Autoregulation of serotonin neurons: Role in antidepressant drug action. *Pharmacol Rev* 51:533–91.

Pollin, W., Cardon, P. V., and Kety, S. S. 1961. Effects of amino acid feedings in schizophrenic patients treated with iproniazid. *Science* 133:104–5.

Popa, D., Léna, C., Alexandre, C., and Adrien, J. 2008. Lasting syndrome of depression produced by reduction in serotonin uptake during postnatal development: Evidence from sleep, stress, and behavior. *J Neurosci* 28:3546–54.

Popa, D., Cerdan, J., Repérant, C., et al. 2010. A longitudinal study of 5-HT outflow during chronic fluoxetine treatment using a new technique of chronic microdialysis in a highly emotional mouse strain. *Eur J Pharmacol* 628:83–90.

Ramage, A. G., and Villalon, C. M. 2008. 5-Hydroxytryptamine and cardiovascular regulation. *Trends Pharmacol Sci* 29:472–81.

Ramboz, S., Oosting, R., Amara, D. A., et al. 1998. Serotonin receptor 1A knockout: An animal model of anxiety-related disorder. *Proc Natl Acad Sci USA* 95:14476–81.

Rapport, M. M., Green, A. A., and Page, I. H. 1948. Crystalline serotonin. *Science* 108:329–30.

Rausch, J. L., Johnson, M. E., Li, J., et al. 2005. Serotonin transport kinetics correlated between human platelets and brain synaptosomes. *Psychopharmacology (Berl)* 180:391–8.

Remick, R. A., Froese, C., and Keller, F. D. 1989. Common side effects associated with monoamine oxidase inhibitors. *Prog Neuropsychopharmacol Biol Psychiatry* 13:497–504.

Richardson-Jones, J. W., Craige, C. P., Guiard, B. P., et al. 2010. 5-HT1A autoreceptor levels determine vulnerability to stress and response to antidepressants. *Neuron* 65:40–52.

Robinson, O. J., and Sahakian, B. J. 2009. Acute tryptophan depletion evokes negative mood in healthy females who have previously experienced concurrent negative mood and tryptophan depletion. *Psychopharmacology (Berl)* 205:227–35.

Roiser, J. P., Blackwell, A. D., Cools, R., et al. 2006. Serotonin transporter polymorphism mediates vulnerability to loss of incentive motivation following acute tryptophan depletion. *Neuropsychopharmacology* 31:2264–72.

Rose, W. C., Haines, W. J., and Warner, D. T. 1954. The amino acid requirements of man. V. The role of lysine, arginine, and tryptophan. *J Biol Chem* 206:421–30.

Rossi, D. V., Valdez, M., Gould, G. G., and Hensler, J. G. 2006. Chronic administration of venlafaxine fails to attenuate 5-HT1A receptor function at the level of receptor–G protein interaction. *Int J Neuropsychopharmacol* 9:393–406.

Rossi, D. V., Burke, T. F., McCasland, M., and Hensler, J. G. 2008. Serotonin-1A receptor function in the dorsal raphe nucleus following chronic administration of the selective serotonin reuptake inhibitor sertraline. *J Neurochem* 105:1091–9.

Roubein, I. F., and Embree, L. J. 1979. Post mortem stability of catecholamines in discrete regions of rat brain. *Res Commun Chem Pathol Pharmacol* 23:143–53.

Roy-Byrne, P., Post, R. M., Rubinow, D. R., Linnoila, M., Savard, R., and Davis, D. 1983. CSF 5HIAA and personal and family history of suicide in affectively ill patients; A negative study. *Psychiatry Res* 10:263–74.

Rudnick, G. 2006. Structure/function relationships in serotonin transporter: New insights from the structure of a bacterial transporter. *Handb Exp Pharmacol* 175:59–73.

Ruhé, H. G., Mason, N. S., and Schene, A. H. 2007. Mood is indirectly related to serotonin, norepinephrine and dopamine levels in humans: A meta-analysis of monoamine depletion studies. *Mol Psychiatry* 12:331–59.

Rush, A. J. 2007. Limitations in efficacy of antidepressant monotherapy. *J Clin Psychiatry* 68 Suppl 10:8–10.

Sakowski, S. A., Geddes, T. J., and Kuhn, D. M. 2006. Mouse tryptophan hydroxylase isoform 2 and the role of proline 447 in enzyme function. *J Neurochem* 96:758–65.

Samuelsson, M., Jokinen, J., Nordström, A.-L., and Nordström, P. 2006. CSF 5-HIAA, suicide intent and hopelessness in the prediction of early suicide in male high-risk suicide attempters. *Acta Psychiatr Scand* 113:44–7.

Sanders-Bush, E., Fentress, H., and Hazelwood, L. 2003. Serotonin 5-HT2 receptors: Molecular and genomic diversity. *Mol Interv* 3:319–30.

Sanz, E. J., De-las-Cuevas, C., Kiuru, A., Bate, A., and Edwards, R. 2005. Selective serotonin reuptake inhibitors in pregnant women and neonatal withdrawal syndrome: A database analysis. *Lancet* 365:482–7.

Saunders, J. C., Radinger, N., Rochlin, D., and Kline, N. S. 1959. Treatment of depressed and regressed patients with iproniazid and reserpine. *Dis Nerv Syst* 20:31–9.

Savelieva, K. V., Zhao, S., Pogorelov, V. M., et al. 2008. Genetic disruption of both tryptophan hydroxylase genes dramatically reduces serotonin and affects behavior in models sensitive to antidepressants. *PLoS ONE* 3:e3301.

Secunda, S. K., Cross, C. K., Koslow, S., et al. 1986. Biochemistry and suicidal behavior in depressed patients. *Biol Psychiatry* 21:756–67.

Sen, G., and Bose, K. 1931. *Rauwolfia serpentina*, a new Indian drug for insanity and high blood pressure. *Ind Med World* 2:194–201.

Sen, S., Duman, R., and Sanacora, G. 2008. Serum brain-derived neurotrophic factor, depression, and antidepressant medications: Meta-analyses and implications. *Biol Psychiatry* 64:527–32.

Shaw, D. M., Camps, F. E., and Eccleston, E. G. 1967. 5-Hydroxytryptamine in the hind-brain of depressive suicides. *Br J Psychiatry* 113:1407–11.

Sheline, Y. I., Mintun, M. A., Barch, D. M., Wilkins, C., Snyder, A. Z., and Moerlein, S. M. 2004. Decreased hippocampal 5-HT(2A) receptor binding in older depressed patients using [^{18}F]altanserin positron emission tomography. *Neuropsychopharmacology* 29:2235–41.

Sibille, E., and Lewis, D. A. 2006. SERT-ainly involved in depression, but when? *Am J Psychiatry* 163:8–11.

Sibille, E., Wang, Y., Joeyen-Waldorf, J., et al. 2009. A molecular signature of depression in the amygdala. *Am J Psychiatry* 166:1011–24.

Siddiqui, A., Clark, J. S., and Gilmore, D. P. 1990. Post-mortem changes of neurotransmitter concentrations in the rat brain regions. *Acta Physiol Hung* 75:179–85.

Singh, Y. S., Sawarynski, L. E., Michael, H. M., et al. 2010. Boron-doped diamond microelectrodes reveal reduced serotonin uptake rates in lymphocytes from adult rhesus monkeys carrying the short allele of the *5-HTTLPR*. *ACS Chem Neurosci* 1:49–64.

Sinyor, M., Schaffer, A., and Levitt, A. 2010. The sequenced treatment alternatives to relieve depression (STAR*D) trial: A review. *Can J Psychiatry* 55:126–35.

Spiller, R. 2008. Serotonin and GI clinical disorders. *Neuropharmacology* 55:1072–80.

Stanley, M., and Mann, J. J. 1983. Increased serotonin-2 binding sites in frontal cortex of suicide victims. *Lancet* 1:214–6.

Stanley, M., Traskman-Bendz, L., and Dorovini-Zis, K. 1985. Correlations between aminergic metabolites simultaneously obtained from human CSF and brain. *Life Sci* 37:1279–86.

Stockmeier, C. A., Shapiro, L. A., Dilley, G. E., Kolli, T. N., Friedman, L., and Rajkowska, G. 1998. Increase in serotonin-1A autoreceptors in the midbrain of suicide victims with major depression—postmortem evidence for decreased serotonin activity. *J Neurosci* 18:7394–401.

Sullivan, G. M., Oquendo, M. A., Huang, Y.-Y., and Mann, J. J. 2006. Elevated cerebrospinal fluid 5-hydroxyindoleacetic acid levels in women with comorbid depression and panic disorder. *Int J Neuropsychopharmacol* 9:547–56.

Tagliamonte, A., Biggio, G., Vargiu, L., and Gessa, G. L. 1973. Free tryptophan in serum controls brain tryptophan level and serotonin synthesis. *Life Sci* II 12:277–87.

Tanis, K. Q., Newton, S. S., and Duman, R. S. 2007. Targeting neurotrophic/growth factor expression and signaling for antidepressant drug development. *CNS Neurol Disord Drug Targets* 6:151–60.

Thomas, A. J., Hendriksen, M., Piggott, M., et al. 2006. A study of the serotonin transporter in the prefrontal cortex in late-life depression and Alzheimer's disease with and without depression. *Neuropathol Appl Neurobiol* 32:296–303.

Tjurmina, O. A., Armando, I., Saavedra, J. M., Goldstein, D. S., and Murphy, D. L. 2002. Exaggerated adrenomedullary response to immobilization in mice with targeted disruption of the serotonin transporter gene. *Endocrinology* 143:4520–6.

Träskman, L., Asberg, M., Bertilsson, L., and Sjöstrand, L. 1981. Monoamine metabolites in CSF and suicidal behavior. *Arch Gen Psychiatry* 38:631–6.

Träskman-Bendz, L., Asberg, M., Bertilsson, L., and Thorén, P. 1984. CSF monoamine metabolites of depressed patients during illness and after recovery. *Acta Psychiatr Scand* 69:333–42.

Trowbridge, S., Narboux-Neme, N., and Gaspar, P. 2010. Genetic models of serotonin (5-HT) depletion: What do they tell us about the developmental role of 5-HT? *Anat Rec (Hoboken)* PMID: 20818612.

Twarog, B. M., and Page, I. H. 1953. Serotonin content of some mammalian tissues and urine and a method for its determination. *Am J Physiol* 175:157–61.

Uebelhack, R., Franke, L., Herold, N., Plotkin, M., Amthauer, H., and Felix, R. 2006. Brain and platelet serotonin transporter in humans—correlation between [^{123}I]-ADAM SPECT and serotonergic measurements in platelets. *Neurosci Lett* 406:153–8.

Uher, R., and McGuffin, P. 2008. The moderation by the serotonin transporter gene of environmental adversity in the aetiology of mental illness: Review and methodological analysis. *Mol Psychiatry* 13:131–46.

Vakil, R. J. 1949. A clinical trial of *Rauwolfia serpentina* in essential hypertension. *Br Heart J* 11:350–5.

van Dyck, C. H., Malison, R. T., Staley, J. K., et al. 2004. Central serotonin transporter availability measured with [^{123}I]beta-CIT SPECT in relation to serotonin transporter genotype. *Am J Psychiatry* 161:525–31.

Velazquez-Moctezuma, J., and Diaz Ruiz, O. 1992. Neonatal treatment with clomipramine increased immobility in the forced swim test: An attribute of animal models of depression. *Pharmacol Biochem Behav* 42:737–9.

Verney, C., Lebrand, C., and Gaspar, P. 2002. Changing distribution of monoaminergic markers in the developing human cerebral cortex with special emphasis on the serotonin transporter. *Anat Rec* 267:87–93.

Vialli, M., and Erspamer, V. 1937. Ricerche sul secreto delle cellule enterocromaffini. IX Intorno alla natura chimica della sostanza specifica. *Boll Soc Med-Chir Pavia* 51:1111–6.

Vialli, M., and Erspamer, V. 1940. Ricerche di caratterizzazione chimica delle sostanze fenoliche presenti negli estratti acetonici di ghiandole salivari posteriori di *Octopus vulgaris*. *Arch Fisiol* 40:293–302.

Vitalis, T., Cases, O., Callebert, J., et al. 1998. Effects of monoamine oxidase A inhibition on barrel formation in the mouse somatosensory cortex: Determination of a sensitive developmental period. *J Comp Neurol* 393:169–84.

Vitalis, T., Fouquet, C., Alvarez, C., et al. 2002. Developmental expression of monoamine oxidases A and B in the central and peripheral nervous systems of the mouse. *J Comp Neurol* 442:331–47.

Vogel, G., Neill, D., Hagler, M., and Kors, D. 1990. A new animal model of endogenous depression: A summary of present findings. *Neurosci Biobehav Rev* 14:85–91.

Walderhaug, E., Magnusson, A., Neumeister, A., et al. 2007. Interactive effects of sex and *5-HTTLPR* on mood and impulsivity during tryptophan depletion in healthy people. *Biol Psychiatry* 62:593–9.

Walther, D. J., Peter, J.-U., Bashammakh, S., et al. 2003. Synthesis of serotonin by a second tryptophan hydroxylase isoform. *Science* 299:76.

Wendland, J. R., Moya, P. R., Kruse, M. R., et al. 2008. A novel, putative gain-of-function haplotype at SLC6A4 associates with obsessive–compulsive disorder. *Hum Mol Genet* 17:717–23.

White, K. J., Walline, C. C., and Barker, E. L. 2005. Serotonin transporters: Implications for antidepressant drug development. *AAPS J* 7:E421–33.

Wilkins, R. W., and Judson, W. E. 1953. The use of *Rauwolfia serpentina* in hypertensive patients. *N Engl J Med* 248:48–53.

Willeit, M., Stastny, J., Pirker, W., et al. 2001. No evidence for in vivo regulation of midbrain serotonin transporter availability by serotonin transporter promoter gene polymorphism. *Biol Psychiatry* 50:8–12.

Yang, G. B., Qiu, C. L., Zhao, H., Liu, Q., and Shao, Y. 2006. Expression of mRNA for multiple serotonin (5-HT) receptor types/subtypes by the peripheral blood mononuclear cells of rhesus macaques. *J Neuroimmunol* 178:24–9.

Yang, G. B., Qiu, C. L., Aye, P., Shao, Y., and Lackner, A. A. 2007. Expression of serotonin transporters by peripheral blood mononuclear cells of rhesus monkeys (*Macaca mulatta*). *Cell Immunol* 248:69–76.

Yates, M., Leake, A., Candy, J. M., Fairbairn, A. F., McKeith, I. G., and Ferrier, I. N. 1990. 5HT2 receptor changes in major depression. *Biol Psychiatry* 27:489–96.

Zalsman, G., Huang, Y. Y., Oquendo, M. A., et al. 2006. Association of a triallelic serotonin transporter gene promoter region (*5-HTTLPR*) polymorphism with stressful life events and severity of depression. *Am J Psychiatry* 163:1588–93.

Zhang, X., Beaulieu, J.-M., Sotnikova, T. D., Gainetdinov, R. R., and Caron, M. G. 2004. Tryptophan hydroxylase-2 controls brain serotonin synthesis. *Science* 305:217.

Zhang, X., Gainetdinov, R. R., Beaulieu, J.-M., et al. 2005. Loss-of-function mutation in tryptophan hydroxylase-2 identified in unipolar major depression. *Neuron* 45:11–6.

Zhang, X., Nicholls, P. J., Laje, G., et al. 2010. A functional alternative splicing mutation in human tryptophan hydroxylase-2. *Mol Psychiatry* doi: 10.1038/mp.2010.99.

Zipursky, R. B., Meyer, J. H., and Verhoeff, N. P. 2007. PET and SPECT imaging in psychiatric disorders. *Can J Psychiatry* 52:146–57.

8 Noradrenergic System in Depression

J. Javier Meana, Luis F. Callado, and Jesús A. García-Sevilla

CONTENTS

8.1 MONOAMINES AND DEPRESSION

The monoaminergic hypothesis of depression states that major depression could be related to a deficiency of brain monoaminergic activity as depression responds to drugs that increase monoamine concentration in the synapses (Bunney and Davis 1965; Schildkraut 1965; Coppen 1967). Thus, treatment with monoamine reuptake inhibitors or monoamine oxidase (MAO) inhibitors induces clinical antidepressant activity, whereas treatment with the monoamine vesicular depletor reserpine produces a depressive syndrome.

Several controlled studies have tried to clarify the role of serotonin and catecholamines such as noradrenaline in individual mood state. A general reduction in brain serotonin levels can be achieved via the tryptophan depletion technique. In these experiments, subjects are given a diet from which tryptophan has been omitted. Since tryptophan is the precursor amino acid for serotonin, synthesis of serotonin is reduced. On the other hand, α-methyl-p-tyrosine induces the inhibition of catecholamine synthesis (dopamine and noradrenaline in the central nervous system [CNS]) via the inhibition of the rate-limiting step enzyme tyrosine hydroxylase (TH) (Heninger et al. 1996). The mood-altering effect of tryptophan depletion was more evident in recently remitted patients, who displayed a remarkably rapid return of depressed mood and the cognitive and neurovegetative symptoms of depression. Moreover, patients treated with selective serotonin reuptake inhibitors are much more sensitive to tryptophan depletion manipulation than to catecholamine depletion, and they are more sensitive to tryptophan depletion than patients treated with noradrenergic active antidepressants (Delgado et al. 1990; Miller et al. 1996). In the same way, depressed subjects who responded to antidepressant treatment with noradrenaline reuptake inhibitors are much more sensitive to catecholamine depletion than to tryptophan depletion (Delgado et al. 1993; Miller et al. 1996). These results suggest that the beneficial effects of antidepressant treatment are directly related to the increase in the concentration of neurotransmitters they produce but do not clarify if other changes in the CNS are necessary. Indeed, reports about depressed mood after tryptophan depletion in subjects without any psychiatric diagnosis have been inconsistent (Young et al. 1985; Delgado et al. 1990; Moreno et al. 1999). Similarly, catecholamine depletion does not produce depressed mood in subjects without psychiatric diagnosis (Miller et al. 1996; Ruhé et al. 2007). The occurrence of depression after monoamine depletion occurs more prominently in subjects without psychiatric diagnosis who have first-degree relatives suffering from major depression than in subjects with no family history of major depression (Benkelfat et al. 1994; Ellenbogen et al. 1996; Ruhé et al. 2007), providing support for a genetic component in depression.

8.2 CATECHOLAMINES IN DEPRESSED SUBJECTS

Numerous studies have tried to analyze noradrenergic and serotonergic neurotransmission directly in the CNS of depressed subjects (Table 8.1). However, ethical and methodological aspects have limited these studies.

Essentially, monoamine metabolites have been determined in the cerebrospinal fluid (CSF) of depressed subjects: 3-methoxy-4-hydroxyphenylglycol (MHPG) and homovanillic acid for noradrenaline and dopamine, respectively, and 5-hydroxyindolacetic acid (5-HIAA) for serotonin. Despite the many studies, it has not been possible to establish a general metabolic pattern for all depressed subjects (van Praag 1982). Interestingly, one of the most robust findings is the correlation between low 5-HIAA CSF concentrations and suicidal behavior (Åsberg et al. 1976; Mann et al.

TABLE 8.1
Opportunities and Pitfalls in the Study of Noradrenergic System in Depression

Biological Marker	Advantages	Disadvantages
Catecholamines and Metabolites in Plasma and Urine	Easy to obtain	Peripheral nervous system activity superimposed
Metabolites in the CSF	Indirect access to CNS in vivo	Confounding variables: aggressive behavior, etc.
		Methodological variability
TH inhibition	In vivo painless test	Marker of response to noradrenergic antidepressants
Monoamine oxidase enzyme	In vivo PET available	Confounding variables: smoking, genetic polymorphisms influencing activity
	Peripheral expression in platelets (MAO-B)	
Adrenergic receptor densities	Peripheral expression in blood cells	Methodological variability
	Short-term stability	In vivo PET unavailable
		Sensitive to treatment: state marker
Hormone response to drugs	Functional in vivo test	Low sensitivity and specificity
Neuropathological studies	Accessible to molecular and cellular studies	Limitations of postmortem tissues
Genetics	Accessibility	Confounding variables: endophenotypes
	Approach to etiology of the disorder: risk factor	Environmental influence on gene activity
-omics	High-throughput screening	Large number of samples to obtain reliable findings
	New hypothesis	

Note: CSF, cerebrospinal fluid; PET, positron emission tomography.

1992). However, evaluation of neurotransmitter and metabolite concentrations in the CSF has been controversial. First, it is difficult to extrapolate values at cerebral or spinal level to the whole CNS, and second, one isolated measure does not reflect the metabolic rhythm of the system.

Plasma noradrenaline is increased in unipolar depressives with melancholia, indicating a higher peripheral sympathetic nervous activity. The measurement of the plasma MHPG levels, as a representative metabolite of CNS noradrenergic activity, has yielded highly variable results (Potter and Manji 1994). Studies performed with urinary MHPG have shown decreased MHPG concentration in depressed subjects (Maas et al. 1972) as indicative of low central noradrenergic activity. Nevertheless, this MHPG index has been questioned as sport, food, and stress can importantly modify urinary MHPG concentration.

Several studies have analyzed noradrenaline concentration in *postmortem* human brains of depressed suicide victims. Unaltered concentrations of noradrenaline or its metabolites were found (Bourne et al. 1968; Beskow et al. 1976). One study reported a decreased noradrenaline concentration (Birkmayer and Riederer 1975), whereas another found an increased concentration (Arango et al. 1993). The confounding effect of suicide is common in most of these studies. Because TH is the rate-limiting enzyme in noradrenaline synthesis, some authors have evaluated its expression in the nucleus locus coeruleus, which is the main noradrenergic source in the CNS. In these studies, an association was found between increased levels of TH and major depression (Ordway 1997; Zhu et al. 1999).

The enzyme MAO plays an important role in the inactivation of monoamines in the CNS. Recent findings have demonstrated by in vivo positron emission tomography neuroimaging that the isoform A of MAO is increased in the brains of subjects with major depressive disorder and normalized after antidepressant treatment (Meyer et al. 2009). In contrast, postmortem brain studies do not reflect this increase of MAO-A (Ordway et al. 1999). The MAO-B isoform is expressed by astrocytes and platelets. MAO-B density has been found increased in postmortem brains of suicide victims (Ballesteros et al. 2008), but in platelets the findings are more controversial (for a review, see Ballesteros et al. 2008). Therefore, although inconclusive, the actions of MAO-A and MAO-B on the metabolism of the different monoamines support the view that hyperactivity of these enzymes may be a contributing factor in the decreased neurotransmitter availability in depressive disorders.

The monoaminergic hypothesis of depression, however, does not explain several facts. First, there are drugs that increase brain monoaminergic activity (e.g., cocaine and amphetamine) but are not clinically effective as antidepressants. Conversely, some antidepressants do not increase local monoamine concentrations in the synapses but act as postsynaptic receptor antagonists (e.g., trazodone). Second, not all depressed patients respond equally to the same antidepressant. Third, and most importantly, changes in the synaptic monoamine concentrations take place within hours after the administration of the antidepressant, but the therapeutic response requires the continuous administration of antidepressant drugs for weeks (Baldessarini 1989). There is also considerable evidence that antidepressants exert effects on monoaminergic pathways both on presynaptic and postsynaptic locations, acting at somatodendritic and nerve terminal areas. This fact represents a more complex regulation of final in vivo responses to antidepressants and considerably expands the potential malfunctioning targets in the brains of depressed subjects.

8.3 HYPOTHESIS OF RECEPTOR SENSITIVITY IN DEPRESSION

The abovementioned limitations of the monoaminergic hypothesis displaced the investigation to the study of monoamine receptors and coupled mechanisms. The time lag observed between the antidepressant administration and the antidepressant effect suggests that adaptations that occur over time are required for the therapeutic action of antidepressants. Such adaptations could also explain why the maintenance of antidepressant treatment is dependent on the presence of monoamines. The mechanisms underlying this neuronal adaptation could involve neurotransmitter receptors, receptor-coupled intracellular signal, transduction pathways, and neuronal gene expression. In this context, monoamine receptor sensitivity hypothesis emerged (Charney et al. 1981a).

Release of serotonin and noradrenaline to the synapse leads to the activation of pre- and post-synaptic receptors. The sustained release and/or elevation of monoamines by antidepressant treatment results in continued activation of the receptors, which could finally lead to desensitization and downregulation of the receptor-dependent functions. Cross-desensitization between receptors is possible, which indicates the existence of common regulatory signaling processes between receptors and/or the influence on gene regulation activities.

Initially, β-adrenoceptors and serotonin 5-HT$_2$ receptors were the most studied receptor systems (Sulser et al. 1978; Peroutka and Snyder 1980). Several years later, and consistent with the suspected presynaptic dysfunction in the brains of depressed subjects, hypersensitivity was also studied for presynaptic inhibitory receptors such as α$_2$-adrenoceptors.

8.3.1 BRAIN α$_1$-ADRENOCEPTORS IN DEPRESSION

Studies on the status of α$_1$-adrenoceptors in depression show controversial results. A significant decrease in the density of these receptors in several areas of prefrontal, temporal, and parietal cortices in depressed suicide victims has been reported, whereas other brain regions, such as the hippocampus, showed no changes (Gross-Isseroff et al. 1990). On the other hand, other studies found no difference (De Paermentier et al. 1997) or even increase (Arango et al. 1993) in the α$_1$-adrenoceptor density in postmortem brain. In some of these studies, the inclusion of subjects under chronic treatment with tricyclic antidepressants, which possess α$_1$-adrenoceptor antagonist properties, could explain the apparent discrepancies and even the obtained findings.

8.3.2 EFFECT OF ANTIDEPRESSANT TREATMENT ON BRAIN α$_1$-ADRENOCEPTORS

Several studies on the effect of antidepressant treatment on brain α$_1$-adrenoceptors indirectly support the notion of an impaired function of these receptors in depression. In this regard, it has been reported that chronic treatment with antidepressants or electroconvulsive shock increases the number, the affinity, or the responsiveness of the brain α$_1$-adrenoceptors (Stone et al. 2003; Elhwuegi 2004).

8.3.3 α$_2$-ADRENOCEPTORS IN DEPRESSION

Numerous studies have analyzed the density of α$_2$-adrenoceptors in the postmortem brains of depressed subjects, many of whom died by suicide. These studies have repeatedly reported an increased density of α$_2$-adrenoceptors (Meana et al. 1992; González et al. 1994; De Paermentier et al. 1997; Callado et al. 1998; García-Sevilla et al. 1999; Ordway et al. 2003). A similar finding has been also described in platelets of depressed patients (García-Sevilla et al. 1981, 1986, 2004; Piletz and Halaris 1988; Piletz et al. 1990; Gurguis et al. 1999b). Platelets represent a peripheral model of CNS cells with some similar properties including serotonin reuptake machinery, MAO-B expression, presence of brain-derived neurotrophic factor (BDNF) and functional activities coupled to monoaminergic receptors. Therefore, blood platelet parameters can be used as peripheral biomarkers to follow CNS responses to antidepressant treatments.

The increased α$_2$-adrenoceptor density reported in brains of depressed suicide victims seems to be selective for the α$_{2A}$-adrenoceptor subtype (Callado et al. 1998), although a more exhaustive evaluation of α$_{2B}$- and α$_{2C}$-adrenoceptors in postmortem brains of depressed subjects is still lacking. This increased α$_2$-adrenoceptor density in postmortem brains of depressed subjects is preferentially observed when agonist radioligands such as [^{3}H]clonidine, [^{3}H]UK14304, or [^{125}I]p-iodoclonidine are used (Meana et al. 1992; González et al. 1994; Callado et al. 1998; Ordway et al. 2003). Conversely, when antagonist radioligands have been used ([^{3}H]idazoxan, [^{3}H]rauwolscine, or [^{3}H] RX821002), no significant differences in α$_2$-adrenoceptor density in depressed subjects compared with controls have been observed (Crow et al. 1984; Ferrier et al. 1986; Meana et al. 1992; Callado et al. 1998). The high-affinity state of the receptor is better labeled by agonist radioligands, whereas antagonists bound both high- and low-affinity states (García-Sevilla et al. 1981; Kenakin 2004). Thus, the disparity of the results with agonists and antagonists might be ascribed to an imbalance between high- and low-affinity states of the α_2-adrenoceptor (Callado et al. 1998). These studies suggest that total α_2-adrenoceptor density is not altered in the postmortem brains of depressed subjects but that α_2-adrenoceptor density in high-affinity state, the functional G-protein-coupled state, is increased. However, some discrepancies have also been reported. Thus, in postmortem brains of depressed subjects, Klimek et al. (1999) reported unaltered α_2-adrenoceptor density with the agonist radioligand [^{125}I]p-iodoclonidine, and De Paermentier et al. (1997) described increased α_2-adrenoceptor density with antagonist radioligands. Moreover, α_2-adrenoceptor mRNA (Escribá et al. 2004) and immunoreactive levels of the receptor protein (García-Sevilla et al. 1999) have been described to be increased in the prefrontal cortex of depressed subjects. Therefore, an increased synthesis of α_2-adrenoceptors cannot be discarded. In vivo neuroimaging of α_2-adrenoceptors by using the radiotracer [^{11}C]mirtazapine has recently shown a decreased binding potential in cortical areas of nonresponder depressed subjects (Smith et al. 2009).

Among the mechanisms that may underlie the increased α_2-adrenoceptors in high-affinity state, altered α_2-adrenoceptor coupling to $G_{i/o}$ proteins has been suggested. Indeed, when the functional α_2-adrenoceptor response was evaluated by agonist-stimulated [^{35}S]GTPγS binding studies, α_2-adrenoceptor supersensitivity was observed in the brains of depressed subjects (González-Maeso et al. 2002). Another possible mechanism accounting for the increased α_2-adrenoceptor density in high-affinity state might be an increased availability of $G_{i/o}$ heterotrimeric proteins. Although the results on G proteins densities in the brains of depressed subjects are controversial (González-Maeso and Meana 2006), G protein activity and receptor coupling could be misregulated by different intracellular factors. In this sense, the proteins termed *regulator of G-protein signaling* modulate G protein activity in a negative manner and could thereby interfere in the G proteins coupling to α_2-adrenoceptors. Therefore, the evaluation of the α_2-adrenoceptor precoupling to $G_{i/o}$ proteins and the study of the regulator of G-protein signaling expression in postmortem brains of depressed subjects would provide another approach to the study of the α_2-adrenoceptor supersensitivity in depression. On the other hand, desensitization of monoamine receptors is dependent of G-protein-coupled receptor kinase (GRK) proteins and arrestins. GRK phosphorylates G-protein-coupled receptors and bind to a family of proteins termed *arrestins*, which induce uncoupling of receptor from G proteins (Claing et al. 2002). GRK2, but not other GRK proteins, has been described altered in the brains and blood cells of subjects with depression, acting as surrogate marker of response to antidepressant treatment (García-Sevilla et al. 1999, 2004, 2010; Grange-Midroit et al. 2003; Matuzany-Ruban et al. 2010). Whether GRK2 alterations contribute to α_2-adrenoceptor hyperactivity in depressed subjects remains to be elucidated.

Receptor-coupled signaling is a difficult functional methodology to perform in postmortem tissue. Therefore, α_2-adrenoceptor-dependent signaling has been evaluated in platelets of depressed patients (Kafka and Paul 1986; García-Sevilla et al. 1990). A seminal work (González-Maeso et al. 2002) demonstrated the functional hyperactive α_2-adrenoceptor in postmortem brains of depressed subjects. More recently, a new study has provided additional evidence about the functional status of this increased α_2-adrenoceptor-mediated signaling in the pathophysiology of depression. Thus, a dual modification in the postmortem frontal cortex brain of depressed subjects is observed with an increase in the α_{2A}-adrenoceptor-mediated G-protein coupling associated to desensitization in the inhibition of the adenylyl cyclase activity produced by these receptors (Valdizán et al. 2010).

Pharmacological challenge strategies have been used to test the functional responses of α_2-adrenoceptors in depressive states. Thus, the growth hormone shows a blunted response to acute administration of clonidine (Potter and Manji 1994), suggesting subsensitive postsynaptic α_2-adrenoceptors at the hypothalamic level. Changes in plasma MHPG, blood pressure, heart rate, sedation, and cortisol after administration of α_2-adrenoceptor agonists or antagonists have been extensively studied. The interpretation of these studies directed to evaluate the functionality of

α_2-adrenoceptors in major depression is sometimes complex because of the simultaneous presynaptic and postsynaptic location of these receptor. Thus, a higher sensitivity to the increase in cerebral blood flow induced by the agonist clonidine has been observed, suggesting a supersensitivity of postsynaptic cortical receptors (Fu et al. 2001). Similarly, melatonin response to clonidine administration used as presynaptic α_2-adrenoceptor indicator is hyperactive (Paparrigopoulos et al. 2001). Postsynaptic cortisol responses to the antagonist yohimbine are also increased in depressed subjects without changes in the plasma levels of the presynaptic indicator MHPG (Price et al. 1986).

8.3.4 EFFECT OF ANTIDEPRESSANT TREATMENT ON α_2-ADRENOCEPTORS

Presynaptically located α_2-adrenoceptors negatively control noradrenaline and serotonin release as auto- or heteroreceptors, respectively (Starke 1987). Therefore, it would be expected that desensitization of α_2-adrenoceptors increased the availability of both noradrenaline and serotonin at the synapse (Mongeau et al. 1994; Mateo et al. 2001).

In human peripheral cells, decreased density and/or sensitivity of platelet α_2-adrenoceptors after antidepressant treatment have been reported (García-Sevilla et al. 1981, 1986, 1990, 2004; Gurguis et al. 1999b). Moreover, imipramine treatment induced a decrease in density and coupling of platelet α_2-adrenoceptors only in treatment responders but not in treatment nonresponders (Gurguis et al. 1999b). These results suggest that increased α_2-adrenoceptor density after antidepressant treatment could represent a marker of lack of response to antidepressant treatment. On the other hand, other authors have reported that α_2-adrenoceptor binding parameters remained unchanged in platelets of patients with major depression after antidepressant treatment (Pandey et al. 1989; Marazziti et al. 2010). Nevertheless, information about the effect of antidepressant treatment on α_2-adrenoceptors in postmortem human brain or in vivo neuroimaging of depressed subjects is very limited. Most of these studies have been performed with samples of suicide victims with unknown psychiatric diagnosis, antidepressant treatment information, and toxicological data, or the experimental design and conditions were not controlled for that purpose. In a study on suicide victims with unknown psychiatric diagnosis, De Paermentier et al. (1997) found that α_2-adrenoceptor density was higher in the temporal cortex of suicide victims who had not received antidepressant treatment. In contrast, in suicide victims who were receiving antidepressant treatment at the time of death, α_2-adrenoceptor density was similar to controls in the temporal cortex and even lower in the occipital cortex and hippocampus.

8.3.5 β-ADRENOCEPTORS IN DEPRESSION

There are very few reports with regard to brain β-adrenoceptors in major depression. Also in this case, the main part of the studies corresponds to suicide victims with unknown psychiatric diagnosis, and these have resulted in controversial data (Brunello et al. 2002). No alterations in frontal cortical β-adrenoceptor population have been reported in subjects with major depression who died by natural causes (Ferrier et al. 1986) and in suicide victims with (De Paermentier et al. 1991) or without (De Paermentier et al. 1990) antidepressant treatment. Nevertheless, De Paermentier et al. (1992) observed lower density of total β- and $\beta1$-adrenoceptors in other cortical areas and also in the frontal cortex of antidepressant-free suicide victims who died by violent means. The use of the radioligand [^{3}H]dihydroalprenolol in most of these binding studies represents a problem because serotonin receptors may be labeled in addition to β-adrenoceptors. β_1-Adrenoceptor subsensitivity has also been described in the frontal cortex of depressed suicide victims by using cAMP accumulation response to agonist drugs (Valdizán et al. 2003).

Mononuclear blood cells express β-adrenoceptors that correspond to the β_2-adrenoceptor subtype. The density of β-adrenoceptors in lymphocytes and leukocytes of untreated patients has yielded conflicting results with decreases, increases, or no changes. Methodological conditions and the source of patients could explain the apparent controversy (Werstiuk et al. 1990). Results of studies on functional responses to β-adrenoceptor stimulation in depressive disorders are more consistent. A decreased β-adrenoceptor-mediated stimulation of adenylyl cyclase activity has been repeatedly observed (Mann et al. 1997). In human fibroblasts from depressed patients, a reduced signaling of β-adrenoceptor-linked protein kinase A activity has also been found (Manier et al. 1996).

8.3.6 EFFECT OF ANTIDEPRESSANT TREATMENT ON β-ADRENOCEPTORS

According to desensitization and/or downregulation mechanisms, chronic exposure to an agonist such as noradrenaline will result in a decrease in the number and/or sensitivity of postsynaptic receptors. This prediction has been consistently reported to be true for β-adrenoceptors. It has been reported that chronic antidepressant treatment decreased β-adrenoceptor number in rat brain (Ordway et al. 1991; Harkin et al. 2000). Furthermore, β-adrenoceptor function is also decreased after antidepressant treatment as evaluated at adenylyl cyclase activation level (Gillespie et al. 1979). However, the changes observed in peripheral β_2-adrenoceptor density and/or function are less consistent (Gurguis et al. 1999a). These results led to the hypothesis that the therapeutic action of antidepressant treatment could be mediated by downregulation of β-adrenoceptor sites and that upregulation of these receptors may underlie the depressive state (Sulser et al. 1978; Charney et al. 1981b). However, there are several problems with this hypothesis. First, not all antidepressants downregulate β-adrenoceptors (Charney et al. 1981b). This could mean that the action of different antidepressants is mediated by different receptors or that other receptor sites are more relevant to the action of antidepressants treatments. Second, the downregulation of β-adrenoceptors is more rapid than the onset of the therapeutic response to these treatments (Riva and Creese 1989). Third, inhibition of β-adrenoceptors by treatment with a selective antagonist is not an effective treatment for depression. In fact, β-adrenoceptor antagonists are reported to produce depression in some individuals (Paykel et al. 1982; Avorn et al. 1986) and β-adrenoceptor agonists have antidepressant effects in behavioral models of depression. Moreover, activation or facilitation of β-adrenoceptor function by thyroid hormone or a specific receptor agonist can have antidepressant efficacy in some patients (Goodwin et al. 1982). Taken together, these findings indicate that β-adrenoceptor downregulation does not mediate the therapeutic action of antidepressants. In fact, results are more consistent with the possibility that β-adrenoceptor downregulation may actually lead to depression (Duman 1999).

8.4 NEUROPATHOLOGICAL AND NEUROCHEMICAL FINDINGS IN NUCLEUS LOCUS COERULEUS OF DEPRESSED SUBJECTS

The locus coeruleus is the main source of noradrenergic cells in the CNS. Projections from the locus coeruleus innervate the brain cortex, subcortical areas, and spinal cord. As a result, there have been different studies on depression looking at neuropathological and neurochemical alterations in the locus coeruleus. The vast majority of the samples represent suicide victims, making it difficult to differentiate between the findings relating to suicide and depression.

A reduced density of noradrenergic neurons in depressed suicide victims has been suggested, although other studies did not reproduce this finding (see review by Ressler and Nemeroff 1999). In contrast, expression of the TH enzyme in the area is increased in the brains of depressed subjects (Ordway 1997; Zhu et al. 1999) together with a decreased level of the noradrenaline reuptake transporter (NET) (Klimek et al. 1997). These observations led to the hypothesis of dysregulated locus coeruleus noradrenergic neurons in depressive disorders with subsequent adaptive changes in the neurochemical processes of noradrenaline synthesis, reuptake, and density of α_2-adrenoceptors. In this context, the immunoreactive presence of neuronal nitric oxide synthase, an intracellular mediator of glutamate receptor activation, seems to reflect altered excitatory input to the locus coeruleus in major depression (Karolewicz et al. 2004). Therefore, whether intrinsically or extrinsically induced alterations of locus coeruleus are present in depression is still not sufficiently clarified.

8.5 BRAIN NEUROGENESIS RELATED TO ACTIVITY OF THE NORADRENERGIC SYSTEM

The decreased volume observed in the hippocampus of depressed subjects has given support to the neurotrophic hypothesis of depression (Duman and Monteggia 2006). The hypothesis suggests a decreased neural plasticity in the adult brain involving decrements in neurotrophic factors, especially the BDNF. Elevated concentrations of glucocorticoids reduce neurogenesis, and this was proposed as the mechanism for the decreased size of the hippocampus. In animals, chronic antidepressant treatment enhances neurogenesis and the BDNF-mediated signaling, whereas blockade of hippocampal neurogenesis inhibits the behavioral effects of most antidepressants. A similar increase of neural proliferative cells after antidepressant use has been demonstrated in the postmortem human brain (Boldrini et al. 2009). In line with these findings, the addition of α_2-adrenoceptor antagonist drugs to antidepressants accelerates the neurogenic and neurotrophic activity of chronic antidepressant treatment in rodent models (Yanpallewar et al. 2010), highlighting the importance of this receptor as a target for antidepressant activity.

8.6 GENETIC STUDIES OF THE NORADRENERGIC SYSTEM IN DEPRESSION

The genetic influence on development of depressive disorders is well established. The heritability of major depression is estimated to be 31% to 42%. However, consistent findings implicating specific genes or polymorphisms are scarce. It is probable that the polygenic component of depression and the important influence of nonheritable factors (environmental influence, epigenetics, etc.) are critical factors in the knowledge of the pathogenesis of depression.

Almost all confirmed findings on single nucleotide polymorphism alterations in depression have implicated the serotonergic system. Polymorphism of the serotonin reuptake transporter and the synthesis enzyme tryptophan hydroxylase has demonstrated functional significance. However, clear associations of gene polymorphisms in the noradrenergic systems with major depression have not been identified.

The synthesis enzyme TH was proposed to show gene variants in depressive symptoms (Serretti et al. 1998; Furlong et al. 1999), but the findings have not been replicated (López-León et al. 2008). In the case of NET, the influence of genetic polymorphisms is more at the pharmacogenetic response to antidepressants than at the pathogenesis of depression. The most recent genome-wide association studies ascribe to NET a potential linkage with major depressive disorders, although the possibility of false-positive findings is latent (Bosker et al. 2011).

The enzyme catechol-O-methyltransferase (COMT) contributes to the metabolism of catecholamines and formation of MHPG. An important variant polymorphism termed *Val/Met* that determines loss of functional COMT activity has been described (Meyer-Lindenberg et al. 2005). This gene polymorphism plays an important role in the cognitive symptoms of schizophrenia because it exerts influence on dopamine metabolism in the prefrontal cortex. However, the influence on potential cognitive alterations in the context of depressive disorders has not been demonstrated. It is probable that the influence of COMT on noradrenaline metabolism is lower than that exerted on cortical dopamine. In the frontal cortex, synaptic noradrenaline is mainly reduced through reuptake by NET, whereas dopamine reuptake transporter is expressed in very low density, forcing dopamine to be metabolized by COMT.

Among adrenoceptors, different gene polymorphisms have been associated with depressive disorder in the α_{2A}- and α_{2C}-adrenoceptors (Sequeira et al. 2004; Martín-Guerrero et al. 2006; Neumeister et al. 2006; Perroud et al. 2009). Interestingly, a C to G transversion at nucleotide 753 that induces an Asn to Lys change at amino acid 251 or N251K, and that results in a gain-of-function phenotype of the α_{2A}-adrenoceptor gene, has been tested in the context of the previously demonstrated hyperactivity of this adrenoceptor in depression (Sequeira et al. 2004; Martín-Guerrero et al. 2006). Unfortunately, no clear association with depression or suicide was observed.

In the past decade, important high-throughput studies have been performed in peripheral blood cells and postmortem human brains of depressed subjects by using microarrays of DNA. Such observations have provided evidence for the absence of altered mRNA expression of the genes influencing noradrenergic transmission. Despite these pitfalls, the microarray approaches have enabled researchers to formulate and support a new hypothesis in relation to glial dysfunction and altered neuroplasticity in the brains of depressed subjects (Pittenger and Duman 2008).

REFERENCES

Arango, V., Ernsberger, P., Sved, A. F., and Mann, J. J. 1993. Quantitative autoradiography of alpha 1- and alpha 2-adrenergic receptors in the cerebral cortex of controls and suicide victims. *Brain Res* 630:271–82.

Åsberg, M., Träskman, L., and Thoren, P. 1976. 5-HIAA in the cerebrospinal fluid. A biochemical suicide predictor? *Arch Gen Psychiatry* 33:1193–7.

Avorn, J., Everitt, D. E., and Weiss, S. 1986. Increased antidepressant use in patients prescribed beta-blockers. *JAMA* 255:357–60.

Baldessarini, R. J. 1989. Current status of antidepressants: Clinical pharmacology and therapy. *J Clin Psychiatry* 50:117–26.

Ballesteros, J., Maeztu, A. I., Callado, L. F., Meana, J. J., and Gutiérrez, M. 2008. Specific binding of [^{3}H]Ro 19-6327 (lazabemide) to monoamine oxidase B is increased in frontal cortex of suicide victims after controlling for age at death. *Eur Neuropsychopharmacol* 18:55–61.

Benkelfat, C., Ellenbogen, M. A., Dean, P., Palmour, R. M., and Young, S. N. 1994. Mood lowering effect of tryptophan depletion. Enhanced susceptibility in young men at genetic risk for major affective disorders. *Arch Gen Psychiatry* 51:687–97.

Beskow, J., Gottfries, C. G., Roos, B. E., and Winblad, B. 1976. Determination of monoamine and monoamine metabolites in the human brain: Post mortem studies in a group of suicides and in a control group. *Acta Psychiatr Scand* 53:7–20.

Birkmayer, W., and Riederer, P. 1975. Biochemical post-mortem findings in depressed patients. *J Neural Transm* 37:95–109.

Boldrini, M., Underwood, M. D., Hen, R., et al. 2009. Antidepressants increase neural progenitor cells in the human hippocampus. *Neuropsychopharmacology* 34:2376–89.

Bosker, F. J., Hartman, C. A., Nolte, I. M., et al. 2011. Poor replication of candidate genes for major depressive disorder using genome-wide association data. *Mol Psychiatry* 16:516–32.

Bourne, M. R., Bunney, W. E., Colburn, R. W., et al. 1968. Noradrenaline, 5-hydroxytryptamine and 5-hydroxy-indolacetic acid in hindbrains of suicidal patients. *Lancet* 2:805–8.

Brunello, N., Mendlewicz, J., Kasper, S., et al. 2002. The role of noradrenaline and selective noradrenaline reuptake inhibition in depression. *Eur Neuropsychopharmacol* 12:461–75.

Bunney, W. E., and Davis, J. M. 1965. Norepinephrine in depressive reactions. *Arch Gen Psychiatry* 13:483–94.

Callado, L. F., Meana, J. J., Grijalba, B., Pazos, A., Sastre, M., and García-Sevilla, J. A. 1998. Selective increase of alpha2A-adrenoceptor agonist binding sites in brains of depressed suicide victims. *J Neurochem* 70:1114–23.

Charney, D. S., Heninger, G. R., Sternberg, D. E., et al. 1981a. Presynaptic adrenergic receptor sensitivity in depression. The effect of long-term desipramine treatment. *Arch Gen Psychiatry* 38:1334–40.

Charney, D. S., Menkes, D. B., and Heninger, G. R. 1981b. Receptor sensitivity and the mechanism of action of antidepressant treatment. Implications for the etiology and therapy of depression. *Arch Gen Psychiatry* 38:1160–80.

Claing, A., Laporte, S. A., Caron, M. G., and Lefkowitz, R. J. 2002. Endocytosis of G protein-coupled receptors: Roles of G protein-coupled receptor kinases and beta-arrestin proteins. *Prog Neurobiol* 66:61–79.

Coppen, A. 1967. The biochemistry of affective disorders. *Br J Psychiatry* 113:1407–11.

Crow, T. J., Cross, A. J., Cooper, S. J., et al. 1984. Neurotransmitter receptors and monoamine metabolites in the brains of patients with Alzheimer-type dementia and depression, and suicides. *Neuropharmacology* 23:1561–9.

De Paermentier, F., Cheetham, S. C., Crompton, M. R., Katona, C. L., and Horton, R. W. 1990. Brain beta-adrenoceptor binding sites in antidepressant-free depressed suicide victims. *Brain Res* 525:71–7.

De Paermentier, F., Cheetham, S. C., Crompton, M. R., Katona, C. L., and Horton, R. W. 1991. Brain beta-adrenoceptor binding sites in depressed suicide victims: Effects of antidepressant treatment. *Psychopharmacology (Berl)* 105:283–8.

De Paermentier, F., Crompton, M. R., Katona, C. L., and Horton, R. W. 1992. Beta-adrenoceptors in brain and pineal from depressed suicide victims. *Pharmacol Toxicol* 71 (Suppl. 1):86–95.

De Paermentier, F., Mauger, J. M., Lowther, S., Crompton, M. R., Katona, C. L., and Horton, R. W. 1997. Brain alpha-adrenoceptors in depressed suicides. *Brain Res* 757:60–8.

Delgado, P. L., Charney, D. S., Price, L. H., Aghajanian, G. K., Landis, H., and Heninger, G. R. 1990. Serotonin function and the mechanism of antidepressant action. Reversal of antidepressant-induced remission by rapid depletion of plasma tryptophan. *Arch Gen Psychiatry* 47:411–8.

Delgado, P. L., Miller, H. L., Salomon, R. M., et al. 1993. Monoamines and the mechanism of antidepressant action: Effects of catecholamine depletion on mood of patients treated with antidepressants. *Psychopharmacol Bull* 29:389–96.

Duman, R. 1999. The neurochemistry of mood disorders: Preclinical studies. In *Neurobiology of Mental Illness*, 1st edn, ed. D. S. Charney, E. J. Nestler, and B. S. Bunney, 333–47. New York: Oxford University Press.

Duman, R. S., and Monteggia, L. M. 2006. A neurotrophic model for stress-related mood disorders. *Biol Psychiatry* 59:1116–27.

Elhwuegi, A. S. 2004. Central monoamines and their role in major depression. *Prog Neuropsychopharmacol Biol Psychiatry* 28:435–51.

Ellenbogen, M. A., Young, S. N., Dean, P., Palmour, R. M., and Benkelfat, C. 1996. Mood response to acute tryptophan depletion in healthy volunteers: Sex differences and temporal stability. *Neuropsychopharmacology* 15:465–74.

Escribá, P. V., Ozaita, A., and García-Sevilla, J. A. 2004. Increased mRNA expression of alpha2A-adrenoceptors, serotonin receptors and mu-opioid receptors in the brains of suicide victims. *Neuropsychopharmacology* 29:1512–21.

Ferrier, I. N., McKeith, I. G., Cross, A. J., Perry, E. K., Candy, J. M., and Perry, R. H. 1986. Postmortem neurochemical studies in depression. *Ann N Y Acad Sci* 487:128–42.

Fu, C. H. Y., Reed, L. J., Meyer, J. H., et al. 2001. Noradrenergic dysfunction in the prefrontal cortex in depression: An [^{15}O] H$_2$O PET study of the neuromodulatory effects of clonidine. *Biol Psychiatry* 49:317–25.

Furlong, R. A., Rubinsztein, J. S., Ho, L., et al. 1999. Analysis and metaanalysis of two polymorphisms within the tyrosine hydroxylase gene in bipolar and unipolar affective disorders. *Am J Med Genet* 88:88–94.

García-Sevilla, J. A., Zis, A. P., Hollingsworth, P. J., Greden, J. F., and Smith, C. B. 1981. Platelet alpha 2-adrenergic receptors in major depressive disorder. Binding of tritiated clonidine before and after tricyclic antidepressant drug treatment. *Arch Gen Psychiatry* 38:1327–33.

García-Sevilla, J. A., Guimon, J., García-Vallejo, P., and Fuster, M. J. 1986. Biochemical and functional evidence of supersensitive platelet alpha 2-adrenoceptors in major affective disorder. Effect of long-term lithium carbonate treatment. *Arch Gen Psychiatry* 43:51–7.

García-Sevilla, J. A., Padro, D., Giralt, M. T., Guimon, J., and Areso, P. 1990. Alpha 2-adrenoceptor-mediated inhibition of platelet adenylate cyclase and induction of aggregation in major depression. Effect of long-term cyclic antidepressant drug treatment. *Arch Gen Psychiatry* 47:125–32.

García-Sevilla, J. A., Escribá, P. V., Ozaita, A., et al. 1999. Up-regulation of immunolabeled alpha2A-adrenoceptors, Gi coupling proteins, and regulatory receptor kinases in the prefrontal cortex of depressed suicides. *J Neurochem* 72:282–91.

García-Sevilla, J. A., Ventayol, P., Pérez, V., et al. 2004. Regulation of platelet alpha 2A-adrenoceptors, Gi proteins and receptor kinases in major depression: Effects of mirtazapine treatment. *Neuropsychopharmacology* 29:580–8.

García-Sevilla, J. A., Alvaro-Bartolomé, M., Diez-Alarcia, R., et al. 2010. Reduced platelet G protein–coupled receptor kinase 2 in major depressive disorder: Antidepressant treatment-induced upregulation of GRK2 protein discriminates between responder and non-responder patients. *Eur Neuropsychopharmacol* 20: 721–30.

Gillespie, D. D., Manier, D. H., and Sulser, F. 1979. Electroconvulsive treatment: Rapid subsensitivity of the norepinephrine receptor coupled adenylate cyclase system in brain linked to down regulation of beta-adrenergic receptors. *Commun Psychopharmacol* 3:191–5.

González, A. M., Pascual, J., Meana, J. J., et al. 1994. Autoradiographic demonstration of increased alpha 2-adrenoceptor agonist binding sites in the hippocampus and frontal cortex of depressed suicide victims. *J Neurochem* 63:256–65.

González-Maeso, J., and Meana, J. J. 2006. Heterotrimeric G proteins: Insights into the neurobiology of mood disorders. *Curr Neuropharmacol* 4:127–38.

González-Maeso, J., Rodríguez-Puertas, R., Meana, J. J., García-Sevilla, J. A., and Guimón, J. 2002. Neurotransmitter receptor-mediated activation of G-proteins in brains of suicide victims with mood disorders: Selective supersensitivity of alpha(2A)-adrenoceptors. *Mol Psychiatry* 7:755–67.

Goodwin, F. K., Prange, A. J. Jr., Post, R. M., Muscettola, G., and Lipton, M. A. 1982. Potentiation of antide-pressant effects by L-triiodothyronine in tricyclic nonresponders. *Am J Psychiatry* 139, 34–8.

Grange-Midroit, M., García-Sevilla, J. A., Ferrer-Alcón, M., La Harpe, R., Huguelet, P., and Guimón, J. 2003. Regulation of GRK2 and 6, beta-arrestin-2 and associated proteins in the prefrontal cortex of drug-free and antidepressant drug-treated subjects with major depression. *Brain Res Mol Brain Res* 111:31–41.

Gross-Isseroff, R., Dillon, K. A., Fieldust, S. J., and Biegon, A. 1990. Autoradiographic analysis of alpha 1-noradrenergic receptors in the human brain postmortem. Effect of suicide. *Arch Gen Psychiatry* 47:1049–53.

Gurguis, G. N., Vo, S. P., Griffith, J. M., and Rush, A. J. 1999a. Neutrophil beta(2)-adrenoceptor function in major depression: G(s) coupling, effects of imipramine and relationship to treatment outcome. *Eur J Pharmacol* 386:135–44.

Gurguis, G. N., Vo, S. P., Griffith, J. M., and Rush, A. J. 1999b. Platelet alpha2A-adrenoceptor function in major depression: Gi coupling, effects of imipramine and relationship to treatment outcome. *Psychiatry Res* 89:73–95.

Harkin, A., Nally, R., Kelly, J. P., and Leonard, B. E. 2000. Effects of reboxetine and sertraline treatments alone and in combination on the binding properties of cortical NMDA and beta1-adrenergic receptors in an animal model of depression. *J Neural Transm* 107:1213–27.

Heninger, G. R., Delgado, P. L., and Charney, D. S. 1996. The revised monoamine theory of depression: A modulatory role for monoamines, based on new findings from monoamine depletion experiments in humans. *Pharmacopsychiatry* 29:2–11.

Kafka, M. S., and Paul, S. M. 1986. Platelet α_2-adrenergic receptors in depression. *Arch Gen Psychiatry* 43:91–5.

Karolewicz, B., Szebeni, K., Stockmeier, C. A., et al. 2004. Low nNOS protein in the locus coeruleus in major depression. *J Neurochem* 91:1057–66.

Kenakin, T. 2004. Principles: Receptor theory in pharmacology. *Trends Pharmacol Sci* 25:186–92.

Klimek, V., Stockmeier, C., Overholser, J., et al. 1997. Reduced levels of norepinephrine transporters in the locus coeruleus in major depression. *J Neurosci* 17:8451–8.

Klimek, V., Rajkowska, G., Luker, S. N., et al. 1999. Brain noradrenergic receptors in major depression and schizophrenia. *Neuropsychopharmacology* 21:69–81.

López-León, S., Janssens, A. C., González-Zuloeta Ladd, A. M., et al. 2008. Meta-analyses of genetic studies on major depressive disorder. *Mol Psychiatry* 13:772–85.

Maas, J. W., Fawcett, J. A., and Dehirmenjian, H. 1972. Catecholamine metabolism, depressive illness and drug response. *Arch Gen Psychiatry* 26:252–62.

Manier, D. H., Eiring, A., Shelton, R. C., and Sulser, F. 1996. β-Adrenoceptor-linked protein kinase A (PKA) activity in human fibroblasts from normal subjects and form patients with major depression. *Neuropsychopharmacology* 15:555–61.

Mann, J. J., McBride, P. A., Brown, R. P., et al. 1992. Relationship between central and peripheral serotonin indexes in depressed and suicidal psychiatric inpatients. *Arch Gen Psychiatry* 49:442–6.

Mann, J. J., Halper, J. P., Wilner, P. J., et al. 1997. Subsensitivity of adenylyl cyclase-coupled receptors on mononuclear leukocytes from drug-free inpatients with a major depressive episode. *Biol Psychiatry* 42:859–70.

Marazziti, D., Consoli, G., Golia, F., et al. 2010. Trazodone effects on [^{3}H]-paroxetine and α_2-adrenoreceptors in platelets of patients with major depression. *Neuropsychiat Dis Treat* 6:255–9.

Martín-Guerrero, I., Callado, L. F., Saitua, K., Rivero, G., García-Orad, A., and Meana, J. J. 2006. The N251K functional polymorphism in the α2A-adrenoceptor gene is not associated with depression: A study in suicide completers. *Psychopharmacology (Berl)* 184:82–6.

Mateo, Y., Fernández-Pastor, B., and Meana, J. J. 2001. Acute and chronic effects of desipramine and clorgyline on alpha(2)-adrenoceptors regulating noradrenergic transmission in the rat brain: A dual-probe micro-dialysis study. *Br J Pharmacol* 133:1362–70.

Matuzany-Ruban, A., Golan, M., Miroshnik, N., Schreiber, G., and Avissar, S. 2010. Normalization of GRK2 protein and mRNA measures in patients with depression predict response to antidepressants. *Int J Neuropsychopharmacol* 13:83–91.

Meana, J. J., Barturen, F., and García-Sevilla, J. A. 1992. Alpha 2-adrenoceptors in the brain of suicide victims: Increased receptor density associated with major depression. *Biol Psychiatry* 31: 471–90.

Meyer, J. H., Wilson, A. A., Sagrati, S., et al. 2009. Brain monoamine oxidase A binding in major depressive disorder: Relationship to selective serotonin reuptake inhibitor treatment, recovery, and recurrence. *Arch Gen Psychiatry* 66:1304–12.

Meyer-Lindenberg, A., Kohn, P. D., Kolachana, B., et al. 2005. Midbrain dopamine and prefrontal function in humans: Interaction and modulation by COMT genotype. *Nat Neurosci* 8:594–6.

Miller, H. L., Delgado, P. L., Salomon, R. M., et al. 1996. Clinical and biochemical effects of catecholamine depletion on antidepressant-induced remission of depression. *Arch Gen Psychiatry* 53:117–28.

Mongeau, R., de Montigny, C., and Blier, P. 1994. Electrophysiologic evidence for desensitization of alpha 2-adrenoceptors on serotonin terminals following long-term treatment with drugs increasing norepinephrine synaptic concentration. *Neuropsychopharmacology* 10:41–51.

Moreno, F. A., Gelenberg, A. J., Heninger, G. R., et al. 1999. Tryptophan depletion and depressive vulnerability. *Biol Psychiatry* 46:498–505.

Neumeister, A., Drevets, W. C., Belfer, I., et al. 2006. Effects of a α2C-adrenoreceptor gene polymorphism on neural responses to facial expressions in depression. *Neuropsychopharmacology* 31:1750–6.

Ordway, G. A. 1997. Pathophysiology of the locus coeruleus in suicide. *Ann N Y Acad Sci* 836:233–52.

Ordway, G. A., Gambarana, C., Tejani-Butt, S. M., Areso, P., Hauptmann, M., and Frazer, A. 1991. Preferential reduction of binding of 125I-iodopindolol to beta-1 adrenoceptors in the amygdala of rat after antidepressant treatments. *J Pharmacol Exp Ther* 257:681–90.

Ordway, G. A., Farley, J. T., Dilley, G. E., et al. 1999. Quantitative distribution of monoamine oxidase A in brainstem monoamine nuclei is normal in major depression. *Brain Res* 847:71–9.

Ordway, G. A., Schenk, J., Stockmeier, C. A., May, W., and Klimek, V. 2003. Elevated agonist binding to alpha2-adrenoceptors in the locus coeruleus in major depression. *Biol Psychiatry* 53: 315–23.

Pandey, G. N., Janicak, P. G., Javaid, J. I., and Davis, J. M. 1989. Increased 3H-clonidine binding in the platelets of patients with depressive and schizophrenic disorders. *Psychiatry Res* 28:73–88.

Paparrigopoulos, T., Psarros, C., Bergiannaki, J.-D., Varsou, E., Dafni, U., and Stefanis, C. 2001. Melatonin response to clonidine administration in depression: Indication of presynaptic α_2-adrenoceptor dysfunction. *J Affect Disord* 65:307–13.

Paykel, E. S., Fleminger, R., and Watson, J. P. 1982. Psychiatric side effects of antihypertensive drugs other than reserpine. *J Clin Psychopharmacol* 2:14–39.

Peroutka, S. J., and Snyder, S. H. 1980. Long-term antidepressant treatment decreases spiroperidol-labeled serotonin receptor binding. *Science* 210:88–90.

Perroud, N., Aitchison, K. J., Uher, R., et al. 2009. Genetic predictors of increase in suicidal ideation during antidepressant treatment in the GENDEP project. *Neuropsychopharmacology* 34:2517–28.

Piletz, J. E., and Halaris, A. 1988. Super high affinity ^{3}H-*para*-aminoclonidine binding to platelet adrenoceptors in depression. *Prog Neuropsychopharmacol Biol Psychiatry* 12:541–53.

Piletz, J. E., Halaris, A., Saran, A., and Marler, M. 1990. Elevated ^{3}H-*para*-aminoclonidine binding to platelet purified plasma membranes from depressed patients. *Neuropsychopharmacology* 3:201–10.

Pittenger, C., and Duman, R. S. 2008. Stress, depression, and neuroplasticity: A convergence of mechanisms. *Neuropsychopharmacology* 33:88–109.

Potter, W. Z., and Manji, H. K. 1994. Catecholamines in depression: An update. *Clin Chem* 40:279–87.

Price, L. H., Charney, D. S., Rubin, L., and Heninger, G. R. 1986. α2-Adrenergic receptor function in depression. The cortisol response to yohimbine. *Arch Gen Psychiatry* 43:849–58.

Ressler, K. J., and Nemeroff, C. B. 1999. Role of norepinephrine in the pathophysiology and treatment of mood disorders. *Biol Psychiatry* 46:1219–33.

Riva, M. A., and Creese, I. 1989. Reevaluation of the regulation of beta-adrenergic receptor binding by desipramine treatment. *Mol Pharmacol* 36:211–8.

Ruhé, H. G., Mason, N. S., and Schene, A. H. 2007. Mood is indirectly related to serotonin, norepinephrine and dopamine levels in humans: A meta-analysis of monoamine depletion studies. *Mol Psychiatry* 12:331–59.

Schildkraut, J. J. 1965. The catecholamine hypothesis of affective disorders: A review of supporting evidence. *Am J Psychiatry* 122:509–22.

Sequeira, A., Mamdani, F., Lalovic, A., et al. 2004. Alpha 2A adrenergic receptor gene and suicide. *Psychiatry Res* 125:87–93.

Serretti, A., Macciardi, F., Verga, M., Cusin, C., Pedrini, S., and Smeraldi, E. 1998. Tyrosine hydroxylase gene associated with depressive symptomatology in mood disorder. *Am J Med Genet* 81:127–30.

Smith, D. F., Stork, B. S., Wegener, G., et al. 2009. [^{11}C]Mirtazapine binding in depressed antidepressant nonresponders studies by PET neuroimaging. *Psychopharmacology (Berl)* 206:133–40.

Starke, K. 1987. Presynaptic alpha-autoreceptors. *Rev Physiol Biochem Pharmacol* 107:73–146.

Stone, E. A., Lin, Y., Rosengarten, H., Kramer, H. K., and Quartermain, D. 2003. Emerging evidence for a central epinephrine-innervated alpha 1-adrenergic system that regulates behavioral activation and is impaired in depression. *Neuropsychopharmacology* 28:1387–99.

Sulser, F., Vetulani, J., and Mobley, P. L. 1978. Mode of action of antidepressant drugs. *Biochem Pharmacol* 27:257–61.

Valdizán, E., Díez-Alarcia, R., González-Maeso, J., et al. 2010. α_2-Adrenoceptor functionality in postmortem frontal cortex of depressed suicide victims. *Biol Psychiatry* 68:869–72.

Valdizán, E. M., Gutiérrez, O., and Pazos, A. 2003. Adenylate cyclase activity in postmortem brain of suicide subjects: Reduced response to beta-adrenergic stimulation. *Biol Psychiatry* 54:1457–64.

van Praag, H. M. 1982. Serotonin precursors in the treatment of depression. *Adv Biochem Psychopharmacol* 34:259–86.

Werstiuk, E. S., Steiner, M., and Burns, T. 1990. Studies on leukocyte beta-adrenergic receptors in depression: A critical appraisal. *Life Sci* 47:85–105.

Yanpallewar, S. U., Fernandes, K., Marathe, S. V., et al. 2010. Alpha2-adrenoceptor blockade accelerates the neurogenic, neurotrophic, and behavioral effects of chronic antidepressant treatment. *J Neurosci* 30:1096–109.

Young, S. N., Smith, S. E., Pihl, R. O., and Ervin, F. R. 1985. Tryptophan depletion causes a rapid lowering of mood in normal males. *Psychopharmacology (Berl)* 87:173–7.

Zhu, M. Y., Klimek, V., Dilley, G. E., et al. 1999. Elevated levels of tyrosine hydroxylase in the locus coeruleus in major depression. *Biol Psychiatry* 46:1275–86.

9 Glutamatergic System and Mood Disorders

Rodrigo Machado-Vieira and Carlos A. Zarate Jr.

CONTENTS

9.1 INTRODUCTION

Mood disorders such as major depressive disorder (MDD) and bipolar disorder (BPD) are among the most severe of all psychiatric illnesses. These chronic, often incapacitating disorders affect the lives of millions worldwide. Indeed, the World Health Organization projected that MDD would be the second leading cause of disability worldwide by 2020 (Murray and Lopez 1996).

For most patients with BPD, monotherapy with mood stabilizers such as lithium and valproate is often insufficient to treat acute manic episodes, and combination treatment is required (Zarate and Quiroz 2003). For long-term prophylaxis, far fewer agents are available. These include lithium, aripiprazole, and lamotrigine; of these, only lamotrigine has been shown to help prevent depressive relapses (McElroy et al. 2004). In addition, some patients are unable to tolerate existing therapies for BPD, which leads to either frequent changes in medications or to nonadherence (Sajatovic et al. 2006; Zarate et al. 1999; Zarate and Tohen 2004). The situation is similarly challenging for treating MDD. Available antidepressants are based on theories developed and popularized more than 40 years ago (Schildkraut 1965), and drug development in mood disorder research noticeably lags behind other areas in medicine (Insel and Scolnick 2006; Manji and Zarate 2002). Despite a variety of currently available treatments, many patients do not respond early enough in the course of a major depressive episode, and this therapeutic lag is associated with considerable negative consequences. Furthermore, currently available therapeutics are still ineffective for a large percentage of patients are associated with increased recurrence rates and persistent subsyndromal depressive symptoms, functional impairment, cognitive deficits, and disability (Fagiolini et al. 2005; Judd et al. 2002; Tohen et al. 2000). Notably, the Sequenced Treatment Alternatives to Relieve Depression (STAR*D) study found that only roughly one-third of patients with MDD experienced remission after an adequate trial (i.e., dose, duration) of a standard antidepressant after 10 to 14 weeks of treatment (Trivedi et al. 2006).

Thus, drugs that are mechanistically distinct from previous ones are still not available for either MDD or BPD and are urgently needed (Agid et al. 2007). Such treatments would be expected to be more effective for more patients, be better tolerated, and act more rapidly than currently available therapeutics. In this context, considerable research has taken place over the past decade on the role of the glutamatergic system in the pathophysiology and therapeutics of mood disorders. Findings from diverse studies suggest the relevance of this neurotransmitter system for mood disorders. In this chapter, we highlight the promising nature of this work.

9.2 FUNCTIONAL GLUTAMATERGIC SYSTEM

Glutamate is the most abundant excitatory neurotransmitter in the brain and acts in three different cell compartments—the pre- and postsynaptic neurons and glia—which collectively characterize the "tripartite glutamatergic synapse" (Machado-Vieira et al. 2009a) (see Figure 9.1). Physiological activity in the glial–neuronal glutamate–glutamine cycle includes the uptake and inactivation of glutamate after its actions as a neurotransmitter have ended to prevent toxic effects secondary to overexposure to high glutamate levels (Sanacora et al. 2008).

Glutamate is produced from α-ketoglutarate, an intermediate in the Krebs cycle, and is packaged into secretory vesicles in the presynaptic neuron by glutamate transporters. After being released in an activity-dependent process through interactions with soluble N-ethylmaleimide-sensitive factor attachment receptor proteins and sodium/calcium channels (Takamori 2006), glutamate is taken up by astrocytes and converted to glutamine by the enzyme glutamine synthetase. Glutamine released by glial cell is transported back to presynaptic neurons, oxidized back into glutamate by the enzyme glutaminase, and repackaged.

Glutamate activates diverse ionotropic and metabotropic receptors involved in synaptic plasticity, learning, behavior, and memory (Collingridge and Bliss 1995). A variety of glutamate ionotropic receptors and their respective subunits have been identified: N-methyl-D-aspartate (NMDA; NR1, NR2, NR2B, NR2C, NR2D, NR3A, and NR3B subunits), α-amino-3-hydroxy-5-methyl-4-isoxazolepropionic acid (AMPA; GluR1, GluR2, GluR3, GluR4), and kainate (GluR5, GluR6, GluR7, KA1, and KA2). Eight G-protein-coupled metabotropic receptors (mGluRs) have also been identified and characterized based on the signaling transduction pathway that they stimulate (mGluR1–8). These are divided into three subgroups: group I [mGluR1 (a, b, c, d) and mGluR5 (a, b)], group II (mGluR2 and mGluR3), and group III (mGluR4, mGluR6, mGluR7, and mGluR8).

The NMDA channel comprises two sites: the "s" site and the phencyclidine (PCP) site. Glutamate's binding sites are mostly expressed in the NR2A and NR2B subunits, which are both highly expressed in brain areas implicated in mood regulation (Magnusson et al. 2002; Sah and Lopez De Armentia 2003). The NR1 subunit is placed on the site for its coagonist, glycine [for a complete overview of the distribution and functional effects of NMDA receptors (NMDARs), see Cull-Candy et al. 2001].

AMPA receptors (AMPARs) are activated in the presence of glutamate, thus inducing a fast excitatory synaptic signal involved in early glutamatergic effects in the synapse. These effects are crucial to calcium metabolism, synaptic strength, and oxidative stress (Machado-Vieira et al.

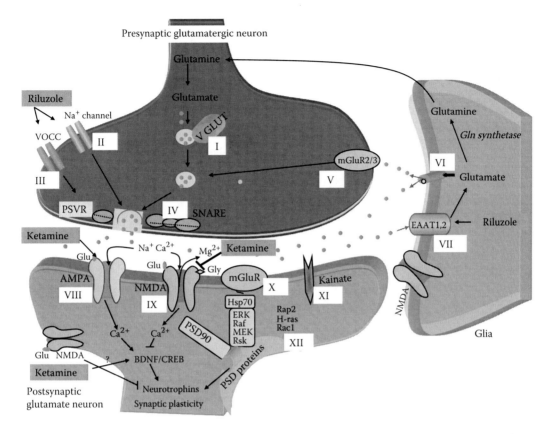

FIGURE 9.1 Pathophysiological basis and potential therapeutic targets for mood disorders involving glutamatergic neurotransmission. Ketamine preferentially targets postsynaptic AMPA/NMDA receptors, whereas riluzole's antidepressant effects occur through direct regulation, mostly at presynaptic voltage-operated channels and glia. cAMP, cyclic adenosine monophosphate; GABA, alpha-aminobutyric acid; Glu, glutamate; Gly, glycine; KA, kainate; PSVR, presynaptic voltage-operated release; SNARE, soluble N-ethylmaleimide-sensitive factor attachment receptor. (Reproduced from Machado-Vieira, R., et al., *Neuroscientist*, 15, 525–39, 2009. With permission.)

2009b; Zarate et al. 2003). In fact, AMPAR activation opens the pore, permitting the inward flow of sodium, thereby resulting in the depolarization of the neuronal membrane. This change in intracellular charge liberates the magnesium cation from the NMDAR, permitting the entrance of calcium through that pore. At mature synapses, AMPARs can be coexpressed with NMDARs, contributing to synaptic plasticity and neuroprotection (Barria and Malinow 2002). However, AMPARs have a lower affinity for glutamate than the NMDARs, allowing for more rapid dissociation of glutamate and a fast deactivation of the AMPAR (reviewed by Zarate et al. 2003).

Kainate receptors (KARs) participate in excitatory neurotransmission by both activating postsynaptic receptors and by inhibiting neurotransmission that occurs via the regulation of gamma aminobutyric acid (GABA) release. KARs have limited distribution in the brain and are believed to affect synaptic signaling and plasticity less than AMPARs (for a review, see Huettner 2003).

With regard to mGluRs, group I mGluRs are coupled to the phospholipase C signal transduction pathway and are located in both the pre- and postsynaptic membranes, whereas both group II and III mGluRs are coupled in an inhibitory manner to the adenylyl cyclase pathway and are involved in the regulation of glutamate and GABA release (for a review, see Witkin et al. 2007; Zarate et al. 2003); presynaptic mGluR2/3s also limit the release of glutamate.

In addition to ionotropic receptors and mGluRs, cytoplasmic postsynaptic density (PSD)–enriched molecules, excitatory amino acid transporters (EAATs or GLASTs), and vesicular glutamate transporters (VGLUTs) are also directly involved in synaptic and extrasynaptic glutamate brain levels and may represent potential therapeutic targets. VGLUT1 is the most abundant isoform in the cerebral cortex and hippocampus, selectively located onto synaptic vesicles of glutamatergic terminals. Notably, variations in VGLUT1 expression levels critically regulate the efficacy of glutamate synaptic transmission (reviewed by Machado-Vieira et al. 2009b; Sanacora et al. 2008). Glial cells also regulate pre- and postsynaptic activity by directly releasing and uptaking glutamate through a number of EAATs (mostly subtypes 1 and 2) (Danbolt 2001; Oliet et al. 2001; O'Shea 2002). Furthermore, cytoplasmic PSD-enriched molecules (such as PSD95) interact with glutamate receptors (particularly NMDARs and AMPARs) to regulate signal transduction (Dingledine et al. 1999; Sheng and Sala 2001), as well as synchronize information from several neurotransmitter systems. For instance, NR1 subunits interact with NL-L and Yotiao, whereas the NR2 subunit acts with several PSD proteins such as PSD95, PSD93, SAP102, CIPP, and Densin-180 (Bleakman and Lodge 1998; Hollmann and Heinemann 1994; Nakanishi 1992).

9.3 DYSFUNCTIONAL TRIPARTITE GLUTAMATERGIC SYNAPSE IN MOOD DISORDERS

Synaptic levels of glutamate can rise to excitotoxic concentrations rapidly after an insult (e.g., trauma, ischemia) or when glutamate transporter function is altered, which may involve direct changes in packaging, release, and reuptake of glutamate. Diverse pathophysiological mechanisms have been described. For example, inhibition of glial reuptake of glutamate rapidly decreases glutamate uptake, leading to hyperglutamatergic states and neural toxicity due to increased extrasynaptic glutamate (Jabaudon et al. 1999; Soriano and Hardingham 2007) (see Figure 9.1).

9.3.1 Transgenic Animal Models of Dysfunctional Glutamatergic Neurotransmission in Mood Disorders

Several animal studies have demonstrated a potential role for the glutamatergic system in the pathophysiology of mood disorders (discussed here as well as later in the chapter). For instance, NR2A knockout mice were found to exhibit a highly robust anxiolytic- and antidepressant-like phenotype (Boyce-Rustay and Holmes 2006), whereas GluR1 knockout mice showed increased learned helplessness (Chourbaji et al. 2008). Mice showing reduced VGLUT1 expression displayed increased anxiety- and depressive-like behaviors, as well as impaired recognition memory (Tordera et al. 2007). A recent study of KAR subunit GluR6 knockout mice described a potential functional role for this receptor in modulating manic-like behavior. These mice displayed increased risk taking, aggressive behavior, motor activity in response to amphetamine, and less despair-like behaviors; these manifestations were decreased after chronic lithium treatment (Shaltiel et al. 2008).

9.3.2 Human Studies

Pathophysiological glutamatergic neurotransmission has been described in subjects with mood disorders. Broadly, altered glutamate levels have been observed in plasma, serum, and cerebrospinal fluid of individuals with BPD (reviewed by Sanacora et al. 2008). Similarly, postmortem studies have shown increased glutamate levels in diverse brain areas in individuals with mood disorders (Hashimoto et al. 2007; Scarr et al. 2003). Imaging studies describe increased levels of glutamate and related metabolites in the occipital cortex and decreased levels in the anterior cingulate cortex (ACC) of individuals with both BPD and MDD (Hasler et al. 2007; Sanacora et al. 2004; Yildiz-

Yesiloglu and Ankerst 2006); these appear to be the most consistent findings in patients with mood disorders.

As regards specific glutamate receptors and subunits, reduced NMDAR binding and expression have also been found in individuals with MDD (Beneyto et al. 2007; Choudary et al. 2005; Law and Deakin 2001; McCullumsmith et al. 2007; Nudmamud-Thanoi and Reynolds 2004; Toro and Deakin 2005). Similar decreases in NR1 and NR2A expression have been observed in individuals with BPD (McCullumsmith et al. 2007). In addition, polymorphisms of *GRIN1*, *GRIN2A*, and *GRIN2B*—the genes encoding for each NMDAR subunit—appear to confer susceptibility to BPD (Itokawa et al. 2003; Martucci et al. 2006; Mundo et al. 2003). Interestingly, decreased NMDAR expression has been associated with PSD signaling proteins in BPD subjects (Clinton and Meador-Woodruff 2004). For instance, decreased PSD95 levels were observed in the dentate of individuals with BPD (Toro and Deakin 2005). A similar decrease in subjects with MDD was described in SAP102 levels (which primarily interacts with the NR2B subunit), and this decrease was correlated with decreased NR1 and NR2A expression in the hippocampus, striatum, and thalamus of patients with BPD and MDD (Clinton and Meador-Woodruff 2004; Kristiansen and Meador-Woodruff 2005; McCullumsmith et al. 2007). Recently, NR2A and PSD95 protein levels were found to be significantly increased in the lateral amygdala in MDD subjects compared with healthy controls (Karolewicz et al. 2009).

Similar findings were described in AMPAR regulation. Decreased GluR2 and GluR3 levels have been reported in the prefrontal cortex of subjects with mood disorders (Beneyto and Meador-Woodruff 2006; Hashimoto et al. 2007; Scarr et al. 2003). A selective decrease in striatal GluR1 expression has also been described in BPD subjects (Meador-Woodruff et al. 2001). Abnormal mGluR3 expression in suicidal subjects with BPD has been reported, but this finding was not subsequently replicated (Devon et al. 2001; Marti et al. 2002).

With regard to KARs, a recent, large, family-based association study evaluating the *GRIK3* gene (GluR7) described linkage disequilibrium in MDD (Schiffer and Heinemann 2007). Likewise, elevated GRIK3 DNA copy numbers were observed in BPD subjects (Wilson et al. 2006). In addition, a common variant in the 3′UTR *GRIK4* gene was found to have a protective effect in BPD (Pickard et al. 2006). Interestingly, the STAR*D and Munich Antidepressant Response Signature projects described an association between treatment-emergent suicidal ideation and the glutamate system via the involvement of the *GRIA3* and *GRIK2* genes (Laje et al. 2007; Menke et al. 2008).

Reduced expression of EAAT1, EAAT2, EAAT3, EAAT4, and glutamine synthetase was also observed in postmortem studies of subjects with mood disorders (Choudary et al. 2005; McCullumsmith and Meador-Woodruff 2002). Taken together, these findings lend further insight to the role that the glutamatergic system plays in developing promising therapeutic targets for the treatment of mood disorders.

9.4 THERAPEUTIC TARGETS AND NEW AGENTS ACTING THROUGH THE TRIPARTITE GLUTAMATERGIC SYNAPSE FOR THE TREATMENT OF MOOD DISORDERS

As early as 1959, studies were conducted evaluating glutamate as a neurotransmitter. Since then, many studies have expanded our understanding of the role that the glutamatergic system plays in the pathophysiology and therapeutics of psychiatric and neurological illnesses. Diverse glutamate-modulating agents have been tested in preclinical models and in patients with mood disorders (Sanacora et al. 2008; Zarate et al. 2002). In particular, riluzole and ketamine—described in greater detail below—are prototypic proof-of-concept glutamatergic agents in mood disorders. Here we describe findings pertaining to these two agents, as well as other agents targeting NMDARs, AMPARs, KARs, mGluRs, and the complex dynamics of VGLUTs, EAATs, and PSD. The effects of standard antidepressants and mood stabilizers are also described.

9.4.1 Currently Available Treatments that Target the Glutamatergic System

Several preclinical and human studies have shown that currently available antidepressants have both short- and long-term regulatory effects on the glutamatergic system. Elegant studies conducted over a decade ago with tricyclic antidepressants found that the NMDA receptor was a final common pathway of antidepressant action and could possibly be involved in faster onset of antidepressant effects (Nowak et al. 1993; Paul et al. 1994). More recently, chronic treatment with standard antidepressants was shown to directly regulate AMPAR expression and phosphorylation (Du et al. 2004, 2007; Paul and Skolnick 2003). Specifically, selective serotonin reuptake inhibitors and serotonin–norepinephrine reuptake inhibitors were all found to downregulate NMDA and activate AMPA (Bleakman and Lodge 1998; Skolnick et al. 1996). Fluoxetine treatment increased phosphorylation of GluR1 at Ser831 and Ser845, and chronic fluoxetine treatment targeted GluR1 Ser845 (Svenningsson et al. 2002). Chronic imipramine increased p845 GluR1 (associated with increased GluR1 insertion) (Maeng et al. 2008). Mostly, however, these effects were often subtle and delayed after administration.

In preclinical studies, fluoxetine, paroxetine, and desipramine all increased VGLUT mRNA levels in the frontal, orbital, and cingulate cortices as well as in the hippocampus (Tordera et al. 2005, 2007). Paroxetine enhanced the interaction of GluR1 with Rab4A, whereas desipramine markedly increased the interaction of GluR1 with SAP97 (Song et al. 1998). Chronic fluoxetine treatment also affected GluR2, GluR5, and GluR6 expression in the hippocampus (GluR5 and GluR6 are kainate subunits). Finally, the atypical antidepressant tianeptine prevented or reversed stress-associated structural and cellular brain changes related to normalization of altered glutamatergic neurotransmission (Kasper and McEwen 2008; Svenningsson et al. 2007).

Interestingly, the *GRIK4* gene polymorphism (rs1954787), which codes for the kainic acid–type glutamate receptor KA1, was found to regulate antidepressant response to citalopram in patients with MDD (Paddock et al. 2007). Chronic fluoxetine treatment similarly affected GluR5 and GluR6 levels in the hippocampus (Barbon et al. 2006). With regard to mGluRs, chronic imipramine reduced the inhibitory effects of group II receptors (Palucha et al. 2007).

Chronic treatment with the mood stabilizers lithium and valproate has also been found to affect the glutamatergic system. Both agents decreased hippocampal GluR1 expression after chronic treatment (Du et al. 2004). Chronic lithium treatment also increased glutamate synaptosomal uptake, thus protecting against glutamate-induced excitotoxicity (Hashimoto et al. 2002). Lithium further enhanced VGLUT1 mRNA expression in neurons of the cerebral cortex (Moutsimilli et al. 2005). Chronic treatment with valproate increased EAAT1 levels and decreased EAAT2 levels in the hippocampus (Hassel et al. 2001; Ueda and Willmore 2000). Lamotrigine, which has been recently approved by the U.S. Food and Drug Administration (FDA) for relapse prevention in BPD, also has antidepressant-like effects in animal models (Kaster et al. 2007). Notably, lamotrigine was found to limit glutamate release by blocking voltage-operated sodium channels, thus stabilizing the neuronal membrane (Prica et al. 2008). Lamotrigine also increased AMPAR activity by upregulating the surface expression of GluR1 and GluR2 phosphorylation and expression in hippocampal neurons (Bhagwagar et al. 2007; Du et al. 2007).

9.4.2 Glutamatergic Agents in Mood Disorders: Clinical and Preclinical Studies

Studies have demonstrated that NMDA antagonists produce rapid antidepressant effects in diverse paradigms (Maeng et al. 2008; Moryl et al. 1993; Papp and Moryl 1994; Trullas and Skolnick 1990; Zarate et al. 2006a). One preclinical study by our group showed that a subunit selective NR2B antagonist exerted antidepressant-like effects in rodents (Maeng et al. 2008). Other brain-penetrant NR2B antagonists currently under development include indole-2-carboxamides, benzimidazole-2-carboxamides, and HON0001 (Borza et al. 2007; Suetake-Koga et al. 2006). In humans, a double-blind, randomized, placebo-controlled study reported that the NR2B subunit-selective antagonist

CP-101,606 had significant and relatively rapid antidepressant efficacy in patients with treatment-resistant MDD (Preskorn et al. 2008).

AMPAR potentiators are a new class of pharmacological agents being tested in mood disorders. These agents decrease the AMPAR rate and/or deactivation in the presence of an agonist (e.g., AMPA and glutamate) (see Black 2005; Bleakman and Lodge 1998). The compounds are classified based on their effects on the biophysical processes of desensitization and deactivation, and include benzothiazides (e.g., cyclothiazide), benzoyliperidines (e.g., CX-516), and birylpropyl-sulfonamides (e.g., LY392098) (Miu et al. 2001; Quiroz et al. 2004). These agents also play a key role in modulating activity-dependent synaptic strength and behavioral plasticity (Sanacora et al. 2008). In preclinical paradigms, several AMPAR potentiators showed significant antidepressant-like effects and also improved cognitive function (reviewed by Black 2005; Lynch 2004; Miu et al. 2001; O'Neill et al. 2004). Whereas AMPAR potentiators are associated with antidepressant-like properties, AMPAR antagonists such as talampanel may have antimanic properties (reviewed by Rogawski 2006).

To date, no KAR modulators have been tested in the treatment of mood disorders. Notably, however, one recent study found that individuals with MDD who had a *GRIK4* gene polymorphism (rs1954787) were more likely to respond to treatment with the antidepressant citalopram than those who did not have this allele (Paddock et al. 2007).

Several mGluR-modulating agents have also been tested in preclinical and clinical studies in mood disorders. Diverse agents that target mGluRs have been found to induce anxiolytic, antidepressant, and neuroprotective effects in preclinical models, especially group II and III mGluR agonists (Chaki et al. 2004; Cosford et al. 2003; Cryan et al. 2003; Gasparini et al. 1999; Maiese et al. 2000; Palucha et al. 2004; Schoepp et al. 2003). Specifically, group II mGluR antagonists have been shown to induce antidepressant-like effects in animal models; for instance, the group II mGluR antagonist MGS-0039 induced antidepressant-like and neuroprotective effects (Yoshimizu and Chaki 2004; Yoshimizu et al. 2006). In addition, group III mGluR agonists exerted antidepressant-like properties in the behavioral despair test (Palucha et al. 2004; Palucha and Pilc 2007), a well-known animal model of depression. The selective group III mGluR agonists (ACPT-I, [1S,3R,4S]-1-aminocyclo-pentane-1,3,4-tricarboxylic acid) and an mGluR8 agonist (RS-PPG, [RS]-4-phosphonophenylglycine) also had antidepressant-like effects (Gasparini et al. 1999). Interestingly, AMPAR activation critically regulates the antidepressant-like effects of the group II mGluR antagonist MGS0039 (Karasawa et al. 2005). Finally, group I mGluR antagonists have also shown potential therapeutic effects in mood disorders. The group I mGluR5 antagonists MPEP (2-methyl-6-[phenylethynyl]-pyridine) and MTEP ([(2-methyl-1,3-thiazol-4-yl)ethynyl]pyridine) have antidepressant-like activity in animal models (Li et al. 2006; Wieronska et al. 2002). The potent and selective mGluR5 antagonist fenobam has been associated with significant psychostimulant effects (Porter et al. 2005). Many agents that modulate the mGluRs are under development, but no clinical studies on mood disorders have yet been published.

PSD-enriched molecules, EAATs, and VGLUTs also represent potential therapeutic targets in mood disorders, although few data are available. PSD regulation by glutamatergic turnover is important as a therapeutic target in mood disorders, but currently no agent targeting PSD proteins has been developed. Similarly, the study of VGLUTs in mood disorders is in its infancy but appears promising. Diverse compounds targeting VGLUTs are under development and may provide insight into the relevance of glutamate transporters in the pathophysiology of mood disorders and their treatment (Thompson et al. 2005). Similarly, increased EAAT2 expression could result in antidepressant-like effects (Mineur et al. 2007) and, preclinically, chronic blockade of glutamate uptake by a glial/neuronal transporter antagonist was found to decrease social exploratory behavior and alter circadian activity (Lee et al. 2007). Thus, its increased expression may represent a potential target. Notably, this is one of the mechanisms by which riluzole is believed to regulate excess synaptic glutamate (Frizzo et al. 2004) (described below).

9.5 RILUZOLE AND KETAMINE: PROTOTYPES FOR NEW, IMPROVED TREATMENTS IN MOOD DISORDERS

9.5.1 RILUZOLE

Riluzole (2-amino-6-trifluoromethoxy benzothiazole), a glutamatergic modulator with neuropro-tective and anticonvulsant properties, was approved by the FDA for the treatment of amyotrophic lateral sclerosis. Although riluzole has no known direct effects on NMDARs or KARs (Debono et al. 1993), it inhibits glutamate release (by inhibiting voltage-dependent sodium channels in neurons, its best known mechanism), enhances AMPA trafficking and membrane insertion of GluR1 and GluR2, and has glutamate reuptake properties (Frizzo et al. 2004). It also stimulates nerve growth factor, brain-derived neurotrophic factor (BDNF), and the synthesis of other neurotrophic factors in cultured astrocytes (Mizuta et al. 2001). Repeated injections of riluzole have been found to induce a prolonged elevation of hippocampal BDNF and neurogenesis (Katoh-Semba et al. 2002). Riluzole also protected glial cells against glutamate excitotoxicity (Dagci et al. 2007) and prevented the overstimulation of extrasynaptic glutamate receptors and consequent excitotoxicity; this mechanism of action is believed to occur through EAATs (Azbill et al. 2000; Frizzo et al. 2004; Fumagalli et al. 2008; Hardingham 2006). In rats, increased ^{13}C glucose metabolism was observed in the hippocampus and prefrontal cortex after 21 days of treatment with riluzole, suggesting increased glutamatergic metabolism rather than decreased glutamate release in these areas (Chowdhury et al. 2008). Furthermore, long-term administration of riluzole in animal models was recently found to induce antidepressant- and antimanic-like behaviors (Banasr and Duman 2008; Banasr et al. 2010; Lourenco Da Silva et al. 2003).

With regard to clinical studies, riluzole had antidepressant effects in an open-label study of patients with treatment-resistant MDD; 13 patients (68%) completed the trial and all showed a sig-nificant improvement at week 6 (Zarate et al. 2004). Similar results were obtained in another study that used riluzole as add-on therapy for MDD. In that study, antidepressant effects were noted within 1 week of treatment, and a significant decrease (36%) in Hamilton Depression Rating Scale scores was observed among completers (Sanacora et al. 2007). Riluzole's efficacy in treating bipolar depression was subsequently assessed. Riluzole was used adjunctively with lithium in 14 patients with bipolar depression and was associated with a 60% overall decrease in Montgomery Asberg Depression rating scale (MADRS) scores across the 8 weeks of treatment; a significant improve-ment on MADRS scores was noted by week 5. Two open-label trials of riluzole have also been con-ducted for the treatment of generalized anxiety disorder and compulsive disorder, with significant therapeutic effects noted in both studies (Coric et al. 2005; Mathew et al. 2005). There is no indica-tion that riluzole acts more rapidly than existing antidepressants; thus, it may be a valuable option predominantly in treatment-resistant cases.

9.5.2 KETAMINE

The noncompetitive, high-affinity NMDA antagonist ketamine is a derivative of phencycli-dine. It prevents excess calcium influx and cellular damage by antagonizing NMDARs. In vitro, ketamine increases the firing rate of glutamatergic neurons as well as presynaptic glutamate release (Moghaddam et al. 1997). Some of these properties are believed to be involved in ke-tamine's antidepressant effects. Interestingly, AMPAR activation critically regulates ketamine's antidepressant-like effects (reviewed below) (Maeng et al. 2008), suggesting that enhanced AMPA transmission may represent a common mechanism for the antidepressant action of this agent.

Ketamine's antidepressant effects have also been assessed in clinical studies. One prelimi-nary investigation in seven subjects with treatment-resistant MDD found that patients' depres-sive symptoms improved within 72 h of ketamine infusion (Berman et al. 2000). Subsequently, a double-blind, placebo-controlled, crossover study in patients with treatment-resistant MDD showed that a single infusion of ketamine had rapid (within the first 2 h after infusion) and relatively sustained antidepressant effects (lasting 1–2 weeks) (Zarate et al. 2006a). Notably, more than 70% of patients responded at 24 h after infusion, and 35% had a sustained response at the end of week 1. The response rates with ketamine after 24 h (71%) were similar to those described after 6 to 8 weeks of treatment with traditional monoaminergic-based antidepressants (65%) (Entsuah et al. 2001; Thase et al. 2005). This finding was replicated in a different cohort involving 26 subjects (Phelps et al. 2009); the magnitude and time frame of response to ketamine in this study were similar to those in the previous finding. Another small, recent trial evaluated 10 patients with treatment-resistant MDD who received repeated ketamine infusions—six infusions over 12 days (aan het Rot et al. 2010). Response criteria were met by nine patients after the first through sixth infusions, suggesting that repeated NMDA blocking is a feasible approach for treating acute, treatment-resistant MDD.

Ketamine's effects in bipolar depression have also been evaluated (Diaz-Granados et al. 2010a). Subjects with bipolar depression who received ketamine showed significant improvement in depressive symptoms compared with patients who received placebo within 40 min after injection; these effects remained significant through day 3. Seventy-one percent of all subjects responded to ketamine at some point during the study. Ketamine has also been associated with significant antisuicidal effects (Diaz-Granados et al. 2010b; Price et al. 2009). In the study by Diaz-Granados and colleagues, 33 subjects with treatment-resistant MDD received a single open-label infusion of ketamine and were rated at baseline and at 40, 80, 120, and 230 min after infusion with the Scale for Suicide Ideation (SSI) as well as several depression rating scales. Suicidal ideation scores significantly decreased on the SSI and suicidality items of depression rating scales within 40 min after infusion, and this effect remained significant through the first 4 h after infusion. Measures of depression, anxiety, and hopelessness were significantly improved at all time points (Diaz-Granados et al. 2010b).

Furthermore, ketamine infusion has also displayed antidepressant effects in depressed patients during pre- and postoperative states, and in patients with MDD comorbid with pain syndrome and/ or alcohol dependence (Goforth and Holsinger 2007; Kudoh et al. 2002; Liebrenz et al. 2007; Ostroff et al. 2005). However, ketamine's sedative and psychotomimetic side effects will probably continue to limit its clinical use in larger samples and is a problem that will need to be addressed by the next generation of novel glutamatergic agents.

9.5.3 Biological Correlates of Antidepressant Response to Ketamine

9.5.3.1 Alcohol Dependence

Previous studies found that alcohol-dependent individuals showed marked reductions to the subjective intoxicating effects of ketamine compared with healthy controls (Krystal et al. 2003) and that healthy individuals with a positive family history of alcohol dependence had fewer perceptual alterations and lower dysphoric mood after receiving ketamine (Petrakis et al. 2004) (see Figure 9.2). In line with these findings, a recent study found that family history of alcohol dependence in individuals with MDD was associated with better short-term outcome after ketamine infusion (Phelps et al. 2009). The precise reasons underlying this improved response in patients with treatment-resistant MDD with a positive family history of alcohol dependence are essentially unknown. However, emerging data suggest that genetically determined alterations in NMDAR subunits may be associated with alcohol dependence. Ketamine acts as a partial NR2A antagonist (Petrenko et al. 2004)—notably, NR2A expression was shown to be regulated by alcohol in the amygdala and hippocampus (Boyce-Rustay and Holmes 2006). Thus, differences in NR2A sensitivity may account for ketamine's differential antidepressant effects in individuals with MDD with or without a family history of alcohol dependence (see Figure 9.2).

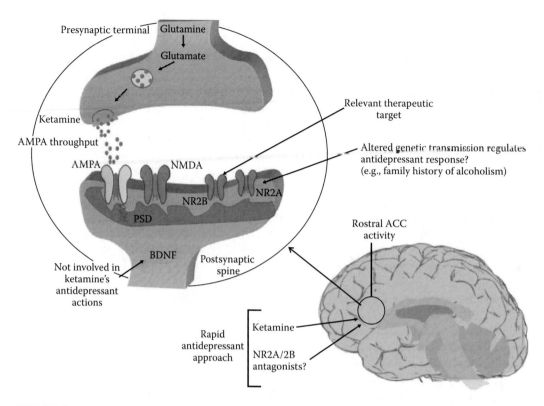

FIGURE 9.2 BDNF does not appear to be involved in ketamine's rapid antidepressant effects; however, AMPA relative to NMDA throughput and its potential effects targeting at PSD may represent a relevant mechanism by which ketamine induces these therapeutic effects. In particular, NR2A and NR2B receptors are believed to mediate these rapid antidepressant effects. Rostral ACC activity has been implicated in rapid antidepressant response to ketamine infusion. Family history of alcohol dependence in subjects with MDD was associated with better short-term outcome after ketamine infusion. BDNF, brain-derived neurotrophic factor. (Reproduced from Zarate, C. A. Jr., et al., *Harvard Rev Psychiatry*, 18, 293–303, 2010. With permission.)

9.5.3.2 Anterior Cingulate Cortex Activity

Studies have shown that ACC activation can predict improved antidepressant response to standard antidepressants, electroconvulsive therapy, transcranial magnetic stimulation, and deep brain stimulation (Chen et al. 2007; Langguth et al. 2007; Mayberg et al. 1997; McCormick et al. 2007; Pizzagalli et al. 2001). Similar effects have been also associated with sleep deprivation (Saxena et al. 2003; Wu et al. 1999). With regard to ketamine, one recent magnetoencephalography study found that increased pretreatment rostral ACC activity positively correlated with rapid antidepressant response to ketamine infusion in 11 patients with MDD versus healthy controls (Salvadore et al. 2009) (see Figure 9.2). Interestingly, another study observed that orbitofrontal and subgenual ACC blood oxygenation level–dependent activity was directly regulated by ketamine in healthy individuals.

9.5.3.3 Brain-Derived Neurotrophic Factor

Changes in BDNF levels were not associated with ketamine's initial antidepressant effects (Figure 9.2). BDNF levels showed no change from baseline after ketamine infusion and up to 230 min after infusion (a time point when antidepressant effects and response were present). However, this clinical study confirmed ketamine's efficacy as a fast-acting antidepressant (Machado-Vieira et al. 2009c).

9.5.3.4 AMPA Relative to NMDA Throughput

The net effect of ketamine's antidepressant actions on a cellular level is increased glutamatergic throughput. One study found that these effects were selectively abolished by using an AMPA antagonist (NBQX) before ketamine infusion (Maeng et al. 2008). The authors proposed that ketamine's effects occurred mostly via AMPAR activation and not critically through NMDAR antagonism, thus inducing rapid AMPA-mediated synaptic potentiation; in contrast, traditional antidepressants induce delayed intracellular signaling changes (Sanacora et al. 2008), which may explain their differential time of onset. Ketamine appears to enhance synaptic efficacy in the amygdala–accumbens pathway (Kessal et al. 2005). Therefore, it is possible that increased glutamatergic throughput of AMPARs relative to NMDARs after ketamine treatment may enhance synaptic potentiation and activate early neuroplastic genes (see Figure 9.2).

9.5.4 OTHER GLUTAMATE MODULATORS

Cytidine, a pyrimidine component of RNA that regulates dysfunctional neuronal–glial glutamate cycling, has been tested in bipolar depression. In a recent double-blind, placebo-controlled study evaluating 35 subjects with BPD experiencing a depressive episode, valproate plus cytidine improved depressive symptoms earlier than valproate plus placebo; the antidepressant effects observed in the cytidine group were positively associated with decreased midfrontal glutamate/glutamine levels (Yoon et al. 2009). This suggests that cytidine supplementation may have therapeutic effects in bipolar depression by decreasing cerebral glutamate/glutamine levels.

Memantine is a moderately selective noncompetitive NMDA antagonist approved for the treatment of Alzheimer's disease. When used as monotherapy or with imipramine in rodent models of depression, memantine displayed antidepressant-like effects. However, a double-blind, placebo-controlled clinical trial showed that memantine had no significant antidepressant effects in MDD (Zarate et al. 2006b). There is only one report of memantine use in bipolar depression. In this case series evaluating two patients, memantine as add-on therapy to mood stabilizers improved depressive symptoms and cognitive performance in patients with BPD (Teng and Demetrio 2006).

9.6 FINAL REMARKS

Glutamate is the major excitatory neurotransmitter controlling synaptic excitability and plasticity in most brain circuits, including limbic pathways involved in mood disorders. In recent years, there has been a growing appreciation that drugs targeting glutamate neuronal transmission may offer novel approaches to treating MDD and BPD.

As this chapter has shown, clear glutamatergic system abnormalities exist in subjects with mood disorders, but the magnitude and extent of the abnormalities require further clarification. In addition, biological parameters are necessary to determine the clinical relevance of glutamatergic modulators that target specific receptors/subunits. Existing effective antidepressants and mood stabilizers also modulate different components of the glutamate system, and these effects may also be relevant to developing improved therapeutics for mood disorders. An increasing number of proof-of-principle studies have attempted to identify relevant therapeutic targets involving the glutamatergic system; however, to date, riluzole and ketamine are still the main proof-of-concept agents.

Mood disorders are common, chronic, recurrent mental illness that affect the lives and functioning of millions of individuals worldwide, and are a major public health concern. Treatment for these disorders often takes days to weeks to take effect, with damaging consequences for ill patients. The work described above—particularly the ketamine studies—provides a direct example that rapid response in mood disorders is possible. Most significantly, pharmacological treatments that exert a rapid and sustained antidepressant effect within hours or even a few days could significantly impact care for our patients as well as public health worldwide.

ACKNOWLEDGMENTS

We would like to acknowledge the support of the Intramural Research Program of the National Institute of Mental Heath, National Institutes of Health, Department of Health and Human Services (IRP-NIMH-NIH-DHHS). Dr. Zarate is listed among the inventors on a patent application submitted for the use of ketamine in depression. He has assigned his rights on the patent to the U.S. government. Ioline Henter (NIMH) provided invaluable editorial assistance.

REFERENCES

aan het Rot, M., Collins, K. A., Murrough, J. W., et al. 2010. Safety and efficacy of repeated-dose intravenous ketamine for treatment-resistant depression. *Biol Psychiatry* 67:139–45.

Agid, Y., Buzsaki, G., Diamond, D. M., et al. 2007. How can drug discovery for psychiatric disorders be improved? *Nat Rev Drug Discov* 6:189–201.

Azbill, R. D., Mu, X., and Springer, J. E. 2000. Riluzole increases high-affinity glutamate uptake in rat spinal cord synaptosomes. *Brain Res* 871:175–80.

Banasr, M., and Duman, R. S. 2008. Glial loss in the prefrontal cortex is sufficient to induce depressive-like behaviors. *Biol Psychiatry* 64:863–70.

Banasr, M., Chowdhury, G. M., Terwilliger, R., et al. 2010. Glial pathology in an animal model of depression: Reversal of stress-induced cellular, metabolic and behavioral deficits by the glutamate-modulating drug riluzole. *Mol Psychiatry* 15:501–11.

Barbon, A., Popoli, M., La Via, L., et al. 2006. Regulation of editing and expression of glutamate alpha-amino-propionic-acid (AMPA)/kainate receptors by antidepressant drugs. *Biol Psychiatry* 59:713–20.

Barria, A., and Malinow, R. 2002. Subunit-specific NMDA receptor trafficking to synapses. *Neuron* 35:345–53.

Beneyto, M., and Meador-Woodruff, J. H. 2006. Lamina-specific abnormalities of AMPA receptor trafficking and signaling molecule transcripts in the prefrontal cortex in schizophrenia. *Synapse* 60:585–98.

Beneyto, M., Kristiansen, L. V., Oni-Orisan, A., McCullumsmith, R. E., and Meador-Woodruff, J. H. 2007. Abnormal glutamate receptor expression in the medial temporal lobe in schizophrenia and mood disorders. *Neuropsychopharmacology* 32:1888–902.

Berman, R. M., Cappiello, A., Anand, A., et al. 2000. Antidepressant effects of ketamine in depressed patients. *Biol Psychiatry* 47:351–4.

Bhagwagar, Z., Wylezinska, M., Jezzard, P., et al. 2007. Reduction in occipital cortex gamma-aminobutyric acid concentrations in medication-free recovered unipolar depressed and bipolar subjects. *Biol Psychiatry* 61:806–12.

Black, M. D. 2005. Therapeutic potential of positive AMPA modulators and their relationship to AMPA receptor subunits. A review of preclinical data. *Psychopharmacology (Berl)* 179:154–63.

Bleakman, D., and Lodge, D. 1998. Neuropharmacology of AMPA and kainate receptors. *Neuropharmacology* 37:1187–204.

Borza, I., Bozo, E., Barta-Szalai, G., et al. 2007. Selective NR1/2B *N*-methyl-D-aspartate receptor antagonists among indole-2-carboxamides and benzimidazole-2-carboxamides. *J Med Chem* 50:901–14.

Boyce-Rustay, J. M., and Holmes, A. 2006. Genetic inactivation of the NMDA receptor NR2A subunit has anxiolytic- and antidepressant-like effects in mice. *Neuropsychopharmacology* 31:2405–14.

Chaki, S., Yoshikawa, R., Hirota, S., et al. 2004. MGS0039: A potent and selective group II metabotropic glutamate receptor antagonist with antidepressant-like activity. *Neuropharmacology* 46:457–67.

Chen, C. H., Ridler, K., Suckling, J., et al. 2007. Brain imaging correlates of depressive symptom severity and predictors of symptom improvement after antidepressant treatment. *Biol Psychiatry* 62:407–14.

Choudary, P. V., Molnar, M., Evans, S. J., et al. 2005. Altered cortical glutamatergic and GABAergic signal transmission with glial involvement in depression. *Proc Natl Acad Sci USA* 102:15653–8.

Chourbaji, S., Vogt, M. A., Fumagalli, F., et al. 2008. AMPA receptor subunit 1 (GluR-A) knockout mice model the glutamate hypothesis of depression. *FASEB J* 22:3129–34.

Chowdhury, G. M., Banasr, M., de Graaf, R. A., et al. 2008. Chronic riluzole treatment increases glucose metabolism in rat prefrontal cortex and hippocampus. *J Cereb Blood Flow Metab* 28:1892–7.

Clinton, S. M., and Meador-Woodruff, J. H. 2004. Abnormalities of the NMDA receptor and associated intracellular molecules in the thalamus in schizophrenia and bipolar disorder. *Neuropsychopharmacology* 29:1353–62.

Collingridge, G. L., Bliss, T. V., 1995. Memories of NMDA receptors and LTP. *Trends Neurosci* 18:54–6.

Coric, V., Taskiran, S., Pittenger, C., et al. 2005. Riluzole augmentation in treatment-resistant obsessive-compulsive disorder: An open-label trial. *Biol Psychiatry* 58:424–8.

Cosford, N. D., Tehrani, L., Roppe, J., et al. 2003. 3-[(2-Methyl-1,3-thiazol-4-yl)ethynyl]-pyridine: A potent and highly selective metabotropic glutamate subtype 5 receptor antagonist with anxiolytic activity. *J Med Chem* 46:204–6.

Cryan, J. F., Kelly, P. H., Neijt, H. C., et al. 2003. Antidepressant and anxiolytic-like effects in mice lacking the group III metabotropic glutamate receptor mGluR7. *Eur J Neurosci* 17:2409–17.

Cull-Candy, S., Brickley, S., and Farrant, M. 2001. NMDA receptor subunits: Diversity, development and disease. *Curr Opin Neurobiol* 11:327–35.

Dagci, T., Yilmaz, O., Taskiran, D., and Peker, G. 2007. Neuroprotective agents: Is effective on toxicity in glial cells? *Cell Mol Neurobiol* 27:171–7.

Danbolt, N. C. 2001. Glutamate uptake. *Prog Neurobiol* 65:1–105.

Debono, M. W., Le Guern, J., Canton, T., Doble, A., and Pradier, L. 1993. Inhibition by riluzole of electrophysiological responses mediated by rat kainate and NMDA receptors expressed in *Xenopus* oocytes. *Eur J Pharmacol* 235:283–9.

Devon, R. S., Anderson, S., Teague, P. W., et al. 2001. Identification of polymorphisms within Disrupted in Schizophrenia 1 and Disrupted in Schizophrenia 2, and an investigation of their association with schizophrenia and bipolar affective disorder. *Psychiatr Genet* 11:71–8.

Diaz-Granados, N., Ibrahim, L., Brutsche, N. E., et al. 2010a. A randomized add-on trial of an *N*-methyl-D-aspartate antagonist in treatment-resistant bipolar depression. *Arch Gen Psychiatry* 67:793–802.

Diaz-Granados, N., Ibrahim, L. A., Brutsche, N. E., et al. 2010b. Rapid resolution of suicidal ideation after a single infusion of an *N*-methyl-D-aspartate antagonist in patients with treatment-resistant major depressive disorder. *J Clin Psychiatry* doi: 10.4088/JCP.09m05327blu.

Dingledine, R., Borges, K., Bowie, D., and Traynelis, S. F. 1999. The glutamate receptor ion channels. *Pharmacol Rev* 51:7–61.

Du, J., Gray, N. A., Falke, C. A., et al. 2004. Modulation of synaptic plasticity by antimanic agents: The role of AMPA glutamate receptor subunit 1 synaptic expression. *J Neurosci* 24:6578–89.

Du, J., Suzuki, K., Wei, Y., et al. 2007. The anticonvulsants lamotrigine, riluzole, and valproate differentially regulate AMPA receptor membrane localization: Relationship to clinical effects in mood disorders. *Neuropsychopharmacology* 32:793–802.

Entsuah, A. R., Huang, H., and Thase, M. E. 2001. Response and remission rates in different subpopulations with major depressive disorder administered venlafaxine, selective serotonin reuptake inhibitors, or placebo. *J Clin Psychiatry* 62:869–77.

Fagiolini, A., Kupfer, D. J., Masalehdan, A., et al. 2005. Functional impairment in the remission phase of bipolar disorder. *Bipolar Disord* 7:281–5.

Frizzo, M. E., Dall'Onder, L. P., Dalcin, K. B., and Souza, D. O. 2004. Riluzole enhances glutamate uptake in rat astrocyte cultures. *Cell Mol Neurobiol* 24:123–8.

Fumagalli, E., Funicello, M., Rauen, T., Gobbi, M., and Mennini, T. 2008. Riluzole enhances the activity of glutamate transporters GLAST, GLT1 and EAAC1. *Eur J Pharmacol* 578:171–6.

Gasparini, F., Bruno, V., Battaglia, G., et al. 1999. (R,S)-4-Phosphonophenylglycine, a potent and selective group III metabotropic glutamate receptor agonist, is anticonvulsive and neuroprotective in vivo. *J Pharmacol Exp Ther* 289:1678–87.

Goforth, H. W., and Holsinger, T. 2007. Rapid relief of severe major depressive disorder by use of preoperative ketamine and electroconvulsive therapy. *J ECT* 23:23–5.

Hardingham, G. E. 2006. Pro-survival signalling from the NMDA receptor. *Biochem Soc Trans* 34:936–8.

Hashimoto, K., Sawa, A., and Iyo, M. 2007. Increased levels of glutamate in brains from patients with mood disorders. *Biol Psychiatry* 62:1310–6.

Hashimoto, R., Hough, C., Nakazawa, T., Yamamoto, T., and Chuang, D. M. 2002. Lithium protection against glutamate excitotoxicity in rat cerebral cortical neurons: Involvement of NMDA receptor inhibition possibly by decreasing NR2B tyrosine phosphorylation. *J Neurochem* 80:589–97.

Hasler, G., van der Veen, J. W., Tumonis, T., et al. 2007. Reduced prefrontal glutamate/glutamine and gamma-aminobutyric acid levels in major depression determined using proton magnetic resonance spectroscopy. *Arch Gen Psychiatry* 64:193–200.

Hassel, B., Iversen, E. G., Gjerstad, L., and Tauboll, E. 2001. Up-regulation of hippocampal glutamate transport during chronic treatment with sodium valproate. *J Neurochem* 77:1285–92.

Hollmann, M., and Heinemann, S. 1994. Cloned glutamate receptors. *Annu Rev Neurosci* 17:31–108.

Huettner, J. E. 2003. Kainate receptors and synaptic transmission. *Prog Neurobiol* 70:387–407.

Insel, T. R., and Scolnick, E. M. 2006. Cure therapeutics and strategic prevention: Raising the bar for mental health research. *Mol Psychiatry* 11:11–7.

Itokawa, M., Yamada, K., Iwayama-Shigeno, Y., et al. 2003. Genetic analysis of a functional GRIN2A promoter (GT)n repeat in bipolar disorder pedigrees in humans. *Neurosci Lett* 345:53–6.

Jabaudon, D., Shimamoto, K., Yasuda-Kamatani, Y., et al. 1999. Inhibition of uptake unmasks rapid extracellular turnover of glutamate of nonvesicular origin. *Proc Natl Acad Sci USA* 96:8733–8.

Judd, L. L., Akiskal, H. S., Schettler, P. J., et al. 2002. The long-term natural history of the weekly symptomatic status of bipolar I disorder. *Arch Gen Psychiatry* 59:530–7.

Karasawa, J., Shimazaki, T., Kawashima, N., and Chaki, S. 2005. AMPA receptor stimulation mediates the anti-depressant-like effect of a group II metabotropic glutamate receptor antagonist. *Brain Res* 1042.92–8.

Karolewicz, B., Szebeni, K., Gilmore, T., et al. 2009. Elevated levels of NR2A and PSD-95 in the lateral amygdala in depression. *Int J Neuropsychopharmacol* 12:143–53.

Kasper, S., and McEwen, B. S. 2008. Neurobiological and clinical effects of the antidepressant tianeptine. *CNS Drugs* 22:15–26.

Kaster, M. P., Raupp, I., Binfare, R. W., Andreatini, R., and Rodrigues, A. L. 2007. Antidepressant-like effect of lamotrigine in the mouse forced swimming test: Evidence for the involvement of the noradrenergic system. *Eur J Pharmacol* 565:119–24.

Katoh-Semba, R., Asano, T., Ueda, H., et al. 2002. Riluzole enhances expression of brain-derived neurotrophic factor with consequent proliferation of granule precursor cells in the rat hippocampus. *FASEB J* 16: 1328–30.

Kessal, K., Chessel, A., Spennato, G., and Garcia, R. 2005. Ketamine and amphetamine both enhance synaptic transmission in the amygdala-nucleus accumbens pathway but with different time-courses. *Synapse* 57:61–5.

Kristiansen, L. V., and Meador-Woodruff, J. H. 2005. Abnormal striatal expression of transcripts encoding NMDA interacting PSD proteins in schizophrenia, bipolar disorder and major depression. *Schizophr Res* 78:87–93.

Krystal, J. H., Petrakis, I. L., Krupitsky, E., et al. 2003. NMDA receptor antagonism and the ethanol intoxication signal: From alcoholism risk to pharmacotherapy. *Ann N Y Acad Sci* 1003:176–84.

Kudoh, A., Takahira, Y., Katagai, H., and Takazawa, T. 2002. Small-dose ketamine improves the postoperative state of depressed patients. *Anesth Analg* 95:114–8.

Laje, G., Paddock, S., Manji, H., et al. 2007. Genetic markers of suicidal ideation emerging during citalopram treatment of major depression. *Am J Psychiatry* 164:1530–8.

Langguth, B., Wiegand, R., Kharraz, A., et al. 2007. Pre-treatment anterior cingulate activity as a predictor of antidepressant response to repetitive transcranial magnetic stimulation (rTMS). *Neuro Endocrinol Lett* 28:633–8.

Law, A. J., and Deakin, J. F. 2001. Asymmetrical reductions of hippocampal NMDAR1 glutamate receptor mRNA in the psychoses. *Neuroreport* 12:2971–4.

Lee, Y., Gaskins, D., Anand, A., and Shekhar, A. 2007. Glia mechanisms in mood regulation: A novel model of mood disorders. *Psychopharmacology (Berl)* 191:55–65.

Li, X., Need, A. B., Baez, M., and Witkin, J. M. 2006. Metabotropic glutamate 5 receptor antagonism is associated with antidepressant-like effects in mice. *J Pharmacol Exp Ther* 319:254–9.

Liebrenz, M., Borgeat, A., Leisinger, R., and Stohler, R. 2007. Intravenous ketamine therapy in a patient with a treatment-resistant major depression. *Swiss Med Wkly* 137:234–6.

Lourenco Da Silva, A., Hoffmann, A., Dietrich, M. O., et al. 2003. Effect of riluzole on MK-801 and amphetamine-induced hyperlocomotion. *Neuropsychobiology* 48:27–30.

Lynch, G. 2004. AMPA receptor modulators as cognitive enhancers. *Curr Opin Pharmacol* 4:4–11.

Machado-Vieira, R., Manji, H. K., and Zarate, C. A. 2009a. The role of the tripartite glutamatergic synapse in the pathophysiology and therapeutics of mood disorders. *Neuroscientist* 15:525–39.

Machado-Vieira, R., Salvadore, G., Ibrahim, L. A., Diaz-Granados, N., and Zarate, C. A., Jr. 2009b. Targeting glutamatergic signaling for the development of novel therapeutics for mood disorders. *Curr Pharm Des* 15:1595–611.

Machado-Vieira, R., Yuan, P., Brutsche, N., et al. 2009c. Brain-derived neurotrophic factor and initial antidepressant response to an *N*-methyl-D-aspartate antagonist. *J Clin Psychiatry* 70:1662–6.

Maeng, S., Zarate, C. A., Jr., Du, J., et al. 2008. Cellular mechanisms underlying the antidepressant effects of ketamine: Role of alpha-amino-3-hydroxy-5-methylisoxazole-4-propionic acid receptors. *Biol Psychiatry* 63:349–52.

Magnusson, K. R., Nelson, S. E., and Young, A. B. 2002. Age-related changes in the protein expression of subunits of the NMDA receptor. *Brain Res Mol Brain Res* 99:40–5.

Maiese, K., Vincent, A., Lin, S. H., and Shaw, T. 2000. Group I and group III metabotropic glutamate receptor subtypes provide enhanced neuroprotection. *J Neurosci Res* 62:257–72.

Manji, H. K., and Zarate, C. A. 2002. Molecular and cellular mechanisms underlying mood stabilization in bipolar disorder: Implications for the development of improved therapeutics. *Mol Psychiatry* 7 (Suppl 1): S1–7.

Marti, S. B., Cichon, S., Propping, P., and Nothen, M. 2002. Metabotropic glutamate receptor 3 (GRM3) gene variation is not associated with schizophrenia or bipolar affective disorder in the German population. *Am J Med Genet* 114:46–50.

Martucci, L., Wong, A. H., De Luca, V., et al. 2006. *N*-methyl-D-aspartate receptor NR2B subunit gene GRIN2B in schizophrenia and bipolar disorder: Polymorphisms and mRNA levels. *Schizophr Res* 84:214–21.

Mathew, S. J., Amiel, J. M., Coplan, J. D., et al. 2005. Open-label trial of riluzole in generalized anxiety disorder. *Am J Psychiatry* 162:2379–81.

Mayberg, H. S., Brannan, S. K., Mahurin, R. K., et al. 1997. Cingulate function in depression: A potential predictor of treatment response. *Neuroreport* 8:1057–61.

McCormick, L. M., Boles Ponto, L. L., Pierson, R. K., et al. 2007. Metabolic correlates of antidepressant and antipsychotic response in patients with psychotic depression undergoing electroconvulsive therapy. *J ECT* 23:265–73.

McCullumsmith, R. E., and Meador-Woodruff, J. H. 2002. Striatal excitatory amino acid transporter transcript expression in schizophrenia, bipolar disorder, and major depressive disorder. *Neuropsychopharmacology* 26:368–75.

McCullumsmith, R. E., Kristiansen, L. V., Beneyto, M., et al. 2007. Decreased NR1, NR2A, and SAP102 transcript expression in the hippocampus in bipolar disorder. *Brain Res* 1127:108–18.

McElroy, S. L., Zarate, C. A., Cookson, J., et al. 2004. A 52-week, open-label continuation study of lamotrigine in the treatment of bipolar depression. *J Clin Psychiatry* 65:204–10.

Meador-Woodruff, J. H., Hogg, A. J., Jr., and Smith, R. E. 2001. Striatal ionotropic glutamate receptor expression in schizophrenia, bipolar disorder, and major depressive disorder. *Brain Res Bull* 55:631–40.

Menke, A., Lucae, S., Kloiber, S., et al. 2008. Genetic markers within glutamate receptors associated with antidepressant treatment-emergent suicidal ideation. *Am J Psychiatry* 165:917–8.

Mineur, Y. S., Picciotto, M. R., and Sanacora, G. 2007. Antidepressant-like effects of ceftriaxone in male C57BL/6J mice. *Biol Psychiatry* 61:250–2.

Miu, P., Jarvie, K. R., Radhakrishnan, V., et al. 2001. Novel AMPA receptor potentiators LY392098 and LY404187: Effects on recombinant human AMPA receptors in vitro. *Neuropharmacology* 40:976–83.

Mizuta, I., Ohta, M., Ohta, K., et al. 2001. Riluzole stimulates nerve growth factor, brain-derived neurotrophic factor and glial cell line–derived neurotrophic factor synthesis in cultured mouse astrocytes. *Neurosci Lett* 310:117–20.

Moghaddam, B., Adams, B., Verma, A., and Daly, D. 1997. Activation of glutamatergic neurotransmission by ketamine: A novel step in the pathway from NMDA receptor blockade to dopaminergic and cognitive disruptions associated with the prefrontal cortex. *J Neurosci* 17:2921–7.

Moryl, E., Danysz, W., and Quack, G. 1993. Potential antidepressive properties of amantadine, memantine and bifemelane. *Pharmacol Toxicol* 72:394–7.

Moutsimilli, L., Farley, S., Dumas, S., et al. 2005. Selective cortical VGLUT1 increase as a marker for antidepressant activity. *Neuropharmacology* 49:890–900.

Mundo, E., Tharmalingham, S., Neves-Pereira, M., et al. 2003. Evidence that the *N*-methyl-D-aspartate subunit 1 receptor gene (GRIN1) confers susceptibility to bipolar disorder. *Mol Psychiatry* 8:241–5.

Murray, C. J., and Lopez, A. D. 1996. Evidence-based health policy—Lessons from the Global Burden of Disease Study. *Science* 274:740–3.

Nakanishi, S. 1992. Molecular diversity of glutamate receptors and implications for brain function. *Science* 258:597–603.

Nowak, G., Trullas, R., Layer, R. T., Skolnick, P., and Paul, I. A. 1993. Adaptive changes in the *N*-methyl-D-aspartate receptor complex after chronic treatment with imipramine and 1-aminocyclopropanecarboxylic acid. *J Pharmacol Exp Ther* 265:1380–6.

Nudmamud-Thanoi, S., and Reynolds, G. P. 2004. The NR1 subunit of the glutamate/NMDA receptor in the superior temporal cortex in schizophrenia and affective disorders. *Neurosci Lett* 372:173–7.

O'Neill, M. J., Bleakman, D., Zimmerman, D. M., and Nisenbaum, E. S. 2004. AMPA receptor potentiators for the treatment of CNS disorders. *Curr Drug Targets CNS Neurol Disord* 3:181–94.

O'Shea, R. D. 2002. Roles and regulation of glutamate transporters in the central nervous system. *Clin Exp Pharmacol Physiol* 29:1018–23.

Oliet, S. H., Piet, R., and Poulain, D. A. 2001. Control of glutamate clearance and synaptic efficacy by glial coverage of neurons. *Science* 292:923–6.

Ostroff, R., Gonzales, M., and Sanacora, G. 2005. Antidepressant effect of ketamine during ECT. *Am J Psychiatry* 162:1385–6.

Paddock, S., Laje, G., Charney, D., et al. 2007. Association of GRIK4 with outcome of antidepressant treatment in the STAR*D cohort. *Am J Psychiatry* 164:1181–8.

Palucha, A., and Pilc, A. 2007. Metabotropic glutamate receptor ligands as possible anxiolytic and antidepressant drugs. *Pharmacol Ther* 115:116–47.

Palucha, A., Tatarczynska, E., Branski, P., et al. 2004. Group III mGlu receptor agonists produce anxiolytic- and antidepressant-like effects after central administration in rats. *Neuropharmacology* 46:151–9.

Palucha, A., Branski, P., Klak, K., and Sowa, M. 2007. Chronic imipramine treatment reduces inhibitory properties of group II mGlu receptors without affecting their density or affinity. *Pharmacol Rep* 59:525–30.

Papp, M., and Moryl, E. 1994. Antidepressant activity of non-competitive and competitive NMDA receptor antagonists in a chronic mild stress model of depression. *Eur J Pharmacol* 263:1–7.

Paul, I. A., and Skolnick, P. 2003. Glutamate and depression: Clinical and preclinical studies. *Ann N Y Acad Sci* 1003:250–72.

Paul, I. A., Nowak, G., Layer, R. T., Popik, P., and Skolnick, P. 1994. Adaptation of the *N*-methyl-D-aspartate receptor complex following chronic antidepressant treatments. *J Pharmacol Exp Ther* 269:95–102.

Petrakis, I. L., Limoncelli, D., Gueorguieva, R., et al. 2004. Altered NMDA glutamate receptor antagonist response in individuals with a family vulnerability to alcoholism. *Am J Psychiatry* 161:1776–82.

Petrenko, A. B., Yamakura, T., Fujiwara, N., et al. 2004. Reduced sensitivity to ketamine and pentobarbital in mice lacking the *N*-methyl-D-aspartate receptor GluRepsilon1 subunit. *Anesth Analg* 99:1136–40.

Phelps, L. E., Brutsche, N., Moral, J. R., et al. 2009. Family history of alcohol dependence and initial antidepressant response to an *N*-methyl-D-aspartate antagonist. *Biol Psychiatry* 65:181–4.

Pickard, B. S., Malloy, M. P., Christoforou, A., et al. 2006. Cytogenetic and genetic evidence supports a role for the kainate-type glutamate receptor gene, GRIK4, in schizophrenia and bipolar disorder. *Mol Psychiatry* 11:847–57.

Pizzagalli, D., Pascual-Marqui, R. D., Nitschke, J. B., et al. 2001. Anterior cingulate activity as a predictor of degree of treatment response in major depression: Evidence from brain electrical tomography analysis. *Am J Psychiatry* 158:405–15.

Porter, R. H., Jaeschke, G., Spooren, W., et al. 2005. Fenobam: A clinically validated nonbenzodiazepine anxiolytic is a potent, selective, and noncompetitive mGlu5 receptor antagonist with inverse agonist activity. *J Pharmacol Exp Ther* 315:711–21.

Preskorn, S. H., Baker, B., Kolluri, S., et al. 2008. An innovative design to establish proof of concept of the antidepressant effects of the NR2B subunit selective *N*-methyl-D-aspartate antagonist, CP-101,606, in patients with treatment-refractory major depressive disorder. *J Clin Psychopharmacol* 28:631–7.

Prica, C., Hascoet, M., and Bourin, M. 2008. Antidepressant-like effect of lamotrigine is reversed by veratrine: A possible role of sodium channels in bipolar depression. *Behav Brain Res* 191:49–54.

Price, R. B., Nock, M. K., Charney, D., and Mathew, S. J. 2009. Effects of intravenous ketamine on explicit and implicit measures of suicidality in treatment-resistant depression. *Biol Psychiatry* 66:522–9.

Quiroz, J. A., Gould, T. D., and Manji, H. K. 2004. Molecular effects of lithium. *Mol Interv* 4:259–72.

Rogawski, M. A., 2006. Diverse mechanisms of antiepileptic drugs in the development pipeline. *Epilepsy Res* 69:273–94.

Sah, P., and Lopez De Armentia, M. 2003. Excitatory synaptic transmission in the lateral and central amygdala. *Ann N Y Acad Sci* 985:67–77.

Sajatovic, M., Valenstein, M., Blow, F. C., Ganoczy, D., and Ignacio, R. V. 2006. Treatment adherence with antipsychotic medications in bipolar disorder. *Bipolar Disord* 8:232–41.

Salvadore, G., Cornwell, B. R., Colon-Rosario, V., et al. 2009. Increased anterior cingulate cortical activity in response to fearful faces: A neurophysiological biomarker that predicts rapid antidepressant response to ketamine. *Biol Psychiatry* 65:289–95.

Sanacora, G., Gueorguieva, R., Epperson, C. N., et al. 2004. Subtype-specific alterations of GABA and glutamate in major depression. *Arch Gen Psychiatry* 61:705–13.

Sanacora, G., Kendell, S. F., Levin, Y., et al. 2007. Preliminary evidence of riluzole efficacy in antidepressant-treated patients with residual depressive symptoms. *Biol Psychiatry* 61:822–5.

Sanacora, G., Zarate, C. A., Krystal, J. H., and Manji, H. K. 2008. Targeting the glutamatergic system to develop novel, improved therapeutics for mood disorders. *Nat Rev Drug Discov* 7:426–37.

Saxena, S., Brody, A. L., Ho, M. L., et al. 2003. Differential brain metabolic predictors of response to parox-etine in obsessive–compulsive disorder versus major depression. *Am J Psychiatry* 160:522–32.

Scarr, E., Pavey, G., Sundram, S., MacKinnon, A., and Dean, B. 2003. Decreased hippocampal NMDA, but not kainate or AMPA receptors in bipolar disorder. *Bipolar Disord* 5:257–64.

Schiffer, H. H., and Heinemann, S. F. 2007. Association of the human kainate receptor GluR7 gene (GRIK3) with recurrent major depressive disorder. *Am J Med Genet B Neuropsychiatr Genet* 144B:20–6.

Schildkraut, J. J. 1965. The catecholamine hypothesis of affective disorders: A review of supporting evidence. *Am J Psychiatry* 122:509–22.

Schoepp, D. D., Wright, R. A., Levine, L. R., Gaydos, B., and Potter, W. Z. 2003. LY354740, an mGlu2/3 receptor agonist as a novel approach to treat anxiety/stress. *Stress* 6:189–97.

Shaltiel, G., Maeng, S., Malkesman, O., et al. 2008. Evidence for the involvement of the kainate receptor subunit GluR6 (GRIK2) in mediating behavioral displays related to behavioral symptoms of mania. *Mol Psychiatry* 13:858–72.

Sheng, M., and Sala, C. 2001. PDZ domains and the organization of supramolecular complexes. *Annu Rev Neurosci* 24:1–29.

Skolnick, P., Layer, R. T., Popik, P., et al. 1996. Adaptation of *N*-methyl-D-aspartate (NMDA) receptors follow-ing antidepressant treatment: Implications for the pharmacotherapy of depression. *Pharmacopsychiatry* 29:23–6.

Song, I., Kamboj, S., Xia, J., et al. 1998. Interaction of the *N*-ethylmaleimide-sensitive factor with AMPA receptors. *Neuron* 21:393–400.

Soriano, F. X., and Hardingham, G. E. 2007. Compartmentalized NMDA receptor signalling to survival and death. *J Physiol* 584 (Pt. 2):381–7.

Suetake-Koga, S., Shimazaki, T., Takamori, K., et al. 2006. In vitro and antinociceptive profile of HON0001, an orally active NMDA receptor NR2B subunit antagonist. *Pharmacol Biochem Behav* 84:134–41.

Svenningsson, P., Tzavara, E. T., Witkin, J. M., et al. 2002. Involvement of striatal and extrastriatal DARPP-32 in biochemical and behavioral effects of fluoxetine (Prozac). *Proc Natl Acad Sci USA* 99:3182–7.

Svenningsson, P., Bateup, H., Qi, H., et al. 2007. Involvement of AMPA receptor phosphorylation in antide-pressant actions with special reference to tianeptine. *Eur J Neurosci* 26:3509–17.

Takamori, S. 2006. VGLUTs: 'Exciting' times for glutamatergic research? *Neurosci Res* 55:343–51.

Teng, C. T., and Demetrio, F. N. 2006. Memantine may acutely improve cognition and have a mood stabilizing effect in treatment-resistant bipolar disorder. *Rev Bras Psiquiatr* 28:252–4.

Thase, M. E., Haight, B. R., Richard, N., et al. 2005. Remission rates following antidepressant therapy with bupropion or selective serotonin reuptake inhibitors: A meta-analysis of original data from 7 randomized controlled trials. *J Clin Psychiatry* 66:974–81.

Thompson, C. M., Davis, E., Carrigan, C. N., et al. 2005. Inhibitor of the glutamate vesicular transporter (VGLUT). *Curr Med Chem* 12:2041–56.

Tohen, M., Strakowski, S. M., Zarate, C., Jr., et al. 2000. The McLean-Harvard first-episode project: 6-Month symptomatic and functional outcome in affective and nonaffective psychosis. *Biol Psychiatry* 48:467–76.

Tordera, R. M., Pei, Q., and Sharp, T. 2005. Evidence for increased expression of the vesicular glutamate trans-porter, VGLUT1, by a course of antidepressant treatment. *J Neurochem* 94:875–83.

Tordera, R. M., Totterdell, S., Wojcik, S. M., et al. 2007. Enhanced anxiety, depressive-like behaviour and impaired recognition memory in mice with reduced expression of the vesicular glutamate transporter 1 (VGLUT1). *Eur J Neurosci* 25:281–90.

Toro, C., Deakin, J. F. 2005. NMDA receptor subunit NRI and postsynaptic protein PSD-95 in hippocampus and orbitofrontal cortex in schizophrenia and mood disorder. *Schizophr Res* 80:323–30.

Trivedi, M. H., Rush, A. J., Wisniewski, S. R., et al. 2006. Evaluation of outcomes with citalopram for depres-sion using measurement-based care in STAR*D: Implications for clinical practice. *Am J Psychiatry* 163:28–40.

Trullas, R., and Skolnick, P. 1990. Functional antagonists at the NMDA receptor complex exhibit antidepres-sant actions. *Eur J Pharmacol* 185:1–10.

Ueda, Y., and Willmore, L. J. 2000. Molecular regulation of glutamate and GABA transporter proteins by val-proic acid in rat hippocampus during epileptogenesis. *Exp Brain Res* 133:334–9.

Wieronska, J. M., Szewczyk, B., Branski, P., Palucha, A., and Pilc, A. 2002. Antidepressant-like effect of MPEP, a potent, selective and systemically active mGlu5 receptor antagonist in the olfactory bulbecto-mized rats. *Amino Acids* 23:213–6.

Wilson, G. M., Flibotte, S., Chopra, V., et al. 2006. DNA copy-number analysis in bipolar disorder and schizo-phrenia reveals aberrations in genes involved in glutamate signaling. *Hum Mol Genet* 15:743–9.

Witkin, J. M., Marek, G. J., Johnson, B. G., and Schoepp, D. D. 2007. Metabotropic glutamate receptors in the control of mood disorders. *CNS Neurol Disord Drug Targets* 6:87–100.

Wu, J., Buchsbaum, M. S., Gillin, J. C., et al. 1999. Prediction of antidepressant effects of sleep deprivation by metabolic rates in the ventral anterior cingulate and medial prefrontal cortex. *Am J Psychiatry* 156:1149–58.

Yildiz-Yesiloglu, A., and Ankerst, D. P. 2006. Neurochemical alterations of the brain in bipolar disorder and their implications for pathophysiology: A systematic review of the in vivo proton magnetic resonance spectroscopy findings. *Prog Neuropsychopharmacol Biol Psychiatry* 30:969–95.

Yoon, S. J., Lyoo, I. K., Haws, C., et al. 2009. Decreased glutamate/glutamine levels may mediate cytidine's efficacy in treating bipolar depression: A longitudinal proton magnetic resonance spectroscopy study. *Neuropsychopharmacology* 34:1810–18.

Yoshimizu, T., and Chaki, S. 2004. Increased cell proliferation in the adult mouse hippocampus following chronic administration of group II metabotropic glutamate receptor antagonist, MGS0039. *Biochem Biophys Res Commun* 315:493–6.

Yoshimizu, T., Shimazaki, T., Ito, A., and Chaki, S. 2006. An mGluR2/3 antagonist, MGS0039, exerts antidepressant and anxiolytic effects in behavioral models in rats. *Psychopharmacology (Berl)* 186:587–93.

Zarate, C. A., Jr., and Quiroz, J. A. 2003. Combination treatment in bipolar disorder: A review of controlled trials. *Bipolar Disord* 5:217–25.

Zarate, C. A., Jr., and Tohen, M. 2004. Double-blind comparison of the continued use of antipsychotic treatment versus its discontinuation in remitted manic patients. *Am J Psychiatry* 161:169–71.

Zarate, C. A., Jr., Tohen, M., Narendran, R., et al. 1999. The adverse effect profile and efficacy of divalproex sodium compared with valproic acid: A pharmacoepidemiology study. *J Clin Psychiatry* 60:232–6.

Zarate, C. A., Quiroz, J., Payne, J., and Manji, H. K. 2002. Modulators of the glutamatergic system: Implications for the development of improved therapeutics in mood disorders. *Psychopharmacol Bull* 36:35–83.

Zarate, C. A., Jr., Du, J., Quiroz, J., et al. 2003. Regulation of cellular plasticity cascades in the pathophysiology and treatment of mood disorders: Role of the glutamatergic system. *Ann N Y Acad Sci* 1003:273–91.

Zarate, C. A., Jr., Payne, J. L., Quiroz, J., et al. 2004. An open-label trial of riluzole in patients with treatment-resistant major depression. *Am J Psychiatry* 161:171–4.

Zarate, C. A., Jr., Singh, J. B., Carlson, P. J., et al. 2006a. A randomized trial of an *N*-methyl-D-aspartate antagonist in treatment-resistant major depression. *Arch Gen Psychiatry* 63:856–64.

Zarate, C. A., Jr., Singh, J. B., Quiroz, J. A., et al. 2006b. A double-blind, placebo-controlled study of memantine in the treatment of major depression. *Am J Psychiatry* 163:153–5.

Zarate, C. A. Jr., Machado-Vieira, R., Henter, I., et al. 2010. Glutamatergic modulators: The future of treating mood disorders? *Harvard Rev Psychiatry* 18:293–303.

10 Role of the Endocannabinoid System in Etiopathology of Major Depression

Ryan J. McLaughlin, Francis Rodriguez Bambico, and Gabriella Gobbi

CONTENTS

10.1 INTRODUCTION

For centuries, different cultures around the world have used the herb *Cannabis sativa* for its therapeutic and mood-elevating properties. After the discovery and classification of the endocannabinoid system in the early 1990s, a growing body of research has unequivocally demonstrated that deficits in this neuromodulatory system may result in the pathological development of behavioral, physiological, cognitive, and endocrine symptoms of major depression. The aim of this chapter is to summarize this current state of knowledge regarding the role of the endocannabinoid system in the etiology of major depression, derived from preclinical animal models of depression as well as from genetic, clinical, and postmortem reports in human populations. Furthermore, this review will present research indicating that endocannabinoid signaling is critically involved in the proper functioning of monoamine neural systems, activation and termination of the neuroendocrine stress axis, and a fundamental component of synaptic remodeling and neuroplasticity in the brain—processes that are often impaired in clinically depressed populations. Lastly, evidence will be described suggesting that various pharmacological and somatic treatment regimens for major depression elicit neurobiological changes in endocannabinoid signaling that may be necessary for the positive clinical response evoked by these treatments. Collectively, this body of literature suggests that the endocannabinoid system is functionally implicated in the pathology of major depression, whereas pharmacological facilitation of this system may represent an appealing antidepressant target for combating symptoms in clinical populations.

10.2 ENDOCANNABINOID SYSTEM

The discovery and characterization of Δ^9-tetrahydrocannabinol, the primary psychoactive constituent of marijuana, in 1964, along with the synthesis of biologically active analogs (collectively termed *cannabinoids*) has incited extensive research focused on elucidating the pharmacological and biochemical properties of these compounds and how they produce their physiological and behavioral effects. Significant progress was made in the early 1990s with the identification and cloning of the cannabinoid CB_1 receptor in rodent brain tissue (Matsuda et al. 1990) and the later characterization of the CB_2 receptor in spleen macrophages (Munro et al. 1993). Both receptors are inhibitory G-protein-coupled receptors with the CB_1 receptor coupling to $G_{i/o}$ proteins, whereas CB_2 receptors couple exclusively to G_i proteins (Howlett 2002). CB_1 receptors represent the most abundant class of G-protein-coupled receptors in the brain, as neuroanatomical studies have confirmed widespread expression throughout the forebrain, basal ganglia, and limbic system suggestive of a ubiquitous neuromodulatory role for this receptor subtype in both humans (Glass et al. 1997; Mato and Pazos 2004) and rodents (Herkenham et al. 1990, 1991) (see Figure 10.1). In rodents, intense CB_1 receptor staining has been observed in the neocortex, hippocampus, striatum, substantia nigra, and the cerebellum, whereas moderate CB_1 receptor immunoreactivity has been detected in the cingulate, entorhinal and piriform cortical areas, olfactory bulbs, amygdala, and nucleus accumbens (Herkenham et al. 1990). CB_1 receptor density is much lower in the diencephalon (thalamus and hypothalamus) and midbrain, and essentially null in the medulla (Herkenham et al. 1990; Tsou et al. 1998).

Activation of CB_1 receptors inhibits adenylyl cyclase activity, leading to a subsequent reduction in the cyclic adenosine monophosphate cascade, augmentation of inward-rectifying potassium channels, and inhibition of subsequent calcium influx via voltage-gated calcium channels (Howlett 1995; Freund et al. 2003). Furthermore, CB_1 receptors are located on presynaptic axon terminals of glutamatergic principal neurons as well as on a subpopulation of noncalbindin and cholecystokinin-positive GABAergic basket cells (Herkenham et al. 1990; Tsou et al. 1998; Freund et al. 2003); thus, CB_1 receptors are ideally positioned to modulate the balance of excitation and inhibition within a given neural circuit. Recent evidence also indicates that CB_1 receptors are present on serotonergic (5-HT) (Morales and Bäckman 2002; Häring et al. 2007) and noradrenergic (NA) (Oropeza et al. 2007) neurons, suggesting that these receptors also directly modulate monoamine efflux as well.

Investigation into cannabinoid receptor pharmacology has provided the foundation for a relatively new field of research focused on the structural and functional characterization of naturally occurring endogenous ligands that bind to these receptors. In 1992, Devane and coworkers identified these endogenous cannabinoids (or "endocannabinoids") as the arachidonate-derived lipophilic molecules *N*-arachidonylethanolamide (anandamide; AEA) and 2-arachidonylglycerol (2-AG) (Devane et al. 1992; Sugiura et al. 1995). Several other putative endocannabinoids have been isolated; however, greater attention has been given to AEA and 2-AG because of their potent agonistic activity at CB_1 and CB_2 receptors. Both AEA and 2-AG are formed postsynaptically by activity-dependent cleavage of phospholipid head groups via activation of specific enzymes (see Ahn et al. 2008 for a review of these biosynthetic pathways). Endocannabinoids are different from traditional neurotransmitters in that they are not stored in vesicles but are instead cleaved from the phospholipid membrane and synthesized on demand in postsynaptic cells after postsynaptic membrane depolarization (Freund et al. 2003). They are then released into the synapse where they travel in a retrograde manner to activate CB_1 receptors on the presynaptic membrane, hyperpolarizing it and

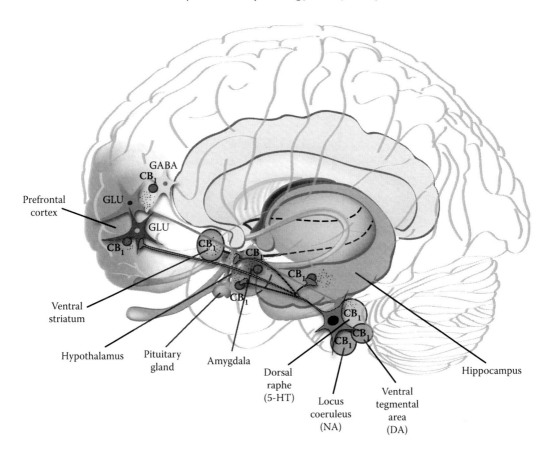

FIGURE 10.1 **(See color insert.)** Localization of CB_1 receptors in brain regions known to regulate mood. CB_1 receptors are located on inhibitory (GABA) and excitatory (GLU) neurons throughout cortical, limbic, hypothalamic, and midbrain monoaminergic structures known to be involved in regulation of mood. 5-HT, NA, and dopaminergic (DA) neurons are located in dorsal raphe (DR), locus coeruleus (LC), and ventral tegmental area (VTA), respectively. Under physiological conditions, these neurons exhibit spontaneous electrical activity, the modulation of which may lead to burst firing associated with enhanced release of their respective monoamines at terminal areas. Glutamatergic pyramidal neurons of prefrontal cortex that are under tonic CB_1 receptor regulation send their axons to DR 5-HT, LC NA, and VTA DA neurons. In turn, 5-HT and NA neurons innervate amygdala, prefrontal cortex, and hippocampus, brain regions that are critically implicated in etiopathology of major depression, whereas DA neurons innervate the ventral striatum, which is likely involved in motivation and hedonic responding. Hippocampus and prefrontal cortex exert negative feedback over HPA axis, which is hyperactive in individuals suffering from melancholic depression. Disturbances in endocannabinoid signaling in any of aforementioned regions may lead to perturbations in monoaminergic signaling, HPA axis negative feedback, and reward processing, as well as deficits in synaptic plasticity in these regions, and thus may contribute to a multitude of behavioral and physiological symptoms present in individuals suffering from major depression. (Adapted from Hill et al. *Trends Pharmacol Sci*, 30, 484–93, 2009.)

thereby reducing postsynaptic currents and depressing subsequent neurotransmitter release (Freund et al. 2003; Di Marzo 1999). Termination of AEA and 2-AG signaling begins with transport across the plasma membrane followed by enzymatic hydrolysis into arachidonic acid and ethanolamine or glycerol, respectively (Ahn et al. 2008). This is accomplished via their respective hydrolytic enzymes; fatty acid amide hydrolase (FAAH) is the catabolic enzyme for AEA, whereas 2-AG is primarily metabolized by monoacylglycerol lipase (Bisogno 2008) (see Figure 10.2).

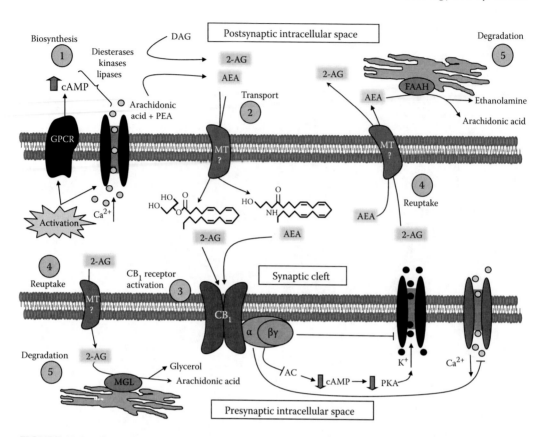

FIGURE 10.2 **(See color insert.)** Simplified schematic diagram of the life cycle of endocannabinoids anand-amide and 2-arachidonylglycerol. (1) AEA and 2-AG are synthesized on demand through the action of multiple diesterases, kinases, and lipases. Stimulation of postsynaptic G-protein-coupled receptors (GPCR) and calcium (Ca^{2+}) channels increases adenylyl cyclase (AC) activity and cyclic adenosine monophosphate (cAMP) produc-tion, which facilitates conversion of endocannabinoid precursors N-arachidonoylphosphatidylethanolamine [from arachidonic acid and phosphatidyl-ethanolamine (PEA)] and diacylglycerol (DAG) into AEA and 2-AG, respectively. Other putative biosynthetic pathways may also be involved (not shown). (2) AEA and 2-AG are transported into synaptic cleft through nonvesicular trafficking that involves a yet uncharacterized membrane transporter (MT). (3) AEA and 2-AG bind to presynaptically located metabotropic CB$_1$ receptors (CB$_1$). CB$_1$ receptor activation hyperpolarizes the presynaptic membrane by activating inward rectifying potassium (K$^+$) channels, reducing protein kinase A (PKA) phosphorylation, and inhibiting L-, N-, and P/Q-type calcium channels (through G$_{i/o}$ protein α subunit). (4) AEA and 2-AG are transported back into cell via a yet unchar-acterized endocannabinoid membrane transporter. (5) In the cell, AEA and 2-AG are metabolized through different catabolic pathways. AEA is degraded in postsynapse by FAAH (located predominantly on mem-brane surface of cytoplasmic organelles) into arachidonic acid and ethanolamine. 2-AG is likely transported into presynaptic cell via a putative membrane transporter and is degraded by monoacylglycerol lipase (MGL) into arachidonic acid and glycerol, although other postsynaptic 2-AG hydrolases may also participate in this process (not shown). ↑, stimulation; ⊤ or ↓, inhibition.

It is a matter of speculation why two endogenous ligands exist for the same receptor, but these molecules do exhibit slight pharmacokinetic differences that may produce differential signaling patterns. For instance, AEA exhibits a high affinity for the CB$_1$ receptor (approximately 50–100 nM) but has poor efficacy as an agonist at inducing intracellular signal transduction (Hillard 2000). In contrast, 2-AG has a lower affinity for CB$_1$ (approximately 1–10 µM) but induces a robust intra-cellular response (Hillard 2000). Thus, 2-AG is thought to induce a rapid and robust CB$_1$ receptor response required for modulation of activity-induced synaptic plasticity, whereas AEA evokes more tonic, mild CB_1 receptor stimulation that may have greater implications for proper behavioral function (Gorzalka et al. 2008).

10.3 EFFECTS OF EXOGENOUS CANNABINOIDS ON MOOD AND EMOTIONALITY

Anecdotal reports from healthy cannabis smokers characterize its acute effects as relaxation, reduced anxiety, stress relief, euphoria, increased well-being and sociability, heightened sensory experience and imagination, distortion of time perception, and feelings of depersonalization (Bambico et al. 2009). In recent years, there has been a growing interest in the therapeutic potential of cannabinergic compounds for various mood disorders. This interest has been inspired by several reports documenting their capacity to improve mood in healthy individuals and mitigate symptoms of major depression in patients being treated for other illnesses such as multiple sclerosis, chronic pain, and human immunodeficiency virus (see Bambico et al. 2009 for a review). For instance, a recent controlled trial examining the effects of cannabis consumption on chronic neuropathic pain revealed significant improvements in measures of anxiety and depression, in addition to reduced pain and better quality of sleep (Ware et al. 2010).

Despite these encouraging findings, no large-scale double-blind study to date has directly tested the anxiolytic and antidepressant effects of cannabis or cannabis-derived drugs on patients suffering from mood disorders. This may be because of the diversity of experiences associated with cannabis intoxication, which are dependent on many factors including baseline emotional states, genetic background, personality and expectations of the user, environmental setting, and the dose of the drug ingested (Bambico et al. 2009). For example, in contrast to low-dose cannabinoid administration, high-dose cannabinoid exposure has been shown to elicit anxiety, panic, and psychotomimetic effects (Curran et al. 2002), which may partially be attributed to the bidirectional effects of low and high doses on the modulation of 5-HT neuronal activity (Bambico et al. 2007). Therefore, the variable and complex effects of cannabinergic drugs on emotional states may present an unstable therapeutic window that is not particularly conducive to the treatment of mood disorders.

Another major obstacle in the implementation of cannabis-derived drugs for mood disorders is that long-term heavy cannabis use, particularly during adolescence, is associated with increased risk of contracting depressive-like symptoms that persist into adulthood (Bovasso 2001; Patton et al. 2002). A constellation of maladaptive behaviors including diminished drive and ambition, increased apathy, dysphoria, decreased ability to carry out long-term plans, and a difficulty dealing with frustration—referred to as amotivational syndrome—is thought to be induced by long-term cannabis abuse (Campbell 1976). Notably, these component symptoms overlap with those of major depression, suggesting that chronic adolescent cannabis use could have a detrimental impact on the development of mood, motivation, and reward processing pathways in the brain (Campbell 1976; Musty and Kaback 1995). Recent preclinical research supports this notion, as chronic CB_1 receptor agonist administration in adolescent rats promotes depressogenic responding along with dysregulated monoaminergic neural firing that persists into adulthood (Bambico et al. 2010b). Given the detrimental effects of chronic cannabinoid exposure during this developmental phase, it is conceivable to speculate that the pathogenesis of major depression could partially be due to alterations in the endocannabinoid system, particularly within monoamine projections to corticolimbic brain regions that regulate emotional behavior.

10.4 DISRUPTION OF ENDOCANNABINOID SIGNALING AND IMPLICATIONS FOR EMOTIONAL BEHAVIOR

The generation of transgenic mice lacking the CB_1 receptor has offered tremendous insight into the role of this signaling system in the regulation of emotional states. CB_1 receptor knockout mice exhibit increased depressive-like passive coping responses (i.e., immobility) in the forced swim test and tail suspension test (Steiner et al. 2008c; Aso et al. 2008), two commonly implemented paradigms used to assess the antidepressant potential of novel pharmacotherapeutic compounds. Similarly, these mice are particularly susceptible to the anhedonic effects of chronic stress (Martin et al. 2002) and exhibit reduced responsiveness to rewarding stimuli such as ethanol (Poncelet et al. 2003) and sucrose (Sanchis-Segura et al. 2004; Cota et al. 2003), in addition to reductions in food intake and weight gain (Haller et al. 2002). CB_1 receptor–deficient mice also display an increase in anxiogenic traits in tests such as the light–dark box (Martin et al. 2002) and elevated plus maze (Haller et al. 2002, 2004a, 2004b), and increased aggression in the resident–intruder test (Martin et al. 2002). Moreover, these animals exhibit strongly impaired short-term and long-term extinction in fear-conditioning tests (Marsicano et al. 2002) but show no differences during extinction of appetitively motivated tasks (Hölter et al. 2005). Furthermore, increases in both basal and stress-induced activity of the hypothalamic–pituitary–adrenal (HPA) axis have been documented in the form of pronounced corticotrophin-releasing hormone, adrenocorticotropin hormone, and corticosterone release (Cota et al. 2003; Haller et al. 2004a; Barna et al. 2004). Lastly, these mice exhibit impaired hippocampal neurogenesis (Jin et al. 2004) and reduced brain-derived neurotrophic factor release in response to neurotoxic insults (Khaspekov et al. 2004). Therefore, CB_1 receptor–deficient mice exhibit a phenotype that is strikingly reminiscent of the symptomatic profile of melancholic depression (Hill and Gorzalka 2005).

Although it has been unanimously demonstrated that CB_1 receptor knockout mice display all the characteristic phenotypic traits of individuals suffering from major depression, the effects of pharmacological CB_1 receptor blockade in preclinical animal models of emotionality have been equivocal. For instance, chronic treatment with the CB_1 receptor antagonist rimonabant has been shown to mimic the effects of chronic unpredictable stress (CUS), increasing immobility time in the forced swim test and reducing consumption of sucrose-sweetened water (Beyer et al. 2010). Also, chronic rimonabant administration decreases 5-HT levels in the frontal cortex; reduces hippocampal cell proliferation, survival, and brain-derived neurotrophic factor levels; and increases concentrations of proinflammatory cytokines (Beyer et al. 2010). However, CB_1 receptor antagonism has also been shown to exert antidepressant (Shearman et al. 2003; Griebel et al. 2005; Steiner et al. 2008a), anxiolytic (Griebel et al. 2005), anxiogenic (Navarro et al. 1997), or even null effects (Adamczyk et al. 2008). These equivocal results may be due to many factors, including differences in species, strain, dose, test conditions, or off-target effects of the agents employed, such as effects at transient receptor potential vanilloid type 1 receptors (Hill and Gorzalka 2009b).

10.5   ENDOCANNABINOID SIGNALING IN ANIMAL MODELS OF DEPRESSION

Our current understanding of the pathophysiological development of major depression largely stems from research using preclinical paradigms that model the progressive development of a depressive-like phenotype in rodents that closely mirrors the symptom profile of major depression in humans. The CUS (Willner 2005) and olfactory bulbectomy (OBX) (Song and Leonard 2005) models of depression are two such paradigms that have been extensively implemented for this purpose. Studies using chronic psychosocial stressors (e.g., social isolation and social conflict paradigms) have also offered considerable insight into the neurobiological underpinnings of major depression (Anisman and Matheson 2005). In recent years, researchers have begun to examine neurobiological alterations in the endocannabinoid system that occur after exposure to these preclinical paradigms with some rather intriguing results. Other models that assess the antidepressant efficacy of novel therapeutic compounds, such as the forced swim test and tail suspension test, have also yielded interesting findings regarding the antidepressant potential of compounds that facilitate endocannabinoid signaling; however, these paradigms do not necessarily speak to the development of depressive symptoms per se and are thus beyond the scope of the current review (see Hill et al. 2009a for a discussion of this literature). The following section will review the biochemical and physiological changes in endocannabinoid signaling parameters induced by regimens that model the pathological development of depressive-like symptoms and discuss their implications in the context of major depression.

10.5.1 Chronic Unpredictable Stress

The CUS model of depression is often regarded as one of the strongest animal models of depression because of its relatively high levels of face validity (symptom profile), construct validity (theoretical rationale), and predictive validity (pharmacological profile) (Willner 2005). CUS exposure induces hypersecretion of glucocorticoid stress hormones, accompanied by profound alterations in hedonic reactivity, reductions in body weight, decreased grooming behaviors, and, notably, a lack of behavioral habituation, all of which are ameliorated by chronic antidepressant treatments in a manner consistent with the time course observed in clinically depressed individuals (Willner 2005). With respect to endocannabinoid signaling, subjecting rodents to this stress regimen has been shown to induce a ubiquitous downregulation in AEA content, as well as a reduction in the binding site density of CB_1 receptors in subcortical limbic structures such as the hippocampus, hypothalamus, and ventral striatum, while exerting an opposite pattern on CB_1 receptors in the PFC (Hill et al. 2005, 2008b; Hillard et al. 2006). Similarly, CUS exposure decreases CB_1 receptor mRNA transcription (Hillard et al. 2006), CB_1 receptor–mediated GTPγS signaling (Perez-Rial et al. 2004), and 2-AG content in the hippocampus (Hill et al. 2005), while increasing CB_1 receptor mRNA transcription in the PFC (Hillard et al. 2006). With respect to the downregulation of CB_1 receptor binding in the hippocampus, it should be noted that a recent study has replicated these effects in male CUS-exposed rats but demonstrated a different profile in CUS-exposed female rats, revealing *enhanced* CB_1 receptor binding in the dorsal hippocampus (Reich et al. 2009).

Wang et al. (2010) have recently measured CB_1 receptor–mediated physiological responses in ventral striatum slices taken from animals exposed to CUS in an effort to capture the dynamic nature of endocannabinoid/CB_1 receptor signaling. In addition to the behavioral changes induced by CUS, this regimen caused persistent downregulation of endocannabinoid-mediated depolarization-induced suppression of excitation, long-term depression, and CB_1 receptor agonist–induced depression of field excitatory postsynaptic potentials in the nucleus accumbens core, which were reversed by chronic antidepressant administration (Wang et al. 2010). This suggests that CUS exposure induces both biochemical and physiological changes in endocannabinoid signaling within key components of the reward circuitry, which may account for anhedonia, lack of motivation, and other behavioral symptoms of major depression. It is also worthwhile to mention that chronic treatment with the FAAH inhibitor URB597 has been shown to reverse CUS-induced anhedonia and weight loss in a manner comparable to a conventional antidepressant (Bortolato et al. 2007), thus highlighting the therapeutic utility of such compounds in depressed populations.

10.5.2 Olfactory Bulbectomy

The OBX model of depression involves bilateral removal of the olfactory bulbs, which produces behavioral, structural, and neurochemical changes that are akin to those observed in the clinical population (Song and Leonard 2005). Furthermore, these alterations are normalized by chronic (but not acute) administration of antidepressant drugs in a time frame that closely resembles the temporal dynamics of these compounds in the clinic (Song and Leonard 2005). Mounting evidence suggests that endocannabinoids may be functionally implicated in the development of these depressogenic and anxiogenic effects. For instance, OBX produces significant increases in CB_1 receptor binding and CB_1 receptor–mediated GTPγS binding in the PFC in a manner similar to that observed after CUS exposure (Rodríguez-Gaztelumendi et al. 2009). Moreover, chronic antidepressant treatment prevents OBX-induced hyperactivity and upregulated CB_1 receptor signaling in the PFC (Rodríguez-Gaztelumendi et al. 2009). OBX animals also exhibit reduced levels of AEA and 2-AG in the ventral striatum, a brain region deafferented by OBX, with no significant changes in CB_1 receptor densities (Eisenstein et al. 2010). Furthermore, 2-AG levels in OBX rats correlated with distance traveled in the habituation phase of open field exposure, whereas CB_1 receptor blockade further increased the distance traveled during this phase (Eisenstein et al. 2010). Thus, dysregulation of the endocannabinoid system (and 2-AG in particular) may be implicated in the anxiogenic response induced by OBX.

10.5.3 SOCIAL ISOLATION

Rearing rats in isolation postweaning is an animal model of social deprivation that recapitulates features of corticolimbic-based psychopathology in humans (Sciolino et al. 2010). Social isolation produces long-term changes characteristic of emotional disorders including anxiety, neophobia, cognitive rigidity, aggression, hypofunction of the mesocortical dopaminergic system, reduced prefrontal cortical volume, and decreased cortical and hippocampal synaptic plasticity and 5-HT function (Fone and Porkess 2008). Immunohistochemical examinations have shown that isolation-reared rats have reduced CB_1 receptor expression in the caudate putamen and amygdala, along with increased FAAH expression in the caudate putamen and ventral striatum (Malone et al. 2008). However, recent biochemical analyses have revealed *increased* CB_1 receptor binding in the caudate putamen, dorsal and ventral striatum, hypothalamus, and thalamus, in addition to significantly increased 2-AG levels in the PFC of isolation-reared rats (Sciolino et al. 2010).

It is interesting to note that the increase in prefrontal 2-AG content observed in isolation-reared rats mirrors the increase in 2-AG found in the medial PFC of rodents exposed to chronic restraint stress (Patel et al. 2005), which is characterized by a habituated behavioral and neuroendocrine stress response that typically occurs after repeated episodes. Similarly, Sciolino et al. (2010) were able to show increased prefrontal 2-AG content in handled vs. nonhandled rats. Handling could also be considered an initially stressful stimulus that loses its aversive quality over time; thus, the increase in prefrontal 2-AG content could be of particular relevance to stress habituation processes, skills that are conspicuously absent in CUS-exposed rodents and characteristic of individuals inflicted with major depression.

10.5.4 SOCIAL DEFEAT

Repeated exposure to intermale confrontations where defeat ensues has also been reliably shown to promote the persistent development of anxiogenic and depressive-like neurobiological and behavioral symptoms in rodents (Avgustinovich et al. 2005). With respect to endocannabinoid signaling, Rossi et al. (2008) have demonstrated that chronic social defeat stress progressively alters CB_1 receptor–mediated control of GABAergic synaptic transmission in the striatum in a glucocorticoid receptor–dependent manner. Moreover, recovery of these synaptic deficits was encouraged when stressed rats were given access to rewarding stimuli such as a running wheel, sucrose, or an injection of cocaine (Rossi et al. 2008). In a follow-up study, this group further revealed that genetic deletion of FAAH or repeated administration of an FAAH inhibitor prevented the anxious phenotype of mice exposed to social defeat stress in a CB_1 receptor–dependent fashion (Rossi et al. 2010). Remarkably, this effect was also associated with preserved activity of CB_1 receptors regulating GABA activity in the striatum (Rossi et al. 2010). Collectively, these findings argue that alterations in striatal endocannabinoid signaling induced by chronic psychosocial stress may have functional consequences that manifest as disturbances in motor, cognitive, and emotional function. Furthermore, inhibiting FAAH-mediated degradation of AEA prevents these disturbances by preserving CB_1 receptor–mediated GABA control in the striatum.

10.5.5 INTERIM SUMMARY

The aforementioned findings cumulatively argue that aberrant endocannabinoid signaling is a common feature present in animal paradigms that model the development of core symptoms of major depression (see Table 10.1 for a summary of these findings). These studies generally

TABLE 10.1
Alterations in Endocannabinoid Signaling after Exposure to Various Animal Models of Depression and Concurrent Antidepressant Administration

Animal Model	Brain Region	AEA	2-AG	CB_1R	Reversal by Antidepressants	Reference
CUS	Prefrontal cortex	↓	↔	↑	CB_1—yes AEA—no	Hillard et al. 2006; Hill et al. 2008b
	Hippocampus	↓	↓	↓♂ ↑♀	CB_1 ♂—no AEA—no	Hill et al. 2005, 2008b; Perez-Rial et al. 2004; Reich et al. 2009
	Hypothalamus	↓	↑	↓	CB_1—yes AEA, 2-AG—no	Hill et al. 2008b
	Amygdala	↓	↔	↔	AEA—no	Hill et al. 2008b
	Midbrain	↓	↑	↔	AEA, 2-AG—no	Hill et al. 2008b
	Ventral striatum	↓	↔	↓	CB_1—yes AEA—no	Hill et al. 2008b; Wang et al. 2010
OBX	Prefrontal cortex	n.d.	n.d.	↑	CB_1—yes	Rodríguez-Gaztelumendi et al. 2009
	Hippocampus	↔	↔	↔	–	Eisenstein et al. 2010
	Amygdala	↓	↔	↔	n.d.	Eisenstein et al. 2010
	Ventral striatum	↓	↓	↔	n.d.	Eisenstein et al. 2010
SI	Prefrontal cortex	↔	↑	↔	n.d.	Sciolino et al. 2010; Malone et al. 2008
	Hippocampus	↔	↔	↔	–	Sciolino et al. 2010; Malone et al. 2008
	Thalamus	↔	↔	↓	n.d.	Sciolino et al. 2010
	Amygdala	↔	↔	↓	n.d.	Sciolino et al. 2010; Malone et al. 2008
	Ventral striatum	↔	↔	↑, ↓	n.d.	Sciolino et al. 2010; Malone et al. 2008
	Caudate putamen	↔	↔	↑, ↓	n.d.	Sciolino et al. 2010; Malone et al. 2008
SD	Striatum	n.d.	n.d.	↓	CB_1—yes (via FAAH inhibition)	Rossi et al. 2008, 2010

Note: CB_1R, cannabinoid CB_1 receptor; SI, social isolation; SD, social defeat; ↑, increase; ↓, decrease; ↔, no change; n.d., not determined; –, not applicable.

demonstrate a reduction in CB_1 receptor activity, AEA, and/or 2-AG signaling in subcortical structures such as the hippocampus, hypothalamus, and striatum, with a paradoxical upregulation of CB_1 receptor activity in the PFC. The robust decrease in subcortical endocannabinoid signaling in these preclinical models of depression is in agreement with studies demonstrating that genetic deletion or chronic pharmacological blockade of the CB_1 receptor promotes behavioral and physiological disturbances that closely mirror the symptoms of major depression. Conversely, the increase in cortical CB_1 receptor activity may represent an adaptive compensatory response engaged to dampen the adverse effects of chronic stress, although this hypothesis has not yet been experimentally validated. However, it is nonetheless apparent that alterations in the endocannabinoid system are functionally implicated in the development of depressive-like symptoms in preclinical models.

10.6 ENDOCANNABINOID DYSFUNCTION IN HUMAN DEPRESSED POPULATIONS

Accumulating evidence from clinical, postmortem, and genetic studies complements the preclinical data and further supports the notion that endocannabinoid signaling may be compromised in humans suffering from major depression. Hill et al. (2008d, 2009c) have shown that circulating levels of endocannabinoids are significantly reduced in two independent populations of depressed women. Postmortem analyses have provided evidence for aberrant receptor density associated with depressive disorders, particularly in prefrontocortical subregions. For instance, Koethe et al. (2007) found a significant decrease in glial CB_1 receptor density in the anterior cingulate cortex of patients with major depression. In the dorsolateral PFC, Hungund et al. (2004) have reported elevated CB_1 receptor density and functionality in suicide victims compared with healthy controls. These observations were also replicated in chronic alcoholics who died of suicide in comparison with matched alcoholic controls who died of other causes (Vinod et al. 2005). Moreover, studies on schizophrenics, whose negative symptoms overlap with melancholic depression, have also shown increased CB_1 receptor density in the dorsolateral PFC (Dean et al. 2001) as well as in anterior (Zavitsanou et al. 2004) and posterior (Newell et al. 2006) cingulate cortices. Notably, this increase in CB_1 receptor parameters in the PFC is in agreement with preclinical reports on rodents exposed to CUS (Hill et al. 2008b) and OBX (Rodríguez-Gaztelumendi et al. 2009).

Multiple converging genetic studies have further revealed that individuals with different variants of the CB_1 receptor gene are characterized by higher levels of neuroticism and low agreeableness, and display increased vulnerability to depression after adverse life events (Juhasz et al. 2009). Moreover, individuals suffering from recurrent major depressive episodes exhibit a significantly higher frequency of the mutant allele of the CB_1 receptor gene compared with healthy controls (Monteleone et al. 2010). These individuals also exhibit resistance to antidepressant treatment and display weaker striatal and thalamic activation in response to emotional stimuli (Domschke et al. 2008). Similarly, patients with Parkinson's disease who possess long alleles in the CB_1 receptor gene may in fact be less susceptible to depression (Barrero et al. 2005).

Intriguing results from clinical trials implementing the CB_1 receptor antagonist rimonabant for the treatment of obesity have repeatedly revealed that a significant proportion of individuals taking the drug spontaneously developed increased anxiety, adverse depressive-like symptoms, and suicidal ideations that inevitably led to the suspension of clinical trials in both North America and Europe (Christensen et al. 2007; Hill and Gorzalka 2009a). Moreover, these effects were also observed in clinical trials for rimonabant for the treatment of atherosclerosis (Strategy to Reduce Atherosclerosis Development Involving Administration of Rimonabant—The Intravascular Ultrasound Study) (Nissen et al. 2008). In these trials, 43% of individuals on rimonabant developed adverse mood and anxiety responses compared with 28% in the placebo condition (Nissen et al. 2008). In fact, the emergence of these symptoms was sufficient for 1 in 13 individuals to discontinue use of rimonabant compared with 1 in 47 individuals who discontinued placebo treatment for the same reasons (Nissen et al. 2008). A likely explanation for the strikingly high incidence of adverse depressive-like symptoms observed in these trials is the nonexclusion of patients with prior psychiatric disorders. This inevitably resulted in a less selected study population that more closely reflected the risks of depression and anxiety with rimonabant treatment in routine clinical practice (Rumsfeld and Nallamothu 2008). The neural substrates responsible for the increased incidence of depressive symptoms after rimonabant administration are still a subject for speculation but may be attributable to interactions with 5-HT systems. Under conditions of enhanced 5-HT tonic activity, electrophysiology studies have shown that a low dose of rimonabant (1 mg/kg) typically reduces 5-HT firing rates (Bambico et al. 2007; Gobbi et al. 2005), but when 5-HT activity is within the normal range, high doses of rimonabant (>3 mg/kg) further decrease 5-HT firing rates to below optimal levels (Bambico and Gobbi, unpublished results), which could have implications for the spontaneous development of depressive-like symptoms. Regardless of the mechanism responsible, it is evident that pharmacological or genetic disturbances in CB_1 receptor signaling have profound consequences for the instantiation of major depression in clinical populations, thus converging with the described preclinical data.

10.7 ENDOCANNABINOID INTERACTIONS WITH MONOAMINERGIC NEURAL SYSTEMS

Although the pathogenesis of major depression is complex and far from being completely understood, preclinical and clinical evidence suggests that a dysfunction of monoamine systems is critically implicated in the neurobiology of this mood disorder. Accordingly, the mechanism of action for the majority of antidepressant drugs currently on the market is via modulation of 5-HT and/or NA signaling in the brain. Accumulating evidence now suggests that activation of CB_1 receptors augments the release of 5-HT and NA from the dorsal raphe and locus coeruleus, respectively, thereby increasing monoaminergic output to limbic structures implicated in emotionality and mood disorders (see Bambico et al. 2009 and Bambico and Gobbi 2008 for a review).

With respect to 5-HT, anatomical studies have shown that CB_1 receptors are present on 5-HT axon terminals in the dorsal raphe that project to the amygdala and hippocampus (Herkenham et al. 1991; Tsou et al. 1998; Häring et al. 2007). CB_1-deficient mice display increased basal 5-HT extracellular levels and attenuated fluoxetine-induced promotion of 5-HT extracellular levels in the PFC (Aso et al. 2009). These mice also exhibit a reduction in the binding site density of the 5-HT transporter in the PFC and hippocampus; functional desensitization of 5-HT_{1A} autoreceptors in the dorsal raphe; and downregulation of 5-HT_{2C} receptor expression in the dorsal raphe, nucleus accumbens, and paraventricular hypothalamic nucleus (Aso et al. 2009). Therefore, 5-HT negative feedback is severely impaired in mice lacking CB_1 receptors, thus emphasizing the role of endocannabinoids in proper 5-HT function. Conversely, genetic deletion of FAAH produces a marked increase in dorsal raphe 5-HT firing, desensitization of prefrontocortical $5\text{-HT}_{2A/2C}$ receptors, and enhancement of hippocampal 5-HT_{1A} receptor activity, which are all hallmarks of antidepressant activity (Bambico et al. 2010a). Similarly, pharmacological inhibition of FAAH augments the firing activity of dorsal raphe 5-HT neurons, accompanied by potent antidepressant-like responses in the forced swim and tail suspension tests (Gobbi et al. 2005). CB_1 receptor activation with the selective agonist WIN 55212-2 increases 5-HT firing activity at low doses, but decreases it at higher doses (Bambico et al. 2007), with the increase mediated by CB_1 receptors and the decrease likely mediated by transient receptor potential vanilloid type 1 receptors.

The increase in 5-HT neural activity in the dorsal raphe may occur via direct actions of cannabinoids locally within the dorsal raphe or via modulation of upstream projections to the dorsal raphe from the medial PFC. For instance, CB_1 receptor blockade decreases dorsal raphe 5-HT firing rates and can be prevented by coapplication of a $GABA_A$ receptor antagonist, suggesting that endocannabinoids may enhance dorsal raphe 5-HT firing by inhibiting GABAergic control in this region (Mendiguren and Pineda 2009). Furthermore, systemic or intramedial PFC injections of WIN-55212-2 were shown to dose-dependently enhance dorsal raphe 5-HT neuronal activity and antidepressant-like responding in a CB_1 receptor–dependent fashion (Bambico et al. 2007). These effects were abolished by medial PFC transection, thus stressing the importance of CB_1 receptor–mediated medial PFC projections in modulating dorsal raphe 5-HT output (Bambico et al. 2007). Collectively, this evidence suggests that endocannabinoids tonically regulate 5-HT activity through CB_1 receptor–mediated inhibition of GABAergic synaptic transmission locally within the dorsal raphe and/or via disinhibition of excitatory projections from the PFC that modulate dorsal raphe 5-HT firing. It is also noteworthy that chronic administration of a CB_1 receptor agonist in adolescence (but not in adulthood) leads to a long-term decrease in 5-HT firing activity. This is likely due to interactions between the CB_1 receptor and 5-HT neurons that are still under development during adolescence (Spear 2000).

NA signaling also appears to be modulated by endocannabinoid signaling. For instance, CB_1 receptor stimulation increases c-*fos* expression in the locus coeruleus (Oropeza et al. 2005). Moreover, systemic (Oropeza et al. 2005) and local PFC (Page et al. 2008) activation of CB_1 receptors stimulates NA release and increases the firing rate of noradrenergic neurons in the coeruleofrontal pathway. In line with these findings, inhibition of FAAH was shown to robustly increase locus coeruleus NA firing activity and was accompanied by antidepressant-like responding in preclinical paradigms (Gobbi et al. 2005). Notably, CB_1 receptor activation also produces a prolonged decrease in α_{2A} adrenoreceptors in the nucleus accumbens (Carvalho et al. 2010), which are typically increased in patients suffering from major depression and normalized after various antidepressant treatment regimens (Delgado and Moreno 2000). Given this complex association between endocannabinoid/CB_1 receptor signaling and monoaminergic networks, it is conceivable to speculate that dysregulated endocannabinoid signaling may be at the crux of monoamine dysfunction observed in clinically depressed populations.

10.8 ENDOCANNABINOIDS AND HPA AXIS

The melancholic subtype depression is characterized by dysfunctional HPA axis activation and disrupted negative feedback processes that have long-term consequences for emotionality and proper stress coping responses (Gold and Chrousos 2002). Accumulating evidence suggests that endocannabinoid signaling in the brain tightly regulates basal HPA axis activity as well as the neuroendocrine and behavioral responses to stressful stimuli. CB_1 receptors are present on glutamatergic neurons that serve to activate corticotrophin-releasing hormone neurosecretory cells in the parvocellular region of the paraventricular nucleus of the hypothalamus (PVN), and thus are ideally positioned to gate the excitatory drive of the HPA axis (Di et al. 2003). Consequently, disruption of endocannabinoid signaling has profound implications for proper neuroendocrine functioning. For example, transgenic mice lacking the CB_1 receptor show increased CRF mRNA in the PVN, decreased glucocorticoid receptor mRNA in the CA1 region of the hippocampus, and elevated corticosterone concentrations at the onset of the dark cycle, and cultured pituitary cells from these mice display exaggerated adrenocorticotropin hormone secretion in response to CRF or forskolin challenge (Cota et al. 2007). In response to acute stress, genetic or pharmacological CB_1 receptor blockade produces pronounced basal and stress-induced adrenocorticotropin hormone and corticosterone secretion (Steiner et al. 2008c; Aso et al. 2008; Haller et al. 2004a; Barna et al. 2004). Furthermore, glucocorticoid signaling in the PVN leads to CB_1 receptor–mediated fast-feedback inhibition of the HPA axis response to restraint stress (Evanson et al. 2010). Hypothalamic 2-AG content is also elevated by restraint stress, consistent with endocannabinoid action on feedback processes (Evanson et al. 2010). In mice, however, decreases in 2-AG in response to acute restraint stress have also been reported (Patel et al. 2004), although this discrepancy could be attributed to species differences or subtle differences in the area of the hypothalamus dissected. Nevertheless, Patel et al. (2004) showed an enhancement of hypothalamic 2-AG content by the fifth exposure session, along with an attenuated corticosterone response reflecting habituation, thus providing further support that endocannabinoid signaling in the hypothalamus negatively modulates HPA axis activation. Moreover, repeated immobilization stress in juvenile rats (that show a lack of habituation) causes a corticosterone-dependent functional downregulation of CB_1 receptors in the hypothalamic PVN that impairs both activity- and receptor-dependent endocannabinoid signaling at glutamatergic synapses on neuroendocrine cells in the PVN (Wamsteeker et al. 2010). Endocannabinoid signaling in extrahypothalamic regions also plays an important role in stress-induced HPA axis activation. Acute restraint exposure enhances FAAH-mediated hydrolysis of AEA within the amygdala, which results in a suppression of tonic AEA/CB_1 receptor signaling in this region, thus determining the magnitude of HPA axis activation (Hill et al. 2009b). Acute administration of corticosterone also significantly reduces hippocampal AEA content, whereas prolonged glucocorticoid administration downregulates CB_1 receptor expression in the hippocampus

(Hill et al. 2008a). Furthermore, conditional mutant mice lacking CB_1 receptors specifically in principal forebrain neurons display exaggerated neuroendocrine stress responses compared with wild-type mice (Steiner et al. 2008b).

Collectively, these data suggest that endocannabinoid signaling tonically regulates hypothalamic and extrahypothalamic activation under basal conditions, thus constraining the persistent drive of the HPA axis. Upon exposure to stress, endocannabinoid content experiences a decline in these stress-responsive brain regions, likely via an enhancement of enzymatic hydrolysis, thus allowing for the cascade of neuroendocrine and behavioral responses induced by stress. If endocannabinoid levels are maintained before stress induction, then this inhibition is maintained and HPA axis activation is attenuated; but if endocannabinoid signaling is disrupted, then HPA axis activation and neuroendocrine output become potentiated (Hill et al. 2009b). This reciprocal cross-talk between endocannabinoids and the HPA axis is particularly relevant for understanding the pathogenesis of major depression, especially given the regulatory role for endocannabinoid signaling in constraining HPA axis hyperactivation. Over time, prolonged exposure to stress downregulates corticolimbic AEA/CB_1 receptor signaling, allowing for improper HPA axis responses, hypersecretion of glucocorticoids, and maladaptive stress coping responses, which are all hallmark symptoms of the melancholic subtype of depression.

10.9 ENDOCANNABINOIDS PROTECT AGAINST NEUROPLASTIC CHANGES IN MAJOR DEPRESSION

10.9.1 HIPPOCAMPUS

Mounting evidence has demonstrated that neuroplasticity, a fundamental mechanism of neuronal adaptation, is disrupted in major depression and animal models of chronic stress (Pittenger and Duman 2008). Volumetric studies have revealed significant atrophy of the hippocampus in patients suffering from major depression (McKinnon et al. 2009). Accordingly, rodents exposed to chronic stress or glucocorticoids also exhibit deficits in synaptic (long-term potentiation) and morphological (apical dendrite retraction) plasticity in the hippocampus, coupled with impairments in hippocampal-dependent behavioral tasks (Sapolsky 2000). Furthermore, chronic stress or excessive glucocorticoid secretion also reduces the genesis and proliferation of new neurons (neurogenesis) in the rodent hippocampus (Duman 2004a; Dranovsky and Hen 2006). It is now well established that all classes of antidepressants increase neurogenesis (Malberg et al. 2000) and the expression of neurotrophins in the hippocampus (Duman 2004b). This has led to the hypothesis that antidepressants curb the neurodegenerative aspects of depression by supporting neurogenic factors and cellular resilience in the hippocampus. There is evidence that endocannabinoid signaling promotes both neurotrophin expression and hippocampal neurogenesis. Exogenous CB_1 receptor activation promotes brain-derived neurotrophic factor expression and encourages neurogenesis (Jiang et al. 2005), whereas FAAH-deficient mice show enhanced cell proliferation in the hippocampus (Aguado et al. 2005). Accordingly, deficiencies in endocannabinoid signaling are associated with impaired cell proliferation, neurogenesis, and brain-derived neurotrophic factor levels (Aso et al. 2008; Jin et al. 2004; Aguado et al. 2007). Therefore, CB_1 receptor facilitation mimics the effects of conventional antidepressants on neurotrophic signaling and neurogenesis in the hippocampus, suggesting that this system may serve as an endogenous mechanism protecting from neurodegeneration in depressed populations.

10.9.2 PREFRONTAL CORTEX

Another neurobiological correlate of major depression is a reduction in the activity of the PFC, referred to as "hypofrontality," which has been argued to contribute to the depressed mood, amotivational syndrome, and deficits in working memory commonly observed in the depressed population (Drevets et al. 2008). Neuroimaging and postmortem studies have demonstrated reduced PFC volume (Drevets et al. 1997) and altered metabolism in depressed individuals (Rajkowska 1997; Rajkowska et al. 1999). Accordingly, rodents exposed to chronic stress display significant regression of apical dendrites in medial PFC pyramidal cells (Cook and Wellman 2004; Radley et al. 2006), coupled with impairments in PFC-dependent behavioral tasks (Liston et al. 2006). Interestingly, this phenomenon has recently been extended to humans, where chronic psychosocial stress was shown to produce impairments in behavioral and functional magnetic resonance imaging measures of PFC function (Liston et al. 2009). Furthermore, CUS exposure reduces the proliferation of glial cells in the medial PFC, which can be abrogated by concurrent antidepressant administration (Banasr et al. 2007). This finding mirrors the reductions in glia observed in prefrontal subregions in clinically depressed populations (Ongür et al. 1998; Cotter et al. 2001, 2002). These studies suggest that exposure to chronic stress produces deficits in prefrontal synaptic plasticity and glial function that may contribute to the hypofrontality extensively documented in depressed individuals.

Augmented endocannabinoid signaling in the medial PFC has been argued to represent a compensatory mechanism in response to chronic stress (Sciolino et al. 2010; Patel et al. 2005), and increased signaling at CB_1 receptors in this region may circumvent morphological changes after these stress regimens. In support of this theory, chronic exposure to low doses of Δ^9-tetrahydrocannabinol increases the length and branching of dendrites in the medial PFC of rodents (Kolb et al. 2006). With respect to glial cell function, it is interesting to note that endocannabinoid signaling actively promotes biochemical signals resulting in a prosurvival fate for these cells while inducing a selective death in glia-derived tumor cells (Massi et al. 2008). Moreover, under neuropathological conditions, glial cells release an increased amount of endocannabinoids and overexpress cannabinoid receptors, which may constitute an endogenous defense mechanism that abrogates further cell damage (Massi et al. 2008). This research persuasively argues that compromised endocannabinoid signaling in the hippocampus and PFC, two brain regions that are particularly susceptible to the neurodegenerative effects of chronic stress, may have implications for the progression of stress-related pathologies such as major depression. Furthermore, augmenting endocannabinoid signaling in these regions may represent an effective neuroprotective strategy, thus preventing the propagation of adverse changes present in this debilitating disease.

10.10   ENDOCANNABINOID ALTERATIONS INDUCED BY ANTIDEPRESSANT REGIMENS

It is also worth noting that varying classes of chemical antidepressants have been found to increase the binding site density of the CB_1 receptor in the hippocampus, hypothalamus, and amygdala of rodents (Hill et al. 2006, 2008c). Repeated electroconvulsive shock treatment in rats also increases CB_1 receptor signaling in the amygdala (Hill et al. 2007), whereas sleep deprivation, which is also known to exert fast-onset antidepressant effects, increases endocannabinoid ligand content and CB_1 receptor signaling in the hippocampus (Chen and Bazan 2005). Voluntary exercise is yet another antidepressant intervention that is capable of boosting mood, increasing hippocampal neurogenesis (Dunn et al. 2005), and effectively combating symptoms of depression in humans (van Praag 2008). Interestingly, free access to a running wheel has been shown to increase CB_1 receptor density, functionality, and AEA content in the hippocampus, and increase hippocampal cell proliferation in a CB_1 receptor–dependent manner (Hill et al. 2010). These findings are particularly interesting in light of the research demonstrating that moderately intense exercise, in addition to its antidepressant-like effects, also significantly increases plasma AEA levels in human participants (Sparling et al. 2003). These data indicate that a multitude of antidepressant regimens elicit biochemical changes in limbic endocannabinoid signaling that may be necessary for some of the neuroadaptive changes evoked by these treatments.

10.11 SUMMARY AND CONCLUDING REMARKS

The endocannabinoid system is a neuromodulatory system that is intricately involved in regulation of excitatory, inhibitory, and monoaminergic networks in the brain. The research described herein persuasively argues that deficits in corticolimbic endocannabinoid signaling, either through downregulation of endogenous ligands and/or alterations in CB_1 receptor functionality, may have implications for the pathological development of major depression. Anecdotal evidence suggests that consumption of cannabinergic compounds has profound effects on mood, whereas preclinical reports have demonstrated that genetic or pharmacological antagonism of the endocannabinoid system produces behavioral, neurochemical, and neuroplastic changes that closely resemble the phenotype of major depression. Conversely, facilitation of endocannabinoid signaling by inhibiting the degradation of AEA has proven effective at ameliorating the detrimental behavioral and physiological effects of various preclinical stress regimens (Hill et al. 2009a). These stress regimens, which effectively model the progression of core major depressive symptoms, elicit biochemical changes in endocannabinoid parameters that may be relevant to understanding the neurobiological underpinnings of major depression. In agreement with this notion, the endocannabinoid system has been demonstrated to be critical in the functionality of monoaminergic neural systems, a potent modulator of the HPA axis, and a fundamental component of synaptic remodeling and neuroplastic processes in the hippocampus and PFC. Moreover, both pharmacotherapeutic and somatic treatment regimens for depression elicit adaptations in endocannabinoid signaling that may be necessary for the positive clinical response evoked by these treatments. This body of literature argues that enhancement of the endocannabinoid system is a key component involved in the etiopathology of major depression and may represent a novel therapeutic target for the treatment of this debilitating mental illness.

REFERENCES

Adamczyk, P., Gołda, A., McCreary, A. C., Filip, M., and Przegaliński, E. 2008. Activation of endocannabinoid transmission induces antidepressant-like effects in rats. *J Physiol Pharmacol* 59:217–28.

Aguado, T., Monory, K., Palazuelos, J., et al. 2005. The endocannabinoid system drives neural progenitor proliferation. *FASEB J* 19:1704–6.

Aguado, T., Romero, E., Monory, K., et al. 2007. The CB_1 cannabinoid receptor mediates excitotoxicity-induced neural progenitor proliferation and neurogenesis. *J Biol Chem* 282:23892–8.

Ahn, K., McKinney, M. K., and Cravatt, B. F. 2008. Enzymatic pathways that regulate endocannabinoid signaling in the nervous system. *Chem Rev* 108:1687–707.

Anisman, H., and Matheson, K. 2005. Stress, depression, and anhedonia: Caveats concerning animal models. *Neurosci Biobehav Rev* 29:525–46.

Aso, E., Ozaita, A., Valdizán, E. M., et al. 2008. BDNF impairment in the hippocampus is related to enhanced despair behavior in CB_1 knockout mice. *J Neurochem* 103:2111–20.

Aso, E., Renoir, T., Mengod, G., et al. 2009. Lack of CB_1 receptor activity impairs serotonergic negative feedback. *J Neurochem* 109:935–44.

Avgustinovich, D. F., Kovalenko, I. L., and Kudryavtseva, N. N. 2005. A model of anxious depression: Persistence of behavioral pathology. *Neurosci Behav Physiol* 35:917–24.

Bambico, F. R., and Gobbi, G. 2008. The cannabinoid CB_1 receptor and the endocannabinoid anandamide: Possible antidepressant targets. *Expert Opin Ther Targets* 12:1347–66.

Bambico, F. R., Katz, N., Debonnel, G., and Gobbi, G. 2007. Cannabinoids elicit antidepressant-like behavior and activate serotonergic neurons through the medial prefrontal cortex. *J Neurosci* 27:11700–11.

Bambico, F. R., Duranti, A., Tontini, A., Tarzia, G., and Gobbi, G. 2009. Endocannabinoids in the treatment of mood disorders: Evidence from animal models. *Curr Pharm Des* 15:1623–46.

Bambico, F. R., Cassano, T., Dominguez-Lopez, S., et al. 2010a. Genetic deletion of fatty acid amide hydrolase alters emotional behavior and serotonergic transmission in the dorsal raphe, prefrontal cortex, and hippocampus. *Neuropsychopharmacology* 35:2083–100.

Bambico, F. R., Nguyen, N. T., Katz, N., and Gobbi, G. 2010b. Chronic exposure to cannabinoids during adolescence but not during adulthood impairs emotional behaviour and monoaminergic neurotransmission. *Neurobiol Dis* 37:641–55.

Banasr, M., Valentine, G. W., Li, X. Y., Gourley, S. L., Taylor, J. R., and Duman, R. S. 2007. Chronic unpredictable stress decreases cell proliferation in the cerebral cortex of the adult rat. *Biol Psychiatry* 62:496–504.

Barna, I., Zelena, D., Arszovszki, A. C., and Ledent, C. 2004. The role of endogenous cannabinoids in the hypothalamo–pituitary–adrenal axis regulation: In vivo and in vitro studies in CB$_1$ receptor knockout mice. *Life Sci* 75:2959–70.

Barrero, F. J., Ampuero, I., Morales, B., et al. 2005. Depression in Parkinson's disease is related to a genetic polymorphism of the cannabinoid receptor gene (CNR1). *Pharmacogenomics J* 5:135–41.

Beyer, C. E., Dwyer, J. M., Piesla, M. J., et al. 2010. Depression-like phenotype following chronic CB$_1$ receptor antagonism. *Neurobiol Dis* 39:148–55.

Bisogno, T. 2008. Endogenous cannabinoids: Structure and metabolism. *J Neuroendocrinol* 20:1–9.

Bortolato, M., Mangieri, R. A., Fu, J., et al. 2007. Antidepressant-like activity of the fatty acid amide hydrolase inhibitor URB597 in a rat model of chronic mild stress. *Biol Psychiatry* 62:1103–10.

Bovasso, G. B. 2001. Cannabis abuse as a risk factor for depressive symptoms. *Am J Psychiatry* 158:2033–7.

Campbell, I. 1976. The amotivational syndrome and cannabis use with emphasis on the Canadian scene. *Ann N Y Acad Sci.* 282:33–6.

Carvalho, A. F., Mackie, K., and Van Bockstaele, E. J. 2010. Cannabinoid modulation of limbic forebrain noradrenergic circuitry. *Eur J Neurosci* 31:286–301.

Chen, C., and Bazan, N. G. 2005. Lipid signaling: Sleep, synaptic plasticity, and neuroprotection. *Prostaglandins Other Lipid Mediat* 77:65–76.

Christensen, R., Kristensen, P. K., Bartels, E. M., Bliddal, H., and Astrup, A. 2007. Efficacy and safety of the weight-loss drug rimonabant: A meta-analysis of randomised trials. *Lancet* 370:1706–13.

Cook, S. C., and Wellman, C. L. 2004. Chronic stress alters dendritic morphology in rat medial prefrontal cortex. *J Neurobiol* 60:236–48.

Cota, D., Marsicano, G., Tschöp, M., et al. 2003. The endogenous cannabinoid system affects energy balance via central orexigenic drive and peripheral lipogenesis. *J Clin Invest* 112:423–31.

Cota, D., Steiner, M. A., Marsicano, G., et al. 2007. Requirement of cannabinoid receptor type 1 for the basal modulation of hypothalamic–pituitary–adrenal axis function. *Endocrinology* 148:1574–81.

Cotter, D., Mackay, D., Landau, S., Kerwin, R., and Everall, I. 2001. Reduced glial cell density and neuronal size in the anterior cingulate cortex in major depressive disorder. *Arch Gen Psychiatry* 58:545–53.

Cotter, D., Mackay, D., Chana, G., Beasley, C., Landau, S., and Everall, I. P. 2002. Reduced neuronal size and glial cell density in area 9 of the dorsolateral prefrontal cortex in subjects with major depressive disorder. *Cereb Cortex* 12:386–94.

Curran, H. V., Brignell, C., Fletcher, S., Middleton, P., and Henry, J. 2002. Cognitive and subjective dose–response effects of acute oral delta(9)-tetrahydrocannabinol (THC) in infrequent cannabis users *Psychopharmacology* 164:61–70.

Dean, B., Sundram, S., Bradbury, R., Scarr, E., and Copolov, D. 2001. Studies on [^{3}H]CP-55940 binding in the human central nervous system: Regional specific changes in density of cannabinoid-1 receptors associated with schizophrenia and cannabis use. *Neuroscience* 103:9–15.

Delgado, P., and Moreno, F. 2000. Role of norepinephrine in depression. *J Clin Psychiatry* 61:5–12.

Devane, W. A., Hanus, L., Breuer, A., et al. 1992. Isolation and structure of a brain constituent that binds to the cannabinoid receptor. *Science* 258:1946–9.

Di Marzo, V. 1999. Biosynthesis and inactivation of endocannabinoids: Relevance to their proposed role as neuromodulators. *Life Sci* 65:645–55.

Di, S., Malcher-Lopes, R., Halmos, K. C., and Tasker, J. G. 2003. Nongenomic glucocorticoid inhibition via endocannabinoid release in the hypothalamus: A fast feedback mechanism. *J Neurosci* 23:4850–7.

Domschke, K., Dannlowski, U., Ohrmann, P., et al. 2008. Cannabinoid receptor 1 (*CNR1*) gene: Impact on antidepressant treatment response and emotion processing in major depression. *Eur Neuropsychopharmacol* 18:751–9.

Dranovsky, A., and Hen, R. 2006. Hippocampal neurogenesis: Regulation by stress and antidepressants. *Biol Psychiatry* 59:1136–43.

Drevets, W. C., Price, J. L., Simpson, J. R., et al. 1997. Subgenual prefrontal cortex abnormalities in mood disorders. *Nature* 386:824–7.

Drevets, W. C., Price, J. L., and Furey, M. L. 2008. Brain structural and functional abnormalities in mood disorders: Implications for neurocircuitry models of depression. *Brain Struct Funct* 213:93–118.

Duman, R. S. 2004a. Depression: A case of neuronal life and death? *Biol Psychiatry* 56:140–5.

Duman, R. S. 2004b. Role of neurotrophic factors in the etiology and treatment of mood disorders. *Neuromolecular Med* 5:11–25.

Dunn, A. L., Trivedi, M. H., Kampert, J. B., Clark, C. G., and Chambliss, H. O. 2005. Exercise treatment for depression: Efficacy and dose response. *Am J Prev Med.* 28:1–8.

Eisenstein, S. A., Clapper, J. R., Holmes, P. V., Piomelli, D., and Hohmann, A. G. 2010. A role for 2-arachidonoylglycerol and endocannabinoid signaling in the locomotor response to novelty induced by olfactory bulbectomy. *Pharmacol Res* 61:419–29.

Evanson, N. K., Tasker, J. G., Hill, M. N., Hillard, C. J., and Herman, J. P. 2010. Fast feedback inhibition of the HPA axis by glucocorticoids is mediated by endocannabinoid signaling. *Endocrinology* 151:4811–9.

Fone, K. C., and Porkess, M. V. 2008. Behavioral and neurochemical effects of post-weaning social isolation in rodents—Relevance to developmental neuropsychiatric disorders. *Neurosci Biobehav Rev* 32:1087–102.

Freund, T. F., Katona, I., and Piomelli, D. 2003. Role of endogenous cannabinoids in synaptic signaling. *Physiol Rev* 83:1017–66.

Glass, M., Dragunow, M., and Faull, R. L. 1997. Cannabinoid receptors in the human brain: A detailed anatomical and quantitative autoradiographic study in the fetal, neonatal and adult human brain. *Neuroscience* 77:299–318.

Gobbi, G., Bambico, F. R., Mangieri, R., et al. 2005. Antidepressant-like activity and modulation of brain monoaminergic transmission by blockade of anandamide hydrolysis. *Proc Natl Acad Sci U S A* 102:18620–5.

Gold, P. W., and Chrousos, G. P. 2002. Organization of the stress system and its dysregulation in melancholic and atypical depression: High vs low CRH/NE states. *Mol Psychiatry* 7:254–75.

Gorzalka, B. B., Hill, M. N., and Hillard, C. J. 2008. Regulation of endocannabinoid signaling by stress: Implications for stress-related affective disorders. *Neurosci Biobehav Rev* 32:1152–60.

Griebel, G., Stemmelin, J., and Scatton, B. 2005. Effects of the cannabinoid CB1 receptor antagonist rimonabant in models of emotional reactivity in rodents. *Biol Psychiatry* 57:261–7.

Haller, J., Bakos, N., Szirmay, M., Ledent, C., and Freund, T. F. 2002. The effects of genetic and pharmacological blockade of the CB1 cannabinoid receptor on anxiety. *Eur J Neurosci* 16:1395–8.

Haller, J., Varga, B., Ledent, C., Barna, I., and Freund, T. F. 2004a. Context-dependent effects of CB_1 cannabinoid gene disruption on anxiety-like and social behaviour in mice. *Eur J Neurosci* 19:1906–12.

Haller, J., Varga, B., Ledent, C., and Freund, T. F. 2004b. CB_1 cannabinoid receptors mediate anxiolytic effects: Convergent genetic and pharmacological evidence with CB_1-specific agents. *Behav Pharmacol* 15:299–304.

Häring, M., Marsicano, G., Lutz, B., and Monory, K. 2007. Identification of the cannabinoid receptor type 1 in serotonergic cells of raphe nuclei in mice. *Neuroscience* 146:1212–9.

Herkenham, M., Lynn, A. B., Johnson, M. R., Melvin, L. S., de Costa, B. R., and Rice, K. C. 1991. Characterization and localization of cannabinoid receptors in rat brain: A quantitative in vitro autoradiographic study. *J Neurosci* 11:563–83.

Herkenham, M., Lynn, A. B., Little, M. D., et al. 1990. Cannabinoid receptor localization in brain. *Proc Natl Acad Sci U S A* 87:1932–6.

Hill, M. N. and Gorzalka, B. B. 2005. Is there a role for the endocannabinoid system in the etiology and treatment of melancholic depression? *Behav Pharmacol* 16:333–52.

Hill, M. N., and Gorzalka, B. B. 2009a. Impairments in endocannabinoid signaling and depressive illness. *JAMA* 301:1165–6.

Hill, M. N., and Gorzalka, B. B. 2009b. The endocannabinoid system and the treatment of mood and anxiety disorders. *CNS Neurol Disord Drug Targets* 8:451–8.

Hill, M. N., Patel, S., Carrier, E. J., et al. 2005. Downregulation of endocannabinoid signaling in the hippocampus following chronic unpredictable stress. *Neuropsychopharmacology* 30:508–13.

Hill, M. N., Ho, W. S., Sinopoli, K. J., et al. 2006. Involvement of the endocannabinoid system in the ability of long-term tricyclic antidepressant treatment to suppress stress-induced activation of the hypothalamic–pituitary–adrenal axis. *Neuropsychopharmacology* 31:2591–9.

Hill, M. N., Barr, A. M., Ho, W. S., et al. 2007. Electroconvulsive shock treatment differentially modulates cortical and subcortical endocannabinoid activity. *J Neurochem* 103:47–56.

Hill, M. N., Carrier, E. J., Ho, W. S., et al. 2008a. Prolonged glucocorticoid treatment decreases cannabinoid CB_1 receptor density in the hippocampus. *Hippocampus* 18:221–6.

Hill, M. N., Carrier, E. J., McLaughlin, R. J., et al. 2008b. Regional alterations in the endocannabinoid system in an animal model of depression: Effects of concurrent antidepressant treatment. *J Neurochem* 106:2322–36.

Hill, M. N., Ho, W. S., Hillard, C. J., and Gorzalka, B. B. 2008c. Differential effects of the antidepressants tranylcypromine and fluoxetine on limbic cannabinoid receptor binding and endocannabinoid contents. *J Neural Transm* 115:1673–9.

Hill, M. N., Miller, G. E., Ho, W. S., Gorzalka, B. B., and Hillard, C. J. 2008d. Serum endocannabinoid content is altered in females with depressive disorders: A preliminary report. *Pharmacopsychiatry* 41:48–53.

Hill, M. N., Hillard, C. J., Bambico, F. R., et al. 2009a. The therapeutic potential of the endocannabinoid system for the development of a novel class of antidepressants. *Trends Pharmacol Sci* 30:484–93.

Hill, M. N., McLaughlin, R. J., Morrish, A. C., et al. 2009b. Suppression of amygdalar endocannabinoid signaling by stress contributes to activation of the hypothalamic–pituitary–adrenal axis. *Neuropsychopharmacology* 34:2733–45.

Hill, M. N., Miller, G. E., Carrier, E. J., Gorzalka, B. B., and Hillard, C. J. 2009c. Circulating endocannabinoids and *N*-acyl ethanolamines are differentially regulated in major depression and following exposure to social stress. *Psychoneuroendocrinology* 34:1257–62.

Hill, M. N., Titterness, A. K., Morrish, A. C., et al. 2010. Endogenous cannabinoid signaling is required for voluntary exercise-induced enhancement of progenitor cell proliferation in the hippocampus. *Hippocampus* 20:513–23.

Hillard, C. J. 2000. Biochemistry and pharmacology of the endocannabinoids arachidonylethanolamide and 2-arachidonylglycerol. *Prostaglandins Other Lipid Mediat* 61:3–18.

Hillard, C. J., Hill, M. N., Carrier, E. J., Shi, L., Cullinan, W. E., and Gorzalka, B. B. 2006. Regulation of cannabinoid receptor expression by chronic unpredictable stress in rats and mice. *Soc Neurosci Abstr*: 746.19.

Hölter, S. M., Kallnik, M., Wurst, W., et al. 2005. Cannabinoid CB_1 receptor is dispensable for memory extinction in an appetitively-motivated learning task. *Eur J Pharmacol* 510:69–74.

Howlett, A. C. 1995. Pharmacology of cannabinoid receptors. *Annu Rev Pharmacol Toxicol* 35:607–34.

Howlett, A. C. 2002. The cannabinoid receptors. *Prostaglandins Other Lipid Mediat* 68–69:619–31.

Hungund, B. L., Vinod, K. Y., Kassir, S. A., et al. 2004. Upregulation of CB_1 receptors and agonist-stimulated $[^{35}S]GTP\gamma S$ binding in the prefrontal cortex of depressed suicide victims. *Mol Psychiatry* 9:184–90.

Jiang, W., Zhang, Y., Xiao, L., et al. 2005. Cannabinoids promote embryonic and adult hippocampus neurogenesis and produce anxiolytic- and antidepressant-like effects. *J Clin Invest* 115:3104–16.

Jin, K., Xie, L., Kim, S. H., et al. 2004. Defective adult neurogenesis in CB_1 cannabinoid receptor knockout mice. *Mol Pharmacol* 66:204–8.

Juhasz, G., Chase, D., Pegg, E., et al. 2009. CNR1 gene is associated with high neuroticism and low agreeableness and interacts with recent negative life events to predict current depressive symptoms. *Neuropsychopharmacology* 34:2019–27.

Khaspekov, L. G., Brenz Verca, M. S., Frumkina, L. E., et al. 2004. Involvement of brain-derived neurotrophic factor in cannabinoid receptor-dependent protection against excitotoxicity. *Eur J Neurosci* 19:1691–8.

Koethe, D., Llenos, I. C., Dulay, J. R., et al. 2007. Expression of CB_1 cannabinoid receptor in the anterior cingulate cortex in schizophrenia, bipolar disorder, and major depression. *J Neural Transm* 114: 1055–63.

Kolb, B., Gorny, G., Limebeer, C. L., and Parker, L. A. 2006. Chronic treatment with Delta-9-tetrahydrocannabinol alters the structure of neurons in the nucleus accumbens shell and medial prefrontal cortex of rats. *Synapse* 60:429–36.

Liston, C., Miller, M. M., and Goldwater, D. S., et al. 2006. Stress-induced alterations in prefrontal cortical dendritic morphology predict selective impairments in perceptual attentional set-shifting. *J Neurosci* 26:7870–4.

Liston, C., McEwen, B. S., and Casey, B. J. 2009. Psychosocial stress reversibly disrupts prefrontal processing and attentional control. *Proc Natl Acad Sci U S A* 106:912–7.

Malberg, J. E., Eisch, A. J., Nestler, E. J., and Duman, R. S. 2000. Chronic antidepressant treatment increases neurogenesis in adult rat hippocampus. *J Neurosci* 20:9104–10.

Malone, D. T., Kearn, C. S., Chongue, L., Mackie, K., and Taylor, D. A. 2008. Effect of social isolation on CB_1 and D_2 receptor and fatty acid amide hydrolase expression in rats. *Neuroscience* 152:265–72.

Marsicano, G., Wotjak, C. T., Azad, S. C., et al. 2002. The endogenous cannabinoid system controls extinction of aversive memories. *Nature* 418: 530–4.

Martin, M., Ledent, C., Parmentier, M., Maldonado, R., and Valverde, O. 2002. Involvement of CB_1 cannabinoid receptors in emotional behaviour. *Psychopharmacology* 159:379–87.

Massi, P., Valenti, M., Bolognini, D., and Parolaro, D. 2008. Expression and function of the endocannabinoid system in glial cells. *Curr Pharm Des* 14:2289–98.

Mato, S., and Pazos, A. 2004. Influence of age, postmortem delay and freezing storage period on cannabinoid receptor density and functionality in human brain. *Neuropharmacology* 46:716–26.

Matsuda, L. A., Lolait, S. J., Brownstein, M. J., Young, A. C., and Bonner, T. I. 1990. Structure of a cannabinoid receptor and functional expression of the cloned cDNA. *Nature* 346:561–4.

McKinnon, M. C., Yucel, K., Nazarov, A., and MacQueen, G. M. 2009. A meta-analysis examining clinical predictors of hippocampal volume in patients with major depressive disorder. *J Psychiatry Neurosci* 34:41–54.

Mendiguren, A., and Pineda, J. 2009. Effect of the CB_1 receptor antagonists rimonabant and AM251 on the firing rate of dorsal raphe nucleus neurons in rat brain slices. *Br J Pharmacol* 158:1579–87.

Monteleone, P., Bifulco, M., Maina, G., et al. 2010. Investigation of CNR1 and FAAH endocannabinoid gene polymorphisms in bipolar disorder and major depression. *Pharmacol Res* 61:400–4.

Morales, M., and Bäckman, C. 2002. Coexistence of serotonin 3 ($5\text{-}HT_3$) and CB_1 cannabinoid receptors in interneurons of hippocampus and dentate gyrus. *Hippocampus* 12:756–64.

Munro, S., Thomas, K. L., and Abu-Shaar, M. 1993. Molecular characterization of a peripheral receptor for cannabinoids. *Nature* 365:61–5.

Musty, R. R., and Kaback, L. 1995. Relationships between motivation and depression in chronic marijuana users. *Life Sci* 56:2151–8.

Navarro, M., Hernández, E., Muñoz, R. M., et al. 1997. Acute administration of the CB1 cannabinoid receptor antagonist SR 141716A induces anxiety-like responses in the rat. *Neuroreport* 8:491–6.

Newell, K. A., Deng, C., and Huang, X. F. 2006. Increased cannabinoid receptor density in the posterior cingulate cortex in schizophrenia. *Exp Brain Res* 172:556–60.

Nissen, S. E., Nicholls, S. J., Wolski, K., et al. 2008. Effect of rimonabant on progression of atherosclerosis in patients with abdominal obesity and coronary artery disease: The STRADIVARIUS randomized controlled trial. *JAMA* 299:1547–60.

Ongür, D., Drevets, W. C., and Price, J. L. 1998. Glial reduction in the subgenual prefrontal cortex in mood disorders. *Proc Natl Acad Sci U S A* 95:13290–5.

Oropeza, V. C., Page, M. E., and Van Bockstaele, E. J. 2005. Systemic administration of WIN 55,212-2 increases norepinephrine release in the rat frontal cortex. *Brain Res* 1046:45–54.

Oropeza, V. C., Mackie, K., and Van Bockstaele, E. J. 2007. Cannabinoid receptors are localized to noradrenergic axon terminals in the rat frontal cortex. *Brain Res* 1127:36–44.

Page, M. E., Oropeza, V. C., and Van Bockstaele, E. J. 2008. Local administration of a cannabinoid agonist alters norepinephrine efflux in the rat frontal cortex. *Neurosci Lett* 431:1–5.

Patel, S., Roelke, C. T., Rademacher, D. J., Cullinan, W. E., and Hillard, C. J. 2004. Endocannabinoid signaling negatively modulates stress-induced activation of the hypothalamic–pituitary–adrenal axis. *Endocrinology* 145:5431–8.

Patel, S., Roelke, C. T., Rademacher, D. J., and Hillard, C. J. 2005. Inhibition of restraint stress-induced neural and behavioural activation by endogenous cannabinoid signaling. *Eur J Neurosci* 21:1057–69.

Patton, G. C., Coffey, C., Carlin, J. B., Degenhardt, L., Lynskey, M., and Hall, W. 2002. Cannabis use and mental health in young people: Cohort study. *BMJ* 325:1195–8.

Perez-Rial, S., Uriguen, L., Palomo, T., and Manzaneres, J. 2004. Acute and chronic stress alter cannabinoid CB_1 receptor function and POMC gene expression in the rat brain. *Eur Neuropsychopharmacol* 14:S318.

Pittenger, C., and Duman, R. S. 2008. Stress, depression, and neuroplasticity: A convergence of mechanisms. *Neuropsychopharmacology* 33:88–109.

Poncelet, M., Maruani, J., Calassi, R., and Soubrié, P. 2003. Overeating, alcohol and sucrose consumption decrease in CB1 receptor deleted mice. *Neurosci Lett* 343:216–8.

Radley, J. J., Rocher, A. B., Miller, M., et al. 2006. Repeated stress induces dendritic spine loss in the rat medial prefrontal cortex. *Cereb Cortex* 16:313–20.

Rajkowska, G. 1997. Morphometric methods for studying the prefrontal cortex in suicide victims and psychiatric patients. *Ann N Y Acad Sci* 836:253–68.

Rajkowska, G., Miguel-Hidalgo, J. J., and Wei, J., 1999. Morphometric evidence for neuronal and glial prefrontal cell pathology in major depression. *Biol Psychiatry* 45:1085–98.

Reich, C. G., Taylor, M. E., and McCarthy, M. M. 2009. Differential effects of chronic unpredictable stress on hippocampal CB_1 receptors in male and female rats. *Behav Brain Res* 203:264–9.

Rodríguez-Gaztelumendi, A., Rojo, M. L., Pazos, A., and Díaz, A. 2009. Altered CB receptor-signaling in prefrontal cortex from an animal model of depression is reversed by chronic fluoxetine. *J Neurochem* 108:1423–43.

Rossi, S., De Chiara, V., Musella, A., et al. 2008. Chronic psychoemotional stress impairs cannabinoid-receptor-mediated control of GABA transmission in the striatum. *J Neurosci* 28:7284–92.

Rossi, S., De Chiara, V., Musella, A., et al. 2010. Preservation of striatal cannabinoid CB$_1$ receptor function correlates with the anti-anxiety effects of fatty acid amide hydrolase inhibition. *Mol Pharmacol* 78:260–8.

Rumsfeld, J. S., and Nallamothu, B. K. 2008. The hope and fear of rimonabant. *JAMA* 299:1601–2.

Sanchis-Segura, C., Cline, B. H., Marsicano, G., Lutz, B., and Spanagel, R. 2004. Reduced sensitivity to reward in CB$_1$ knockout mice. *Psychopharmacology* 176:223–32.

Sapolsky, R. M. 2000. Glucocorticoids and hippocampal atrophy in neuropsychiatric disorders. *Arch Gen Psychiatry* 57:925–35.

Sciolino, N. R., Bortolato, M., Eisenstein, S. A., et al. 2010. Social isolation and chronic handling alter endocannabinoid signaling and behavioral reactivity to context in adult rats. *Neuroscience* 168:371–86.

Shearman, L. P., Rosko, K. M., Fleischer, R., et al. 2003. Antidepressant-like and anorectic effects of the cannabinoid receptor inverse agonist AM251 in mice. *Behav Phamacol* 14:573–82.

Song, C., and Leonard, B. E. 2005. The olfactory bulbectomized rat as a model of depression. *Neurosci Biobehav Rev* 29:627–47.

Sparling, P. B., Giuffrida, A., Piomelli, D., Rosskopf, L., and Dietrich, A. 2003. Exercise activates the endocannabinoid system. *Neuroreport* 14:2209–11.

Spear, L. P. 2000. The adolescent brain and age-related behavioral manifestations. *Neurosci Biobehav Rev* 24:417–63.

Steiner, M. A., Marsicano, G., Nestler, E. J., Holsboer, F., Lutz, B., and Wotjak, C. T. 2008a. Antidepressant-like behavioral effects of impaired cannabinoid receptor type 1 signaling coincide with exaggerated corticosterone secretion in mice. *Psychoneuroendocrinology* 33:54–67.

Steiner, M. A., Marsicano, G., Wotjak, C. T., and Lutz, B. 2008b. Conditional cannabinoid receptor type 1 mutants reveal neuron subpopulation-specific effects on behavioral and neuroendocrine stress responses. *Psychoneuroendocrinology* 33:1165–70.

Steiner, M. A., Wanisch, K., Monory, K., et al. 2008c. Impaired cannabinoid receptor type 1 signaling interferes with stress coping behavior in mice. *Pharmacogenomics J* 8:196–208.

Sugiura, T., Kondo, S., Sukagawa, A., et al. 1995. 2-Arachidonylglycerol: A possible endogenous cannabinoid receptor ligand in brain. *Biochem Biophys Res Commun* 215:89–97.

Tsou, K., Brown, S., Sañudo-Peña, M. C., Mackie, K., and Walker, J. M. 1998. Immunohistochemical distribution of cannabinoid CB$_1$ receptors in the rat central nervous system. *Neuroscience* 83:393–411.

van Praag, H. 2008. Neurogenesis and exercise: Past and future directions. *Neuromolecular Med* 10:128–40.

Vinod, K. Y., Arango, V., Xie, S., et al. 2005. Elevated levels of endocannabinoids and CB$_1$ receptor–mediated G-protein signaling in the prefrontal cortex of alcoholic suicide victims. *Biol Psychiatry* 57:480–6.

Wamsteeker, J. I., Kuzmiski, J. B., and Bains, J. S. 2010. Repeated stress impairs endocannabinoid signaling in the paraventricular nucleus of the hypothalamus. *J Neurosci* 30:11188–96.

Wang, W., Sun, D., Pan, B., et al. 2010. Deficiency in endocannabinoid signaling in the nucleus accumbens induced by chronic unpredictable stress. *Neuropsychopharmacology* 35:2249–61.

Ware, M. A., Wang, T., Shapiro, S., et al. 2010. Smoked cannabis for chronic neuropathic pain: A randomized controlled trial. *CMAJ* 182(14):E694–701.

Willner, P. 2005. Chronic mild stress (CMS) revisited: Consistency and behavioral–neurobiological concordance in the effects of CMS. *Neuropsychobiology* 52:90–110.

Zavitsanou, K., Garrick, T., and Huang, X. F. 2004. Selective antagonist [H-3]SR141716A binding to cannabinoid CB$_1$ receptors is increased in the anterior cingulate cortex in schizophrenia. *Prog Neuropsychopharmacol Biol Psychiatry* 28:355–60.

11 Opioid System and Depression

Cristina Alba-Delgado, Pilar Sánchez-Blázquez,
Esther Berrocoso, Javier Garzón, and Juan Antonio Micó

CONTENTS

11.1 INTRODUCTION

The opioid system constitutes an important neuropeptidergic system in the central nervous system. There are many central and peripheral biological functions that are regulated by the opioid peptides. Among these functions, the opioid system is able to regulate the emotional status. There are a number of findings that prove that the opioid system is able to modulate mood in depression, a devastating illness that affects millions of individuals worldwide. In this chapter, we review the current knowledge about the implication of the opioid system in depression. Before we proceed, however, as a mode of introduction, we first review the main aspects of the biology of this system.

11.2 BRIEF INTRODUCTION TO OPIOID SYSTEM

The discovery, in the 1970s, of an opiate receptor and the linked discovery of a previously unknown group of endogenous neuropeptides with opioid activity was undoubtedly a breakthrough that still influences our view on the nature of pain and how to treat pain. Histologic analysis showed the presence of enkephalin peptides at supraspinal levels in areas relevant to pain. Moreover, it was established that opiates are not just pain-reducing substances, but are psychotropic agents with widespread effects on emotional (tranquilizing) and motivational states. Today, it is accepted that opioid pharmacology is rather complex because of the existence of multiple mu, delta, and kappa receptor types. Gene cloning led to the characterization of three receptor genes only: *Oprm1, Oprd1,*

223

and *Oprk1*. These genes encode mu, delta, and kappa receptors (MOR, DOR, and KOR), respectively. Alternative splicing has been reported, but it has been difficult to establish the biological relevance of these alternative transcripts in vivo and to correlate their existence with the multiple opioid receptors that were described earlier based on pharmacological aspects. Moreover, the multiplicity of receptor-regulated signaling proteins may also account for the wide diversity of receptor subtypes. Thus, we must now envisage opioid receptors as dynamic multicomponent units, rather than single protein entities, and that ligands establish and stabilize different receptor-containing protein complexes.

11.2.1 ENDOGENOUS OPIOIDS

The sensibility of the stimulation-produced analgesia to the opiate antagonist naloxone and the demonstration of opioid receptors strongly pointed to the existence of an endogenous opioid substance in the brain. Such a substance would be released upon electrical stimulation of certain brain regions to produce analgesia and would be the natural ligand for the opiate receptor. In 1975, Hughes et al. identified an endogenous, opiate-like factor that they termed *enkephalin*. Surprisingly, this factor represented not a single morphine-like molecule but two structurally related pentapeptides: Tyr–Gly–Gly–Phe–Met and Tyr–Gly–Gly–Phe–Leu. These peptides were named *methionine-enkephalin* (*Met-enkephalin*) and *leucine-enkephalin* (*Leu-enkephalin*). A period of intense scientific activity led to the discovery of a dozen of peptides that became part of the growing family of "endorphins" or endogenous opioids (Akil et al. 1984). The amino-terminal sequence of all opioid peptides contain Tyr–Gly–Gly–Phe–[Met/Leu], later known as the opioid motif, followed by various carboxy-terminal extensions that result in peptides that vary in length from 5 to 31. Two interesting members of the endorphin family are β-endorphin, an extremely potent opioid analgesic with long-lasting effects (Loh et al. 1976), and dynorphin A, a 17-amino-acid peptide with high potency and a distinctive pharmacological profile (Goldstein et al. 1979). Anatomical mapping studies revealed that β-endorphin and dynorphin A have different anatomical distributions, suggesting that they are synthesized and expressed separately (Watson et al. 1981). The analysis of the affinities for MOR, DOR, and KOR of the endogenous opioid peptides and related sequences revealed that the K_i for DOR changed very little when the sequence of the enkephalins was enlarged; on the contrary, the MOR and especially the KOR activity were highly dependent not only on the specificity of the sequence but also on the length of the peptide (Garzon et al. 1983).

Although there were suggestions of precursor–product relationships between the various opioid peptides, the confusion about the molecular relationship was clarified when their three precursors were cloned between 1979 and 1982 (Table 11.1). The first to be characterized were proopiomelanocortin, the common protein precursor for β-endorphin, as well as the stress adrenocorticotropic hormone (Nakanishi et al. 1979). The observation that a given precursor can give rise to multiple active peptides held true for the other two opioid precursors. Proenkephalin encodes multiple copies of Met-enkephalin, including two extended forms of Met-enkephalin, a heptapeptide and an octapeptide, and one copy of Leu-enkephalin (Comb et al. 1982). Prodynorphin encodes three opioid peptides of various lengths that all begin with the Leu-enkephalin sequence: dynorphin A, dynorphin B, and neo-endorphin (Kakidani et al. 1982). Within a decade, the field went from uncertainty about the existence of an endogenous opiate-like system to having demonstrated the existence of opiate receptors at the functional and binding levels and having identified the three genes encoding precursors that could give rise to a wide range of endogenous opioid ligands through complex post-translational processing.

Finally, the search for high-affinity/high-selectivity endogenous ligands for the MOR led to the discovery of novel endogenous opioids, the endomorphins (Zadina et al. 1997). Endomorphin-1 and endomorphin-2 are tetrapeptides with the sequence Tyr–Pro–Trp–Phe and Tyr–Pro–Phe–Phe, respectively. These novel peptides violate the canonical opioid core, but they nevertheless bind the mu receptor with very high affinity and selectivity.

TABLE 11.1
Opioid Peptides, Their Precursor Proteins, and Targeted Receptor

Precursor	Typical Peptide	Targeted Receptor
Proenkephalin	Leu-enkephalin	MOR, DOR
	Met-enkephalin	
Proopiomelanocortin[a]	β-Endorphin	MOR, DOR
Prodynorphin	Dynorphins A and B	KOR
Unknown	Endomorphins 1 and 2	MOR

Note: DOR, MOR, and KOR indicate delta-, mu-, and kappa-opioid receptors, respectively.

[a] Also precursor to adrenocorticotropic hormone and melanocyte-stimulating hormone acting on non-opioid receptors.

11.2.2 OPIOID RECEPTORS

In 1973, independent teams showed that opiates bind to membrane receptors in the brain (Pert and Snyder 1973; Terenius 1973). A few years later, the opioid binding sites were named *MOR*, *DOR*, and *KOR* (Martin et al. 1976). Gene cloning and characterization was a necessary step to develop our understanding of opioid receptors as proteins that operate in the nervous system and control nociceptive, hedonic, emotional, as well as autonomic, neuroendocrine, and immune responses in vivo. The first opioid receptor gene isolated by expression cloning in 1992 was the delta subtype (Evans et al. 1992; Kieffer et al. 1992). Because of strong sequence homology across receptors, the entire opioid receptor gene family was readily cloned and characterized at the molecular level in the following 2 years (Kieffer 1995). The cloning enabled the creation of mice lacking opioid receptors that could dissect out the role of each receptor in exogenous and endogenous opioid-mediated behaviors and characterize requirements for ligand trafficking, signaling, and selectivity for the different opioid receptors.

The extensive clinical experience with a wide range of opioids raised the question of receptor heterogeneity. Furthermore, binding studies suggest multiple opioid receptors subtypes. The availability of receptor DNA sequences has been essential to correlate the cloned receptors with those mediating the pharmacological effects in vivo. Thus, MOR-mediated actions involve mu1 or mu2 receptor subtypes. MOR heterogeneity was also implied by the unmatched profile of the morphine metabolite, morphine-6β-glucuronide (Pasternak et al. 1987). This agent is far more potent than morphine when given centrally. Furthermore, clinical studies confirmed the accumulation of M6G in serum of patients, often to levels greater than those of morphine. Pharmacological evidence suggests the existence of two subtypes of delta receptors (Jiang et al. 1991), whereas binding studies suggest four subtypes for KOR (Pasternak 2004).

As noted earlier, the initial identification of a single gene for each MOR, DOR, and KOR left open the question of their relationship to the pharmacologically defined receptor subtypes. The development of approaches that use antisense oligodeoxynucleotides and the posterior cloning of a host of MOR-1 splice variants in mice, rats, and humans (Garzon et al. 2000; Pasternak and Standifer 1995) have provided a means of assessing the functional significance of the cloned receptors and underscore the complexity of the opioid system. In these studies, antisense probes against all the exons within MOR-1 blocked mu-opioid analgesia (Garzon et al. 2000). However, the sensitivity profile of morphine and M6G differed. Whereas probes against exons 1 and 4 blocked morphine but not M6G, other probes against exons 2 and 3 blocked M6G but not morphine (Pasternak and Standifer 1995). This work shows differences in the analgesic mechanisms for M6G and morphine analgesia, raising the possibility that MOR-1 splice variants might be involved in their effects. In fact, they have been described as a variety of variants that encode truncated receptors in the mouse (Pasternak

2004). Indeed, there are far more receptors identified at the molecular level than were predicted pharmacologically, although the functional significance of truncated receptors is not clear.

All receptors in the opioid superfamily are G-protein-coupled receptors (GPCRs) and cause hyperpolarization (inhibition) of neurons (Figure 11.1b). As for all GPCRs, opioid receptors convey extracellular signals within the cell by activating different subtypes of heterotrimeric G proteins (Garzon et al. 2000). KOR is also G protein coupled but differs from the others in that it activates a separate signal transduction pathway. Moreover, KOR agonists are behaviorally different; they are aversive and not subject to self-administration. Overall, MOR, DOR, and KOR show 60% amino acid sequence identity. Extracellular domains, including three extracellular loops and the N-terminal domain, determine receptor selectivity, whereas intracellular loops of the receptor form a large part of the receptor–G protein interface. These intracellular receptor domains are almost identical across MOR, DOR, and KOR, consistent with the fact that all three receptors interact with inhibitory G proteins of the $G_i/G_z/G_o$ type (Figure 11.1b).

11.2.3 PHARMACOLOGICAL DIVERSITY

The variable combinations of the receptor with G protein subunits and the nature of associated signaling networks necessarily generate neuron-specific or even neuron-compartment-specific responses. Upon agonist binding, G protein subunits dissociate from the activated receptor and, in turn, modulate intracellular G-protein-associated signaling pathways (Garzon et al. 2005). Activation of opioid receptors and of other classes of GPCRs triggers the phosphorylation of specific residues by different kinases that indirectly modified receptor signaling by altering the availability of the receptors and providing a new dimension to the relationship between opioid receptors and transduction. Also, the coexistence of ligand-regulated metabotropic and ionotropic receptors in the postsynapse raises the possibility of mutual interactions triggered by the relative abundance of their presynaptic mediators. Among the ligand-gated ionotropic receptors, the glutamate N-methyl-D-aspartate receptors (NMDARs) have received particular attention because of their crucial roles in excitatory synaptic transmission, plasticity, neurodegeneration, and mood. Among the opioid receptors, the mu subtype is probably one of the GPCRs in which this relation with NMDARs has been better characterized. The mu-receptor signaling can be modulated by the NMDA/nitric oxide cascade and, indeed, the development of morphine-induced desensitization has been related to the activity of glutamate NMDARs. The interaction between these receptors is bidirectional since MOR signaling in the brainstem increases the activity of NMDARs. The glutamatergic regulation of MORs supports the attenuation of opioid efficacy in states of persistent pain where there is an increased function of NMDARs and provides the molecular basis to understand the clinical efficacy of NMDAR antagonists on opioid-induced acute tolerance.

Interestingly, pharmacological studies indicate a critical role for NMDAR in depression. Thus, depression seems to be related to hyperfunction of NMDARs, and the administration of NMDAR antagonists produces antidepressant effects in patients with treatment-resistant major depression (Maeng and Zarate 2007) and in animal models of depression (Garcia et al. 2008). Indeed, ketamine, an antagonist of these glutamate receptors, displays rapid antidepressant properties (Mathew et al. 2005). Thus, efficacy of NMDAR antagonists as fast-acting antidepressants can be explained by their capacity to preserve the signaling stamina of associated GPCR. In accordance with this role for NMDARs in mood disorders, molecular studies have revealed alterations in NMDAR subunit expression and that of postsynaptic associated proteins in brain areas of patients with depression (Kristiansen and Meador-Woodruff 2005; Mueller and Meador-Woodruff 2004). The functional relationship between MOR and NMDAR has several physiological and pharmacological implications. Since depression has been related to NMDAR hyperactivity, opioid antidepressive activity may be the result of their capacity to regulate NMDAR function. In this respect, in depression, the NMDAR hyperactivity correlates with an increased PKC activity (Battaini 2001; Maeng and

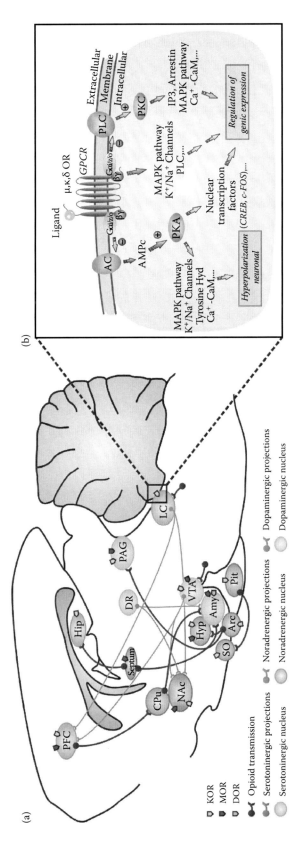

FIGURE 11.1 **(See color insert.)** Opioid system in emotional control areas in rat brain. (a) Potential opioid projections between brain nuclei implicated in mood disorders. Opioid system components have been localized in structures implicated in regulation of mood and motivation. Such areas include dopaminergic, serotonergic, and noradrenergic systems, in addition to amygdala (Amy), hippocampus (Hip), and hypothalamus (Hyp). Usually, anatomical distribution of endogenous opioid peptides agrees with localization of opioid receptor (OR). Since opioid system is present in brain regions containing monoamine neurotransmitters (serotonin, dopamine, and noradrenaline), it is possible that this system could play a key role in physiopathology of depressive disorders. (b) Cellular signaling general pathways of opioid receptors. Opioid receptors are coupled to heterotrimeric G proteins (GPCR) and their activation by opioid agonists triggers an intracellular signaling cascade, which may derive in neuronal inhibition (hyperpolarization) and in expression of specific genes. AC, adenylate cyclase; NAc, nucleus accumbens; AMPc, cyclic adenosine monophosphate; Arc, arcuate hypothalamic nucleus; CaM, calmodulin; CPu, caudate putamen; CREB, AMPc response element-binding; DR, dorsal raphe nucleus; Gα/βγ, G protein subunits; IP3, inositol trisphosphate; LC, locus coeruleus nucleus; MAPK, mitogen-activated protein kinases; PAG, periaqueductal gray; PFC, prefrontal cortex; Pit, pituitary gland; PKA, protein kinase A; PKC, protein kinase C; PLC, phospholipase C; SO, supraoptic nucleus; Tyrosine hyd, tyrosin hydroxylase; VTA, ventral tegmental area.

Zarate 2007). Mu-acting opioids such as morphine raise PKCγ activation threshold by reducing its sensitivity to local concentrations of DAG (Rodriguez-Munoz et al. 2008).

Thus, the existence of multiple opioid receptor subtypes may help explain the range of responses clinically seen among patients for the various opioid drugs. It is important to consider that patient characteristics and structural differences between opioids contribute to differences in opioid metabolism and thereby to the variability of the efficacy, safety, and tolerability of specific opioids in individual patients and diverse patient populations. This variability is also influenced by a series of genetic factors, such as allelic variants, that dictate the complement of opioid receptors and subtle differences in the receptor-binding profiles of opioids. The rate and pathways of opioid metabolism may also be influenced by genetic factors, race, and liver or kidney diseases. Most opioids are metabolized via CYP-mediated oxidation and results in the production of both inactive and active metabolites. Moreover, the risk of drug interactions with an opioid is determined largely by which enzyme systems metabolize the opioid. Understanding this variability would greatly enhance our ability to adequately select a particular opioid to treat patients appropriately.

11.3 STUDY OF OPIOID SYSTEM BY PRECLINICAL MODELS OF DEPRESSION

Opioid compounds are essentially used as analgesic drugs for pain treatment. However, substantial evidence supports the possible implication of the endogenous opioid system in the modulation and regulation of the emotional states as well in the pathophysiology of various mental disorders including depression (Agren et al. 1982; Fink et al. 1970; Tejedor-Real et al. 1998).

These studies have been supported by neurochemical and neurobehavioral findings, including the use of well-validated preclinical models in which the experimental animals, mainly rodents, are subjected to acute or chronic stress inducing depressive-like behaviors. Recently, the number of these behavioral paradigms proposed as animal models of depression has considerably increased.

The most commonly used tests in the study of opioid system within the context of stress and depressive disorders include the forced swim test (FST), the tail suspension test (TST), the learned helplessness (LH) test, the chronic mild stress, and the place conditioning test. These tests are sensitive to treatments that cause aversion (dysphoria) or reduced sensitivity to rewarding stimuli (anhedonia), and the intensity of these signs can be quantified. Collectively, these behavioral tests establish a useful and accurate tool for studying the mechanisms of action and effect of treatment of the opioid compounds as antidepressant potentials.

11.3.1 Use of Endogenous or Synthetic Opioid Agonists

As noted above, the endogenous opioid system comprises three equivalent receptor subtypes: MOR, DOR, and KOR. Certain effects mediated through KOR are similar to those mediated via MOR and DOR, whereas others are unique and thus are specific to each receptor. Proof of this is the opposite effects on mood states after activation of these receptors: MOR or DOR activation elevates mood (Filliol et al. 2000; Shippenberg et al. 2008), whereas KOR activation produces prodepressive-like behaviors in rodents (including anhedonia, aversion, and anxiety), all typical signs of depressive disorders (Bals-Kubik et al. 1993; Carlezon et al. 2009; Todtenkopf et al. 2004; Tomasiewicz et al. 2008).

The activation of these opioid receptors takes place in endogenous ligands, such as δ-selective enkephalins, κ-selective dynorphins, and nonselective β-endorphins, and may be modulated by two different strategies. The classical approach is to use exogenous agonists, synthetic or not, that ubiquitously stimulate the receptor(s).

Several studies have shown that the administration of endogenous opioids, and subsequent activation of specific receptor, decreases the occurrence of behavioral despair observed in animal models of depression, such as FST (Borsini and Meli 1988; Porsolt et al. 1978) and TST (Steru et al. 1985) (Table 11.2). Both tests elicit in rats and mice a profound state of behavioral inhibition known as

TABLE 11.2
Behavioral Effects of the Administration of Opioid Agonists, Antagonists and Enkephalinase Inhibitors in Animal Models of Mood Disorders

Behavioral Paradigm	Species, Strain	Treatment	Target	Doses	Behavioral Effects	References
			Opioid agonists			
FST	Mice, CD1	Endorphin-1	Nonselective	0.3–30 μg i.c.v.	↓ Immobility	Fichna et al. (2007)
	Mice, CD1	Endorphin-2	Nonselective	0.3–30 μg i.c.v.	↓ Immobility	Fichna et al. (2007)
	Rats, SD	SNC-80	DOR	32–100 mg/kg s.c.	↓ Immobility	Broom et al. (2002)
	Rats, SD	(+)BW373U86	DOR	3.2–10 mg/kg s.c.	↓ Immobility	Broom et al. (2003)
	Mice, Swiss	UFP-512	DOR	0.01–1 nmol i.c.v.	↓ Immobility	Vergura et al. (2006, 2009)
	Mice, Swiss	UFP-512	DOR	0.1–0.3 mg/kg i.p.	↓ Immobility	Vergura et al. (2009)
	Rats, SD	UFP-512	DOR	0.3–1 mg/kg i.p.	↓ Immobility	Vergura et al. (2008)
	Rats, SD	Deltorphin II	DOR	0.03–0.1 nmol i.c.v.	↓ Immobility	Torregrossa et al. (2006)
	Rats, SD	Jom-13	DOR	32 mg/kg i.v.	↓ Immobility	Torregrossa et al. (2007)
	Rats, SD	DPDPE	DOR	155 nmol i.c.v.	↓ Immobility	Torregrossa et al. (2008)
	Mice, Swiss	DPDPE	DOR	0.1–1 nmol i.c.v.	↓ Immobility	Vergura et al. (2006)
	Rats, SD	U-69593	KOR	0.3–10 mg/kg i.p.	↓ Immobility	Mague et al. (2003)
TST	Mice, CD1	Endorphin-1	Nonselective	1–30 μg i.c.v.	↓ Immobility	Fichna et al. (2007)
	Mice, CD1	Endorphin-2	Nonselective	1–30 μg i.c.v.	↓ Immobility	Fichna et al. (2007)
LH	Rats, Wistar	Met-enkephalin	DOR	50 μg/day i.c.v.	↓ Escape failures	Tejedor-Real et al. (1995)
	Rats, Wistar	Leu-enkephalin	DOR	50 μg/day i.c.v.	↓ Escape failures	Tejedor-Real et al. (1995)
	Rats, Wistar	BUBU	DOR	2 mg/kg i.v.	↓ Escape failures	Tejedor-Real et al. (1998)
Conditioned place preference	Rats, SD	DANGO	MOR	0.05–1 μg intra-VTA	Place aversions	Bals-Kubik et al. (1989)
	Rats, SD	DANGO	MOR	0.1–0.5 μg intra-NAC	No effect	Bals-Kubik et al. (1989)
	Rats, SD	DANGO	MOR	0.05–0.5 μg intra-MCF	No effect	Bals-Kubik et al. (1989)
	Rats, SD	DANGO	MOR	0.1–0.5 μg intra-LH	No effect	Bals-Kubik et al. (1989)
	Rats, SD	U-50,488H	KOR	0.3–1 μg intra-VTA	Place aversions	Bals-Kubik et al. (1989)
	Rats, SD	U-50,488H	KOR	10 μg intra-NAC	Place aversions	Bals-Kubik et al. (1989)
	Rats, SD	U-50,488H	KOR	1–3.3 μg intra-MCF	Place aversions	Bals-Kubik et al. (1989)
	Rats, SD	U-50,488H	KOR	3.3 μg intra-Hip	Place aversions	Bals-Kubik et al. (1989)

(continued)

TABLE 11.2 (Continued)
Behavioral Effects of the Administration of Opioid Agonists, Antagonists and Enkephalinase Inhibitors in Animal Models of Mood Disorders

Behavioral Paradigm	Species, Strain	Treatment	Target	Doses	Behavioral Effects	References
	Rats, SD	U-50,488H	KOR	10 mg/kg i.p.	Place aversions	Suzuki et al. (1992)
	Rats, Wistar	U-50,488H	KOR	10 nmol intra-PAG	Place aversions	Sante et al. (2000)
	Rats, SD	E-2078	KOR	0.1–0.3 μg intra-VTA	Place aversions	Bals-Kubik et al. (1989)
	Rats, SD	E-2079	KOR	0.3–1 μg intra-NAC	Place aversions	Bals-Kubik et al. (1989)
	Rats, SD	E-2080	KOR	3.3 μg intra-MFC	Place aversions	Bals-Kubik et al. (1989)
	Rats, SD	E-2081	KOR	0.3–1 μg intra-LH	Place aversions	Bals-Kubik et al. (1989)
Opioid antagonists						
FST	Mice, Swiss	Naltrindole	DOR	0.1–3 mg/kg s.c.	No effect	Vergura et al. (2006, 2009); Baamonde et al. (1992)
	Rats, SD	norBNI	KOR	20 μg/animal i.c.v.	↓ Immobility	Mague et al. (2003); Pliakas et al. (2001)
	Mice, G57BL/6	norBNI	KOR	10 mg/kg i.p.	↓ Immobility	McLaughlin et al. (2003)
	Rats, SD	GNTI	KOR	10–20 μg/animal i.c.v.	↓ Immobility	Mague et al. (2003)
	Rats, SD	GNTI	KOR	1–10 mg/kg i.p.	No effect	Mague et al. (2003)
	Rats, SD	ANTI	KOR	0.3–3 mg/kg i.p.	↓ Immobility	Mague et al. (2003)
	Mice, C57BL/6	Naloxone	Nonselective	2.5–40 mg/kg i.p.	↓ Immobility	Amir (1982)
	Mice, BALB/C	Naloxone	Nonselective	10–40 mg/kg i.p.	↓ Immobility	Amir (1983)
LH	Rats, Wistar	Naltrindole	DOR	0.01–1 mg/kg i.p.	No effect	Tejedor-Real et al. (1998)
	Rats, SD	norBNI	KOR	0.25–2.5 μg intra-Hip	↓ Escape failures	Shirayama et al. (2004)
	Rats, SD	norBNI	KOR	0.25–2.5 μg intra-NAC	↓ Escape failures	Shirayama et al. (2005)
	Rats, Wistar	Naloxone	Nonselective	2.5 mg/kg s.c.	↓ Escape failures	Tejedor-Real et al. (1995, 1998)

Test	Species, strain	Compound	Type	Dose	Effect	Reference
Conditioned place preference	Mice, C57Bl/6	norBNI	KOR	10 mg/kg i.p.	No effect	McLaughlin et al. (2006)
	Rats, Wistar	norBNI	KOR	2 mg/kg i.p.	No effect	Sante et al. (2000)
Conditioned suppression of motility test	Mice, Swiss	Naltrindole	DOR	0.1 mg/kg s.c.	No effect	Baamonde et al. (1992)
Chronic mild stress	Rats, Wistar	Naltrexone	Nonselective	2 mg/kg i.p.	↑ Sucrose consumption	Zurita et al. (2000)
Enkephalinase inhibitors						
FST	Mice, Swiss	Acetorphan	Neutral endopeptidase	50 mg/kg i.p.	↓ Immobility	Lecomte et al. (1986)
	Mice, Swiss	RB 101	Dual	5–10 mg/kg i.v.	↓ Immobility	Baamonde et al. (1992)
	Rats, Wistar	Opiorphin	Dual	1–2 mg/kg i.v.	↓ Immobility	Javelot et al. (2010)
	Mice, ddY	BL-2401	Neutral endopeptidase	100 mg/kg oral	↓ Immobility	Kita et al. (1997)
LH	Rats, Wistar	RB 101	Dual	5 mg/kg i.p. or i.v.	↓ Escape failures	Tejedor-Real et al. (1998)
	Rats, Wistar	RB 38A	Dual	6 µg i.c.v.	↓ Escape failures	Tejedor-Real et al. (1993, 1995, 1998)
	Rats, Wistar	RB 38B	Neutral endopeptidase	30 µg i.c.v.	↓ Escape failures	Tejedor-Real et al. (1993, 1995, 1998)
Conditioned suppression of motility test	Mice, Swiss	RB 101	Dual	5–10 mg/kg i.v.	↓ Immobility	Baamonde et al. 1992; Smadja et al. (1995)
	Rats, Wistar	RB 101	Dual	2.5–5 mg/kg i.v.	↓ Immobility	Smadja et al. (1997)

Note: DOR, MOR, and KOR indicate delta-, mu-, and kappa-opioid receptors, respectively; i.c.v., intracerebroventricular; i.p., intraperitoneal; i.v., intravenous; s.c., subcutaneous; Hip, hippocampus; LH, lateral hypothalamus; MCF, medial prefrontal cortex; NAC, nucleus accumbens; PAG, periaqueductal gray; SD, Sprague–Dawley; VTA, ventral tegmental area; GNTI, 5′-guanidinonaltrindole; ANTI, 5′-acetamidinoethylnaltrindole. ↑ or ↓ indicates significant increase or decrease in behavior, respectively. Dual indicates neutral endopeptidase and aminopeptidase N inhibitor.

behavioral despair reflecting a state of lowered mood, which is suppressed by antidepressants such as imipramine, desipramine, or amitryptiline (Kameyama et al. 1985; Porsolt et al. 1979). Thus, intracerebroventricular injection of two endogenous opioid peptides isolated from mammalian brain (Zadina et al. 1997), endomorphin-1 and endomorphin-2, decreased the immobility time in FST and TST in mice (Fichna et al. 2007). In both tests, the duration of immobility was interpreted as an expression of the above-mentioned behavioral despair, demonstrating that endomorphin-1 and endomorphin-2 present antidepressant-like effects. Moreover, antagonist administrations, such as naloxone (nonspecific opioid antagonist) and β-funaltrexamine (MOR selective antagonist), reversed the antidepressant-like effect of these endomorphins (Fichna et al. 2007).

Other examples of endogenous opioid peptides are Met- and Leu-enkephalins. In general, enkephalins have been proposed as endogenous ligands for DOR, eliciting an affinity for these receptors 10- to 20-fold higher than for MOR (Simon 1991). Their exogenous administration in rats showed antidepressant-like action in the LH assay (Tejedor-Real et al. 1995) (Table 11.2). The LH paradigm, initially described by Overmier and Seligman (1967), consists in the exposition of the rodents to inescapable stress, for example, foot shocks. Untreated animals often make no attempt to escape, deriving in a behavioral phenotype validated as model of depression (Thiébot et al. 1992; Willner 1984). This phenotype is reversed by antidepressants and by Met- and Leu-enkephalin treatment (Tejedor-Real et al. 1995).

In respect to synthetic agonists, opioid receptor activation by administration of these drugs has been extensively characterized in rodents. Such is the case of nonpeptidic or pseudopeptidic DOR agonists: pen-enkephalin, SNC-80, (+)BW373U86, UFP-502, and UFP-512 produce antidepressant-like effects in the FST of mice and rats, decreasing the immobility time (Table 11.2) and mimicking the effects of classical antidepressants such as desipramine and fluoxetine (Broom et al. 2002; Torregrossa et al. 2006; Vergura et al. 2006, 2008). BUBU, another selective DOR agonist, reverses escape deficits in rats previously subjected to inescapable shocks (Tejedor-Real et al. 1998), similar to what happens after the enkephalin administration in the LH test (Tejedor-Real et al. 1995).

In contrast, the administration of KOR agonist produces opposite effects (Table 11.2). As has been noted, KOR activation mediates some endogenous neurobiological aspects of aversion and also comprises an important part of the response to environmental stressors (Bals-Kubik et al. 1993; Iwamoto 1985; McLaughlin et al. 2003). These depressive-like behaviors have been extensively characterized in rodents by a conditioned place preference procedure. This test evaluates preferences for environmental stimuli that have been associated with a positive or a negative reward. The amount of time spent in the compartments previously associated with the positive stimulus serves as an indicator of preference and a measure of reward learning. Indeed, intravenous administration of agonists U-50,488H and U-69,593 had aversive effects as measured by this test (Bals-Kubik et al. 1993; Suzuki et al. 1992), and U-69,593 increased immobility in the rat FST (Mague et al. 2003). In addition, microinjections of KOR agonist U-50488 and the dynorphin derivative E-2078 into the ventral tegmental area, nucleus accumbens, medial prefrontal cortex, and lateral hypothalamus produced place aversions (Bals-Kubik et al. 1993). Sante et al. (2000) demonstrated, in the same manner, that U-50488 injection in dorsal periaqueductal gray matter also produced place aversion in the corral method, a similar procedure to a conditioned place preference test.

Although current data place these compounds as potential antidepressant drugs, much remains to be investigated about their intrinsic mechanism of action. Indeed, some studies have showed that a single administration of (+)BW373U86 increases brain-derived neurotrophic factor mRNA expression in rat frontal cortex, hippocampus, and basolateral amygdala (Torregrossa et al. 2004). Increases in brain-derived neurotrophic factor have been suggested to be responsible for the clinical efficacy of antidepressant treatment (Duman 2002; Vaidya and Duman 2001).

In addition, the agonist use of opioid receptors can be associated with serious negative effects related to overstimulation of receptors (e.g., morphine). Moreover, several assays have demonstrated that use of opioid receptors also induce convulsions in a number of species, thus potentially limiting their therapeutic utility. In this respect, the convulsive effects of opioid peptides as well as their ability to modulate models of convulsions have been debated for some time (Broom et al. 2002; Comer et al. 1993; Jutkiewicz et al. 2004; Pakarinen et al. 1995).

11.3.2 USE OF ENKEPHALINASE INHIBITORS

A second approach—possibly more physiological—is to modulate the extracellular concentrations of endogenous peptide effectors by inhibiting their metabolizing enzymes. Endogenous opioid peptides are rapidly inactivated by peptidases, called enkephalinases, which increase their extracellular levels in brain structures where they are tonically or phasically released (Roques and Fournie-Zaluski 1986), minimizing any detectable effects. Thus, by inhibiting the action of these enkephalinases, primarily neutral endopeptidase and aminopeptidase N (zinc metallopeptidases), the action of endogenous opioid peptides could be prolonged and produce significant behavioral and potential clinical effects.

In this respect, administration of synthetic enkephalinase inhibitors has been widely used for the treatment of pain (Carenzi et al. 1983; Dickenson et al. 1987; Fournie-Zaluski et al. 1986; Nieto et al. 2001), representing a promising means to develop "physiological" analgesics devoid of morphine side effects. Previous studies have reported that, addition to their analgesic effect, these inhibitors also have antidepressant-like properties through enkephalin-related activation of DOR. Administration of enkephalinase inhibitors, such as acetorphan, thiorphan, or opiorphin, produces antidepressant-like effects in FST in rodents (Javelot et al. 2010; Lecomte et al. 1986; Nabeshima et al. 1987) (Table 11.2). These enkephalinase inhibitors produce an attenuation of the conditioned suppression of motility in the FST similar to those obtained with antidepressants (Rojas-Corrales et al. 2004), suggesting a potential role of endogenous enkephalins in depressive syndromes.

However, one of the main limitations of these compounds is their hydrophilic structure and incapacity to cross the blood–brain barrier, which limits their therapeutic use (Maldonado et al. 1989). Based on these findings, a number of selective peptidase inhibitors with improved lipophilic profiles were developed to enhance penetration in the central nervous system (Fournie-Zaluski et al. 1992). This is the case of inhibitor RB 101, a lipophilic "prodrug" that, associated with an N-(mercaptoacyl) amino acid, acts as a neutral endopeptidase inhibitor, and, associated with an h-mercaptoalkylamine, acts as an aminopeptidase N inhibitor (Fournie-Zaluski et al. 1992; Noble et al. 1992). This dual inhibitor increased the extracellular concentrations of Met-enkephalins in brains of freely moving rats (Dauge et al. 1996) and reversed the escape deficit after inescapable shock in the LH assay in a dose-dependent manner (Tejedor-Real et al. 1998) (Table 11.2). This antidepressant-like action of RB 101 could be reversed by DOR antagonist naltrindole, implicating these opioid receptors in its mechanism of action (Tejedor-Real et al. 1998). A similar effect has also been reported after RB 101 administration in the conditioned suppression of motility test in mice (Baamonde et al. 1992; Smadja et al. 1995) and rats (Smadja et al. 1997). In this animal model of depression, the rodent reacts to the adverse situation (electric foot shocks) by motor immobility, which is reduced by antidepressants such as imipramine, desipramine, and amitryptiline (Kameyama and Nagasaka 1982; Kameyama et al. 1985).

The same thing occurs with RB 38A (a mixed enkephalinase inhibitor) and RB 38B (a selective inhibitor of neutral endopeptidase). Both inhibitors cross the blood–brain barrier and reduced the helpless behavior in rat, as illustrated by the decrease in the number of escape failures (Table 11.2). RB 38A, however, produced a complete inhibition of enkephalin metabolism, resulting in a greater response than that obtained with partial inhibitor RB 38B (Tejedor-Real et al. 1993; Xie et al. 1989). Another example is BL-2401. Oral administration of BL-2401 decreased the duration of mice immobility in the FST, and the decrease was antagonized by naloxone (Kita et al. 1997).

In addition, another disadvantage of enkephalinase inhibitors compared with exogenous agonists of opioid receptors could be their lower pharmacological potency. Nevertheless, this disadvantage is compensated for by their more specific effects that correlate with the phasic release of opioid peptide in brain structures recruited by a particular stimulus (pain, stress, or emotion) and the absence or low change in either the secretion of the peptide or the expression of its targets (metabolizing enzymes or receptors).

11.3.3 USE OF OPIOID RECEPTOR ANTAGONISTS

In contrast to these two strategies is the use of antagonist of opioid receptors. It is important to note that during receptor blockade by an antagonist, the circulating levels of endogenous opioid peptides remain unchanged or even increased by a feedback mechanism. Indeed, as we have already noted, the KOR system is implicated in depression and other mood disorders and their activation derives in prodepressive-like behavior in preclinical models (Table 11.2). Thus, KOR antagonists have been investigated as potential antidepressant treatments. The most studied is nor-binaltorphimine (norBNI), an irreversible, long-lasting, and selective KOR antagonist (Spanagel and Shippenberg 1993). Recent evidence suggests that injections of norBNI into the nucleus accumbens, dentate gyrus, or CA3 regions of the hippocampus are sufficient to cause an antidepressant-like effect in the LH paradigm, resulting in a decrease in the number of escape failures and latency to escape (Newton et al. 2002; Shirayama et al. 2004), equivalent to subchronic administration of antidepressant desipramine (Shirayama et al. 2004). In addition, intracerebroventricularly administered norBNI dose-dependently decreased immobility in the FST, supporting the view that this agent has antidepressant-like effects (Mague et al. 2003; Pliakas et al. 2001).

Similarly, intracerebroventricular administration of another KOR-antagonist, 5'-guanidinonaltrindole, had similar antidepressant-like effects (Mague et al. 2003). Previous reports have compared both antagonists (norBNI and 5'-guanidinonaltrindole), concluding that 5'-guanidinonaltrindole presents the most potency (Jones and Portoghese 2000; Negus and Mello 2002). The fact that structurally different KOR antagonists have similar actions could suggest that their antidepressant-like effects are attributable to a specific blockade of receptor rather than to nonspecific side effects. Another KOR antagonist is 5'-acetamidinoethylnaltrindole. This compound is a potent and selective KOR antagonist that crosses the blood–brain barrier since it contains three additional hydrophobic groups (two methylenes and one methyl) that increase its lipophilic character. Like norBNI and 5'-guanidinonaltrindole, repeated intraperitoneal administration of 5'-acetamidinoethylnaltrindole reduced immobility time in the FST in rats (Mague et al. 2003), providing additional evidence that antidepressant-like effects may be a general attribute of KOR antagonists.

Nevertheless, the mechanisms by which these KOR antagonists regulate behavior in the depression model are unknown. Several studies suggest that dopaminergic system may be involved. Accordingly, one possibility is that behavioral despair is mediated by KOR located on terminals of mesolimbic dopaminergic neurons that project to the nucleus accumbens (Svingos et al. 1999). Administration of KOR antagonists directly into this nucleus increases dopamine extracellular concentrations (Maisonneuve and Kreek 1994), an increase that may be sufficient to cause antidepressant-like effects without stimulating locomotor activity. However, other possibilities exist, such as activation of CREB protein, which regulates dynorphin expression (Carlezon et al. 1998), or blocking dynorphin-mediated reductions in neurotransmitter extracellular concentrations (Pliakas et al. 2001).

Many studies have been carried out using these KOR antagonists and others not mentioned in various animal models of depression (Table 11.2). In general, taken together, these compounds present potent antidepressant-like effects and could be used as potential antidepressant drugs. However, other nonselective antagonists of opioid system exist. Such is the case of naloxone. Administration of this antagonist in mice produces an attenuation of immobility time in the FST (Amir 1982) and facilitates the induction of LH (Tejedor-Real et al. 1993). It is notable that the naloxone dose reported to decrease the duration of tonic immobility vary between species and even among strains of the same species. For example, in rabbits, the dose of naloxone effectively blocking of behavioral despair is approximately 15 mg/kg (Carli et al. 1981), whereas in mice, the dose used is <10 mg/kg (Amir 1982). These variations may be due to differences in naloxone sensitivity, although the particular opiate mechanism(s) interacting with these behavioral effects may also differ. Despite this variability in dosing, naloxone is the universal opioid antagonist and has been used in several preclinical studies of depression.

Naltrexone is another nonselective antagonist. Zurita et al. (2000) demonstrated that rats subjected to the chronic mild stress paradigm and pretreated with naltrexone (2 mg/kg, i.p.) showed sucrose consumption values similar to those in control animals (Table 11.2). This paradigm is a model that simulates the "anhedonic behavior" associated with major depression. Animals are exposed to a chronic regimen of unpredictable stressors (food or water deprivation, cage tilt, paired housing, continuous lighting, or soiled cage) and given sucrose solution to measure anhedonia (Miller et al. 2005; Muscat and Willner 1992; Papp et al. 1996).

11.3.4 DEVELOPMENT OF OPIOID RECEPTOR KNOCKOUT MICE

Another option in the study of the relationship between opioid system and mood disorders is the use of a genetically selected line of rodents, which displays behavioral characteristics of depression or anxiety in validated tests. Several such lines have been developed, but again must be interpreted with caution, as they are not known to model all of the physiological and behavioral characteristics of the illness. In recent years, the availability of selective ligands for opioid receptors together with the generation of knockout mice for these receptors has permitted researchers to explore novel biological actions modulated by opioidergic signaling, which are implicated in the pathophysiology of depression. Moreover, the use of knockout mice could provide novel theories on the molecular mechanisms underlying the modulation of depression-like behavior.

Consequently, knockout mice for DOR, KOR, and MOR have been tested in the several experimental models of depression mentioned above (Table 11.3). Indeed, DOR gene knockout mice exhibited a depressive-like phenotype in the FST, displaying a significant increase in immobility time, which was interpreted as a sign of increased to depression-related behaviors (Filliol et al. 2000). In the case of KOR−/− mice, many groups have studied the behavioral responses in these animals together with those that lack dynorphins. Thus, the deletion of KOR or dynorphin (PDyn) genes resulted in a decrease of immobility in the FST (McLaughlin et al. 2003, 2006) and in an absence of aversive behavior after administration of the KOR agonist U-50,488H (Simonin et al. 1998), demonstrating that the aversive effects of KOR agonists require intact KOR signaling.

The development of knockout mice for the MOR gene has revealed a central role of these receptors in the various effects of opioids, including analgesia (Raynor et al. 1994). In addition, as will be discussed in the following section, the opioid system has been implicated in the mechanism of action of antidepressants. Therefore, recent studies have investigated the role of MOR in the effects of antidepressants such as venlafaxine (serotonin–noradrenaline reuptake inhibitor), in the FST using MOR knockout mice (Ide et al. 2010). In this report, venlafaxine reduced immobility time in wild-type mice, but not in the absence of the MOR gene, supporting the view that these receptors play an important role in the antidepressant-like effects of venlafaxine.

On the other hand, mice with deletion of the preproenkephalin gene (Penk1−/−) have also been developed. These animals (lacking enkephalin) showed no depression-related phenotype in FST and TST (Bilkei-Gorzo et al. 2007; McLaughlin et al. 2003). Additionally, Penk1−/− mice had a lower frequency of depression-related behavior in stress-induced hypoactivity as model of depression, similar to animals treated with the antidepressants imipramine and fluoxetine. Bilkei-Gorzo et al. (2007) also demonstrated that the dual enkephalinase inhibitor, RB 101, still had an antidepressant effect in double-knockout animals for preproenkephalin and preprodynorphin genes (Penk1−/−/ Pdyn+/+), which conclusively demonstrates that this drug effect cannot be exclusively dependent on the elevation of enkephalin levels.

TABLE 11.3
Behavioral Effects of Opioid Gene Ablation in Animal Models of Depression

Behavioral Paradigm	Species, Strain	Genotype	Encoded Molecule	Behavioral Effects	References
FST	Mice, hybrid 129SV/C57BL/6	Oprd1–/–	DOR	↑ Immobility	Filliol et al. (2000)
	Mice, hybrid 129SV/C57BL/6	Oprk1–/–	KOR	No effect	Filliol et al. (2000)
	Mice, hybrid 129SV/C57BL/6	Oprm–/–	MOR	↓ Immobility	Filliol et al. (2000)
	Mice, C57BL/6	Oprm–/–	MOR	No effect	Ide et al. (2010)
	Mice, C57BL/6	PDyn–/–	Prodynorphin	↓ Immobility	McLaughlin et al. (2002)
	Mice, C57BL/7	Penk1–/–	Preproenkephalin	No effect	Bilkei-Gorzo et al. (2007)
	Mice, DBA/2J	Penk1–/–	Preproenkephalin	No effect	Bilkei-Gorzo et al. (2007)
TST	Mice, C57Bl/7	Penk1–/–	Preproenkephalin	No effect	Bilkei-Gorzo et al. (2007)
Conditioned place preference	Mice, C57Bl/6	PDyn–/–	Prodynorphin	No effect	McLaughlin et al. (2006)
Conditioned suppression of	Mice, hybrid 129SV/C57BL/6	Oprd1–/–	DOR	No effect	Filliol et al. (2000)
motility test	Mice, hybrid 129SV/C57BL/6	Oprk1–/–	KOR	No effect	Filliol et al. (2000)
	Mice, hybrid 129SV/C57BL/6	Oprm–/–	MOR	↓ Immobility	Filliol et al. (2000)

Note: DOR, MOR, and KOR indicate delta-, mu-, and kappa-opioid receptors, respectively. ↑ or ↓ indicates significant increase or decrease in behavior, respectively.

11.4 INTERACTION OF ANTIDEPRESSANTS AND OPIOID SYSTEM IN DEPRESSION PRECLINICAL MODELS

Several preclinical studies have showed a functional relationship between endogenous opioid peptides and antidepressants drugs. In line with this observation, previous reports have implicated the analgesic effect of antidepressants to the opioid system. For example, naloxone or norBNI administrations in models of acute and chronic pain antagonize the analgesic effect of several tricyclic antidepressants and monoamine reuptake inhibitors (Ardid and Guilbaud 1992; Schreiber et al. 1999; Valverde et al. 1994). Considering that opioid and monoaminergic systems seem to share common molecular mechanisms in the control of nociception, opioid compounds are frequently coadministered with antidepressants to relieve pain. However, this therapeutic strategy has not been well explored in the treatment of depression. Studies to date show that the opioid doses required to produce antidepressant-like effect are higher than those needed to produce analgesic effect, suggesting that the mechanisms underlying analgesic and antidepressant effects could be different.

That putative interaction has been studied in two directions: evidence shows that antidepressant drugs have an influence on opioid system, for example, modifying the receptor–opioid binding (Reisine and Soubrie 1982) or the endogenous enkephalin level (Hamon et al. 1987); and vice versa, opiates are also able to modify the effect of antidepressants.

11.4.1 INFLUENCE OF OPIOID SYSTEM ON EFFECT OF ANTIDEPRESSANTS

Devoize et al. (1984) demonstrated that the nonselective opioid antagonist naloxone (2 mg/kg) significantly reduced the behavioral effect of two tricyclic antidepressants, clomipramine (20 and 30 mg/kg) and desipramine (20 and 30 mg/kg), in the mice FST. This antagonism was consistent with later results for tricyclic and nontricyclic antidepressants. For example, Baamonde et al. (1992) established that the opioid antagonist naltrindole administered subcutaneously 30 min before the FST antagonized the imipramine-induced reduction in immobility time. Accordingly, naloxone suppressed the antidepressant effect of imipramine in the LH paradigm (Besson et al. 1999; Tejedor-Real et al. 1995) and antagonized the immobility time reduction by venlafaxine (10 mg/kg) in the mice FST (Berrocoso et al. 2004). However, opioid antagonist administration not always reverses the antidepressant effect produced by antidepressants. Indeed, naloxone did not significantly reduce the activity of other antidepressants, such as nomifensine (a noradrenaline–dopamine reuptake inhibitor), mianserin (a tetracyclic antidepressant) (Berrocoso et al. 2004), or amineptine (a dopamine reuptake inhibitor) (Besson et al. 1999), suggesting that the opioid antagonist–antidepressant combination is not always satisfactory.

In addition, previous reports have indicated a pharmacological interaction by combining opioid compounds with serotonin reuptake inhibitors in the TST, which does not seem to be affected by any locomotor impairment. Berrocoso et al. (2009) showed that the coadministration of noneffective doses of codeine together with fluoxetine or citalopram produces a significant reduction in immobility time in this animal model of depression. Altogether, this seems to suggest that the combination of opioid and serotonergic mechanisms might be a new strategy for the development of antidepressant drugs.

Another notable example of an opioid–monoaminergic combination of putative utility in depression is (i)-tramadol. Tramadol is a weak agonist of MOR but, like antidepressant drugs, is able to inhibit the reuptake of serotonin and noradrenaline. Rojas-Corrales et al. (1998) demonstrated that tramadol and its enantiomers reduced the behavioral despair in the FST in mice. Four years later, the same group proved that tramadol decreased the number of failures to avoid or escape aversive stimulus in the helpless rat (Rojas-Corrales et al. 2004).

On the other hand, the potentiation of antidepressant effects by enkephalinase inhibitors had been suggested. de Felipe et al. (1989) studied the implication of opioid peptides in the antidepressant action of imipramine and iprindole by using the mice FST. Effective doses of these tricyclic antidepressants induced a reduction of immobility time in this test. However, behavioral despair was again decreased after intracerebroventricular injection of thiorphan or bestatin together with subeffective doses of imipramine or iprindole. In line with this result, administration of RB 101 was also efficient in inhibiting the effect of imipramine in this depression paradigm (Baamonde et al. 1992).

11.4.2 Effect of Antidepressant Administration on Opioid System

As noted earlier, a putative interaction between antidepressant drugs and the opioid system could also be studied in another direction. There is evidence that antidepressants have an influence on opioid system, but this influence seems to occur differentially, being region-specific. In addition, discrepancies exist about the effect of antidepressant treatment on opioid receptors. Stengaard-Pedersen and Schou (1986) did not observe any change in either region after the same treatment. However, there are studies that affirm a downregulation of opioid receptors in whole rat brain or in specific regions, such as the cortex, hypothalamus, and thalamus, after chronic treatment (14–28 days) with desipramine, amoxapine, paroxetine, or amitriptyline (Benkelfat et al. 1989; Hamon et al. 1987; Reisine and Soubrie 1982; Vilpoux et al. 2002).

In contrast, other studies claim that antidepressant treatment elevates the densities of opioid receptor binding sites. Antkiewicz-Michaluk et al. (1984) showed that citalopram administered for 24 days elevated naloxone binding in cortical membranes, and de Gandarias et al. (1998, 1999) described an increase of neural cell density immunostained for MOR in the frontal cortex, lateral septum, dentate gyrus, and caudate putamen in the rat brain after 14 days of imipramine and fluoxetine treatment. Similarly, after 21 days of paroxetine or reboxetine treatment, MOR binding site density was increased in the cingulate cortex, hippocampus, and thalamus (Vilpoux et al. 2002). Also, densities of DOR and MOR binding sites were increased in the spinal cord after amoxapine and amitriptyline chronic administrations (Hamon et al. 1987). All these data, in one direction or the other, suggest that opioid receptors displayed dissimilar patterns of adaptations in response to distinct antidepressant treatments. These patterns of adaptations seem to depend on various parameters such as action mechanism of antidepressant, treatment duration, doses, brain regions, and so on.

Otherwise, the chronic treatment with antidepressants has been found to increase or decrease endogenous opioid peptide levels in numerous brain regions. Thus, amoxapine or amitriptyline (14 days) did not affect the levels of dynorphin but markedly enhanced the levels of enkephalins in the cerebral cortex, spinal cord, and hypothalamus (Hamon et al. 1987). Accordingly, imipramine treatment has a profound effect on the levels of enkephalins in nucleus accumbens and ventral tegmentum of the brain rat (Dziedzicka-Wasylewska and Papp 1996; Dziedzicka-Wasylewska and Rogoz 1995; Przewlocki et al. 1997). In the nucleus accumbens, this increase was accompanied by an increase in proenkephalin mRNA, indicating the enhancement of biosynthesis of enkephalinergic peptides after antidepressant treatment (Dziedzicka-Wasylewska and Rogoz 1995). However, these changes in the encephalin synthesis could depend on the treatment duration and the brain region studied. For example, in situ hybridization showed that single imipramine injection (10 mg/kg, i.p.) decreased the level of proenkephalin and prodynorphin mRNA to the same extent in rat nucleus accumbens and striatum, but chronic treatment with this tricyclic antidepressant (10 days) had no effect on proenkephalin mRNA level. With respect to prodynorphin levels, the gene expression was regulated differently and differed depending on the time point of study (Przewlocki et al. 1997).

11.5 DISTRIBUTION OF OPIOID SYSTEM IN BRAIN NUCLEI IMPLICATED IN DEPRESSION

In previous sections, we have provided evidence on the participation of the opioid system in the neurochemical processes underlying depression disorders by using preclinical models. Thus, these studies have showed this implication through various approaches: exogenous administration of opioid agonist and antagonist, enkephalinase inhibitors, antidepressant treatments, knockout animals, and combinations thereof. However, other studies have focused on localization of opioid system components in structures implicated in the regulation of mood and motivation. Such areas include the mesocorticolimbic dopaminergic system (comprising ventral tegmental area, nucleus accumbens, and prefrontal cortex), the serotonergic and noradrenergic systems (comprising dorsal raphe and locus coeruleus nuclei, respectively), the basolateral amygdala, the hippocampus, and the hypothalamus (Figure 11.1a) (Hurd 1996; Mansour et al. 1995; Nestler and Carlezon 2006; Peckys and Landwehrmeyer 1999; Schwarzer 2009; Shuster et al. 2000).

As noted earlier, the endogenous opioid peptides have antidepressant-like effects after their exogenous administration and are involved in action of antidepressant drugs. Accordingly, endomorphins have been found in limbic system structures (septum, nucleus accumbens, and amygdala), thalamic nuclei, and locus coeruleus (Martin-Schild et al. 1999; Schreff et al. 1998; Zadina 2002), regions involved in mood disorders, such as anxiety and depression (Drevets and Raichle 1992; Sheline 2000). Moreover, they are present in the brain regions containing monoamine neurotransmitters (serotonin, dopamine, and noradrenaline), which play a key role in the physiopathology of depressive disorders (Figure 11.1a) (Chen et al. 2001; Hung et al. 2003). The anatomical distribution of these endomorphins agrees with MOR localization.

Other opioid peptides are the dynorphins. Both dynorphins and their targets (KOR) are expressed throughout the mesocorticolimbic and nigrostriatal areas implicated in the pathophysiology of depression and anxiety disorders. This distribution suggests that KOR activation may regulate dopamine transmission in the nucleus accumbens and that the aversive effects of KOR agonists may result from dopamine decreased in this region (Carlezon and Thomas 2009; Fallon and Leslie 1986; Margolis et al. 2003). Moreover, it was speculated that these dopaminergic regions are connected with enkephalin pathways (Figure 11.1a) (Maldonado et al. 1990; Petit et al. 1986; Stinus et al. 1980).

Opioid receptors are also expressed at moderate to high levels in the hippocampus (Clarke et al. 2001; Drake et al. 2007), another structure often implicated in the neurobiology of stress. As previously noted, microinfusions of KOR antagonist in this structure have antidepressant-like effects in the LH test in rats (Shirayama et al. 2004).

11.6 CONCLUSION

Depression is a very multifaceted illness with diverse symptoms and signs that probably respond to a complex neurobiology. Several neurotransmitters and neuropeptides have been implicated in the biological origin of this illness. The monoaminergic system appears to be the principal pathway affected by depression. However, we and others have demonstrated that among neuropeptides, opioids play a relevant role in regulating mood in depression. It is a challenge to find a way to tackle depression with psychotherapeutics that minimizes the side effects of opioids, which might help patients with this mental disorder.

ACKNOWLEDGMENTS

This study was supported by grants from the Spanish Ministry of Health, CIBERSAM G09 and G18, "Fondo de Investigación Sanitaria" (PI070687, PI10/01221, and PI080417), "Junta de Andalucía; Consejería de Innovación, Ciencia y Empresa" (CTS-510 and CTS-4303), "Cátedra del Dolor Fundación Grünenthal-Universidad de Cádiz," and fellowship FPU (AP2007-02397).

REFERENCES

Agren, H., Terenius, L., and Wahlstrom, A. 1982. Depressive phenomenology and levels of cerebrospinal fluid endorphins. *Ann N Y Acad Sci* 398:388–98.
Akil, H., Watson, S. J., Young, E., et al. 1984. Endogenous opioids: Biology and function. *Annu Rev Neurosci* 7:223–55.

Amir, S. 1982. Involvement of endogenous opioids with forced swimming-induced immobility in mice. *Physiol Behav* 28:249–51.

Antkiewicz-Michaluk, L., Rokosz-Pelc, A., and Vetulani, J. 1984. Increase in rat cortical [³H]naloxone binding site density after chronic administration of antidepressant agents. *Eur J Pharmacol* 102:179–81.

Ardid, D., and Guilbaud, G. 1992. Antinociceptive effects of acute and 'chronic' injections of tricyclic antidepressant drugs in a new model of mononeuropathy in rats. *Pain* 49:279–87.

Baamonde, A., Dauge, V., Ruiz-Gayo, M., et al. 1992. Antidepressant-type effects of endogenous enkephalins protected by systemic RB 101 are mediated by opioid delta and dopamine D1 receptor stimulation. *Eur J Pharmacol* 216:157–66.

Bals-Kubik, R., Ableitner, A., Herz, A., and Shippenberg, T. S. 1993. Neuroanatomical sites mediating the motivational effects of opioids as mapped by the conditioned place preference paradigm in rats. *J Pharmacol Exp Ther* 264:489–95.

Battaini, F. 2001. Protein kinase C isoforms as therapeutic targets in nervous system disease states. *Pharmacol Res* 44:353–61.

Benkelfat, C., Aulakh, C. S., Bykov, V., et al. 1989. Apparent down-regulation of rat brain mu- and kappa-opioid binding sites labelled with [³H]cycloFOXY following chronic administration of the potent 5-hydroxytryptamine reuptake blocker, clomipramine. *J Pharm Pharmacol* 41:865–7.

Berrocoso, E., Rojas-Corrales, M. O., and Mico, J. A. 2004. Non-selective opioid receptor antagonism of the antidepressant-like effect of venlafaxine in the forced swimming test in mice. *Neurosci Lett* 363:25–8.

Berrocoso, E., Sanchez-Blazquez, P., Garzon, J., and Mico, J. A. 2009. Opiates as antidepressants. *Curr Pharm Des* 15:1612–22.

Besson, A., Privat, A. M., Eschalier, A., and Fialip, J. 1999. Dopaminergic and opioidergic mediations of tricyclic antidepressants in the learned helplessness paradigm. *Pharmacol Biochem Behav* 64:541–8.

Bilkei-Gorzo, A., Michel, K., Noble, F., Roques, B. P., and Zimmer, A. 2007. Preproenkephalin knockout mice show no depression-related phenotype. *Neuropsychopharmacology* 32:2330–7.

Borsini, F., and Meli, A. 1988. Is the forced swimming test a suitable model for revealing antidepressant activity? *Psychopharmacology (Berl)* 94:147–60.

Broom, D. C., Jutkiewicz, E. M., Folk, J. E., et al. 2002. Nonpeptidic delta-opioid receptor agonists reduce immobility in the forced swim assay in rats. *Neuropsychopharmacology* 26:744–55.

Carenzi, A., Frigeni, V., Reggiani, A., and Della Bella, D. 1983. Effect of inhibition of neuropeptidases on the pain threshold of mice and rats. *Neuropharmacology* 22:1315–9.

Carlezon, W. A., Jr., Beguin, C., Knoll, A. T. and Cohen, B. M. 2009. Kappa-opioid ligands in the study and treatment of mood disorders. *Pharmacol Ther* 123:334–43.

Carlezon, W. A., Jr., and Thomas, M. J. 2009. Biological substrates of reward and aversion: A nucleus accumbens activity hypothesis. *Neuropharmacology* 56 (Suppl 1):122–32.

Carlezon, W. A., Jr., Thome, J., Olson, V. G., et al. 1998. Regulation of cocaine reward by CREB. *Science* 282:2272–5.

Carli, G., Farabollini, F., and Fontani, G. 1981. Effects of pain, morphine and naloxone on the duration of animal hypnosis. *Behav Brain Res* 2:373–85.

Clarke, R. W., Bhandari, R. N., and Leggett, J. 2001. Opioid and GABA receptors involved in mediation and modulation of tonic and stimulus-evoked inhibition of a spinal reflex in the decerebrated and spinalized rabbit. *Neuropharmacology* 41:311–20.

Comb, M., Seeburg, P. H., Adelman, J., Eiden, L., and Herbert, E. 1982. Primary structure of the human Met- and Leu-enkephalin precursor and its mRNA. *Nature* 295:663–6.

Comer, S. D., Hoenicke, E. M., Sable, A. I., et al. 1993. Convulsive effects of systemic administration of the delta opioid agonist BW373U86 in mice. *J Pharmacol Exp Ther* 267:888–95.

Chen, B., Dowlatshahi, D., MacQueen, G. M., Wang, J. F., and Young, L. T. 2001. Increased hippocampal BDNF immunoreactivity in subjects treated with antidepressant medication. *Biol Psychiatry* 50:260–5.

Dauge, V., Mauborgne, A., Cesselin, F., Fournie-Zaluski, M. C., and Roques, B. P. 1996. The dual peptidase inhibitor RB101 induces a long-lasting increase in the extracellular level of Met-enkephalin-like material in the nucleus accumbens of freely moving rats. *J Neurochem* 67:1301–8.

de Felipe, M. C., Jimenez, I., Castro, A., and Fuentes, J. A. 1989. Antidepressant action of imipramine and iprindole in mice is enhanced by inhibitors of enkephalin-degrading peptidases. *Eur J Pharmacol* 159:175–80.

de Gandarias, J. M., Echevarria, E., Acebes, I., Silio, M., and Casis, L. 1998. Effects of imipramine administration on mu-opioid receptor immunostaining in the rat forebrain. *Arzneimittelforschung* 48:717–9.

de Gandarias, J. M., Echevarria, E., Acebes, I., et al. 1999. Effects of fluoxetine administration on mu-opioid receptor immunostaining in the rat forebrain. *Brain Res* 817:236–40.

Devoize, J. L., Rigal, F., Eschalier, A., Trolese, J. F., and Renoux, M. 1984. Influence of naloxone on antidepressant drug effects in the forced swimming test in mice. *Psychopharmacology (Berl)* 84:71–5.

Dickenson, A. H., Sullivan, A. F., Fournie-Zaluski, M. C., and Roques, B. P. 1987. Prevention of degradation of endogenous enkephalins produces inhibition of nociceptive neurones in rat spinal cord. *Brain Res* 408:185–91.

Drake, C. T., Chavkin, C., and Milner, T. A. 2007. Opioid systems in the dentate gyrus. *Prog Brain Res* 163:245–63.

Drevets, W. C., and Raichle, M. E. 1992. Neuroanatomical circuits in depression: Implications for treatment mechanisms. *Psychopharmacol Bull* 28:261–74.

Duman, R. S. 2002. Pathophysiology of depression: The concept of synaptic plasticity. *Eur Psychiatry* 17 (Suppl 3):306–10.

Dziedzicka-Wasylewska, M., and Papp, M. 1996. Effect of chronic mild stress and prolonged treatment with imipramine on the levels of endogenous Met-enkephalin in the rat dopaminergic mesolimbic system. *Pol J Pharmacol* 48:53–6.

Dziedzicka-Wasylewska, M., and Rogoz, R. 1995. The effect of prolonged treatment with imipramine and electroconvulsive shock on the levels of endogenous enkephalins in the nucleus accumbens and the ventral tegmentum of the rat. *J Neural Transm Gen Sect* 102:221–8.

Evans, C. J., Keith, D. E., Jr., Morrison, H., Magendzo, K., and Edwards, R. H. 1992. Cloning of a delta opioid receptor by functional expression. *Science* 258:1952–5.

Fallon, J. H., and Leslie, F. M. 1986. Distribution of dynorphin and enkephalin peptides in the rat brain. *J Comp Neurol* 249:293–336.

Fichna, J., Janecka, A., Piestrzeniewicz, M., Costentin, J., and do Rego, J. C. 2007. Antidepressant-like effect of endomorphin-1 and endomorphin-2 in mice. *Neuropsychopharmacology* 32:813–21.

Filliol, D., Ghozland, S., Chluba, J., et al. 2000. Mice deficient for delta- and mu-opioid receptors exhibit opposing alterations of emotional responses. *Nat Genet* 25:195–200.

Fink, M., Simeon, J., Itil, T. M., and Freedman, A. M. 1970. Clinical antidepressant activity of cyclazocine—A narcotic antagonist. *Clin Pharmacol Ther* 11:41–8.

Fournie-Zaluski, M. C., Bourgoin, S., Cesselin, F., et al. 1986. Increase in Met-enkephalin level and antinociceptive effects induced by kelatorphan in the rat spinal cord. *NIDA Res Monogr* 75:454–6.

Fournie-Zaluski, M. C., Soleilhac, J. M., Turcaud, S., et al. 1992. Development of [125I]RB104, a potent inhibitor of neutral endopeptidase 24.11, and its use in detecting nanogram quantities of the enzyme by "inhibitor gel electrophoresis". *Proc Natl Acad Sci U S A* 89:6388–92.

Garcia, L. S., Comim, C. M., Valvassori, S. S., et al. 2008. Acute administration of ketamine induces antidepressant-like effects in the forced swimming test and increases BDNF levels in the rat hippocampus. *Prog Neuropsychopharmacol Biol Psychiatry* 32:140–4.

Garzon, J., de Antonio, I., and Sanchez-Blazquez, P. 2000. In vivo modulation of G proteins and opioid receptor function by antisense oligodeoxynucleotides. *Methods Enzymol* 314:3–20.

Garzon, J., Rodriguez-Munoz, M., de la Torre-Madrid, E., and Sanchez-Blazquez, P. 2005. Effector antagonism by the regulators of G protein signalling (RGS) proteins causes desensitization of mu-opioid receptors in the CNS. *Psychopharmacology (Berl)* 180:1–11.

Garzon, J., Sanchez-Blazquez, P., Hollt, V., Lee, N. M., and Loh, H. H. 1983. Endogenous opioid peptides: Comparative evaluation of their receptor affinities in the mouse brain. *Life Sci* 33 (Suppl 1):291–4.

Goldstein, A., Tachibana, S., Lowney, L. I., Hunkapiller, M., and Hood, L. 1979. Dynorphin-(1–13), an extraordinarily potent opioid peptide. *Proc Natl Acad Sci U S A* 76:6666–70.

Hamon, M., Gozlan, H., Bourgoin, S., et al. 1987. Opioid receptors and neuropeptides in the CNS in rats treated chronically with amoxapine or amitriptyline. *Neuropharmacology* 26:531–9.

Hughes, J., Smith, T. W., Kosterlitz, H. W., et al. 1975. Identification of two related pentapeptides from the brain with potent opiate agonist activity. *Nature* 258:577–80.

Hung, K. C., Wu, H. E., Mizoguchi, H., Leitermann, R., and Tseng, L. F. 2003. Intrathecal treatment with 6-hydroxydopamine or 5,7-dihydroxytryptamine blocks the antinociception induced by endomorphin-1 and endomorphin-2 given intracerebroventricularly in the mouse. *J Pharmacol Sci* 93:299–306.

Hurd, Y. L. 1996. Differential messenger RNA expression of prodynorphin and proenkephalin in the human brain. *Neuroscience* 72:767–83.

Ide, S., Fujiwara, S., Fujiwara, M., et al. 2010. Antidepressant-like effect of venlafaxine is abolished in mu-opioid receptor-knockout mice. *J Pharmacol Sci* 114:107–10.

Iwamoto, E. T. 1985. Place-conditioning properties of mu, kappa, and sigma opioid agonists. *Alcohol Drug Res* 6:327–39.

Javelot, H., Messaoudi, M., Garnier, S., and Rougeot, C. 2010. Human opiorphin is a naturally occurring antidepressant acting selectively on enkephalin-dependent delta-opioid pathways. *J Physiol Pharmacol* 61:355–62.

Jiang, Q., Takemori, A. E., Sultana, M., et al. 1991. Differential antagonism of opioid delta antinociception by [D-Ala2,Leu5,Cys6]enkephalin and naltrindole 5′-isothiocyanate: Evidence for delta receptor subtypes. *J Pharmacol Exp Ther* 257:1069–75.

Jones, R. M., and Portoghese, P. S. 2000. 5′-Guanidinonaltrindole, a highly selective and potent kappa-opioid receptor antagonist. *Eur J Pharmacol* 396:49–52.

Jutkiewicz, E. M., Eller, E. B., Folk, J. E., et al. 2004. Delta-opioid agonists: Differential efficacy and potency of SNC80, its 3-OH (SNC86) and 3-desoxy (SNC162) derivatives in Sprague-Dawley rats. *J Pharmacol Exp Ther* 309:173–81.

Kakidani, H., Furutani, Y., Takahashi, H., et al. 1982. Cloning and sequence analysis of cDNA for porcine beta-neo-endorphin/dynorphin precursor. *Nature* 298:245–9.

Kameyama, T., and Nagasaka, M. 1982. Effects of apomorphine and diazepam on a quickly learned conditioned suppression in rats. *Pharmacol Biochem Behav* 17:59–63.

Kameyama, T., Nagasaka, M., and Yamada, K. 1985. Effects of antidepressant drugs on a quickly-learned conditioned-suppression response in mice. *Neuropharmacology* 24:285–90.

Kieffer, B. L. 1995. Recent advances in molecular recognition and signal transduction of active peptides: Receptors for opioid peptides. *Cell Mol Neurobiol* 15:615–35.

Kieffer, B. L., Befort, K., Gaveriaux-Ruff, C., and Hirth, C. G. 1992. The delta-opioid receptor: Isolation of a cDNA by expression cloning and pharmacological characterization. *Proc Natl Acad Sci U S A* 89:12048–52.

Kita, A., Imano, K., Seto, Y., et al. 1997. Antinociceptive and antidepressant-like profiles of BL-2401, a novel enkephalinase inhibitor, in mice and rats. *Jpn J Pharmacol* 75:337–46.

Kristiansen, L. V., and Meador-Woodruff, J. H. 2005. Abnormal striatal expression of transcripts encoding NMDA interacting PSD proteins in schizophrenia, bipolar disorder and major depression. *Schizophr Res* 78:87–93.

Lecomte, J. M., Costentin, J., Vlaiculescu, A., et al. 1986. Pharmacological properties of acetorphan, a parenterally active "enkephalinase" inhibitor. *J Pharmacol Exp Ther* 237:937–44.

Loh, H. H., Tseng, L. F., Wei, E., and Li, C. H. 1976. Beta-endorphin is a potent analgesic agent. *Proc Natl Acad Sci U S A* 73:2895–8.

Maeng, S., and Zarate, C. A., Jr. 2007. The role of glutamate in mood disorders: Results from the ketamine in major depression study and the presumed cellular mechanism underlying its antidepressant effects. *Curr Psychiatry Rep* 9:467–74.

Mague, S. D., Pliakas, A. M., Todtenkopf, M. S., et al. 2003. Antidepressant-like effects of kappa-opioid receptor antagonists in the forced swim test in rats. *J Pharmacol Exp Ther* 305:323–30.

Maisonneuve, I. M., and Kreek, M. J. 1994. Acute tolerance to the dopamine response induced by a binge pattern of cocaine administration in male rats: An in vivo microdialysis study. *J Pharmacol Exp Ther* 268:916–21.

Maldonado, R., Dauge, V., Callebert, J., et al. 1989. Comparison of selective and complete inhibitors of enkephalin-degrading enzymes on morphine withdrawal syndrome. *Eur J Pharmacol* 165:199–207.

Maldonado, R., Dauge, V., Feger, J., and Roques, B. P. 1990. Chronic blockade of D2 but not D1 dopamine receptors facilitates behavioural responses to endogenous enkephalins, protected by kelatorphan, administered in the accumbens in rats. *Neuropharmacology* 29:215–23.

Mansour, A., Fox, C. A., Burke, S., Akil, H., and Watson, S. J. 1995. Immunohistochemical localization of the cloned mu opioid receptor in the rat CNS. *J Chem Neuroanat* 8:283–305.

Margolis, E. B., Hjelmstad, G. O., Bonci, A., and Fields, H. L. 2003. Kappa-opioid agonists directly inhibit midbrain dopaminergic neurons. *J Neurosci* 23:9981–6.

Martin, W. R., Eades, C. G., Thompson, J. A., Huppler, R. E., and Gilbert, P. E. 1976. The effects of morphine- and nalorphine-like drugs in the nondependent and morphine-dependent chronic spinal dog. *J Pharmacol Exp Ther* 197:517–32.

Martin-Schild, S., Gerall, A. A., Kastin, A. J., and Zadina, J. E. 1999. Differential distribution of endomorphin 1- and endomorphin 2-like immunoreactivities in the CNS of the rodent. *J Comp Neurol* 405:450–71.

Mathew, S. J., Keegan, K., and Smith, L. 2005. Glutamate modulators as novel interventions for mood disorders. *Rev Bras Psiquiatr* 27:243–8.

McLaughlin, J. P., Land, B. B., Li, S., Pintar, J. E., and Chavkin, C. 2006. Prior activation of kappa opioid receptors by U50,488 mimics repeated forced swim stress to potentiate cocaine place preference conditioning. *Neuropsychopharmacology* 31:787–94.

McLaughlin, J. P., Marton-Popovici, M., and Chavkin, C. 2003. Kappa opioid receptor antagonism and pro-dynorphin gene disruption block stress-induced behavioral responses. *J Neurosci* 23:5674–83.

Miller, P. D., Roux, C., Boonen, S., et al. 2005. Safety and efficacy of risedronate in patients with age-related reduced renal function as estimated by the Cockcroft and Gault method: A pooled analysis of nine clinical trials. *J Bone Miner Res* 20:2105–15.

Mueller, H. T., and Meador-Woodruff, J. H. 2004. NR3A NMDA receptor subunit mRNA expression in schizophrenia, depression and bipolar disorder. *Schizophr Res* 71:361–70.

Muscat, R., and Willner, P. 1992. Suppression of sucrose drinking by chronic mild unpredictable stress: A methodological analysis. *Neurosci Biobehav Rev* 16:507–17.

Nabeshima, T., Yamaguchi, K., Ishikawa, K., Furukawa, H., and Kameyama, T. 1987. Potentiation in phencyclidine-induced serotonin-mediated behaviors after intracerebroventricular administration of 5,7-dihydroxytryptamine in rats. *J Pharmacol Exp Ther* 243:1139–46.

Nakanishi, S., Inoue, A., Kita, T., et al. 1979. Nucleotide sequence of cloned cDNA for bovine corticotropin-beta-lipotropin precursor. *Nature* 278:423–7.

Negus, S. S., and Mello, N. K. 2002. Effects of mu-opioid agonists on cocaine- and food-maintained responding and cocaine discrimination in rhesus monkeys: Role of mu-agonist efficacy. *J Pharmacol Exp Ther* 300:1111–21.

Nestler, E. J., and Carlezon, W. A., Jr. 2006. The mesolimbic dopamine reward circuit in depression. *Biol Psychiatry* 59:1151–9.

Newton, S. S., Thome, J., Wallace, T. L., et al. 2002. Inhibition of cAMP response element–binding protein or dynorphin in the nucleus accumbens produces an antidepressant-like effect. *J Neurosci* 22:10883–90.

Nieto, M. M., Wilson, J., Walker, J., et al. 2001. Facilitation of enkephalins catabolism inhibitor-induced antinociception by drugs classically used in pain management. *Neuropharmacology* 41:496–506.

Noble, F., Soleilhac, J. M., Soroca-Lucas, E., et al. 1992. Inhibition of the enkephalin-metabolizing enzymes by the first systemically active mixed inhibitor prodrug RB 101 induces potent analgesic responses in mice and rats. *J Pharmacol Exp Ther* 261:181–90.

Overmier, J. B., Seligman, M. E. 1967. Effects of inescapable shock upon subsequent escape and avoidance responding. *J Comp Physiol Psychol* 63(1):28–33.

Pakarinen, E. D., Woods, J. H., and Moerschbaecher, J. M. 1995. Repeated acquisition of behavioral chains in squirrel monkeys: Comparisons of a mu, kappa and delta opioid agonist. *J Pharmacol Exp Ther* 272:552–9.

Papp, M., Moryl, E., and Willner, P. 1996. Pharmacological validation of the chronic mild stress model of depression. *Eur J Pharmacol* 296:129–36.

Pasternak, G. W. 2004. Multiple opiate receptors: Déjà vu all over again. *Neuropharmacology* 47 (Suppl 1): 312–23.

Pasternak, G. W., Bodnar, R. J., Clark, J. A., and Inturrisi, C. E. 1987. Morphine-6-glucuronide, a potent mu agonist. *Life Sci* 41:2845–9.

Pasternak, G. W., and Standifer, K. M. 1995. Mapping of opioid receptors using antisense oligodeoxynucleotides: Correlating their molecular biology and pharmacology. *Trends Pharmacol Sci* 16:344–50.

Peckys, D., and Landwehrmeyer, G. B. 1999. Expression of mu, kappa, and delta opioid receptor messenger RNA in the human CNS: A 33P in situ hybridization study. *Neuroscience* 88:1093–135.

Pert, C. B., and Snyder, S. H. 1973. Opiate receptor: Demonstration in nervous tissue. *Science* 179:1011–4.

Petit, F., Hamon, M., Fournie-Zaluski, M. C., Roques, B. P., and Glowinski, J. 1986. Further evidence for a role of delta-opiate receptors in the presynaptic regulation of newly synthesized dopamine release. *Eur J Pharmacol* 126:1–9.

Pliakas, A. M., Carlson, R. R., Neve, R. L., et al. 2001. Altered responsiveness to cocaine and increased immobility in the forced swim test associated with elevated cAMP response element–binding protein expression in nucleus accumbens. *J Neurosci* 21:7397–403.

Porsolt, R. D., Anton, G., Blavet, N., and Jalfre, M. 1978. Behavioural despair in rats: A new model sensitive to antidepressant treatments. *Eur J Pharmacol* 47:379–91.

Porsolt, R. D., Deniel, M., and Jalfre, M. 1979. Forced swimming in rats: Hypothermia, immobility and the effects of imipramine. *Eur J Pharmacol* 57:431–6.

Przewlocki, R., Lason, W., Turchan, J., and Przewlocka, B. 1997. Imipramine induces alterations in proenkephalin and prodynorphin mRNAs level in the nucleus accumbens and striatum in the rat. *Pol J Pharmacol* 49:351–5.

Raynor, K., Kong, H., Hines, J., et al. 1994. Molecular mechanisms of agonist-induced desensitization of the cloned mouse kappa opioid receptor. *J Pharmacol Exp Ther* 270:1381–6.

Reisine, T., and Soubrie, P. 1982. Loss of rat cerebral cortical opiate receptors following chronic desimipramine treatment. *Eur J Pharmacol* 77:39–44.

Rodriguez-Munoz, M., de la Torre-Madrid, E., Sanchez-Blazquez, P., Wang, J. B., and Garzon, J. 2008. NMDAR-nNOS generated zinc recruits PKCgamma to the HINT1-RGS17 complex bound to the C terminus of mu-opioid receptors. *Cell Signal* 20:1855–64.

Rojas-Corrales, M. O., Berrocoso, E., Gibert-Rahola, J., and Micó, J. A. 2004. Antidepressant-like effect of tramadol and its enantiomers in reserpinized mice: Comparative study with desipramine, fluvoxamine, venlafaxine and opiates. *J Psychopharmacol* 18:404–11.

Rojas-Corrales, M. O., Gibert-Rahola, J., and Mico, J. A. 1998. Tramadol induces antidepressant-type effects in mice. *Life Sci* 63:PL175–80.

Roques, B. P., and Fournie-Zaluski, M. C. 1986. Enkephalin degrading enzyme inhibitors: A physiological way to new analgesics and psychoactive agents. *NIDA Res Monogr* 70:128–54.

Sante, A. B., Nobre, M. J., and Brandao, M. L. 2000. Place aversion induced by blockade of mu or activation of kappa opioid receptors in the dorsal periaqueductal gray matter. *Behav Pharmacol* 11:583–9.

Schreff, M., Schulz, S., Wiborny, D., and Hollt, V. 1998. Immunofluorescent identification of endomorphin-2-containing nerve fibers and terminals in the rat brain and spinal cord. *Neuroreport* 9:1031–4.

Schreiber, S., Backer, M. M., and Pick, C. G. 1999. The antinociceptive effect of venlafaxine in mice is mediated through opioid and adrenergic mechanisms. *Neurosci Lett* 273:85–8.

Schwarzer, C. 2009. 30 years of dynorphins—New insights on their functions in neuropsychiatric diseases. *Pharmacol Ther* 123:353–70.

Sheline, Y. I. 2000. 3D MRI studies of neuroanatomic changes in unipolar major depression: The role of stress and medical comorbidity. *Biol Psychiatry* 48:791–800.

Shippenberg, T. S., LeFevour, A., and Chefer, V. I. 2008. Targeting endogenous mu- and delta-opioid receptor systems for the treatment of drug addiction. *CNS Neurol Disord Drug Targets* 7:442–53.

Shirayama, Y., Ishida, H., Iwata, M., et al. 2004. Stress increases dynorphin immunoreactivity in limbic brain regions and dynorphin antagonism produces antidepressant-like effects. *J Neurochem* 90:1258–68.

Shuster, S. J., Riedl, M., Li, X., Vulchanova, L., and Elde, R. 2000. The kappa opioid receptor and dynorphin co-localize in vasopressin magnocellular neurosecretory neurons in guinea-pig hypothalamus. *Neuroscience* 96:373–83.

Simon, E. J. 1991. Opioid receptors and endogenous opioid peptides. *Med Res Rev* 11:357–4.

Simonin, F., Valverde, O., Smadja, C., et al. 1998. Disruption of the kappa-opioid receptor gene in mice enhances sensitivity to chemical visceral pain, impairs pharmacological actions of the selective kappa-agonist U-50,488H and attenuates morphine withdrawal. *EMBO J* 17:886–97.

Smadja, C., Maldonado, R., Turcaud, S., Fournie-Zaluski, M. C., and Roques, B. P. 1995. Opposite role of CCKA and CCKB receptors in the modulation of endogenous enkephalin antidepressant-like effects. *Psychopharmacology (Berl)* 120:400–8.

Smadja, C., Ruiz, F., Coric, P., et al. 1997. CCK-B receptors in the limbic system modulate the antidepressant-like effects induced by endogenous enkephalins. *Psychopharmacology (Berl)* 132:227–36.

Spanagel, R., and Shippenberg, T. S. 1993. Modulation of morphine-induced sensitization by endogenous kappa opioid systems in the rat. *Neurosci Lett* 153:232–6.

Stengaard-Pedersen, K., and Schou, M. 1986. Opioid receptors in the brain of the rat following chronic treatment with desipramine and electroconvulsive shock. *Neuropharmacology* 25:1365–71.

Steru, L., Chermat, R., Thierry, B., and Simon, P. 1985. The tail suspension test: A new method for screening antidepressants in mice. *Psychopharmacology (Berl)* 85:367–70.

Stinus, L., Koob, G. F., Ling, N., Bloom, F. E., and Le Moal, M. 1980. Locomotor activation induced by infusion of endorphins into the ventral tegmental area: Evidence for opiate–dopamine interactions. *Proc Natl Acad Sci U S A* 77:2323–7.

Suzuki, T., Shiozaki, Y., Masukawa, Y., Misawa, M., and Nagase, H. 1992. The role of mu- and kappa-opioid receptors in cocaine-induced conditioned place preference. *Jpn J Pharmacol* 58:435–42.

Svingos, A. L., Colago, E. E., and Pickel, V. M. 1999. Cellular sites for dynorphin activation of kappa-opioid receptors in the rat nucleus accumbens shell. *J Neurosci* 19:1804–13.

Tejedor-Real, P., Mico, J. A., Maldonado, R., Roques, B. P., and Gibert-Rahola, J. 1993. Effect of mixed (RB 38A) and selective (RB 38B) inhibitors of enkephalin degrading enzymes on a model of depression in the rat. *Biol Psychiatry* 34:100–7.

Tejedor-Real, P., Mico, J. A., Maldonado, R., Roques, B. P., and Gibert-Rahola, J. 1995. Implication of endogenous opioid system in the learned helplessness model of depression. *Pharmacol Biochem Behav* 52:145–52.

Tejedor-Real, P., Mico, J. A., Smadja, C., et al. 1998. Involvement of delta-opioid receptors in the effects induced by endogenous enkephalins on learned helplessness model. *Eur J Pharmacol* 354:1–7.

Terenius, L. 1973. Stereospecific interaction between narcotic analgesics and a synaptic plasma membrane fraction of rat cerebral cortex. *Acta Pharmacol Toxicol (Copenh)* 32:317–20.

Thiébot, M. H., Martin, P., Puech, A. J. 1992. Animal behavioural studies in the evaluation of antidepressant drugs. *Br J Psychiatry Suppl.* (15):44–50. Review.

Todtenkopf, M. S., Marcus, J. F., Portoghese, P. S., and Carlezon, W. A., Jr. 2004. Effects of kappa-opioid receptor ligands on intracranial self-stimulation in rats. *Psychopharmacology (Berl)* 172:463–70.

Tomasiewicz, H. C., Todtenkopf, M. S., Chartoff, E. H., Cohen, B. M., and Carlezon, W. A., Jr. 2008. The kappa-opioid agonist U69,593 blocks cocaine-induced enhancement of brain stimulation reward. *Biol Psychiatry* 64:982–8.

Torregrossa, M. M., Isgor, C., Folk, J. E., et al. 2004. The delta-opioid receptor agonist (+)BW373U86 regulates BDNF mRNA expression in rats. *Neuropsychopharmacology* 29:649–59.

Torregrossa, M. M., Jutkiewicz, E. M., Mosberg, H. I., et al. 2006. Peptidic delta opioid receptor agonists produce antidepressant-like effects in the forced swim test and regulate BDNF mRNA expression in rats. *Brain Res* 1069:172–81.

Vaidya, V. A., and Duman, R. S. 2001. Depression—Emerging insights from neurobiology. *Br Med Bull* 57:61–79.

Valverde, O., Mico, J. A., Maldonado, R., Mellado, M., and Gibert-Rahola, J. 1994. Participation of opioid and monoaminergic mechanisms on the antinociceptive effect induced by tricyclic antidepressants in two behavioural pain tests in mice. *Prog Neuropsychopharmacol Biol Psychiatry* 18:1073–92.

Vergura, R., Balboni, G., Spagnolo, B., et al. 2008. Anxiolytic- and antidepressant-like activities of H-Dmt-Tic-NH-CH(CH$_2$-COOH)-Bid (UFP-512), a novel selective delta opioid receptor agonist. *Peptides* 29:93–103.

Vergura, R., Valenti, E., Hebbes, C. P., et al. 2006. Dmt-Tic-NH-CH$_2$-Bid (UFP-502), a potent DOP receptor agonist: In vitro and in vivo studies. *Peptides* 27:3322–30.

Vilpoux, C., Carpentier, C., Leroux-Nicollet, I., Naudon, L., and Costentin, J. 2002. Differential effects of chronic antidepressant treatments on micro- and delta-opioid receptors in rat brain. *Eur J Pharmacol* 443:85–93.

Watson, S. J., Akil, H., Ghazarossian, V. E., and Goldstein, A. 1981. Dynorphin immunocytochemical localization in brain and peripheral nervous system: Preliminary studies. *Proc Natl Acad Sci U S A* 78:1260–3.

Willner, P. 1984. The validity of animal models of depression. *Psychopharmacology* 83(1):1–16. Review.

Xie, J., Soleilhac, J. M., Renwart, N., et al. 1989. Inhibitors of the enkephalin degrading enzymes. Modulation of activity of hydroxamate containing compounds by modifications of the C-terminal residue. *Int J Pept Protein Res* 34:246–55.

Zadina, J. E. 2002. Isolation and distribution of endomorphins in the central nervous system. *Jpn J Pharmacol* 89:203–8.

Zadina, J. E., Hackler, L., Ge, L. J., and Kastin, A. J. 1997. A potent and selective endogenous agonist for the mu-opiate receptor. *Nature* 386:499–502.

Zurita, A., Martijena, I., Cuadra, G., Brandao, M. L., and Molina, V. 2000. Early exposure to chronic variable stress facilitates the occurrence of anhedonia and enhanced emotional reactions to novel stressors: Reversal by naltrexone pretreatment. *Behav Brain Res* 117:163–71.

12 Chronobiological Rhythms, Melatonergic System, and Depression

Cecilio Álamo and Francisco López-Muñoz

CONTENTS

12.1 INTRODUCTION

The knowledge that the numerous biological functions and physiological activities—in human and many mammals (hibernation, etc.), birds (migratory behavior, etc.), and plants (flowering, autumn fall leaves, etc.)—are subject to a temporary rhythmicity is as old as man himself. All living organisms are characterized by the presence of a cyclic rhythmicity in a wide variety of biological and behavioral processes. "Biological rhythms" are known as the recurrence of phenomenon within a biological system at regular intervals. These rhythms involve an adaptation to the environment.

In 1729, Jean Jacques d'Ortous de Mairan performed the first experiment that revealed the existence of an endogenous circadian rhythm. This astronomer had a *Mimosa pudica* in his observatory, whose leaves remains outstretched during the day and retracts during the night, a phenomenon that persisted even if the mimosa was deprived of light. This demonstrated that circadian rhythms were endogenous and could persist in the absence of external temporary signals. The experiment's text was prepared by his friend Jean Marchant and was published in *Histoire de l'Academie Royale des Sciences* (Paris, 1729). A century later, in 1814, Julien-Joseph Virey, phamacist-in-chief at the Val-de-Grâce Military Hospital in Paris, presented his doctoral thesis in medicine at the University of Paris, entitled "Éphémérides de la vie humaine, ou recherches sur la révolution journalière et la périodicité de ses phénomènes dans la santé et les maladies," which was the first published scientific work on biological rhythmicity. Virey, considered the founding father of chronobiology, proposed that biological rhythms in human beings can be explained not only by external changes in the environment. However, Virey was harshly criticized for his idea on the endogenous origin of rhythmicity. In 1825, an anonymous article entitled *Périodicité* was published ridiculing his idea, which disaccredited his career and his reputation as a scientist (Reinberg et al. 2001).

A turning point in this area of research was the year 1925, when the Canadian zoologist William Rowan demonstrated in certain types of migratory birds such as finches that gonadal function was regulated by the ambient light during seasonal cycles; testes and ovaries had an increased size in the spring and a decreased size in the autumn. However, the exposure of these animals to artificial light during the autumn avoided these involutive chronobiological alterations (Rowan 1925). Later research carried out on bacteria, insects, and rodents were determining factors for the current knowledge on biological clocks. The paradigms discovered in these species led to the study in human beings. Indeed, systematic studies on chronobiology began around 1950, from the work of Franz Halberg on circadian variations in body temperature or in the amount of circulating eosinophils, culminating in the publication, in 1958, of the famous book entitled *Die physiologische Uhr* (*The Physiological Clock*) by Erwin Bünning. A new discipline was developed, called *chronobiology*, with Jürgen Aschoff and Colin S. Pittendrich as its modern founders, by fostering chronobiological studies in animals and human beings.

Chronobiology development and the study of the biological rhythms have since come a long way with advances in the knowledge of the physiology of melatonin, a hormone synthesized and released in the pineal gland. As part of their studies to identify the factor responsible for the darkening of the skin in amphibians, Aaron B. Lerner, from the Yale University, together Yoshiyata Takahashi, an internist at the University of Tokyo, isolated a very small amount (100 μg) of indolamine (*N*-acetyl-5-methoxytryptamine) extracted from 250,000 bovine pineal glands (100 kg of material), which was named *melatonin* (from the Greek word *melas* meaning "black" or "dark") (Lerner et al. 1958). Subsequently, they confirmed that it was a derivative of serotonin (Lerner 1961). Over the following years, several authors, basically from the groups of Julius Axelrod (from the Department of Clinical Pharmacology at the National Institutes of Health in Bethesda) and Virginia M. Fiske (from Wellesley College), have shown that the synthesis of melatonin was regulated in mammals by ambient light (Axelrod and Weissbach 1960; Fiske et al. 1960; Wurtman et al. 1961, 1963, 1964; Quay 1963) through a neural pathway that, starting from the retina, ended in the sympathetic neurons of the superior cervical ganglion (Ariëns-Kappers 1960). Melatonin, in turn, would act as a powerful neurotransmitter in the central nervous system, making the pineal body a "neuroendocrine transducer" and an essential gear of the "biological clock" (Wurtman and Axelrod 1965). At the beginning of the 1970s, the existence of a circadian oscillator located in the suprachiasmatic nucleus (SCN) of the hypothalamus was proven, which would also control the synthesis of melatonin in the pineal body depending on the activity of the enzyme serotonin *N*-acetyltransferase (NAT) (Klein et al. 1971; Deguchi and Axelrod 1972; Moore and Klein 1974) and which, in turn, would be regulated by melatonin, given the high density of melatonergic receptors present in this nucleus. Therefore, the melatonergic system constitutes an essential part of neurobiological circuits responsible for the adaptation of the biological rhythms to periodic changes in the environment (Webb and Puig-Domingo 1995).

The molecular basis of endogenous clocks has been discovered, at least partially, in the past decade. Today, it seems evident that alterations in endogenous rhythms and abnormal secretion of melatonin can manifest through clinical disorders in the areas of sleep, cardiovascular system, endocrinometabolic system, neurology, and psychiatry (Schulz and Steimer 2009), including affective disorders. In the present chapter, we mainly focus on the description of circadian rhythms and melatonergic system, the desynchronization of the biological clock, and their role in the pathophysiology of depression.

12.2 BIOLOGICAL RHYTHMS

All organisms, including human beings, have developed a system for measuring time, which allows them to adapt to environmental phenomena that are repeated at a constant and regular fashion. These daily cycles, light–dark, temperature, and so on, force the subject to adapt to environmental changes. The circadian rhythms of the different biological variables (temperature, sleep–wake, hormones, etc.) can be described by various parameters of a sinusoidal curve. "Period" is the time interval between two identical points of the rhythm and measures the cycle duration. "Mesor" is the arithmetic mean of all the values obtained within a cycle. It is the value around which the variable oscillates. "Amplitude" is the difference between the maximum or minimum values and the mean value of an oscillation. "Phase" is the value of a variable at a determined time, and "acrophase" would be the moment at which the variable reached its maximum value (Figure 12.1).

Biological rhythms can be arbitrarily divided into different categories. Thus, we have "high-frequency rhythms" (freely occurring period <30 min), such as the movements through ionic membrane channels or some enzymatic reactions, which occur in nanoseconds, and heartbeat, breathing rate, or electroencephalographic activity, which lasts seconds or minutes. "Medium-frequency rhythms" (freely occurring period between 30 min and 6 days) can be divided into ultradian, circadian, and infradian rhythms. Ultradian rhythms (between 30 min and 20 h) are, for example, the sleep phases, and circadian rhythms (20–28 h) are undoubtedly the most popular since they affect the sleep–wake cycle, motor activity, release of melatonin, blood pressure, and temperature, among other functions. Infradian rhythms (between 28 h and 6 days) can also exist as, for example, the release of adrenal hormones. Finally, there are "low-frequency rhythms" (freely occurring period >6 days), which can be circalunar (approximately 29 days); the most common representation in humans would be

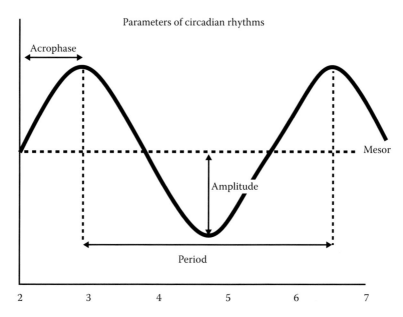

FIGURE 12.1 Phase map: temporal relation between different psychological processes that appear with a characteristic sequence within a cycle.

menstrual cycle. Other rhythms may occur annually, called circannual, such as migrations and reproduction in some animals, or even in periods of more than 1 year (Márquez 2004).

Focusing on the circadian rhythms, a good adaptation to the environment requires being able to anticipate these daily predictable cycles. It is for this reason that a subject shows cycles that affect its physiological and behavioral functions that are best adapted to the alternation of activity and sleep functions. Circadian rhythms affect many physiological parameters, including heartbeat, blood pressure, body temperature, the release of hormones [such as melatonin, cortisol (Figure 12.2), thyrotropin, or insulin among others], renal elimination, and even the enzymatic activity of cytochrome P-450. From an experimental standpoint, the modifications that occur during the sleep–wake cycle, cortisol and melatonin levels, body temperature, and motor activity are typically assessed. Thus, it is possible to observe peaks of hormone release, such as melatonin levels during the night and cortisol during the early morning, followed by reduction in the levels at other hours (Figure 12.2) (Arendt 1995).

In relation to circadian rhythms, we can speak of situations of phase advance or phase delay when the cyclic variable of measurement shows an acrophase, which occurs before (advance) or after (delay) the normal. The biological functions that are typically used as clinical markers of the endogenous rhythms in mammals are the sleep–wake cycle, body temperature, and melatonin secretion. When these parameters are within typical values, the individual has a normal chronobiological configuration and may be classified as chronotype or pattern (Chandrashekaran 1998; Schulz and Steimer 2009). However, the circadian rhythms do not only affect physiological processes but also various pathologies and diseases in human beings. However, the frontiers between the chronotype subjects compared with individuals with altered rhythms must be assessed in large populations to determine the role of alterations in the rhythm in some symptoms or in the predisposition to different diseases (Roenneberg et al. 2007).

Under normal conditions, the circadian rhythms are endogenous (temperature, sleep phases, hormones, enzyme synthesis, etc.) and they are synchronized in days and nights. Nevertheless, these circadian rhythms are influenced by timers, better known by the German term *zeitgebers* (time markers), which are external elements or environmental signals that act repetitively and which use the organism as time references to be able to guide their rhythms. The main *zeitgeber* is light, although other elements exist that may act as such in determined circumstances and animal species,

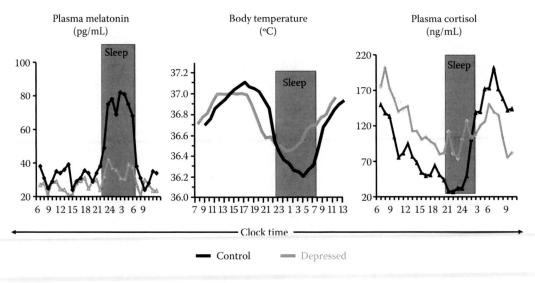

FIGURE 12.2   Circadian rhythms in physiological parameters (body temperature and plasma levels of melatonin and cortisol) and its alteration in depressive patients. (Modified with permission from Souetre E., et al., *Biol Psychiatry*, 24, 336–40, 1988.)

for example, social contact with beings of the same species, the availability of food, and motor activity. Metaphorically, the biological clock would be like an old clock that typically runs slow or fast and that has to be set to the right time every so often. This setting to the right time or synchronization mainly occurs because of light. When the *zeitgeber* or external timer does not exist, in cases of total darkness and light, the endogenous period of circadian rhythms of the individual is expressed, which generally differs from 24 h and is called the free rhythm period. This period differs depending on how strict is the isolation. Hence, when we speak of the period of a circadian rhythm, we must consider if it occurs in the presence of *zeitgebers* or in free rhythm (Mieda et al. 2006; Schulz and Steimer 2009).

12.3 ANATOMIC–FUNCTIONAL AND GENETIC CONTROL MECHANISMS OF CIRCADIAN RHYTHMS

The circadian control is formed by three essential components: a biological clock or master pacemaker, located in the SCN of the anterior hypothalamus; afferent pathways that inform the SCN of the outer world; and efferent pathways that are responsible for coupling the biological clock with the expression of the changes of functions or circadian rhythms. A simplified scheme is shown in Figure 12.3 (Álamo et al. 2008; Schulz and Steimer 2009).

This entire anatomic–functional structure, governed by the master pacemaker located in the SCN of the anterior hypothalamus, will be responsible for generation of the circadian rhythms. It should be borne in mind that this genetically determined individual endogenous periodicity is slightly different to the 24-h circadian cycle and needs the daily synchronization by *zeitgebers* or environmental timers (Monk and Welsh 2003). The main *zeitgeber* of the SCN is light, which reaches this nucleus from the photoreceptors of the retina ganglionic cells. Other nonphotic *zeitgebers* (exercise,

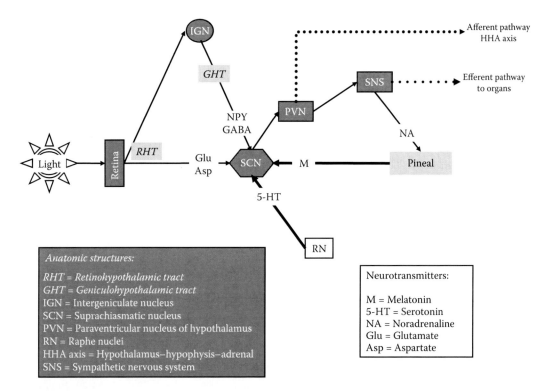

FIGURE 12.3 Schematic representation of the anatomic–functional complex that controls biological rhythms.

perhaps sleep or darkness, and nutrients) have been less researched and are probably weaker or less potent timers than light.

In relation to the presence of circadian rhythms outside the brain, it must be pointed out that it has been observed for years that the synthesis of hepatic enzymes was governed by circadian rhythms (Clarke et al. 1984). The existence of peripheral circadian clocks was later confirmed in many tissues, both at a hepatic level and a nervous level, which control the innervation of the gastrointestinal and respiratory systems, in the adipose tissue, the thyroid gland, and the adrenal gland (Schulz and Steimer 2009). The SCN can influence these peripheral pacemakers, but, against that postulated until recent years, there are peripheral structures, such as fibroblast cultures, that may be totally autonomous and independent of the SCN (Levi and Schibler 2007).

12.3.1 CENTRAL BIOLOGICAL CLOCK

The SCN is a pair nucleus located in the anteroventral region of the hypothalamus, just in front of the optic chiasma, formed by about 10,000 neurons, which largely contain γ-aminobutyric acid (GABA) and, in a lesser proportion, vasopressin, vasoactive intestinal polypeptide, encephalins, substance P, and other peptides (Ángeles et al. 2007). The importance of this nucleus in the control of circadian rhythm is fundamental and is revealed by two fundamental facts.

On one hand, the SCN has a spontaneous circadian rhythm, which is revealed both in the use of glucose and in the electrical activity of its neurons, being that this rhythmicity is autonomous, such that it is maintained when this nucleus is isolated from the other cerebral areas. We should highlight that although some neurons isolated from the SCN have a period of about 20 h, others maintain periods of up to 30 h. However, when they act together, the period is approximately 24 h, a fact indicative of the existence of a paracrine communication between the said neurons within the SCN (Levi and Schibler 2007).

On the other hand, lesion of this nucleus causes a loss in circadian rhythms (Stephan and Zucker 1972), such as motor activity, sleep–wake cycle, intake of food and drink, temperature, and hormonal secretion, in several species of mammals, including man, and these rhythms are recovered after transplanting the SCN tissue (Drucker-Colin et al. 1984).

12.3.2 AFFERENT PATHWAYS

There are three main afferent pathways that reach the SCN (Márquez 2004). The first comes from the retina photoreceptors that project their axons directly to the SCN and is called the retinohypothalamic tract (RHT). This pathway, which carries the light information, uses the excitatory amino acids glutamate and aspartate as neurotransmitters and facilitates synchronization of SCN activity with the outside environment.

A second afferent pathway comes from the RHT and sends collaterals to the intergeniculate leaflet nucleus, which also communicates with the SCN through the geniculohypothalamic tract (GHT). The intergeniculate nucleus, through the GHT, releases neuropeptide Y and GABA on the SCN. It has been suggested that the intergeniculate leaflet nucleus integrates light and nonlight information for the circadian synchronization system, which modulates SCN function.

Furthermore, the SCN is controlled by serotonergic pathways that arise in neurons of the mesencephalic nuclei of the raphe and that act on 5-HT$_2$ receptors located in SCN neurons. The activity of these neurons depends on the waking state of the individual, the time when the subject is functioning normally. However, serotonergic activity is slow during the slow sleep phases and is silent during the rapid eye movement (REM) sleep phases. It has been suggested that this pathway would act in the SCN, modulating the synchronization produced by the light information (Moore-Ede 1999).

The SCN further receives afferences from the cerebral cortex, basal telencephalon, thalamus, hypothalamus, periventricular zones, and brain stem that transmit nonlight information, mainly related to the internal state the organism finds itself in (Márquez 2004).

12.3.3 Efferent Pathways

The efferences that exit the SCN have a dual component as they are nervous and hormonal. The main nerve efferences from the SCN are projected to the anterior hypothalamus and tuberal hypothalamus, which also launch projections to the effector organs. Furthermore, by a multisynaptic pathway, the paraventricular nucleus of the anterior hypothalamus sends its axons to the sympathetic nervous system, reaching the superior cervical ganglion. From here, the information continues by internal carotid nerve and finally by the *nervii conarii*, which innervates the pineal gland and controls melatonin synthesis (Figures 12.3 and 12.4). This pathway, which is in fact a hormonal pathway, becomes activated in darkness and releases noradrenaline to the pineal gland, which stimulates the nighttime synthesis of melatonin. Melatonin, which does not have the capacity of being stored, is secreted in a circadian manner to the bloodstream and synchronizes the rhythms of the entire organism including its own SCN (Schulz and Steimer 2009).

12.3.4 Melatonergic System

12.3.4.1 Pineal Gland

A key anatomic structure of the melatonergic system is the pineal gland, a small neuroendocrine organ similar in size to a grain of rice (5–9 mm long, 15 mm wide, and 3–5 mm thick), which in mammals embryologically comes from a neuroepithelial out-pouching of the dorsal diencephalic roof region. The pineal gland is located just rostrodorsal to the superior colliculus and behind and beneath the stria medullaris, between the laterally positioned thalamic bodies.

The pineal gland of mammals is an organ that develops a high biochemical activity, emphasizing the synthesis and secretion of melatonin in response to environmental light changes. But this is not the only amine in the pineal body because there are also other amines, such as serotonin, noradrenaline, or histamine, in addition to multiple peptidergic substances (vasopressin, vasoactive intestinal polypeptide, oxytocin, neuropeptide Y, somatostatin, substance P, etc.) and other hormones (luteinizing hormone, follicle-stimulating hormone, thyrotropin-releasing hormone, adrenocorticotropic hormone, prolactin, etc.) (Korf et al. 1998; Macchi and Bruce 2004). Thus, we can talk of pineal gland as a neuroendocrine organ capable of synthesizing and releasing active substances, which exert their hormonal action on a variable number of target organs and tissues (Wurtman et al. 1968), including the hypothalamus, pituitary, gonads, thyroid, etc. Also, the pineal gland can be catalogued, without risk of error, as an important component of photoneuroendocrine systems (Oksche and Hartwig 1979).

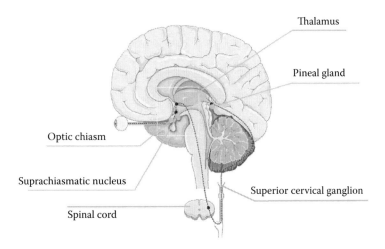

FIGURE 12.4 Anatomic location of biological clock (retina–SCN–pineal gland neural circuit).

12.3.4.2 Melatonin

The biosynthesis of all pineal indolamines (Figure 12.5) (Smith 1983) begins with the uptake of circulating tryptophan by the main pineal functional cell, the pinealocyte. Captured tryptophan is hydroxylated by tryptophan hydroxylase producing 5-hydroxytryptophan, which is quickly decarboxylated by 3,4-dihydroxyphenylalanine decarboxylase to form 5-hydroxytryptamine or serotonin. Serotonin is, in turn, converted into N-acetylserotonin under the action of the NAT. Finally, the resulting compound is methylated by hydroxy-indole-O-methyl-transferase to form N-acetyl-5-methoxytryptamine or melatonin (Macchi and Bruce 2004). After its release to blood capillaries, melatonin is widely distributed by the organism, joined by 70% to albumin. Besides in blood, saliva, and urine, this hormone is detectable in cerebrospinal fluid as well as in other types of fluids related to reproductive activity, such as semen, amniotic fluid, breast milk, etc. (Cagnacci 1996). Melatonin is metabolized initially at the liver level, by 6-hydroxylation, and then, at the kidney level, resulting in sulfate of 6-hydroxymelatonin (90%) and 6-hydroxymelatonin-glucuronide (10%), although other minor metabolites have also been described (Arendt 1995; Macchi and Bruce 2004).

Conversion of serotonin in melatonin is influenced, as all the glandular physiological responses, by the light–dark cycle. Whereas NAT activity is very low during luminous periods, resulting in an increase in serotonin levels and a decrease in the amount of melatonin, quite the opposite occurs in the dark phases. The hydroxy-indole-O-methyl-transferase variations are much less marked, so it can be considered in NAT as the enzyme that limits the synthesis of melatonin (Klein and Weller 1973; Klein et al. 1981).

We must bear in mind that although the pineal gland is the main melatonin-producing organ, this hormone is also synthesized, although in much less amount, in other organs, such as the retina,

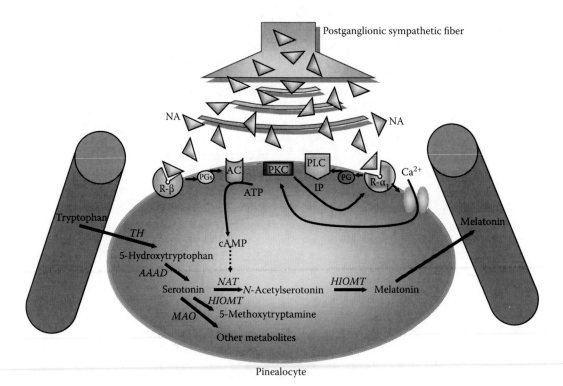

FIGURE 12.5 Illustration of melatonin biosynthesis and effect of interaction between noradrenaline (NA), released from postganglionic sympathetic fibers of superior cervical ganglion, and adrenergic receptors situated on membrane of pinealocyte. This interaction triggers a series of biochemical cascades within the pinealocyte that culminates with production of melatonin.

gastrointestinal tract, Harderian organ, or platelets. Similarly, the production of melatonin is not specific to mammals because it is also synthesized by other nonmammal vertebrates, some invertebrates, and numerous plants (Hardeland et al. 1996).

12.3.4.3 Melatonergic Receptors

Melatonin exerts its physiological actions by initial binding to their corresponding receptors. So far, three subtypes of melatonin receptors have been cloned in mammalian species, of which the so-called MT_1 and MT_2 (previously called Mel_{1a} or MTNR1A and Mel_{1b} or MTNR1B, respectively) are of high affinity, whereas the MT_3 is of low affinity. These melatonin receptors belong to a differential group of metabotropic receptors of the G-protein-coupled receptor family (Sugden et al. 2004; Guardiola-Lemaitre 2005). Another receptor, called Mel_{1c} (MTNR1C), was also described and identified in amphibians and birds and not included in the G-protein-coupled receptor family (Sugden et al. 2004). Characterization of these receptors and their pharmacological profile definition became possible in the late 1980s, because of the use of the 2-[^{125}I]iodomelatonin radioligand, a high-affinity agonist (Dubocovich and Takahashi 1987; Dubocovich 1988).

MT_1 receptor stimulation activates the signal pathways mediated by G protein ($G\alpha_i$ and $G\alpha_o$), resulting in the inhibition of adenylate cyclase and possibly in the stimulation of inositol phosphate pathway (Gordon and Reppert 1997), modifying, therefore, all the biochemical transduction cascade associated with the cAMP response element–binding, which concludes with the inhibition of immediate-early gene expressions, such as *c-fos* and *jun-B*. However, studies conducted with cell cultures that express the recombinant subtypes MT_1 and MT_2 have shown a potential interaction with other systems of signal transduction, such as phospholipase C receptorial activation, activation of protein kinase C, or inhibition of cGMP pathway (von Gall et al. 2002; Guardiola-Lemaitre 2005). Similarly, melatonin is also able to directly bind with the calmodulin and proteins of the cytoskeleton (Benítez-King et al. 1996). Moreover, coexpression of the membrane receptor type MT_1 with the nuclear receptor $ROR\alpha1$ (Tomás-Zapico et al. 2005) has been found in the pineal gland of Syrian hamsters. This data set suggests the existence of a large number of cellular responses induced by melatonin after binding on their high-affinity receptors (Guardiola-Lemaitre 2005).

The density and location of melatonergic receptors vary considerably between different species of mammals and have not only been located in the central nervous system but also in other peripheral organs such as lymphocytes, platelets, prostatic epithelium, granulosa cells of the preovulatory follicles, sperm cells (Krause and Dubocovich 1990; Morgan et al. 1994; Dubocovich 1995; Reppert et al. 1996). This wide distribution allows the explanation of the role of melatonin on cardiovascular rhythms and on gastrointestinal, endocrine, and immunological functions. In general, whereas MT_2 receptor is found primarily in the retina, MT_1 receptor is present in the SCN and the pars tuberalis of the pituitary gland (Sugden et al. 2004). They have been characterized as independent biochemical entities through their chemical structure and chromosomal location [4q35 (MT_1) and 11q21–22 (MT_2)]. In humans, MT_2 receptor is also expressed, in lower density, in the SCN, the regulatory center of rhythmical melatonin synthesis in the pineal gland. At this level, the role of both receptors is different; when melatonin acts on the MT_1 receptor, an inhibitory effect on activated neurons of SCN is observed, as well as the release of melatonin in the pineal gland, whereas the MT_2 receptor plays its role by controlling the sleep–wake cycle, being responsible for circadian phase changes of neuronal discharges (Kennedy 2007).

12.3.5 Circadian Rhythms and Melatonergic System

As previously noted, in terms of the integrity of the retina–SCN–pineal neural circuit, melatonin exhibits a typical rhythmic pattern of synthesis and secretion of form wherein daytime hormone plasma concentrations are low (10–20 pg/ml), whereas overnight concentrations show a significant increase (80–120 pg/ml), with a marked peak between midnight and 3:00 a.m. (Arendt 1995) (Figure 12.2). Similarly, a circadian pattern in the capacity of melatonin binding to its receptors in the central nervous system has been confirmed, at an experimental level, with a significant increase during the daylight period (Benloucif et al. 1997). More recently, it has also been demonstrated that other biochemical structures involved in the physiological mechanism of melatonin action are also subject to circadian oscillations; such is the case, for example, of mRNA levels for ML_{1A} melatonin receptor subtype in SCN (Neu and Niles 1997).

But melatonin is not the only biochemical pineal entity subject to a rhythmic pattern of secretion; other amines present in the gland, such as serotonin and noradrenaline, are also subject to this pattern. Serotonin synthesized by pinealocyte can be released into the extracellular space and be incorporated to the nerve terminal or serve as substrate to the formation of other pineal indoles, such as 5-methoxytryptamine (Miller and Maickel 1970) or 5-methoxytryptophol (McIsaac et al. 1964). But serotonin is also essential for the maintenance of sleep cycles in the rat, and the pineal gland regulates the release of serotonin in cycles of 12 h, which would mark the so-called circadian rhythms (Snyder and Axelrod 1964). Furthermore, noradrenaline contained in the granular vesicles of the sympathetic fibers that innervate the pineal gland also suffers a variation in its levels, in the same sense that melatonin does with the circadian rhythm (Wurtman and Axelrod 1966), which has spurred some authors (Klein 1969; Klein et al. 1971; Klein and Weller 1973) to ensure that the noradrenaline level in these fibers could stimulate the melatonin synthesis.

The effect of noradrenaline on pineal melatonin and serotonin synthesis is mediated by β-adrenergic receptors present in the cellular membrane of pinealocytes (Klein and Weller 1970; Wurtman et al. 1971; Deguchi and Axelrod 1972; Brownstein et al. 1973); so after administration of β-adrenergic antagonists, such as propranolol, glandular response to external stimuli (especially lighting) is unchanged (Deguchi 1973; Parfitt et al. 1976). Activation of β-adrenergic (and α-adrenergic) receptors increases intracellular concentration of cAMP, which induces an increase in the activity of NAT (Figure 12.5) (Klein et al. 1981; Reiter 1991). However, the presence of α-adrenergic receptors in the pineal gland, as well as a central innervation of peptidergic type, further complicates the exact knowledge of the biochemical mechanisms involved in the synthesis of melatonin and its circadian regulation (Claustrat et al. 2005).

12.3.6 MOLECULAR BIOLOGY OF CIRCADIAN RHYTHMS

For more than one decade, physiological studies, which have been essential for studying the phenotypic manifestations of most circadian rhythms, have led to the knowledge, at least in part, of the molecular mechanisms of hereditary transmission of the circadian rhythms in plants, *Drosophila*, mammals, and human beings (Ashkenazi et al. 1993).

Circadian rhythm oscillations are generated by the existence of intracellular feedback pathways, with positive and negative regulation elements, which affect the genetic transcription and posttranscriptional mechanisms (Figure 12.6). The positive regulation elements are three transcription factors: BMAL1 (brain and muscle ARNT-like protein 1), CLOCK (circadian locomotor output cycles kaput), and NPAS2 (neuronal-PAS-domain-containing protein 2; closely related to the CLOCK). The negative regulatory elements are the PER and CRY proteins, which are produced as a consequence of the activation of their *per* (period: Per1, Per2, Per3) and *cry* (cryptochrom: Cry1, Cry2) genes by the BMAL1–CLOCK or BMAL1–NPAS2 heterodimers. When they accumulate in excess, the PER and CRY proteins form heterotypic aggregates (PER/CRY), which are capable of inactivating the transactivation capacity of the BMAL1–CLOCK or BMAL1–NPAS2 heterodimers; and as a result, they self-repress their own *per* and *cry* genes. This leads to a deactivation of PER and CRY protein synthesis such that when these proteins reach levels lower than the necessary concentration so that the self-repression occurs, a new transcription cycle starts led by the positive elements BMAL1, CLOCK, and NPAS2 (Reppert and Weaver 2002; Levi and Schibler 2007).

The same positive transcription regulators (CLOCK, BMAL1, and NPAS2) and negative transcription regulators (CRY and PER proteins) also control the circadian expression of the orphan

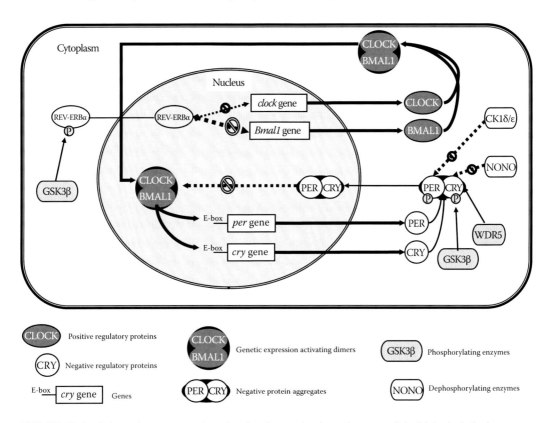

FIGURE 12.6 Schematic representation of molecular mechanisms that control the biological clock.

receptor REV-ERBα, which is a strong transcription repressor induced by the *Bmal1* gene and moderate repressor of the *clock* gene. This receptor alternates antiphasic cycles of transcription, both for negative and positive elements, which, despite not being indispensible, contribute to the solidity of the phase changes within the circadian rhythms. Other additional components seem necessary for the correct functioning of the biological clock, such as the WD-repeat-containing protein 5 platform, which serves as a link platform for histone methyltransferase; the NONO protein for binding to DNA and to RNA, which affects generation of the circadian rhythm in an unknown manner; and the casein kinases CK1ε/δ and CK2, which modulate stability and activity of the positive and negative elements. To date, *Bmal1* is the only gene whose inactivation is translated automatically because of a loss of circadian rhythms. All the other genetic components of the biological clock are represented in the genome by at least two isoforms, so certain alterations or mutations can functionally be replaced by the action of other genes (Levi and Schibler 2007).

These molecular events are also regulated by other mechanisms. Thus, the PER and CRY proteins are phosphorylated by various enzymes, among which we can highlight glycogen synthase kinase 3β, which influences its stability and the entry rate in the nucleus. Furthermore, the rate of formation and degradation of the BMAL/CLOCK heterodimer may be modified by various enzymes. These molecular regulators of the circadian system involve more than 20 genes and mutations that may influence the parameters of the circadian period. The multiple genes that are expressed in relation to the circadian rhythms, both in the SCN and at a level of peripheral clocks, seem to be regulated in a different manner, which permits specific tissue responses to endogenous or exogenous fluctuations. This may explain not only the presence of rhythmic alterations in certain pathological conditions but also the possibility of acting therapeutically on certain targets, such as glycogen synthase kinase 3β, which seems to be the case in lithium and other mood stimulators. Thus, in an animal model, a mutation in the *clock* gene is translated by behavior similar to mania, with hyperactivity, insomnia, and greater tendency to consume cocaine. Treatment with lithium restored normality in these animals (Roybal et al. 2007; Schulz and Steimer 2009).

12.4 ALTERATIONS OF CIRCADIAN RHYTHMS IN DEPRESSION

There are many studies centered on clarifying the role of circadian rhythms, from various perspectives, in depression. Below, we will discuss some approaches from the genetic perspective and others from the phenotypic perspective.

12.4.1 ALTERATIONS OF GENES RELATED TO CIRCADIAN RHYTHMS

In some disorders characterized by modifications of the circadian rhythms, for example, in conditions related to depression such as sleep disorders, it has been possible to detect genetic alterations of different characteristics. Thus, in a familiar condition characterized by a phase advance in sleep rhythm, it has been observed that there are different genetic mutations that alter the phosphorylation of the *per* gene by the casein kinases CKIε and CKIδ, or mutations of the actual *per* gene, which are translated phenotypically by a tendency to advance the start of sleep and waking earlier in the mornings (Mendlewicz 2009). Genetic mutations have likewise been described in conditions that include a phase delay in the sleep–wake rhythm. Ebisawa et al. (2001) observed that alterations in the *per3* gene, characterized by the repetition of four alleles (H4 haplotype), are accompanied by a late start in sleep together with waking up preferably during the day. More recently, Gourin et al. (2010) have shown that patients with a history of depression had higher Clock, Per1, and Bmal1 mRNA levels compared with nondepressed participants in this study.

However, at present, different disturbances of the genes have been described that can be associated with the susceptibility, recurrence, and typically circadian phenotypes observed in some affective conditions, including depression. Thus, a single nucleotide polymorphism of the *clock* gene has been detected that has been related to a greater rate of recurrence throughout life. In this work, it is speculated that the greatest recurrence is due to a circadian alteration characterized by a greater state of alertness in individuals on waking up (Benedetti et al. 2003). Likewise, other alterations of the CLOCK proteins have been described that individually does not justify the appearance of depression. However, when three or more of these alterations appear in a subject, it seems that the susceptibility to develop depressive conditions increases (Shi et al. 2008). Similarly, the *cry2* gene has been significantly associated with winter depression (Lavebratt et al. 2010b), and genetic variation in *per2* has been also associated with depression vulnerability (Lavebratt et al. 2010a). In addition, this genetic risk did not seem to require exposure to other risk factors for depression, such as sleep disturbance factors or negative life events.

12.4.2 PHENOTYPIC ALTERATIONS OF CIRCADIAN RHYTHMS IN DEPRESSION

Many of the diseases affecting human beings have time components, meaning that the vulnerability, the appearance of its signs and symptoms, as well as their recurrence may be manifested in a circadian manner. Thus, peaks of blood pressure have been described between noon and 6:00 P.M.; asthmatic crises are more frequent between midnight and 6:00 A.M.; precordial pain, the painful peak of rheumatoid arthritis, and the crisis of migraines are more frequent between 6:00 A.M. and noon (Smolensky 2001).

Depression can be considered a paradigmatic pathology of a relation between symptomatology and circadian disturbances. In fact, the alteration of mood and sleep rhythms is an integral part of affective disorders, both of major depressive disorder (MDD) and of bipolar disorder (Westrich and Sprouse 2010). There is evidence on the participation of this dysregulation in the genesis of depressive disorders (Pandi-Perumal et al. 2006; Srinivasan et al. 2006). Furthermore, circadian alterations in body temperature, cortisol secretion melatonin, thyrotropin, and monoamines have been detected in depressive patients in relation to healthy subjects.

In this sense, the cardinal symptom of depression, depressive mood, usually has a regular change throughout the day in depressed patients, which are usually accompanied in parallel with modifications in activity, attention capacity, level of anxiety, and alterations in the sleep rhythm (Albrecht 2010). It is normal that, in most patients with melancholic or typical depressive condition, there exist mood deterioration in the morning and improvement in the evening. However, in some subjects who have an "atypical" depression, frequently associated with a neurotic condition, anxiety, and atypical symptoms, the mood would deteriorate in the evening. These atypical patients respond more poorly to light therapy than those who show a morning deterioration (Germain and Kupfer 2008; Soria and Urretavizcaya 2009).

Likewise, it is known that positive emotions, such as enthusiasm, interest, and satisfaction, determined by an electronic outpatient control in real time throughout the day, have a phase delay in depressed patients. In methodological conditions similar to negative emotions, manifested by symptoms of anxiety, nervousness, tension, and culpability, there is a more pronounced daily rhythm and which is more variable than that observed in healthy subjects (Peeters et al. 2006).

From a functional perspective, it has been determined that patients with depression have different glucose consumption patterns in cerebral areas throughout the day in comparison with healthy individuals, who have an increase in cerebral glucose metabolism in the afternoon. The early evening improvement of mood, observed in patients with depression, seems to be associated with an increase in the functional activity of neuronal networks involved in affective control. Furthermore, patients with depression showed a sustained activity in the brain stem and hypothalamic regions, areas involved in the maintenance of the alert state throughout the day (Germain and Kupfer 2008).

In addition to mood, the sleep–wake rhythm is also affected in depression—the association of insomnia and depression not being arguable from a clinical standpoint—since subjects with insomnia are more likely to suffer from depression and depressive patients have, in more than 80% of cases, insomnia as a symptom (Armitage 2007; Ohayon 2007). In depressive patients, sleep is abnormal (Figure 12.7), with a shortened latency and an accumulation of REM in the initial phases of the sleep, with fragmentation of sleep, and with advanced morning wakening (Nutt et al. 2008; Schulz and Steimer 2009). These data are interpreted as a phase advance of the circadian system. Furthermore, the presence of these sleep disorders is a prodromic factor and risk factor of the appearance of recurrences in subjects with MDD (Soria and Urretavizcaya 2009). Likewise, in

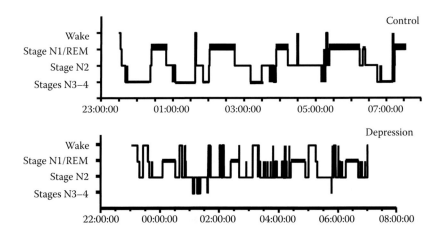

FIGURE 12.7 Circadian alterations (polysomnographic measures) of sleep–wake rhythm in depressive patients vs. control.

depression, there is also a tendency that a phase advance of other biological rhythms occurs, such as body temperature, which is high at night in depressive patients compared with the flattening out in healthy subjects (Azorin and Kaladjian 2009). Furthermore, variations in the secretion rhythm of cortisol and noradrenaline have been described, characterized by a phase advance, in some subtypes of depression (Linkowski et al. 1987; Koenigsberg et al. 2004).

However, in depression, a nighttime flattening out of the thyrotropin and growth hormone secretion has been observed (Soria and Urretavizcaya 2009). Furthermore, the most consistent finding that links melatonin with depression is the nighttime decrease in the concentration of the said hormone in these patients, as well as the alterations in its secretion rhythm (Claustrat et al. 1998; Crasson et al. 2004; Srinivasan et al. 2006; Van Reeth and Maccari 2007). Similarly, it has been confirmed that nocturnal melatonin levels are significantly lower in patients with chronic primary insomnia (Hajak et al. 1995), as well as urinary 6-sulfatoxymelatonin concentrations in elderly patients with insomnia (Haimov et al. 1994). Although these findings have not always been confirmed in other studies, it is probable that this is due to the complexity and subtypes of depressed patients.

12.5 CIRCADIAN AND MELATONERGIC HYPOTHESES OF DEPRESSION

Based on these observations, several hypotheses have been proposed that involve circadian rhythms and melatonin in depression. The hypotheses of depression phase displacement indicate that mood disturbances occur as a consequence of an advance or delay in the phase of the pacemaker or central biological clock, which would alter the circadian rhythms that regulate temperature, cortisol secretion, melatonin secretion, and the rhythm of REM sleep, in relation to other circadian rhythms, and with an important phase displacement of the sleep–wake rhythm.

In this sense, there are data indicating a phase advance of some parameters in patients with depression, such as early morning wakening, early appearance of REM sleep phases with respect to the start of nonparadoxical sleep, as well as changes in melatonin secretion, in comparison with nondepressed individuals. These modifications appear more evident in seasonal depressions (Germain and Kupfer 2008). An alternative model, called internal phase coincidence, postulates that depression occurs due to the coincidence of wakening with a sensitive phase of the circadian period (Borbély and Wirz-Justice 1982). However, probably as a consequence of the complexity of depression, not all studies support these theories. Nevertheless, the hypotheses of phase advance, independent of the therapeutic results of light therapies, have stimulated the development and research of interventions based on circadian principles (Germain and Kupfer 2008).

Another circadian hypothesis of depression is based on observation of the reduction in the latency of REM sleep in depression, together with an improvement in the depressive condition associated to the suppression of REM sleep, whether pharmacologically or behaviorally. However, the suppression of REM sleep is not always necessary to obtain a mood improvement (Argyropoulos and Wilson 2005).

However, it has been suggested that the apparent intensification of REM sleep in depression is due to a deficiency and less activity of the slow sleep waves. In accordance with this hypothesis, the antidepressant effects of sleep deprivation may be attributed to an improvement in the short wave process, whereas the relapse of depression after the recovery of sleep would be due to a return to initial abnormal levels of short waves (Borbély and Wirtz-Justice 1982). However, against this hypothesis is the fact that antidepressant drugs do not improve and may even further reduce the characteristic waves of slow sleep (Germain and Kupfer 2008).

The hypothesis that places emphasis on the interruption of social rhythms and the associated physiological changes in the etiology of depression suggests that vulnerable people show more severe disturbances of their circadian rhythms and of the sleep rhythm with the interruption of their social rhythms. According to this hypothesis, greater importance is usually given to the interruption of nonphotic *zeitgebers*, which normally modulate certain physiological circadian rhythms, as triggers of depressive episodes. In fact, different studies have demonstrated that in patients with mood and anxiety disorders, as well as in persons with stressful life events, the social rhythms are interrupted or have less regularity. However, a greater regularity of social rhythms is associated with a greater amount of sleep and reduces the severity of depressive symptoms. However, despite these data, trials reporting a direct relation between interruption of social rhythms and disturbance of physiological rhythms in depression are limited (Germain and Kupfer 2008).

12.5.1 Melatonergic Hypotheses of Depression

With respect to melatonin, the observation that this hormone was capable of increasing the levels of serotonin in the pineal gland allowed consideration of the hypothesis of its possible usefulness in the treatment of affective disorders, as the serotonergic theory of depressions had already been put forward at the end of the 1960s, postulating a deficit in this indolamine in the synaptic gap in the serotonergic pathways of patients with mood disorders (Lapin and Oxenkrug 1969). Similarly, some authors felt, as previously noted, that melatonin secretion may be considered as an indicator of noradrenergic activity in depressive patients, a consistent event with the so-called catecholaminergic hypothesis of depression, proposed by Joseph J. Schildkraut in 1965 (Schildkraut 1965).

Anton-Tay et al. (1971) suggested that this hormone might have an antidepressive effect, an argument reinforced by the observation of the euphoric effect caused by melatonin after administration to healthy volunteers. Continuing with this research line, by the end of the 1970s, Wetterberg et al. (1979) found a reduction in the plasma levels of melatonin in depressive subjects (Figure 12.2) and a specific reduction of nighttime values in subjects with suicidal ideas, as well as normalization after spontaneous recovery or after antidepressant treatment, whereas Mendlewicz et al. (1979) described the rhythms for the synthesis of melatonin in depressive patients, highlighting the absence of the nocturnal peak in melatonin characteristic of normal subjects (Figure 12.2). A hypothesis known as "low melatonin syndrome" was even proposed, whereby a low secretion of melatonin would constitute a biological marker for greater susceptibility to suffer a depressive disorder (Wetterberg et al. 1979). However, against this hypothesis, other authors have described an increase in nighttime levels of melatonin in patients with MDD (Rubin et al. 1992; Sekula et al. 1997; Szymanska et al. 2001). In any case, alterations in the normal rhythm of melatonin in almost all affective disorders have been observed, as in MDD (Beck-Friis 1985), in affective seasonal disorder (Teicher et al. 1997), premenstrual dysphoric disorder (Parry 1997), or bipolar disorder (Kennedy et al. 1996). In relation to bipolar disorder, Lewy et al. (1979) observed how low levels of melatonin during depressive bouts in bipolar patients changed to an increase in the hormone's secretion during the manic episodes, which has led some authors to consider whether this kind of affective disorder might be characterized by an internal desynchronization of biological rhythms (Srinivasan et al. 2006), as an alteration in hormonal secretion was also verified in other neuroendocrine systems, particularly cortisol, and an alteration in body temperature and, of course, in the sleep (Figure 12.2).

From the neuroendocrinological perspective, the melatonin/cortisol ratio is an interesting indicator in depressive disorders, insofar as the reduction in melatonin levels is associated with an increase in cortisol levels, suggesting a close interrelationship between the hypothalamus–hypophysis–adrenal axis and the melatonergic system in the etiopathogeny of depression. This connection further reaffirmed the clinical hypotheses that were put forward since the 1970s with respect to the link between affective disorders and different endocrine pathologies (Whybrow and Hurwitz 1976).

Additionally, Wirz-Justice and Arendt (1980) confirmed experimentally that antidepressant drugs were capable of modifying the levels of pineal hormone, which gave further support to the idea that the melatonergic system might play a role in the etiology of mood disorders. The use of antidepressants, among other factors, modifies the sensitivity of β-adrenergic receptors, suppressing nocturnal release of melatonin (Heydorn 1982; Thompson 1985). An increase in melatonin secretion has been seen with the chronic administration of tricyclic antidepressants, such as imipramine (Venkoba et al. 1983) or demethylimipramine (Thompson et al. 1985). Similarly, an increase in urinary excretion of 6-sulfatoxymelatonin has been also demonstrated in patients treated with this type of antidepressants, such as imipramine (Sack and Lewy 1986) or desipramine (Golden et al. 1988). In the case of monoamine oxidase inhibitors (MAOIs), it has been found that these agents can also cause an increase in the levels of melatonin, to increase NAT activity (Lewis et al. 1990).

Giving even more support to the implication of the melatonergic system in the genesis of depression, experimental trials with *knockout* mice for MT_1 receptor have confirmed that these animals, which do not have this receptor, behave similarly to those subjected to experimental models of depression (Weil et al. 2006).

Bearing these facts in mind, it should be noted that the therapeutic use of melatonin has been considered (Srinivasan et al. 2006; Sánchez-Barceló et al. 2010). In fact, some studies provide data in favor of the efficacy of melatonin in sleep disorders in adult patients (Zhdanova et al. 1995), in the elderly (Haimov et al. 1995), as well as in children with different neurological disorders (Jan and O'Donnell 1996). However, other studies reveal the lack of consistent results on the amount of total sleep or the decrease in its latency. Likewise, the effects of the exogenous administration of melatonin have been studied in different affective disorders, fundamentally in MDD, in seasonal affective disorder, and in bipolar disorder, although the results have also been contradictory (Pacchierotti et al. 2001; Srinivasan et al. 2006). With regard to the seasonal affective disorder, the exogenous melatonin administration has no effect on depressive symptoms (Wirz-Justice et al. 1990) or even cause worsening of symptoms, suggesting the existence of other relevant factors in the specific etiopathogenesis of this disorder, such as a dysfunction of various neurotransmitters or an increase in the photosensitivity of the retina (Pacchierotti et al. 2001). Furthermore, with respect to bipolar disorder, Leibenluft et al. (1997) noted that the administration, in the morning, of 5 to 10 mg of exogenous melatonin stabilized the endogenous rhythm of melatonin secretion but has no therapeutic effects in the patients studied.

Looking at all the factors discussed, we can conclude that the hypothesis of the origin of depression, only in a melatonergic dysfunction, seems clearly a simplistic and reductionist approach. For example, Paparrigopoulos et al. (2001) reported that the administration of clonidine, an α_2-adrenergic partial agonist, results in a decrease in the levels of melatonin in depressive subjects but not in healthy volunteers. This observation suggests that depressive symptoms might be due to a phenomenon of hypersensitivity of these receptors, which would result in a reduction of the release of noradrenaline and, secondarily, of melatonin at the pineal level. In this sense, some authors, bearing in mind the variability of these pathophysiological events as well as the variable response to therapeutic use of melatonin, estimate that alterations of the melatonin secretion rhythm observed in subjects with mental disorders can essentially be an epiphenomenon associated with a concomitant dysfunction of different neurotransmission systems (noradrenergic and/or serotonergic systems) in the central nervous system, since the synthesis and secretion of melatonin in the pineal gland is controlled, as mentioned, by several neurotransmitters (Pacchierotti et al. 2001).

12.6 RESYNCHRONIZATION OF CIRCADIAN RHYTHMS IN THERAPY OF DEPRESSION

In accordance with alterations in the biological rhythms described in depression, it seems acceptable, with the logical methodological and clinical limitations, to consider the circadian system to be an important therapeutic target in the therapeutic approach to depression. In this sense, various therapeutic strategies have been developed, designed to resynchronize the biological clock, to adjust the circadian rhythms that appear in depression back to a normal cycle, with the consideration that this would also improve the depressive condition.

Thus, different nonpharmacological interventions have been developed as an alternative treatment or as a support therapy to the use of psychotropic drugs. Among these therapies, we can highlight sleep deprivation and light therapy, together with interpersonal therapies aimed at improving social rhythm. From a pharmacological perspective, it is very well known that antidepressants, irrespective of their mechanisms of action, modify the sleep parameters. Thus, it has been observed

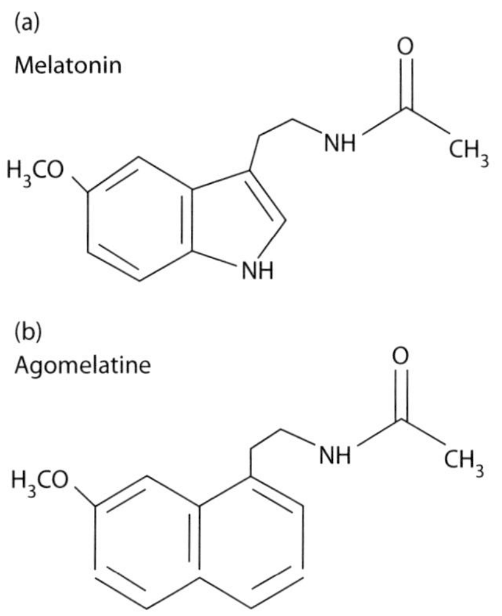

FIGURE 12.8 Comparative chemical structures between (a) melatonin and (b) agomelatine.

that tricyclic antidepressants generally shorten sleep latency and, although they may improve continuity of sleep in patients with depression, they usually cause daytime sleepiness. In this sense, most tricyclics increase the latency of REM sleep and reduce its percentage, tending to normalize the architecture of altered sleep in patients with depression. However, both with MAOIs and with selective serotonin reuptake inhibitors, or with the dual serotonin and noradrenaline reuptake inhibitors, insomnia production is presented as a common characteristic. Although on occasions these sleep modifications are beneficial, in others these are considered adverse effects (Álamo et al. 2005). Generally, it should be pointed out that there is little evidence on the fact that the circadian processes mediate the antidepressant effects of these agents (Germain and Kupfer 2008).

However, it seems that changes may occur in the next few years. A milestone in this sense may come from melatonergic agonists, such as agomelatine (Figure 12.8). This agent, initially named S-20098, is the first antidepressant with proven clinical efficacy, which tackles the treatment of depression from a different pharmacological perspective to that of the other pharmacological agents used to date. The primary mechanism of this agent is its agonist action on melatonin receptors MT_1 and MT_2, and it also has the capacity of resynchronizing the "biological clock" (Álamo et al. 2008).

12.7 MELATONERGIC AGONIST DRUGS AND DEPRESSION

There is much theoretical interest in melatonin from the physiological standpoint. However, from the therapeutic perspective, there is little interest because of several factors, such as its short half-life, its fast metabolization, which gives it low bioavailability by oral route, its high interindividual variability in the response, as well as its lack of receptorial selectivity (Guardiola-Lemaitre 2005). Finally, it must be pointed out that melatonin is a natural substance, not subject to commercial patent.

For these reasons, and considering the interest in the mechanism of action of melatonin in the treatment of depressed patients, analogues of this hormone have been sought for the treatment of depressed patients, which may act through this innovative pathway (Pandi-Perumal et al. 2007). The first drug of this kind, with exclusively melatonergic activity, that has reached clinical practice was ramelteon (a molecule synthesized by Takeda Pharmaceuticals of North America in 1996), a powerful agonist specifically for MT_1 and MT_2 receptors that was authorized by the Food and Drug Administration in 2005 for therapeutic use in primary insomnia. Agomelatine (Figure 12.8), an agonist of MT_1/MT_2 receptors and a selective antagonist of $5\text{-}HT_{2C}$ receptors (Dubocovich and Markowska 2005; Pandi-Perumal et al. 2006; Álamo et al. 2008), was synthesized in 1991 by Adir & Co., a subsidiary of Laboratoires Servier. It has since been marketed as an antidepressant drug with the authorization of the European Medicines Agency since 2008 and has confirmed its antidepressive efficacy in clinical practice through both short-term and long-term studies (Loo et al. 2002; Álamo et al. 2008; Kennedy and Rizvi 2010). This antidepressive efficacy has been shown independently of the equally demonstrable efficacy of the drug for controlling the sleep alterations, fundamentally insomnia, that has been frequently observed in these patients (Álamo et al. 2008).

Agomelatine is a powerful agonist of the high-affinity melatonin receptors MT_1 and MT_2 ($K_i = 6.15 \times 10^{-11}$ and 2.68×10^{-10} M, respectively); its capacity for binding to the receptor is comparable to that of melatonin ($K_i = 8.52 \times 10^{-11}$ and 2.63×10^{-10} M, respectively) (Ying et al. 1996; Álamo et al. 2008). Stimulation of melatonergic receptors, belonging to the superfamily of G-protein-coupled receptor, by melatonin or agomelatine, produces a decrease in cAMP and cAMP response element–binding, which modifies the expression of early genes such as *c-fos* and *jun-B*. Furthermore, it inhibits the activity of protein kinases activated by various mitogenic factors (MAPkinase, $ERK_{1/2}$, $MEK_{1/2}$) and activates phospholipase C, which stimulates protein kinase C, facilitating the intracellular entry of calcium. Furthermore, the stimulation of MT_1 receptors causes the stimulation of hyperpolarizing potassium currents through Kir3 channels coupled to the PGi. However, MT_2 receptors inhibit the accumulation of cyclic guanosine monophosphate. This data set suggests the existence of a large number of cellular responses induced by melatonin, hence imitated by agomelatine, after acting on their high-affinity receptors (Guardiola-Lemaitre 2005; Álamo et al. 2008). The second pharmacological characteristic of agomelatine is its capacity for blocking $5\text{-}HT_{2C}$ receptors. This blockade eliminates the serotonergic brake, facilitating the release of dopamine and noradrenaline in the prefrontal cortex. This increase in monoamines in the prefrontal cortex constitutes an important characteristic that also contributes to the antidepressant effect (Millan et al. 2000) and may be related to its anxiolytic profile (Stein et al. 2007; Millan et al. 2005).

Agomelatine has demonstrated efficacy in different animal models of depression (Pandi-Perumal et al. 2006; Zupancic and Guilleminault 2006; Popoli 2009), such as chronic stress, forced swimming, or learned helplessness models, including tests for circadian rhythms and of stress/anxiety. Whereas melatonin has a positive action only in models that involve circadian rhythms and antagonists of $5\text{-}HT_{2C}$ receptors have action only on anxiety models, agomelatine offers positive results in all model types. Furthermore, in many of the experimental models of depression, agomelatine is effective irrespective of the time when it is administered, which indicates that it has an antidepressant effect regardless of resynchronization of biological rhythms (Popoli 2009).

12.7.1 Agomelatine: Resynchronizer of Biological Rhythms

Because of its properties as a melatonergic receptor agonist, agomelatine can be considered an agent capable of synchronizing distorted rhythms. Thus, in rats subjected to forced darkness, the administration of agomelatine half an hour before the darkness phase restores the nocturnal activity rhythm of the animal (Martinet et al. 1996). This effect is similar to that obtained with melatonin (Armstrong et al. 1993). Likewise, agomelatine behaved similarly to melatonin in the recovery of circadian rhythm, body temperature, and locomotor activity in different experimental models in rats (Redman et al. 1995; Martinet et al. 1996; Pitrosky et al. 1999). However, both melatonin and agomelatine, administered after 18 h, advance the circadian rhythm of nighttime body temperature and heartbeat in young humans. This phase advance is maintained the day after treatment (Kräuchi et al. 1997).

The administration of agomelatine to hamsters of advanced age allows them to recover the phase change before dark, in a similar manner as in young animals (Van Reeth et al. 2001). Likewise, in the elderly, administering agomelatine at nightfall produces an advance in rhythm of approximately 2 h, similar to that observed in young subjects (Leproult et al. 2005). These experiments confirm the chronobiotic capacity of agomelatine in young animals and may explain their use in the depressed elderly (Popoli 2009).

As a whole, these studies demonstrate the capacity of agomelatine to resynchronize altered rhythms, as well as its property to produce a cycle advance. This chronobiotic effect is produced by agomelatine via its agonist action on the melatonergic receptors, both MT_1 and MT_2, located in the SCN (Zupancic and Guilleminault 2006), and its antagonist action of the 5-HT_{2C} receptor in this nucleus (Millan et al. 2003). The first effect is more prolonged than that observed with melatonin, which attributes a greater agonist effect of the drug on the MT_1 receptors (Ying et al. 1996). These MT_1 receptors have a very high expression in the SCN and are related to enhancing the GABAergic activity, which decreases neuronal activity in this nucleus (Pandi-Perumal et al. 2006). Some authors have proposed the term *rhythm stabilizing antidepressants* to designate these pharmacological agents able to restore synchronization of the various body rhythms while not inducing or worsening desynchronization (Fountoulakis 2010).

12.7.2 Agomelatine and Neuroplasticity as Mechanism of Antidepressant Action

For some years now, new theories have been elaborated that attempt to explain how the antidepressant agents act beyond their synaptic actions. In this sense, it is considered that the initial actions of antidepressants do no more than act on switches that activate more complex intracellular and extrasynaptic mechanisms that, if we can use an electronic simile, place the "hard disk" in charge of the antidepressant effect into action.

In this way, as we have discussed, agomelatine, "switching on" the melatonergic receptors, modifies cAMP response element–binding levels, inhibits the activity of protein kinases sensitive to various mitogenic agents, activates phospholipase C, and increases intracellular levels of calcium, while it stimulates hyperpolarizing currents of potassium. In contrast, "switching off" the 5-HT_{2C}, since it acts as an antagonist, cancels out the activity of the Gq/11 and Gi_3 proteins and inhibits phosphoinositol depletion. We can consider that changes in the neuronal hard disk may influence the physiopathology of depression.

Indeed, in the physiopathology of depression, especially in those that progress with continued stress, three factors stand out, which have greater significance (Figure 12.9). On one hand are the hyperfunction of the hypothalamus–hypophysis–adrenal axis, which maintains high cortisol levels, and the increase of excitatory amino acid release, which would behave as neurotoxic factors. On the other hand is the deficit of neurotrophins, particularly the brain-derived neurotrophic factor (BDNF), which disappears as a protective factor. Therefore, in the physiopathology of depression, the two first factors—cortisol and excitatory amino acids—would act in a harmful manner on neuronal survival and neuroplasticity, whereas the deficit of neurotrophins would not counteract the first (Álamo et al. 2009).

Bearing these paradigms in mind, it is interesting to highlight the effects of agomelatine. In this sense, psychosocial stress in subordinate howler monkeys produces, in addition to a desynchronization in the circadian rhythms, a significant increase in urinary cortisol. The administration of agomelatine in prolonged treatment (4 weeks), despite maintaining the stress situation, means that the cortisol levels normalize and return to a level before the stressful situation. In parallel, agomelatine resynchronizes body temperature in these animals. We should highlight that an antagonist of 5-HT_{2C} receptors (S32006) had a tendency to decrease cortisol levels without modifying temperature alterations. These facts seem to indicate that the efficacy of agomelatine in the resynchronization of biological rhythms, therefore in their antidepressant properties, is due to its actions on melatonin and 5-HT_{2C} receptors (Fuchs et al. 2008).

The second aggressive factor discussed would be the neurotoxicity produced by excitatory amino acids, particularly due to the action of glutamate mediated by *N*-methyl-d-aspartate (NMDA) receptors. Agomelatine lacks the capacity to block these receptors (Gressens et al. 2008), although it could decrease glutamate release up to 30%, which would translate into less excitotoxic effect (Bonanno

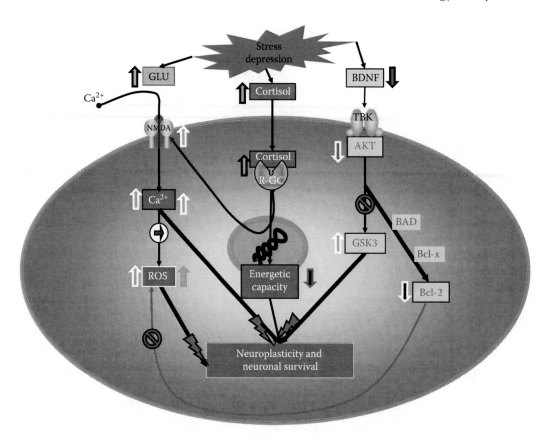

FIGURE 12.9 Schematic representation of cellular and molecular mechanisms involved in physiopathology of stress/depression.

et al. 2005). Recent studies indicate that the administration of agomelatine (40 mg/kg, 2 weeks) completely inhibits the release of glutamate, induced by electric stress, in the frontal and prefrontal cortices of rats. This inhibitory effect may be due to a decrease in complexin, a protein necessary for the fusion of the synaptic vesicle and, as a consequence, glutamate release. Likewise, this effect was not observed with the administration of said 5-HT$_{2C}$ agonist (S32006) nor with melatonin, which seems to indicate that the inhibitory action of the excitotoxity induced by agomelatine is attributable to its joint properties on both types of melatonergic and serotonergic receptors, which would be translated by a decrease in stress-induced glutamate release (Mallei et al. 2009).

The influence of agomelatine (40 mg/kg, 3 weeks) has been studied in rats on different cellular and molecular parameters related to neurogenesis. In these conditions, agomelatine significantly increased BDNF levels in the hippocampus by 20%. Furthermore, agomelatine increased the ERK$_{1/2}$ (protein kinase activated by mitogens), Akt (protein kinase B), and GSK3β (glycogen synthetase kinase-3β) signaling pathways by 91%, 45%, and 45%, respectively. It should be highlighted that these signal transduction pathways are involved in the control of neuronal proliferation and survival (Figure 12.9) and are also modulated by antidepressant and mood-stabilizing drugs. Furthermore, it is important to indicate that these molecular changes go hand in hand with an increase in cell proliferation of 39% in the ventral hippocampus, induced by agomelatine administration. These effects, as occurred in the experimental models of antidepressants, are due to the joint melatonergic and 5-HT$_{2C}$ antagonist properties exhibited by the agomelatine, since the melatonin or 5-HT$_{2C}$ agonists lacked the said effects (Soumier et al. 2008).

Calabrese et al. (2009) studied the influence of acute administration of agomelatine on the genetic expression of mRNA of BDNF and the gene related to the protein-associated cytoskeleton activity

(*Arc*). More effects were observed in the prefrontal cortex, in which a 46% decrease in the mRNA gene of BDNF was observed in the controls, an effect that was not observed in the animals treated with agomelatine, neither for *BDNF* gene nor for *Arc*. Again, these modifications in the neurological factors observed with agomelatine seem due to its joint properties on melatonergic and serotonergic receptors.

The results seem to reveal that in the antidepressant activity of agomelatine, its receptorial action as MT_1 and MT_2 receptor agonist is fundamental, as is its $5\text{-}HT_{2C}$ receptor antagonist capacity. Likewise, it is interesting to highlight that the capacity of joint action sets in motion transynaptic mechanisms responsible for a neuroprotector effect, thanks to its capacity to decrease the negative effects of cortisol and glutamate and to enhance the neuroprotector effects, fundamentally via BDNF.

12.8   CONCLUSIONS

The control of circadian rhythms is complex and involves the participation of multiple molecular, biochemical, physiological, and behavioral mechanisms. This complexity is necessary for the adaptability and survival of human beings in different environments, but it is also the basis of potential internal conflicts. In today's society, where artificial light is available at all times and the working hours and social habits take place throughout the day, the sleep and wake periods and other circadian rhythms are extremely irregular, which may make them vulnerable. There are multiple arguments, some of which are dealt with in this chapter, that converge and support the hypothesis that disorders of different circadian rhythms, particularly those relating to sleep and hormonal secretions, related to melatonergic system, may play a decisive role in the physiopathology of depression (and other psychopathological conditions).

Recent progress in the understanding of molecular and cellular mechanisms responsible for control of the biological clock open new pathways to research to clarify the basis of the relations between circadian rhythm disorders and depression. From the pharmacological perspective, we know that antidepressant agents that exclusively act through monoaminergic functionalism lack efficacy in a substantial part of depressive patients. In this sense, melatonergic agonists, such as agomelatine, which combines a potent activity as a selective agonist of MT_1/MT_2 receptors with its antagonist capacity of $5\text{-}HT_{2C}$ serotonergic receptors, have shown a great antidepressive efficacy. Numerous experimental studies indicate that the synergy between actions on both types of receptors is vital in antidepressants, resynchronizing of biological rhythms, and neuroprotective activities of agomelatine. Moreover, the results of preclinical studies on animal models show that other melatonin receptor agonists can be considered promising agents for the treatment of depression (Mor et al. 2010). All these events reinforce the hypothesis of direct involvement of melatonergic system in the etiology of depressive disorders.

REFERENCES

Álamo, C., E. Cuenca, and F. López-Muñoz. 2005. Yatrogenia medicamentosa: Reacciones adversas a los psicofármacos. In *Tratado de Psiquiatría*, ed. J. Vallejo, and C. Leal, 1875–900. Barcelona: Ars Médica.

Álamo, C., F. López-Muñoz, and M.J. Armada. 2008. Agomelatina: Un nuevo enfoque farmacológico en el tratamiento de la depresión con traducción clínica. *Psiq Biol* 15:125–39.

Álamo, C., F. López-Muñoz, and P. García-García. 2009. Trastornos del estado de ánimo. In *Principios de fisiopatología para la atención farmacéutica. Modulo IV. Plan Nacional de Formación Continuada*, 97–138. Madrid: Consejo General de Colegios Oficiales de Farmacéuticos.

Albrecht, U. 2010. Circadian clocks in mood-related behaviors. *Ann Med* 42:241–51.

Ángeles, M., K. Rodríguez, R. Salgado, and C. Escobar. 2007. Anatomía de un reloj (anatomía del sistema circadiano). *Arch Mex Anat* 2:15–20.

Anton-Tay, F., J. Diaz, and A. Fernández. 1971. On the effect of melatonin upon human brain. Its possible therapeutic implications. *Life Sci* 10:841.

Arendt, A. 1995. *Melatonin and the Mammalian Pineal Gland*. London: Chapman & Hall.

Argyropoulos, S. V., and S. J. Wilson. 2005. Sleep disturbances in depression and the effects of antidepressants. *Int Rev Psychiatry* 17:237–45.

Ariëns-Kappers, J. A. 1960. The development, topographical relations and innervation of the epiphysis cerebri in the albino rat. *Z Zellforsch* 52:163–215.

Armitage, R. 2007. Sleep and circadian rhythms in mood disorders. *Acta Psychiatr Scand Suppl* 433:104–15.

Armstrong, S. M., O. M. McNulty, B. Guardiola-Lemaitre, and J. R. Redman. 1993. Properties of the melatonin agonist S 20098 (agomelatine) and melatonin in an animal model of delayed sleep-phase syndrome (DSPS). *Pharmacol Biochem Behav* 46:45–9.

Ashkenazi, I. E., A. Reinberg, A. Bicakova-Rocher, and A. Ticher. 1993. The genetic background of individual variations of circadian rhythm periods in healthy human adults. *Am J Hum Genet* 52:1250–9.

Axelrod, J., and H. Weissbach. 1960. Enzymatic *O*-methylation of *N*-acetylserotonin to melatonin. *Science* 131:1312.

Azorin, J. M., and A. Kaladjian. 2009. Depression et rythmes circadiens. *L'Encephale* 35:S68–S71.

Beck-Friis, J. 1985. Serum melatonin in relation to clinical variables in patient with major depressive disorder and a hypothesis of a low melatonin syndrome. *Acta Psychiatr Scand* 71:319–30.

Benedetti, F., A. Serretti, A., C. Colombo, et al. 2003. Influence of CLOCK gene polymorphism on circadian mood fluctuation and illness recurrence in bipolar depression. *Am J Med Genet B Neuropsychiatr Genet* 123B:23–6.

Benitez-King, G., A. Ríos, A. Martínez, and F. Antón-Tay. 1996. In vitro inhibition of Ca^{2+}/calmodulin-dependent kinase II activity by melatonin. *Biochim Biophys Acta* 1290:191–6.

Benloucif, S., M. Masane, M. Dubocovich. 1997. Responsiveness to melatonin and its receptor expression in the aging circadian clock of mice. *Am J Physiol* 273:1855–60.

Bonanno, G., R. Giambelli, L. Raiteri, et al. 2005. Chronic antidepressants reduce depolarization-evoked glutamate release and protein interactions favoring formation of SNARE complex in hippocampus. *J Neurosci* 25:3270–9.

Borbély, A. A., and A. Wirz-Justice. 1982. Sleep, sleep deprivation and depression. A hypothesis derived from a model of sleep regulation. *Hum Neurobiol* 1:205–10.

Brownstein, M. J., J. Saavedra, R. Holz, and J. Axelrod. 1973. The control of indole metabolism in the rat pineal gland by a β-adrenergic receptor. *Fed Proc* 32:695.

Cagnacci, A. 1996. Melatonin in relation to physiology in adult humans. *J Pineal Res* 21:200–13.

Calabrese, F., R. Molteni, S. Pisoni, et al. 2009. Synergic mechanisms in the modulation of BDNF and Arc following agomelatine administration. *Eur Neuropsychopharmacol* 19(3):S442.

Chandrashekaran, M. K. 1998. Biological rhythms research: A personal account. *J Biosci* 23:545–55.

Clarke, C. F., A. M. Fogelman, P. A. Edwards. 1984. Diurnal rhythm of rat liver mRNAs encoding 3-hydroxy-3-methylglutaryl coenzyme A reductase. Correlation of functional and total mRNA levels with enzyme activity and protein. *J Biol Chem* 259:10439–47.

Claustrat, B., J. Brun, and G. Chazot. 2005. The basic physiology and pathophysiology of melatonin. *Sleep Med Rev* 9:11–24.

Claustrat, B., J. Brun, M. Geoffriau, and G. Chazot. 1998. Melatonin: From the hormone to the drug? *Restor Neurol Neurosci* 12:151–7.

Crasson, M., S. Kjiri, A. Colin, et al. 2004. Serum melatonin and urinary 6-sulfatoxymelatonin in major depression. *Psychoneuroendocrinology* 29:1–12.

Deguchi, T. 1973. Role of beta adrenergic receptor in the elevation of adenosine cyclic 3′,5′-monophosphate and induction of serotonin *N*-acetyltransferase in rat pineal glands. *Mol Pharmacol* 9:184–90.

Deguchi, T., and J. Axelrod. 1972. Control of circadian change of serotonin *N*-acetyltransferase activity in the pineal organ by the β-adrenergic receptor. *Proc Natl Acad Sci U S A* 68:3106–9.

Drucker-Colin, R., R. Aguilar-Roblero, F. Garcia-Hernandez, F. Fernandez-Cancino, and F. Bermudez-Rattoni. 1984. Fetal suprachiasmatic nucleus transplants; diurnal rhythm recovery of lesioned rats. *Brain Res* 311:353–7.

Dubocovich, M. L. 1988. Pharmacology and function of melatonin receptors. *FASEB J* 2:2765–73.

Dubocovich, M. L. 1995. Melatonin receptors: Are there multiple subtypes? *Trends Pharmacol Sci* 16:50–6.

Dubocovich, M. L., and M. Markowska. 2005. Functional MT1 and MT2 melatonin receptors in mammals. *Endocrine* 27:101–10.

Dubocovich, M., and J. Takahashi. 1987. Use of 2-[^{125}I]-iodomelatonin to characterize melatonin binding sites in chicken retina. *Proc Natl Acad Sci U S A* 84:3916–20.

Ebisawa, T., M. Uchiyama, N. Kajimura, et al. 2001. Association of structural polymorphisms in the human period3 gene with delayed sleep phase syndrome. *EMBO Rep* 2:342–6.

Fiske, V. M., G. K. Bryant, and J. Putnam. 1960. Effect of light in the weight of the pineal in the rat. *Endocrinology* 66:489–91.

Fountoulakis, K. N. 2010. Disruption of biological rhythms as a core problem and therapeutic target in mood disorders: The emerging concept of 'rhythm regulators'. *Ann Gen Psychiatry* 9:3. doi: 10.1186/1744-859X-9-3.

Fuchs, E., S. Corbach-Söhle, B. Schmelting, et al. 2008. Effects of agomelatine and S32006, a selective 5HT2C receptor antagonist, in chronically-stressed tree shrews. *Eur Neuropsychopharmacol* 18(4):S348.

Germain, A., and D. J. Kupfer. 2008. Circadian rhythm disturbances in depression. *Hum Psychopharmacol* 23:571–85.

Golden, R., S. Markey, E. Risby, M. V. Rudorfer, R. W. Cowdry, and W. Z. Potter. 1988. Antidepressants reduce whole-body norepinephrine turnover while enhancing 5-hydroxymelatonin output. *Arch Gen Psychiatry* 45:150–4.

Gordon, C., and S. M. Reppert. 1997. The Mel1a melatonin receptor is coupled to parallel signal transduction pathways. *Endocrinology* 138:397–404.

Gourin, J. P., J. Connors, J. K. Kiecolt-Glaser, et al. 2010. Altered expression of circadian rhythm genes among individuals with a history of depression. *J Affect Disord* 126:161–6.

Gressens, P., L. Schwendimann, I. Husson, et al. 2008. Agomelatine, a melatonin receptor agonist with 5-HT(2C) receptor antagonist properties, protects the developing murine white matter against excitotoxicity. *Eur J Pharmacol* 588:58–63.

Guardiola-Lemaitre, B. 2005. Agonistes et antagonistes des récepteurs mélatoninergiques: Effets pharmacologiques et perspectives thérapeutiques. *Ann Pharm Fr* 63:385–400.

Haimov, I., M. Laudon, N. Zisapel, et al. 1994. Sleep disorders and melatonin rhythms in elderly people. *Br Med J* 309:167.

Haimov, I., P. Lavie, M. Laudon, P. Herer, C. Vigder, and N. Zisapel. 1995. Melatonin replacement therapy of elderly insomniacs. *Sleep* 18:598–603.

Hajak, G., A. Rodenbeck, J. Staedt, B. Bandelow, G. Huether, and E. Riither. 1995. Nocturnal plasma melatonin levels in patients suffering from chronic primary insomnia. *J Pineal Res* 19:116–22.

Hardeland, R., I. Balzer, B. Fuhrberg, and G. Behrmann. 1996. Melatonin in unicellular organisms and plants. *Front Horm Res* 21:1–6.

Heydorn, W. 1982. Effect of treatment of rats with antidepressants on melatonin concentrations in the pineal gland and serum. *J Pharmacol Exp Ther* 222:534–43.

Jan, J., and M. O'Donnell. 1996. Use of melatonin in the treatment of paediatric sleep disorders. *J Pineal Res* 21:193–99.

Kennedy, S. H. 2007. Agomelatine: An antidepressant with a novel mechanism of action. *Future Neurol* 2:145–51.

Kennedy, S., S. Kutcher, E. Ralevski, and G. M. Brown. 1996. Nocturnal melatonin and 24-h 6-sulphatoxymelatonin levels in various phases of bipolar affective disorder. *Psychiatr Res* 63:219–22.

Kennedy, S. H., and S. J. Rizvi. 2010. Agomelatine in the treatment of major depressive disorder: Potential for clinical effectiveness. *CNS Drugs* 24:479–99.

Klein, D. C. 1969. Pineal gland metabolism. The relationship between hydroxyindole-*O*-methyl transferase, melatonin production and secretion as stimulated by norepinephrine. *Fed Proc* 28:734 Abs.

Klein, D. C., D. A. Auerbach, M. A. A. Nambodiri, and C. H. T. Wheler. 1981. Indole metabolism in the mammalian pineal gland. In *The Pineal Gland. Vol 1, Anatomy and Biochemistry,* ed. R. J. Reiter, 199–228. Boca Raton, FL: CRC Press.

Klein, D. C., and J. L. Weller. 1970. Indole metabolism in the pineal gland: A circadian rhythm in *N*-acetyltransferase. *Science* 169:1093–5.

Klein, D. C., and J. L. Weller. 1973. Adrenergic-adenosine 3,-5, monophosphate regulation of serotonin *N*-acetyltransferase activity to synthesis of ³H-*N*-acetylserotonin and ³H-melatonin in the cultured rat pineal gland. *J Pharmacol Exp Ther* 186:516–27.

Klein, D. C., J. L. Weller, and R. Y. Moore. 1971. Melatonin metabolism: Neural regulation of pineal serotonin: Acetyl coenzyme A *N*-acetyltransferase activity. *Proc Natl Acad Sci Wash* 68:3107–10.

Koenigsberg, H. W., M. H. Teicher, V. Mitropoulou, et al. 2004. 24-h monitoring of plasma norepinephrine, MHPG, cortisol, growth hormone and prolactin in depression. *J Psychiatr Res* 38:503–11.

Korf, H., C. Schomerus, and J. Stehle. 1998. *The Pineal Organ, Its Hormone Melatonin, and the Photoneuroendocrine System*. Berlin: Springer.

Kräuchi, K., C. Cajochen, D. Möri, P. Graw, and A. Wirz-Justice. 1997. Early evening melatonin and S-20098 advance circadian phase and nocturnal regulation of core body temperature. *Am J Physiol Integr Comp Psysiol* 41:R1176–R1186.

Krause, D. N., and M. L. Dubocovich. 1990. Regulatory sites in the melatonin system of mammals. *Trends Neurosci* 3:464–70.

Lapin, J. P., and G. F. Oxenkrug. 1969. Intensification of the central serotonergic processes as a possible determinal of the thymoleptic effect. *Lancet* 1:132–6.

Lavebratt, C., L. K. Sjöholm, T. Partonen, M. Schalling, and Y. Forsell. 2010a. PER2 variation is associated with depression vulnerability. *Am J Med Gen Part B Neuropsychiatr Gen* 153:570–81.

Lavebratt, C., L. K. Sjöholm, P. Soronen, et al. 2010b. CRY2 is associated with depression. *PLoS One* 5:e9407. doi: 10.1371/journal.pone.0009407.

Leibenluft, E., S. Feldman-Naim, E. Turner, T. A. Wehr, and N. E. Rosenthal. 1997. Effects of exogenous melatonin administration and withdrawal in five patients with rapid-cycling bipolar disorder. *J Clin Psychiatry* 58:383 8.

Leproult, R., A. Van Onderbergen, M. L'Hermite-Baleriaux, E. Van Cauter, and G. Copinschi. 2005. Phase-shifts on 24-h rhythms of hormonal release and body temperature following early evening administration of the melatonin agonist agomelatine in healthy older men. *Clin Endocrinol* 63:298–304.

Lerner, A. 1961. Hormones and skin color. *Scientif Am* July:98–108.

Lerner, A. B., J. D. Case, Y. Takahashi, T. H. Lee, and. W. Mori. 1958. Isolation of melatonin, the pineal gland factor that lightens melanocytes. *J Am Chem Soc* 80:2587.

Levi, F., and U. Schibler. 2007. Circadian rhythms: Mechanisms and therapeutic implications. *Annu Rev Pharmacol Toxicol* 47:593–628.

Lewis, A., N. Kerenyi, and G. Feuer. 1990. Neuropharmacology of pineal secretions. *Drug Metab Interact* 8:247–312.

Lewy, A., T. A. Wehr, P. W. Gold, and F. K. Goodwin. 1979. Plasma melatonin in manic-depressive illness. In *Catecholamines: Basic and Clinical Frontiers*, ed. E. Usdin, I. Kopin, and J. Barchas, 1173–5. New York: Pergamon.

Linkowski, P., J. Mendlewicz, M. Kerkhofs, et al. 1987. 24-hour profiles of adrenocorticotropin, cortisol, and growth hormone in major depressive illness: Effect of antidepressant treatment. *J Clin Endocrinol Metab* 65:141–52.

Loo, H., A. Hale, and H. D'Haenen. 2002. Determination of the dose of agomelatine, a melatoninergic agonist and selective 5-HT$_{2C}$ antagonist, in the treatment of major depressive disorder: A placebo-controlled dose range study. *Int Clin Psychopharmacol* 17:239–47.

Macchi, M. M., and J. N. Bruce. 2004. Human pineal physiology and functional significance of melatonin. *Front Neuroendocrinol* 25:177–95.

Mallei, M., S. Zappettini, L. Musazzi, et al. 2009. Agomelatine reduces glutamate release induced by acute stress, possible synergism between melatonin and 5-HT2C properties. *Eur Neuropsychopharmacol* 19(3):S441.

Márquez, B. 2004. Ritmos circadianos y neurotransmisores: Estudios en la corteza prefrontal de la rata. Memoria presentada para optar al grado de Doctor. Universidad Complutense, Madrid.

Martinet, L., B. Guardiola-Lemaitre, and E. Mocaer. 1996. Entrainment of circadian rhythms by S-20098, a melatonin agonist, is dose and plasma concentration dependent. *Pharmacol Biochem Behav* 54:713–8.

McIsaac, W. M., R. G. Taborshy, and G. Farrell. 1964. 5-Methoxytryptophol: Effect on estrus and ovarian weight. *Science* 145:63–4.

Mendlewicz, J. 2009. Disruption of the circadian timing systems. Molecular mechanisms in mood disorders. *CNS Drugs* 23:15–26.

Mendlewicz, J., P. Linkowski, L. Branchey, U. Weinberg, E. D. Weitzman, and M. Branchey. 1979. Abnormal 24-hour pattern of melatonin secretion in depression. *Lancet* 2:1362.

Mieda, M., S. C. Williams, J. A. Richardson, K. Tanaka, and M. Yanagisawa. 2006. The dorsomedial hypothalamic nucleus as a putative food-entrainable circadian pacemaker. *Proc Natl Acad Sci U S A* 103:12150–5.

Millan, M. J., F. Lejeune, and A. Gobert. 2000. Reciprocal autoreceptor and heteroreceptor control of serotonergic dopaminergic and noradrenergic transmission in the frontal cortex: Relevance to the actions of antidepressant agents. *J Psychopharmacol* 14:114–38.

Millan, M. J., A. Gobert, F. Lejeune, et al. 2003. The novel melatonin agonist agomelatine (S20098) is an antagonist at 5-hydroxy-tryptamine 2C receptors, blockade of which enhances the activity of frontocortical dopaminergic and adrenergic pathways. *J Pharmacol Exp Ther* 306:954–64.

Millan, M. J., M. Brocco, A. Gobert, and A. Dekeyne. 2005. Anxiolytic properties of agomelatine, an antidepressant with melatonergic and serotonergic properties: Role of 5HT2C receptor blockade. *Psychopharmacology* 177:1–12.

Miller, F. P., and R. P. Maickel. 1970. Fluorimetric determination of indole derivatives. *Life Sci* 9(13):747–52.

Monk, T. H., and D. K. Welsh. 2003. The role of chronobiology in sleep disorders medicine. *Sleep Med Rev* 6:455–73.

Moore, R. Y., and D. C. Klein. 1974. Visual pathways and the central neural control of a circadian rhythm in pineal serotonin *N*-acetyltransferase activity. *Brain Res* 71:17–33.

Moore-Ede, R. Y. 1999. Circadian timing. In *Fundamental Neuroscience*, ed. M. J. Zigmond, F. E. Bloom, S. C. Landis, J. L. Roberts, and L. S. Squire, 1189–206. San Diego: Academic Press.

Mor, M., S. Rivara, D. Pala, A. Bedini, G. Spadoni, and G. Tarzia. 2010. Recent advances in the development of melatonin MT1 and MT2 receptor agonists. *Exp Opin Ther Patents* 20:1059–77.

Morgan, P. J., P. Barrett, H. E. Howell, and R. Helliwell. 1994. Melatonin receptors: Localization, molecular pharmacology and physiological significance. *Neurochem Int* 24:101–46.

Neu, J., and L. Niles. 1997. A marked diurnal rhythm of melatonin ML1A receptor mRNA expression in the suprachiasmatic nucleus. *Brain Res Mol Brain Res* 49:303–6.

Nutt, D., S. Wilson, and L. Paterson. 2008. Sleep disorders as core symptoms of depression. *Dial Clin Neurosci* 10:329–36.

Ohayon, M. M. 2007. Epidemiology of circadian rhythm disorders in depression. *Medicographia* 29:10–16.

Oksche, A., and H. G. Hartwig. 1979. Pineal sense organs components of photoneuroendocrine systems. *Prog Brain Res* 52:113–30.

Pacchierotti, C., S. Iapichino, L. Bossini, F. Pieraccini, and P. Castrogiovanni. 2001. Melatonin in psychiatric disorders: A review on the melatonin involvement in psychiatry. *Front Neuroendocrinol* 22:18–32.

Pandi-Perumal, S. R., V. Srinivasan, D. P. Cardinali, and J. M. Monti. 2006. Could agomelatine be the ideal antidepressant? *Expert Rev Neurother* 6:1595–608.

Pandi-Perumal, S. R., V. Srinivasan, B. Poeggeler, R. Hardeland, and D. P. Cardinali. 2007. Drug insight: The use of melatonergic agonists for the treatment of insomnia. Focus on ramelteon. *Nat Clin Pract Neurol* 3:221–8.

Paparrigopoulos, T., C. Psarros, J. Bergiannaki, E. Varsou, U. Dafni, and C. Stefanis. 2001. Melatonin response to clonidine administration in depression: Indication of presynaptic alpha$_2$-adrenoceptor dysfunction. *J Affect Disord* 65:307–13.

Parfitt, A., J. L. Weller, and D. C. Klein. 1976. Beta-adrenergic-blockers decrease adrenergically stimulated *N*-acetyltransferase activity in pineal glands in organ culture. *Neuropharmacology* 15:353–8.

Parry, B. 1997. Psychobiology of premenstrual dysphoric disorder. *Semin Reprod Endocrinol* 15:55–68.

Peeters, F., J. Berkhof, P. Delespaul, J. Rottenberg, and N. A. Nicolson. 2006. Diurnal mood variation in major depressive disorder. *Emotion* 6:383–91.

Pitrosky, B., R. Kirsch, A. Malan, E. Mocaer, and P. Pevet. 1999. Organisation of rat circadian rhythms during daily infusion of melatonin or S20098, a melatonin agonist. *Am J Physiol* 277:R812-R828.

Popoli, M. 2009. Agomelatine innovative pharmacological approach in depression. *CNS Drugs* 23(Suppl. 2): 27–34.

Quay, W. B. 1963. Circadian rhythm in rat pineal serotonin and its modifications by estrous cycle and photoperiod. *Gen Comp Endocrinol* 3:473–9.

Redman, J. R., B. Guardiola-Lemaitre, M. Brown, P. Delagrange, and S. M. Armstrong. 1995. Dose dependent effects of S-20098, a melatonin agonist, on direction of re-entrainment of rat circadian activity rhythms. *Psychopharmacology Berl* 118:385–90.

Reinberg, A. E., H. Lewy, and M. Smolensky. 2001. The birth of chronobiology: Julien Joseph Virey 1814. *Chronobiol Int* 18:173–186.

Reiter, R. J. 1991. Pineal melatonin: Cell biology of its synthesis and of its physiological interactions. *Endocr Rev* 12:151–80.

Reppert, S. M., and D. R. Weaver. 2002. Coordination of circadian timing in mammals. *Nature* 418(6901):935–41.

Reppert, S. M., D. R. Weaver, and C. Godson. 1996. Melatonin receptors step into the light: Cloning and classification of subtypes. *Trends Pharmacol Sci* 17:100–2.

Roenneberg, T., T. Kuehnle, M. Juda, et al. 2007. Epidemiology of the human circadian clock. *Sleep Med Rev* 11:429–38.

Rowan, W. 1925. Relation of light and bird migration and developmental changes. *Nature* 115:494–5.

Roybal, K., D. Theobold, A. Graham, et al. 2007. Mania-like behavior induced by disruption of CLOCK. *Proc Natl Acad Sci U S A* 104:6406–11.

Rubin, R., E. Helst, S. McGeoy, et al. 1992. Neuroendocrine aspects of primary endogenous depression: XI. Serum melatonin measures in patients and matched control subjects. *Arch Gen Psychiatry* 49:558–67.

Sack, R., and A. Lewy. 1986. Desmethylimipramine treatment increases melatonin production in humans. *Biol Psychiatry* 21:406–10.

Sánchez-Barceló, E. J., M. D. Mediavilla, D. X. Tan, and R. J. Reiter. 2010. Clinical uses of melatonin: Evaluation of human trials. *Curr Med Chem* 17:2070–95.

Schildkraut, J. J. 1965. The catecholamine hypothesis of affective disorders: A review of supporting evidence. *Am J Psychiat* 122:509–22.

Schulz, P., and T. Steimer. 2009. Neurobiology of circadian systems. *CNS Drugs* 23:3–13.

Sekula, L., J. Lucke, E. Heist, R. K. Czambel, and R. T. Rubin. 1997. Neuroendocrine aspects of primary endogenous depression: XV. Mathematical modeling of nocturnal melatonin secretion in major depressives and normal controls. *Psychiatry Res* 69:143–53.

Shi, J., J. K. Wittke-Thompson, J. A. Badner, et al. 2008. Clock genes may influence bipolar disorder susceptibility and dysfunctional circadian rhythm. *Am J Med Genet B Neuropsychiatr Genet* 147B:1047–55.

Smith, I. 1983. Indoles pineal origin: Biochemical and physiological status. *Psychoneuroendocrinology* 8:41–60.

Smolensky, M. H. 2001. Circadian rhythms in medicine. *CNS Spectr* 6:467–74.

Snyder, S. H., and J. Axelrod. 1964. Influence of the light and the sympathetic nervous system on 5-hydroxytryptophan decarboxylase (5-HTPD) activity in the pineal gland. *Fed Proc* 23:206.

Soria, V., and M. Urretavizcaya. 2009. Circadian rhythms and depression. *Actas Esp Psiquiatr* 37:222–32.

Souetre, E., E. Salvati, H. Rix, et al. 1988. Effect of recovery on the cortisol circadian rhythm of depressed patients. *Biol Psychiatry* 24:336–40.

Soumier, A., S. Lortet, C. Gabriel, et al. 2008. Cellular and molecular mechanisms underlying increased adult hippocampal neurogenesis induced by agomelatine. *Eur Neuropsychopharmacol* 18(4):S350.

Srinivasan, V., M. Smits, W. Spencer, et al. 2006. Melatonin in mood disorders. *World J Biol Psychiatry* 7:138–51.

Stein, D. J., A. Ahokas, and A. Fabiano. 2007. Agomelatine in generalized anxiety disorder: A randomized, placebo-controlled study. Poster presented at 20th European College of Neuropsychopharmacology (ECNP) Congress, 13–17, Vienna.

Stephan, F. K., and I. Zucker. 1972. Circadian rhythms in drinking behaviour and locomotor activity of rats are eliminated by hypothalamic lesions. *Proc Natl Acad Sci U S A* 69:1583–6.

Sugden, D., K. Davidson, K. A. Hough, and M. T. Teh. 2004. Melatonin, melatonin receptors and melanophores: A moving story. *Pigment Cell Res* 17:454–60.

Szymanska, A., J. Rabe-Jablonska, and M. Karasek. 2001. Diurnal profile of melatonin concentrations in patients with major depression: Relationship to the clinical manifestation and antidepressant treatment. *Neuroendocrinol Lett* 22:192–98.

Teicher, M., C. Glod, and E. Magnus. 1997. Circadian rest–activity disturbances in seasonal affective disorder. *Arch Gen Psychiatr* 54:124–30.

Thompson, C. 1985. The effect of desipramine upon melatonin and cortisol secretion in depressed and normal subjects. *Br J Psychiatr* 147:389–93.

Thompson, C., G. Mezey, T. Corn, et al. 1985. The effect of desipramine upon melatonin and cortisol secretion in depressed and normal subjects. *Br J Psychiatry* 147:389–93.

Tomás-Zapico, C., J. A. Boga, B. Caballero, et al. 2005. Coexpression of MT_1 and $ROR\alpha_1$ melatonin receptors in the Syrian hamster Harderian gland. *J Pineal Res* 39:21–6.

Van Reeth, O., and S. Maccari. 2007. Biology of circadian rhythms: Possible links to the pathophysiology of human depression. *Medicographia* 29:17–21.

Van Reeth, O., L. Weibel, E. Olivares, S. Maccari, E. Mocaer, and F. W. Turek. 2001. Melatonin or a melatonin agonist corrects age-related changes in circadian response to environmental stimulus. *Am J Physiol Regul Integr Comp Physiol* 280:R1582–91.

Venkoba, A., S. Parvathi, and V. Srinivasan. 1983. Urinary melatonin in depression. *Indian J Psychiatry* 25:167–72.

Von Gall, C., J. Stehle, and D. Weaver. 2002. Mammalian melatonin receptors: Molecular biology and signal transduction. *Cell Tissue Res* 309:151–62.

Webb, S., and M. Puig-Domingo. 1995. Role of melatonin in health and disease. *Clin Endocrinol* 42:221–34.

Weil, Z. M., A. K. Hotchkiss, M. L. Gatien, S. Pieke-Dahl, and R. J. Nelson. 2006. Melatonin receptor (MT1) knockout mice display depression-like behaviors and deficits in sensorimotor gating. *Brain Res Bull* 68:425–9.

Westrich, L., and J. Sprouse. 2010. Circadian rhythm dysregulation in bipolar disorder. *Curr Opin Invest Drugs* 11:779–87.

Wetterberg, L., J. Beckfriis, B. Aperia, and U. Petterson. 1979. Melatonin cortisol ratio in depression. *Lancet* 2:1361.

Whybrow, P., and T. Hurwitz. 1976. Psychological disturbances associated with endocrine disease and hormone therapy. In *Hormones, Behavior, and Psychopathology*, ed. E. J. Sachar, 125–43. New York, NY: Raven Press.

Wirz-Justice, A., and J. Arendt. 1980. Plasma melatonin and antidepressant drugs. *Lancet* 8165:425.

Wirz-Justice, A., P. Graw, K. Krauchi, et al. 1990. Morning or night-time melatonin is ineffective in seasonal affective disorder. *J Psychiatry Res* 24:129–37.

Wurtman, R. J., and J. Axelrod. 1965. The pineal gland. *Sci Am* 213:50–60.

Wurtman, R. J., and J. Axelrod. 1966. A 24-hour rhythm in the content of norepinephrine in the pineal and salivary glands of the rat. *Life Sci* 5:655–69.

Wurtman, R. J., W. Roth, M. D. Altschule, and J. J. Wurtman. 1961. Interactions of the pineal and exposure to continuous light on organ weights of female rats. *Acta Endocrinol Kbh* 36:617–24.

Wurtman, R. J., J. Axelrod, and L. S. Philips. 1963. Melatonin synthesis in the pineal gland: Control by light. *Science* 142:1071–3.

Wurtman, R. J., J. Axelrod, and J. E. Fischer. 1964. Melatonin synthesis in the pineal gland: Effect of light mediated by the sympathetic nervous system. *Science* 143:1328–30.

Wurtman, R. J., J. Axelrod, and D. E. Kelly. 1968. *The Pineal*. New York: Academic Press.

Wurtman, R. J., H. M. Shein, and F. Larin. 1971. Mediation by β-adrenergic receptors of effect of norepinephrine on pineal synthesis of ^{14}C-serotonin and ^{14}C-melatonin. *J Neurochem* 18:1683–7.

Ying, S. W., B. Rusak, P. Delagrange, E. Mocaër, P. Renard, and B. Guardiola-Lemaître. 1996. Melatonin analogues as agonists and antagonists in the circadian system and other brain areas. *Eur J Pharmacol* 296:33–4.

Zhdanova, I., R. Wurtman, and D. Schomer. 1995. Sleep-inducing effects of low doses of melatonin in the evening. *Clin Pharmacol Ther* 57:552–8.

Zupancic, M., and C. Guilleminault. 2006. Agomelatine: A preliminary review of a new antidepressant. *CNS Drugs* 20:981–92.

13 Corticotropin-Releasing Factor and Hypothalamic–Pituitary–Adrenal Axis Regulation of Behavioral Stress Response and Depression

Glenn R. Valdez

CONTENTS

13.1 INTRODUCTION

Depression is a highly prevalent form of mental illness, with a lifetime occurrence of 8–12% in all individuals (Andrade et al. 2003). This condition is often considered a stress-related disorder because some form of stressful life event frequently triggers depressive symptoms. Depression also has a high incidence of comorbidity with anxiety disorders (Lenze et al. 2000, 2001, 2005), which occur in approximately 30% of all adults (Kessler et al. 2005). Both of these conditions place a considerable economic burden on society (Andrade et al. 2003; DuPont et al. 1996). The body's normal stress response is thought to be essential to survival because it allows for adaptation to environmental demands. Chronic exposure to stress, however, has been hypothesized to lead to long-term alterations of the physiological systems mediating the body's responses to these demands.

One of the key components to understanding the biological foundations of depression may be the interaction between brain corticotropin-releasing factor (CRF) systems and the hypothalamic–pituitary–adrenal (HPA) axis. CRF is a 41 amino acid neuropeptide that has long been considered one of the body's major regulators of the stress response. It is involved in mediating the neuroendocrine HPA axis response to stress (Vale et al. 1981), as well as autonomic (Dunn and Berridge 1990; Vale et al. 1983) and behavioral responses to environmental demands (Koob et al. 1994; Koob and Heinrichs 1999). Results from clinical studies suggest that normal functioning of the CRF systems

275

and the HPA axis are altered in depression. For example, clinically depressed individuals show increased levels of CRF in cerebrospinal fluid (Nemeroff et al. 1984) and plasma (Galard et al. 2002). Postmortem studies have demonstrated that individuals who were known to have depression show decreased CRF binding in the frontal cortex (Nemeroff et al. 1988).

Two genes encoding distinct G-protein-coupled CRF receptors have been identified. The CRF_1 receptor is found mainly in the pituitary, amygdala, hippocampus, cerebellum, and cortex and is generally associated with increases in anxiety-like behavior (Koob and Heinrichs 1999). The CRF_2 receptor is found mainly in the lateral septum, ventromedial hypothalamus, and choroid plexus and exists in three splice variants: CRF_{2a}, CRF_{2b}, and CRF_{2c} receptors (Chalmers et al. 1995; Perrin et al. 1995). Originally, the belief was that activation of the CRF system itself would simply lead to increases in the stress response. Recent characterization of the CRF receptor subtypes, however, suggests that there may be a differential regulation of stress within this system. One hypothesis is that activation of the CRF_1 receptor leads to increases in the stress response, whereas the CRF_2 receptor represents a compensatory coping mechanism to oppose this action (Valdez et al. 2005).

Preclinical models show strong evidence for increased CRF_1 receptor activity in the regulation of the activational component of the stress response, and a number of nonpeptide CRF_1 receptor antagonists have recently been developed with the hope that these drugs may be of therapeutic value in the treatment of stress-related psychiatric illnesses (Valdez 2006; Zorrilla and Koob 2004). Although the role of CRF_2 receptors remains unclear in depression, preclinical evidence suggests that underactivation of this receptor may be involved in the regulation of increased depression-like behavior in animals.

Disruptions in HPA axis function are also observed during depression. Patients diagnosed with major depression often exhibit consistent hyperactivity of the HPA, which is likely due to increased CRF expression in the paraventricular nucleus of the hypothalamus (Raadsheer et al. 1994), leading to increases in the glucocorticoid hormone cortisol (Holsboer 2000; Pariante 2003). When given dexamethasone, a synthetic glucocorticoid, depressed patients fail to suppress plasma adrenocorticotropic hormone (ACTH) and cortisol, suggesting an impairment of the negative feedback loop of normal HPA axis function (Galard et al. 2002).

Although disturbances in HPA axis function are regulated in part by central CRF receptors, disruption of brain glucocorticoid receptor function may also contribute to the manifestation of depressive symptoms. Glucocorticoids exert their effects through two receptor subtypes: mineralocorticoid receptors and glucocorticoid receptors (Evans and Arriza 1989). Whereas glucocorticoids bind with high affinity to mineralocorticoid receptors, they display a low binding affinity for glucocorticoid receptors (de Kloet et al. 2000; Meijer and de Kloet 1998). This particular binding profile suggests a role for the glucocorticoid receptor in the regulation of depression because biologically active concentrations of cortisol at these receptors may only be present during the sustained high levels of HPA axis activity observed in depressed patients.

This chapter will review the role of CRF-related ligands, CRF receptors, HPA axis activity, and glucocorticoid receptors in the regulation of the stress response and depression. Understanding the interactions between these systems may provide further insight into the biological foundations of major depression, as well as stress-related psychiatric conditions.

13.2 CRF AND HPA AXIS

The neuroanatomical substrates associated with the mediation of the stress response by CRF include the central nucleus of the amygdala, locus coeruleus, medulla oblongata, and hypothalamus (Menzaghi et al. 1994a; Valentino and Wehby 1988). CRF in the medulla oblongata can activate the sympathetic nervous system, stimulating the release of adrenaline from the adrenal medulla and other parts of the sympathetic nervous system (Dunn and Berridge 1990; Vale et al. 1983). In order to stimulate HPA axis activity, CRF neurons are activated in the paraventricular nucleus of the hypothalamus, from where they send axon terminals to the median eminence, when the body is faced with a stressor. From the median eminence, CRF is released into the portal blood system and carried to the anterior lobe of the pituitary gland, where it stimulates the synthesis and release of ACTH. ACTH subsequently stimulates the release of glucocorticoids by the outer shell of the adrenals (Vale et al. 1981).

Glucocorticoid hormones, such as cortisol in humans and nonhuman primates and corticosterone in rodents, are involved in the regulation of various physiological and behavioral processes, including stress responses, energy metabolism, immune function, and learning and memory (Stratakis and Chrousos 1995). The effects of glucocorticoid hormones appear to be mediated by two receptor subtypes: mineralocorticoid receptors and glucocorticoid receptors (Evans and Arriza 1989). Mineralocorticoid receptors are found mainly in the kidneys, heart, and intestines, as well as limbic brain regions, such as the hippocampus (Funder 1992; Jacobson and Sapolsky 1991). Glucocorticoid receptors are expressed in all body tissues, including the frontal cortex, hippocampus, and hypothalamus (de Kloet et al. 1998; Jacobson and Sapolsky 1991). Glucocorticoids are also the final effectors of HPA axis activity (Dunn and Berridge 1990; Vale et al. 1981, 1983). The termination of the neuroendocrine stress response appears to be regulated by negative feedback effects exerted by glucocorticoids in the hypothalamus, pituitary, and hippocampus (de Kloet 1995; Vale et al. 1983) (Figure 13.1).

Chronic elevations in CRF and the hormones involved in the neuroendocrine stress response can result in various detrimental physiological effects. Increased CRF levels can lead to the decreased toxicity of natural killer cells in the immune system (Irwin et al. 1990) and induce stress-like changes in gastrointestinal, cardiovascular, and metabolic functions (Brown 1991; Fisher 1993; Taché et al. 1989) via activation of the HPA axis and sympathetic nervous system. ACTH secretion directly influences immune function and acts within the brain to regulate sleep (Chastrette et al. 1990; Rivier 1996). Hypersecretion of glucocorticoids can result in infection via suppressed

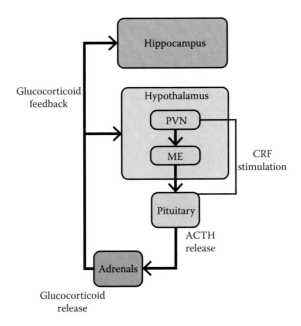

FIGURE 13.1 CRF regulation of HPA axis response. CRF neurons are activated in paraventricular nucleus (PVN) of hypothalamus and stimulate the median eminence (ME). CRF is released into portal blood system and carried to anterior lobe of pituitary gland, where it stimulates synthesis and release of ACTH. Outer shell of adrenals is then stimulated by ACTH, resulting in release of glucocorticoids. Termination of neuroendocrine stress response is regulated by negative feedback effects of glucocorticoids to the brain.

immune function, whereas low levels of these hormones can result in inflammatory conditions in laboratory animals (Rivier 1996).

Although the neuroendocrine stress response is important in the regulation of physiological responses to stress, the behavioral response to stress may also occur independently of the HPA axis. Hypophysectomy and blockade of the HPA axis response via dexamethasone suppression do not alter the behavioral response to stress produced by central administration of CRF (Britton et al. 1986; Eaves et al. 1985). Although it appears that a central site of action is likely responsible for coordinating stress-related behavior, HPA axis function is clearly impaired in patients with major depression. Although these findings may suggest that the HPA axis may not be necessary in the regulation of behaviors seen in stress-related disorders, its dysregulation likely still contributes to some of the symptoms associated with depression, a hypothesis that will be discussed later in this chapter.

13.3   CRF AND BEHAVIORAL STRESS RESPONSE

Central administration of CRF mimics the behavioral response to stress in laboratory animals. Many of the data regarding CRF itself and nonselective CRF receptor antagonists focus more on animal models of anxiety rather than depression with regard to stress-related behaviors. These results should be strongly considered when examining the role of CRF in depression, however, given the high incidence of comorbidity between these two disorders (Lenze et al. 2000, 2001, 2005). CRF injected into unstressed animals under familiar conditions leads to increased locomotor activation (Koob et al. 1984; Sutton et al. 1982), whereas in unfamiliar settings, centrally administered CRF can lead to behavioral suppression (Sutton et al. 1982). In addition, CRF administration can suppress behavior during the conflict test, as animals show an even greater reduction in responding when the response is accompanied by a shock (Britton et al. 1985). Other examples of stress-related behavior induced by central administration of CRF include an enhanced acoustic startle response (Swerdlow et al. 1986), increases in the conditioned fear response (Cole and Koob 1988), and decreased appetite (Arase et al. 1988; Krahn et al. 1986). Furthermore, transgenic mice overproducing CRF show decreased exploration of a novel environment compared with wild types, an effect potentiated by exposure to social defeat stress (Stenzel-Poore et al. 1994).

Further evidence that brain CRF systems play an important role in the regulation of the behavioral response to stress comes from studies using competitive CRF receptor antagonists, as these antagonists attenuate both CRF- and stress-induced behavioral changes. α-Helical CRF$_{(9-41)}$, a CRF receptor antagonist, decreases the CRF-enhanced acoustic startle response in rats when centrally injected (Swerdlow et al. 1989). CRF-overproducing mice also exhibit increased exploratory behavior in the elevated plus maze after injection with α-helical CRF$_{(9-41)}$ compared with those receiving vehicle (Stenzel-Poore et al. 1994). A second CRF receptor antagonist, D-Phe-CRF$_{(12-41)}$, reduces both CRF- and stress-induced increases in locomotor activation (Menzaghi et al. 1994b). Astressin, a third CRF receptor antagonist, decreases CRF-induced locomotor activation and closed arm exploration in the elevated plus maze (Spina et al. 2000).

CRF receptor antagonists have also been shown to reduce stress-induced behavioral changes. In the elevated plus maze, administration of α-helical CRF$_{(9-41)}$ leads to increased exploration of the open arms when subjected to restraint stress, swim stress, or social conflict stress (Heinrichs et al. 1994). Astressin also increases open arm exploration in the elevated plus maze in rats subjected to social conflict stress (Spina et al. 2000). D-Phe-CRF$_{(12-41)}$ attenuates stress-induced increases in locomotor activation and decreases in exploration of the open arms in the elevated plus maze (Menzaghi et al. 1994b).

With regard to animal models of depression, D-Phe-CRF$_{(12-41)}$ reverses increases in intracranial self-stimulation reward thresholds induced by central CRF administration (Macey et al. 2000). The forced swim test is an animal model of depression in which immobility during a period of swim stress has been proposed as a depressive-like behavior (Porsolt et al. 1977). Injections of CRF have also shown to increase immobility in rats observed in this model (Dunn and Swiergiel 2008; Swiergiel et al. 2007). These results suggest that increased CRF activity may be a key component in regulating depressive-like behaviors in animals.

13.4   CRF$_1$ RECEPTORS, STRESS-RELATED BEHAVIOR, AND ANIMAL MODELS OF DEPRESSION

Two genes encoding distinct G-protein-coupled CRF receptors have been identified. The CRF$_1$ receptor is found mainly in the pituitary, amygdala, hippocampus, cerebellum, and cortex, and its activation is generally associated with increases in stress-related behavior (Koob and Heinrichs 1999). As described above, CRF, which preferentially binds to the CRF$_1$ receptor, leads to increases in the behavioral stress response. For example, ovine CRF, which shows an 80-fold higher affinity in binding to the CRF$_1$ receptor (Behan et al. 1996), produces increased motor activation, and decreases open arm exploration in the elevated plus maze (Valdez et al. 2002). CRF$_1$ receptor knockout mice also show decreased responsiveness to stressful stimuli (Contarino et al. 1999; Smith et al. 1998; Timpl et al. 1998) and less spontaneous motor activity (Contarino et al. 2000). These data demonstrating that activation of the CRF$_1$ receptor increases the behavioral response to stress suggest that the CRF$_1$ receptor may be a key biological component in regulating stress-related psychiatric disorders.

A number of CRF$_1$ receptor antagonists have been examined in animal models of depression. CP-154,526, one of the earliest nonpeptide CRF$_1$ receptor antagonists to be developed (Schulz et al. 1996), reversed escape deficit in rats exposed to inescapable shock without affecting the behavior of controls when tested in the learned helplessness task (Mansbach et al. 1997). During this task, animals were exposed to a series of inescapable shocks and were then given the opportunity to escape shock after a period of time. Decreased escape behavior has been proposed to indicate a depressive-like state (Maier and Seligman 1976). When administered 60 min before a test session, acute and chronic injections of CP-154,526 and CRA1000, another nonpeptide CRF$_1$ receptor antagonist, reduced escape failure in rats in a manner comparable to chronic treatment with the tricyclic antidepressant imipramine (Takamori et al. 2001a, 2001b). Acute and chronic treatment with the CRF$_1$ receptor antagonist R278995/CRA0450 also reduced escape failures (Chaki et al. 2004). However, the CRF$_1$ receptor antagonists DMP696 and DMP904 were not effective in the learned helplessness task (Li et al. 2005).

The forced swim test has also yielded mixed results regarding the effectiveness of CRF$_1$ antagonists in animal models of depression. For example, CP-154,526, R121919 and antalarmin, a CRF$_1$ antagonist that is structurally similar to CP-154,526 (Webster et al. 1996), have been found to be ineffective in reversing swim stress–induced immobility in the rat (Jutkiewicz et al. 2005). Similar results have also been found using R278995/CRA0450 (Chaki et al. 2004), DMP696 and DMP904 (Li et al. 2005). In mice, antalarmin, DMP696, DMP904, and R121919 were also unable to reverse immobility in the forced swim test (Nielsen et al. 2004). In contrast, the CRF$_1$ receptor antagonist LWH234 reduced immobility but did not affect stress-induced increases in ACTH (Jutkiewicz et al. 2005). Another study has found that SSR125543A and antalarmin can also attenuate immobility due to swim stress (Griebel et al. 2002).

In the tail suspension test, subchronic dosing of R121919 and DMP696 has been shown to decrease immobility in mice in a similar manner to that of the selective serotonin reuptake inhibitors (SSRI) fluoxetine and paroxetine or the selective norepinephrine reuptake inhibitor reboxetine (Nielsen et al. 2004). In contrast, antalarmin, DMP904 (Nielsen et al. 2004), CP-154,526 (Yamano et al. 2000), and R278995/CRA0450 (Chaki et al. 2004) were found to be ineffective in the tail suspension test.

These conflicting findings may be explained, in part, by the baseline behavior of the animals. Recently, a model of depression has been developed while examining the Flinders Sensitive Line

(FSL) rats. These animals show a higher baseline level of immobility compared with other rat strains (Overstreet et al. 2005). FSL rats receiving chronic treatment with SSR125543A showed decreases in immobility in the forced swim test comparable to that observed following chronic injections of fluoxetine and the tricyclic antidepressant desipramine. These drugs did not affect immobility in Flinders Resistant Line (FRL) rats (Overstreet and Griebel 2004). Similar results have also been observed in FSL and FRL rats receiving chronic treatment with CP-154,526, imipramine, and the SSRI citalopram (Overstreet et al. 2004). Given these results, it appears that CRF_1 receptor antagonists are most effective in animal models of depression when tested in animals that show behaviors indicative of a depressive-like state. CRF_1 receptor antagonists appear to be silent in unstressed animals.

More consistent results have been seen in animals examined in the chronic mild stress model. BALB/c mice exposed to a series of mild stressors, including restraint, food restriction, and changes in housing, show a deteriorated physical state and decreased body weight. Chronic treatment with antalarmin or fluoxetine reverses the decrease in physical state, but only fluoxetine increased body weight (Ducottet et al. 2003). Another study has shown that chronic treatment with SSR125543A or fluoxetine can also reverse deterioration of physical state in mice. In addition, both of these drugs attenuated the stress-induced reduction of cell proliferation in the dentate gyrus (Alonso et al. 2004). Interestingly, fluoxetine treatment led to an increase in cell proliferation in the dentate gyrus in nonstressed mice, whereas SSR125543A had no effects on cell proliferation in these animals (Alonso et al. 2004). This result further supports the hypothesis that CRF_1 antagonists may only be effective in altering depressive-like behaviors related to stressful conditions.

13.5 CRF₂ RECEPTORS, STRESS-RELATED BEHAVIOR, AND ANIMAL MODELS OF DEPRESSION

The CRF_2 receptor is found mainly in the lateral septum, ventromedial hypothalamus, and choroid plexus (Chalmers et al. 1995; Perrin et al. 1995). Activation of the CRF_2 receptor is most strongly associated with alterations in feeding behavior (Pelleymounter et al. 2000; Spina et al. 1996). Because of conflicting experimental findings, there is currently no general consensus regarding the role of CRF_2 receptors in the behavioral stress response. CRF_2 receptor knockout mice show an anxiogenic-like phenotype when examined in the elevated plus maze, open field test, and light–dark emergence test (Bale et al. 2000; Kishimoto et al. 2000). Central infusion of CRF_2 receptor antisense decreases stress-coping behaviors and induces an anxiogenic-like response in rats (Isogawa et al. 2003). Amidated 38 amino acid synthetic peptides encoded by the urocortin 2 (Ucn 2) (Reyes et al. 2001) and urocortin 3 (Ucn 3) (Lewis et al. 2001) genes have been identified as selective CRF_2 receptor agonists, and these ligands produce distinct behavioral profiles compared with that of CRF. Central injections of Ucn 2 and Ucn 3 lead to suppressed motor activity in the locomotor activity test and an increase in open arm exploration in the elevated plus maze (Valdez et al. 2002, 2003). Ucn 3 also increases exploratory behavior in mice examined in the open field test (Venihaki et al. 2004).

In contrast, there also have been findings that suggest an anxiogenic role for the CRF_2 receptor. Contrary to the findings discussed above, studies have found that intracerebroventricular injections of antisauvagine-30 produced anxiolytic-like effects in the rat (Takahashi et al. 2001) and the mouse (Pelleymounter et al. 2002). In addition, antisense inhibition of CRF_2 receptors in the lateral septum also attenuates fear conditioning (Ho et al. 2001). Ucn 2 has also been shown to produce a decrease in open arm exploration in the elevated plus maze in mice (Pelleymounter et al. 2002). These seemingly contradictory findings may be the result of site-specific actions. Astressin, a nonselective CRF receptor antagonist, but not antisauvagine-30, impairs CRF-enhanced fear conditioning when injected into the hippocampus. Both of these antagonists, however, attenuate fear conditioning when injected into the lateral septum (Radulovic et al. 1999). Lesions of the lateral septum also produce anxiolytic-like effects in rat models of anxiety (Menard and Treit 1996; Yadin et al. 1993).

Another possibility is that CRF_2 receptors may regulate specific aspects of the stress-coping response, such as sensory information. In C57BL/6J and 129S6/SvEvTac mice, both the CRF_1 receptor antagonist NBI-03775 and antisauvagine-30 attenuated enhancement of the acoustic startle response by CRF. In addition, Ucn 2 also increased the acoustic startle response but with less efficacy than CRF (Risbrough et al. 2003). Further investigation, however, showed that although NBI-30775 and antisauvagine-30 reduced CRF-induced increases in startle and CRF-induced deficits in prepulse inhibition, CRF_2 receptor activation via Ucn 2 and Ucn 3 injections enhanced prepulse inhibition of the acoustic startle response (Risbrough et al. 2004). These data suggest that although it is possible that CRF_2 receptors have some stress-inducing properties, they appear to be a critical component tempering the stress response.

A third potential explanation is that the CRF_2 receptor may act as a presynaptic autoreceptor, inhibiting the release of CRF, which would in turn lead to decreased activation of the CRF_1 receptor. The neuropeptide Y system within the brain has been shown to have anxiety-reducing properties (Britton et al. 1997; Heilig et al. 1992, 1993; Heilig and Murison 1987). Specific activation of the Y_2 receptor, however, has been shown to increase the behavioral stress response in animal models of anxiety (Britton et al. 2000; Heilig et al. 1992, 1993; Heilig and Murison 1987). There are data indicating that the Y_2 receptor functions as a presynaptic autoreceptor that inhibits the release of neuropeptide Y (Malmstrom et al. 2002; Smith-White et al. 2001), which would subsequently lead to increases in stress-related behaviors. It is possible that activation of the CRF_2 receptor may work in a similar manner to inhibit the release of CRF, resulting in a decreased stress response.

With regard to depressive-like behaviors specifically, the role of CRF_2 receptors remains unclear. When observed in the forced swim test, CRF_2 receptor knockout mice displayed increased depression-like behavior as indicated by increased immobility (Bale and Vale 2003). In contrast, female, but not male, Ucn 2 knockout mice showed less immobility time in the tail suspension and forced swim tests (Chen et al. 2006). The difference in behavioral profiles between these strains of mice may have been due to the body-wide absence of CRF_2 receptors and a different compensation in CRF_1 signaling in the CRF_2 receptor knockout mice. Ucn 2 knockout mice still have CRF_2 receptors available for Ucn 1 and Ucn 3 to bind and, in fact, show upregulated CRF_2 receptor expression in the dorsal raphe. Conflicting data have also been found with regard to CRF_2 receptor–specific ligands on depressive-like behaviors. Ucn 2 injected into the dorsal raphe has been shown to potentiate learned helplessness behavior in response to inescapable shock, possibly due to interactions with the serotonergic system (Hammack et al. 2003). A more recent study, however, showed that Ucn 2 and Ucn 3 decreased immobility and increased climbing and swimming in mice examined in the forced swim test (Tanaka and Telegdy 2008). Clearly, further work is needed to fully understand the role of CRF_2 receptors in depression.

13.6 HPA AXIS AND ANIMAL MODELS OF DEPRESSION

As discussed previously, the behavioral response to stressors associated with depression may occur independently of HPA axis activation. However, HPA axis function is clearly impaired during depression. Therefore, while the HPA axis may not be necessary in the behavioral response to external stressors, its dysregulation likely still contributes to depressive symptoms.

Repeated glucocorticoid injections have been found to result in behavioral changes that are often associated with animal models of depression. Rats receiving 40 mg/kg corticosterone injections for 21 consecutive days displayed increases in immobility when tested using the forced swim test model (Gregus et al. 2005; Marks et al. 2009). Similar results were observed in both male and female rats tested using similar procedures, although male rats were shown to spend an even greater percentage of time immobile (Kalynchuk et al. 2004). In addition, repeated injections of corticosterone suppressed endogenous glucocorticoid release after exposure to a novel stressor (Johnson et al. 2006).

With regard to external stressors, rats subjected to forced swim stress displayed robust increases in corticosterone secretion for up to 2 h after exposure (Connor et al. 1997). Unavoidable chronic stress can also lead to alterations in HPA axis function and increases in depression-like behaviors. Rats exposed to 4 weeks of restraint stress and inescapable footshock showed significantly fewer escapes in the learned helplessness test compared with naïve rats (Raone et al. 2007). Stressed rats also showed higher basal corticosterone levels and a decreased response to dexamethasone (Raone et al. 2007).

Exposure to stressors during prenatal or early postnatal development has also been studied in order to examine the HPA axis in the pathogenesis of depression. For example, prenatally stressed rats displayed increased immobility in the forced swim test and decreased time in the illuminated portion of the open field test at 4 months postnatal (Szymanska et al. 2009). These rats also showed significantly elevated levels of corticosterone after swim stress compared with controls (Szymanska et al. 2009). Congenital learned helpless (cLH) rats, which are selectively bred to be susceptible to learned helplessness behavior as a model of depression, also showed disruptions in HPA axis activity when exposed to early postnatal stress (King et al. 1993). At postnatal day 14, cLH rats showed a significant increase in ACTH secretion after exposure to cold stress compared with control rats. However, a blunted ACTH response was observed in response to this stressor at postnatal day 21 (King et al. 1993). A similar pattern has been observed with regard to corticosterone levels in cLH rats exposed to similar conditions (King and Edwards 1999). These alterations in HPA axis function appear to be long lasting, as enhanced ACTH secretion and blunted corticosterone responses were observed in adult cLH rats after footshock, an effect that was particularly pronounced in cLH rats exposed to early maternal deprivation (King and Edwards 1999).

One hypothesis regarding the contribution of elevated glucocorticoid levels and the development of depression is that high concentrations of glucocorticoids lead to impaired glucocorticoid receptor signaling and decreased neurogenesis, which in turn correlate with the induction of depressive symptoms. As previously discussed, glucocorticoids exert their effects through two receptor subtypes: mineralocorticoid receptors and glucocorticoid receptors (Evans and Arriza 1989). Whereas glucocorticoids bind with high affinity to mineralocorticoid receptors, glucocorticoid receptors display a low binding affinity for endogenous glucocorticoids (de Kloet et al. 1998; Meijer and de Kloet 1998). This particular binding profile suggests a role for the glucocorticoid receptor in the regulation of depression because biologically active concentrations of cortisol at these receptors may only be present during the sustained high levels of HPA axis activity observed in depressed patients. Exposure to chronic stress leads to decreased glucocorticoid receptor expression in the hippocampus, hypothalamus, medial prefrontal cortex, and pituitary in rats, in addition to increases in depression-like behaviors (Raone et al. 2007). Reductions in hippocampal volume are also observed after chronic glucocorticoid treatment (Sapolsky et al. 1985). Stressed rat pups show increased corticosterone levels, which correlate with decreased hippocampal neurogenesis (Tanapat et al. 1998), suggesting that these elevated glucocorticoid levels may impair negative feedback control of the HPA axis.

Recently, glucocorticoid receptor transgenic mice have been developed to further examine the role of HPA axis activity in depression (Urani and Gass 2003). Glucocorticoid receptor heterozygous (GR$^{+/-}$) mice with a 50% reduction in glucocorticoid receptor gene done showed increased escape latency in the learned helplessness test and altered HPA axis activity (Ridder et al. 2005). More specifically, GR$^{+/-}$ mice failed to suppress corticosterone after injections of dexamethasone and showed increased glucocorticoid release in response to a CRF challenge compared with wild-type littermates (Ridder et al. 2005). GR$^{+/-}$ mice also showed decreased hippocampal neurogenesis when exposed to restraint stress (Kronenberg et al. 2009). Interestingly, mice with a specific deletion of the glucocorticoid receptor in the hippocampus showed increases in depression-like behaviors, as well as HPA axis hyperactivity and impaired negative feedback regulation (Boyle et al. 2005), suggesting a particular role for the hippocampus in the regulation of depressive behaviors related to HPA axis impairment.

Further support for the role of HPA axis dysregulation, glucocorticoid receptors, and decreased neurogenesis in depression comes from studies examining the effects of classic antidepressants.

Whereas stress has been shown to decrease glucocorticoid receptor expression in various brain regions and increase depression-related behaviors in rats, imipramine treatment reverses these behavioral and physiological stress-induced modifications (Raone et al. 2007). Injections of imipramine and desipramine also increase hippocampal and hypothalamic glucocorticoid receptor mRNA (Peiffer et al. 1991). Additionally, desipramine decreases immobility in the forced swim test and attenuates stress-induced increases in corticosterone levels (Connor et al. 2000). Collectively, these data indicate that disrupted HPA axis function and alterations in glucocorticoid receptor expression, as well as the accompanied decrease in neurogenesis in the hippocampus, likely contribute to symptoms associated with depression.

13.7 HPA AXIS AND CLINICAL STUDIES ON DEPRESSION

Patients diagnosed with major depression often exhibit consistent hyperactivity of the HPA axis (Holsboer 2000; Pariante 2003), as evidenced by increased pituitary and adrenal gland volume and chronic high levels of glucocorticoids in saliva, cerebrospinal fluid, blood plasma, and urine (Nemeroff et al. 1992; Pariante 2009). Although this is a common finding in clinical studies of depression, the specific role of the HPA axis in the mediation of depressive symptoms remains unclear.

The dexamethasone suppression test has often been used to examine impaired negative feedback regulation of HPA axis activity. Being a synthetic glucocorticoid, dexamethasone binds to glucocorticoid receptors and activates HPA axis feedback inhibition in healthy subjects. Depressed patients, however, fail to suppress plasma ACTH and cortisol after dexamethasone administration (Evans and Nemeroff 1984; Galard et al. 2002; Nemeroff and Evans 1984), suggesting an impairment of the negative feedback loop of normal HPA axis function.

It is possible that the high levels of glucocorticoids resulting in this impairment of negative feedback contribute to the symptoms of depression through reduced neurogenesis. One hypothesis is that loss in hippocampal volume may contribute to mood and especially memory disturbances that are observed during depression (Sahay and Hen 2007). Data to support this hypothesis, however, have been conflicting. Although some studies have found hippocampal atrophy in depressed patients (Boldrini et al. 2009; Sheline et al. 1996), others have found no changes in neurogenesis (Reif et al. 2006). Clinical trials examining the glucocorticoid receptor antagonist RU-486 (mifepristone) have shown that reducing glucocorticoid activity can reverse neurocognitive impairments and alleviate depressive symptoms in bipolar depressed patients (Young et al. 2004). However, conflicting evidence has been observed in trials examining patients with psychotic major depression. Whereas improvements in the symptoms of depression have been found in one study (Belanoff et al. 2002), other studies have found that RU-486 does not improve symptoms of patients diagnosed with psychotic major depression (DeBattista and Belanoff 2006; Flores et al. 2006). These studies suggest that although glucocorticoids may play a role in regulating the cognitive impairments associated with major depression, further study is needed to fully examine their role in depressive symptoms.

Given the regulatory role of CRF, it is possible that many of the effects of impaired HPA axis function may be a secondary result of impaired CRF activity. For instance, depressed patients with increased plasma cortisol levels, a blunted ACTH response to CRF, and adrenal hypertrophy show normalized HPA axis activity after the resolution of clinical symptoms (Amsterdam et al. 1988; Rubin et al. 1995). As discussed previously, although the neuroendocrine stress response is important in the regulation of physiological responses to stress, the behavioral response to stress can occur independently of the HPA axis. These clinical studies, along with preclinical data, suggest that the HPA axis may not be necessary in the regulation of depressive symptoms. Therefore, a central CRF-related mechanism may be a more viable target in the overall treatment of depression. However, disruption of HPA axis function does appear to contribute to the cognitive and memory impairments associated with depression. As such, pharmacotherapies that target regulation of the HPA axis, such as glucocorticoid receptor antagonists, may be a valuable adjunct to the treatment of certain major depressive symptoms.

13.8   CRF$_1$ ANTAGONISTS IN CLINICAL TRIALS

Clinical trials have been performed at the Max Planck Institute of Psychiatry in Munich, Germany, examining the CRF$_1$ antagonist NBI-30775. In an open label trial designed to assess the safety of NBI-30775, patients with major depression did not show significant alterations in liver enzymes or heart rate. Moreover, NBI-30775 did not alter the normal neuroendocrine response to intravenous administration of CRF. These patients also showed reduced indications of anxiety and depression, which returned after discontinuation of treatment (Zobel et al. 2000).

A subsequent trial has shown that a 30-day treatment of low- or high-dose regimens of NBI-30775 did not affect various endocrine measures, including HPA axis; hypothalamic–pituitary–gonadal axis and plasma renin activity; and aldosterone, human growth hormone, insulin-like growth factor vasopressin, and thyroid hormone levels. In addition, reports of adverse side effects such as headache, nausea, and dizziness that were observed during the trial did not appear to be the direct result of the experimental compound (Kunzel et al. 2003). NBI-30775 has also been shown to normalize electroencephalogram sleep patterns in patients diagnosed with major depression (Held et al. 2004). Finally, subsequent analysis of plasma leptin levels and body weight of the patients examined by Zobel et al. (2000) showed that neither of these measures was affected by NBI-30775 treatment (Kunzel et al. 2005). Whereas this study showed NBI-30775 to be relatively safe, an unpublished study in the United Kingdom found elevated liver enzyme levels in two patients, leading to the termination of the development of NBI-30775, according to a media release from Janssen published on April 7, 2000.

Phase I clinical trials have recently been conducted to assess the effects of another CRF$_1$ receptor antagonist, NBI-34041, on neuroendocrine function (Ising et al. 2007). For 14 days, 24 healthy male subjects received either NBI-34041 or placebo, and the HPA axis response to CRF and psychosocial stress was examined. NBI-34041 reversed the increases in ACTH and cortisol due to intravenous CRF or psychosocial stress but did not impair normal HPA axis function. Although these initial results demonstrate that NBI-34041 is effective in reversing physiological responses to stress without affecting basal hormone levels, further work is clearly needed to asses the safety and efficacy of this drug in the treatment of depression.

13.9   CONCLUSIONS

The data presented are clear evidence that the CRF system and HPA axis are important in mediating heightened stress, which in turn may lead to stress-related psychiatric disorders. These behavioral changes appear to be the result of an imbalance of CRF$_1$ and CRF$_2$ receptor activity, rather than simply CRF activation. The increases in CRF typically observed during the stress response appear to lead to an overactivation of the CRF$_1$ receptor given the binding profile of this neuropeptide (Behan et al. 1996). This hypothesis is further supported by the data indicating an antistress function for CRF$_2$ receptor activation (Valdez et al. 2002, 2003; Venihaki et al. 2004), leading to the idea that the CRF$_1$ receptor is responsible for coordinating the activational components of stress whereas the CRF$_2$ receptor acts as a coping mechanism to compensate for this action.

In addition, alterations in normal HPA axis function are often observed in patients with major depression, suggesting that this impairment may contribute to depressive symptoms. Patients diagnosed with major depression often exhibit consistent hyperactivity of the HPA axis, as evidenced by sustained levels of increased cortisol (Holsboer 2000; Nemeroff et al. 1992; Pariante 2003, 2009). In addition, depressed patients fail to suppress plasma ACTH and cortisol after dexamethasone administration (Evans and Nemeroff 1984; Galard et al. 2002; Nemeroff and Evans 1984), suggesting an impairment of negative feedback regulation of the HPA axis. It is possible that the resulting high levels of glucocorticoids contribute to the cognitive symptoms of depression through reduced hippocampal neurogenesis (Sahay and Hen 2007). Clinical studies support a role for HPA axis impairment in neurocognitive symptoms but yield conflicting evidence regarding depressive symptoms (Belanoff et al. 2002; DeBattista and Belanoff 2006; Flores et al. 2006; Young et al. 2004). When considered together, data regarding CRF and HPA axis alterations in major depression suggest that complementary approaches targeting these particular systems may provide a valuable approach in the treatment of depression.

Currently, the most promising approach appears to be the development of small-molecule CRF_1 antagonists. Observations from both clinical and preclinical studies suggest that CRF_1 receptor antagonists are potentially effective candidate medications in the treatment of stress-related psychiatric illnesses. Patients diagnosed with affective disorders have shown increased CRF levels in cerebrospinal fluid and the hypothalamus, decreased CRF binding in the frontal cortex, and impaired HPA axis function (Galard et al. 2002; Nemeroff et al. 1984, 1988; Raadsheer et al. 1994). Preclinical evidence demonstrates that reducing brain CRF activity via antagonism of CRF_1 receptors can attenuate the behavioral changes observed in animal models of depression.

There are some issues that still need to be considered in the development of CRF_1 receptor antagonists. One concern is that most structures are excessively lipophilic and have poor aqueous solubility and pharmacokinetic properties. One highly encouraging characteristic of these drugs is their specificity in altering behavior only under stressed conditions, a feature that bodes well for side-effect liability. Although major side effects were not observed in clinical studies examining NBI-30775, animal studies have suggested that some CRF_1 antagonists may have slight sedative effects (Chaki et al. 2004), endocrine disruptive actions (Bornstein et al. 1998), and transient abuse-related potential (Broadbear et al. 2002). Further research is clearly warranted to fully investigate the efficacy and safety profiles of these candidate medications. The results of preclinical experiments and clinical trials examining CRF_1 receptor antagonists, however, demonstrate that further development of these compounds may one day lead to new pharmacotherapies for the treatment of depression.

An alternative approach to targeting the CRF system in the development of new antidepressant medications may be to increase CRF_2 receptor activity. Although there is still no general consensus regarding the role of the CRF_2 receptor in the behavioral stress response, much of the data regarding the behavioral profiles of CRF-related neuropeptides and CRF-receptor knockout mice have led to a hypothesis that the CRF receptor subtypes have an opposing role in the regulation of the stress-related behaviors. It appears that activation of the CRF_1 receptor increases the activational component of the stress response, whereas the CRF_2 receptor acts as a compensatory coping mechanism to oppose this action. For example, CRF and urocortin 1 (Ucn 1), a CRF-related neuropeptide with equal affinity for the CRF_1 and CRF_2 receptors (Vaughan et al. 1995), have been shown to increase overall behavioral activity in rats in a differential manner compared with Ucn 2 and Ucn 3. For example, de Groote et al. (2005) found that CRF and Ucn 1 increase measures of exploratory behavior and grooming, whereas Ucn 2 and Ucn 3 only increase exploratory behavior.

In addition, CRF_2 receptor–specific ligands have been shown to reverse the effects of CRF. Ucn 2 and Ucn 3 have been shown to decrease the stimulatory effect of CRF on locomotor activity in a familiar environment. Furthermore, Ucn 2 injections attenuated CRF-induced decreases in exploratory behavior in the open field (Ohata and Shibasaki 2004). Although the precise mechanism of this action remains unclear, one hypothesis is that activation of the CRF_2 receptor may simply lead to a functional antagonism of the behaviors produced by CRF_1 receptor activation. Activation of the CRF_2 receptor may oppose the stress-inducing actions that result from CRF_1 receptor activation. Although this hypothesis must be further tested, the development of small-molecule CRF_2 receptor agonists may also provide a novel approach for the treatment of depression.

Both preclinical and clinical data suggest that glucocorticoid receptor antagonism may be a mechanism of particular interest in the treatment of the neurocognitive impairments associated with depression. Rat pups exposed to stress showed increased corticosterone levels and decreased hippocampal neurogenesis (Tanapat et al. 1998), and chronic glucocorticoid treatment has been found to reduce hippocampal volume (Sapolsky et al. 1985). $GR^{+/-}$ mice also showed decreased hippocampal neurogenesis when exposed to stress (Kronenberg et al. 2009). Hippocampal atrophy has also been observed in depressed patients (Boldrini et al. 2009; Sheline et al. 1996). In clinical trials, RU-486 can reverse neurocognitive impairments (Young et al. 2004). Although the hypothesis regarding the role of glucocorticoids in depression needs to be further examined, glucocorticoid receptors appear to be a promising potential target in the neurocognitive impairments associated with this disorder.

The experiments discussed in this chapter suggest that the underlying mechanisms for increases in depressive behaviors may be the overactivation of CRF_1 receptors, possible underactivation of CRF_2 receptors, and impaired HPA axis function. Studies describing the effects of CRF_1 receptor antagonists imply that antagonizing the effects associated with activation of this receptor can attenuate the behavioral stress response. These compounds have also shown promising results when tested in preclinical models of depression and clinical trials. Although data regarding CRF_2 receptor regulation of depressive behaviors are less clear, there is evidence that CRF_2 receptor ligands oppose the effects of CRF_1 receptor activation. Thus, the development of CRF_2 receptor agonists may also be worth exploring as potential novel antidepressants. Finally, glucocorticoid receptor antagonism has been shown to reduce the neurocognitive impairments associated with depression. In conclusion, the data presented regarding CRF, its receptors and related ligands, and HPA axis dysfunction strongly suggest that targeting the CRF system and HPA axis has the potential to generate novel pharmacotherapies for depression.

ACKNOWLEDGMENTS

The author was supported by Academic Research Enhancement Award AA18213 from the National Institute on Alcohol Abuse and Alcoholism.

REFERENCES

Alonso, R., G. Griebel, G. Pavone, J. Stemmelin, G. Le Fur, and P. Soubrie. 2004. Blockade of CRF(1) or V(1b) receptors reverses stress-induced suppression of neurogenesis in a mouse model of depression. *Mol Psychiatry* 9:278–86.

Amsterdam, J. D., G. Maislin, A. Winokur, N. Berwish, M. Kling, and P. Gold. 1988. The oCRH stimulation test before and after clinical recovery from depression. *J Affect Disord* 14:213–22.

Andrade, L., J. J. Caraveo-Anduaga, P. Berglund, et al. 2003. The epidemiology of major depressive episodes: Results from the international consortium of psychiatric epidemiology (ICPE) surveys. *Int J Methods Psychiatr Res* 12:3–21.

Arase, K., D. A. York, H. Shimizu, N. Shargill, and G. A. Bray. 1988. Effects of corticotropin-releasing factor on food intake and brown adipose tissue thermogenesis in rats. *Am J Physiol* 255:E255–9.

Bale, T. L., and W. W. Vale. 2003. Increased depression-like behaviors in corticotropin-releasing factor receptor-2-deficient mice: Sexually dichotomous responses. *J Neurosci* 23:5295–301.

Bale, T. L., A. Contarino, G. W. Smith, et al. 2000. Mice deficient for corticotropin-releasing hormone receptor-2 display anxiety-like behaviour and are hypersensitive to stress. *Nat Genet* 24:410–14.

Behan, D. P., D. E. Grigoriadis, T. Lovenberg, et al. 1996. Neurobiology of corticotropin releasing factor (CRF) receptors and CRF-binding protein: Implications for the treatment of CNS disorders. *Mol Psychiatry* 1:265–77.

Belanoff, J. K., J. Jurik, L. D. Schatzberg, C. DeBattista, and A. F. Schatzberg. 2002. Slowing the progression of cognitive decline in Alzheimer's disease using mifepristone. *J Mol Neurosci* 19:201–6.

Boldrini, M., M. D. Underwood, R. Hen, et al. 2009. Antidepressants increase neural progenitor cells in the human hippocampus. *Neuropsychopharmacology* 34:2376–89.

Bornstein, S. R., E. L. Webster, D. J. Torpy, et al. 1998. Chronic effects of a nonpeptide corticotropin-releasing hormone type I receptor antagonist on pituitary–adrenal function, body weight, and metabolic regulation. *Endocrinology* 139:1546–55.

Boyle, M. P., J. A. Brewer, M. Funatsu, et al. 2005. Acquired deficit of forebrain glucocorticoid receptor produces depression-like changes in adrenal axis regulation and behavior. *Proc Natl Acad Sci USA* 102:473–8.

Britton, K. T., J. Morgan, J. Rivier, W. Vale, and G. F. Koob. 1985. Chlordiazepoxide attenuates response suppression induced by corticotropin-releasing factor in the conflict test. *Psychopharmacology (Berl)* 86:170–4.

Britton, K. T., G. Lee, R. Dana, S. C. Risch, and G. F. Koob. 1986. Activating and 'anxiogenic' effects of corticotropin releasing factor are not inhibited by blockade of the pituitary–adrenal system with dexamethasone. *Life Sci* 39:1281–6.

Britton, K. T., S. Southerland, E. Van Uden, D. Kirby, J. Rivier, and G. Koob. 1997. Anxiolytic activity of NPY receptor agonists in the conflict test. *Psychopharmacology* (*Berl*) 132:6–13.

Britton, K. T., Y. Akwa, M. G. Spina, and G. F. Koob. 2000. Neuropeptide Y blocks anxiogenic-like behavioral action of corticotropin-releasing factor in an operant conflict test and elevated plus maze. *Peptides* 21:37–44.

Broadbear, J. H., G. Winger, K. C. Rice, and J. H. Woods. 2002. Antalarmin, a putative CRH-RI antagonist, has transient reinforcing effects in rhesus monkeys. *Psychopharmacology* (*Berl*) 164:268–76.

Brown, M. R. 1991. Brain peptide regulation of autonomic nervous and neuroendocrine functions. In *Neurobiology and Neuroendocrinology of Stress*, ed. M. R. Brown, C. Rivier, and G. F. Koob, 193–216. New York: Marcel Dekker.

Chaki, S., A. Nakazato, L. Kennis, et al. 2004. Anxiolytic- and antidepressant-like profile of a new CRF1 receptor antagonist, R278995/CRA0450. *Eur J Pharmacol* 485:145–58.

Chalmers, D. T., T. W. Lovenberg, and E. B. De Souza. 1995. Localization of novel corticotropin-releasing factor receptor (CRF2) mRNA expression to specific subcortical nuclei in rat brain: Comparison with CRF1 receptor mRNA expression. *J Neurosci* 15:6340–50.

Chastrette, N., R. Cespublio, and M. Jouvet. 1990. Proopiomelanocortin (POMC)-derived peptides and sleep in the rat. Part 1. Hypnogenic properties of ACTH derivatives. *Neuropeptides* 15:61–74.

Chen, A., E. Zorrilla, S. Smith, et al. 2006. Urocortin 2-deficient mice exhibit gender-specific alterations in circadian hypothalamus–pituitary–adrenal axis and depressive-like behavior. *J Neurosci* 26:5500–10.

Cole, B. J., and G. F. Koob. 1988. Propranolol antagonizes the enhanced conditioned fear produced by corticotropin releasing factor. *J Pharmacol Exp Ther* 247:902–10.

Connor, T. J., J. P. Kelly, and B. E. Leonard. 1997. Forced swim test-induced neurochemical endocrine, and immune changes in the rat. *Pharmacol Biochem Behav* 58:961–7.

Connor, T. J., P. Kelliher, Y. Shen, A. Harkin, J. P. Kelly, and B. E. Leonard. 2000. Effect of subchronic antidepressant treatments on behavioral, neurochemical, and endocrine changes in the forced-swim test. *Pharmacol Biochem Behav* 65:591–7.

Contarino, A., F. Dellu, G. F. Koob, et al. 1999. Reduced anxiety-like and cognitive performance in mice lacking the corticotropin-releasing factor receptor 1. *Brain Res* 835:1–9.

Contarino, A., F. Dellu, G. F. Koob, et al. 2000. Dissociation of locomotor activation and suppression of food intake induced by CRF in CRFR1-deficient mice. *Endocrinology* 141:2698–702.

de Groote, L., R. G. Penalva, C. Flachskamm, J. M. Reul, and A. C. Linthorst. 2005. Differential monoaminergic, neuroendocrine and behavioural responses after central administration of corticotropin-releasing factor receptor type 1 and type 2 agonists. *J Neurochem* 94:45–56.

de Kloet, E. R. 1995. Steroids, stability and stress. *Front Neuroendocrinol* 16:416–25.

de Kloet, E. R., E. Vreugdenhil, M. S. Oitzl, and M. Joels. 1998. Brain corticosteroid receptor balance in health and disease. *Endocr Rev* 19:269–301.

de Kloet, E. R., S. A. Van Acker, R. M. Sibug, et al. 2000. Brain mineralocorticoid receptors and centrally regulated functions. *Kidney Int* 57:1329–36.

DeBattista, C., and J. Belanoff. 2006. The use of mifepristone in the treatment of neuropsychiatric disorders. *Trends Endocrinol Metab* 17:117–21.

Ducottet, C., G. Griebel, and C. Belzung. 2003. Effects of the selective nonpeptide corticotropin-releasing factor receptor 1 antagonist antalarmin in the chronic mild stress model of depression in mice. *Prog Neuropsychopharmacol Biol Psychiatry* 27:625–31.

Dunn, A. J., and C. W. Berridge. 1990. Physiological and behavioral responses to corticotropin-releasing factor administration: Is CRF a mediator of anxiety or stress responses? *Brain Res Rev* 15:71–100.

Dunn, A. J., and A. H. Swiergiel. 2008. Effects of acute and chronic stressors and CRF in rat and mouse tests for depression. *Ann N Y Acad Sci* 1148:118–26.

DuPont, R. L., D. P. Rice, L. S. Miller, S. S. Shiraki, C. R. Rowland, and H. J. Harwood. 1996. Economic costs of anxiety disorders. *Anxiety* 2:167–72.

Eaves, M., K. Thatcher-Britton, J. Rivier, W. Vale, and G. F. Koob. 1985. Effects of corticotropin releasing factor on locomotor activity in hypophysectomized rats. *Peptides* 6:923–6.

Evans, D. L., and C. B. Nemeroff. 1984. The dexamethasone suppression test in organic affective syndrome. *Am J Psychiatry* 141:1465–7.

Evans, R. M., and J. L. Arriza. 1989. A molecular framework for the actions of glucocorticoid hormones in the nervous system. *Neuron* 2:1105–12.

Fisher, L. A. 1993. Central actions of corticotropin-releasing factor on autonomic nervous activity and cardio-vascular functioning. In *Corticotropin-Releasing Factor*, ed. D. J. Chadwick, J. Marsh, and K. Ackrill, 243–257. New York: John Wiley and Sons.

Flores, B. H., H. Kenna, J. Keller, H. B. Solvason, and A. F. Schatzberg. 2006. Clinical and biological effects of mifepristone treatment for psychotic depression. *Neuropsychopharmacology* 31:628–36.

Funder, J. W. 1992. Glucocorticoid receptors. *J Steroid Biochem Mol Biol* 43:389–94.

Galard, R., R. Catalan, J. M. Castellanos, and J. M. Gallart. 2002. Plasma corticotropin-releasing factor in depressed patients before and after the dexamethasone suppression test. *Biol Psychiatry* 51:463–8.

Gregus, A., A. J. Wintink, A. C. Davis, and L. E. Kalynchuk. 2005. Effect of repeated corticosterone injections and restraint stress on anxiety and depression-like behavior in male rats. *Behav Brain Res* 156: 105–14.

Griebel, G., J. Simiand, R. Steinberg, et al. 2002. 4-(2-Chloro-4-methoxy-5-methylphenyl)-N-[(1S)-2-cyclopropyl-1-(3-fluoro-4-methylphenyl)ethyl]5-methyl-N-(2-propynyl)-1, 3-thiazol-2-amine hydro-chloride (SSR125543A), a potent and selective corticotrophin-releasing factor(1) receptor antagonist: II. Characterization in rodent models of stress-related disorders. *J Pharmacol Exp Ther* 301:333–45.

Hammack, S. E., M. J. Schmid, M. L. LoPresti, et al. 2003. Corticotropin releasing hormone type 2 receptors in the dorsal raphe nucleus mediate the behavioral consequences of uncontrollable stress. *J Neurosci* 23:1019–25.

Heilig, M., and R. Murison. 1987. Intracerebroventricular neuropeptide Y suppresses open field and home cage activity in the rat. *Regul Pept* 19:221–31.

Heilig, M., S. McLeod, G. K. Koob, and K. T. Britton. 1992. Anxiolytic-like effect of neuropeptide Y (NPY), but not other peptides in an operant conflict test. *Regul Pept* 41:61–9.

Heilig, M., S. McLeod, M. Brot, et al. 1993. Anxiolytic-like action of neuropeptide Y: Mediation by Y1 receptors in amygdala, and dissociation from food intake effects. *Neuropsychopharmacology* 8:357–63.

Heinrichs, S. C., F. Menzaghi, E. M. Pich, et al. 1994. Anti-stress action of a corticotropin-releasing factor antagonist on behavioral reactivity to stressors of varying type and intensity. *Neuropsychopharmacology* 11:179–86.

Held, K., H. Kunzel, M. Ising, et al. 2004. Treatment with the CRH1-receptor-antagonist R121919 improves sleep-EEG in patients with depression. *J Psychiatr Res* 38:129–36.

Ho, S. P., L. K. Takahashi, V. Livanov, et al. 2001. Attenuation of fear conditioning by antisense inhibition of brain corticotropin releasing factor-2 receptor. *Brain Res Mol Brain Res* 89:29–40.

Holsboer, F. 2000. The corticosteroid receptor hypothesis of depression. *Neuropsychopharmacology* 23:477–501.

Irwin, M., W. Vale, and C. Rivier. 1990. Central corticotropin-releasing factor mediates the suppressive effect of stress on natural killer cytotoxicity. *Endocrinology* 126:2837–44.

Ising, M., U. S. Zimmermann, H. E. Kunzel, et al. 2007. High-affinity CRF1 receptor antagonist NBI-34041: Preclinical and clinical data suggest safety and efficacy in attenuating elevated stress response. *Neuropsychopharmacology* 32:1941–9.

Isogawa, K., J. Akiyoshi, T. Tsutsumi, K. Kodama, Y. Horinouti, and H. Nagayama. 2003. Anxiogenic-like effect of corticotropin-releasing factor receptor 2 antisense oligonucleotides infused into rat brain. *J Psychopharmacol* 17:409–13.

Jacobson, L., and R. Sapolsky. 1991. The role of the hippocampus in feedback regulation of the hypothalamic–pituitary–adrenocortical axis. *Endocr Rev* 12:118–34.

Johnson, S. A., N. M. Fournier, and L. E. Kalynchuk. 2006. Effect of different doses of corticosterone on depression-like behavior and HPA axis responses to a novel stressor. *Behav Brain Res* 168:280–8.

Jutkiewicz, E. M., S. K. Wood, H. Houshyar, L. W. Hsin, K. C. Rice, and J. H. Woods. 2005. The effects of CRF antagonists, antalarmin, CP154,526, LWH234, and R121919, in the forced swim test and on swim-induced increases in adrenocorticotropin in rats. *Psychopharmacology (Berl)* 180:215–23.

Kalynchuk, L. E., A. Gregus, D. Boudreau, and T. S. Perrot-Sinal. 2004. Corticosterone increases depression-like behavior, with some effects on predator odor-induced defensive behavior, in male and female rats. *Behav Neurosci* 118:1365–77.

Kessler, R. C., O. Demler, R. G. Frank, et al. 2005. Prevalence and treatment of mental disorders, 1990 to 2003. *N Engl J Med* 352:2515–23.

King, J. A., and E. Edwards. 1999. Early stress and genetic influences on hypothalamic–pituitary–adrenal axis functioning in adulthood. *Horm Behav* 36:79–85.

King, J. A., D. Campbell, and E. Edwards. 1993. Differential development of the stress response in congenital learned helplessness. *Int J Dev Neurosci* 11:435–42.

Kishimoto, T., J. Radulovic, M. Radulovic, et al. 2000. Deletion of crhr2 reveals an anxiolytic role for corticotropin-releasing hormone receptor-2. *Nat Genet* 24:415–9.

Koob, G. F., and S. C. Heinrichs. 1999. A role for corticotropin releasing factor and urocortin in behavioral responses to stressors. *Brain Res* 848:141–52.

Koob, G. F., N. Swerdlow, M. Seeligson, et al. 1984. Effects of alpha-flupenthixol and naloxone on CRF-induced locomotor activation. *Neuroendocrinology* 39:459–64.

Koob, G. F., S. C. Heinrichs, F. Menzaghi, E. M. Pich, and K. T. Britton. 1994. Corticotropin releasing factor, stress, and behavior. *Sem Neurosci* 6:221–9.

Krahn, D. D., B. A. Gosnell, M. Grace, and A. S. Levine. 1986. CRF antagonist partially reverses CRF- and stress-induced effects on feeding. *Brain Res Bull* 17:285–9.

Kronenberg, G., I. Kirste, D. Inta, et al. 2009. Reduced hippocampal neurogenesis in the GR(+/-) genetic mouse model of depression. *Eur Arch Psychiatry Clin Neurosci* 259:499–504.

Kunzel, H. E., A. W. Zobel, T. Nickel, et al. 2003. Treatment of depression with the CRH-1-receptor antagonist R121919: Endocrine changes and side effects. *J Psychiatr Res* 37:525–33.

Kunzel, H. E., M. Ising, A. W. Zobel, et al. 2005. Treatment with a CRH-1-receptor antagonist (R121919) does not affect weight or plasma leptin concentration in patients with major depression. *J Psychiatr Res* 39:173–7.

Lenze, E. J., B. H. Mulsant, M. K. Shear, et al. 2000. Comorbid anxiety disorders in depressed elderly patients. *Am J Psychiatry* 157:722–8.

Lenze, E. J., B. H. Mulsant, M. K. Shear, G. S. Alexopoulos, E. Frank, and C. F. Reynolds III. 2001. Comorbidity of depression and anxiety disorders in later life. *Depress Anxiety* 14:86–93.

Lenze, E. J., B. H. Mulsant, J. Mohlman, et al. 2005. Generalized anxiety disorder in late life: Lifetime course and comorbidity with major depressive disorder. *Am J Geriatr Psychiatry* 13:77–80.

Lewis, K., C. Li, M. H. Perrin, et al. 2001. Identification of urocortin III, an additional member of the corticotropin-releasing factor (CRF) family with high affinity for the CRF2 receptor. *Proc Natl Acad Sci U S A* 98:7570–5.

Li, Y. W., L. Fitzgerald, H. Wong, et al. 2005. The pharmacology of DMP696 and DMP904, non-peptidergic CRF1 receptor antagonists. *CNS Drug Rev* 11:21–52.

Macey, D. J., G. F. Koob, and A. Markou. 2000. CRF and urocortin decreased brain stimulation reward in the rat: Reversal by a CRF receptor antagonist. *Brain Res* 866:82–91.

Maier, S. F., and M. E. P. Seligman. 1976. Learned helplessness: Theory and evidence. *J Exp Psychol Gen* 105:3–46.

Malmstrom, R. E., J. O. Lundberg, and E. Weitzberg. 2002. Autoinhibitory function of the sympathetic prejunctional neuropeptide Y Y(2) receptor evidenced by BIIE0246. *Eur J Pharmacol* 439:113–19.

Mansbach, R. S., E. N. Brooks, and Y. L. Chen. 1997. Antidepressant-like effects of CP-154,526, a selective CRF1 receptor antagonist. *Eur J Pharmacol* 323:21–6.

Marks, W., N. M. Fournier, and L. E. Kalynchuk. 2009. Repeated exposure to corticosterone increases depression-like behavior in two different versions of the forced swim test without altering nonspecific locomotor activity or muscle strength. *Physiol Behav* 98:67–72.

Meijer, O. C., and E. R. de Kloet. 1998. Corticosterone and serotonergic neurotransmission in the hippocampus: Functional implications of central corticosteroid receptor diversity. *Crit Rev Neurobiol* 12:1–20.

Menard, J., and D. Treit. 1996. Lateral and medial septal lesions reduce anxiety in the plus-maze and probe-burying tests. *Physiol Behav* 60:845–53.

Menzaghi, F., S. C. Heinrichs, E. Merlo-Pich, F. J. Tilders, and G. F. Koob. 1994a. Involvement of hypothalamic corticotropin-releasing factor neurons in behavioral responses to novelty in rats. *Neurosci Lett* 168:139–42.

Menzaghi, F., R. L. Howard, S. C. Heinrichs, W. Vale, J. Rivier, and G. F. Koob. 1994b. Characterization of a novel and potent corticotropin-releasing factor antagonist in rats. *J Pharmacol Exp Ther* 269:564–72.

Nemeroff, C. B., and D. L. Evans. 1984. Correlation between the dexamethasone suppression test in depressed patients and clinical response. *Am J Psychiatry* 141:247–9.

Nemeroff, C. B., E. Widerlov, G. Bissette, et al. 1984. Elevated concentrations of CSF corticotropin-releasing factor-like immunoreactivity in depressed patients. *Science* 226:1342–4.

Nemeroff, C. B., M. J. Owens, G. Bissette, A. C. Andorn, and M. Stanley. 1988. Reduced corticotropin releasing factor binding sites in the frontal cortex of suicide victims. *Arch Gen Psychiatry* 45:577–9.

Nemeroff, C. B., K. R. Krishnan, D. Reed, R. Leder, C. Beam, and N. R. Dunnick. 1992. Adrenal gland enlargement in major depression. A computed tomographic study. *Arch Gen Psychiatry* 49:384–7.

Nielsen, D. M., G. J. Carey, and L. H. Gold. 2004. Antidepressant-like activity of corticotropin-releasing factor type-1 receptor antagonists in mice. *Eur J Pharmacol* 499:135–46.

Ohata, H., and T. Shibasaki. 2004. Effects of urocortin 2 and 3 on motor activity and food intake in rats. *Peptides* 25:1703–9.

Overstreet, D. H., and G. Griebel. 2004. Antidepressant-like effects of CRF1 receptor antagonist SSR125543 in an animal model of depression. *Eur J Pharmacol* 497:49–53.

Overstreet, D. H., A. Keeney, and S. Hogg. 2004. Antidepressant effects of citalopram and CRF receptor antagonist CP-154,526 in a rat model of depression. *Eur J Pharmacol* 492:195–201.

Overstreet, D. H., E. Friedman, A. A. Mathe, and G. Yadid. 2005. The Flinders Sensitive Line rat: A selectively bred putative animal model of depression. *Neurosci Biobehav Rev* 29:739–59.

Pariante, C. M. 2003. Depression, stress and the adrenal axis. *J Neuroendocrinol* 15:811–12.

Pariante, C. M. 2009. Risk factors for development of depression and psychosis. Glucocorticoid receptors and pituitary implications for treatment with antidepressant and glucocorticoids. *Ann N Y Acad Sci* 1179:144–52.

Peiffer, A., S. Veilleux, and N. Barden. 1991. Antidepressant and other centrally acting drugs regulate glucocorticoid receptor messenger RNA levels in rat brain. *Psychoneuroendocrinology* 16:505–15.

Pelleymounter, M. A., M. Joppa, M. Carmouche, et al. 2000. Role of corticotropin-releasing factor (CRF) receptors in the anorexic syndrome induced by CRF. *J Pharmacol Exp Ther* 293:799–806.

Pelleymounter, M. A., M. Joppa, N. Ling, and A. C. Foster. 2002. Pharmacological evidence supporting a role for central corticotropin-releasing factor(2) receptors in behavioral, but not endocrine, response to environmental stress. *J Pharmacol Exp Ther* 302:145–52.

Perrin, M., C. Donaldson, R. Chen, et al. 1995. Identification of a second corticotropin-releasing factor receptor gene and characterization of a cDNA expressed in heart. *Proc Natl Acad Sci USA* 92:2969–73.

Porsolt, R. D., M. Le Pichon, and M. Jalfre. 1977. Depression: A new animal model sensitive to antidepressant treatments. *Nature* 266:730–2.

Raadsheer, F. C., W. J. Hoogendijk, F. C. Stam, F. J. Tilders, and D. F. Swaab. 1994. Increased numbers of corticotropin-releasing hormone expressing neurons in the hypothalamic paraventricular nucleus of depressed patients. *Neuroendocrinology* 60:436–44.

Radulovic, J., A. Ruhmann, T. Liepold, and J. Spiess. 1999. Modulation of learning and anxiety by corticotropin-releasing factor (CRF) and stress: Differential roles of CRF receptors 1 and 2. *J Neurosci* 19:5016–25.

Raone, A., A. Cassanelli, S. Scheggi, R. Rauggi, B. Danielli, and M. G. De Montis. 2007. Hypothalamus–pituitary–adrenal modifications consequent to chronic stress exposure in an experimental model of depression in rats. *Neuroscience* 146:1734–42.

Reif, A., S. Fritzen, M. Finger, et al. 2006. Neural stem cell proliferation is decreased in schizophrenia, but not in depression. *Mol Psychiatry* 11:514–22.

Reyes, T. M., K. Lewis, M. H. Perrin, et al. 2001. Urocortin II: A member of the corticotropin-releasing factor (CRF) neuropeptide family that is selectively bound by type 2 CRF receptors. *Proc Natl Acad Sci U S A* 98:2843–8.

Ridder, S., S. Chourbaji, R. Hellweg, et al. 2005. Mice with genetically altered glucocorticoid receptor expression show altered sensitivity for stress-induced depressive reactions. *J Neurosci* 25:6243–50.

Risbrough, V. B., R. L. Hauger, M. A. Pelleymounter, and M. A. Geyer. 2003. Role of corticotropin releasing factor (CRF) receptors 1 and 2 in CRF-potentiated acoustic startle in mice. *Psychopharmacology (Berl)* 170:178–87.

Risbrough, V. B., R. L. Hauger, A. L. Roberts, W. W. Vale, and M. A. Geyer. 2004. Corticotropin-releasing factor receptors CRF1 and CRF2 exert both additive and opposing influences on defensive startle behavior. *J Neurosci* 24:6545–52.

Rivier, C. 1996. Alcohol stimulates ACTH secretion in the rat: Mechanisms of action and interactions with other stimuli. *Clin Exp Res* 20:240–54.

Rubin, R. T., J. J. Phillips, T. F. Sadow, and J. T. McCracken. 1995. Adrenal gland volume in major depression. increase during the depressive episode and decrease with successful treatment. *Arch Gen Psychiatry* 52:213–18.

Sahay, A., and R. Hen. 2007. Adult hippocampal neurogenesis in depression. *Nat Neurosci* 10:1110–15.

Sapolsky, R. M., M. J. Meaney, and B. S. McEwen. 1985. The development of the glucocorticoid receptor system in the rat limbic brain: III. Negative-feedback regulation. *Brain Res* 350:169–73.

Schulz, D. W., R. S. Mansbach, J. Sprouse, et al. 1996. CP-154,526: A potent and selective nonpeptide antagonist of corticotropin releasing factor receptors. *Proc Natl Acad Sci U S A* 93:10477–82.

Sheline, Y. I., P. W. Wang, M. H. Gado, J. G. Csernansky, and M. W. Vannier. 1996. Hippocampal atrophy in recurrent major depression. *Proc Natl Acad Sci U S A* 93:3908–13.

Smith, G. W., J. M. Aubry, F. Dellu, et al. 1998. Corticotropin releasing factor receptor 1-deficient mice display decreased anxiety, impaired stress response, and aberrant neuroendocrine development. *Neuron* 20:1093–102.

Smith-White, M. A., T. A. Hardy, J. A. Brock, and E. K. Potter. 2001. Effects of a selective neuropeptide Y Y2 receptor antagonist, BIIE0246, on Y2 receptors at peripheral neuroeffector junctions. *Br J Pharmacol* 132:861–8.

Spina, M., E. Merlo-Pich, R. K. Chan, et al. 1996. Appetite-suppressing effects of urocortin, a CRF-related neuropeptide. *Science* 273:1561–4.

Spina, M. G., A. M. Basso, E. P. Zorrilla, et al. 2000. Behavioral effects of central administration of the novel CRF antagonist astressin in rats. *Neuropsychopharmacology* 22:230–9.

Stenzel-Poore, M. P., S. C. Heinrichs, S. Rivest, G. F. Koob, and W. W. Vale. 1994. Overproduction of corticotropin-releasing factor in transgenic mice: A genetic model of anxiogenic behavior. *J Neurosci* 14:2579–84.

Stratakis, C. A., and G. P. Chrousos. 1995. Neuroendocrinology and pathophysiology of the stress system. *Ann N Y Acad Sci* 771:1–18.

Sutton, R. E., G. F. Koob, M. Le Moal, J. Rivier, and W. Vale. 1982. Corticotropin releasing factor produces behavioural activation in rats. *Nature* 297: 331–3.

Swerdlow, N. R., M. A. Geyer, W. W. Vale, and G. F. Koob. 1986. Corticotropin-releasing factor potentiates acoustic startle in rats: Blockade by chlordiazepoxide. *Psychopharmacology (Berl)* 88:147–52.

Swerdlow, N. R., K. T. Britton, and G. F. Koob. 1989. Potentiation of acoustic startle by corticotropin-releasing factor (CRF) and by fear are both reversed by alpha-helical CRF(9–41). *Neuropsychopharmacology* 2:285–92.

Swiergiel, A. H., Y. Zhou, and A. J. Dunn. 2007. Effects of chronic footshock, restraint and corticotropin-releasing factor on freezing, ultrasonic vocalization and forced swim behavior in rats. *Behav Brain Res* 183:178–87.

Szymanska, M., B. Budziszewska, L. Jaworska-Feil, et al. 2009. The effect of antidepressant drugs on the HPA axis activity, glucocorticoid receptor level and FKBP51 concentration in prenatally stressed rats. *Psychoneuroendocrinology* 34:822–32.

Taché, Y., M. Gunion, and R. Stephens. 1989. CRF: Central nervous system action to influence gastrointestinal function and role in the gastrointestinal response to stress. In *Corticotropin-Releasing Factor: Basic and Clinical Studies of a Neuropeptide*, ed. E. D. Souza and C. Nemeroff, 299–307. Boca Raton, FL: CRC.

Takahashi, L. K., S. P. Ho, V. Livanov, N. Graciani, and S. P. Arneric. 2001. Antagonism of CRF(2) receptors produces anxiolytic behavior in animal models of anxiety. *Brain Res* 902:135–42.

Takamori, K., N. Kawashima, S. Chaki, A. Nakazato, and K. Kameo. 2001a. Involvement of corticotropin-releasing factor subtype 1 receptor in the acquisition phase of learned helplessness in rats. *Life Sci* 69:1241–8.

Takamori, K., N. Kawashima, S. Chaki, A. Nakazato, and K. Kameo. 2001b. Involvement of the hypothalamus-pituitary-adrenal axis in antidepressant activity of corticotropin-releasing factor subtype 1 receptor antagonists in the rat learned helplessness test. *Pharmacol Biochem Behav* 69:445–9.

Tanaka, M., and G. Telegdy. 2008. Antidepressant-like effects of the CRF family peptides, urocortin 1, urocortin 2 and urocortin 3 in a modified forced swimming test in mice. *Brain Res Bull* 75:509–12.

Tanapat, P., L. A. Galea, and E. Gould. 1998. Stress inhibits the proliferation of granule cell precursors in the developing dentate gyrus. *Int J Dev Neurosci* 16:235–9.

Timpl, P., R. Spanagel, I. Sillaber, et al. 1998. Impaired stress response and reduced anxiety in mice lacking a functional corticotropin-releasing hormone receptor. *Nat Genet* 19:162–6.

Urani, A., and P. Gass. 2003. Corticosteroid receptor transgenic mice: Models for depression? *Ann N Y Acad Sci* 1007:379–93.

Valdez, G. R. 2006. Development of CRF1 receptor antagonists as antidepressants and anxiolytics: Progress to date. *CNS Drugs* 20:887–96.

Valdez, G. R., K. Inoue, G. F. Koob, J. Rivier, W. W. Vale, and E. P. Zorrilla. 2002. Human urocortin II: Mild locomotor suppressive and delayed anxiolytic-like effects of a novel corticotropin-releasing factor related peptide. *Brain Res* 943:142–50.

Valdez, G. R., E. P. Zorrilla, J. Rivier, W. W. Vale, and G. F. Koob. 2003. Locomotor suppressive and anxiolytic-like effects of urocortin 3, a highly selective type 2 corticotropin-releasing factor agonist. *Brain Res* 980:206–12.

Valdez, G. R., E. P. Zorrilla, and G. F. Koob. 2005. Homeostasis within the corticotropin-releasing factor system via CRF2 receptor activation: A novel approach for the treatment of anxiety. *Drug Dev Res* 65:205–15.

Vale, W., J. Spiess, C. Rivier, and J. Rivier. 1981. Characterization of a 41-residue ovine hypothalamic peptide that stimulates secretion of corticotropin and beta-endorphin. *Science* 213:1394–7.

Vale, W., C. Rivier, M. R. Brown, et al. 1983. Chemical and biological characterization of corticotropin releasing factor. *Recent Prog Horm Res* 39:245–70.

Valentino, R. J., and R. G. Wehby. 1988. Corticotropin-releasing factor: Evidence for a neurotransmitter role in the locus coeruleus during hemodynamic stress. *Neuroendocrinology* 48:674–7.

Vaughan, J., C. Donaldson, J. Bittencourt, et al. 1995. Urocortin, a mammalian neuropeptide related to fish pretension I and to corticotropin-releasing factor. *Nature* 378:287–92.

Venihaki, M., S. Sakharov, S. Subramanian, et al. 2004. Urocortin III, a brain neuropeptide of the corticotropin-releasing hormone family: Modulation by stress and attenuation of some anxiety-like behaviors. *J Neuroendocrinol* 16:411–22.

Webster, E. I., D. B. Lewis, D. J. Torpy, E. K. Zachman, K. C. Rice, and G. P. Chrousos. 1996. In vivo and in vitro characterization of antalarmin, a nonpeptide corticotropin-releasing hormone (CRH) receptor antagonist: Suppression of pituitary ACTH release and peripheral inflammation. *Endocrinology* 137:5747–50.

Yadin, E., E. Thomas, H. L. Gresham, and C. E. Strickland. 1993. The role of the lateral septum in anxiolytic. *Physiol Behav* 53:1071–83.

Yamano, M., H. Yuki, S. Yasuda, and K. Miyata. 2000. Corticotropin-releasing hormone receptors mediate consensus interferon-alpha YM643-induced depression-like behavior in mice. *J Pharmacol Exp Ther* 292:181–7.

Young, A. H., P. Gallagher, S. Watson, D. Del-Festal, B. M. Owen, and I. N. Ferrier. 2004. Improvements in neurocognitive function and mood following adjunctive treatment with mifepristone (RU-486) in bipolar disorder. *Neuropsychopharmacology* 29:1538–45.

Zobel, A. W., T. Nickel, H. E. Kunzel, et al. 2000. Effects of the high-affinity corticotropin-releasing hormone receptor 1 antagonist R121919 in major depression: The first 20 patients treated. *J Psychiatr Res* 34:171–81.

Zorrilla, E. P., and G. F. Koob. 2004. The therapeutic potential of CRF1 antagonists for anxiety. *Expert Opin Investig Drugs* 13:799–828.

14 Depression and Cytokine-Regulated Pathways

Bernhard T. Baune

CONTENTS

14.1 CYTOKINE EFFECTS IN THE CENTRAL NERVOUS SYSTEM

Emerging from recent research has been the understanding that depression involves a complex and bidirectional interaction between the immune system and the central nervous system (CNS) (Gold and Irwin 2006). It is now generally accepted that immune dysregulation plays an important role in the etiology and pathogenesis of major depression. There is growing evidence that the increased production of proinflammatory cytokines during stress, immune, or inflammatory responses can exert powerful influences on the CNS.

The term *cytokine* encompasses a large and diverse family of signaling molecules that primarily have immune modulating activity and are produced widely throughout the body by cells of diverse embryological origin. Cytokines play a role in the pathogenesis of numerous disorders, including infection, autoimmune disease, stroke, trauma, and neurodegenerative disease (Rothwell and Loddick 2002; Ransohoff and Benveniste 2006). Cytokines have also been implicated in the pathogenesis of a number of neuropsychiatric disorders, including depression (Capuron and Dantzer 2003; Irwin and Miller 2007; Raison et al. 2006; Hickie and Lloyd 1995; Kronfol and Remick 2000). There have been numerous publications on the association between depression and immune and cytokine function. More specifically, cytokines may exert direct or indirect effects on complex behavioral patterns, emotional processing, sleep patterns, feeding behavior, and complex cognitive function, including memory and learning. The description of pathways by which cytokines may modulate molecular mechanisms of complex depression-like behavior and symptoms forms the major part of this chapter.

14.1.1 Depression and Immunity

Although this chapter focuses on the involvement of cytokines in the pathophysiology of symptom clusters of depression, such as emotion processing, cognitive performance (including learning and memory), appetite, and sleep behavior, it is emphasized that the corresponding literature forms part of a larger theoretical model pertaining to the role of the immune system in depression. The introduction of the macrophage theory in the early 1990s (Smith 1991) is often seen as a seminal advance in the conceptualization of depression and its relationship to immune function and to cytokines in particular. Over recent years, researchers have sought to clarify and better understand the role that the immune system plays in the pathogenesis of depression.

Quantitative markers of immune function, such as leukocyte, lymphocyte, neutrophil, monocyte, basophil, eosinophil, granulocyte, and CD4/CD8 cell numbers, are often affected in depression (Müller et al. 1993, who were the first to describe the increased CD4/CD8 ratio in depression), but findings have been partly inconsistent (Herbert and Cohen 1993a). Many aspects of cellular immunity, including phagocytosis, appear impaired or reduced in depression (Schlatter et al. 2004b). Natural killer cytotoxicity (NKCA) and lymphocyte proliferative response have also been shown to be reduced or impaired in depression (Herbert and Cohen 1993a; Schleifer et al. 1999). Depressed patients have impaired or reduced immune cell function, as measured by reduced lymphocyte proliferation (Zorrilla et al. 2001) and lower NKCA (Zorrilla et al. 2001; Herbert and Cohen 1993a; Schleifer et al. 1999). Early studies reviewing CD4/CD8 ratio in depression have been mixed with Herbert and Cohen's (Schleifer et al. 1999) finding a decreased ratio, whereas Müller et al. (1993) and Zorrilla et al. (2001) detected an increase. Since 2001, however, all the studies we have located demonstrated an increase in CD4 levels and CD4/CD8 ratios in depression (Schlatter et al. 2004a).

Immune markers such as haptoglobin, α1-antitrypsin, and IgM are all consistently elevated in depressed cohorts, indicating that this illness is associated with inflammatory activation similar to an acute phase response (Zorrilla et al. 2001). The main inflammatory cytokines IL-1, IL-6, and tumor necrosis factor (TNF) appear elevated in depressive states or in response to psychological stress (O'Brien et al. 2004). The direct and indirect effects of cytokines in the pathophysiology of depression have led to the formulation of a model called cytokine-induced depression (Dantzer et al. 2008). The model has important implications for understanding the role cytokines play in sickness behavior relevant to depression.

14.1.2 Cytokine—Depression Hypothesis

There is growing evidence to suggest that proinflammatory cytokines play a major role in the pathophysiology of depression. Medical illness can induce a variety of biological and behavioral responses, including fever, anorexia, weight loss, fatigue, sleep disturbances, motor retardation, dysphoria, anhedonia, impaired cognition, and depressed mood, which are all seen in major depression. Administration of proinflammatory cytokines can induce such symptoms in animals (Kent et al. 1992) and humans (McDonald et al. 1987; Capuron and Ravaud 1999; Rosenstein et al. 1999).

Such symptoms are sometimes referred to as *sickness behavior*. It is postulated that antidepressants may prevent the onset of such sickness behavior by modulating proinflammatory cytokines (Flanigan et al. 1992; Yirmiya 1996; Shen et al. 1999; Connor et al. 2000; Castanon et al. 2001). Depressed patients who are otherwise healthy have increased levels of proinflammatory cytokines (IL-1, IL-6, TNF-α, etc.), acute phase proteins, chemokines, and adhesion molecules.

The biological pathways by which cytokines may mediate depression are poorly understood, although a number of mechanisms have been proposed:

1. The proinflammatory cytokines affect serotonin (5-HT) metabolism by reducing tryptophan levels. Cytokines appear to activate indoleamine-2-3-dioxygenase, an enzyme that metabolizes tryptophan, thereby reducing serotonin levels. Furthermore, inflammatory cytokines, such as IL-1β, may reduce extracellular 5-HT levels via activation of the serotonin transporter mechanisms.
2. Proinflammatory cytokines have a potent direct effect on the hypothalamic–pituitary–adrenal (HPA) axis. Cytokines, including IL-1, IL-6, TNF-α, and IFN-α, have been shown to increase inflammatory responses by disrupting the function of glucocorticoid receptors.

Infection and tissue damage lead to increased local cytokine production within and outside of the CNS. Cytokines are large, hydrophilic molecules, which do not easily cross the blood–brain barrier under normal conditions. Peripheral cytokines may, however, communicate with the cerebrum through the afferent sensory fibers of the vagus nerve. Peripheral cytokines have also been shown to enter the CNS. This can occur through passive diffusion at areas where the blood–brain barrier is deficient or by active transport of cytokines stimulated by the central noradrenergic system. Peripheral cytokines may also be activated through neural afferents, leading to synthesis of IL-6 within the brain by microglia and endothelial cells (Pollmacher et al. 2002; O'Brien et al 2004; Szelenyi and Vizi 2007).

In the CNS, cytokines may also exert their effect by activating the HPA axis. Proinflammatory cytokines induce gene expression and synthesis of corticotrophin-releasing factor (CRF), which stimulates adrenocorticotropic hormone release and causes glucocorticoid secretion (Song 2002). An activated HPA axis may lead to a further rise in proinflammatory cytokines, through a complex positive feedback loop. Stress can lead to increased cytokine levels and induction of catecholamines via an activated HPA axis, which may further increase proinflammatory cytokines (Szelenyi and Vizi 2007). Cytokines may also directly affect higher cognitive and emotional functions (McAfoose and Baune 2009), possibly leading to depression and the associated cognitive dysfunction.

14.1.3 BIOLOGICAL PROPERTIES OF CYTOKINES

Cytokines are a group of signaling molecules, primarily protein peptides or glycoproteins, used extensively in immune modulation. They are secreted by a wide variety of cells in the immune system. Traditionally, they can be classified as lymphokines, interleukins (ILs), and chemokines, depending on their presumed function, target, or cells of origin. The term *interleukins* initially indicated that these compounds primarily targeted leukocytes; however, this notion is outdated, because it is clear that the vast majority is produced by T-helper cells. Most new cytokines are designated as ILs. Chemokines are a specific type of cytokine, thought to stimulate chemoattraction.

A classification method that is sometimes used is to divide immunological cytokines into those that enhance cytokine responses, type 1 (such as IFN-γ and transforming growth factor [TGF]-beta [β]) and type 2 (such as IL-4 and IL-10), which favor antibody responses. A complete discussion of cytokine functioning and classification is beyond the scope of this review. Generally, cytokines, especially IL-1, IL-6, and TNF, enhance inflammatory responses. Some cytokines, such as IL-10, can be considered, broadly speaking, "anti-inflammatory", although this distinction of pro- and anti-inflammatory cytokines is an oversimplification.

Cytokines must bind on specific receptors, which are broadly distributed on cells of the immune system. Cytokine receptors have attracted much attention in recent years because of their remarkable characteristics. Currently, five cytokine receptor families are known: (1) immunoglobulin, (2) hematopoietin (class I), (3) interferon (IFN; class II), (4) TNF, and (5) chemokine. It appears from research to date that IL-1, TNF-α, and IL-6 receptors play a significant role in depression. The distribution of these receptors has important significance with regard to their actions.

14.1.4 Cytokines and Their Receptors in CNS

Cytokines were initially identified as a group of soluble mediators within the immune system; they are a heterogeneous group of polypeptide hormones (Connor et al. 1998) involved in notably inflammatory reactions. Mostly they are produced within inflammation cells, such as mast cells and macrophages, and are secreted into the surrounding tissues or fluids; when attached to the cell surface, they can also function as membrane-bound complexes (John et al. 2003). Recently, cytokines and their receptors have been located in other tissues as well, other than the immune system, under which are the peripheral nervous system and the CNS (Rothwell and Hopkins 1995; Schobitz et al. 1994). They appear to be involved in specific neuronal and behavioral processes and different neuropathologies (Dunn 2006). There is also evidence of their effect on neuronal degeneration and neurogenesis (Rothwell and Hopkins 1995), suggesting that they act on the CNS either by inhibiting or reducing neuropathology (John et al. 2003).

To understand the effects of different cytokines on the nervous system, and on the amygdala in particular, the action of cytokines and their receptors will be discussed briefly. Focus will be on the cytokines TNF-α, IL-1β, IL-6, TGF, and IFNs, because the influence of these cytokines on the amygdala is most widely studied.

14.1.4.1 Classification of Cytokines and Their Receptors

Data about cytokines crossing the blood–brain barrier are contradictory. Kaur and Salm (2008) pointed out that cytokines easily cross the blood–brain barrier, whereas Dunn and Quan (Dunn 2002, 2006; Quan and Herkenham 2002) mentioned earlier that although cytokines are typically small proteins, they are too large to readily pass the blood–brain barrier. However, Dunn (2006) named different mechanisms through which cytokines can enter the CNS, without passing the blood–brain barrier. Mechanisms by which cytokines or their effects could enter the CNS are as follows:

- They act on one or other brain regions lacking a blood–brain barrier (so-called circumventricular organs) (Dunn 2006; Plata-Salaman 2001; Turrin and Plata-Salaman 2000; Connor et al. 1998).
- They use specific uptake systems (retrograde axonal cytokine transport) to cross the blood–brain barrier (Plata-Salaman 2001; Turrin and Plata-Salaman 2000); this mechanism has been proven for IL-1β, IL-6, and TNF-α and their soluble receptors (Dunn 2006).
- They use periphery-to-brain communication through the vagus nerve and other neural afferents (Turrin and Plata-Salaman 2000; Dunn 2006; Plata-Salaman 2001).
- They act on tissues outside the CNS that can secrete mediators that are able to cross the blood–brain barrier (cytokine-induced generation of chemical mediators) (Dunn 2006; Plata-Salaman 2001; Turrin and Plata-Salaman 2000).
- They may be synthesized by immune cells that infiltrate the CNS (Pawlak et al. 2005).
- They are produced by neurons or glial cells within the CNS (Connor et al. 1998); this mechanism has been proven for IL-1β and TNF-α (Plata-Salaman 2001). It has to be noted that brain production of cytokines has been demonstrated as a response to several diseases such as peripheral cancer and inflammation (Plata-Salaman 2001) and that production under normal circumstances has not been proven yet.

In order to maintain physiological levels and act on processes in the CNS, it is very likely that the aforementioned mechanisms operate in parallel (Dunn 2006); however, different cytokines might use different mechanisms. When they enter the CNS, cytokines have effects on diverse processes within the CNS and the amygdala and they might act in parallel with or counteract each other.

14.1.4.2 TGF-β

TGF-β has been shown to mainly act as a neuroprotective, anti-inflammatory cytokine (Kaur and Salm 2008). TGF-β plays a key role in neural development, disease, and tissue repair (Gomes et al. 2005), partly by inhibiting the expression of other proinflammatory cytokines, such as TNF-α and IL-1β, hence supporting strong immunosuppressive effects. TGF-β also blocks the cytokine induction of adhesion molecules (Benveniste et al. 2001; Le et al. 2004).

TGF-β is expressed in the brain under normal conditions (Kaur and Salm 2008) and can act at two different receptors, TGFR1 and TGFR2, which activate transcription factors after binding to TGF-β. These transcription factors translocate to the nucleus, subsequently regulating transcriptional responses to TGF-β (Gomes et al. 2005). Because very few data are available about the underlying mechanism, the exact intracellular pathway essential for the effect of TGF-β in the CNS remains unknown.

14.1.4.3 Interleukins

ILs may constitute one of the largest cytokine families, and this family is still growing; IL-1, IL-2, IL-3, IL-4, IL-6, IL-8, and IL-1Ra (a soluble IL-1 receptor antagonist in vivo; Rothwell and Hopkins 1995), just to name a few, are the most widely studied in the past decade. Here, the focus will be on IL-1 (mainly subtype IL-1β) and IL-6; IL-1Ra will be mentioned briefly.

14.1.4.3.1 IL-1β and IL-1 Receptor Antagonist

IL-1β is mainly released from macrophages, acts as a proinflammatory cytokine on the CNS (Connor et al. 1998), and is said to be by far the most potent cytokine (Dunn 2006). IL-1β is able to bind on two different IL-1 receptors. However, the type 1 receptor seems to be the only active receptor; the type 2 receptor has no intracellular signaling effect and is therefore called a "decoy" receptor (Dunn 2006). The naturally occurring receptor antagonist for IL-1β, IL-1Ra, is able to inhibit the effect of IL-1β by blocking the IL-1 receptor. The balance between the levels of IL-1β and IL-1Ra may establish the intensity of neurodegeneration in the CNS (Apte et al. 1998). Moreover, the ability of IL-1β to exert either neurotrophic or neurotoxic effects depends mainly on the IL-1β concentration in the CNS (Rothwell and Relton 1993; Araujo and Cotman 1993).

The CNS (and therefore the amygdala) appears to have a respectable sensitivity to IL-1β; nevertheless, it contains very few receptors for IL-1β. IL-1 receptors have been found in the capillary endothelium and the choroid plexus (Dunn 2006), suggesting that IL-1β might use these receptors to activate mediators, which in turn act on the amygdala.

14.1.4.3.2 Interleukin-6

IL-6 is shown to mainly have an influence on the repair and regulation of inflammation processes, therefore promoting neuronal survival and neurogenesis. On the other hand, it has been shown that IL-6 also has a proinflammatory capacity (John et al. 2003), principally during development of the CNS, and consequently increases neurodegeneration. The ability of IL-6 to exert either neurotrophic or neurotoxic effects depends mainly on the IL-6 concentration in the CNS (Rothwell and Relton 1993; Araujo and Cotman 1993). Whereas a low central IL-6 concentration does not have notable influences on neuronal vitality, higher concentrations showed a remarkable improvement (Peng et al. 2005).

The mechanism through which IL-6 affects the CNS and influences neuronal functions is very diverse, complicated, and, for the biggest part, still unknown. Peng et al. (2005) suggested that the influence on neuronal repair of IL-6 may be via the activation of the gp130 receptor, leading to the protection against glutamate-induced neurotoxicity. Previous research has pointed out that both human IL-1β and human IL-6 are active on their receptors in rodents; as a consequence, it can be concluded that human and rodent IL-1β and IL-6 are similar (Dunn 2006). This is of remarkable importance for the interpretation of the outcome of animal studies and the extrapolation to human studies.

14.1.4.3.3 Tumor Necrosis Factor-Alpha

TNF-α is widely known as a proinflammatory cytokine, which is involved in the regulation of both inflammatory as well as survival components (Saha et al. 2006), hence affecting physiological and pathophysiological processes in the CNS. The implementation of neurodegeneration by TNF-α is a result of either the silencing or the induction of survival signals in neurons (Venters et al. 1999). The effect of TNF-α on glial cells, on the other hand, will increase the production of proinflammatory mediators, which might predominate the survival post of these neurons (if survival signals were induced), hence causing neurodegeneration (Saha et al. 2006). The aforementioned is suggesting that, one way or the other, the proinflammatory effect of TNF-α will prevail over the anti-inflammatory effect. This is not always the case: although TNF-α on itself acts neurodegenerative, it can also contribute to a neuroprotective process by activating neurotrophins. It is therefore of importance that proinflammatory products in the CNS are balanced with the expression of neurotrophins, to enable TNF-α to protect the CNS from neurodegeneration (Saha et al. 2006). The ability of TNF-α to exert either neurotrophic or neurotoxic effects depends mainly on the TNF-α concentration in the CNS (Rothwell and Relton 1993; Araujo and Cotman 1993), as well as on the receptor to which it binds. Recently, two TNF-α receptors have been identified: TNF receptor 1 (TNFR1; p55) and TNF receptor 2 (TNFR2; p75) (John et al. 2003). Interaction between TNF-α and the TNFR2 receptor leads to cell activation, neurogenesis, and neuronal repair (Dopp et al. 1997). Increased TNF-α was found to enhance the extent of toxicity (John et al. 2003), suggesting that an altered level of TNF-α will also activate TNFR1, which is believed to mediate cell death (John et al. 2003). TNF-α is able to upregulate the TNF receptors, which in turn induce the levels of this cytokine (Nadeau and Rivest 2000).

Conclusively, it can be stated that a low TNF-α density in the CNS is required for normal brain functioning and neuroprotection; however, when levels rise, due to either infection or external stimuli, TNF-α has the ability to cause cell death and is mainly acting as a neurodegenerative agent.

14.1.4.3.4 Interferons

The family of IFNs consists of the three members, IFN-α, IFN-β, and IFN-γ, which all act differently on the CNS. Neurons have been found to express IFN-α outside of pathological circumstances (Yamada et al. 1994). Although there is so far no evidence for the expression of IFN-β and IFN-γ in normal brain functioning, their presence is not denied either. As a consequence of their ability to bind different receptors, a distinction can be made between type 1 IFNs, which include IFN-α and IFN-β, and type 2 IFNs, which include only IFN-γ. Type 1 IFNs share a common receptor, which consists of two subunits, IFN alpha receptor-1 (IFNAR-1) and IFNAR-2; type 2 IFNs act on IFN gamma receptor (IFN GR) (Dunn 2006). IFNs have a mainly proinflammatory character (Hurlock 2001). The ability of IFNs to exert either neurotrophic or neurotoxic effects depends mainly on the IFN concentration in the CNS (Rothwell and Relton 1993; Araujo and Cotman 1993) and on the receptor activated by the IFNs; therefore, type 1 IFNs (IFN-α and IFN-β) are more likely to act in neurodegeneration, whereas type 2 IFNs (IFN-γ) are more likely to have a neuroprotective effect (DiProspero et al. 1997). The positive effect of IFN-γ after binding to IFN GR is thought to be due to the expression of cell-adhesion molecules such as laminin, fibronectin, and tenacin, which are known to be involved in neuronal development (DiProspero et al. 1997).

In the limbic system, of which the amygdala is a section, IFNs and IFN receptors are widespread. Even though their levels rise principally during CNS infections, it is proven that they also have an important role in normal CNS functioning. Influences on the communication between the CNS and the immune system, as well as their role in neuronal development, have been shown for IFN-α and IFN-γ (Hurlock 2001). A study with IFN null mice has shown a general hypertrophy of emotions and behavior (Hurlock 2001), hence suggesting that physiologic IFN levels in the amygdala are important in evolving basic emotions.

It can be concluded that a low IFN density in the amygdala is required for normal functioning and the evolvement of emotions; moreover, higher levels of IFN-γ are better tolerated than IFN-α and IFN-β levels. However, when levels rise, due to either CNS infection or external stimuli, IFNs develop the ability to cause cell death and mainly act as a neurodegenerative agent.

14.1.4.3.5   Cytokine Balance

Because most cytokines can act as a ligand for more than one receptor, their effect and functional properties depend mostly on the receptor, which is the one expressed instead of the cytokine itself. However, as the modulation of these receptors depends on cytokines, the cytokine levels of the environment, as well as the balance between the different cytokines, influence the physiological effects of these cytokines to a large extent (John et al. 2003). Therefore, the most obvious mechanism of the contribution of cytokines to the origin of neurodegenerative and psychiatric disorders is a dysregulation of the balance of proinflammatory and anti-inflammatory cytokines (O'Brien et al. 2004). A shift in the cytokine balance in the CNS could ultimately lead to deleterious amplification cycles of cellular activation cytotoxicity (Plata-Salaman 2001).

14.2   INTRACELLULAR SIGNALING PATHWAYS OF CYTOKINES

Before the proposed mechanisms by which cytokines exert their multiple functions are described, the intracellular signaling pathways implicated in the cytokine activity are reported. Cytokines must bind with specific receptors in order for signal transduction to take place. The formation of a cytokine receptor complex leads to an intricate chain of downstream effects, ultimately culminating in the activation of transcription and the modulation of cellular activity. These receptors are linked to tyrosine kinases, a group of enzymes that phosphorylate target molecules. Phosphorylation of the cytokine receptor complex activates it, leading to further phosphorylation and activation of other proteins. This process initiates a complex signaling cascade, which amplifies and modulates the actions of the cytokine molecules.

To date, two families of tyrosine kinases, the Janus kinase (JAK) and Src kinases, have been implicated in cytokine signal transduction. Importantly, because most cytokine receptors lack intrinsic catalytic activity (i.e., tyrosine kinase activities), signal transduction by cytokine receptors relies on JAK and Src kinases (Ihle 1995; Yeh and Pellegrini 1999). These kinases primarily serve to amplify intracellular signaling pathways.

Three major intracellular cytokine signaling pathways have been described in more detail: (1) the JAK-STAT (signal transducer and activator of transcription) pathway (Ihle 1995; Yeh and Pellegrini 1999; O'Shea et al. 2002), (2) the Ras/MAPK (mitogen-activated protein kinase pathway, and (3) the phosphoinositide-3-kinase (PI-3-kinase) pathway. These pathways are unique and complex and beyond the scope of this paper to discuss in detail. A final action of the signal transduction, however, occurs through the effect on gene expression via an effect on transcription. An important and commonly implicated transcription pathway occurs through nuclear factor κB (NF-κB), a pleiotrophic transcription factor with broad inflammatory actions.

14.2.1   NF-κB PATHWAY

NF-κB is a protein complex that acts as a transcription factor in the nucleus and is found in almost all mammal cell types. NF-κB has long since been known to be a regulator protein in apoptosis and cell death and operates by the induction of genes mediating cell survival (Mattson and Meffert 2006). Recently, the action of NF-κB in the CNS has been clarified. Within the developing brain,

NF-κB is responsible for neurogenesis; in the forebrain, the amygdala and the hippocampus particularly (Yeh et al. 2002), NF-κB remains elevated in adults (Bhakar et al. 2002). In these brain regions, NF-κB is mostly available in synaptic terminals, dendrites, and axons, and activates as a response to mediators such as cytokines and neurotrophins (Mattson and Meffert 2006). NF-κB is also found in the blood–brain barrier, where its activation by mediators, which are not capable of crossing the blood–brain barrier, is crucial for neural activation and development (Nadjar et al. 2005; Bhakar et al. 2002).

NF-κB is a dimer that operates as a DNA-binding factor (Siebenlist et al. 1994). In the cytoplasm, NF-κB is present in an inactive form, bound to an inhibitory protein called IκB (Baldwin 1996; Yeh et al. 2004). For NF-κB to become activated, phosphorylation of IκB by the enzyme IκB kinase (IKK) is required; this phosphorylation by IKK requires activation of IP-3, MAPK, and N-methyl-D-aspartic acid (NMDA) receptors, which will be discussed in the following paragraphs. After activation by NF-κB of the NF-κB binding sites, a rapid resynthesis of IκB will follow (Yeh et al. 2002; Bhakar et al. 2002). Next to phosphorylation of IκB, the NF-κB complex enters the nucleus and causes gene transcription and translation of specific genes that have binding sites for NF-κB (Yeh et al. 2004; Hayden and Ghosh 2004), resulting in the production of cytokines, cytokine receptors, chemokines, and prostaglandins (Verma et al. 1995). Activation of the NF-κB signaling pathway is essential for the induction of proinflammatory cytokines in the amygdala, such as TNF-α and IL-1β (John et al. 2003).

Activation of NF-κB in the amygdala is a requirement for synaptic plasticity, memory consolidation, and the evolvement of fear cognition (Mattson and Meffert 2006); it has been proven that stimuli generating long-term potentiation (LTP) in animals activate NF-κB in the amygdala (Yeh et al. 2002). Yeh et al. (2002) have also described the possibility of increased NF-κB activity in the amygdala after fear-conditioned training and the impairment of fear memory after disruption of this signaling pathway.

NF-κB activation is now proven for IL-1β, TNF-α, TGF-β, and nerve growth factor (NGF); by doing so, they cause neuronal survival and neurogenesis as well as LTP (Mattson and Meffert 2006; Hayden and Ghosh 2004; Le et al. 2004). Cytokines and neurotrophins are able to influence each other through NF-κB signaling as well. For example, TNF-α has been demonstrated to cause an expression of brain-derived neurotrophic factor (BDNF) through the activation of NF-κB by inducing one of the four BDNF transcripts (Saha et al. 2006).

The involvement of several cytokine and neurotrophin receptors in the activation of NF-κB has also been demonstrated; TrkA, p75, TNFR1, and TNFR2 signaling has been linked to either cell-survival (TrkA and TNFR2) or cell-death (TNFR1) functions, or both (p75) (Marchetti et al. 2004; Carter et al. 1996; Mattson and Meffert 2006). In addition, NF-κB appears to have different subunits (Castagne et al. 2001); depending on the subunit that becomes activated, NF-κB is involved in either neuroprotection or neurodegeneration. Other studies suggest that, besides depending on the receptor, NF-κB-mediated actions may also depend on the location of NF-κB activation. When NF-κB is activated in glial cells, it may induce cell death; when NF-κB is activated in neurons, it may serve as a neuroprotective pathway (Mattson and Meffert 2006; Bhakar et al. 2002). However, the factors that influence either the neuroprotective or neurodegenerative actions of NF-κB are poorly understood (Mattson and Meffert 2006).

14.2.2 MAPK ACTIVATION BY PI-3 KINASE

MAPK activation by PI-3 kinase is required for the phosphorylation of IκB by IKK and therefore the release of NF-κB into the nucleus (Yeh et al. 2002). MAPK is an intracellular protein kinase that responds to extracellular stimuli and regulates various cellular activities, including NF-κB activation. The requirement of MAPK in LTP in the amygdala and the evolvement of emotions such as fear memory formation, the extinction of fear, and taste-aversion memory has been demonstrated (Berman et al. 1998; Blum et al. 1999; Lu et al. 2001; Huang et al. 2000), hence suggesting that

MAPK activation is a crucial factor in the NF-κB cascade. Further evidence demonstrated that MAPK acts on the NF-κB cascade by binding to IKK proteins, therefore regulating the activation of NF-κB (Yin et al. 2001).

To be able to bind IKK proteins, MAPK has to be activated by PI-3 kinase, with PI-3 kinase being an early intermediate in the MAPK-signaling pathway (Lin et al. 2001). The requirement for MAPK activation by PI-3 kinase in fear cognition and LTP in the amygdala has been demonstrated by Lin et al. (2001) and Huang et al. (2000); they showed an induction in PI-3 kinase in the amygdala as a result of fear conditioning and LTP. Activation of PI-3 kinase is thought to be the first step in the NF-κB cascade. PI-3 kinase becomes activated by two mechanisms acting in parallel: signaling through cytokine or neurotrophin receptors on one side and by Ca^{2+} influx through the NMDA receptor on the other (Rattiner et al. 2004). BDNF has been shown to play a major role in the translocation and activation of MAPK, by signaling through the TrkB receptor (Patterson et al. 2001), therefore providing evidence that BDNF is required for normal amygdala functioning. BDNF most likely triggers MAPK phosphorylation by PI-3 kinase activation within the amygdala (Levine et al. 1998). Several growth factors, such as NGF, have also been shown to be involved in PI-3 kinase activation (Miller et al. 1997).

14.2.3   Involvement of NMDA Receptor in NF-κB Cascade

The influx of Ca^{2+} through the NMDA receptor, which is triggered by glutamate, is thought to be involved in the translocation of both MAPK and PI-3 kinase to the nucleus (Weisskopf et al. 1999), where they can act on IKK proteins. TrkB signaling has been proven to act in parallel with NMDA signaling on the activation of PI-3 kinase and, as a result, the activation of MAPK (Gewirtz et al. 2002). The blocking of one of these two mechanisms will prevent the activation of PI-3 kinase and MAPK, hence preventing NF-κB from entering the nucleus. It has been demonstrated that a prevention of Ca^{2+} influx through NMDA receptors will decrease synaptic transmission in the basolateral nucleus (BLA) (Weisskopf et al. 1999), thus proving that NMDA signaling is crucial for processes evolved by the BLA, for example, LTP (Ren and Dubner 2007). The NMDA receptor has also been shown to be involved in gene transcription and protein synthesis, therefore also influencing another step in the NF-κB cascade (Nader et al. 2000).

Whereas cytokines can influence only one side of the NF-κB cascade, neurotrophins can act on both. They can either activate PI-3 kinase directly by signaling through their receptors (e.g., p75) or act on the NMDA receptor, hence inducing Ca^{2+} influx, which also leads to PI-3 kinase activation (Ren and Dubner 2007). BDNF-TrkB signaling occurs through the intracellular cascade that involves IP-3 kinase activation, as previously described; another effect of BDNF involves the phosphorylation of NMDA receptors, hence initiating a descending facilitatory drive (Ren and Dubner 2007), which is also described to occur by signaling through TrkB receptors (Rattiner et al. 2005; Suen et al. 1997; Ren and Dubner 2007).

14.3   EFFECTS OF CYTOKINES ON DEPRESSION-LIKE SYMPTOM CLUSTERS

The effects of cytokines on the response of an individual to biologically relevant stimuli have been extensively studied in the past few decades. The most important result in this category is the ability of cytokines to affect cognitive performance and emotional processing and to induce sickness behavior in mammals (Anisman et al. 2002), which includes anhedonia, feelings of depression and guilt, decreased sexual behavior, and a decreased appetite. Centrally administered IL-1β has been shown to inhibit food intake and social behavior (Nadjar et al. 2005; Brebner et al. 2000; Connor et al. 1998). Centrally administered TNF-α demonstrated equal results and also increased when the subject is lying down (compared to standing) (Sakumoto et al. 2003; Brebner et al. 2000). Other cytokines such as IFN (Hurlock 2001) and IL-6 (Connor et al. 1998) have also been shown to affect the response to biologically relevant behavior. However, none of these studies have focused especially on the role of the amygdala in these processes. Because the amygdala is involved in the response to biologically relevant stimuli (Dougherty and Rauch 2001) and cytokines have been shown to affect the amygdala (Kaur and Salm 2008), it is to be expected that cytokines may act on this response by acting on the amygdala. However, the available evidence so far has not proven this hypothesis. Until now, cytokines have been shown to act on the response to biologically relevant stimuli mainly by acting on other brain regions, either by affecting the HPA axis (Ericsson et al. 1994) or by the upregulation of several neurotransmitters (Dunn 2006). Although the role of the amygdala in the response to biologically relevant stimuli as a result of administration of cytokines is not ruled out, further research on these processes is required.

14.3.1 Cognitive Deficits in Depression

The symptomatology of depression is rarely limited to the affective symptoms that need to be present for a diagnosis. Empirical and clinical studies have shown that depression does not exist in isolation but is rather linked intrinsically to cognitive dysfunction (Ravnkilde et al. 2002; Austin et al. 1992). Early research of neuropsychological performance in depressed patients demonstrated that there were deficits in memory in these patients as compared to normal subjects (Cronholm and Ottosson 1961; Golinkoff and Sweeney 1989; Blaney 1986). More recent studies continue to support the conclusion that cognitive impairments are a part of the symptomatology of depression (Dupont et al. 1995; Ravnkilde et al. 2002; Rose and Ebmeier 2006). The recognized cognitive deficits that have been noted in patients with depression have resulted in research in neuropsychiatric (Andreasen 1997) and neuropsychological investigation (Veiel 1997) over the past few decades (Ravnkilde et al. 2002). Neuropsychological tests have reported multiple domains of cognitive dysfunction, which include spatial learning, memory, and digit span (Gruzelier et al. 1988); explicit, declarative memory (Danion et al. 1991); selective attention and set shifting (Austin et al. 1999); attention, executive function, and visuospatial learning and memory (Porter et al. 2003); free recall (Ilsley et al. 1995); and frontal executive functions (Fossati et al. 1999, 2003).

14.3.2 Molecular Mechanisms Underlying Cytokine Effects on Cognition

It is now believed that cytokine-mediated pathophysiological processes underlie the cognitive impairments associated with several neuropsychiatric diseases, including depression, making cytokines ideal targets for therapeutic intervention (Reichenberg et al. 2001; Tobinick 2007; Tweedie et al. 2007; Wilson et al. 2002). Despite these pathological implications (Bitsch et al. 2000; He et al. 2007), cytokines have also been shown to exert physiological (Blatteis 1990; Sei et al. 1995; Vitkovic et al. 2000) and even neuroprotective (Pan and Kastin 2001; Schwartz et al. 1991, 1994) functions.

Recent findings suggest that in healthy elderly humans, cytokines, in particular IL-8, under physiological conditions, are possibly involved in cognitive processes such as memory, perceptual speed, and motor function (Baune et al. 2008a). Moreover, the associations between genetic variants of cytokines (IL-1β, IL-6, and TNF) and cognitive function in healthy elderly humans in the general population (Baune et al. 2008b) have recently been reported. More specifically, the results suggest that genetic variants of TNF may have protective effects in cognitive functions such as perceptual speed (Baune et al. 2008b). In an animal model of cytokine-mediated cognitive functions such as memory and learning, we were able to demonstrate that the presence of TNF under immunologically nonchallenged conditions is essential for normal functions of memory and learning (Baune et al. 2008c). This research linking complex cognitive phenotypes and cytokine effects in the CNS is well based on previous evidence suggesting that cytokines play a role in normal CNS function at a cellular and molecular level (see Jankowsky and Patterson 1999 for review). For instance, during unchallenged healthy conditions, both IL-1 and TNF have been shown to act as neuromodulators, as well as proinflammatory factors (Blatteis 1990; Sei et al. 1995; Vitkovic et al. 2000).

Although these detrimental effects on cognitive function in both overexpressing and cytokine-deficient models suggest that cytokines play an important physiological role in the CNS at the molecular and cognitive levels, it still remains to be fully understood as to how cytokines participate in the molecular and cellular mechanisms subserving complex CNS functions such as learning, memory, and cognition. Interestingly, growing evidence suggests that the cytokines IL-1, IL-6, and TNF, in particular, are involved in the molecular and cellular mechanisms subserving complex cognitive processes (Pickering and O'Connor 2007; Vitkovic et al. 2000; Viviani et al. 2007). Whereas direct and indirect evidence shows that IL-1, IL-6, and TNF may play a major role in synaptic plasticity, LTP, neurogenesis, and memory consolidation, the evidence for involvement with cognitive function is less conclusive for other cytokines such as IFN-γ (Baron et al. 2008), alpha(1)-antichymotrypsin (Dik et al. 2005; Nilsson et al. 2004; McIlroy et al. 2000), and IL-2 (Beck et al. 2002, 2005).

14.3.3 EMOTION PROCESSING IN AMYGDALA RELEVANT TO DEPRESSION

The amygdala is part of the telencephalon (Murray 2007) and appears as an almond-shaped neuronal core. It lies deep within the anterior temporal lobe (Murray 2007) and is part of the limbic system. The limbic system is a principal region involved in social behaviors such as affiliation, affection, and emotional intelligence (Killgore and Yurgelun-Todd 2004). The amygdala is, when it comes to emotion regulation, a key region of the limbic system. Yet, it receives signals from most cortical fields (Stefanacci and Amaral 2000; Ghashghaei and Barbas 2002) as well as brain sections outside the limbic system, predominantly ventral forebrain structures (Phillips et al. 2003a, 2003b), and usually returns them.

The amygdala consists anatomically of two different parts: the BLA and the central nucleus (CeA). Whereas the BLA is characteristically considered as the sensory interface of the amygdala, the CeA is seen as the output region to other brain areas (Pare et al. 2004). The BLA can physiologically be subdivided into two groups of cells: one is involved in memory storage and the other in initial learning (Radwanska et al. 2002). The BLA encodes the visionary relationship between impulses and primary reinforcers, such as food and fluids. In contrast to what was thought in the past, positive influences are as often encoded as negative ones (sad vs. happy faces) (Murray 2007).

Although it has been known for quite some time that the amygdala is involved in emotion processing (Weiskrantz 1956), most recently accomplished research is based on its role in fear conditioning, which is a model of emotional learning in which psychological responses are being associated with an aversive stimulus (Davis and Whalen 2001; Bailey et al. 1999; Lin et al. 2001). However, the role of the amygdala in emotion processing and implicit learning is not limited to fear (Phelps and LeDoux 2005) and includes other emotional conditioning processes as well. The amygdala is involved in the modulation and habituation of spatial memories, and it may even favor stimuli that are predisposed, leading to an emotional reaction (Phelps and LeDoux 2005). However, it does not appear to be required in the modulation of memories of fear conditioning (Lee et al. 2001). There are many other brain systems, including the limbic system, involved in the encoding and expression of this form of emotional learning (Phelps and LeDoux 2005).

A large range of emotions is recognized to be related to processes in the amygdala, such as anger, happiness, sadness, pride, fear, anxiety, relief, and shame (Ramos and Mormede 1998; Phelps and LeDoux 2005), all relevant to depression. More complicated affective processes, including emotional learning, emotional memory, reward, rage moderation, emotional influences on attention and perception, sexuality, and social behavior, have also been linked to the amygdala (McMillian et al. 1995; Phelps and LeDoux 2005; Murray 2007). Previous research is suggesting that the amygdala not only mediates unconscious biases and preferences but also influences feelings such as hopes, dreams, ideas, beliefs, and dreads (Murray 2007).

14.3.4 EFFECTS OF CYTOKINES ON NEUROBIOLOGY OF AMYGDALA

Cytokines are acting on the neurobiology of the amygdala through the onset of several mechanisms: the induction of other mediators such as chemokines, adhesion molecules, extracellular proteins, growth factors, and other cytokines, and through the induction of cell death (John et al. 2003).

Churchill et al. (2006) have found that locally induced changes in the cytokine levels in the amygdala can be a result of the effect of peripheral cytokines. In this research, it was clarified that peripheral cytokines induce brain cytokine messenger RNA, which is evident for the transcription and translation of cytokines in the amygdala. Churchill also defined a relationship between peripheral cytokines and the cytokines that will ultimately be produced in the amygdala: IL-1β increases brain IL-1β messenger RNA, and TNF-α increases brain TNF-α messenger RNA.

Certain cytokines will, after interaction with their receptor, induce a pathway of signal transduction within the amygdala. For TNF-α and IL-1β, it has been proven that both the NF-κB pathway and the MAPK pathway become activated (Saha et al. 2006). It is very likely that these pathways will also be activated by other cytokines (John et al. 2003).

14.3.5 Emotion Processing in Depression

A recent study in major depression applying a combined pharmacogenetic and imaging approach suggests possible pathways linking inflammatory genes, regulation of emotions, and pharmacoresponse in major depression disorder (Baune et al. 2010). One of those pathways might be the anterior cingulate cortex (ACC), which plays a pivotal role in the experience and regulation of emotions (Phillips et al. 2003a, 2003b) and has been strongly implicated in the pathophysiology of major depression (Phillips et al. 2003b). Several studies have demonstrated that pre- and subgenual ACC activity at rest (Mayberg et al. 1997; Pizzagalli et al. 2001), as well as reactivity to emotional faces (Chen et al. 2008) and pictures (Pizzagalli et al. 2001), predicts response to antidepressant medication in depression (also see DeRubeis et al. 2008 for review). Along the same line, two recent studies demonstrated that responsiveness of the amygdala to emotional pictures (Canli et al. 2005) and words (Siegle et al. 2006) is associated with a beneficial treatment outcome.

New data suggest a consistent finding that the same genetic variants of IL-1β associated with nonremission after antidepressant treatment are also associated with decreased amygdala and ACC functions, which seem to represent neurobiological markers of a more severe course of disease and worse antidepressant treatment outcome (Baune et al. 2010). Interestingly, this finding parallels the results of other imaging genetics studies describing a relative uncoupling of the amygdala and ACC in carriers of potential risk alleles for unfavorable treatment response in depression (5-HTTLPR S allele (Pezawas et al. 2005) and MAO-A H alleles (Dannlowski et al. 2009)). Furthermore, these findings are in line with a study by Harrison et al. showing that a reduced amygdala–subgenual ACC coupling is associated with inflammation-induced mood decrease (Harrison et al. 2009).

14.3.6 Anxiety and Fear Processing

In the literature, there is a distinction between fear and anxiety. Whereas anxiety is known as the emotional expectation of an aversive position, which is difficult to control or predict and which is likely to occur, it has been suggested that fear is the reaction of an individual to a dangerous situation that is already there and that is real (Ramos and Mormede 1998). Much of our knowledge about how the amygdala links memory and emotions has come from studies of Pavlovian fear conditioning (Bailey et al. 1999) and neuroimaging (Killgore and Yurgelun-Todd 2004). It has been shown that the amygdala is a possible site of neuronal plasticity, which forms the basis of the storage of fear memories and the modulation of fear learning (Rattiner et al. 2005; Phelps and LeDoux 2005). Killgore and Yurgelun-Todd (2004) have described several studies showing that the activation of the amygdala in response to fearful faces is below the threshold of visual perception, which means that the amygdala is capable of evolving emotions below the level of conscious awareness. Neuroimaging studies have suggested that there are two distinct neural systems for emotional perception: one is operating at the conscious level and the other is operating in the subconsciousness (de Gelder et al. 1999; Driver et al. 2001).

Conditioned fear responses are thought to be acquired in the BLA of the amygdala, which appears as a critical site of plasticity when it comes to fear conditioning (Walker and Davis 2002; Bailey et al. 1999). These conditional fear responses occur through exiting neurons of the CeA of the amygdala by the BLA. The CeA then projects stimuli to the brainstem and hypothalamic sites, hence creating a fear response (Davis and Shi 2000).

Recent studies have reported a relationship between fear conditioning and LTP in the amygdala (Samson and Pare 2005). LTP is the formation of changes in associative encoding in amygdala neurons, which will make the connection between two synapses stronger, mediated by specific alterations in gene expression, hence increasing the efficiency of the synapse (Murray 2007; Yeh et al. 2002). LTP in the amygdala is required for the origination of new fear-related memories (Yeh et al. 2002). The amplification of fear memory also requires new protein synthesis (Bailey et al. 1999). Yeh et al. (2002) have demonstrated that biochemical changes observed during LTP in vitro are similar to those observed in the amygdala during fear conditioning, hence providing evidence for the requirement of LTP in the amygdala to acquire fear memory.

There are many forms of LTP in the amygdala, and they are depending partly on NMDA receptors for their induction of LTP. The critical NMDA receptors were found to have mostly postsynaptic locations (Tsvetkov et al. 2004). Involvement of the NMDA receptor in LTP in the amygdala was proven by Fanselow and Kim (1994) by applying specific NMDA-receptor antagonists to the amygdala; in doing so, they prevented common forms of LTP and induced the acquisition of fear conditioning, thus concluding that NMDA receptors are essential to the induction of LTP. The NF-κB pathway on its own has also been proven to be involved in fear conditioning (Yeh et al. 2002). Fear learning has been demonstrated to increase the acetylation on NF-κB in the amygdala, which led to an increase in DNA-binding activity (Yeh et al. 2004). The maintenance of LTP (and consequently the storage of memory in the amygdala) seems to require the synthesis of messenger RNA and, subsequently, protein synthesis (Kang and Schuman 1996) in causing persistent changes in the connection and function of neurons (Bailey et al. 1999; Phelps and LeDoux 2005).

In conclusion, plasticity in amygdala neurons is required for fear conditioning. A pathological appearance of fear, social fear, is also modified in the amygdala. Studies that describe social fear and autism are those of Phelps and LeDoux (2005) and Baron-Cohen et al. (2000).

14.3.7 Effects of Cytokines on Anxiety and Fear Processing

As mentioned above, one of the basic emotions is fear. Pawlak et al. (2005) have described that there is evidence for the involvement of cytokines IL-1β, IL-6, and TNF-α in the evolvement of fear and fear learning; it has been proven that administration of these cytokines can elicit a fear-like response (Connor et al. 1998). IFN-γ has also been shown to play an important role in moderating fear and anxiety (Hurlock 2001). However, the exact underlying mechanism remains unknown. A probability is the involvement of HPA-axis activation (mechanism outside the amygdala). However, this has only been shown for IL-1β and is said to be not the only mechanism responsible for fear expression (Connor et al. 1998), which suggests that other cytokines, and IL-1β partly, induce the anxiogenic response through another mechanism, which is indeed most likely by acting on the amygdala.

The complexity of fear processing is suggested by Swiergiel and Dunn (2006), who showed that IL-6 deficiency did not induce any change in fear expression (less fear was expected), although IL-6 has been shown to be the less potent inductor of anxiogenic behavior (Connor et al. 1998). Hence, it remains to be clarified whether the deficiency of more potent cytokines such as IL-1β or TNF-α will show a decrease in fear response. The effect of cytokines on the development of fear memory seems more distinct because LTP in the amygdala is inhibited by IL-1β, TNF-α, IFN-α, and IFN-β (Rothwell and Hopkins 1995). The role of the amygdala in fear and fear learning is essential and is based mainly on the initiation of cytokine activation as a response to stimuli such as stress and auditory fear-potentiated startle (Kaur and Salm 2008; Yeh et al. 2002), hence inducing LTP. Thus, by inhibiting LTP in the amygdala, IL-1β, TNF-α, IFN-α, and IFN-β are restricting the evolvement of fear memory. Other mechanisms by which cytokines act on the evolvement of fear through the amygdala have also been studied: IFN-α and IFN-γ increase vasopressin and CRF, substances related to aggression, pair bonding, and learning. As a consequence, a potent influence of these cytokines on learned associations of stimuli with fear was demonstrated (Hurlock 2001).

In conclusion, the data suggest that increasing levels of IL-1β, IL-6, and TNF-α have the ability to induce a fear response, and increasing IL-1β, TNF-α, IFN-α, and IFN-β can also inhibit fear learning by decreasing LTP development in the amygdala. So far, no conclusions can be reached about the effects of reducing levels of IL-1β, IL-6, and TNF-α on fear sensations and fear memory.

14.3.8 DEPRESSION-LIKE BEHAVIOR: SLEEP, APPETITE, AND CLINICAL DEPRESSION

Mammal behaviors such as appetite, avoidance behavior, and sleep are a result of processes within the limbic system, particularly the amygdala. Both the CeA and the BLA of the amygdala are involved in certain expressions of behavior (Calvo and Fernandez-Guardiola 1984; Parker and Coscina 2001). The most frequently studied effects of cytokines on behavior involve sleep and feeding.

14.3.8.1 Sleep

A normal human sleep episode lasts approximately 8 h and can be divided into slow-wave sleep and rapid eye movement (REM) sleep. The amygdala appears to influence both phases of sleep, and both the CeA and the BLA are involved and work in parallel with other brain regions (Calvo and Fernandez-Guardiola 1984). Calvo et al. (1987) have demonstrated that stimulation of the amygdala affects the number of sleep episodes, mean duration of sleep, and general sleep organization, thereby proving its requirement in normal sleep rhythm.

14.3.8.2 Appetite and Control of Feeding

The mammalian urge to feed themselves is mainly regulated by the BLA; the CeA plays a less important role (Holland and Gallagher 2003). Although feeding seems to be a stimulus–reward process, it appears to be more complicated because not only food intake (reward) but also the control of eating is as important in this process. Whereas the urge to eat is an inborn feature, the control of feeding has to be learned by motivational cues. "Feed learning" is a form of cognition that may require LTP in the amygdala (Holland and Gallagher 2003).

The BLA is responsible for the control of food intake by the individual and mediates these actions by affecting the hypothalamus (Petrovich and Gallagher 2003). Holland and Gallagher (2003) have demonstrated that the CeA does not contribute to the control of feeding, by creating lesions in this part of the amygdala in rats. However, lesions in the BLA did affect feeding (Petrovich and Gallagher 2003), resulting in overeating and weight gain (Parker and Coscina 2001). The mechanism as to why lesions in the BLA did not influence appetite but affected the control of feeding has to be further examined.

14.3.8.3 Effects of Cytokines on Sleep, Appetite, and Clinical Depression

The fact that cytokines have an effect on behavior, for example, on sleep and appetite, has been a certainty for quite some time (Rothwell and Hopkins 1995), especially with the role of IL-1β having been studied widely. It is proven that different IL-1 receptors are involved in behavioral responses (Rothwell and Hopkins 1995) and that central administration of IL-1β reduces exploration behavior in mice (Spadaro and Dunn 1990). IL-6 seems to have no influence on exploration behavior, because no changes have been reported using IL-6 deficient mice (Swiergiel and Dunn 2006).

14.3.8.3.1 Effects of Cytokines on Sleep

Both peripheral and central cytokines are able to modulate sleep (Churchill et al. 2006). After central injection of IL-1β, IFN-α, or TNF-α in rat, both slow-wave sleep and non–rapid eye movement sleep (NREMS) increased (Obal et al. 2003; Hansen and Krueger 1997). Injection of IL-6 showed no effect on slow-wave sleep or NREMS (Rothwell and Hopkins 1995). IFN-α and IFN-γ have a somnolent effect when administered to humans. IFN-γ is proven to act on sleep rhythms in normal amygdala functioning (Lundkvist et al. 1998), although it is not clear whether IFN-γ affects the amygdala directly or indirectly via other brain regions. The effect of cytokines on sleep induction mediated by the amygdala depends probably on vagal afferents. Churchill et al. (2006) speculated about the induction of the homeostatic regulation of central growth factors through the activation of the cytokine network within the amygdala being involved in sleep regulation in normal brain functioning. However, Churchill et al. concluded that although cytokines do act on the amygdala and the amygdala has been proven to be involved in sleep induction, the effects of cytokines on sleep are not through the amygdala. Cytokines act on other brain regions (such as the hypothalamus and the hippocampus) to induce either slow-wave sleep or NREMS. A connection between these brain regions and the amygdala clarifies the involvement of the amygdala in sleep and suggests an indirect response of the amygdala to most cytokines (Churchill et al. 2006).

IL-6 appears to be the only cytokine to have a role in affecting normal sleep by acting on the gp130 receptor in the amygdala. IL-6 levels in the amygdala increased by 70% at the end of sleep, while not influencing IL-6-producing monocytes or gp130 concentrations (Dimitrov et al. 2006). The authors concluded that sleep widens the profile of IL-6 actions; however, both the consequences and the underlying mechanism remain unknown.

14.3.8.3.2 Effects of Cytokines on Appetite and Feeding

Both peripheral and central cytokines are capable of influencing appetite in mammals (Churchill et al. 2006), mainly by decreasing food intake (Bray 2000). Most cytokines seem to have an effect on the control of feeding, including IL-1β, IL-6, TNF-α, and IFN-α, probably through activation of the gp130 receptor in the amygdala (Plata-Salaman 2001). Central administration of IL-β and TNF-α has been proven to reduce food intake in rats (Kent et al. 1994; Connor et al. 1998).

Again, the effect of IL-6 seems questionable: whereas feeding is not modified in IL-6-deficient mice (Swiergiel and Dunn 2006), mice that hypersecrete IL-6 show a reduction of sucrose consumption in a study by Sakic et al. (1997). However, Connor et al. (1998) showed no effect on food consumption after central administration of IL-6. Although the effects of increased cytokine levels in the amygdala on declined food intake have been proven, the physiological mechanism by which cytokines influence appetite in normal amygdala functioning remains unclear.

14.3.8.3.3 Cytokines in Clinical Depression

Commensurate with animal studies showing a relationship between stress, depression, and immunity (Herbert and Cohen 1993a, 1993b), major depression in humans is associated with alterations of various aspects of the immune response, including a reduction of mitogen-stimulated lymphocyte proliferation, as well as reduced natural killer cell activity (Herbert and Cohen 1993a, 1993b; Maes et al. 1999). These effects are most pronounced in severely depressed patients (i.e., those exhibiting melancholia) (Maes et al. 1999), and the altered immunity may be attenuated with symptom remission (Irwin et al. 1992). In contrast to the assumption that depression was associated with the suppression of nonspecific immunity, it has been argued that affective disturbances may actually be secondary to an initial immune activation (Maes et al. 1999; Licinio and Wong 1999) possibly due to the altered HPA functioning and central monoamine activity. Indeed, depressed patients were found to display signs of immune activation, reminiscent of an acute phase response, including increased plasma concentrations of complement proteins, C3 and C4, and IgM, as well as positive acute phase proteins, haptoglobin, α1-antitrypsin, and α1- and α2-macroglobulin, coupled with reduced levels of negative acute phase proteins (Maes et al. 1999; Sluzewska 1999). Major depressive illness was also accompanied by an increased number of activated T cells (CD25β and HLA-DRβ), secretion of neopterin, prostaglandin E_2, and thromboxane (Maes et al. 1999).

In addition to these factors, it has been reported that depression was accompanied by increased levels of circulating cytokines or their soluble receptors, including IL-2, soluble IL-2 receptors, IL-1β, IL-1Ra, IL-6, soluble IL-6 receptor, and c-interferon (IFN) (Berk et al. 1997; Frommberger et al. 1997; Maes et al. 1993; Müller and Ackenheil 1998; Nässberger and Träskman-Bendz 1993; Słuzewska et al. 1995). In addition, increased mitogen-elicited production of the proinflammatory cytokines IL-1β, IL-6, and TNF-α has been reported (Anisman et al. 1999; Maes et al. 1993). Although the elevated levels of IL-1β, IL-6, and α1-acid glycoprotein normalized with antidepressant medication (Frommberger et al. 1997; Słuzewska et al. 1995), such treatment did not affect the upregulated production of soluble IL-2 receptor, IL-6, and soluble IL-6 receptor in major depression (Maes et al. 1999) or that of IL-1 in patients with chronic low-grade depression (dysthymia) (Anisman et al. 1999). Likewise, serum levels of IL-6, as well as the anti-inflammatory cytokines IL-10 and IL-1Ra, were moderately elevated in depressed patients, but once again, successful pharmacotherapy was not associated with normalization of these cytokines (Kubera et al. 2000). Thus, these cytokines may be trait markers of the illness but do not play an etiological role in depression (Anisman et al. 1999; Maes et al. 1999).

The data currently available concerning cytokine elevations in depressive illness are largely correlational. Thus, it is unclear whether the cytokine elevations are secondary to the illness (i.e., being directly or indirectly brought on by the depression) or contribute to the provocation of the disorder. Yet, high doses of IL-2, IFN-α, and TNF-α in humans undergoing immunotherapy induce neuropsychiatric symptoms, including depression, and these effects were related to the cytokine treatment rather than to the primary illness (Maes et al. 2001; Denicoff et al. 1987; Musselman et al. 2001). Interestingly, it was reported (Maes et al. 2001) that IL-2, alone or in combination with IFN-α, induced a decrease in serum dipepridyl peptidase IV, a membrane-bound serine protease that acts to catalyze the cleavage of at least some cytokines and peptides, hence affecting cytokine production and immune activity. At 3 to 5 days after treatment, depression scores were elevated in patients being treated for metastatic cancer, and the extent of the increase was inversely related to dipepridyl peptidase IV levels. In addition, the treatment resulted in elevated levels of IL-6 and IL-2R, which were inversely related to dipepridyl peptidase IV levels. Thus, these data were taken to suggest that cytokines contributed to the provocation of depressive symptoms. In a more recent report (Musselman et al. 2001), it was observed that depressive symptoms were provoked in humans treated with IFN-α and that these symptoms could be attenuated by treatment with the selective serotonin reuptake inhibitor, paroxetine. Interestingly, by attenuating the side effects of the immunotherapy, the positive effects of the treatment were also maximized. In effect, although the bulk of the data concerning the cytokine–depression relationship is essentially correlational, there is certainly good reason to suspect an etiological role for cytokines in depressive illness. At the same time, however, it needs to be remembered that the populations being assessed in studies where cytokines are being administered are relatively unique and that the participants are undergoing considerable strain. Thus, the effects of the cytokine treatments may well represent the interactive and combined effects of a number of factors, beyond simply that of immune activation.

14.4 CONCLUSIONS

Cytokines as part of an immune response have been implicated in the etiology, pathophysiology, and clinical course of major depression. More recently, research suggests cytokine-specific effects on various symptom clusters of depression, such as sickness behavior, cognitive impairment, and emotion processing. Basic research has been instrumental in understanding the underlying molecular mechanisms and signaling pathways by which cytokines may affect complex depression-like symptoms and behavior. Most importantly, cytokines appear to play an important role in the CNS under immune-unchallenged as well as under challenged conditions, which indicates significant implications for a variety of neuropsychiatric disorders, not only for depression. Future research may help enhance our understanding of these complex effects of cytokines in the CNS, aiming at the development of cytokine-specific pharmacological interventions in depression.

REFERENCES

Andreasen, N. C. 1997. Linking mind and brain in the study of mental illnesses: A project for a scientific psychopathology. *Science* 275 (5306):1586–93.

Anisman, H., A. V. Ravindran, J. Griffiths, and Z. Merali. 1999. Interleukin-1 beta production in dysthymia before and after pharmacotherapy. *Biol Psychiatry* 46 (12):1649–55.

Anisman, H., L. Kokkinidis, and Z. Merali. 2002. Further evidence for the depressive effects of cytokines: Anhedonia and neurochemical changes. *Brain Behav Immun* 16 (5):544–56.

Apte, R. S., D. Sinha, E. Mayhew, G. J. Wistow, and J. Y. Niederkorn. 1998. Cutting edge: Role of macrophage migration inhibitory factor in inhibiting NK cell activity and preserving immune privilege. *J Immunol* 160 (12):5693–6.

Araujo, D. M., and C. W. Cotman. 1993. Trophic effects of interleukin-4, -7 and -8 on hippocampal neuronal cultures: Potential involvement of glial-derived factors. *Brain Res* 600 (1):49–55.

Austin, M. P., M. Ross, C. Murray, R. E. O'Carroll, K. P. Ebmeier, and G. M. Goodwin. 1992. Cognitive function in major depression. *J Affect Disord* 25 (1):21–9.

Austin, M. P., P. Mitchell, K. Wilhelm, et al. 1999. Cognitive function in depression: A distinct pattern of frontal impairment in melancholia? *Psychol Med* 29 (1):73–85.

Bailey, D. J., J. J. Kim, W. Sun, R. F. Thompson, and F. J. Helmstetter. 1999. Acquisition of fear conditioning in rats requires the synthesis of mRNA in the amygdala. *Behav Neurosci* 113 (2):276–82.

Baldwin, A. S., Jr. 1996. The NF-kappa B and I kappa B proteins: New discoveries and insights. *Annu Rev Immunol* 14:649–83.

Baron, R., A. Nemirovsky, I. Harpaz, H. Cohen, T. Owens, and A. Monsonego. 2008. IFN-gamma enhances neurogenesis in wild-type mice and in a mouse model of Alzheimer's disease. *FASEB J* 22:2843–52.

Baron-Cohen, S., H. A. Ring, E. T. Bullmore, S. Wheelwright, C. Ashwin, and S. C. Williams. 2000. The amygdala theory of autism. *Neurosci Biobehav Rev* 24 (3):355–64.

Baune, B. T., G. Ponath, J. Golledge, et al. 2008a. Association between IL-8 cytokine and cognitive performance in an elderly general population—The MEMO-Study. *Neurobiol Aging* 29 (6):937–44.

Baune, B. T., G. Ponath, M. Rothermundt, O. Riess, H. Funke, and K. Berger. 2008b. Association between genetic variants of IL-1beta, IL-6 and TNF-alpha cytokines and cognitive performance in the elderly general population of the MEMO-study. *Psychoneuroendocrinology* 33 (1):68–76.

Baune, B. T., F. Wiede, A. Braun, J. Golledge, V. Arolt, and H. Koerner. 2008c. Cognitive dysfunction in mice deficient for TNF- and its receptors. *Am J Med Genet B Neuropsychiatr Genet* 147B:1056–64.

Baune, B. T., U. Dannlowski, K. Domschke, et al. 2010. The interleukin 1 beta (IL1B) gene is associated with failure to achieve remission and impaired emotion processing in major depression. *Biol Psychiatry* 67 (6):543–9.

Beck, R. D., Jr., M. A. King, Z. Huang, and J. M. Petitto. 2002. Alterations in septohippocampal cholinergic neurons resulting from interleukin-2 gene knockout. *Brain Res* 955:16–23.

Beck, R. D., Jr., C. Wasserfall, G. K. Ha, J. D. Cushman, Z. Huang, M. A. Atkinson, and J. M. Petitto. 2005. Changes in hippocampal IL-15, related cytokines, and neurogenesis in IL-2 deficient mice. *Brain Res* 1041:223–30.

Benveniste, E. N., V. T. Nguyen, and G. M. O'Keefe. 2001. Immunological aspects of microglia: Relevance to Alzheimer's disease. *Neurochem Int* 39 (5–6):381–91.

Berk, M., A. A. Wadee, R. H. Kuschke, and A. O'Neill-Kerr. 1997. Acute phase proteins in major depression. *J Psychosom Res* 43 (5):529–34.

Berman, D. E., S. Hazvi, K. Rosenblum, R. Seger, and Y. Dudai. 1998. Specific and differential activation of mitogen-activated protein kinase cascades by unfamiliar taste in the insular cortex of the behaving rat. *J Neurosci* 18 (23):10037–44.

Bhakar, A. L., L. L. Tannis, C. Zeindler, et al. 2002. Constitutive nuclear factor-kappa B activity is required for central neuron survival. *J Neurosci* 22 (19):8466–75.

Bitsch, A., T. Kuhlmann, et al. 2000. Tumour necrosis factor alpha mRNA expression in early multiple sclerosis lesions: Correlation with demyelinating activity and oligodendrocyte pathology. *Glia* 29: 366–75.

Blaney, P. H. 1986. Affect and memory: A review. *Psychol Bull* 99 (2):229–46.

Blatteis, C. M. 1990. Neuromodulative actions of cytokines. *Yale J Biol Med* 63 (2):133–46.

Blum, S., A. N. Moore, F. Adams, and P. K. Dash. 1999. A mitogen-activated protein kinase cascade in the CA1/CA2 subfield of the dorsal hippocampus is essential for long-term spatial memory. *J Neurosci* 19 (9):3535–44.

Bray, G. A. 2000. Afferent signals regulating food intake. *Proc Nutr Soc* 59 (3):373–84.

Brebner, K., S. Hayley, R. Zacharko, Z. Merali, and H. Anisman. 2000. Synergistic effects of interleukin-1beta, interleukin-6, and tumor necrosis factor-alpha: Central monoamine, corticosterone, and behavioral variations. *Neuropsychopharmacology* 22 (6):566–80.

Calvo, J. M., and A. Fernandez-Guardiola. 1984. Phasic activity of the basolateral amygdala, cingulate gyrus, and hippocampus during REM sleep in the cat. *Sleep* 7 (3):202–10.

Calvo, J. M., S. Badillo, M. Morales-Ramirez, and P. Palacios-Salas. 1987. The role of the temporal lobe amygdala in ponto-geniculo-occipital activity and sleep organization in cats. *Brain Res* 403 (1):22–30.

Canli, T., R. E. Cooney, P. Goldin, et al. 2005. Amygdala reactivity to emotional faces predicts improvement in major depression. *Neuroreport* 16 (12):1267–70.

Capuron, L., and R. Dantzer. 2003. Cytokines and depression: The need for a new paradigm. *Brain Behav Immun* 17 (Suppl 1):S119–24.

Capuron, L., and A. Ravaud. 1999. Prediction of the depressive effects of interferon alfa therapy by the patient's initial affective state. *N Engl J Med* 340 (17):1370.

Carter, B. D., C. Kaltschmidt, B. Kaltschmidt, et al. 1996. Selective activation of NF-kappa B by nerve growth factor through the neurotrophin receptor p75. *Science* 272 (5261):542–5.

Castagne, V., K. Lefevre, and P. G. Clarke. 2001. Dual role of the NF-kappaB transcription factor in the death of immature neurons. *Neuroscience* 108 (3):517–26.

Castanon, N., R. M. Bluthe, and R. Dantzer. 2001. Chronic treatment with the atypical antidepressant tianeptine attenuates sickness behavior induced by peripheral but not central lipopolysaccharide and interleukin-1-beta in the rat. *Psychopharmacology (Berl)* 154 (1):50–60.

Chen, C. H., J. Suckling, C. Ooi, et al. 2008. Functional coupling of the amygdala in depressed patients treated with antidepressant medication. *Neuropsychopharmacology* 33 (8):1909–18.

Churchill, L., P. Taishi, M. Wang, et al. 2006. Brain distribution of cytokine mRNA induced by systemic administration of interleukin-1beta or tumor necrosis factor alpha. *Brain Res* 1120 (1):64–73.

Connor, T. J., C. Song, B. E. Leonard, Z. Merali, and H. Anisman. 1998. An assessment of the effects of central interleukin-1beta, -2, -6, and tumor necrosis factor-alpha administration on some behavioural, neurochemical, endocrine and immune parameters in the rat. *Neuroscience* 84 (3):923–33.

Connor, T. J., A. Harkin, J. P. Kelly, and B. E. Leonard. 2000. Olfactory bulbectomy provokes a suppression of interleukin-1beta and tumour necrosis factor-alpha production in response to an in vivo challenge with lipopolysaccharide: Effect of chronic desipramine treatment. *Neuroimmunomodulation* 7 (1):27–35.

Cronholm, B., and J. O. Ottosson. 1961. Memory functions in endogenous depression before and after electroconvulsive therapy. *Arch Gen Psychiatry* 5:193–9.

Danion, J. M., D. Willard-Schroeder, M. A. Zimmermann, D. Grange, J. L. Schlienger, and L. Singer. 1991. Explicit memory and repetition priming in depression. Preliminary findings. *Arch Gen Psychiatry* 48 (8):707–11.

Dannlowski, U., P. Ohrmann, C. Konrad, et al. 2009. Reduced amygdala–prefrontal coupling in major depression: Association with MAOA genotype and illness severity. *Int J Neuropsychopharmacol* 12 (1):11–22.

Dantzer, R., J. C. O'Connor, G. G. Freund, R. W. Johnson, and K. W. Kelley. 2008. From inflammation to sickness and depression: When the immune system subjugates the brain. *Nat Rev Neurosci* 9 (1):46–56.

Davis, M., and C. Shi. 2000. The amygdala. *Curr Biol* 10 (4):R131.

Davis, M., and P. J. Whalen. 2001. The amygdala: Vigilance and emotion. *Mol Psychiatry* 6 (1):13–34.

de Gelder, B., J. Vroomen, G. Pourtois, and L. Weiskrantz. 1999. Non-conscious recognition of affect in the absence of striate cortex. *Neuroreport* 10 (18):3759–63.

Denicoff, KD., D. R. Rubinow, M. Z. Papa, et al. 1987. The neuropsychiatric effects of treatment with interleukin-2 and lymphokine-activated killer cells. *Ann Intern Med* 107 (3):293–300.

DeRubeis, R. J., G. J. Siegle, and S. D. Hollon. 2008. Cognitive therapy versus medication for depression: Treatment outcomes and neural mechanisms. *Nat Rev Neurosci* 9 (10):788–96.

Dik, M. G., C. Jonker, C. E. Hack, J. H. Smit, H. C. Comijs, and P. Eikelenboom. 2005. Serum inflammatory proteins and cognitive decline in older persons. *Neurology* 64 (8):1371–7.

Dimitrov, S., T. Lange, C. Benedict, et al. 2006. Sleep enhances IL-6 trans-signaling in humans. *FASEB J* 20 (12):2174–6.

DiProspero, N. A., S. Meiners, and H. M. Geller. 1997. Inflammatory cytokines interact to modulate extracellular matrix and astrocytic support of neurite outgrowth. *Exp Neurol* 148 (2):628–39.

Dopp, J. M., A. Mackenzie-Graham, G. C. Otero, and J. E. Merrill. 1997. Differential expression, cytokine modulation, and specific functions of type-1 and type-2 tumor necrosis factor receptors in rat glia. *J Neuroimmunol* 75 (1–2):104–12.

Dougherty, DD, and SL Rauch, eds. 2001. *Psychiatric Neuroimaging Research; Contemporary Strategies*, 1st ed. Washington, DC: American Psychiatric Press, Inc.

Driver, J., P. Vuilleumier, M. Eimer, and G. Rees. 2001. Functional magnetic resonance imaging and evoked potential correlates of conscious and unconscious vision in parietal extinction patients. *Neuroimage* 14 (1 Pt 2):S68–75.

Dunn, A. J. 2002. Mechanisms by which cytokines signal the brain. *Int Rev Neurobiol* 52:43–65.

Dunn, A. J. 2006. Effects of cytokines and infections on brain neurochemistry. *Clin Neurosci Res* 6 (1–2):52–68.

Dupont, R. M., T. L. Jernigan, W. Heindel, et al. 1995. Magnetic resonance imaging and mood disorders. Localization of white matter and other subcortical abnormalities. *Arch Gen Psychiatry* 52 (9):747–55.

Ericsson, A., K. J. Kovacs, and P. E. Sawchenko. 1994. A functional anatomical analysis of central pathways subserving the effects of interleukin-1 on stress-related neuroendocrine neurons. *J Neurosci* 14 (2):897–913.

Fanselow, M. S., and J. J. Kim. 1994. Acquisition of contextual Pavlovian fear conditioning is blocked by application of an NMDA receptor antagonist D,L-2-amino-5-phosphonovaleric acid to the basolateral amygdala. *Behav Neurosci* 108 (1):210–12.

Flanigan, M. J., Q. Accone, and H. P. Laburn. 1992. Amitriptyline attenuates the febrile response to a pyrogen in rabbits. *J Basic Clin Physiol Pharmacol* 3 (1):19–32.

Fossati, P., G. Amar, N. Raoux, A. M. Ergis, and J. F. Allilaire. 1999. Executive functioning and verbal memory in young patients with unipolar depression and schizophrenia. *Psychiatry Res* 89 (3):171–87.

Fossati, P., B. Guillaume le, A. M. Ergis, and J. F. Allilaire. 2003. Qualitative analysis of verbal fluency in depression. *Psychiatry Res* 117 (1):17–24.

Frommberger, UH., J. Bauer, P. Haselbauer, A. Fräulin, D. Riemann, and M. Berger. 1997. Interleukin-6-(IL-6) plasma levels in depression and schizophrenia: Comparison between the acute state and after remission. *Eur Arch Psychiatry Clin Neurosci* 247 (4):228–33.

Gewirtz, J. C., A. C. Chen, R. Terwilliger, R. C. Duman, and G. J. Marek. 2002. Modulation of DOI-induced increases in cortical BDNF expression by group II mGlu receptors. *Pharmacol Biochem Behav* 73 (2):317–26.

Ghashghaei, H. T., and H. Barbas. 2002. Pathways for emotion: Interactions of prefrontal and anterior temporal pathways in the amygdala of the rhesus monkey. *Neuroscience* 115 (4):1261–79.

Gold, S. M., and M. R. Irwin. 2006. Depression and immunity: Inflammation and depressive symptoms in multiple sclerosis. *Neurol Clin* 24 (3):507–19.

Golinkoff, M., and J. A. Sweeney. 1989. Cognitive impairments in depression. *J Affect Disord* 17 (2):105–12.

Gomes, F. C., O. Sousa Vde, and L. Romao. 2005. Emerging roles for TGF-beta1 in nervous system development. *Int J Dev Neurosci* 23 (5):413–24.

Gruzelier, J., K. Seymour, L. Wilson, A. Jolley, and S. Hirsch. 1988. Impairments on neuropsychologic tests of temporohippocampal and frontohippocampal functions and word fluency in remitting schizophrenia and affective disorders. *Arch Gen Psychiatry* 45 (7):623–9.

Hansen, M. K., and J. M. Krueger. 1997. Subdiaphragmatic vagotomy blocks the sleep- and fever-promoting effects of interleukin-1beta. *Am J Physiol* 273 (4 Pt 2):R1246–53.

Harrison, N. A., L. Brydon, C. Walker, M. A. Gray, A. Steptoe, and H. D. Critchley. 2009. Inflammation causes mood changes through alterations in subgenual cingulate activity and mesolimbic connectivity. *Biol Psychiatry* 66 (5):407–14

Hayden, M. S., and S. Ghosh. 2004. Signaling to NF-kappaB. *Genes Dev* 18 (18):2195–224.

He, P., Z. Zhong, et al. 2007. Deletion of tumor necrosis factor death receptor inhibits amyloid {beta} generation and prevents learning and memory deficits in Alzheimer's mice. *J Cell Biol* 178 (5):829–41.

Herbert, T. B., and S. Cohen. 1993a. Depression and immunity: A meta-analytic review. *Psychol Bull* 113 (3):472–86.

Herbert, TB., and S. Cohen. 1993b. Stress and immunity in humans: A meta-analytic review. *Psychosom Med* 55 (4):364–79.

Hickie, I., and A. Lloyd. 1995. Are cytokines associated with neuropsychiatric syndromes in humans? *Int J Immunopharmacol* 17 (8):677–83.

Holland, P. C., and M. Gallagher. 2003. Double dissociation of the effects of lesions of basolateral and central amygdala on conditioned stimulus-potentiated feeding and Pavlovian-instrumental transfer. *Eur J Neurosci* 17 (8):1680–94.

Huang, Y. Y., K. C. Martin, and E. R. Kandel. 2000. Both protein kinase A and mitogen-activated protein kinase are required in the amygdala for the macromolecular synthesis-dependent late phase of long-term potentiation. *J Neurosci* 20 (17):6317–25.

Hurlock, E. C. 2001. Interferons: Potential roles in affect. *Med Hypotheses* 56 (5):558–66.

Ihle, J. N. 1995. The Janus protein tyrosine kinase family and its role in cytokine signaling. *Adv Immunol* 60:1–35.

Ilsley, J. E., A. P. Moffoot, and R. E. O'Carroll. 1995. An analysis of memory dysfunction in major depression. *J Affect Disord* 35 (1–2):1–9.

Irwin, M., T. L. Smith, and J. C. Gillin. 1992. Electroencephalographic sleep and natural killer activity in depressed patients and control subjects. *Psychosom Med* 54 (1):10–21.

Irwin, M. R., and A. H. Miller. 2007. Depressive disorders and immunity: 20 years of progress and discovery. *Brain Behav Immun* 21 (4):374–83.

Jankowsky, J. L., and P. H. Patterson. 1999. Cytokine and growth factor involvement in long-term potentiation. *Mol Cell Neurosci* 14 (6):273–86.

John, G. R., S. C. Lee, and C. F. Brosnan. 2003. Cytokines. Powerful regulators of glial cell activation. *Neuroscientist* 9 (1):10–22.

Kang, H., and E. M. Schuman. 1996. A requirement for local protein synthesis in neurotrophin-induced hippocampal synaptic plasticity. *Science* 273 (5280):1402–6.

Kaur, G., and A. K. Salm. 2008. Blunted amygdalar anti-inflammatory cytokine effector response to postnatal stress in prenatally stressed rats. *Brain Res* 1196:1–12.

Kent, S., R. M. Bluthe, K. W. Kelley, and R. Dantzer. 1992. Sickness behavior as a new target for drug development. *Trends Pharmacol Sci* 13 (1):24–8.

Kent, S., F. Rodriguez, K. W. Kelley, and R. Dantzer. 1994. Reduction in food and water intake induced by microinjection of interleukin-1 beta in the ventromedial hypothalamus of the rat. *Physiol Behav* 56 (5):1031–6.

Killgore, W. D., and D. A. Yurgelun-Todd. 2004. Activation of the amygdala and anterior cingulate during nonconscious processing of sad versus happy faces. *Neuroimage* 21 (4):1215–23.

Kronfol, Z., and D. G. Remick. 2000. Cytokines and the brain: Implications for clinical psychiatry. *Am J Psychiatry* 157 (5):683–94.

Kubera, M., G. Kenis, E. Bosmans, et al. 2000. Suppressive effect of TRH and imipramine on human interferon-gamma and interleukin-10 production in vitro. *Pol J Pharmacol* 52 (6):481–6.

Le, Y., P. Iribarren, W. Gong, Y. Cui, X. Zhang, and J. M. Wang. 2004. TGF-beta1 disrupts endotoxin signaling in microglial cells through Smad3 and MAPK pathways. *J Immunol* 173 (2):962–8.

Lee, H. J., J. S. Choi, T. H. Brown, and J. J. Kim. 2001. Amygdalar NMDA receptors are critical for the expression of multiple conditioned fear responses. *J Neurosci* 21 (11):4116–24.

Levine, E. S., R. A. Crozier, I. B. Black, and M. R. Plummer. 1998. Brain-derived neurotrophic factor modulates hippocampal synaptic transmission by increasing *N*-methyl-D-aspartic acid receptor activity. *Proc Natl Acad Sci U S A* 95 (17):10235–9.

Licinio, J., and M. L. Wong. 1999. The role of inflammatory mediators in the biology of major depression: Central nervous system cytokines modulate the biological substrate of depressive symptoms, regulate stress-responsive systems, and contribute to neurotoxicity and neuroprotection. *Mol Psychiatry* 4 (4):317–27.

Lin, C. H., S. H. Yeh, C. H. Lin, et al. 2001. A role for the PI-3 kinase signaling pathway in fear conditioning and synaptic plasticity in the amygdala. *Neuron* 31 (5):841–51.

Lu, K. T., D. L. Walker, and M. Davis. 2001. Mitogen-activated protein kinase cascade in the basolateral nucleus of amygdala is involved in extinction of fear-potentiated startle. *J Neurosci* 21 (16):RC162.

Lundkvist, G. B., B. Robertson, J. D. Mhlanga, M. E. Rottenberg, and K. Kristensson. 1998. Expression of an oscillating interferon-gamma receptor in the suprachiasmatic nuclei. *Neuroreport* 9 (6):1059–63.

Maes, M., E. Bosmans, H. Y. Meltzer, S. Scharpé, and E. Suy. 1993. Interleukin-1 beta: A putative mediator of HPA axis hyperactivity in major depression? *Am J Psychiatry* 150 (8):1189–93.

Maes, M., N. De Vos, P. Demedts, A. Wauters, and H. Neels. 1999. Lower serum zinc in major depression in relation to changes in serum acute phase proteins. *J Affect Disord* 56 (2–3):189–94.

Maes, M., L. Capuron, A. Ravaud, et al. 2001. Lowered serum dipeptidyl peptidase IV activity is associated with depressive symptoms and cytokine production in cancer patients receiving interleukin-2-based immunotherapy. *Neuropsychopharmacology* 24 (2):130–40.

Marchetti, L., M. Klein, K. Schlett, K. Pfizenmaier, and U. L. Eisel. 2004. Tumor necrosis factor (TNF)-mediated neuroprotection against glutamate-induced excitotoxicity is enhanced by *N*-methyl-D-aspartate receptor activation. Essential role of a TNF receptor 2–mediated phosphatidylinositol 3-kinase-dependent NF-kappa B pathway. *J Biol Chem* 279 (31):32869–81.

Mattson, M. P., and M. K. Meffert. 2006. Roles for NF-kappaB in nerve cell survival, plasticity, and disease. *Cell Death Differ* 13 (5):852–60.

Mayberg, H. S., S. K. Brannan, R. K. Mahurin, et al. 1997. Cingulate function in depression: A potential predictor of treatment response. *Neuroreport* 8 (4):1057–61.

McAfoose, J., and B. T. Baune. 2009. Evidence for a cytokine model of cognitive function. *Neurosci Biobehav Rev* 33 (3):355–66.

McDonald, E. M., A. H. Mann, and H. C. Thomas. 1987. Interferons as mediators of psychiatric morbidity. An investigation in a trial of recombinant alpha-interferon in hepatitis-B carriers. *Lancet* 2 (8569):1175–8.

McIlroy, S. P., M. D. Vahidassr, D. A. Savage, et al. 2000. Association of serum AACT levels and AACT signal polymorphism with late-onset Alzheimer's disease in Northern Ireland. *Int J Geriatr Psychiatry* 15 (3):260–6.

McMillian, M., L. Y. Kong, S. M. Sawin, et al. 1995. Selective killing of cholinergic neurons by microglial activation in basal forebrain mixed neuronal/glial cultures. *Biochem Biophys Res Commun* 215 (2):572–7.

Miller, T. M., M. G. Tansey, E. M. Johnson, Jr., and D. J. Creedon. 1997. Inhibition of phosphatidylinositol 3-kinase activity blocks depolarization- and insulin-like growth factor I-mediated survival of cerebellar granule cells. *J Biol Chem* 272 (15):9847–53.

Müller, N., and M. Ackenheil. 1998. Psychoneuroimmunology and the cytokine action in the CNS: Implications for psychiatric disorders. *Prog Neuropsychopharmacol Biol Psychiatry* 22 (2):1–33.

Müller, N., E. Hofschuster, M. Ackenheil, W. Mempel, and R. Eckstein. 1993. Investigations of the cellular immunity during depression and the free interval: Evidence for an immune activation in affective psychosis. *Prog Neuropsychopharmacol Biol Psychiatry* 17 (5):713–30.

Murray, E. A. 2007. The amygdala, reward and emotion. *Trends Cogn Sci* 11 (11):489–97.

Musselman, DL., A. H. Miller, M. R. Porter, et al. 2001. Higher than normal plasma interleukin-6 concentrations in cancer patients with depression: Preliminary findings. *Am J Psychiatry* 158 (8):1252–7.

Nadeau, S., and S. Rivest. 2000. Role of microglial-derived tumor necrosis factor in mediating CD14 transcription and nuclear factor kappa B activity in the brain during endotoxemia. *J Neurosci* 20 (9):3456–68.

Nader, K., G. E. Schafe, and J. E. Le Doux. 2000. Fear memories require protein synthesis in the amygdala for reconsolidation after retrieval. *Nature* 406 (6797):722–6.

Nadjar, A., R. M. Bluthe, M. J. May, R. Dantzer, and P. Parnet. 2005. Inactivation of the cerebral NFkappaB pathway inhibits interleukin-1beta-induced sickness behavior and c-Fos expression in various brain nuclei. *Neuropsychopharmacology* 30 (8):1492–9.

Nässberger, L., and L. Träskman-Bendz. 1993. Increased soluble interleukin-2 receptor concentrations in suicide attempters. *Acta Psychiatr Scand* 88 (1):48–52.

Nilsson, L. N., G. W. Arendash, R. E. Leighty, et al. 2004. Cognitive impairment in PDAPP mice depends on ApoE and ACT-catalyzed amyloid formation. *Neurobiol Aging* 25 (9):1153–67.

O'Brien, S. M., L. V. Scott, and T. G. Dinan. 2004. Cytokines: Abnormalities in major depression and implications for pharmacological treatment. *Hum Psychopharmacol* 19 (6):397–403.

O'Shea, J. J., M. Gadina, and R. D. Schreiber. 2002. Cytokine signaling in 2002: New surprises in the Jak/STAT pathway. *Cell* 109 (Suppl):S121–31.

Obal, F., Jr., J. Alt, P. Taishi, J. Gardi, and J. M. Krueger. 2003. Sleep in mice with nonfunctional growth hormone-releasing hormone receptors. *Am J Physiol Regul Integr Comp Physiol* 284 (1):R131–9.

Pan, W., and A. Kastin. 2001. Upregulation of the transport system for TNFa at the blood–brain barrier. *Arch Physiol Biochem* 109 (4):350–3.

Pare, D., G. J. Quirk, and J. E. Ledoux. 2004. New vistas on amygdala networks in conditioned fear. *J Neurophysiol* 92 (1):1–9.

Parker, G. C., and D. V. Coscina. 2001. Lesions of the posterior basolateral amygdala block feeding induced by systemic 8-OH-DPAT. *Pharmacol Biochem Behav* 68 (4):729–34.

Patterson, S. L., C. Pittenger, A. Morozov, et al. 2001. Some forms of cAMP-mediated long-lasting potentiation are associated with release of BDNF and nuclear translocation of phospho-MAP kinase. *Neuron* 32 (z1):123–40.

Pawlak, C. R., R. K. Schwarting, and A. Bauhofer. 2005. Cytokine mRNA levels in brain and peripheral tissues of the rat: Relationships with plus-maze behavior. *Brain Res Mol Brain Res* 137 (1–2):159–65.

Peng, Y. P., Y. H. Qiu, J. H. Lu, and J. J. Wang. 2005. Interleukin-6 protects cultured cerebellar granule neurons against glutamate-induced neurotoxicity. *Neurosci Lett* 374 (3):192–6.

Petrovich, G. D., and M. Gallagher. 2003. Amygdala subsystems and control of feeding behavior by learned cues. *Ann N Y Acad Sci* 985:251–62.

Pezawas, L., A. Meyer-Lindenberg, E. M. Drabant, et al. 2005. 5-HTTLPR polymorphism impacts human cingulate–amygdala interactions: A genetic susceptibility mechanism for depression. *Nat Neurosci* 8 (6):828–34.

Phelps, E. A., and J. E. LeDoux. 2005. Contributions of the amygdala to emotion processing: From animal models to human behavior. *Neuron* 48 (2):175–87.

Phillips, M. L., W. C. Drevets, S. L. Rauch, and R. Lane. 2003a. Neurobiology of emotion perception: I. The neural basis of normal emotion perception. *Biol Psychiatry* 54 (5):504–14.

Phillips, M. L., W. C. Drevets, S. L. Rauch, and R. Lane. 2003b. Neurobiology of emotion perception: II. Implications for major psychiatric disorders. *Biol Psychiatry* 54 (5):515–28.

Pickering, M., and J. J. O'Connor. 2007. Cytokines and their effects in the dentate gyrus. In *The Dentate Gyrus: A Comprehensive Guide to Structure, Function, and Clinical Implications*, ed. H. Scharfman. *Progr Brain Res* 163:339–54.

Pizzagalli, D., R. D. Pascual-Marqui, J. B. Nitschke, et al. 2001. Anterior cingulate activity as a predictor of degree of treatment response in major depression: Evidence from brain electrical tomography analysis. *Am J Psychiatry* 158 (3):405–15.

Plata-Salaman, C. R. 2001. Cytokines and feeding. *Int J Obes Relat Metab Disord* 25 (Suppl 5):S48–52.

Pollmacher, T., M. Haack, A. Schuld, A. Reichenberg, and R. Yirmiya. 2002. Low levels of circulating inflammatory cytokines — Do they affect human brain functions? *Brain Behav Immun* 16 (5):525–32.

Porter, R. J., P. Gallagher, J. M. Thompson, and A. H. Young. 2003. Neurocognitive impairment in drug-free patients with major depressive disorder. *Br J Psychiatry* 182:214–20.

Quan, N., and M. Herkenham. 2002. Connecting cytokines and brain: A review of current issues. *Histol Histopathol* 17 (1):273–88.

Radwanska, K., E. Nikolaev, E. Knapska, and L. Kaczmarek. 2002. Differential response of two subdivisions of lateral amygdala to aversive conditioning as revealed by c-Fos and P-ERK mapping. *Neuroreport* 13 (17):2241–6.

Raison, C. L., L. Capuron, and A. H. Miller. 2006. Cytokines sing the blues: Inflammation and the pathogenesis of depression. *Trends Immunol* 27 (1):24–31.

Ramos, A., and P. Mormede. 1998. Stress and emotionality: A multidimensional and genetic approach. *Neurosci Biobehav Rev* 22 (1):33–57.

Ransohoff, R. M., and E. N. Benveniste, eds. 2006. *Cytokines and the CNS*, 2nd ed. New York: Taylor & Francis Group.

Rattiner, L. M., M. Davis, C. T. French, and K. J. Ressler. 2004. Brain-derived neurotrophic factor and tyrosine kinase receptor B involvement in amygdala-dependent fear conditioning. *J Neurosci* 24 (20):4796–806.

Rattiner, L. M., M. Davis, and K. J. Ressler. 2005. Brain-derived neurotrophic factor in amygdala-dependent learning. *Neuroscientist* 11 (4):323–33.

Ravnkilde, B., P. Videbech, K. Clemmensen, A. Egander, N. A. Rasmussen, and R. Rosenberg. 2002. Cognitive deficits in major depression. *Scand J Psychol* 43 (3):239–51.

Reichenberg, A., R. Yirmiya, et al. 2001. Cytokine-associated emotional and cognitive disturbances in humans. *Arch Gen Psychiatry* 58 (5):445–52.

Ren, K., and R. Dubner. 2007. Pain facilitation and activity-dependent plasticity in pain modulatory circuitry: Role of BDNF-TrkB signaling and NMDA receptors. *Mol Neurobiol* 35 (3):224–35.

Rose, E. J., and K. P. Ebmeier. 2006. Pattern of impaired working memory during major depression. *J Affect Disord* 90 (2–3):149–61.

Rosenstein, D. L., D. Lerner, and J. Cai. 1999. More on the depressive effects of interferon alfa. *N Engl J Med* 341 (11):849–50.

Rothwell, N. J., and S. J. Hopkins. 1995. Cytokines and the nervous system: II. Actions and mechanisms of action. *Trends Neurosci* 18 (3):130–6.

Rothwell, N. J., and S. Loddick, eds. 2002. *Immune and Inflammatory Responses in the Nervous System, 2nd ed., Molecular and Cellular Neurobiology Series*. New York: Oxford University Press.

Rothwell, N. J., and J. K. Relton. 1993. Involvement of cytokines in acute neurodegeneration in the CNS. *Neurosci Biobehav Rev* 17 (2):217–27.

Saha, R. N., X. Liu, and K. Pahan. 2006. Up-regulation of BDNF in astrocytes by TNF-alpha: A case for the neuroprotective role of cytokine. *J Neuroimmune Pharmacol* 1 (3):212–22.

Sakic, B., H. Szechtman, T. Braciak, C. Richards, J. Gauldie, and J. A. Denburg. 1997. Reduced preference for sucrose in autoimmune mice: A possible role of interleukin-6. *Brain Res Bull* 44 (2):155–65.

Sakumoto, R., E. Kasuya, T. Komatsu, and T. Akita. 2003. Central and peripheral concentrations of tumor necrosis factor-alpha in Chinese Meishan pigs stimulated with lipopolysaccharide. *J Anim Sci* 81 (5):1274–80.

Samson, R. D., and D. Pare. 2005. Activity-dependent synaptic plasticity in the central nucleus of the amygdala. *J Neurosci* 25 (7):1847–55.

Schlatter, J., F. Ortuno, and S. Cervera-Enguix. 2004a. Lymphocyte subsets and lymphokine production in patients with melancholic versus nonmelancholic depression. *Psychiatry Res* 128 (3):259–65.

Schlatter, J., F. Ortuno, and S. Cervera-Enguix. 2004b. Monocytic parameters in patients with dysthymia versus major depression. *J Affect Disord* 78 (3):243–7.

Schleifer, S. J., S. E. Keller, and J. A. Bartlett. 1999. Depression and immunity: Clinical factors and therapeutic course. *Psychiatry Res* 85 (1):63–9.

Schobitz, B., E. R. De Kloet, and F. Holsboer. 1994. Gene expression and function of interleukin 1, interleukin 6 and tumor necrosis factor in the brain. *Prog Neurobiol* 44 (4):397–432.

Schwartz, M., A. Solomon, et al. 1991. Tumor necrosis factor facilitates regeneration of injured central nervous system axons. *Brain Res* 545 (1–2):334–8.

Schwartz, M., T. Sivron, et al. 1994. Cytokines and cytokine-related substances regulating glial cell response to injury of the central nervous system. *Prog Brain Res* 103:331–41.

Sei, Y., L. Vitkovic, et al. 1995. Cytokines in the central nervous system: Regulatory roles in neuronal function, cell death and repair. *Neuroimmunomodulation* 2 (3):121–33.

Shen, Y., T. J. Connor, Y. Nolan, J. P. Kelly, and B. E. Leonard. 1999. Differential effect of chronic antidepressant treatments on lipopolysaccharide-induced depressive-like behavioural symptoms in the rat. *Life Sci* 65 (17):1773–86.

Siebenlist, U., G. Franzoso, and K. Brown. 1994. Structure, regulation and function of NF-kappa B. *Annu Rev Cell Biol* 10:405–55.

Siegle, G. J., C. S. Carter, and M. E. Thase. 2006. Use of FMRI to predict recovery from unipolar depression with cognitive behavior therapy. *Am J Psychiatry* 163 (4):735–8.

Sluzewska, A. 1999. Indicators of immune activation in depressed patients. *Adv Exp Med Biol* 461:59–73.

Słuzewska, A., J. K. Rybakowski, M. Laciak, A. Mackiewicz, M. Sobieska, and K. Wiktorowicz. 1995. Interleukin-6 serum levels in depressed patients before and after treatment with fluoxetine. *Ann N Y Acad Sci* 762:474–6.

Smith, R. S. 1991. The macrophage theory of depression. *Med Hypotheses* 35 (4):298–306.

Song, C. 2002. The effect of thymectomy and IL-1 on memory: Implications for the relationship between immunity and depression. *Brain Behav Immun* 16 (5):557–68.

Spadaro, F., and A. J. Dunn. 1990. Intracerebroventricular administration of interleukin-1 to mice alters investigation of stimuli in a novel environment. *Brain Behav Immun* 4 (4):308–22.

Stefanacci, L., and D. G. Amaral. 2000. Topographic organization of cortical inputs to the lateral nucleus of the macaque monkey amygdala: A retrograde tracing study. *J Comp Neurol* 421 (1):52–79.

Suen, P. C., K. Wu, E. S. Levine, et al. 1997. Brain-derived neurotrophic factor rapidly enhances phosphorylation of the postsynaptic *N*-methyl-D-aspartate receptor subunit 1. *Proc Natl Acad Sci U S A* 94 (15):8191–5.

Swiergiel, A. H., and A. J. Dunn. 2006. Feeding, exploratory, anxiety- and depression-related behaviors are not altered in interleukin-6-deficient male mice. *Behav Brain Res* 171 (1):94–108.

Szelenyi, J., and E. S. Vizi. 2007. The catecholamine cytokine balance: Interaction between the brain and the immune system. *Ann N Y Acad Sci* 1113:311–24.

Tobinick, E. 2007. Perispinal etanercept for treatment of Alzheimer's disease. *Curr Alzheimer Res* 4 (5): 550–2.

Tsvetkov, E., R. M. Shin, and V. Y. Bolshakov. 2004. Glutamate uptake determines pathway specificity of long-term potentiation in the neural circuitry of fear conditioning. *Neuron* 41 (1):139–51.

Turrin, N. P., and C. R. Plata-Salaman. 2000. Cytokine–cytokine interactions and the brain. *Brain Res Bull* 51 (1):3–9.

Tweedie, D., K. Sambamurti, et al. 2007. TNF-alpha inhibition as a treatment strategy for neurodegenerative disorders: New drug candidates and targets. *Curr Alzheimer Res* 4 (4):378–5.

Veiel, H. O. 1997. A preliminary profile of neuropsychological deficits associated with major depression. *J Clin Exp Neuropsychol* 19 (4):587–603.

Venters, H. D., Q. Tang, Q. Liu, R. W. VanHoy, R. Dantzer, and K. W. Kelley. 1999. A new mechanism of neurodegeneration: A proinflammatory cytokine inhibits receptor signaling by a survival peptide. *Proc Natl Acad Sci U S A* 96 (17):9879–84.

Verma, I. M., J. K. Stevenson, E. M. Schwarz, D. Van Antwerp, and S. Miyamoto. 1995. Rel/NF-kappa B/I kappa B family: Intimate tales of association and dissociation. *Genes Dev* 9 (22):2723–35.

Vitkovic, L., J. Bockaert, et al. 2000. "Inflammatory" cytokines: Neuromodulators in normal brain? *J Neurochem* 74 (2):457–71.

Walker, D. L., and M. Davis. 2002. The role of amygdala glutamate receptors in fear learning, fear-potentiated startle, and extinction. *Pharmacol Biochem Behav* 71 (3):379–92.

Weiskrantz, L. 1956. Behavioral changes associated with ablation of the amygdaloid complex in monkeys. *J Comp Physiol Psychol* 49 (4):381–91.

Weisskopf, M. G., E. P. Bauer, and J. E. LeDoux. 1999. L-type voltage-gated calcium channels mediate NMDA-independent associative long-term potentiation at thalamic input synapses to the amygdala. *J Neurosci* 19 (23):10512–19.

Wilson, C. J., C. E. Finch, et al. 2002. Cytokines and cognition—the case for a head-totoe inflammatory paradigm. *J Am Geriatr Soc* 50 (12):2041–56.

Yamada, T., M. A. Horisberger, N. Kawaguchi, I. Moroo, and T. Toyoda. 1994. Immunohistochemistry using antibodies to alpha-interferon and its induced protein, MxA, in Alzheimer's and Parkinson's disease brain tissues. *Neurosci Lett* 181 (1–2):61–4.

Yeh, S. H., C. H. Lin, C. F. Lee, and P. W. Gean. 2002. A requirement of nuclear factor-kappaB activation in fear-potentiated startle. *J Biol Chem* 277 (48):46720–9.

Yeh, S. H., C. H. Lin, and P. W. Gean. 2004. Acetylation of nuclear factor-kappaB in rat amygdala improves long-term but not short-term retention of fear memory. *Mol Pharmacol* 65 (5):1286–92.

Yeh, T. C., and S. Pellegrini. 1999. The Janus kinase family of protein tyrosine kinases and their role in signaling. *Cell Mol Life Sci* 55 (12):1523–34.

Yin, L., L. Wu, H. Wesche, et al. 2001. Defective lymphotoxin-beta receptor-induced NF-kappaB transcriptional activity in NIK-deficient mice. *Science* 291 (5511):2162–5.

Yirmiya, R. 1996. Endotoxin produces a depressive-like episode in rats. *Brain Res* 711 (1–2):163–74.

Zorrilla, E. P., L. Luborsky, J. R. McKay, et al. 2001. The relationship of depression and stressors to immunological assays: A meta-analytic review. *Brain Behav Immun* 15 (3):199–226.

15 Tachykinins and Tachykinin Receptor Antagonists in Depression: Therapeutic Implications

Tih-Shih Lee and K. Ranga Krishnan

CONTENTS

15.1 INTRODUCTION

Von Euler and Gaddum described an unidentified substance extracted from horse brain and intestine. This substance, when administered to rabbits, displayed stimulant action on the jejunum and induced hypotension. This was called *P substance*. Substance P (SP) was isolated in pure form from bovine hypothalamus (Von Euler and Gaddum 1931).

Other similar peptides were identified and were called *tachykinins*. They were so named because of their ability to rapidly induce contraction of gut tissue. Three tachykinins have been isolated: SP, neurokinin A (NKA), and neurokinin B (NKB). SP is generally cosynthesized, colocalized, and cosecreted with NKA. The tachykinin family of peptides is characterized by a common C-terminal sequence, Phe-X-Gly-Leu-Met-NH$_2$, where X is an aromatic or an aliphatic amino acid. The tachykinin genes encode precursor proteins called *protachykinins*; these are cleaved into tachykinins by posttranslational processing. Mammalian tachykinins are derived from two preprotachykinin genes: the *PPT-A* gene, which encodes the sequences of SP, NKA, neuropeptide K and neuropeptide-γ, and the *PPT-B* gene, which encodes the sequence of NKB (Nawa et al. 1983; Kotani et al. 1986; Bonner et al. 1987; Sivam et al. 1987).

Neuronal tachykinins are released by a calcium-dependent mechanism (Maggi et al. 1993a, 1993b). There are three tachykinin receptors: NK$_1$, NK$_2$, and NK$_3$. All are members of the 7-transmembrane G-protein-coupled family of receptors and induce the activation of phospholipase C, producing inositol triphosphate. NK$_1$, NK$_2$, and NK$_3$ selectively bind to SP, NKA, and NKB, respectively, and have weaker binding to the other tachykinins (Table 15.1).

TABLE 15.1
Tachykinins and Preferential Binding by Receptors

NK_1	SP > NKA > NKB
NK_2	NKA > NKB > SP
NK_3	NKB > NKA > SP

Source: Page, N. M., *Peptides*, 26, 1356–68, 2005. With permission.

15.2 LOCALIZATION OF TACHYKININS AND TACHYKININ RECEPTORS IN CENTRAL NERVOUS SYSTEM

SP is the most abundant neurokinin in the mammalian central nervous system (Bensaid et al. 2001). In rats, autoradiographic, immunohistochemical staining, and messenger RNA expression studies have shown broad distribution of SP and NK_1 receptors. Staining was particularly intense in the amygdala, locus ceruleus, hypothalamus, and substantia nigra. In the human brain, areas most rich in SP are the subcortical regions, such as the amygdala, caudate, putamen, globus pallidus, hypothalamus, substantia nigra, and locus coeruleus (Gale et al. 1978; Emson et al. 1980; Cooper et al. 1981). Some noradrenaline- and serotonin-containing cell bodies also coexpress SP. Although NK_1 receptors are expressed only by a minority (5–7%) of neurons in the brain and spinal cord, they are found in close association to serotonergic and noradrenergic neurons.

There are relatively much lower concentrations of NK_2 and NK_3 receptors in the brain. Using reverse transcription-polymerase chain reaction, Bensaid et al. (2001) found barely detectable expression of NK_2 receptor mRNA in the caudate, putamen, hippocampus, substantia nigra, and cerebral cortex. The distribution of NK_2 receptor expression is much greater in the frontal and temporal cortices as compared with that in the occipital and parietal areas. NK_3 receptors in humans are expressed in cortical regions, the amygdala, hippocampus, and substantia nigra, often in proximity to dopaminergic pathways (Almeida et al. 2004; Rigby et al. 2005).

The distribution of SP and the other tachykinins provides multiple opportunities for interactions with the convergent noradrenaline and serotonin pathways through which established antidepressant drugs act, suggesting that tachykinin antagonists might have some use in the treatment of psychiatric disorders.

15.3 THERAPEUTIC APPLICATIONS OF NK_1 RECEPTOR ANTAGONISTS

Phylogenetic differences in central nervous system localization of tachykinin receptors and species variants in NK receptor pharmacology have complicated efforts to develop SP antagonists and have limited the usefulness of standard models and preclinical studies (Rupniak and Kramer 1999). Potent tachykinin receptor antagonists have been developed over the past two decades directed at various disorders. In particular, NK_1 receptor antagonists have been proposed for the treatment of pain and emesis, for inflammatory diseases, arthritis, inflammatory and motor diseases, and cystitis (Quartara and Maggi 1998). However, efficacy for most of the conditions of interest remains elusive.

There was initial promise of NK_1 receptor antagonists for analgesia as activation of central SP pathways occurs in response to noxious or aversive stimulation. Radioligand-binding studies showed expression of tachykinin NK_1 and NK_3 receptors in the dorsal horn of the spinal cord (Shults et al. 1984). This was supported by functional studies. However, clinical trials with NK_1 receptor antagonists conducted in patients with various pain conditions, including peripheral neuropathy, migraine, and osteoarthritis, uniformly failed to confirm the analgesic efficacy of these compounds in humans (Suarez et al. 1994).

The one area where NK_1 receptor antagonists have convincingly been shown to be effective appears to be for some types of nausea and vomiting, especially for chemotherapy-induced nausea and vomiting. Chemotherapy regimens (e.g., cisplatin) can be highly emetogenic. Chemotherapy-induced nausea and vomiting is postulated to be centrally mediated via the nucleus of the solitary tract in the brainstem that discharges through the vagal efferent pathways and involves multiple neurotransmitters, including $5\text{-}HT_3$, NK_1, and dopamine (Rupniak and Kramer 1999; Hesketh et al. 2003).

An NK_1 receptor antagonist, aprepitant, was first shown to be efficacious for chemotherapy-induced nausea and vomiting in two multicenter, randomized, parallel, double-blind, placebo-controlled trials involving 1044 patients who received cisplatin. This was confirmed by subsequent studies that included various chemotherapeutic agents and antiemetic combinations. The only adverse effects reported were mild, such as fatigue, hiccups, headache, and diarrhea (Quartara and Maggi 1998).

In 2003, aprepitant was launched as the first tachykinin receptor antagonist, with the trade name Emend® (Merck), for the prevention of chemotherapy-induced nausea and vomiting and subsequently also applied to postoperative nausea and vomiting. A parenterally administered alternative, fosaprepitant, was made available in 2007.

15.4 APPLICATIONS IN PSYCHIATRIC DISORDERS

The rationale for the use in psychiatric disorders derives from observations on the colocalization and location of SP and NK_1 receptors in the brain involved in the regulation of mood, anxiety, and stress response. In addition, there are spatial and functional overlaps between SP-NK_1 receptors and noradrenergic and serotonergic pathways. Some neurons coexpress SP, noradrenaline, and serotonin, suggesting a possible interaction between the different neurotransmitter systems (Krishnan 2002). SP receptors in rats differed significantly from those in humans, and NK_1 antagonist compounds developed for humans did not work in rats. However, they worked in guinea pigs, and these were found to suppress isolation-induced stress vocalization in these guinea pigs, which was similar to that produced by clinically used antidepressant and anxiolytic drugs (Kramer et al. 2004).

It was conceptualized that SP was a neurotransmitter that might play a role in stress responses, such as when an organism is exposed to noxious stimuli (Maggi 1995), and it may be synchronized to survival-type responses (Boyce et al. 2001). The model that was therefore built proposed that excessive release of SP in key fear and mood circuits led to a cascade of signs and symptoms such as those observed in depression or anxiety, and antagonism of dysregulated SP activity might provide novel mechanisms for antidepressant or anxiolytic activity. Preclinical and clinical studies can therefore be used as a putative validator of the model.

15.4.1 DEPRESSION CLINICAL TRIALS

An exciting initial study was the basis for the major interest in SP antagonists in major depression. A randomized, double-blind, placebo-controlled study was conducted to evaluate the safety and efficacy of single daily doses of 300 mg of MK-869 (aprepitant) in comparison with paroxetine (20 mg) or placebo in patients with major depressive disorder (MDD). In this initial study, there was a 4.3-point difference in mean change in total Hamilton Depression Rating Scale (HAMD)-21 score from baseline to week 6 between MK-869 and placebo. This suggested that MK-869 was an effective antidepressant. The effect of MK-869 was similar to that of paroxetine. Like the known antidepressive drugs, MK-869's effect was evident only after 2 to 3 weeks, suggesting the possibility that all antidepressant drugs act via a yet uncharacterized "common pathway" mechanism (Kramer et al. 1998).

This was followed by a second study with another NK_1 antagonist, L-759274. This was investigated in MDD with melancholic features. A randomized, double-blind, placebo-controlled study was carried out. Patients were randomized to oral L-759274 40 mg daily ($n = 66$) or placebo ($n = 62$) for 6 weeks. Patients receiving L-759274 showed improvement in mean HAMD-17 score of 10.7 points, compared with a mean 7.8-point improvement in patients receiving placebo ($p < .009$). Mean scores for item 1 of HAMD-17 (depressed mood) also improved to a greater extent in the active group compared with the placebo group (0.3 points, $p < .058$).

These two studies sparked considerable interest in the mechanism as a distinct pathway for antidepressant activity (Kramer et al. 2004). This further led to a series of studies of NK_1 antagonists for major depression. A major endeavor was a large five-study program with MK-869 (aprepitant). The results from these five randomized, double-blind, phase III studies in more than 2500 patients demonstrated that aprepitant (at doses of 160 and 80 mg) given once daily for 8 weeks were well tolerated. Significantly, however, they did not prove to be more effective than placebo in the treatment of MDD (Keller et al. 2005).

Although high placebo response rates can confound the interpretation of results from clinical trials in depression, this was not an issue with this trial because paroxetine (20 mg) was included as an active comparator in three of the studies and showed significant advantages over placebo on the primary measure and most secondary measures across all three studies at the 8-week end point.

The demonstration that neither dose of aprepitant exhibited antidepressant efficacy was very surprising and perhaps disappointing, given the extensive preclinical evidence implicating SP in depression and the previous clinical finding in a phase II trial of MK-869 that a 300-mg tablet formulation of aprepitant was more effective than placebo and as effective as paroxetine (20 mg) in patients with MDD (Kramer et al. 1998). This also contradicted the finding that the structurally similar SP receptor antagonist L-759274 was more effective than placebo in patients with MDD with melancholic features (Kramer et al. 2004).

It also seemed unlikely that differences in the aprepitant formulation/dose between the later studies and earlier ones could have accounted for this difference. The data from a concurrent positron emission specstroscopy (PET) study indicated that plasma levels achieved in the depression clinical trials were likely to have produced high occupancies that blocked the NK_1 receptor system in the brain continuously throughout the study period (Figure 15.1). This suggests that the right formulation, doses, and dosing interval of aprepitant were selected for the phase III clinical studies.

A possible explanation for the discrepancy in the findings between the initial phase II trial and the phase III studies is that the former study might have recruited a subgroup of patients who were preferentially responsive to SP blockade, although the patients recruited seemed generally comparable with regard to measured characteristics. Another possibility is that supra high saturation of the NK_1 site or off-target effects may have been effective, and that may have been the case in the early studies. However, there is no easy way to identify substantive differences between the phase II and phase III studies. Table 15.2 summarizes the major studies with NK_1 antagonists for MDD.

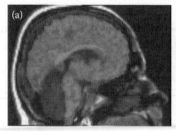 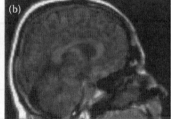

FIGURE 15.1   **(See color insert.)** Positron emission tomography scans of brain neurokinin1 receptors, visualized with [^{18}F] SPA-RQ (substance P antagonist receptor quantifier) overlaid on magnetic resonance images before (a) and after (b) blockade by administration of 160 mg aprepitant for 40 days. Blue indicates low tracer binding; yellow and orange, high tracer binding. (Reproduced with permission from Keller, M., et al., *Biol Psychiatry*, 59, 216–23, 2005.)

TABLE 15.2
Major Studies with NK₁ Antagonists for MDD

Study		Type	Drug	n	Time (weeks)	Comparator	Results
Kramer	1998	Randomized, double-blind, controlled	MK869	213	6	Paroxetine/ placebo	Similar to comparator
	2004		L759274	128	6	Placebo	Significant difference
Keller 2006	059		Aprepitant	468	8	Paroxetine/ placebo	No difference from placebo; inferior to paroxetine
	061			439	8	Paroxetine/ placebo	No difference from placebo; inferior to paroxetine
	062			494	8	Placebo	No difference from placebo
	063			495	8	Placebo	No difference from placebo
	073			324	8	Placebo	No difference from placebo

15.4.2 Anxiety Disorders Trials

Walsh et al. (1995) reported the anxiolytic-like activity of GR159897, an NK₂ receptor antagonist, in rodent and primate models of anxiety. Similar encouraging results were obtained by other researchers based on rat social interaction (Louis et al. 2008). The NK₁ antagonist LY-686017 (Eli Lilly) was tested for social anxiety. After 12 weeks of treatment, the Liebowitz Social Anxiety Score did not differ between the compound and the placebo. Another NK₁ receptor antagonist, vofopitant (GlaxoSmithKline), was investigated in conjunction with citalopram, but efficacy results have thus far not been published (Quartara and Maggi 1998).

An NK₂ antagonist, saredutant (Sanofi-Aventis), underwent phase III trials for MDD and generalized anxiety disorder. Pooled analysis of all saredutant studies showed some positive short-term benefit for patients with MDD in terms of HAMD scores. However, further trials have been discontinued. Similarly, an 8-week treatment with saredutant in adult patients with generalized anxiety disorder was evaluated in three phase III studies. These have been discontinued, and there is no further information available (Quartara and Maggi 1998).

15.4.3 Schizophrenia and NK₃ Receptor Selective Antagonists

Preclinical studies suggested a possible role of NK₃ antagonists for the treatment of schizophrenia (Maggi et al. 1993a). Microdialysis studies demonstrated that acute administration of talnetant (GlaxoSmithKline) produced significant increases in extracellular dopamine and noradrenaline in the medial prefrontal cortex and attenuated haloperidol-induced increase in nucleus accumbens dopamine levels in guinea pigs (Dawson et al. 2008).

In a placebo-controlled trial, schizophrenic subjects ($n = 236$) received oral talnetant or risperidone for 6 weeks. The primary outcome was a change in Positive and Negative Scale for Schizophrenia (PANSS) score. A greater or equal reduction in PANSS score was achieved in 40% of patients who received talnetant, compared with 48% who received risperidone and 30% who received placebo. Notably, patients on talnetant did not gain weight or had increased prolactin levels (Quartara and Maggi 1998).

However, in another placebo-controlled trial ($n = 20$), a greater or equal to 20% reduction of PANSS was seen in 25% of subjects treated with talnetant compared with 49% and 29% of those receiving risperidone or placebo, respectively. As of 2007, the development of talnetant was halted.

Another NK$_3$ receptor antagonist, osanetant (Sanofi-Synthélabo), was tested as part of a multi-trial study group (Meltzer et al. 2004). There was a mean reduction of Clinical Global Impression Scale severity of illness score, the Brief Psychosis Rating Scale, and the PANSS total score that was significantly larger than the placebo. However, there was no significant difference between the efficacy of osanetant and haloperidol. Although it can be postulated that NK$_3$ antagonists may lead to an improvement in psychopathology and possess the advantage of good tolerability, the findings were inconclusive regarding the effect of these compounds to cognition. Moreover, both compounds were limited by their bioavailability—osanetant was found to have cleared too quickly, and talnetant had difficulty passing the blood–brain barrier (Quartara and Maggi 1998). Osanetant was also tested in MDD in a phase II study, but no significant difference was found as compared with placebo and paroxetine. Development of both compounds has been discontinued.

15.5   CONCLUSIONS

There is little doubt that tachykinins play an important role in the nervous system and that their effects are widespread, from the central to the peripheral and enteric nervous systems. Hence, tremendous efforts were expended over the past two decades to discover therapeutic roles for the tachykinins. Despite a promising start, the overall results have thus far been disappointing.

Aprepitant and the parenteral formulation for it are now in use for chemotherapy-induced and postoperative nausea and vomiting. Several drug developments directed at the treatment of depression, anxiety, and schizophrenia have stalled or been curtailed (Table 15.3). Given the good tolerability and mild side-effect profiles of the NK antagonists tested thus far, there remains potential to develop new formulations of NK antagonists with greater bioavailability that may be tested again for depression, anxiety, schizophrenia, and other psychiatric conditions.

But this story has several important lessons:

1. There is limited utility of animal models.
2. New models that are not based simply on what worked with monoamines need to be developed.
3. Dose response has to be addressed rigorously to find adequate, if not optimal, dosing to achieve symptom relief or remission. For example, a radioactive, brain-penetrant tracer with high affinity for NK$_1$ receptor, which permits real-time imaging of occupancy of the receptors in human subjects via positron emission tomography, may be useful in this regard (Krishnan 2002).

TABLE 15.3
Psychiatric Indications for Some NK Receptor Antagonists

INN (Lab Code)	TK Receptor(s)	Clinical Condition	Highest Development Status
Aprepitant	NK$_1$	Alcohol craving/posttraumatic stress disorder	Phase II
Orvepitant	NK$_1$	MDD	Phase II
LY-686017	NK$_1$	Alcohol craving/social anxiety	Phase II
AZD-2624	NK$_1$	Schizophrenia	Phase II
Saredutant	NK$_2$	MDD, generalized anxiety disorder	Phase III discontinued
Osanetant	NK$_3$	Schizophrenia, MDD, panic disorder	Phase II discontinued
Talnetant	NK$_3$	Schizophrenia	Phase II discontinued

Source: Quartara, L., et al., *Expert Opin Invest Drugs*, 18, 1843–64, 2009. With permission.

4. There is enormous variability and heterogeneity in psychiatric disorders. Hence, appropriate patient selection appears to be a key issue for future trials and, ultimately, in clinical practice.
5. This heterogeneity may account for why some study compounds work and others do not, and why some may work only in subsets of the disease phenotype.
6. Personalized medicine based on the genotypic and phenotypic profiles of the patient may be the best way forward.

We are in the early stages of this process, and the lessons learned from previous work with SP and the other tachykinins may be of enormous value in helping to pave the way forward.

REFERENCES

Almeida, T.A., Rojo, J., Nieto, P.M., et al. 2004. Tachykinin and tachykinin receptors: Structure and activity relationships. *Curr Med Chem* 11:2045–81.

Bensaid, M., Faucheux, B.A., Hirsh, E., et al. 2001. Expression of tachykinin NK2 receptor mRNA in human brain. *Neurosci Lett* 303:25–8.

Bonner, T.I., Affolter, H.U., Young, A.C., et al. 1987. A cDNA encoding the precursor of the rat neuropeptide, neurokinin B. *Brain Res* 388:243–9.

Boyce, S., Smith, D., Carlson, E., et al. 2001. Intra-amygdala injection of the substance P (NK1 receptor) Antagonist L-760735 inhibits neonatal vocalization in guinea pigs. *Neuropharmacology* 41:130–7.

Cooper, P.E., Fernstrom, M.H., Rorstad, O.P., et al. 1981. The regional distribution of somatostatin, substance P and neurotensin in human brain. *Brain Res* 218:219–32.

Dawson, L.A., Cato, K.J., Scott, C., et al. 2008. In vitro and in vivo characterization of the non-peptide NK3 receptor antagonist SB-223412 (talnetant): Potential therapeutic utility in the treatment of schizophrenia. *Neuropsychopharmacology* 33:1642–52.

Emson, P.C., Arregul, A., Clement, J.V., et al. 1980. Regional distribution of methionine-enkephalin and substance P-like immunoreactivity in normal human brain and in Huntington's disease. *Brain Res* 199:147–60.

Gale, K., Costa, E., Toffano, G., et al. 1978. Evidence for a role of nigral γ- aminobutyric acid and substance P in the haloperidol-induced activation of striatal tyrosine hydroxylase. *J Pharmacol Exp Ther* 206:29–37.

Hesketh, P.J., Van Belle, S., Aapro, M., et al. 2003. Aprepitant Protocol 052 Study Group. The oral neurokinin-1 antagonist aprepitant for the preventive of chemotherapy-induced nausea and vomiting: A multinational, randomized, double-blind, placebo-controlled trial in patients receiving high-dose cisplatin — The Aprepitant Protocol 052 Study Group. *J Clin Oncol* 21:4112–9.

Keller, M., Montgomery, S., Ball, W., et al. 2005. Lack of efficacy of the substance P (neurokinin₁ receptor) antagonist aprepitant in the treatment of major depressive disorder. *Biol Psychiatry* 59:216–23.

Kotani, H., Hoshimaru, M., Nawa, H., et al. 1986. Structure and gene organization of bovine neuromedin K precursor. *Proc Natl Acad Sci USA* 83:7074–8.

Kramer, M.S., Cutler, N., Feighner, J., et al. 1998. Distinct mechanism for antidepressant activity by blockage of central substance P receptors. *Science* 281:1640–5.

Kramer, M.S., Winokur, A., Kelsey, J., et al. 2004. Demonstration of the efficacy and safety of a novel substance P (NK1) receptor antagonist in major depression. *Neuropsychopharmacology* 29:385–92.

Krishnan, K.R.K. 2002. Clinical experience with substance P receptor (NK₁) antagonists in depression. *J Clin Psychiatry* 63 (Suppl 11):25–9.

Louis, C., Stemmelin, J., Boulay, D., et al. 2008. Additional evidence for anxiolytic- and antidepressant-like activities of saredutant (SR48968), an antagonist at the neurokinin-2 receptor in various rodent-models. *Pharmacol Biochem Behav* 89:36–45.

Maggi, C.A. 1995. The mammalian tachykinin receptor. *Gen Pharmacol* 26:911–44.

Maggi, C.A., Patacchini, R., Rovero, P., et al. 1993a. Tachykinin receptors and tachykinin receptor antagonists. *J Aut Pharmacol* 13:23–93.

Maggi, C.A., Quartara, L., Patacchini, R., et al. 1993b. MEN 10,573 and MEN 10,612, novel cyclic pseudopeptides which are potent tachykinin NK-2 receptor antagonists. *Regul Pept* 47:151–8.

Meltzer, H.Y., Arvanitis, L., Bauer, D., et al. 2004. Placebo-controlled evaluation of four novel compounds for the treatment of schizophrenia and schizoaffective disorder. *Am J Psychiatry* 161:975–84.

Nawa, H., Hirose, T., Takashima, H., et al. 1983. Nucleotide sequences of cloned cDNAs for two types of bovine brain substance P precursor. *Nature* 306 (5938):32–6.

Page, N.M. 2005. New challenges in the study of the mammalian tachykinins. *Peptides* 26:1356–68.

Quartara, L., and Maggi, C.A. 1998. The tachykinin NK 1 receptor: Part II. Distribution and pathophysiological roles. *Neuropeptides* 32:1–49.

Quartara, L., Altamura, M., Evangelista, S., and Maggi, C.A. 2009. Tachykinin receptor antagonists in clinical trials. *Expert Opin Invest Drugs* 18:1843–64.

Rigby, M., O'Donnell, R., and Rupniak, N.M. 2005. Species differences in tachykinin receptor distribution: Further evidence that the substance P (NK1) receptor predominates in human brain. *J Comp Neurol* 490:335–53.

Rupniak, N.M., and Kramer, M.S. 1999. Discovery of the anti-depressant and anti-emetic efficacy of substance P receptor (NK1) antagonists. *Trends Pharmacol Sci* 20:485–90.

Shults, C.W., Quirion, R., Chronwall, B., et al. 1984. A comparison of the anatomical distribution of substance P and substance P receptors in the rat central nervous system. *Peptides* 5:1097–128.

Sivam, S.P., Breese, G.R., Krause, J.E., et al. 1987. Neonatal and adult 6-hydroxydopamine-induced lesions differentially alter tachykinin and enkephalin gene expression. *J Neurochem* 49:1623–33.

Suarez, G.A., Opfer-Gehrking, T.L., MacLean, D.B., and Low, P.A. 1994. Double-blind, placebo-controlled study of the efficacy of a substance P (NK1) receptor antagonist in painful peripheral neuropathy. *Neurology* 44: 373P.

Von Euler, U.S., and Gaddum, J.H. 1931. An unidentified depressor substance in certain tissue extracts. *J Physiol* 72:74–87.

Walsh, D.M., Stratton, S.C., Harvey, F.J., et al. 1995. The anxiolytic-like activity of GR159897, a non-peptide NK2 receptor antagonist, in rodent and primate models of anxiety. *Psychopharmacology (Berl)* 121:186–91.

16 Role of Neuropeptide Y in Depression: Current Status

Julio César Morales-Medina and Rémi Quirion

CONTENTS

16.1 INTRODUCTION

Major depression is a severe and debilitating emotional disorder predicted to become the second leading cause of disability worldwide (Murray and Lopez 1996). In addition, Kessler et al. (2003) observed up to 60% co-occurrence of major depression with anxiety in the U.S. National Comorbidity Survey. This complex emotional dysfunction is often characterized by anhedonia, hopelessness, exacerbated guilt, and memory deficits (Castaneda et al. 2008). Commonly prescribed antidepressants are not curative, and a considerable subpopulation of depressed subjects is resistant to these treatments (Berlim et al. 2008). In recent years, various neuropeptides have been proposed to play significant roles in the etiology of depression in addition to classical neurotransmitters (Alldredge 2010; Rotzinger et al. 2009). Among those peptides, neuropeptide Y (NPY) was shown to be dysregulated in depression and other emotional conditions and to have antidepressant properties (Dumont et al. 2009; Morales-Medina et al. 2010).

NPY is a 36-amino-acid polypeptide first isolated from porcine brain by Tatemoto et al. (1982). This peptide, along with pancreatic polypeptides (PPs) and peptide YY (PYY), shares differential affinities for at least four G-protein-coupled receptors known as the Y_1, Y_2, Y_4, and Y_5 receptors

(Michel et al. 1998). The possible relevance of this peptide in emotional processes is suggested by the finding that NPY is conserved in evolution and across different species (Larhammar and Salaneck 2004). In addition, NPY is widely distributed in the central nervous system, particularly in stress- and emotion-processing regions (Dumont et al. 2000; Dumont and Quirion 2006).

In this chapter, we summarize the current literature on the role of NPY in depression on the basis of results obtained in animal as well as human studies. Where appropriate, data related to anxiety-related behaviors are also discussed.

16.2 ANIMAL STUDIES

Multiple approaches have evaluated the role of NPY and its receptors in depression and other emotional processes. Direct and indirect pieces of evidence on the role of this peptide have been obtained using genetic manipulations and pharmacological treatments in control animals as well as in models of depression-related behaviors.

16.2.1 GENETIC MANIPULATIONS

A summary of data on the role of NPY and its receptors in genetically modified animals is presented in Table 16.1, with details discussed below.

16.2.1.1 Loss of Function Studies

Numerous studies have investigated the role of NPY and Y_1, Y_2, and Y_4 receptor subtypes using germinal knockout (KO) animals. Bannon et al. (2000) observed a decrease in center distance in the open field (OF) in NPY KO mice, suggestive of an anxiogenic-like phenotype. Data on Y_1 KO mice generated behavioral results that are rather controversial. Painsipp et al. (2010) observed that female Y_1 KO mice show reduced immobility in the tail-suspension test (TST) but not in the forced-swim test (FST). Reduced immobility is interpreted as an antidepressant phenotype in both tests (Cryan et al. 2005a, 2005b). In contrast, Karlsson et al. (2008) found increased immobility time in the FST in Y_1 KO mice. However, both male and female mice were used in the latter study. Interestingly, Karl et al. (2006) found that Y_1 KO mice displayed normal behavior in the elevated plus maze (EPM), whereas stress produced an anxiolytic-like effect in these animals. Additionally, Painsipp et al. (2010) observed an antidepressant effect after stress in Y_1 KO mice in various depression-related paradigms. Early on, Thorsell et al. (2000) had suggested that the NPYergic system, especially the Y_1 receptor subtype, is activated to counterbalance the effects of stress.

In contrast to the mild antidepressant effects observed in Y_1 KO mice, accumulated evidence has revealed the strong antidepressant- and anxiolytic-related phenotype of germinal Y_2 KO mice independent of sex, age, or behavioral paradigm (Carvajal et al. 2006; Painsipp et al. 2008a; Redrobe et al. 2003b; Tschenett et al. 2003). In contrast to these reproducible findings, Zambello et al. (2010b) recently found that the deletion of the Y_2 receptor subtype had no effect on any of the studied behavioral paradigms. However, there is a tendency toward a decrease in the immobility in the FST in Y_2 KO compared to wild-type animals 30 min after the injection of a vehicle. The latter study backcrossed the original Y_2 KO mice (C57BL/6-129SvJ) to C57BL/6, for eight generations. In this regard, Lucki et al. (2001) evaluated the effect of genetic background in 11 strains of mice in the FST. This group observed a 10-fold difference in baseline immobility time as well as differential susceptibility to antidepressant treatment. Accordingly, Cryan et al. (2002) suggested exhaustive analysis of inbred and outbred strains in different behavioral tests. Most interestingly, Tasan et al. (2010) just reported increases in time spent in open arms in the EPM in animals having a targeted deletion of Y_2 receptors in the basolateral (BLA) and central (CeA) amygdala. Additionally, a decrease in immobility time in the TST was specifically observed after the deletion of the Y_2 receptor subtype in the CeA. Taken together, these series of studies strongly support the role of the Y_2 receptor subtype in depression-related behaviors.

Recently, it was shown that the Y_4 KO mice display reduced immobility in the FST and TST in male (Tasan et al. 2009) and female (Painsipp et al. 2008a, 2008b) mice. Rather interestingly, Tasan et al. (2009) observed that a double deletion of the Y_2 and Y_4 receptors further increased the movement component of the FST in these animals compared with Y_2 KO mice. This group also suggested that the antidepressant effect of the deletion of the Y_4 receptor was induced peripherally rather than centrally.

Finally, although the role of germinal Y_5 KO mice has been exhaustively evaluated in feeding behavior (Higuchi et al. 2008; Marsh et al. 1998; Raposinho et al. 2004) and seizures (Marsh et al. 1999), data on its role in emotion-related processes are still awaited.

16.2.1.2 Gain of Function Studies

The technique of overexpression of certain genes for the NPY peptide family and its receptors has not been widely used so far. Transgenic rats that overexpress NPY present normal behavior in the EPM when tested for the first time. However, after repeated stress, the time spent in the open arm of the EPM increased independently of the age of the animals, sugestive of an anxiolytic-like effect (Carvajal et al. 2004; Thorsell et al. 2000). Additionally, PP-overexpressing transgenic mice showed a decreased number of entries in the open arm of the EPM, a behavior associated with increased anxiety (Ueno et al. 2007). As PP-related peptides have particularly high affinity for the Y_4 receptor, it may suggest a role for this subtype in the noted behavior.

In summary, the germinal genetic manipulation of NPY, PP, and their receptors suggests that the Y_1 receptor subtype induces an antidepressant-like effect after stressful stimuli, whereas Y_2 KO mice display marked antidepressant and anxiolytic phenotypes. Complementary studies of the Y_4 KO and PP transgenic mice also revealed antidepressant-like phenotypes. Inasmuch as it is now possible to produce inducible KO or transgenic animals, further studies using such approaches to investigate further the role of NPY and related peptides in emotional processes are certainly warranted.

16.2.2 Pharmacological Treatment in Normal Naive Animals

Different groups have administered various agonists and antagonists of Y_1, Y_2, and Y_5 receptor subtypes to investigate their potential antidepressant-related effects in control animals. A summary of data is shown in Table 16.2. Acute intracerebroventricular (ICV) administration of NPY produced an antidepressant-like effect in the FST in naive rats (Redrobe et al. 2005; Stogner and Holmes 2000). Similarly, the Y_1-like agonist, [Leu^{31}Pro^{34}]PYY, when administered, acutely, reduced the immobility time in this paradigm in mice (Redrobe et al. 2002). Additionally, Redrobe et al. (2002) showed that the coadministration of the Y_1 antagonist, BIBO3304, abolished the behavioral effects induced by [Leu^{31}Pro^{34}]PYY. Acute ICV administration of the Y_2 antagonist, BIIE0246, also decreased the immobility time in the FST in mice (Redrobe et al. 2002).

Whereas Y_4 KO mice induced an antidepressant-like phenotype, the overexpression of Y4 receptors, PPs had opposite effects. The intraperitonial administration of PP increased anxiety-related behavior in mice (Asakawa et al. 2003), whereas the ICV administration of this peptide had limited effect on emotion-related behaviors (Asakawa et al. 1999). In this regard, Kastin and Pan (2010) recently highlighted the relevance of the role of peripheral peptides in the brain.

The Y_5 receptor subtype was initially regarded as the "feeding" receptor (Marsh et al. 1998) but has gone through dramatic changes over the years. More recently, various groups have suggested an important role for this receptor in emotional processes, with particular emphasis on anxiety-related behaviors. Acute administration of the Y_5 agonist, [cPP$^{1-7}$, NPY$^{19-23}$, Ala^{31}, Aib^{32}, Gln^{34}] hPP, either in the lateral ventricle (Sorensen et al. 2004) or BLA (Sajdyk et al. 2002), produced an anxiolytic-like effect in control animals. Additionally, Sajdyk et al. (2002) observed that Novartis 1, a Y_5 antagonist, blocked the anxiolytic-related effects of NPY3-36 in the BLA. However, the administration of this Y_5 antagonist by itself had no effect on anxiety-related behaviors (Sajdyk et al. 2002). Interestingly, an acute administration of another Y_5 antagonist, CGP71683A, induced an

TABLE 16.1
Summary of Data on the Role of NPY and Its Receptors in Genetically Modified Animals

Genetically Modified Animal	Type of Deletion	Species	Sex	Strain	Paradigm	Effect	Time of start of the test	Reference
NPY KO	germinal	mice	male	129/SvJ-c57BL/6	OF	↓ Distance center of arena	INA	Bannon et al. 2000
NPY transgenic	germinal	rat	male	Sprague–Dawley	EPM	No change	INA	Carvajal et al. 2004
					EPM/Stress	No change	INA	
					OF	↑ Entries center of arena/↑ center time	INA	
					EPM	NC	INA	Thorsell et al. 2000
					EPM/Stress	NC	INA	
Y1 KO	germinal	mice	female	129/SvJ-c57BL/6	TST	↓ Immobility	INA	Painsipp et al. 2010
					FST	NC	INA	
			male and female		OF	NC	INA	Karlsson et al. 2008
					EPM	NC	INA	
			INA		OF	↑ Distance center of arena	14:00	Karl et al. 2006
Y2 KO	germinal	mice	female	129/SvJ-c57BL/6	TST	↓ Immobility	10:00	Painsipp et al. 2008b
					EPM	↑ Open arm entries/↑ open arm time	10:00	
			INA		OF	↑ Center time	10:00	Carvajal et al. 2006
					FST	↓ Immobility	INA	
					EPM	↑ Open arm entries	INA	
					OF	↑ Distance center of arena	INA	
			female		TST	↓ Immobility	10:30	Painsipp et al. 2008a
					OF	↑ Entries center of arena	10:30	
					EPM	↑ Open arm entries	10:30	
			male		EPM	↑ Open arm entries/↑ open arm time	INA	Redrobe et al. 2003b
					OF	↑ Entries center of arena/↑ time center of arena	INA	

Genotype	Type		Strain	Sex	Test	Result	Time	Reference
				male	FST	↓ Immobility	9:00	Tschenett et al. 2003
					OF	↑ Entries center of arena	9:00	
					EPM	↑ Open arm entries	9:00	
			129/SvJ-c57BL/6 backcrossed 8 generations onto a c57BL/6	male and female	FST	NC	9:00	Zambello et al. 2010b
					EPM	NC	9:00	
Y4 KO	conditional-CeA		129/SvJ-c57BL/6	male	TST	↓ Immobility	8:00	Tasan et al. 2010
					EPM	↑ Open arm entries/↑ open arm time	8:00	
	conditional-BLA				TST	NC	8:00	
					EPM	↑ Open arm time	8:00	
	germinal	mice	129/SvJ-c57BL/6	female	TST	↓ Immobility	10:00	Painsipp et al. 2008b
					EPM	↑ Open arm entries/↑ open arm time	10:00	
					OF	↑ Entries center of arena/↑ center time	10:00	
				male	FST	↓ Immobility	8:00	Tasan et al. 2009
					TST	↓ Immobility	8:00	
					EPM	NC	8:00	
					OF	↑ Entries center of arena/↑ center time	8:00	
PP transgenic	germinal	mice	C57BL/6J	INA	EPM	↓ Open arm entries/↓ open arm time	INA	Ueno et al. 2007
Y2/Y4 KO	germinal	mice	129/SvJ-c57BL/6	male	FST	↓ Immobility	8:00	Tasan et al. 2009
					TST	↓ Immobility	8:00	
					EPM	↑ Open arm entries/↑ open arm time	8:00	
					OF	↑ Entries center of arena/↑ center time	8:00	

Note: INA, information not available; NC, no change.

TABLE 16.2
Effects of NPY- Related Ligands in Control Naive Animals

Ligand	Target	Species	Strain	Dose	Route	Duration	Test	Effect	Reference
NPY	NPY system	Rat	Sprague–Dawley	0.5–10 ng	ICV	Twice/24 h	FST	↓ Immobility	Stogner and Holmes 2000
NPY	NPY system	Mice	CD-one	3.0 nmol	ICV	Acute	FST	↓ Immobility	Redrobe et al. 2005
NPY	NPY system	Mice	CD-one	0.1–3.0 nmol	ICV	Acute	FST	↓ Immobility	Redrobe et al. 2002
[D-His26]NPY	Y_1 agonist	Rat	Wistar	0.8–3.0 nmol	ICV	Acute	EPM	↑ Open arm entries	Sorensen et al. 2004
[Leu[31]Pro[34]]PYY	Y_1/Y_5 agonist	Mice	CD-one	3.0 nmol	ICV	Acute	FST	↓ Immobility	Redrobe et al. 2002
BIBP3226	Y_1 antagonist	Mice	CD-one	3.0 nmol	ICV	Acute	FST	NC	Redrobe et al. 2002
BIBO3304	Y_1 antagonist	Mice	CD-one	0.3 nmol	ICV	Acute	FST	NC	Redrobe et al. 2002
C2-NPY	Y_2 agonist	Rat	Wistar	0.2–3.0 nmol	ICV	Acute	EPM	NC	Sorensen et al. 2004
NPY13-36	Y_2 agonist	Mice	CD-one	0.3–3.0 nmol	ICV	Acute	FST	NC	Redrobe et al. 2002
BIIE0246	Y_2 antagonist	Mice	CD-one	3.0 nmol	ICV	Acute	FST	↓ Immobility	Redrobe et al. 2002
PP	Y_4 agonist	Mice	ddY	3.0 nmol	IP	Twice per day/14 days	EPM	↑ Open arm entries/↑ open arm time	Asakawa et al. 2003
PP	Y_4 agonist	Mice	ddY	0.003–3.0 nmol	ICV	Acute	EPM	NC	Asakawa et al. 1999
[cPP[1-7], NPY[19-23], Ala[31], Aib[32], Gln[34]]hPP	Y_5 agonist	Rat	Wistar	0.8–3.0 nmol	ICV	Acute	EPM	↑ Open arm entries	Sorensen et al. 2004
CGP71683A	Y_5 antagonist	Rat	Wistar	10 mg/kg	IP	Acute	OF	↓ Horizontal activity/↓ vertical activity	Kask et al. 2001
CGP71683A	Y_5 antagonist	Rat	Wistar	10 mg/kg	IP	Acute	EPM	NC	Kask et al. 2001
CGP71683A	Y_5 antagonist	Rat	Wistar	10 mg/kg	IP	Acute	SI	NC	Kask et al. 2001
Lu AA33810	Y_5 antagonist	Rat	Sprague–Dawley	3–10 mg/kg	Oral	Acute	SI	↑ Active social interaction	Walker et al. 2009
Lu AA33810	Y_5 antagonist	Rat	Sprague–Dawley	10 mg/kg	Oral	14 days	SI	↑ Active social interaction	Walker et al. 2009

Note: INA, information not available; NC, no change; IP, intraperitoneal.

anxiolytic-related effect that was task-specific (Kask et al. 2001). Recently, Walker et al. (2009) observed that the novel Y_5 antagonist, Lu AA33810, reduced the [cPP$^{1-7}$, NPY$^{19-23}$, Ala^{31}, Aib^{32}, Gln^{34}]hPP-induced increase in plasma levels of corticosterone and adrenocorticotropic hormone in normal naive rats. In addition, either acute or repeated administration of Lu AA33810 resulted in an anxiolytic-like effect in the social interaction (SI) test in control rats.

This first generation of pharmacological approaches in normal animals added further evidence on the role of the NPYergic system in emotional processes to those obtained using germline KO and transgenic animals. In contrast to Y_1 KO mice, these studies showed that the acute activation of the Y_1 receptor subtype can induce an antidepressant effect. These pharmacological approaches also suggested that the antidepressant effects of NPY are mostly mediated through the Y_1 receptor subtype. In agreement with data obtained in Y_2 KO mice, the pharmacological blockade of Y_2 receptors induced an antidepressant phenotype. In addition, these pharmacological data suggested that the action of the Y_4 receptor subtype could be peripherally mediated. As to the Y_5 receptor, this subtype is apparently mostly associated with anxiety but not depression-like behaviors.

16.2.3 Animal Model Studies

Animal models are powerful tools to help understand different pathologies, including emotional conditions, as well as to test potential therapeutic agents (Cryan and Mombereau 2004). In this regard, data from various animal models have suggested both direct and indirect roles for NPY in depression and anxiety. For example, the NPYergic system is disturbed in various animal models of depression, and the administration of different NPYergic compounds improve behavioral despairs in these models.

16.2.3.1 Antidepressant Treatments

Several studies have investigated the effects of various treatments with antidepressants in animal models of depression-like behavior as well as changes in NPY and its receptors after such treatment. Repeated administration of the selective serotonin reuptake inhibitor (SSRI), fluoxetine, increased NPY-like immunoreactivity in Flinders sensitive line (FSL) rats, a genetic model of depression-related behavior, whereas NPY levels were decreased in the Flinders resistant line (FRL) rats (Caberlotto et al. 1998). A repeated treatment with fluoxetine also increased NPY mRNA levels in the dentate gyrus following psychosocial stress in the rat (Zambello et al. 2010a). Additionally, repeated electroconvulsive therapy (ECT) increased NPY levels in the hippocampus (Stenfors et al. 1989). Recently, running, as well as the combined treatment of running and the SSRI, escitalopram, increased both NPY mRNA and Y_1 receptor mRNA in the hippocampus, particularly in the dentate gyrus in the FSL rat (Bjornebekk et al. 2010). Interestingly, running did not increase the hippocampal levels of NPY in the FRL rats (Bjornebekk et al. 2006). Furthermore, these treatments induced antidepressant-like effects in the FST, and these actions were associated with increased levels of NPY and Y_1 receptors in these animals. Recently, Christiansen et al. (2011) observed that repeated fluoxetine treatment decreased immobility time in the FST in a model of stress-induced depression. This treatment increased the mRNA expression of NPY in the dentate gyrus and amygdala. However, NPY mRNA levels in these regions were not altered in control animals and in animals exposed to repeated stress (Christiansen et al. 2011).

16.2.3.2 Altered Levels of NPY and NPY Receptors in Animal Models of Depression-Related Behaviors

Studies in animal models have suggested that the synthesis of NPY as well as the expression of various subclasses of NPY receptors is altered in key brain regions involved in emotional processing, including the hippocampus, amygdala, and cortex. For example, low levels of NPY have been observed in the dorsal hippocampus of maternally separated adult rat, a model of depression-related behavior (Jimenez-Vasquez et al. 2001). Similarly, NPY-like immunoreactivity is decreased in the dorsal hippocampus of FSL rats (Caberlotto et al. 1999). Likely as a compensatory mechanism, the level of the Y_1 receptor subtype was increased in the hippocampal region in this genetic model of depression (Caberlotto et al. 1999). Another study by Caberlotto et al. (1998) proposed that Y1-mRNA expression was downregulated in various cortical regions as well as in the hippocampus of FSL rats. This group also suggested that the expression of the Y_2 receptor subtype was not affected in the hippocampus in this model (Caberlotto et al. 1998). Additionally, repeated stress induced increases in NPY levels (protein and mRNA) in the amygdala (de Lange et al. 2008; Thorsell et al. 1998, 1999). Recently, Christiansen et al. (2011) replicated these findings in a model of stress-induced depression. This group also observed that the mRNA levels of NPY in the hippocampus were unchanged in their model. Interestingly, after psychosocial stress, rats displayed decreased levels of NPY mRNA in various hippocampal subfields (Zambello et al. 2010a). In the olfactory bulbectomized (OBX) rat, accumulated evidence suggested disturbances in the NPYergic system. The prepro-NPY mRNA levels were found to be increased in the hippocampus and piriform cortex of the OBX rat (Holmes et al. 1998). Additionally, NPY levels were increased in the amygdala (Rutkoski et al. 2002) in this animal model. In the cortex of OBX animals, Uriguen et al. (2008) also observed increased gene expression of NPY, whereas Widerlov et al. (1988a) reported decreased NPY-like immunoreactivity. Our group has recently investigated the effect of OBX lesion on the expression of the Y_1, Y_2, and Y_5 receptor subtypes in the hippocampus and amygdala. We observed that this lesion increased the expression of Y_2 receptors in both the dorsal hippocampus and amygdala, whereas Y_5 receptor expression was changed only in the amygdala (Morales-Medina et al. 2009). No changes in the expression of the Y_1 receptor subtype were found in this particular model.

These studies consistently reported decreased levels of NPY in the hippocampus but its overexpression in the amygdala. However, the expression of the various NPY receptor subtypes was selectively changed depending on the model used. Future studies should aim to evaluate the antidepressant effects of NPY-related molecules and to demonstrate if alterations in NPY levels and receptor expression are correlated with behaviors.

16.2.3.3 Exogenous Administration of NPY-Related Molecules in Animal Models

A variety of well-known animal models of depression-related behaviors has thus far been used to provide additional evidence on the role of NPY and its receptors in emotional processes, as observed in Table 16.3. In the helplessness model of depression-like behaviors, the acute administration of NPY and [Leu^{31}, Pro^{34}]PYY as well as the Y_2 antagonist BIIE0246 in the CA3 subfield of the hippocampus decreased the escape failure in the active avoidance paradigm, which is inferred as antidepressant-like behavior (Ishida et al. 2007). Interestingly, the Y_1 antagonist BIBO3304 blocked the antidepressant-related effect of NPY in this model (Ishida et al. 2007). Our group observed that a continuous ICV administration of either [Leu^{31}Pro^{34}]PYY or BIIE0246 significantly decreased whereas the Y_2 agonist PYY3-36 increased the immobility time in the FST in OBX rats (Morales-Medina et al. 2009). Additionally, Goyal et al. (2009) showed that the Y_1-like agonist [Leu^{31}Pro^{34}] NPY decreased the hyperlocomotion in OF, a disturbed behavior reversed only by administration of antidepressants (Song and Leonard 2005).

Recently, the novel Y_5 antagonist Lu AA33810 was shown to have anxiolytic- and antidepressant-like effects in the FSL rat (Walker et al. 2009). Lu AA33810 increased sucrose consumption compared with vehicle-treated animals, interpreted as an antidepressant-like effect, in the chronic mild stress model (Walker et al. 2009). In contrast, preliminary data from our group suggest that the repeated ICV administration of [cPP$^{1-7}$, NPY$^{19-23}$, Ala^{31}, Aib^{32}, Gln^{34}]hPP reduced hyperlocomotion in OF and increased social contacts in the SI test in the OBX rat (Morales-Medina et al. 2009).

Taken together, these studies confirmed the role of NPY in depression-related behaviors in the challenged brain. The activation of the Y_1 receptor subtype as well as the blockade of the Y_2 receptor consistently induced antidepressant effects in various animal models. However, the role of Y_5 receptor subtype is still controversial. Inasmuch as both a Y_5 agonist and an antagonist improved

TABLE 16.3
Effects of NPY-Related Ligands in Various Animal Models of Depression-Related Behavior

Ligand	Target	Model	Species	Strain	Douse	Route	Duration	Test	Effect	Reference
NPY	NPY system	LH	Rat	Sprague–Dawley	5 ng	CA3	Acute	LH	↓ Escape failure	Ishida et al. 2007
					500 ng	ICV	Acute	LH	↓ Escape failure	Ishida et al. 2007
					0.5–500 ng	DG	Acute	LH	NC	Ishida et al. 2007
					5 ng	CA3	Acute	OF	NC	Ishida et al. 2007
					5 ng	DG	Acute	OF	NC	Ishida et al. 2007
		OBX			5.0–10.0 nmol	ICV	14 days	OF	↓ Hyperlocomotion	Goyal et al. 2009
[Leu^{31}Pro^{34}]PYY	Y$_1$/Y$_5$ agonist	LH			1 ng	CA3	Acute	LH	↓ Escape failure	Ishida et al. 2007
		OBX			0.3–1.0 nmol	ICV	14 days	FST	↓ Immobility	Morales-Medina et al. 2009
					0.3–1.0 nmol	ICV	14 days	OF	↑ Hyperlocomotion	Morales-Medina et al. 2009
					0.1–1.0 nmol	ICV	14 days	SI	↑ Active social interaction	Morales-Medina et al. 2009
[Leu^{31}Pro^{34}]NPY	Y$_1$/Y$_5$ agonist	OBX			0.5–1.0 nmol	ICV	14 days	OF	↓ Hyperlocomotion	Goyal et al. 2009
NPY13-36	Y$_2$ agonist	LH			5–50 ng	CA3	Acute	LH	NC	Ishida et al. 2007
PYY3-36		OBX			1.0 nmol	ICV	14 days	FST	↑ Immobility	Morales-Medina et al. 2009
BIIE0246	Y$_2$ antagonist	LH			10 ng	CA3	Acute	LH	↓ Escape failure	Ishida et al. 2007
		LH			10 ng	DG	Acute	LH	NC	Ishida et al. 2007
		OBX			1.0–10.0 nmol	ICV	14 days	FST	↓ Immobility	Morales-Medina et al. 2009
		OBX			1.0–10.0 nmol	ICV	14 days	OF	NC	Morales-Medina et al. 2009
[cPP$^{1–7}$, NPY$^{19–23}$, Ala^{31}, Aib^{32}, Gln^{34}]hPP	Y$_5$ agonist	OBX			1.0–3.0 nmol	ICV	14 days	OF	↓ Hyperlocomotion	Morales-Medina et al. 2009
		OBX			3.0 nmol	ICV	14 days	SI	↑ Active social interaction	Morales-Medina et al. 2009
Lu AA33810	Y$_5$ antagonist	FSL		FSL/FRL	10 mg/kg	IP	14 days	FST	↓ Immobility	Walker et al. 2009
		FSL			10 mg/kg	IP	14 days	SI	↑ Active social interaction	Walker et al. 2009
		CMS		Wistar	3–10 mg/kg	IP	35 days/twice per day	SPT	↑ in sucrose consumption	Walker et al. 2009

Note: LH, learned helplessness; CMS, chronic mild stress; IP, intraperitoneal; SPT, sucrose preference test; NC, no change.

emotional despairs, their cotreatment will be required to clarify the role of the Y_5 receptor in depression- and anxiety-related behaviors.

16.2.4 MEMORY PROCESSES

Substantial evidence has shown that in addition to the emotional despairs observed in human subjects who suffer from depression, memory and attention deficits are common traits (Castaneda et al. 2008). However, still relatively little is known on the role of NPY and its receptors in memory processes. Acute ICV administration of NPY improved learning in two well-known models of amne sia, the anisomycin and scopolamine treatments in rats (Flood et al. 1987). In contrast, memory deficits were observed in young (Thorsell et al. 2000) but not in old (Carvajal et al. 2004) NPY-overexpressing rats. NPY KO mice showed no changes in memory-related paradigms (Bannon et al. 2000). In the object recognition test, Y_1 KO mice displayed some cognitive impairment (Costoli et al. 2005). Similarly, Y_2 KO mice showed learning deficits in a variety of memory tasks (Greco and Carli 2006; Redrobe et al. 2003a). In contrast, Y_4 KO mice have normal learning abilities compared with wild-type mice (Painsipp et al. 2008b).

Thus, further research is certainly warranted on the role of NPY and its receptors in learning and memory deficits known to occur in depressive states.

16.3 HUMAN STUDIES

Cumulative evidence suggested the existence of a negative correlation between NPY levels and depression. This hypothesis is based on the measurement of peripheral levels of this peptide under stressful conditions and in depressed subjects. Similarly, alterations in NPY levels in suicide attempters and completers who suffered from depression-related disorders and increased NPY levels after chronic antidepressant treatments as well as in genetic association studies in humans suggest a role for NPY in depressive disorders.

16.3.1 NPY LEVELS UNDER STRESSFUL CONDITIONS IN HUMANS

Growing evidence shows that stress is a strong risk factor for depression (Lupien et al. 2009). In humans, NPY administration resulted in reduced cortisol levels in the dark phase in healthy subjects (Antonijevic et al. 2000). Subsequently, Morgan et al. (2000, 2001, 2002) observed that psychological stress increased both cortisol and NPY levels in volunteer subjects. Interestingly, the highest levels of NPY were observed in participants who were apparently less affected by stress (Morgan et al. 2002). Taken together, these findings suggest that stressful conditions can increase NPY levels in humans as observed in animal studies. The subgroup that is most able to manage stressful experiences also expressed the highest level of the peptide, possibly due to genetic predisposition.

16.3.2 ALTERED LEVELS OF NPY IN DEPRESSED SUBJECTS

The measurement of plasma or cerebrospinal fluid (CSF) levels of monoamines and neuropeptides has been widely used as a potential state- and/or trait-dependent marker of various neuropsychiatric conditions, including depression (Raedler and Wiedemann 2006). Numerous studies have suggested that CSF levels of NPY are decreased in depressed subjects (Heilig et al. 2004; Hou et al. 2006; Olsson et al. 2004; Widerlov et al. 1988a, 1988b). Additionally, Hashimoto et al. (1996) observed that plasma NPY levels were decreased in major depression, whereas Kuromitsu et al. (2001) showed that NPY mRNA expression was downregulated in bipolar subjects. In complementary studies, Widdowson et al. (1992) reported that NPY-like immunoreactivity was decreased in the frontal cortex of suicide completers. Interestingly, this alteration in NPY levels was exacerbated in a subgroup of depressed suicide victims. Moreover, NPY Y_2 receptor mRNA level was reported to be increased in the prefrontal cortex of suicide victims (Caberlotto and Hurd 2001). Taken together, these results provide strong evidence that NPY levels are negatively correlated with depression, although few studies have also shown poor association (Gjerris et al. 1992; Nikisch et al. 2005; Ordway et al. 1995; Roy 1993).

16.3.3 ANTIDEPRESSANT TREATMENT AND NPY LEVELS IN HUMAN SUBJECTS

In animal models of depression-related behaviors, clear correlation between antidepressant treatment and increased levels of NPY has been reported. In contrast, human studies have provided conflicting data. For example, long-term treatment with the SSRI citalopram increased CSF NPY levels (Nikisch et al. 2005), with a similar effect being observed after ECT (Nikisch and Mathe 2008). Conversely, treatment with the selective monoamine oxidase A inhibitor amiflamine failed to alter CSF NPY levels (Widerlov et al. 1988b). Olsson et al. (2004) also monitored the levels of NPY in depressed subjects, but failed to observe any correlation between antidepressant treatment and NPY levels. In this study, patients received various antidepressants and at different dose regimens. Accordingly, additional studies are required in order to obtain more definitive conclusions on the relevance of concomitant increases in NPY levels and the alleviation of depressive symptoms observed during antidepressant treatment. It is interesting to note here that in contrast to animal studies, a considerable number of depressed subjects are refractory to any antidepressant treatments (Berlim et al. 2008). Thus, conflicting results may be attributable to the lack of efficacy of antidepressant therapies in a subgroup of patients.

16.3.4 GENETIC COMPONENT OF NPY IN DEPRESSION

Depression is a multifactorial disorder with a role for a large array of susceptibility genes and epigenetic and environmental factors (Bale et al. 2010; Brown and Harris 2008; Lupien et al. 2009). Genetic association studies have attempted to link NPY to emotional disorders, particularly depression. Sebat et al. (2009) hypothesized that single-nucleotide polymorphisms (SNPs) are most likely linked with limited increase in the probability of developing a given disease. In contrast, rare copy number of variations is more strongly associated to the likelihood of developing a certain psychiatric disease. The SNP rs16147, Leu7Pro7 (T/C), a polymorphism observed in the prepro-NPY gene, was first examined in control subjects under an exercise paradigm (Kallio et al. 2001). This study suggested that Leu7Pro7 carriers produced higher levels of plasma NPY compared with Leu7Leu7 and in a much shorter time. Further studies found that alterations in this promoter resulted in higher NPY levels (Buckland et al. 2004; Itokawa et al. 2003; Shah et al. 2009).

Inasmuch as previous findings have suggested that NPY levels are decreased in depression in humans, a working hypothesis was that carriers of this polymorphism had decreased risks to develop depression. Heilig et al. (2004) tested this assumption and observed a negative correlation between depression and this polymorphism in a Swedish population and this finding was subsequently replicated in another study (Sjoholm et al. 2009). In contrast, Lindberg et al. (2006) reported a lack of correlation between this SNP and depression in a Danish population. This polymorphism was further evaluated in regard to a stress response in a Finnish population (Zhou et al. 2008). Recently, Sommer et al. (2010) and Domschke et al. (2010) suggested that variations in the polymorphism contribute to emotional responses in humans. However, Cotton et al. (2009) failed to observe an association between this polymorphism and emotional arousal in a very large population.

In summary, these studies globally suggest that this particular SNP plays a key role in the processing of NPY. However, the association between this polymorphism and emotional processes is still controversial. Further studies will thus be necessary to establish the role of this SNP and its impact on NPYergic systems in various subgroups of patients.

16.4 CONCLUSION

Depression has been associated with underlying deficits in a number of classical neurotransmitters and most recently with neuropeptides, including NPY. *In vivo* studies in rodents have provided strong evidence for a role for this abundant peptide in the pathology of depression as well as its possible antidepressant effects. Additionally, human studies have supported a direct role for NPY in depression and emotional processing.

The Y_1 and the Y_2 receptors are the main transducers of the effects of NPY in affective disorders. However, the dissection of the precise respective contribution of these two receptor subtypes awaits further characterization. Rather surprisingly, data on the Y_4 receptor subtype suggest a role for this receptor in emotional processing. The effects mediated by this receptor seem to occur via the peripheral nervous system rather than centrally. Interestingly, recent studies on the Y_5 receptor subtype suggest a role in anxiety rather than depression. Future studies using well-characterized animal models of depression-related behaviors and that will investigate the levels and activities of NPY and its receptors are awaited to further dissect the role of each receptor subtype in emotional processing. Because disturbances in memory are a confounding factor in depression, those studies should also evaluate the role of NPY and related molecules in cognitive processes.

Finally, recent findings have suggested the possible role of genetic variants of NPY and NPY receptors as potential contributors in the pathology of depression-related disorders in humans. Further studies (SNPs, genome-wide association study) are certainly warranted using large populations of well-characterized individuals to provide clearer data on the relevance of these genetic variants of NPY markers in their role in emotional processing.

ACKNOWLEDGMENTS

This study was supported by grants from the Canadian Institutes of Health Research (RQ). JCMM is a PhD student with fellowship from CONACyT-Mexico. We also thank Mira Thakur for editing and proofreading the text.

REFERENCES

Alldredge, B. 2010. Pathogenic involvement of neuropeptides in anxiety and depression. *Neuropeptides* 44:215–24.

Antonijevic, I.A., Murck, H., Bohlhalter, S., et al. 2000. Neuropeptide Y promotes sleep and inhibits ACTH and cortisol release in young men. *Neuropharmacology* 39:1474–81.

Asakawa, A., Inui, A., Ueno, N., et al. 1999. Mouse pancreatic polypeptide modulates food intake, while not influencing anxiety in mice. *Peptides* 20:1445–8.

Asakawa, A., Inui, A., Yuzuriha, H., et al. 2003. Characterization of the effects of pancreatic polypeptide in the regulation of energy balance. *Gastroenterology* 124:1325–36.

Bale, T.L., Baram, T.Z., Brown, A.S., et al. 2010. Early life programming and neurodevelopmental disorders. *Biol Psychiatry* 68:314–9.

Bannon, A.W., Seda, J., Carmouche, M., et al. 2000. Behavioral characterization of neuropeptide Y knockout mice. *Brain Res* 868:79–87.

Berlim, M.T., Fleck, M.P., and Turecki, G. 2008. Current trends in the assessment and somatic treatment of resistant/refractory major depression: An overview. *Ann Med* 40:149–59.

Bjornebekk, A., Mathe, A.A., and Brene, S. 2006. Running has differential effects on NPY, opiates, and cell proliferation in an animal model of depression and controls. *Neuropsychopharmacology* 31:256–64.

Bjornebekk, A., Mathe, A.A., and Brene, S. 2010. The antidepressant effects of running and escitalopram are associated with levels of hippocampal NPY and Y1 receptor but not cell proliferation in a rat model of depression. *Hippocampus* 20:820–8.

Brown, G.W., and Harris, T.O. 2008. Depression and the serotonin transporter 5-HTTLPR polymorphism: A review and a hypothesis concerning gene–environment interaction. *J Affect Disord* 111:1–12.

Buckland, P.R., Hoogendoorn, B., Guy, C.A., et al. 2004. A high proportion of polymorphisms in the promoters of brain expressed genes influences transcriptional activity. *Biochem Biophys Acta* 1690:238–49.

Caberlotto, L., and Hurd, Y.L. 2001. Neuropeptide Y Y(1) and Y(2) receptor mRNA expression in the prefrontal cortex of psychiatric subjects. Relationship of Y(2) subtype to suicidal behavior. *Neuropsychopharmacology* 25:91–7.

Caberlotto, L., Fuxe, K., Overstreet, D.H., et al. 1998. Alterations in neuropeptide Y and Y1 receptor mRNA expression in brains from an animal model of depression: Region specific adaptation after fluoxetine treatment. *Brain Res Mol Brain Res* 59:58–65.

Caberlotto, L., Jimenez, P., Overstreet, D.H., et al. 1999. Alterations in neuropeptide Y levels and Y1 binding sites in the Flinders sensitive line rats, a genetic animal model of depression. *Neurosci Lett* 265:191–4.

Carvajal, C.C., Vercauteren, F., Dumont, Y., et al. 2004. Aged neuropeptide Y transgenic rats are resistant to acute stress but maintain spatial and non-spatial learning. *Behav Brain Res* 153:471–80.

Carvajal, C., Dumont, Y., and Quirion, R. 2006. Neuropeptide Y: Role in emotion and alcohol dependence. *CNS Neurol Disord Drug Targets* 5:181–95.

Castaneda, A.E., Tuulio-Henriksson, A., Marttunen, M., et al. 2008. A review on cognitive impairments in depressive and anxiety disorders with a focus on young adults. *J Affect Disord* 106:1–27.

Christiansen, S.H., Olesen, M.V., Wortwein, G., et al. 2011. Fluoxetine reverts chronic restraint stress-induced depression-like behaviour and increases neuropeptide Y and galanin expression in mice. *Behav Brain Res* 216:585–91.

Costoli, T., Sgoifo, A., Stilli, D., et al. 2005. Behavioural, neural and cardiovascular adaptations in mice lacking the NPY Y1 receptor. *Neurosci Biobehav Rev* 29:113–23.

Cotton, C.H., Flint, J., and Campbell, T.G. 2009. Is there an association between NPY and neuroticism? *Nature* 458:E6; discussion E7.

Cryan, J.F., and Mombereau, C. 2004. In search of a depressed mouse: Utility of models for studying depression-related behavior in genetically modified mice. *Mol Psychiatry* 9:326–57.

Cryan, J.F., Markou, A., and Lucki, I. 2002. Assessing antidepressant activity in rodents: Recent developments and future needs. *Trends Pharmacol Sci* 23:238–45.

Cryan, J.F., Page, M.E., and Lucki, I. 2005a. Differential behavioral effects of the antidepressants reboxetine, fluoxetine, and moclobemide in a modified forced swim test following chronic treatment. *Psychopharmacology (Berl)* 182:335–44.

Cryan, J.F., Valentino, R.J., and Lucki, I. 2005b. Assessing substrates underlying the behavioral effects of antidepressants using the modified rat forced swimming test. *Neurosci Biobehav Rev* 29:547–69.

de Lange, R.P., Wiegant, V.M., and Stam, R. 2008. Altered neuropeptide Y and neurokinin messenger RNA expression and receptor binding in stress-sensitised rats. *Brain Res* 1212:35–47.

Domschke, K., Dannlowski, U., Hohoff, C., et al. 2010. Neuropeptide Y (NPY) gene: Impact on emotional processing and treatment response in anxious depression. *Eur Neuropsychopharmacol* 20:301–9.

Dumont, Y., and Quirion, R. 2006. An overview of neuropeptide Y: Pharmacology to molecular biology and receptor localization. *Experientia* 95:7–11.

Dumont, Y., Jacques, D., St-Pierre, J., et al. 2000. Neuropeptide Y, peptide YY and pancreatic polypeptide receptor proteins and mRNAs in mammalian brain. In *Peptide Receptors*, Part 1, vol. 16, eds. A. Bjornebekk, and T. Hokfelt, 375–475. Amsterdam: Elsevier Science B.V.

Dumont, Y., Morales-Medina, J.C., and Quirion, R. 2009. Neuropeptide Y and its role in anxiety-related disorders. In *Transmitters and Modulators in Health and Disease*, ed. S. Shioda, I. Homma, and N. Kato, 51–82. Kato Bunmeisha: Springer.

Flood, J.F., Hernandez, E.N., and Morley, J.E. 1987. Modulation of memory processing by neuropeptide Y. *Brain Res* 421:280–90.

Gjerris, A., Widerlov, E., Werdelin, L., et al. 1992. Cerebrospinal fluid concentrations of neuropeptide Y in depressed patients and in controls. *J Psychiatry Neurosci* 17:23–7.

Goyal, S.N., Upadhya, M.A., Kokare, D.M., et al. 2009. Neuropeptide Y modulates the antidepressant activity of imipramine in olfactory bulbectomized rats: Involvement of NPY Y1 receptors. *Brain Res* 1266:45–53.

Greco, B., and Carli, M. 2006. Reduced attention and increased impulsivity in mice lacking NPY Y2 receptors: Relation to anxiolytic-like phenotype. *Behav Brain Res* 169:325–34.

Hashimoto, H., Onishi, H., Koide, S., et al. 1996. Plasma neuropeptide Y in patients with major depressive disorder. *Neurosci Lett* 216:57–60.

Heilig, M., Zachrisson, O., Thorsell, A., et al. 2004. Decreased cerebrospinal fluid neuropeptide Y (NPY) in patients with treatment refractory unipolar major depression: Preliminary evidence for association with preproNPY gene polymorphism. *J Psychiatr Res* 38:113–21.

Higuchi, H., Niki, T., and Shiiya, T. 2008. Feeding behavior and gene expression of appetite-related neuropeptides in mice lacking for neuropeptide Y Y5 receptor subclass. *World J Gastroenterol* 14:6312–7.

Holmes, P.V., Davis, R.C., Masini, C.V., et al. 1998. Effects of olfactory bulbectomy on neuropeptide gene expression in the rat olfactory/limbic system. *Neuroscience* 86:587–96.

Hou, C., Jia, F., Liu, Y., et al. 2006. CSF serotonin, 5-hydroxyindolacetic acid and neuropeptide Y levels in severe major depressive disorder. *Brain Res* 1095:154–8.

Ishida, H., Shirayama, Y., Iwata, M., et al. 2007. Infusion of neuropeptide Y into CA3 region of hippocampus produces antidepressant-like effect via Y1 receptor. *Hippocampus* 17:271–80.

Itokawa, M., Arai, M., Kato, S., et al. 2003. Association between a novel polymorphism in the promoter region of the neuropeptide Y gene and schizophrenia in humans. *Neurosci Lett* 347:202–4.

Jimenez-Vasquez, P.A., Mathe, A.A., Thomas, J.D., et al. 2001. Early maternal separation alters neuropeptide Y concentrations in selected brain regions in adult rats. *Brain Res Dev Brain Res* 131:149–52.

Kallio, J., Pesonen, U., Kaipio, K., et al. 2001. Altered intracellular processing and release of neuropeptide Y due to leucine 7 to proline 7 polymorphism in the signal peptide of preproneuropeptide Y in humans. *FASEB J* 15:1242–4.

Karl, T., Burne, T.H.J., and Herzog, H. 2006. Effect of Y1 receptor deficiency on motor activity, exploration, and anxiety. *Behav Brain Res* 167:87–93.

Karlsson, R.M., Choe, J.S., Cameron, H.A., et al. 2008. The neuropeptide Y Y1 receptor subtype is necessary for the anxiolytic-like effects of neuropeptide Y, but not the antidepressant-like effects of fluoxetine, in mice. *Psychopharmacology (Berl)* 195:547–57.

Kask, A., Vasar, E., Heidmets, L.T., et al. 2001. Neuropeptide Y Y(5) receptor antagonist CGP71683A: The effects on food intake and anxiety-related behavior in the rat. *Eur J Pharmacol* 414:215–24.

Kastin, A.J., and Pan, W. 2010. Concepts for biologically active peptides. *Curr Pharm Des* 16:3390–400.

Kessler, R.C., Berglund, P., Demler, O., et al. 2003. The epidemiology of major depressive disorder: Results from the National Comorbidity Survey Replication (NCS-R). *JAMA* 289:3095–105.

Kuromitsu, J., Yokoi, A., Kawai, T., et al. 2001. Reduced neuropeptide Y mRNA levels in the frontal cortex of people with schizophrenia and bipolar disorder. *Brain Res Gene Expr Patterns* 1:17–21.

Larhammar, D., and Salaneck, E. 2004. Molecular evolution of NPY receptor subtypes. *Neuropeptides* 38: 141–51.

Lindberg, C., Koefoed, P., Hansen, E.S., et al. 2006. No association between the –399 C > T polymorphism of the neuropeptide Y gene and schizophrenia, unipolar depression or panic disorder in a Danish population. *Acta Psychiatr Scand* 113:54–8.

Lucki, I., Dalvi, A., and Mayorga, A.J. 2001. Sensitivity to the effects of pharmacologically selective antidepressants in different strains of mice. *Psychopharmacology (Berl)* 155:315–22.

Lupien, S.J., McEwen, B.S., Gunnar, M.R., et al. 2009. Effects of stress throughout the lifespan on the brain, behaviour and cognition. *Nat Rev Neurosci* 10:434–45.

Marsh, D.J., Hollopeter, G., Kafer, K.E., et al. 1998. Role of the Y5 neuropeptide Y receptor in feeding and obesity. *Nat Med* 4:718–21.

Marsh, D.J., Baraban, S.C., Hollopeter, G., et al. 1999. Role of the Y5 neuropeptide Y receptor in limbic seizures. *Proc Natl Acad Sci U S A* 96:13518–23.

Michel, M.C., Beck-Sickinger, A., Cox, H., et al. 1998. XVI. International Union of Pharmacology recommendations for the nomenclature of neuropeptide Y, peptide YY, and pancreatic polypeptide receptors. *Pharmacol Rev* 50:143–50.

Morales-Medina, J.C., Dumont, Y., Benoit, C.E., et al. 2009. Neuropeptide Y Y1 and Y2 receptors reverse behavioral despairs by increasing neurogenesis. *Abstract of 5th Symposium Peptide Receptors and Kinin*, Quebec, 26–30 June 2009.

Morales-Medina, J.C., Dumont, Y., and Quirion, R. 2010. A possible role of neuropeptide Y in depression and stress. *Brain Res* 1314:194–205.

Morgan, C.A., III, Wang, S., Southwick, S.M., et al. 2000. Plasma neuropeptide-Y concentrations in humans exposed to military survival training. *Biol Psychiatry* 47:902–9.

Morgan, C.A. III, Wang, S., Rasmusson, A., et al. 2001. Relationship among plasma cortisol, catecholamines, neuropeptide Y, and human performance during exposure to uncontrollable stress. *Psychosom Med* 63:412–22.

Morgan, C.A., 3rd, Rasmusson, A.M., Wang, S., et al. 2002. Neuropeptide-Y, cortisol, and subjective distress in humans exposed to acute stress: Replication and extension of previous report. *Biol Psychiatry* 52:136–42.

Murray, C.J., and Lopez, A.D. 1996. Evidence-based health policy—Lessons from the Global Burden of Disease Study. *Science* 274:740–3.

Nikisch, G., and Mathe, A.A. 2008. CSF monoamine metabolites and neuropeptides in depressed patients before and after electroconvulsive therapy. *Eur Psychiatry* 23:356–9.

Nikisch, G., Agren, H., Eap, C.B., et al. 2005. Neuropeptide Y and corticotropin-releasing hormone in CSF mark response to antidepressive treatment with citalopram. *Int J Neuropsychopharmacol* 8:403–10.

Olsson, A., Regnell, G., Traskman-Bendz, L., et al. 2004. Cerebrospinal neuropeptide Y and substance P in suicide attempters during long-term antidepressant treatment. *Eur Neuropsychopharmacol* 14:479–85.

Ordway, G.A., Stockmeier, C.A., Meltzer, H.Y., et al. 1995. Neuropeptide Y in frontal cortex is not altered in major depression. *J Neurochem* 65:1646–50.

Painsipp, E., Herzog, H., and Holzer, P. 2008a. Implication of neuropeptide-Y Y2 receptors in the effects of immune stress on emotional, locomotor and social behavior of mice. *Neuropharmacology* 55:117–26.

Painsipp, E., Wultsch, T., Edelsbrunner, M.E., et al. 2008b. Reduced anxiety-like and depression-related behavior in neuropeptide Y Y4 receptor knockout mice. *Genes Brain Behav* 7:532–42.

Painsipp, E., Sperk, G., Herzog, H., et al. 2010. Delayed stress-induced differences in locomotor and depression-related behaviour in female neuropeptide-Y Y1 receptor knockout mice. *J Psychopharmacol* 24:1541–9.

Raedler, T.J., and Wiedemann, K. 2006. CSF-studies in neuropsychiatric disorders. *Neuro Endocrinol Lett* 27:297–305.

Raposinho, P.D., Pedrazzini, T., White, R.B., et al. 2004. Chronic neuropeptide Y infusion into the lateral ventricle induces sustained feeding and obesity in mice lacking either Npy1r or Npy5r expression. *Endocrinology* 145:304–10.

Redrobe, J.P., Dumont, Y., Fournier, A., et al. 2002. The neuropeptide Y (NPY) Y1 receptor subtype mediates NPY-induced antidepressant-like activity in the mouse forced swimming test. *Neuropsychopharmacology* 26:615–24.

Redrobe, J.P., Dumont, Y., Herzog, H., et al. 2003a. Characterization of neuropeptide Y, Y2 receptor knockout mice in two animal models of learning and memory processing. *J Mol Neurosci* 22:121–8.

Redrobe, J.P., Dumont, Y., Herzog, H., et al. 2003b. Neuropeptide Y (NPY) Y2 receptors mediate behaviour in two animal models of anxiety: Evidence from Y2 receptor knockout mice. *Behav Brain Res* 141:251–5.

Redrobe, J.P., Dumont, Y., Fournier, A., et al. 2005. Role of serotonin (5-HT) in the antidepressant-like properties of neuropeptide Y (NPY) in the mouse forced swim test. *Peptides* 26:1394–400.

Rotzinger, S., Lovejoy, D.A., and Tan, L.A. 2009. Behavioral effects of neuropeptides in rodent models of depression and anxiety. *Peptides* 31:736–56.

Roy, A. 1993. Neuropeptides in relation to suicidal behavior in depression. *Neuropsychobiology* 28:184–6.

Rutkoski, N.J., Lerant, A.A., Nolte, C.M., et al. 2002. Regulation of neuropeptide Y in the rat amygdala following unilateral olfactory bulbectomy. *Brain Res* 951:69–76.

Sajdyk, T.J., Schober, D.A., and Gehlert, D.R. 2002. Neuropeptide Y receptor subtypes in the basolateral nucleus of the amygdala modulate anxiogenic responses in rats. *Neuropharmacology* 43:1165–72.

Sebat, J., Levy, D.L., and McCarthy, S.E. 2009. Rare structural variants in schizophrenia: One disorder, multiple mutations; one mutation, multiple disorders. *Trends Genet* 25:528–35.

Shah, S.H., Freedman, N.J., Zhang, L., et al. 2009. Neuropeptide Y gene polymorphisms confer risk of early-onset atherosclerosis. *PLoS Genet* 5:e1000318.

Sjoholm, L.K., Melas, P.A., Forsell, Y., et al. 2009. PreproNPY Pro7 protects against depression despite exposure to environmental risk factors. *J Affect Disord* 118:124–30.

Sommer, W.H., Lidstrom, J., Sun, H., et al. 2010. Human NPY promoter variation rs16147:T>C as a moderator of prefrontal NPY gene expression and negative affect. *Hum Mutat* 31:E1594–608.

Song, C., and Leonard, B. E. 2005. The olfactory bulbectomised rat as a model of depression. *Neurosci Biobehav Rev* 29:627–47.

Sorensen, G., Lindberg, C., Wortwein, G., et al. 2004. Differential roles for neuropeptide Y Y1 and Y5 receptors in anxiety and sedation. *J Neurosci Res* 77:723–9.

Stenfors, C., Theodorsson, E., and Mathe, A.A. 1989. Effect of repeated electroconvulsive treatment on regional concentrations of tachykinins, neurotensin, vasoactive intestinal polypeptide, neuropeptide Y, and galanin in rat brain. *J Neurosci Res* 24:445–50.

Stogner, K.A., and Holmes, P.V. 2000. Neuropeptide-Y exerts antidepressant-like effects in the forced swim test in rats. *Eur J Pharmacol* 387:R9–10.

Tasan, R.O., Lin, S., Hetzenauer, A., et al. 2009. Increased novelty-induced motor activity and reduced depression-like behavior in neuropeptide Y (NPY)-Y4 receptor knockout mice. *Neuroscience* 158:1717–30.

Tasan, R.O., Nguyen, N.K., Weger, S., et al. 2010. The central and basolateral amygdala are critical sites of neuropeptide Y/Y2 receptor-mediated regulation of anxiety and depression. *J Neurosci* 30:6282–90.

Tatemoto, K., Carlquist, M., and Mutt, V. 1982. Neuropeptide Y—A novel brain peptide with structural similarities to peptide YY and pancreatic polypeptide. *Nature* 296:659–60.

Thorsell, A., Svensson, P., Wiklund, L., et al. 1998. Suppressed neuropeptide Y (NPY) mRNA in rat amygdala following restraint stress. *Regul Pept* 75–76:247–54.

Thorsell, A., Carlsson, K., Ekman, R., et al. 1999. Behavioral and endocrine adaptation, and up-regulation of NPY expression in rat amygdala following repeated restraint stress. *Neuroreport* 10:3003–7.

Thorsell, A., Michalkiewicz, M., Dumont, Y., et al. 2000. Behavioral insensitivity to restraint stress, absent fear suppression of behavior and impaired spatial learning in transgenic rats with hippocampal neuropeptide Y overexpression. *Proc Natl Acad Sci U S A* 97:12852–7.

Tschenett, A., Singewald, N., Carli, M., et al. 2003. Reduced anxiety and improved stress coping ability in mice lacking NPY-Y2 receptors. *Eur J Neurosci* 18:143–8.

Ueno, N., Asakawa, A., Satoh, Y., et al. 2007. Increased circulating cholecystokinin contributes to anorexia and anxiety behavior in mice overexpressing pancreatic polypeptide. *Regul Pept* 141:8–11.

Uriguen, L., Arteta, D., Diez-Alarcia, R., et al. 2008. Gene expression patterns in brain cortex of three different animal models of depression. *Genes Brain Behav* 7:649–58.

Walker, M.W., Wolinsky, T.D., Jubian, V., et al. 2009. The novel neuropeptide Y Y5 receptor antagonist Lu AA33810 [*N*-[[*trans*-4-[(4,5-dihydro[1]benzothiepino[5,4-*d*]thiazol-2-yl)amino]cycloh exyl]methyl]-methanesulfonamide] exerts anxiolytic- and antidepressant-like effects in rat models of stress sensitivity. *J Pharmacol Exp Ther* 328:900–11.

Widdowson, P.S., Ordway, G.A., and Halaris, A.E. 1992. Reduced neuropeptide Y concentrations in suicide brain. *J Neurochem* 59:73–80.

Widerlov, E., Heilig, M., Ekman, R., et al. 1988a. Possible relationship between neuropeptide Y (NPY) and major depression — Evidence from human and animal studies. *Nordic J Psychiatry* 42:131–7.

Widerlov, E., Lindstrom, L.H., Wahlestedt, C., et al. 1988b. Neuropeptide Y and peptide YY as possible cere-brospinal fluid markers for major depression and schizophrenia, respectively. *J Psychiatr Res* 22:69–79.

Zambello, E., Fuchs, E., Abumaria, N., et al. 2010a. Chronic psychosocial stress alters NPY system: Different effects in rat and tree shrew. *Prog Neuropsychopharmacol Biol Psychiatry* 34:122–30.

Zambello, E., Zanetti, L., Hedou, G.F., et al. 2010b. Neuropeptide Y-Y2 receptor knockout mice: Influence of genetic background on anxiety-related behaviors. *Neuroscience* doi: 10.1016/j.neuroscience.2010.10.075.

Zhou, Z., Zhu, G., Hariri, A.R., et al. 2008. Genetic variation in human NPY expression affects stress response and emotion. *Nature* 452:997–1001.

17 Nitric Oxide Signaling in Depression and Antidepressant Action

Gregers Wegener and Aleksander A. Mathé

CONTENTS

17.1 INTRODUCTION

Recent data from Denmark and Europe (Olesen and Leonardi 2003; Olesen et al. 2008) indicate that brain disorders account for 12% of all direct costs in the Danish health system and that 9% of the total drug consumption was used for treatment of brain diseases. Expenses for brain diseases constituted 3% of the country's gross national product, and the total expenses for all investigated brain diseases were DKK37.3 billion. Among brain disorders, affective disorders were among the most costly diseases, and anxiety disorders among the most prevalent.

The pathogenesis of mood disorders remains elusive, but it is evident that multiple factors, genetic and environmental, play a crucial role for adult psychopathology and neurobiology (Caspi et al. 2003). With regard to therapy, a significant proportion of affective disorder patients are partial or nonresponders, and there has been no major breakthrough in finding novel effective drug targets since the introduction of the currently marketed antidepressant drugs in the 1950s to the 1980s, which are all based on mono-aminergic pharmacological effects. Consequently, there exists a pressing need to develop novel treatment strategies—and ultimately understand the etiology and pathophysiology of affective disorders.

Nitric oxide (NO), originally termed *endothelial-derived relaxing factor* (EDRF) before it was discovered that NO and EDRF were the same substance, serve important roles in the cardiovascular system and macrophages (Palmer et al. 1987; Hibbs et al. 1987). In addition, NO has been shown also to have an important role in the nervous system (Bredt and Snyder 1989; Garthwaite et al. 1989), where NO serves as a messenger molecule in a number of physiological processes, including processes being linked to the major psychiatric diseases (Knott and Bossy-Wetzel 2009; Oosthuizen et al. 2005; Reif et al. 2006a, 2009). This chapter will review the general aspects of the NO system in major depressive disorder (MDD), as well as focus on inhibitors of NO production as putative therapeutic agents toward depression.

17.2 GENERAL ASPECTS OF NO

NO, a small molecule (MW = 30 Da), is a colorless gas in vitro and a product from the breakdown of N_2. NO is degraded into nitrites and nitrates, and—depending on the environmental conditions—the half-life ranges from minutes to years (Beckman 1996). In biological systems, the half-life of NO is much shorter and estimated to be about 30 s or less (Beckman 1996). The molecule is uncharged and is therefore freely diffusible across cell membranes and other structures. NO is produced and released by many different cells in multicellular organisms and can thus act as a tool for intercellular communication (Bredt and Snyder 1994; Garthwaite et al. 1988; Kerwin and Heller 1994; Marletta 1993; Moncada et al. 1989; Nathan 1992). These properties are very important, as they form the basis for the understanding of NO as a retrograde transmitter (Arancio et al. 1996), which is proposed to play a major role in the induction of long-term potentiation (LTP) and long-term depression (LTD) (Bliss and Collingridge 1993). LTP and LTD have been observed to be impaired in experimental depression models and in depressive subjects (Normann et al. 2007; Holderbach et al. 2007).

The combination of one atom of N and one atom of O results in the presence of an unpaired electron. However, NO is less reactive than many other free radicals and does not react with itself. Being a free radical, NO has both pro- and antioxidant properties (Radi and Rubbo 2001; Pacher et al. 2007) (Figure 17.1). NO can be protective against oxidative injury, depending on the specific conditions (Radi and Rubbo 2001). An NO radical can both stimulate lipid oxidation and mediate oxidant-protective reactions in membranes (Radi et al. 1991). At high rates of NO production, the pro-oxidant versus antioxidant outcome depends critically on the relative concentrations of the individual reactive species (Rubbo et al. 1994). The pro-oxidant reactions of NO occur with superoxide, whereas the antioxidant effects of NO are consequent to direct reactions with alkoyl and peroxyl radical intermediate during lipid peroxidation, terminating the propagation of lipid radical chain reactions (Rubbo et al. 1994).

NO may therefore limit injury to target molecules or tissues during events associated with excess production of reactive oxygen species. Based on these properties, it is not surprising that disturbance of NO production may greatly affect cell functioning and survival.

17.2.1 NO SYNTHASE ENZYMES

The enzyme responsible for the synthesis of NO, NO synthase (NOS), appears in different isoforms that are constitutive or inducible (see Table 17.1). The activity of the constitutive NOS depends on

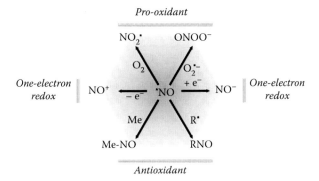

FIGURE 17.1 NO chemistry. One-electron redox (to NO^+ or NO^-), pro-oxidant (to NO_2 or $ONOO^-$), and antioxidant reactions (to Me-NO or RNO) are shown. Me, metal center (e.g., iron or copper); $R^•$, organic radical.

Ca^{2+} and calmodulin, whereas the inducible NOSs are independent from both Ca^{2+} and calmodulin. A distinction of the isoforms is also made based on the tissue where the NOS was identified the first time and primarily located. Of the constitutive isoforms, NOS in endothelial cells is mainly located in the cell membrane and is termed *eNOS*. NOS in neuronal cells is located throughout the cell and is termed *nNOS*. The inducible isoform is located in macrophages and is termed *iNOS*; it consists of soluble and membrane-bound NOS (Nathan 1992; Forstermann et al. 1991). However, exceptions from this rule exist, and nNOS has been found in a variety of nonneuronal cells, whereas eNOS has been demonstrated in some neurons (Forstermann et al. 1995; O'Dell et al. 1991). The NH_2 terminus of nNOS contains a postsynaptic density protein, disks-large, ZO-1 (PDZ) domain, participating in the formation of active nNOS dimers and interacting with many other proteins in specific regions of the cell (Chen et al. 2008; Cui et al. 2007; Lemaire and McPherson 2006). Proteins bearing PDZ domains typically localize to specialized cell compartments and are believed to be important in linking components of signal transduction pathways in multiple complexes (Chanrion et al. 2007; Riefler and Firestein 2001; Saitoh et al. 2004).

TABLE 17.1
NOS Isoforms

	Neural (nNOS, Type I)	Inducible (iNOS, Type II)	Endothelial (eNOS, Type III)
First identified in	Neurons	Macrophages	Endothelium
Other cells expressing	Myocytes	Astrocytes	Neurons
	Astrocytes	Microglia	
Intracellular localization	Soluble or membrane bound	Soluble or membrane bound	Largely membrane bound
Ca^{2+} dependency	Activity depends on elevated Ca^{2+}	Activity is independent of elevated Ca^{2+}	Activity depends on elevated Ca^{2+}
Expression	Constitutive Inducible under certain circumstances, e.g., trauma	Inducible	Constitutive
Amounts of NO released	Small, pulses	Large, continuous	Small, pulses
Proposed function	Regulation	Host defense	Regulation
Activators	Glutamate Noradrenaline	Lipopolysaccharide	Acetylcholine

By anchoring nNOS to membrane or cytosolic protein via direct PDZ–PDZ domain or C-terminal–PDZ interactions, NO signaling is altered. Postsynaptic density protein-95 (PSD95), a multivalent synaptic scaffolding protein and core component of the postsynaptic density, can link nNOS to *N*-methyl-D-aspartate (NMDA) receptor (Christopherson et al. 1999; Mungrue and Bredt 2004) and accounts for the efficient activation of nNOS by NMDA receptor stimulation by coupling NO synthesis by nNOS to the Ca^{2+} influx through NMDA receptors (Sattler et al. 1999). Binding of nNOS to PSD95 is a determinant of postsynaptic targeting of nNOS. nNOS has also been found capable of negatively regulating the NMDA receptor by *S*-nitrosylation (Kim et al. 1999).

17.2.2 Synthesis of NO

NO is synthesized in the brain by NOS from the amino acid L-arginine. In brief, L-arginine is converted to N^{ω}-hydroxy-L-arginine, which is further converted to NO and citrulline by NOS (Figure 17.2). The process is rather complex, and further discussion lies beyond the scope of this chapter. Briefly, the process involves five electrons, three cosubstrates, and five prosthetic groups (Dawson and Snyder 1994; Knowles and Moncada 1994; Nathan 1992).

17.2.3 Localization of NOS in Central Nervous System

The NOS enzymes are widely distributed within the mammalian brain (Blottner et al. 1995; de Vente et al. 1998). The neuronal isoform accounts for the majority of the NOS activity in the brain (Hara et al. 1996), and NOS-positive neurons are located in the hippocampal layers CA1–CA3, the medial amygdaloid nucleus, the olfactory bulb, the layers II–VI in the cerebral cortex, the granular and deep molecular layers of the cerebellum, and, with special interest regarding the serotonin system, the dorsal and medial raphe nuclei (Blottner et al. 1995). Measurements of NOS activity in different brain regions have shown the highest activity in the cerebellum, midbrain, hypothalamus, cortex, striatum, and hippocampus (Barjavel and Bhargava 1995; Salter et al. 1995). Interestingly, NO has been shown to colocalize with several other known transmitters within the same neuron, for example, serotonin (5-HT) in the medial and dorsal raphe nuclei (Johnson and Ma 1993), norepinephrine in the solitarian tract nucleus (Simonian and Herbison 1996), γ-amino-butyric acid in the cerebral cortex (Valtschanoff et al. 1993), and neuropeptide Y and somatostatin in the striatum (Kowall et al. 1987).

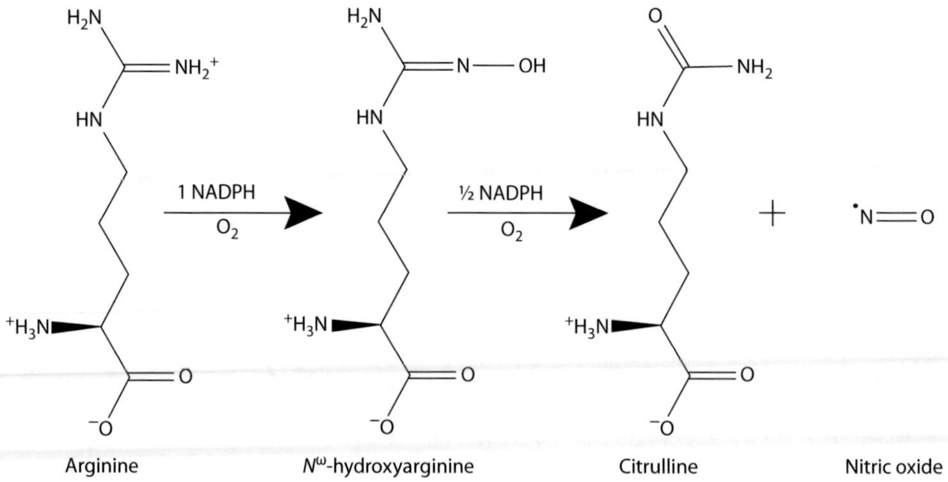

FIGURE 17.2 Synthesis of NO.

17.2.4 REGULATION OF NOS ACTIVITY

Regulation of the NOS enzyme expression has to be clarified in detail. Most of the studies performed have focused on the iNOS isoform. This isoform is not present in the cells under normal circumstances but can be expressed after activation by different cytokines/endotoxins (Nathan and Xie 1994; Schmidt et al. 1993). Less is known about the expression of nNOS and eNOS, but it has become evident that expression of nNOS in the brain and spinal cord during the embryonic and postnatal period can change markedly, which is in line with evidence indicating that NO is implicated in synaptic plasticity in adults and in regulating neurite outgrowth, as exemplified by the finding that NO donors enhance neurotrophin-induced neurite outgrowth through a cyclic guanosine monophosphate (cGMP)–dependent mechanism (Hindley et al. 1997; Contestabile 2000; Hess et al. 1993).

The cofactors and especially the NOS–Ca^{2+}-calmodulin interaction are a primary regulator for NO production. Following an action potential, increases in the intracellular Ca^{2+} environment (about 500 nM [Schmidt et al. 1992]) triggers Ca^{2+}-calmodulin to bind to NOS, activating the NOS enzymatic activity. As the intracellular Ca^{2+} level can rapidly change, the catalytic activity can be turned on and off within a short time. iNOS binds calmodulin very tightly and continues to synthesize NO throughout the life of the enzyme, regardless of the intracellular Ca^{2+} concentration (Nathan 1992). In addition to the cofactor and Ca^{2+} level regulations, phosphorylation is used to regulate the activity, as exemplified by the finding that nNOS phosphorylation by protein kinase C inhibits NO production (Lowenstein and Snyder 1992). NO itself has also been shown to regulate NOS activity (Assreuy et al. 1993; Buga et al. 1993; Rengasamy and Johns 1993). The nature of this inhibition needs to be fully clarified but can be hypothesized to involve nitrosylation (Gaston et al. 2003).

nNOS is connected to NMDA receptors and sharply increases NO production after activation of this receptor (Brenman and Bredt 1997), with the consequence that the level of endogenously produced NO around NMDA synapses reflects the activity of glutamate-mediated neurotransmission (Akyol et al. 2004). However, there is also evidence showing that non-NMDA glutamate receptors [i.e., α-amino-3-hydroxyl-5-methyl-4-isoxazoleproprionic acid (AMPA) and type I metabotropic receptors] may contribute to NO generation (Okada et al. 2004). Besides regulation on the levels above, the synthesis of NO can be inhibited by direct inhibition of NOS (see Section 17.2.5) or by limited availability of the substrate L-arginine.

17.2.5 TARGETS OF NO

NO has multiple targets in the brain, with the soluble form of the guanylate cyclase (sGC) being the most extensively characterized (Denninger and Marletta 1999; Miki et al. 1977; Schmidt et al. 1993). Activation of sGC subsequently increases the production of cGMP, and the level of cGMP in the cerebellum, striatum, and hippocampus has been shown to depend largely on the NOS activity (Laitinen et al. 1994; Luo and Vincent 1994; Vallebuona and Raiteri 1994).

Some physiological effects of NO, however, are independent of sGC activation, and it has been demonstrated that NO induced by NMDA receptor stimulation activates the p21 (ras) pathway of signal transduction with a cascade involving extracellular signal–regulated kinases and phosphoinositide-3-kinase (Yun et al. 1998; Dawson et al. 1998). These pathways are known to be involved in the transmission of signals to the cell nuclei and may therefore form a basis of a generation of long-lasting neuronal responses to NO. Other enzymes that constitute cellular targets for NO are cyclooxygenases, ribonucleotide reductase, some mitochondrial enzymes, and NOS itself (Dawson and Dawson 1995; Garthwaite and Boulton 1995). Finally, NO can nitrosylate proteins and damage the DNA (Dawson et al. 1998; Stamler 1994, 1995; Stamler et al. 2001).

Besides its influence on glutamate, NO is known to have effects on the storage, uptake, and/or release of most other neurotransmitters in the CNS (see Section 17.3.1.3), and because NO is a highly diffusible molecule, it may reach extrasynaptic receptors at some distance from the place of NO synthesis (Wiklund et al. 1997; Kiss and Vizi 2001). NO is thus capable of mediating both synaptic and nonsynaptic communication processes.

17.3 NO AND DEPRESSION

A role of the NO signaling pathway has been established in several different psychiatric disease entities, including schizophrenia, bipolar disorder, and MDD. A detailed overview of all these disorders is beyond the scope of this chapter. The present overview will therefore only relate to unipolar depression, where several lines of evidence support an association between abnormalities in nNOS and mood disorders.

17.3.1 Preclinical Studies

Preclinical studies of nitrergic signaling utilize methods from the concept of transgenic animals to stress models. The studies highlighted in the following section all provide basic evidence for the involvement of nitrergic signaling in affective disease.

17.3.1.1 NO and Behavior

Mice with targeted disruption of the nNOS gene exhibit abnormal behaviors. In a recent published work, it was demonstrated that transgenic mice lacking nNOS showed hyperlocomotor activity in a novel environment, increased social interaction in their home cage, decreased depression-related behavior, and impaired spatial memory retention (Tanda et al. 2009). The hyperlocomotion has also been demonstrated in reports from other groups, although some variances in methodology exist (Weitzdoerfer et al. 2004; Nelson et al. 1995; Bilbo et al. 2003). In a study using adult nNOS-deficient and wild-type (WT) mice for their recognition memory abilities in a social olfactory discrimination paradigm, it was observed that short-term and intermediate-term recognition memory was normal in nNOS-deficient mice, but unlike WT mice, nNOS-deficient mice failed to consolidate an olfactory cued long-term recognition memory (Juch et al. 2009). These data, together with proteomic observations, suggest that NO of nNOS origin is critically involved in the regulation of protein synthesis–dependent olfactory long-term memory consolidation within relevant brain structures, including the olfactory bulb (Juch et al. 2009).

Interestingly, mice lacking the gene encoding nNOS (nNOS$^{-/-}$) are more aggressive than WT mice in standard testing paradigms (Nelson et al. 2006). Testosterone is necessary, but not sufficient, to evoke persistent aggression in these mutants. In male nNOS$^{-/-}$ mice, a dramatic loss of behavioral inhibition reflected by persistent fighting and mounting behavior despite obvious signals of submissiveness or disinterest by their test partners was observed, a finding that was linked to blood testosterone levels (Nelson et al. 1995).

Despite elevated corticosterone concentrations (see section 17.3.1.2), nNOS knockout mice seem to be less "anxious" or "fearful" than WT mice in some studies. Male nNOS$^{-/-}$ mice spend more time in the open field than do WT mice (Nelson 2005) and display less anxiety in the elevated plus maze (EPM) (Zhang et al. 2010). However, in other studies, it has been reported that nNOS transgenic mice have an increased time spent in the closed arm of the EPM (Weitzdoerfer et al. 2004).

As a casual association exists between stressful life events and number of prior depressive episodes (Kendler et al. 2000), a majority of the preclinical characterization of NOS involvement in disease has been carried out in various animal stress paradigms, ranging from models of unipolar depression to models of posttraumatic stress disorder (PTSD). Thus, several preclinical studies have confirmed excessive NOS activation and NO release in the cortex and hippocampus after a protracted stressful event (Harvey et al. 2004, 2005; Madrigal et al. 2001, 2002), as well as concomitant changes in hippocampal NMDA receptor density (Harvey et al. 2004). Moreover, in a recent study using a highly validated psychopathological animal model of depression, it was observed that environmental factors such as stress may augment a predisposed depressed individual to hyperactivity in the nitrergic signaling pathway, as measured by the expression of key genes involved in NMDA-NO response and increased nNOS expression and NOS activity (Wegener et al. 2010) (Figure 17.3).

17.3.1.2 NO and HPA Axis

Accumulating evidence suggests that NO is involved in the tonic control of the hypothalamic–pituitary–adrenal (HPA) axis function. nNOS is widely present in the HPA axis and in closely related anatomical structures. At the hypothalamic level, nNOS immunoreactivity can be found in parvocellular neurosecretory and medullary-projecting preautonomic neurons (Arevalo et al. 1992; Bhat et al. 1996; Ceccatelli et al. 1996; Nylén et al. 2001), and at the lower level, nNOS mRNA and immunoreactivity have been detected in the adrenal cortex (Tsuchiya et al. 1996).

The involvement of NO in neuroendocrine regulation and that it specifically modulates the release of hypothalamic neurohormones were based on the findings that the NO donor molsidomine was able to inhibit the interleukin-1β-induced corticotropin-releasing hormone (CRH) release from the hypothalamus (Costa et al. 1993). Similar experimental evidence on the conclusions on an inhibitory role of NO on CRH release was shown by the augmentation of the IL-1β-stimulated adenocorticotropic hormone (ACTH) release in the presence of the NOS inhibitor L-NAME (see below). Since the authors could not detect a direct influence of NO at the pituitary level, the ACTH release in response to IL-1β was considered to reflect CRH activity. Similarly, the stimulatory action of lipopolysaccharide (LPS) on ACTH and corticosterone secretion was also augmented by inhibition of NO formation (Rivier and Shen 1994). In another study on the activity of the HPA axis in intact and adrenalectomized rats, the effects of two NOS inhibitors were evaluated, with results suggesting that the central NO system exerts a tonic negative influence on the activity of the HPA axis in the presence or absence of circulating glucocorticoids (Givalois et al. 2002).

This is confirmed in another study, where 7-nitroindazole (7-NI) and L-NAME reduced ACTH and corticosterone secretion induced by an alpha-1-adrenergic receptor agonist and also diminished the HPA response to the nonselective beta-adrenergic receptor agonist isoprenaline, suggesting a complex functional relationship between NO and the regulation of HPA axis activity (Gadek-Michalska and Bugajski 2008).

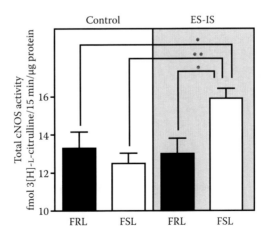

FIGURE 17.3 Hippocampal constitutive NOS (cNOS) activity data under basal conditions (Flinders sensitive line [FSL], $n = 8$; Flinders resistant line [FRL], $n = 7$) and following escapable stress/inescapable stress (ES-IS) (FSL, $n = 8$; FRL, $n = 7$). Following ES-IS, cNOS activity is significantly elevated in FSL rats compared to unstressed FSL controls (**$p < .005$) and versus pre- and poststress FRL animals (*$p < .05$). Pre- and poststress activity levels for FRL rats did not differ from one another. Values shown are means ± SEM. (Reprinted with permission from Wegener, G., et al., *Int J Neuropsychopharmacol*, 13, 461–73, 2010. © Cambridge University Press.)

Comparing transgenic animals lacking nNOS and WT mice, the levels of CRH mRNA were similar, and plasma ACTH levels, under basal conditions and in response to forced swimming, were unchanged in mutant when compared to WT mice (Orlando et al. 2008). These findings suggest that nNOS gene disruption does not significantly impair basal CRH gene expression but may be (although still debated) an important mediator of homeostatic adaptation, as confirmed by the finding that transgenic mice lacking nNOS, single-housed at weaning, showed significantly higher basal corticosterone plasma levels (Bilbo et al. 2003).

Larger animals have also been studied. For example, in a study of the ACTH response in female Yucatan miniature swine, it was observed that chronic NOS inhibition with L-NAME (see below) increased the ACTH response to exercise, indicating that NO modulates the neuroendocrine component of the HPA axis during exercise, despite the possible change in blood flow following L-NAME (Jankord et al. 2009).

Stress has also been found to influence the NO functioning. For example, swim stress increases the number of hypothalamic NOS-containing neurons, confirming the involvement of NOS neurons of the paraventricular nucleus in response to different types of acute stressors (Sanchez et al. 1999).

17.3.1.3 NO and Other Transmitter Systems

Deletion of the nNOS gene not only eliminates nNOS protein, common to many gene deletions, but also affects several downstream processes. For example, serotonin (5-HT) metabolism is altered in male nNOS$^{-/-}$ mice, where the ratio of the metabolite 5-HIAA and 5-HT was significantly reduced in several brain regions, including the cortex, hypothalamus, midbrain, and cerebellum of nNOS$^{-/-}$ in comparison with WT (Chiavegatto et al. 2001). In the same study, norepinephrine, dopamine, and metabolites were generally unaffected (Chiavegatto et al. 2001).

Several in vivo studies have demonstrated that NO can modulate the extracellular level of various neurotransmitters in the central nervous system, for example, 5-HT, dopamine, γ-amino-butyric acid, and glutamate (Kaehler et al. 1999; Lorrain and Hull 1993; Segovia et al. 1994, 1997, 1999; Strasser et al. 1994; Wegener et al. 2000) (Figure 17.4). In addition, NO can inactivate the rate-limiting enzyme in the synthesis of 5-HT, tryptophan hydroxylase (Kuhn and Arthur 1996, 1997), and it has been suggested to stimulate synaptic vesicle release from hippocampal synaptosomes (Meffert et al. 1994, 1996). Furthermore, NO regulates 5-HT reuptake (Pogun et al. 1994a, 1994b;

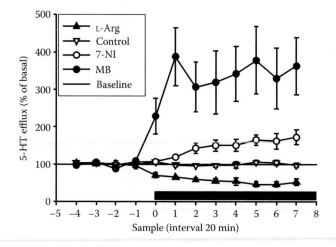

FIGURE 17.4 Effect of local perfusion using microdialysis of L-Arg (2 mM), 7-NI (1 mM), MB (1 mM), or artificial on 5-HT ($n = 8$, $n = 12$, $n = 7$, and $n = 13$, respectively) overflow in ventral hippocampus of rats. Drugs were infused into ventral hippocampus as indicated by bar. Results are expressed as % of basal efflux ± SEM. (Reprinted with permission from Wegener, G., et al., *Br J Pharmacol*, 130, 575–80, 2000. © Macmillan Publishers Ltd.)

Pogun and Kuhar 1994), inhibits uptake of [³H]dopamine by striatal synaptosomes (Lonart et al. 1993; Lonart and Johnson 1994), and transforms 5-HT into an inactive form (Fossier et al. 1999).

More recently, it was demonstrated that a physical interaction between the serotonin transporter and nNOS, via PDZ–PDZ interactions, may underlie the reciprocal modulation of their activity (Chanrion et al. 2007). The connection between NO and 5-HT is substantiated by observations showing that NO and 5-HT are involved in the pathophysiology of migraine (Lassen et al. 1997, 1998; Thomsen 1997; Thomsen and Olesen 1998), as well as in the inverse relationship between NO and 5-HT in peripheral tissue.

17.3.1.4 NO and Neurogenesis

Neurogenesis have attracted significant interest in the past two decades, and it has been suggested that neurogenesis may be linked to recovery from clinical depression (Duman et al. 2001a, 2001b) and that neurogenesis may be a prerequisite for antidepressant response (Santarelli et al. 2003). However, it should be strongly emphasized that evidence for clinical relevant neurogenesis in humans does not exist. In the brain, neurogenesis has been observed in the subventricular zone and in the subgranular zone of the dentate gyrus (DG) (Ehninger and Kempermann 2008). Interestingly, it has been demonstrated that the subventricular zone is surrounded by nNOS-positive neurons (Romero-Grimaldi et al. 2008), and cells expressing nNOS also have been identified in neuronal precursors in the DG (Shariful Islam et al. 2003), suggesting that nNOS could participate in the regulation of neurogenesis.

Moreover, it was demonstrated that inhibition of NO synthesis with 7-NI increases proliferation of neural precursors isolated from the postnatal mouse subventricular zone (Matarredona et al. 2004). However, a recent report has also demonstrated that nNOS inhibition with 7-NI enhanced the proliferation of progenitor cells in the DG (Zhu et al. 2006). The abovementioned results are in line with findings using a null mutant nNOS knockout mouse line, where the number of new cells generated in neurogenic areas of the adult brain, the olfactory subependyma and the DG, was strongly augmented, indicating that division of neural stem cells in the adult brain is negatively controlled by NO (Packer et al. 2003).

In addition, inhibition of nNOS with 7-NI has been shown to reverse the neurogenetic impairment induced by a chronic mild stress regimen (Zhou et al. 2007). Interestingly, contrary to what may be expected based on the findings with the inhibitors, administration of exogenous NO donors have also been shown to stimulate cell proliferation (Hua et al. 2008; Zhang et al. 2001). The neurobiology underlying this phenomenon remains to be established, but it should be mentioned here that several of the pharmacological tools available possess a U-shaped dose–effect pharmacology.

17.3.2 Clinical Studies

17.3.2.1 Postmortem Studies

Postmortem material from patients with major depression has demonstrated a reduced nNOS activity and protein content in various brain regions. In a study of eight patients (including two with schizoaffective diagnosis and two with bipolar depression) diagnosed according to the *Diagnostic and Statistical Manual of Mental Disorders* (*Third Edition, Revised*), a reduced number of NOS-I-containing neurons in the paraventricular hypothalamic nucleus was observed (Bernstein et al. 1998). This finding was later expanded and confirmed in 11 patients and 11 matched controls (Bernstein et al. 2002). In another study, a strong trend ($p < .06$) in decreased activity of the constitutive NOS in prefrontal cortex of 15 patients with unipolar depression diagnosed versus 15 nonpsychiatric controls from the Stanley Consortium was observed (Xing et al. 2002). Moreover, a study examining 12 depressed subjects and 12 psychiatrically normal control subjects, obtained at autopsy at the Coroner's Office of Cuyahoga County, Cleveland, OH, USA, found a significantly lower amount of nNOS in the locus coeruleus of depressed subjects (Karolewicz et al. 2004). However, no changes were observed in the cerebellum (Karolewicz et al. 2004).

Because the hippocampus is a crucial region in the pathophysiology of affective disorders, the possible hippocampal involvement is of great interest. Findings from the CA1 hippocampal area in brains from the Stanley Consortium have reported an increase in NOS-I immunoreactivity in depression and bipolar disorder (Oliveira et al. 2008). In the same study, no changes were observed in brains from schizophrenic patients. However, the study was not a detailed stereological examination, which may limit the overall conclusions. In another postmortem study, preliminary results suggest that the hippocampal expression of NOS-I may be increased in depression (De Oliveira et al. 2000).

17.3.2.2 Peripheral Markers

Several studies have examined peripheral NO metabolism in major depression; however, rather mixed results were obtained. In a study conducted on suicide attempters, increased NO metabolites (NO_2 and NO_3) have been observed (Kim et al. 2006; Lee et al. 2006), indicating a hyperfunction of the nitrergic system. The same finding was reported a few years earlier, where it was found that 17 drug-naïve patients suffering from depression, diagnosed according to the *Diagnostic and Statistical Manual of Mental Disorders* (*Fourth Edition*) (DSM-IV), had elevated nitrite levels (Suzuki et al. 2001). In the same study, treatment with antidepressant normalized the nitrite levels, correlating with clinical response (Suzuki et al. 2001). Finally, in a study including 36 depressed patients, diagnosed according to *DSM-IV*, and 20 healthy subjects, there was no correlation between depressive symptoms and levels of nitrate, but there was a significant effect of antidepressant treatment, lowering the nitrate levels (Herken et al. 2007). In addition, there are some studies demonstrating involvement of NO in some, but not all, forms of interferon-alpha-induced depression (Suzuki et al. 2003).

However, measurement of nitrate in serum will only detect the overall nitrate pool. Indeed, a study by Srivastava et al. (2002) examining 66 cases of depression and 114 controls revealed a 73% decrease in nitrite content in the polymorphonuclear leukocytes. Inasmuch as human polymorphonuclear leukocytes express nNOS-like neurons (Wallerath et al. 1997), this measure may be hypothesized to be more relevant than serum values.

This assumption is also reflected in a study where a decreased platelet NOS activity and plasma NO metabolites in depressed patients were found (Chrapko et al. 2004, 2006).

Several human association studies have been published linking endogenous inhibitors of NOS with disease. The levels of the endogenous inhibitors, N^G-monomethyl-L-arginine (symmetric dimethylarginine [SDMA]) and N^G-dimethyl-L-arginine (asymmetric dimethylarginine [ADMA]) (Boger et al. 2009a, 2009b, 2009c), have been shown to be changed in depression, schizophrenia, and Alzheimer disease (Selley 2004; Das et al. 1996; Arlt et al. 2008). However, it is not clear whether these associations are clinically important.

Taken together, although human studies have been predominantly carried out on peripheral tissue samples (e.g., plasma or serum), a support for the role of the NO system in psychiatric disease exists. Findings from the peripheral tissue seem to suggest an elevated NO metabolism in depressive states. However, it is noteworthy that the measures mentioned here were carried out on specific subcomponents in the blood compartment and that the generalization may therefore be limited.

17.3.2.3 Genetics

Recently, results from of a few laboratories have implicated a role of NOSs polymorphisms in brain diseases such as bipolar disorders, schizophrenia, Alzheimer disease, and autism (Buttenschon et al. 2004; Reif et al. 2006a, 2006b; Galimberti et al. 2008). A few genetic studies have been carried out investigating NOS-I in unipolar depression, and several genetic association studies are currently being carried out; data will likely contribute to the clarification of NOS's roles in these brain disorders.

In a population-based association study investigating NOS-I in unipolar depression, it was tested whether the nNOS C276T polymorphism confers susceptibility to unipolar depression and treatment response to fluoxetine. No association with disease or selective serotonin reuptake inhibitor treatment response was found in 108 Chinese patients (Yu et al. 2003), but because of the restricted design of the study, it is concluded that other variants of the nNOS gene may play a role.

Moreover, in a recent genetic association analysis of case-control samples (325 MDD patients, 154 bipolar patients, and 807 controls) in a Japanese population, using single-nucleotide polymorphism (SNP; rs41279104, also called ex1c), no associations were detected between one marker (rs41279104) in NOS-I and Japanese mood disorder patients, although the sample sizes were probably too small to allow a meaningful test (Okumura et al. 2010). Moreover, the paper did not perform an association analysis based on linkage disequilibrium and a mutation scan of NOS-I (Okumura et al. 2010).

In a large genome-wide association study of 435,291 SNPs genotyped in 1738 MDD cases and 1802 controls selected to be at low liability for MDD, it was reported that an association of NOS-I with the disease was present, although the size of the NOS-I gene makes the authors cautious about the finding (Sullivan et al. 2009).

Interestingly, in a recent study carried out in a group of 181 depressed patients and 149 control subjects of Polish origin, it was examined whether an SNP present in the genes encoding iNOS and nNOS could contribute to the risk of developing recurrent depressive disorder (Galecki et al. 2010). It was shown that both investigated polymorphisms are associated with depression and that the *NOS2A* and *nNOS* genes may confer an increased risk of recurrent depressive disorder (Galecki et al. 2010).

In conclusion, although mixed findings directly related to the NOS-I gene exist and despite the fact that several larger replication studies are needed, the available genetic studies support a role of NO in depressive disorders.

17.3.2.4 Pathways Interacting with NO and Depression

As mentioned above, the glutamatergic input constitutes the major activating transmitter to the nNOS. A detailed review lies beyond the scope of this chapter, but several prominent findings will be highlighted here.

Several studies have demonstrated differences related to glutamate receptors in postmortem brain samples from individuals with major depression. A receptor binding assay and a Western blot study revealed a reduction in [^{3}H]L-689,560 (a potent antagonist at the glycine modulatory site on the NMDA receptor) binding as well as a reduction in NR1 immunoreactivity in the superior temporal cortex in major depression (Nudmamud-Thanoi and Reynolds 2004). Moreover, receptor binding and autoradiography in patients with MDD showed increased binding of [^{3}H]CGP39653 (glutamate site on the NMDA receptors) in the hippocampus but not in the entorhinal or perirhinal cortex (Beneyto et al. 2007). In the same study, in situ hybridization demonstrated that mRNA levels of the NR2A and NR2B subunits of the NMDA receptors and the GluR1, GluR3, and GluR5 subunits of the AMPA receptors in the perirhinal cortex, but not in the hippocampus or entorhinal cortex, were significantly lower than those of a control group (Beneyto et al. 2007). This is in line with reports suggesting that the levels of the NR2A and NR2B subunits of the NMDA receptors are decreased in the prefrontal cortex (Brodmann's area 10) of patients with MDD (Feyissa et al. 2009).

Stress is recognized as a main risk factor for major depression, and it has been demonstrated that acute stress is associated with increased glutamatergic neurotransmission in areas of the forebrain as measured by an in vivo microdialysis technique (Bagley and Moghaddam 1997; Lowy et al. 1995; Reznikov et al. 2007). In this regard, it is of significant interest that after chronic treatment with antidepressants, a reduction of glutamate release under basal conditions can be observed (Bonanno et al. 2005; Michael-Titus et al. 2000; Tokarski et al. 2008). Using a different technique with direct measurement of glutamate release, it was recently reported that depolarization-dependent release of glutamate is selectively upregulated by acute stress relative to γ-amino-butyric acid release, and that chronic treatment with antidepressants (desipramine, fluoxetine, or venlafaxine) completely abolished the stress-induced upregulation of glutamate release (Musazzi et al. 2010). It can be speculated that this inhibition may be a relevant component of the therapeutic action of anti-depressants (Musazzi et al. 2010). Furthermore, it has been reported that 4-week administration of the serotonin reuptake inhibitor fluoxetine caused upregulation of the subunits (e.g., NR2A, GluR1, GluR2) of glutamate receptors in the retrosplenial cortex of the rat brain, and that these changes in the subunit levels were associated with an upregulation of dendritic spine.

Interestingly, the magnesium ion plays an important role in the NMDA receptor functioning, and several studies suggest that other studies demonstrate that depletion of magnesium precipitates depressive symptomatology in both humans and animals (Barragán-Rodríguez et al. 2008; Herzberg and Herzberg 1977; Hojgaard Rasmussen et al. 1989; Murck 2002; Pavlinac et al. 1979; Szewczyk et al. 2008; Poleszak et al. 2007; Singewald et al. 2004). As activation of the NMDA receptor is known to increase the activity of NOS, these symptoms may arise because of nitrergic overactivity. Importantly, the depressive symptoms after magnesium depletion can be ameliorated after antide-pressant treatment with either desipramine or *Hypericum* extract (Singewald et al. 2004).

Finally, it is noteworthy that the clinical studies carried out with the NMDA receptor antagonist ketamine clearly demonstrated a rapid sustained antidepressant effect (Machado-Vieira et al. 2009; Maeng and Zarate 2007; Zarate et al. 2006). However, although ketamine clearly affects the NMDA receptor, it may also have other important targets, such as the rapamycin pathway (Li et al. 2010).

Considering these results together, it seems likely that glutamatergic neurotransmission is impli-cated in the action of antidepressants.

17.4 NO AND ANTIDEPRESSANT TREATMENT

Over the past two decades, a number of preclinical studies have demonstrated that inhibition of NOS produces anxiolytic and antidepressant-like behavioral effects in a variety of animal paradigms, and there also have been studies showing the involvement of NO in the action of classical antidepres-sants. These studies include systemic injections as well as targeted infusions into the brain. Only a few, very limited clinical studies are available, and they are confounded by the nonselectivity of the drug used (see Section 17.4.1.4).

17.4.1 NOS INHIBITION PRODUCES ANTIDEPRESSANT-LIKE EFFECTS

The typical NOS inhibitor is an amino acid that associates with the substrate-binding site for L-argi-nine (Griffith and Kilbourn 1996). The inhibitor will compete with L-arginine, and usually, extra arginine will reverse the NOS inhibition produced by the inhibitor.

17.4.1.1 Depression-Like Behavior

The best investigated inhibiting amino acids are L-NNA (L-N^G-nitro arginine), its methyl ester (L-NAME), L-NMMA (L-N^G-monomethyl-arginine), and N^G-propyl-L-arginine. L-NAME requires hydrolysis of the methyl ester by cellular esterases to become a fully functional inhibitor (Griffith and Kilbourn 1996). Acute antidepressant effects have been found in both rats and mice models. L-NNA and L-NAME have been reported to be effective in both the forced-swim test (FST) and tail-suspension test in mice (da Silva et al. 2000; Harkin et al. 1999), and in the FST in rats (Jefferys and Funder 1996; Harkin et al. 2003). The effect of the drugs seems to display a U-shaped phar-macology, where both low and high doses have no effect (Harkin et al. 1999; da Silva et al. 2000; Karolewicz et al. 2001). Pretreatment with L-Arg counteracts the behavioral effects of L-NAME and L-NNA (Jefferys and Funder 1996; Harkin et al. 1999; da Silva et al. 2000; Inan et al. 2004) but has also been reported in some studies to have an antidepressant-like effect by itself (da Silva et al. 2000).

Similar to the findings with amino acids, antidepressant-like properties have also been demon-strated with the non–amino acid compounds. The primary benefit with the indazoles and imidazole derivates is a potential superiority in selectivity among the different isoforms of the NOS enzymes.

This was clear when 7-NI was discovered (Babbedge et al. 1993), as it did not have a profound effect on blood pressure (Moore et al. 1993), which can be observed with most of the amino acid inhibitors. Studies suggest that 7-NI not only interacts competitively at the substrate binding site in the NOS enzyme (Babbedge et al. 1993) but also acts competitively regarding the cofactor tetra-hydrobiopterin (BH_4) (Mayer et al. 1994). Because 7-NI is also a potent inhibitor of bovine aortic endothelial eNOS in vitro, in spite of the lack of cardiovascular side effects of this compound in vivo (Babbedge et al. 1993) suggesting other factors to be involved, other more selective isoform inhibitors have been screened. One such compound is 1-(2-trifluoromethyphenyl)imidazole, which is described as a potent and relatively selective inhibitor of nNOS both in vitro and in vivo (Handy et al. 1995, 1996). The selectivity of this compound seems to be centered around the cofactor BH_4 and the availability of BH_4 in the tissues (Handy and Moore 1997).

In the FST, acute administration of 7-NI and 1-(2-trifluoromethyphenyl)imidazole has antidepressant-like effects but does not affect locomotion (Yildiz et al. 2000; Harkin et al. 2003; Volke et al. 2003a; Spiacci et al. 2008; Heiberg et al. 2002) (Figure 17.5). Interestingly, the effects of 7-NI are centrally based because intrahippocampal administration of 7-NI causes a dose-dependent antidepressant-like effect in the FST, an effect that could be prevented after intrahippocampal coad-ministration of L-arginine (Joca and Guimaraes 2006).

17.4.1.2 Cognition and Memory

The clinically important features in depression, cognition and memory, have been extensively examined, and a major role for NO in the formation of memory and as a mediator in synaptic plas-ticity has been suggested (Prast and Philippu 2001; Papa et al. 1994). The majority of studies sup-port a facilitatory role of NO in learning processes, and nNOS has been proposed as the principal source of this retrograde messenger during LTP (O'Dell et al. 1991; Schuman and Madison 1991), a highly important process for memory formation (Bliss and Collingridge 1993; Madison et al. 1991; Malenka 1994). However, some controversy about this finding exists, as LTP in the hippocampus and cerebellum was reported to be normal in nNOS transgenic mice (O'Dell et al. 1994; Linden et al. 1995). The involvement of NOS in memory has been confirmed in studies with NOS inhibitors. For example, it was shown that systemic administration of L-NAME and L-NNA impairs acquisi-tion but not retention of spatial learning in rats (Bohme et al. 1993; Chapman et al. 1992; Estall et al. 1993), and L-NA reduces hippocampal mediation of place learning in the rat (Mogensen et al. 1995a, 1995b). Similarly, intrahippocampal administration of L-NAME impairs working memory

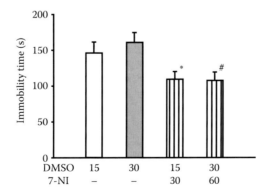

FIGURE 17.5 Effects of tNOS inhibitor 7-NI (30 or 60 mg/kg, $n = 10$ and $n = 8$, respectively) on duration of immobility time (in seconds) in FST recorded for 5 min. Control groups were treated with dimethyl sulfoxide (DMSO) 15% ($n = 7$) or DMSO 30% ($n = 8$). Values shown are means ± SEM. An asterisk (*) indicates signifi-cant difference from DMSO 15% control group ($p < .05$) and a hash (#) indicates difference from DMSO 30% control group ($p < .05$). (Reprinted with permission from Heiberg, I. L., et al., *Behav Brain Res*, 134, 479–84, 2002. © Elsevier BV.)

on a runway task without affecting reference memory (Ohno et al. 1993, 1994, and L-NAME has been shown to disrupt learning of an associative memory task, the conditioned eyeblink response in rabbits (Chapman et al. 1992). However, in a well-learned operant task—a delayed nonmatch-to-position—no effect of L-NAME was found (Wiley and Willmore 2000), and similarly, L-NAME did not affect learning in a Morris water maze paradigm (Bannerman et al. 1994a). In agreement with these observations, central and systemic administration of the NO precursor L-Arginine has been found to significantly prolong the latency time in the passive avoidance test without inhibition of locomotor and exploratory activity (Plech et al. 2003). The interpretation of the overall neurobiological consequences of these findings remains unclear. The early findings with NOS inhibitors do not seem to correspond well with the results published about other clinically relevant antidepressants, such as the serotonin reuptake inhibitors, where cognitive performance in patients has been shown to be unaffected (Siepmann et al. 2003) and independent from clinical recovery (Herrera-Guzman et al. 2009). However, a recent rodent study found that acute administration of imipramine and aroxetine to rats impaired the discrimination of old from the recently presented objects (Naudon et al. 2007). Interestingly, after chronic administration, the imipramine-treated rats were unable to differentiate between the two objects, whereas paroxetine-treated rats spent more time exploring the old object (Naudon et al. 2007). Similarly, it is important to note that the studies with NOS inhibitors and cognitive testing predominantly have been carried out after one acute dose. The relevance for this paradigm related to a clinical context is, as it also is the case with the other depression and anxiety tests, questionable. Only limited information is available concerning chronic administration of NOS inhibitors. However, it has been shown that L-NAME in the drinking water over 14 days impairs working memory in rats (Cobb et al. 1995). On the other hand, it has also been demonstrated that acute, but not chronic, administration of L-NAME impairs LTP formation induced by a weak near-threshold tetanus (Bannerman et al. 1994b) and that chronic L-NAME in the drinking water did not affect working memory in a water-maze working task (Wohlert-Johannsen and Wegener, unpublished data). Further studies must be carried out to elucidate the overall effects of NOS inhibition on cognition.

In the examination of the non–amino acid inhibitors, 7-NI induces amnesia in a passive avoidance task in the chick (Holscher 1994) and impairs learning and memory in different tasks such as the Morris water maze, radial maze, and passive avoidance and EPM tests (Holscher et al. 1996; Mizuno et al. 2000; Yildiz Akar et al. 2007, 2009; Zou et al. 1998). 7-NI also produces taste aversions and enhances the lithium-based taste aversion learning in a conditioned taste aversion paradigm, an effect that was counteracted with simultaneous administration of L-arginine (Wegener et al. 2001).

Some compounds based on hydrazine structure have been extensively studied in relation to cardiovascular (Yang et al. 2004; Ishibashi et al. 2001; Wang et al. 1999; Takano et al. 1998; Gardiner et al. 1996) and endocrinological diseases (Rydgren and Sandler 2002; Suarez-Pinzon et al. 2001; Shimabukuro et al. 1997; Holstad et al. 1996). The compounds are predominantly inhibitors of iNOS, with much less activity on the other isoforms. Aminoguanidine (AG) is a hydrazine derivate and the best characterized compound (Corbett and McDaniel 1996; Hasan et al. 1993; Griffiths et al. 1993), which selectively decreases cGMP levels produced by iNOS (Griffiths et al. 1995). Furthermore, AG has been observed to protect against neurodegeneration produced by chronic stress in rats (Olivenza et al. 2000) and to prevent the impairment of learning behavior and hippocampal LTP following transient cerebral ischemia in rats (Mori et al. 2001). Moreover, intracerebroventricular infusion of AG prevents the depression-like behavior following a chronic unpredictable stress paradigm (Wang et al. 2008). Supporting these findings, a model of PTSD seems to exclusively involve the iNOS isoform, as AG, but not 7-NI, was effective in attenuating neurobiological readouts (Harvey et al. 2004). Together, these findings highlight the possible involvement of inflammatory processes in depression and anxiety, which is plausible considering the significant involvement of stress in those disorders. AG has also recently been demonstrated to display anxiolytic-like effects in EPM, open field test, light/dark test, and social interaction test in stressed mice (Gilhotra and Dhingra 2009). Whether these effects are independent of the presence of stressors remains to be established.

17.4.1.3 Endogenous Inhibitors

Some antagonists of NOS require special attention, as they may be considered endogenous inhibitors. These include L-citrulline, agmatine, NG, ADMA, SDMA, and argininosuccinic acid. Whereas L-citrulline is a very weak inhibitor, a derivate, L-thiocitrulline, is much more powerful (Frey et al. 1994). Agmatine, decarboxylated arginine (Wiesinger 2001), is potentially significant, as there is evidence of antidepressant effects in preclinical animal models of depression (Zomkowski et al. 2002; Aricioglu and Altunbas 2003; Li et al. 2003; Krass et al. 2008), as well as in humans (Halaris et al. 1999; Halaris and Piletz 2001, 2003; Taksande et al. 2009). It is, however, noteworthy that agmatine also has been conceptualized as an endogenous clonidine-displacing substance on imidazoline receptors (Regunathan and Reis 1996; Li et al. 1994) and to have affinity for several transmembrane receptors, such as α2-adrenergic (Olmos et al. 1999), imidazoline I1, and glutamatergic NMDA receptors (Yang and Reis 1999). Therefore, the effects observed in the preclinical studies may be mediated via these pathways and not linked to NOS.

No solid preclinical data exist for the other endogenous inhibitors, although there are reports of their presence in animals (Ogawa et al. 1987).

17.4.1.4 Clinically Tested Compounds

Methylene blue (MB), although not exclusively selective for the NOS, is potentially of special relevance because it is so far the only compound proven to be effective in patients (Naylor et al. 1986, 1987, 1988). MB oxidizes protein-bound heme and non-heme ferrous iron (Salaris et al. 1991), inhibiting the stimulation of sGC by NO and nitrovasodilators (Murad et al. 1978). MB potently inhibits NOS both in vitro (Mayer et al. 1993a, 1993b) and in vivo (Volke et al. 1999). As early as 1899, MB was described to have a calming—probably antipsychotic—effect on patients (Bodoni 1899). However, more recent work has focused on the beneficial effects of MB in manic-depressive disorder, where a response of 63% among 24 lithium refractory patients was found (Narsapur and Naylor 1983). The studies were supplemented and expanded, confirming this action (Naylor et al. 1986, 1987, 1988). At the time of the study, the mechanistic hypotheses were based on changes in the vanadium ion (Naylor and Smith 1982; Naylor 1984; Naylor et al. 1981). Unfortunately, the studies cited above were not fully randomized, but better controlled trials are now being carried out (Alda 2008).

Several preclinical studies confirm a positive effect of MB in the FST and EPM (Eroglu and Caglayan 1997), although with a U-shaped dose–response efficacy curve. MB produces taste aversion in a conditioned taste aversion paradigm, an effect comparable to the effects of 7-NI, which also could be counteracted with simultaneous administration of L-arginine (Wegener et al. 2001). As indicated by the mode of action, MB is expected to be a very nonselective compound. Indeed, MB inhibits not only NOS and sGC but also several other heme-containing enzymes, making it a potent inhibitor of monoamine oxidase (Ehringer et al. 1961; Gillman 2008; Jakubovic and Necina 1963) and various cytochromes. This effect probably accounts for the case reports suggesting a hyperserotonergic state after use of MB (Stanford et al. 2010; Gillman 2008) and can also be an explanation for the clinical efficacy.

As MB affects the NO downstream signaling pathway, including sGC, a few compounds which affect sGC (but not NOS), have been examined. Studies with selective (i.e., non-NOS) inhibitors of NO-dependent cGMP formation with [1H-[1,2,4]oxadiazole[4,3-a]quinoxalin-1-one] have shown antidepressant-like effects in the FST (Heiberg et al. 2002) as well as prevention of prodepressant effect of L-arginine in the FST (Ergun and Ergun 2007). Similarly, [1H-[1,2,4]oxadiazole[4,3-a] quinoxalin-1-one] has been shown to have anxiolytic-like properties, with an increase in the percentage of time spent on the open arm in the EPM after administration of the drug (Spolidorio et al. 2007). These findings are in agreement with other studies showing that an increase in cGMP, after inhibition of phosphodiesterase type V with sildenafil, can produce anxiogenic-like responses in the EPM (Kurt et al. 2004; Volke et al. 2003b). However, the mechanisms have not been fully elucidated, as sildenafil has also been shown to have antidepressant-like effects following central muscarinic receptor blockade (Brink et al. 2008).

17.4.2 Classical Antidepressants and NO

Direct interaction between clinically used antidepressants and nitrergic signaling has been shown in a few studies. In a study with patients with ischemic heart disease and depression, 17 received paroxetine and 14 patients received nortriptyline, and it was observed that serum nitrite and nitrate levels were significantly decreased following paroxetine, but not nortriptyline, treatment (Finkel et al. 1996). In addition, paroxetine was also shown to be a significantly more potent inhibitor of the NOS enzyme activity than nortriptyline (Finkel et al. 1996). Recently, it was established that the antidepressant action of imipramine and venlafaxine involves suppression of nitric oxide synthesis (Krass et al. 2011), and several established antidepressants of distinct chemical classes, including imipramine, paroxetine, citalopram, and tianeptine, have all been shown to inhibit hippocampal NOS activity in vivo when applied locally in the brain in therapeutic-relevant concentrations (Wegener et al. 2003) (Figure 17.6). This effect may involve a substantial pathway through the NMDA complex, and selective inhibition of NOS will produce pronounced behavioral effects (see below).

It has also been reported that the precursor of NO, L-Arg, antagonizes the effects of the classic tricyclic antidepressant, imipramine (Harkin et al. 1999). This observation has led to hypotheses regarding the potential contribution of serotonergic/noradrenergic mechanisms in the observed antidepressant-like effects of the NOS inhibitors (see below).

Subsequently, it has been demonstrated that low and ineffective doses of L-NAME were able to potentiate the behavioral effects of imipramine and fluoxetine, but not reboxetine, a noradrenaline reuptake inhibitor, in the FST (Harkin et al. 2003, 2004). In addition, it was shown that a serotonergic mediation of the antidepressant-like effects of L-NA and 7-NI was present, inasmuch as serotonergic depletion abolished the antidepressant-like effect of the inhibitors (Harkin et al. 2003). Not all inhibitors seem to display this profile, as it also was demonstrated that the effect of agmatine was independent of the 5-HT depletion (Krass et al. 2008). However, as already discussed, agmatine likely has multiple effects on several receptor systems.

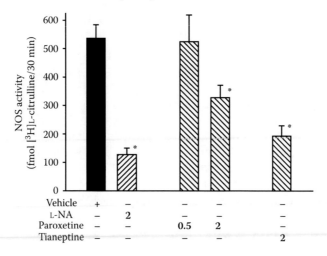

FIGURE 17.6 NOS activity in vivo (microdialysis) after acquisition of steady state following local perfusion with vehicle ($n = 19$), paroxetine (500 nM, $n = 53$; and 2 mM, $n = 11$), and tianeptine (2 mM, $n = 6$). An asterisk (*) indicates statistical difference from control group ($p < .05$). Values shown are means $\pm$ SEM. (Reprinted with permission from Wegener, G., et al., *Brain Res*, 959, 128–34, 2003. © Elsevier BV.)

Finally, NO has also been implicated in the antidepressant role of several other substances such as tramadol (Jesse et al. 2008), bupropion (Dhir and Kulkarni 2007), and lithium (Ghasemi et al 2008).

17.5 CONCLUSION

Although the studies cited in the current chapter use different methodologies, from a preclinical genetic approach to postmortem human material, the physiological role of NOS emerges to be relatively clear. Therefore, the conclusion of this current review is that despite significant challenges in developing compounds that may differentially inhibit the "right" NOS isoform at the right place, the NO system continues to be an interesting novel approach in the future development of antidepressants.

ACKNOWLEDGMENTS

GW was supported by grants from The Danish Medical Research Council (grant 271-08-0768) and the Research Foundation of County Midtjylland. AAM received support from the Swedish Medical Research Council (grant 10414) and the Karolinska Institute.

REFERENCES

Akyol, O., Zoroglu, S. S., Armutcu, F., Sahin, S., and Gurel, A. 2004. Nitric oxide as a physiopathological factor in neuropsychiatric disorders. *In Vivo* 18 (3):377–90.

Alda, M. 2008. Nct00214877: Methylene blue for cognitive dysfunction in bipolar disorder. http://clinicaltrials .gov/show/NCT00214877. Accessed November 8th 2009.

Arancio, O., Lev-Ram, V., Tsien, R. Y., Kandel, E. R., and Hawkins, R. D. 1996. Nitric oxide acts as a retrograde messenger during long-term potentiation in cultured hippocampal neurons. *J Physiol (Paris)* 90 (5–6):321–2.

Arevalo, R., Sanchez, F., Alonso, J. R., Carretero, J., Vazquez, R., and Aijon, J. 1992. NADPH-diaphorase activity in the hypothalamic magnocellular neurosecretory nuclei of the rat. *Brain Res Bull* 28 (4):599–603.

Aricioglu, F., and Altunbas, H. 2003. Is agmatine an endogenous anxiolytic/antidepressant agent? *Ann N Y Acad Sci* 1009:136–40.

Arlt, S., Schulze, F., Eichenlaub, M., et al. 2008. Asymmetrical dimethylarginine is increased in plasma and decreased in cerebrospinal fluid of patients with Alzheimer's disease. *Dement Geriatr Cogn Disord* 26 (1):58–64.

Assreuy, J., Cunha, F. Q., Liew, F. Y., and Moncada, S. 1993. Feedback inhibition of nitric oxide synthase activity by nitric oxide. *Br J Pharmacol* 108 (3):833–7.

Babbedge, R. C., Bland-Ward, P. A., Hart, S. L., and Moore, P. K. 1993. Inhibition of rat cerebellar nitric oxide synthase by 7-nitro indazole and related substituted indazoles. *Br J Pharmacol* 110 (1):225–8.

Bagley, J., and Moghaddam, B. 1997. Temporal dynamics of glutamate efflux in the prefrontal cortex and in the hippocampus following repeated stress: Effects of pretreatment with saline or diazepam. *Neuroscience* 77 (1):65–73.

Bannerman, D. M., Chapman, P. F., Kelly, P. A., Butcher, S. P., and Morris, R. G. 1994a. Inhibition of nitric oxide synthase does not impair spatial learning. *J Neurosci* 14 (12):7404–14.

Bannerman, D. M., Chapman, P. F., Kelly, P. A., Butcher, S. P., and Morris, R. G. 1994b. Inhibition of nitric oxide synthase does not prevent the induction of long-term potentiation in vivo. *J Neurosci* 14 (12):7415–25.

Barjavel, M. J., and Bhargava, H. N. 1995. Nitric oxide synthase activity in brain regions and spinal cord of mice and rats: Kinetic analysis. *Pharmacology* 50 (3):168–74.

Barragán-Rodríguez, L., Rodríguez-Morán, M., and Guerrero-Romero, F. 2008. Efficacy and safety of oral magnesium supplementation in the treatment of depression in the elderly with type 2 diabetes: A randomized, equivalent trial. *Magnes Res* 21 (4):218–23.

Beckman, J. S. 1996. The physiological and pathological chemistry of nitric oxide. In *Nitric Oxide: Principles and Actions*, ed. J. Lancaster, 1–82. San Diego: Academic Press, Inc.

Beneyto, M., Kristiansen, L. V., Oni-Orisan, A., McCullumsmith, R. E., and Meador-Woodruff, J. H. 2007. Abnormal glutamate receptor expression in the medial temporal lobe in schizophrenia and mood disorders. *Neuropsychopharmacology* 32 (9):1888–902.

Bernstein, H. G., Stanarius, A., Baumann, B., et al. 1998. Nitric oxide synthase-containing neurons in the human hypothalamus: Reduced number of immunoreactive cells in the paraventricular nucleus of depressive patients and schizophrenics. *Neuroscience* 83 (3):867–75.

Bernstein, H. G., Heinemann, A., Krell, D., et al. 2002. Further immunohistochemical evidence for impaired NO signaling in the hypothalamus of depressed patients. *Ann N Y Acad Sci* 973:91–3.

Bhat, G., Mahesh, V. B., Aguan, K., and Brann, D. W. 1996. Evidence that brain nitric oxide synthase is the major nitric oxide synthase isoform in the hypothalamus of the adult female rat and that nitric oxide potently regulates hypothalamic cGMP levels. *Neuroendocrinology* 64 (2):93–102.

Bilbo, S. D., Hotchkiss, A. K., Chiavegatto, S., and Nelson, R. J. 2003. Blunted stress responses in delayed type hypersensitivity in mice lacking the neuronal isoform of nitric oxide synthase. *J Neuroimmunol* 140 (1–2):41–8.

Bliss, T. V. P., and Collingridge, G. L. 1993. A synaptic model of memory: Long-term potentiation in the hippocampus. *Nature* 361 (6407):31–9.

Blottner, D., Grozdanovic, Z., and Gossrau, R. 1995. Histochemistry of nitric oxide synthase in the nervous system. *Histochem J* 27 (10):785–811.

Bodoni, P. 1899. Le bleu de méthylène comme calmant chez le aliénés. *Semaine Méd* 7:56.

Boger, R. H., Diemert, A., Schwedhelm, E., Luneburg, N., Maas, R., and Hecher, K. 2009a. The role of nitric oxide synthase inhibition by asymmetric dimethylarginine in the pathophysiology of preeclampsia. *Gynecol Obstet Invest* 69 (1):1–13.

Boger, R. H., Maas, R., Schulze, F., and Schwedhelm, E. 2009b. Asymmetric dimethylarginine (ADMA) as a prospective marker of cardiovascular disease and mortality—An update on patient populations with a wide range of cardiovascular risk. *Pharmacol Res* 60:481–7.

Boger, R. H., Sullivan, L. M., Schwedhelm, E., et al. 2009c. Plasma asymmetric dimethylarginine and incidence of cardiovascular disease and death in the community. *Circulation* 119 (12):1592–600.

Bohme, G. A., Bon, C., Lemaire, M., et al. 1993. Altered synaptic plasticity and memory formation in nitric oxide synthase inhibitor-treated rats. *Proc Natl Acad Sci U S A* 90 (19):9191–4.

Bonanno, G., Giambelli, R., Raiteri, L., et al. 2005. Chronic antidepressants reduce depolarization-evoked glutamate release and protein interactions favoring formation of snare complex in hippocampus. *J Neurosci* 25 (13):3270–9.

Bredt, D. S., and Snyder, S. H. 1989. Nitric oxide mediates glutamate-linked enhancement of cGMP levels in the cerebellum. *Proc Natl Acad Sci U S A* 86 (22):9030–3.

Bredt, D. S., and Snyder, S. H. 1994. Nitric oxide: A physiologic messenger molecule. *Annu Rev Biochem* 63:175–95.

Brenman, J. E., and Bredt, D. S. 1997. Synaptic signaling by nitric oxide. *Curr Opin Neurobiol* 7 (3):374–8.

Brink, C. B., Clapton, J. D., Eagar, B. E., and Harvey, B. H. 2008. Appearance of antidepressant-like effect by sildenafil in rats after central muscarinic receptor blockade: Evidence from behavioural and neuroreceptor studies. *J Neural Transm* 115 (1):117–25.

Buga, G. M., Griscavage, J. M., Rogers, N. E., and Ignarro, L. J. 1993. Negative feedback regulation of endothelial cell function by nitric oxide. *Circ Res* 73 (5):808–12.

Buttenschon, H. N., Mors, O., Ewald, H., et al. 2004. No association between a neuronal nitric oxide synthase (NOS1) gene polymorphism on chromosome 12q24 and bipolar disorder. *Am J Med Genet B Neuropsychiatr Genet* 124 (1):73–5.

Caspi, A., Sugden, K., Moffitt, T. E., et al. 2003. Influence of life stress on depression: Moderation by a polymorphism in the 5-HTT gene. *Science* 301 (5631):386–9.

Ceccatelli, S., Grandison, L., Scott, R. E. M., Pfaff, D. W., and Kow, L. M. 1996. Estradiol regulation of nitric oxide synthase mRNAs in rat hypothalamus. *Neuroendocrinology* 64 (5):357–63.

Chanrion, B., Mannoury La Cour, C., Bertaso, F., et al. 2007. Physical interaction between the serotonin transporter and neuronal nitric oxide synthase underlies reciprocal modulation of their activity. *Proc Natl Acad Sci U S A* 104 (19):8119–24.

Chapman, P. F., Atkins, C. M., Allen, M. T., Haley, J. E., and Steinmetz, J. E. 1992. Inhibition of nitric oxide synthesis impairs two different forms of learning. *Neuroreport* 3 (7):567–70.

Chen, M., Cheng, C., Yan, M., et al. 2008. Involvement of capon and nitric oxide synthases in rat muscle regeneration after peripheral nerve injury. *J Mol Neurosci* 34 (1):89–100.

Chiavegatto, S., Dawson, V. L., Mamounas, L. A., Koliatsos, V. E., Dawson, T. M., and Nelson, R. J. 2001. Brain serotonin dysfunction accounts for aggression in male mice lacking neuronal nitric oxide synthase. *Proc Natl Acad Sci U S A* 98 (3):1277–81.

Chrapko, W. E., Jurasz, P., Radomski, M. W., Lara, N., Archer, S. L., and Le Melledo, J. M. 2004. Decreased platelet nitric oxide synthase activity and plasma nitric oxide metabolites in major depressive disorder. *Biol Psychiatry* 56 (2):129–34.

Chrapko, W., Jurasz, P., Radomski, M. W., et al. 2006. Alteration of decreased plasma NO metabolites and platelet NO synthase activity by paroxetine in depressed patients. *Neuropsychopharmacology* 31 (6):1286–93.

Christopherson, K. S., Hillier, B. J., Lim, W. A., and Bredt, D. S. 1999. PSD-95 assembles a ternary complex with the N-methyl-D-aspartic acid receptor and a bivalent neuronal NO synthase PDZ domain. *J Biol Chem* 274 (39):27467–73.

Cobb, B. L., Ryan, K. L., Frei, M. R., Guel-Gomez, V., and Mickley, G. A. 1995. Chronic administration of L-NAME in drinking water alters working memory in rats. *Brain Res Bull* 38 (2):203–7.

Contestabile, A. 2000. Roles of NMDA receptor activity and nitric oxide production in brain development. *Brain Res Rev* 32 (2–3):476–509.

Corbett, J. A., and McDaniel, M. L. 1996. The use of aminoguanidine, a selective iNOS inhibitor, to evaluate the role of nitric oxide in the development of autoimmune diabetes. *Methods* 10 (1):21–30.

Costa, A., Trainer, P., Besser, M., and Grossman, A. 1993. Nitric oxide modulates the release of corticotropin-releasing hormone from the rat hypothalamus in vitro. *Brain Res* 605 (2):187–92.

Cui, H., Hayashi, A., Sun, H. S., et al. 2007. PDZ protein interactions underlying NMDA receptor-mediated excitotoxicity and neuroprotection by PSD-95 inhibitors. *J Neurosci* 27 (37):9901–15.

da Silva, G. D., Matteussi, A. S., dos Santos, A. R., Calixto, J. B., and Rodrigues, A. L. 2000. Evidence for dual effects of nitric oxide in the forced swimming test and in the tail suspension test in mice. *Neuroreport* 11 (17):3699–702.

Das, I., Khan, N. S., Puri, B. K., and Hirsch, S. R. 1996. Elevated endogenous nitric oxide synthase inhibitor in schizophrenic plasma may reflect abnormalities in brain nitric oxide production. *Neurosci Lett* 215 (3):209–11.

Dawson, T. M., and Dawson, V. L. 1995. ADP-ribosylation as a mechanism for the action of nitric oxide in the nervous system. *New Horiz* 3 (1):85–92.

Dawson, T. M., and Snyder, S. H. 1994. Gases as biological messengers: Nitric oxide and carbon monoxide in the brain. *J Neurosci* 14 (9):5147–59.

Dawson, T. M., Sasaki, M., Gonzalez-Zulueta, M., and Dawson, V. L. 1998. Regulation of neuronal nitric oxide synthase and identification of novel nitric oxide signaling pathways. *Prog Brain Res* 118:3–11.

De Oliveira, R. M. W., Deakin, J. F., and Guimaraes, F. S. 2000. Neuronal nitric oxide synthase (NOS) expression in the hippocampal formation of patients with schizophrenia and affective disorder. *J Psychopharmacol* 14 (Suppl):1–8.

de Vente, J., Hopkins, D. A., Markerink-Van, I. M., Emson, P. C., Schmidt, H. H., and Steinbusch, H. W. 1998. Distribution of nitric oxide synthase and nitric oxide-receptive, cyclic GMP–producing structures in the rat brain. *Neuroscience* 87 (1):207–41.

Denninger, J. W., and Marletta, M. A. 1999. Guanylate cyclase and the NO/cGMP signaling pathway. *Biochim Biophys Acta* 1411 (2–3):334–50.

Dhir, A., and Kulkarni, S. K. 2007. Involvement of nitric oxide (NO) signaling pathway in the antidepressant action of bupropion, a dopamine reuptake inhibitor. *Eur J Pharmacol* 568 (1–3):177–85.

Duman, R. S., Malberg, J., and Nakagawa, S. 2001a. Regulation of adult neurogenesis by psychotropic drugs and stress. *J Pharmacol Exp Ther* 299 (2):401–7.

Duman, R. S., Nakagawa, S., and Malberg, J. 2001b. Regulation of adult neurogenesis by antidepressant treatment. *Neuropsychopharmacology* 25 (6):836–44.

Ehninger, D., and Kempermann, G. 2008. Neurogenesis in the adult hippocampus. *Cell Tissue Res* 331 (1):243–50.

Ehringer, H., Hornykiewicz, O., and Lechner, K. 1961. Die Wirkung von Methylenblau auf die Monoaminoxydase und den Katecholamin-und 5-Hydroxytryptaminstoffwechsel des Gehirnes. *Naunyn Schmiedebergs Arch Exp Pathol Pharmakol* 241:568–82.

Ergun, Y., and Ergun, U. G. 2007. Prevention of pro-depressant effect of L-arginine in the forced swim test by NG-nitro-L-arginine and [1H-[1,2,4]oxadiazole[4,3-a]quinoxalin-1-one]. *Eur J Pharmacol* 554 (2–3):150–4.

Eroglu, L., and Caglayan, B. 1997. Anxiolytic and antidepressant properties of methylene blue in animal models. *Pharmacol Res* 36 (5):381–5.

Estall, L. B., Grant, S. J., and Cicala, G. A. 1993. Inhibition of nitric oxide (NO) production selectively impairs learning and memory in the rat. *Pharmacol Biochem Behav* 46 (4):959–62.

Feyissa, A. M., Chandran, A., Stockmeier, C. A., and Karolewicz, B. 2009. Reduced levels of NR2A and NR2B subunits of NMDA receptor and PAD-95 in the prefrontal cortex in major depression. *Prog Neuropsychopharmacol Biol Psychiatry* 33:70–5.

Finkel, M. S., Laghrissi-Thode, F., Pollock, B. G., and Rong, J. 1996. Paroxetine is a novel nitric oxide synthase inhibitor. *Psychopharmacol Bull* 32 (4):653–8.

Forstermann, U., Schmidt, H. H., Pollock, J. S., et al. 1991. Isoforms of nitric oxide synthase. Characterization and purification from different cell types. *Biochem Pharmacol* 42 (10):1849–57.

Forstermann, U., Gath, I., Schwarz, P., Closs, E. I., and Kleinert, H. 1995. Isoforms of nitric oxide synthase. Properties, cellular distribution and expressional control. *Biochem Pharmacol* 50 (9):1321–32.

Fossier, P., Blanchard, B., Ducrocq, C., Leprince, C., Tauc, L., and Baux, G. 1999. Nitric oxide transforms serotonin into an inactive form and this affects neuromodulation. *Neuroscience* 93 (2):597–603.

Frey, C., Narayanan, K., McMillan, K., Spack, L., Gross, S. S., Masters, B. S., and Griffith, O. W. 1994. L-Thiocitrulline. A stereospecific, heme-binding inhibitor of nitric-oxide synthases. *J Biol Chem* 269 (42):26083–91.

Gadek-Michalska, A., and Bugajski, J. 2008. Nitric oxide in the adrenergic- and CRH-induced activation of hypothalamic–pituitary–adrenal axis. *J Physiol Pharmacol* 59 (2):365–78.

Galecki, P., Maes, M., Florkowski, A., et al. 2011. Association between inducible and neuronal nitric oxide synthase polymorphisms and recurrent depressive disorder. *J Affect Disord* 129(1–3):175–82.

Galimberti, D., Scarpini, E., Venturelli, E., et al. 2008. Association of a NOS1 promoter repeat with Alzheimer's disease. *Neurobiol Aging* 29 (9):1359–65.

Gardiner, S. M., Kemp, P. A., March, J. E., and Bennett, T. 1996. Influence of aminoguanidine and the endothelin antagonist, sb 209670, on the regional haemodynamic effects of endotoxaemia in conscious rats. *Br J Pharmacol* 118 (7):1822–8.

Garthwaite, J., and Boulton, C. L. 1995. Nitric oxide signaling in the central nervous system. *Annu Rev Physiol* 57:683–706.

Garthwaite, J., Charles, S. L., and Chess-Williams, R. 1988. Endothelium-derived relaxing factor release on activation of NMDA receptors suggests role as intercellular messenger in the brain. *Nature* 336 (6197):385–8.

Garthwaite, J., Garthwaite, G., Palmer, R. M., and Moncada, S. 1989. NMDA receptor activation induces nitric oxide synthesis from arginine in rat brain slices. *Eur J Pharmacol* 172 (4–5):413–16.

Gaston, B. M., Carver, J., Doctor, A., and Palmer, L. A. 2003. S-nitrosylation signaling in cell biology. *Mol Interv* 3 (5):253–63.

Ghasemi, M., Sadeghipour, H., Mosleh, A., Sadeghipour, H. R., Mani, A. R., and Dehpour, A. R. 2008. Nitric oxide involvement in the antidepressant-like effects of acute lithium administration in the mouse forced swimming test. *Eur Neuropsychopharmacol* 18 (5):323–32.

Gilhotra, N., and Dhingra, D. 2009. Involvement of NO-cGMP pathway in anti-anxiety effect of aminoguanidine in stressed mice. *Prog Neuropsychopharmacol Biol Psychiatry* 33(8):1502–7.

Gillman, P. K. 2008. Methylene blue is a potent monoamine oxidase inhibitor. *Can J Anaesth* 55 (5):311–12.

Givalois, L., Li, S., and Pelletier, G. 2002. Central nitric oxide regulation of the hypothalamic–pituitary–adrenocortical axis in adult male rats. *Mol Brain Res* 102 (1–2):1–8.

Griffiths, M. J., Messent, M., MacAllister, R. J., and Evans, T. W. 1993. Aminoguanidine selectively inhibits inducible nitric oxide synthase. *Br J Pharmacol* 110 (3):963–8.

Griffiths, M. J., Messent, M., Curzen, N. P., and Evans, T. W. 1995. Aminoguanidine selectively decreases cyclic GMP levels produced by inducible nitric oxide synthase. *Am J Respir Crit Care Med* 152 (5 Pt 1):1599–604.

Griffith, O. W., and Kilbourn, R. G. 1996. Nitric oxide synthase inhibitors: Amino acids. *Methods Enzymol* 268:375–92.

Halaris, A., and Piletz, J. E. 2001. Imidazoline receptors: Possible involvement in the pathophysiology and treatment of depression. *Hum Psychopharmacol Clin Exp* 16 (1):65–9.

Halaris, A., and Piletz, J. E. 2003. Relevance of imidazoline receptors and agmatine to psychiatry: A decade of progress. *Ann N Y Acad Sci* 1009:1–20.

Halaris, A., Zhu, H., Feng, Y., and Piletz, J. E. 1999. Plasma agmatine and platelet imidazoline receptors in depression. *Ann N Y Acad Sci* 881:445–51.

Handy, R. L., Harb, H. L., Wallace, P., Gaffen, Z., Whitehead, K. J., and Moore, P. K. 1996. Inhibition of nitric oxide synthase by 1-(2-trifluoromethylphenyl) imidazole (TRIM) in vitro: Antinociceptive and cardiovascular effects. *Br J Pharmacol* 119 (2):423–31.

Handy, R. L., and Moore, P. K. 1997. Mechanism of the inhibition of neuronal nitric oxide synthase by 1-(2- trifluoromethylphenyl) imidazole (TRIM). *Life Sci* 60 (25):L389–94.

Handy, R. L., Wallace, P., Gaffen, Z. A., Whitehead, K. J., and Moore, P. K. 1995. The antinociceptive effect of 1-(2-trifluoromethylphenyl) imidazole (TRIM), a potent inhibitor of neuronal nitric oxide synthase in vitro, in the mouse. *Br J Pharmacol* 116 (5):2349–50.

Hara, H., Waeber, C., Huang, P. L., Fujii, M., Fishman, M. C., and Moskowitz, M. A. 1996. Brain distribution of nitric oxide synthase in neuronal or endothelial nitric oxide synthase mutant mice using [³H]L-NG-nitro-arginine autoradiography. *Neuroscience* 75 (3):881–90.

Harkin, A. J., Bruce, K. H., Craft, B., and Paul, I. A. 1999. Nitric oxide synthase inhibitors have antidepressant-like properties in mice. 1. Acute treatments are active in the forced swim test. *Eur J Pharmacol* 372 (3):207–13.

Harkin, A., Connor, T. J., Walsh, M., St John, N., and Kelly, J. P. 2003. Serotonergic mediation of the antidepressant-like effects of nitric oxide synthase inhibitors. *Neuropharmacology* 44 (5):616–23.

Harkin, A., Connor, T. J., Burns, M. P., and Kelly, J. P. 2004. Nitric oxide synthase inhibitors augment the effects of serotonin re-uptake inhibitors in the forced swimming test. *Eur Neuropsychopharmacol* 14 (4):274–81.

Harvey, B. H., Oosthuizen, F., Brand, L., Wegener, G., and Stein, D. J. 2004. Stress–restress evokes sustained iNOS activity and altered GABA levels and NMDA receptors in rat hippocampus. *Psychopharmacology* 175 (4):494–502.

Harvey, B. H., Bothma, T., Nel, A., Wegener, G., and Stein, D. J. 2005. Involvement of the NMDA receptor, NO-cyclic GMP and nuclear factor k-beta in an animal model of repeated trauma. *Hum Psychopharmacol Clin Exp* 20 (5):367–73.

Hasan, K., Heesen, B. J., Corbett, J. A., et al. 1993. Inhibition of nitric oxide formation by guanidines. *Eur J Pharmacol* 249 (1):101–6.

Heiberg, I. L., Wegener, G., and Rosenberg, R. 2002. Reduction of cGMP and nitric oxide has antidepressant-like effects in the forced swimming test in rats. *Behav Brain Res* 134 (1–2):479–84.

Herken, H., Gurel, A., Selek, S., et al. 2007. Adenosine deaminase, nitric oxide, superoxide dismutase, and xanthine oxidase in patients with major depression: Impact of antidepressant treatment. *Arch Med Res* 38 (2):247–52.

Herrera-Guzman, I., Gudayol-Ferre, E., Herrera-Guzman, D., Guardia-Olmos, J., Hinojosa-Calvo, E., and Herrera-Abarca, J. F. 2009. Effects of selective serotonin reuptake and dual serotonergic-noradrenergic reuptake treatments on memory and mental processing speed in patients with major depressive disorder. *J Psychiatr Res* 43 (9):855–63.

Herzberg, L., and Herzberg, B. 1977. Mood change and magnesium. A possible interaction between magnesium and lithium? *J Nerv Ment Dis* 165 (6):423–6.

Hess, D. T., Patterson, S. I., Smith, D. S., and Skene, J. H. 1993. Neuronal growth cone collapse and inhibition of protein fatty acylation by nitric oxide. *Nature* 366 (6455):562–5.

Hibbs, J. B. Jr., Taintor, R. R., and Vavrin, Z. 1987. Macrophage cytoxicity: Role for L-arginine deiminase and imino nitrogen oxidation to nitrite. *Science* 235 (4787):473–6.

Hindley, S., Juurlink, B. H., Gysbers, J. W., Middlemiss, P. J., Herman, M. A., and Rathbone, M. P. 1997. Nitric oxide donors enhance neurotrophin-induced neurite outgrowth through a cGMP-dependent mechanism. *J Neurosci Res* 47 (4):427–39.

Hojgaard Rasmussen, H., Bo Mortensen, P., and Jensen, I. W. 1989. Depression and magnesium deficiency. *Int J Psychiatry Med* 19 (1):57–63.

Holderbach, R., Clark, K., Moreau, J. L., Bischofberger, J., and Normann, C. 2007. Enhanced long-term synaptic depression in an animal model of depression. *Biol Psychiatry* 62 (1):92–100.

Holscher, C. 1994. 7-Nitro indazole, a neuron-specific nitric oxide synthase inhibitor, produces amnesia in the chick. *Learn Mem* 1 (4):213–16.

Holscher, C., McGlinchey, L., Anwyl, R., and Rowan, M. J. 1996. 7-Nitro indazole, a selective neuronal nitric oxide synthase inhibitor in vivo, impairs spatial learning in the rat. *Learn Mem* 2 (6):267–78.

Holstad, M., Jansson, L., and Sandler, S. 1996. Effects of aminoguanidine on rat pancreatic islets in culture and on the pancreatic islet blood flow of anaesthetized rats. *Biochem Pharmacol* 51 (12):1711–17.

Hua, Y., Huang, X. Y., Zhou, L., et al. 2008. DETA/NONOate, a nitric oxide donor, produces antidepressant effects by promoting hippocampal neurogenesis. *Psychopharmacology* 200 (2):231–42.

Inan, S. Y., Yalcin, I., and Aksu, F. 2004. Dual effects of nitric oxide in the mouse forced swimming test: Possible contribution of nitric oxide–mediated serotonin release and potassium channel modulation. *Pharmacol Biochem Behav* 77 (3):457–64.

Ishibashi, Y., Shimada, T., Murakami, Y., et al. 2001. An inhibitor of inducible nitric oxide synthase decreases forearm blood flow in patients with congestive heart failure. *J Am Coll Cardiol* 38 (5):1470–6.

Jakubovic, A., and Necina, J. 1963. The effect of methylene blue on the monoamine oxidase activity of the liver and brain of rats after various routes of administration. *Arzneimittelforschung* 13:134–6.

Jankord, R., McAllister, R. M., Ganjam, V. K., and Laughlin, M. H. 2009. Chronic inhibition of nitric oxide synthase augments the ACTH response to exercise. *Am J Physiol Regul Integr Comp Physiol* 296 (3):R728–34.

Jefferys, D., and Funder, J. 1996. Nitric oxide modulates retention of immobility in the forced swimming test in rats. *Eur J Pharmacol* 295 (2–3):131–5.

Jesse, C. R., Bortolatto, C. F., Savegnago, L., Rocha, J. B., and Nogueira, C. W. 2008. Involvement of L-arginine–nitric oxide–cyclic guanosine monophosphate pathway in the antidepressant-like effect of tramadol in the rat forced swimming test. *Prog Neuropsychopharmacol Biol Psychiatry* 32 (8):1838–43.

Joca, S. R., and Guimaraes, F. S. 2006. Inhibition of neuronal nitric oxide synthase in the rat hippocampus induces antidepressant-like effects. *Psychopharmacology (Berl)* 185 (3):298–305.

Johnson, M. D., and Ma, P. M. 1993. Localization of NADPH diaphorase activity in monoaminergic neurons of the rat brain. *J Comp Neurol* 332 (4):391–406.

Juch, M., Smalla, K. H., Kahne, T., et al. 2009. Congenital lack of nNOS impairs long-term social recognition memory and alters the olfactory bulb proteome. *Neurobiol Learn Mem* 92 (4):469–84.

Kaehler, S. T., Singewald, N., Sinner, C., and Philippu, A. 1999. Nitric oxide modulates the release of serotonin in the rat hypothalamus. *Brain Res* 835 (2):346–9.

Karolewicz, B., Paul, I. A., and Antkiewicz-Michaluk, L. 2001. Effect of NOS inhibitor on forced swim test and neurotransmitters turnover in the mouse brain. *Pol J Pharmacol* 53 (6):587–96.

Karolewicz, B., Szebeni, K., Stockmeier, C. A., et al. 2004. Low nNOS protein in the locus coeruleus in major depression. *J Neurochem* 91 (5):1057–66.

Kendler, K. S., Thornton, L. M., and Gardner, C. O. 2000. Stressful life events and previous episodes in the etiology of major depression in women: An evaluation of the 'kindling' hypothesis. *Am J Psychiatry* 157 (8):1243–51.

Kerwin, J. F. Jr., and Heller, M. 1994. The arginine–nitric oxide pathway: A target for new drugs. *Med Res Rev* 14 (1):23–74.

Kim, W. K., Choi, Y. B., Rayudu, P. V., Das, P., Asaad, W., Arnelle, D. R., Stamler, J. S., and Lipton, S. A. 1999. Attenuation of NMDA receptor activity and neurotoxicity by nitroxyl anion, NO⁻. *Neuron* 24 (2):461–9.

Kim, Y. K., Paik, J. W., Lee, S. W., Yoon, D., Han, C., and Lee, B. H. 2006. Increased plasma nitric oxide level associated with suicide attempt in depressive patients. *Prog Neuropsychopharmacol Biol Psychiatry* 30 (6):1091–6.

Kiss, J. P., and Vizi, E. S. 2001. Nitric oxide: A novel link between synaptic and nonsynaptic transmission. *Trends Neurosci* 24 (4):211–15.

Knott, A. B., and Bossy-Wetzel, E. 2009. Nitric oxide in health and disease of the nervous system. *Antioxid Redox Signal* 11 (3):541–54.

Knowles, R. G., and Moncada, S. 1994. Nitric oxide synthases in mammals. *Biochem J* 298 (Pt 2):249–58.

Kowall, N. W., Ferrante, R. J., Beal, M. F., et al. 1987. Neuropeptide Y, somatostatin, and reduced nicotinamide adenine dinucleotide phosphate diaphorase in the human striatum: A combined immunocytochemical and enzyme histochemical study. *Neuroscience* 20 (3):817–28.

Krass, M., Wegener, G., Vasar, E., and Volke, V. 2008. Antidepressant-like effect of agmatine is not mediated by serotonin. *Behav Brain Res* 188 (2):324–8.

Krass, M., Wegener, G., Vasar, E., and Volke, V. 2011. The antidepressant action of imipramine and venlafaxine involves suppression of nitric oxide synthesis. *Behav Brain Res* 218 (1):57–63.

Kuhn, D. M., and Arthur, R. E. Jr. 1996. Inactivation of brain tryptophan hydroxylase by nitric oxide. *J Neurochem* 67 (3):1072–7.

Kuhn, D. M., and Arthur, R. Jr. 1997. Molecular mechanism of the inactivation of tryptophan hydroxylase by nitric oxide: Attack on critical sulfhydryls that spare the enzyme iron center. *J Neurosci* 17 (19):7245–51.

Kurt, M., Bilge, S. S., Aksoz, E., Kukula, O., Celik, S., and Kesim, Y. 2004. Effect of sildenafil on anxiety in the plus-maze test in mice. *Pol J Pharmacol* 56 (3):353–7.

Laitinen, J. T., Laitinen, K. S., Tuomisto, L., and Airaksinen, M. M. 1994. Differential regulation of cyclic GMP levels in the frontal cortex and the cerebellum of anesthetized rats by nitric oxide: An in vivo microdialysis study. *Brain Res* 668 (1–2):117–21.

Lassen, L. H., Ashina, M., Christiansen, I., Ulrich, V., and Olesen, J. 1997. Nitric oxide synthase inhibition in migraine [letter]. *Lancet* 349 (9049):401–2.

Lassen, L. H., Ashina, M., Christiansen, I., et al. 1998. Nitric oxide synthase inhibition: A new principle in the treatment of migraine attacks. *Cephalgia* 18 (1):27–32.

Lee, B. H., Lee, S. W., Yoon, D., et al. 2006. Increased plasma nitric oxide metabolites in suicide attempters. *Neuropsychobiology* 53 (3):127–32.

Lemaire, J. F., and McPherson, P. S. 2006. Binding of vac14 to neuronal nitric oxide synthase: Characterisation of a new internal PDZ-recognition motif. *FEBS Lett* 580 (30):6948–54.

Li, G., Regunathan, S., Barrow, C. J., Eshraghi, J., Cooper, R., and Reis, D. J. 1994. Agmatine: An endogenous clonidine-displacing substance in the brain. *Science* 263 (5149):966–9.

Li, N., Lee, B., Liu, R. J., et al. 2010. MTOR-dependent synapse formation underlies the rapid antidepressant effects of NMDA antagonists. *Science* 329 (5994):959–64.

Li, Y. F., Gong, Z. H., Cao, J. B., Wang, H. L., Luo, Z. P., and Li, J. 2003. Antidepressant-like effect of agmatine and its possible mechanism. *Eur J Pharmacol* 469 (1–3):81–8.

Linden, D. J., Dawson, T. M., and Dawson, V. L. 1995. An evaluation of the nitric oxide/cGMP/cGMP-dependent protein kinase cascade in the induction of cerebellar long-term depression in culture. *J Neurosci* 15 (7 Pt 2):5098–105.

Lonart, G., and Johnson, K. M. 1994. Inhibitory effects of nitric oxide on the uptake of [³H]dopamine and [³H] glutamate by striatal synaptosomes. *J Neurochem* 63 (6):2108–17.

Lonart, G., Cassels, K. L., and Johnson, K. M. 1993. Nitric oxide induces calcium-dependent [3H]dopamine release from striatal slices. *J Neurosci Res* 35 (2):192–8.

Lorrain, D. S., and Hull, E. M. 1993. Nitric oxide increases dopamine and serotonin release in the medial pre-optic area. *Neuroreport* 5 (1):87–9.

Lowenstein, C. J., and Snyder, S. H. 1992. Nitric oxide, a novel biologic messenger. *Cell* 70 (5):705–7.

Lowy, M. T., Wittenberg, L., and Yamamoto, B. K. 1995. Effect of acute stress on hippocampal glutamate levels and spectrin proteolysis in young and aged rats. *J Neurochem* 65 (1):268–74.

Luo, D., and Vincent, S. R. 1994. NMDA-dependent nitric oxide release in the hippocampus in vivo: Inter-actions with noradrenaline. *Neuropharmacology* 33 (11):1345–50.

Machado-Vieira, R., Salvadore, G., Diazgranados, N., and Zarate, C. A. Jr. 2009. Ketamine and the next gen-eration of antidepressants with a rapid onset of action. *Pharmacol Ther* 123 (2):143–50.

Madison, D. V., Malenka, R. C., and Nicoll, R. A. 1991. Mechanisms underlying long-term potentiation of synaptic transmission. *Annu Rev Neurosci* 14:379–97.

Madrigal, J. L., Moro, M. A., Lizasoain, I., et al. 2001. Inducible nitric oxide synthase expression in brain cor-tex after acute restraint stress is regulated by nuclear factor kappaB-mediated mechanisms. *J Neurochem* 76 (2):532–8.

Madrigal, J. L. M., Moro, M. A., Lizasoain, I., Lorenzo, P., and Leza, J. C. 2002. Stress-induced increase in extracellular sucrose space in rats is mediated by nitric oxide. *Brain Res* 938 (1–2):87–91.

Maeng, S., and Zarate, C. A. Jr. 2007. The role of glutamate in mood disorders: Results from the ketamine in major depression study and the presumed cellular mechanism underlying its antidepressant effects. *Curr Psychiatry Rep* 9 (6):467–74.

Malenka, R. C. 1994. Synaptic plasticity in the hippocampus: LTP and LTD. *Cell* 78 (4):535–8.

Marletta, M. A. 1993. Nitric oxide synthase structure and mechanism. *J Biol Chem* 268 (17):12231–4.

Matarredona, E. R., Murillo-Carretero, M., Moreno-López, B., and Estrada, C. 2004. Nitric oxide synthesis inhibition increases proliferation of neural precursors isolated from the postnatal mouse subventricular zone. *Brain Res* 995 (2):274–84.

Mayer, B., Brunner, F., and Schmidt, K. 1993a. Inhibition of nitric oxide synthesis by methylene blue. *Biochem Pharmacol* 45 (2):367–74.

Mayer, B., Brunner, F., and Schmidt, K. 1993b. Novel actions of methylene blue. *Eur Heart J* 14 (Suppl I): 22–6.

Mayer, B., Klatt, P., Werner, E. R., and Schmidt, K. 1994. Molecular mechanisms of inhibition of porcine brain nitric oxide synthase by the antinociceptive drug 7-nitro-indazole. *Neuropharmacology* 33 (11):1253–9.

Meffert, M. K., Premack, B. A., and Schulman, H. 1994. Nitric oxide stimulates Ca(2+)-independent synaptic vesicle release. *Neuron* 12 (6):1235–44.

Meffert, M. K., Calakos, N. C., Scheller, R. H., and Schulman, H. 1996. Nitric oxide modulates synaptic vesicle docking fusion reactions. *Neuron* 16 (6):1229–36.

Michael-Titus, A. T., Bains, S., Jeetle, J., and Whelpton, R. 2000. Imipramine and phenelzine decrease gluta-mate overflow in the prefrontal cortex — A possible mechanism of neuroprotection in major depression? *Neuroscience* 100 (4):681–4.

Miki, N., Kawabe, Y., and Kuriyama, K. 1977. Activation of cerebral guanylate cyclase by nitric oxide. *Biochem Biophys Res Commun* 75 (4):851–6.

Mizuno, M., Yamada, K., Olariu, A., Nawa, H., and Nabeshima, T. 2000. Involvement of brain-derived neu-rotrophic factor in spatial memory formation and maintenance in a radial arm maze test in rats. *J Neurosci* 20 (18):7116–21.

Mogensen, J., Wortwein, G., Gustafson, B., and Ermens, P. 1995a. ʟ-Nitroarginine reduces hippocampal medi-ation of place learning in the rat. *Neurobiol Learn Mem* 64 (1):17–24.

Mogensen, J., Wortwein, G., Hasman, A., Nielsen, P., and Wang, Q. 1995b. Functional and neurochemical pro-file of place learning after ʟ-nitro-arginine in the rat. *Neurobiol Learn Mem* 63 (1):54–65.

Moncada, S., Palmer, R. M., and Higgs, E. A. 1989. Biosynthesis of nitric oxide from L-arginine. A pathway for the regulation of cell function and communication. *Biochem Pharmacol* 38 (11):1709–15.

Moore, P. K., Babbedge, R. C., Wallace, P., Gaffen, Z. A., and Hart, S. L. 1993. 7-nitro indazole, an inhibitor of nitric oxide synthase, exhibits anti-nociceptive activity in the mouse without increasing blood pressure. *Br J Pharmacol* 108 (2):296–7.

Mori, K., Togashi, H., Ueno, K. I., Matsumoto, M., and Yoshioka, M. 2001. Aminoguanidine prevented the impairment of learning behavior and hippocampal long-term potentiation following transient cerebral ischemia. *Behav Brain Res* 120 (2):159–68.

Mungrue, I. N., and Bredt, D. S. 2004. nNOS at a glance: Implications for brain and brawn. *J Cell Sci* 117 (Pt 13):2627–9.

Murad, F., Mittal, C. K., Arnold, W. P., Katsuki, S., and Kimura, H. 1978. Guanylate cyclase: Activation by azide, nitro compounds, nitric oxide, and hydroxyl radical and inhibition by hemoglobin and myoglobin. *Adv Cyclic Nucleotide Res* 9:145–58.

Murck, H. 2002. Magnesium and affective disorders. *Nutr Neurosci* 5 (6):375–89.

Musazzi, L., Milanese, M., Farisello, P., et al. 2010. Acute stress increases depolarization-evoked glutamate release in the rat prefrontal/frontal cortex: The dampening action of antidepressants. *PLoS One* 5 (1): e8566.

Narsapur, S. L., and Naylor, G. J. 1983. Methylene blue. A possible treatment for manic depressive psychosis. *J Affect Disord* 5 (2):155–61.

Nathan, C. 1992. Nitric oxide as a secretory product of mammalian cells. *FASEB J* 6 (12):3051–64.

Nathan, C., and Xie Q. W. 1994. Regulation of biosynthesis of nitric oxide. *J Biol Chem* 269 (19):13725–8.

Naudon, L., Hotte, M., and Jay, T. M. 2007. Effects of acute and chronic antidepressant treatments on memory performance: A comparison between paroxetine and imipramine. *Psychopharmacology (Berl)* 191 (2):353–64.

Naylor, G. G., and Smith, A. H. 1982. Reduction of vanadate, a possible explanation of the effect of phenothiazines in manic-depressive psychosis. *Lancet* 1 (8268):395–6.

Naylor, G. J. 1984. Vanadium and manic depressive psychosis. *Nutr Health* 3 (1–2):79–85.

Naylor, G. J., Dick, D. A., Johnston, B. B., et al. 1981. Possible explanation for therapeutic action of lithium, and a possible substitute (methylene-blue) [letter]. *Lancet* 2 (8256):1175–6.

Naylor, G. J., Martin, B., Hopwood, S. E., and Watson, Y. 1986. A two-year double-blind crossover trial of the prophylactic effect of methylene blue in manic-depressive psychosis. *Biol Psychiatry* 21 (10):915–20.

Naylor, G. J., Smith, A. H., and Connelly, P. 1987. A controlled trial of methylene blue in severe depressive illness. *Biol Psychiatry* 22 (5):657–9.

Naylor, G. J., Smith, A. H., and Connelly P. 1988. Methylene blue in mania [letter]. *Biol Psychiatry* 24 (8):941–2.

Nelson, R. J. 2005. Effects of nitric oxide on the HPA axis and aggression. In *Molecular Mechanisms Influencing Aggressive Behaviours*. Novartis Foundation Symposium 268: 147–70.

Nelson, R. J., Demas, G. E., Huang, P. L., et al. 1995. Behavioural abnormalities in male mice lacking neuronal nitric oxide synthase. *Nature* 378 (6555):383–6.

Nelson, R. J., Trainor, B. C., Chiavegatto, S., and Demas, G. E. 2006. Pleiotropic contributions of nitric oxide to aggressive behavior. *Neurosci Biobehav Rev* 30 (3):346–55.

Normann, C., Schmitz, D., Furmaier, A., Doing, C., and Bach, M. 2007. Long-term plasticity of visually evoked potentials in humans is altered in major depression. *Biol Psychiatry* 62 (5):373–80.

Nudmamud-Thanoi, S., and Reynolds, G. P. 2004. The NR1 subunit of the glutamate/NMDA receptor in the superior temporal cortex in schizophrenia and affective disorders. *Neurosci Lett* 372 (1–2):173–7.

Nylén, A., Skagerberg, G., Alm, P., Larsson, B., Holmqvist, B., and Andersson, K. E. 2001. Nitric oxide synthase in the hypothalamic paraventricular nucleus of the female rat; organization of spinal projections and coexistence with oxytocin or vasopressin. *Brain Res* 908 (1):10–24.

O'Dell, T. J., Hawkins, R. D., Kandel, E. R., and Arancio, O. 1991. Tests of the roles of two diffusible substances in long-term potentiation: Evidence for nitric oxide as a possible early retrograde messenger. *Proc Natl Acad Sci U S A* 88 (24):11285–9.

O'Dell, T. J., Huang, P. L., Dawson, T. M., et al. 1994. Endothelial NOS and the blockade of LTP by NOS inhibitors in mice lacking neuronal NOS. *Science* 265 (5171):542–6.

Ogawa, T., Kimoto, M., Watanabe, H., and Sasaoka, K. 1987. Metabolism of N^G,N^G- and N^G,N'^G-dimethylarginine in rats. *Arch Biochem Biophys* 252 (2):526–37.

Ohno, M., Yamamoto, T., and Watanabe, S. 1993. Deficits in working memory following inhibition of hippocampal nitric oxide synthesis in the rat. *Brain Res* 632 (1–2):36–40.

Ohno, M., Yamamoto, T., and Watanabe, S. 1994. Intrahippocampal administration of the NO synthase inhibitor L-NAME prevents working memory deficits in rats exposed to transient cerebral ischemia. *Brain Res* 634 (1):173–7.

Okada, D., Yap, C. C., Kojima, H., Kikuchi, K., and Nagano, T. 2004. Distinct glutamate receptors govern differential levels of nitric oxide production in a layer-specific manner in the rat cerebellar cortex. *Neuroscience* 125 (2):461–72.

Okumura, T., Kishi, T., Okochi, T., et al. 2010. Genetic association analysis of functional polymorphisms in neuronal nitric oxide synthase 1 gene (NOS1) and mood disorders and fluvoxamine response in major depressive disorder in the Japanese population. *Neuropsychobiology* 61 (2):57–63.

Olesen, J., and Leonardi, M. 2003. The burden of brain diseases in Europe. *Eur J Neurol* 10 (5):471–7.

Olesen, J., Sobscki, P., Truelsen, T., Sestoft, D., and Jonsson, B. 2008. Cost of disorders of the brain in Denmark. *Nord J Psychiatry* 62 (2):114–20.

Oliveira, R. M., Guimaraes, F. S., and Deakin, J. F. 2008. Expression of neuronal nitric oxide synthase in the hippocampal formation in affective disorders. *Braz J Med Biol Res* 41 (4):333–41.

Olivenza, R., Moro, M. A., Lizasoain, I., et al. 2000. Chronic stress induces the expression of inducible nitric oxide synthase in rat brain cortex. *J Neurochem* 74 (2):785–91.

Olmos, G., De Gregorio-Rocasolano, N., Paz Regalado, M., et al. 1999. Protection by imidazol(ine) drugs and agmatine of glutamate-induced neurotoxicity in cultured cerebellar granule cells through blockade of NMDA receptor. *Br J Pharmacol* 127 (6):1317–26.

Oosthuizen, F., Wegener, G., and Harvey, B. H. 2005. Nitric oxide as inflammatory mediator in post-traumatic stress disorder (PTSD): Evidence from an animal model. *Neuropsychiat Dis Treat* 1 (2):109–23.

Orlando, G. F., Langnaese, K., Schulz, C., Wolf, G., and Engelmann, M. 2008. Neuronal nitric oxide synthase gene inactivation reduces the expression of vasopressin in the hypothalamic paraventricular nucleus and of catecholamine biosynthetic enzymes in the adrenal gland of the mouse. *Stress* 11 (1):42–51.

Pacher, P., Beckman, J. S., and Liaudet, L. 2007. Nitric oxide and peroxynitrite in health and disease. *Physiol Rev* 87 (1):315–424.

Packer, M. A., Stasiv, Y., Benraiss, A., et al. 2003. Nitric oxide negatively regulates mammalian adult neurogenesis. *Proc Natl Acad Sci U S A* 100 (16):9566–71.

Palmer, R. M., Ferrige, A. G., and Moncada, S. 1987. Nitric oxide release accounts for the biological activity of endothelium-derived relaxing factor. *Nature* 327 (6122):524–6.

Papa, M., Pellicano, M. P., and Sadile, A. G. 1994. Nitric oxide and long-term habituation to novelty in the rat. *Ann N Y Acad Sci* 738:316–24.

Pavlinac, D., Langer, R., Lenhard, L., and Deftos, L. 1979. Magnesium in affective disorders. *Biol Psychiatry* 14 (4):657–61.

Plech, A., Klimkiewicz, T., and Maksym, B. 2003. Effect of L-arginine on memory in rats. *Pol J Pharmacol* 55 (6):987–92.

Pogun, S., and Kuhar, M. J. 1994. Regulation of neurotransmitter reuptake by nitric oxide. *Ann N Y Acad Sci* 738:305-15:305–15.

Pogun, S., Baumann, M. H., and Kuhar, M. J. 1994a. Nitric oxide inhibits [^{3}H]dopamine uptake. *Brain Res* 641 (1):83–91.

Pogun, S., Dawson, V., and Kuhar, M. J. 1994b. Nitric oxide inhibits ^{3}H-glutamate transport in synaptosomes. *Synapse* 18 (1):21–6.

Poleszak, E., Wlaź, P., Kedzierska, E., et al. 2007. NMDA/glutamate mechanism of antidepressant-like action of magnesium in forced swim test in mice. *Pharmacol Biochem Behav* 88 (2):158–64.

Prast, H., and Philippu, A. 2001. Nitric oxide as modulator of neuronal function. *Prog Neurobiol* 64 (1):51–68.

Radi, R., and Rubbo, H. 2001. Antioxidant properties of nitric oxide. In *Handbook of Antioxidants*, 2nd edn., ed. E. Cadenas and L. Parker, 689–706. New York: Marcel Dekker Inc.

Radi, R., Beckman, J. S., Bush, K. M., and Freeman, B. A. 1991. Peroxynitrite-induced membrane lipid peroxidation: The cytotoxic potential of superoxide and nitric oxide. *Arch Biochem Biophys* 288 (2):481–7.

Regunathan, S., and Reis, D. J. 1996. Imidazoline receptors and their endogenous ligands. *Annu Rev Pharmacol Toxicol* 36:511–44.

Reif, A., Herterich, S., Strobel, A., et al. 2006a. A neuronal nitric oxide synthase (NOS-I) haplotype associated with schizophrenia modifies prefrontal cortex function. *Mol Psychiatry* 11 (3):286–300.

Reif, A., Strobel, A., Jacob, C. P., et al. 2006b. A NOS-III haplotype that includes functional polymorphisms is associated with bipolar disorder. *Int J Neuropsychopharmacol* 9 (1):13–20.

Reif, A., Jacob, C. P., Rujescu, D., et al. 2009. Influence of functional variant of neuronal nitric oxide synthase on impulsive behaviors in humans. *Arch Gen Psychiatry* 66 (1):41–50.

Rengasamy, A., and Johns, R. A. 1993. Regulation of nitric oxide synthase by nitric oxide. *Mol Pharmacol* 44 (1):124–8.

Reznikov, L. R., Grillo, C. A., Piroli, G. G., Pasumarthi, R. K., Reagan, L. P., and Fadel, J. 2007. Acute stress-mediated increases in extracellular glutamate levels in the rat amygdala: Differential effects of antidepressant treatment. *Eur J Neurosci* 25 (10):3109–14.

Riefler, G. M., and Firestein, B. L. 2001. Binding of neuronal nitric-oxide synthase (nNOS) to carboxyl-terminal-binding protein (CTBP) changes the localization of CTBP from the nucleus to the cytosol: A novel function for targeting by the PDZ domain of nNOS. *J Biol Chem* 276 (51):48262–8.

Rivier, C., and Shen, G. H. 1994. In the rat, endogenous nitric oxide modulates the response of the hypothalamic–pituitary–adrenal axis to interleukin-1β, vasopressin, and oxytocin. *J Neurosci* 14 (4):1985–93.

Romero-Grimaldi, C., Moreno-López, B., and Estrada, C. 2008. Age-dependent effect of nitric oxide on subventricular zone and olfactory bulb neural precursor proliferation. *J Comp Neurol* 506 (2):339–46.

Rubbo, H., Radi, R., Trujillo, M., et al. 1994. Nitric oxide regulation of superoxide and peroxynitrite-dependent lipid peroxidation. Formation of novel nitrogen-containing oxidized lipid derivatives. *J Biol Chem* 269 (42):26066–75.

Rydgren, T., and Sandler, S. 2002. Efficacy of 1400 w, a novel inhibitor of inducible nitric oxide synthase, in preventing interleukin-1beta-induced suppression of pancreatic islet function in vitro and multiple low-dose streptozotocin-induced diabetes in vivo. *Eur J Endocrinol* 147 (4):543–51.

Saitoh, F., Tian, Q. B., Okano, A., Sakagami, H., Kondo, H., and Suzuki, T. 2004. NIDD, a novel DHHC-containing protein, targets neuronal nitric-oxide synthase (nNOS) to the synaptic membrane through a PDZ-dependent interaction and regulates nNOS activity. *J Biol Chem* 279 (28):29461–8.

Salaris, S. C., Babbs, C. F., and Voorhees, W. D. III. 1991. Methylene blue as an inhibitor of superoxide generation by xanthine oxidase. A potential new drug for the attenuation of ischemia/reperfusion injury. *Biochem Pharmacol* 42 (3):499–506.

Salter, M., Duffy, C., Garthwaite, J., and Strijbos, P. J. 1995. Substantial regional and hemispheric differences in brain nitric oxide synthase (NOS) inhibition following intracerebroventricular administration of *N* omega-nitro-L-arginine (L-NA) and its methyl ester (L-NAME). *Neuropharmacology* 34 (6):639–49.

Sanchez, F., Moreno, M. N., Vacas, P., Carretero, J., and Vazquez, R. 1999. Swim stress enhances the NADPH-diaphorase histochemical staining in the paraventricular nucleus of the hypothalamus. *Brain Res* 828 (1–2):159–62.

Santarelli, L., Saxe, M., Gross, C., et al. 2003. Requirement of hippocampal neurogenesis for the behavioral effects of antidepressants. *Science* 301 (5634):805–9.

Sattler, R., Xiong, Z., Lu, W. Y., Hafner, M., MacDonald, J. F., and Tymianski, M. 1999. Specific coupling of NMDA receptor activation to nitric oxide neurotoxicity by PSD-95 protein. *Science* 284 (5421): 1845–8.

Schmidt, H. H., Pollock, J. S., Nakane, M., Forstermann, U., and Murad, F. 1992. Ca^{2+}/calmodulin-regulated nitric oxide synthases. *Cell Calcium* 13 (6–7):427–34.

Schmidt, H. H., Lohmann, S. M., and Walter, U. 1993. The nitric oxide and cGMP signal transduction system: Regulation and mechanism of action. *Biochim Biophys Acta* 1178 (2):153–75.

Schuman, E. M., and Madison, D. V. 1991. A requirement for the intercellular messenger nitric oxide in long-term potentiation. *Science* 254 (5037):1503–6.

Segovia, G., Porras, A., and Mora, F. 1994. Effects of a nitric oxide donor on glutamate and GABA release in striatum and hippocampus of the conscious rat. *Neuroreport* 5 (15):1937–40.

Segovia, G., Del Arco, A., and Mora, F. 1997. Endogenous glutamate increases extracellular concentrations of dopamine, GABA, and taurine through NMDA and AMPA/kainate receptors in striatum of the freely moving rat: A microdialysis study. *J Neurochem* 69 (4):1476–83.

Segovia, G., Del Arco, A., and Mora, F. 1999. Role of glutamate receptors and glutamate transporters in the regulation of the glutamate–glutamine cycle in the awake rat. *Neurochem Res* 24 (6):779–83.

Selley, M. L. 2004. Increased (e)-4-hydroxy-2-nonenal and asymmetric dimethylarginine concentrations and decreased nitric oxide concentrations in the plasma of patients with major depression. *J Affect Disord* 80 (2–3):249–56.

Shariful Islam, A. T. M., Kuraoka, A., and Kawabuchi, M. 2003. Morphological basis of nitric oxide production and its correlation with the polysialylated precursor cells in the dentate gyrus of the adult guinea pig hippocampus. *Anat Sci Int* 78 (2):98–103.

Shimabukuro, M., Ohneda, M., Lee, Y., and Unger, R. H. 1997. Role of nitric oxide in obesity-induced beta cell disease. *J Clin Invest* 100 (2):290–5.

Siepmann, M., Grossmann, J., Muck-Weymann, M., and Kirch W. 2003. Effects of sertraline on autonomic and cognitive functions in healthy volunteers. *Psychopharmacology (Berl)* 168 (3):293–8.

Simonian, S. X., and Herbison, A. E. 1996. Localization of neuronal nitric oxide synthase-immunoreactivity within sub-populations of noradrenergic a1 and a2 neurons in the rat. *Brain Res* 732 (1–2):247–52.

Singewald, N., Sinner, C., Hetzenauer, A., Sartori, S. B., and Murck, H. 2004. Magnesium-deficient diet alters depression- and anxiety-related behavior in mice — Influence of desipramine and *Hypericum perforatum* extract. *Neuropharmacology* 47 (8):1189–97.

Spiacci, A. Jr., Kanamaru, F., Guimaraes, F. S., and Oliveira, R. M. 2008. Nitric oxide–mediated anxiolytic-like and antidepressant-like effects in animal models of anxiety and depression. *Pharmacol Biochem Behav* 88 (3):247–55.

Spolidorio, P. C., Echeverry, M. B., Iyomasa, M., Guimaraes, F. S., and Del Bel, E. A. 2007. Anxiolytic effects induced by inhibition of the nitric oxide-cGMP pathway in the rat dorsal hippocampus. *Psychopharmacology (Berl)* 195 (2):183–92.

Srivastava, N., Barthwal, M. K., Dalal, P. K., et al. 2002. A study on nitric oxide, beta-adrenergic receptors and antioxidant status in the polymorphonuclear leukocytes from the patients of depression. *J Affect Disord* 72 (1):45–52.

Stamler. J. S. 1994. Redox signaling: Nitrosylation and related target interactions of nitric oxide. *Cell* 78 (6):931–6.

Stamler, J. S. 1995. *S*-nitrosothiols and the bioregulatory actions of nitrogen oxides through reactions with thiol groups. *Curr Top Microbiol Immunol* 196:19–36.

Stamler, J. S., Lamas, S., and Fang, F. C. 2001. Nitrosylation. The prototypic redox-based signaling mechanism. *Cell* 106 (6):675–83.

Stanford, S. C., Stanford, B. J., and Gillman, P. K. 2010. Risk of severe serotonin toxicity following co-administration of methylene blue and serotonin reuptake inhibitors: An update on a case report of post-operative delirium. *J Psychopharmacol* 24 (10):1433–8.

Strasser, A., McCarron, R. M., Ishii, H., Stanimirovic, D., and Spatz, M. 1994. L-Arginine induces dopamine release from the striatum in vivo. *Neuroreport* 5 (17):2298–300.

Suarez-Pinzon, W. L., Mabley, J. G., Strynadka, K., Power, R. F., Szabo, C., and Rabinovitch, A. 2001. An inhibitor of inducible nitric oxide synthase and scavenger of peroxynitrite prevents diabetes development in nod mice. *J Autoimmun* 16 (4):449–55.

Sullivan, P. F., de Geus, E. J., Willemsen, G., et al. 2009. Genome-wide association for major depressive disorder: A possible role for the presynaptic protein piccolo. *Mol Psychiatry* 14 (4):359–75.

Suzuki, E., Yagi, G., Nakaki, T., Kanba, S., and Asai, M. 2001. Elevated plasma nitrate levels in depressive states. *J Affect Disord* 63 (1–3):221–4.

Suzuki, E., Yoshida, Y., Shibuya, A., and Miyaoka, H. 2003. Nitric oxide involvement in depression during interferon-alpha therapy. *Int J Neuropsychopharmacol* 6 (4):415–9.

Szewczyk, B., Poleszak, E., Sowa-Kućna, M., et al. 2008. Antidepressant activity of zinc and magnesium in view of the current hypotheses of antidepressant action. *Pharmacol Rep* 60 (5):588–99.

Takano, H., Manchikalapudi, S., Tang, X. L., et al. 1998. Nitric oxide synthase is the mediator of late preconditioning against myocardial infarction in conscious rabbits. *Circulation* 98 (5):441–9.

Taksande, B. G., Kotagale, N. R., Tripathi, S. J., Ugale, R. R., and Chopde, C. T. 2009. Antidepressant like effect of selective serotonin reuptake inhibitors involve modulation of imidazoline receptors by agmatine. *Neuropharmacology* 57 (4):415–24.

Tanda, K., Nishi, A., Matsuo, N., et al. 2009. Abnormal social behavior, hyperactivity, impaired remote spatial memory, and increased d1-mediated dopaminergic signaling in neuronal nitric oxide synthase knockout mice. *Mol Brain* 2 (1):19.

Thomsen, L. L. 1997. Investigations into the role of nitric oxide and the large intracranial arteries in migraine headache. *Cephalgia* 17 (8):873–95.

Thomsen, L. L., and Olesen, J. 1998. Nitric oxide theory of migraine. *Clin Neurosci* 5 (1):28–33.

Tokarski, K., Bobula, B., Wabno, J., and Hess, G. 2008. Repeated administration of imipramine attenuates glutamatergic transmission in rat frontal cortex. *Neuroscience* 153 (3):789–95.

Tsuchiya, T., Kishimoto, J., and Nakayama, Y. 1996. Marked increases in neuronal nitric oxide synthase (nNOS) mRNA and NADPH-diaphorase histostaining in adrenal cortex after immobilization stress in rats. *Psychoneuroendocrinology* 21 (3):287–93.

Vallebuona, F., and Raiteri, M. 1994. Extracellular cGMP in the hippocampus of freely moving rats as an index of nitric oxide (NO) synthase activity. *J Neurosci* 14 (1):134–9.

Valtschanoff, J. G., Weinberg, R. J., Kharazia, V. N., Schmidt, H. H., Nakane, M., and Rustioni, A. 1993. Neurons in rat cerebral cortex that synthesize nitric oxide: NADPH diaphorase histochemistry, NOS immunocytochemistry, and colocalization with GABA. *Neurosci Lett* 157 (2):157–61.

Volke, V., Wegener, G., Vasar, E., and Rosenberg, R. 1999. Methylene blue inhibits hippocampal nitric oxide synthase activity in vivo. *Brain Res* 826 (2):303–5.

Volke, V., Wegener, G., Bourin, M., and Vasar, E. 2003a. Antidepressant- and anxiolytic-like effects of selective neuronal NOS inhibitor 1-(2-trifluoromethylphenyl)-imidazole in mice. *Behav Brain Res* 140 (1–2):141–7.

Volke, V., Wegener, G., and Vasar, E. 2003b. Augmentation of the NO-cGMP cascade induces anxiogenic-like effect in mice. *J Physiol Pharmacol* 54 (4):653–60.

Wallerath, T., Gath, I., Aulitzky, W. E., Pollock, J. S., Kleinert, H., and Förstermann, U. 1997. Identification of the NO synthase isoforms expressed in human neutrophil granulocytes, megakaryocytes and platelets. *Thromb Haemost* 77 (1):163–7.

Wang, D., Yang, X. P., Liu, Y. H., Carretero, O. A., and LaPointe, M. C. 1999. Reduction of myocardial infarct size by inhibition of inducible nitric oxide synthase. *Am J Hypertens* 12 (2 Pt 1):174–82.

Wang, D., An, S. C., and Zhang, X. 2008. Prevention of chronic stress-induced depression-like behavior by inducible nitric oxide inhibitor. *Neurosci Lett* 433 (1):59–64.

Wegener, G., Volke, V., and Rosenberg, R. 2000. Endogenous nitric oxide decreases hippocampal levels of serotonin and dopamine in vivo. *Br J Pharmacol* 130 (3):575–80.

Wegener, G., Volke, V., Bandpey, Z., and Rosenberg, R. 2001. Nitric oxide modulates lithium-induced conditioned taste aversion. *Behav Brain Res* 118 (2):195–200.

Wegener, G., Volke, V., Harvey, B. H., and Rosenberg, R. 2003. Local, but not systemic, administration of serotonergic antidepressants decreases hippocampal nitric oxide synthase activity. *Brain Res* 959 (1):128–34.

Wegener, G., Harvey, B. H., Bonefeld, B., et al. 2010. Increased stress-evoked nitric oxide signalling in the flinders sensitive line (FSL) rat: A genetic animal model of depression. *Int J Neuropsychopharmacol* 13 (4):461–73.

Weitzdoerfer, R., Hoeger, H., Engidawork, E., et al. 2004. Neuronal nitric oxide synthase knock-out mice show impaired cognitive performance. *Nitric Oxide* 10 (3):130–40.

Wiesinger, H. 2001. Arginine metabolism and the synthesis of nitric oxide in the nervous system. *Prog Neurobiol* 64 (4):365–91.

Wiklund, N. P., Cellek, S., Leone, A. M., et al. 1997. Visualisation of nitric oxide released by nerve stimulation. *J Neurosci Res* 47 (2):224–32.

Wiley, J. L., and Willmore, C. B. 2000. Effects of nitric oxide synthase inhibitors on timing and short-term memory in rats. *Behav Pharmacol* 11 (5):421–9.

Xing, G., Chavko, M., Zhang, L. X., Yang, S., and Post, R. M. 2002. Decreased calcium-dependent constitutive nitric oxide synthase (cNOS) activity in prefrontal cortex in schizophrenia and depression. *Schizophr Res* 58 (1):21–30.

Yang, B., Larson, D. F., and Watson, R. R. 2004. Modulation of iNOS activity in age-related cardiac dysfunction. *Life Sci* 75 (6):655–67.

Yang, X. C., and Reis, D. J. 1999. Agmatine selectively blocks the *N*-methyl-D-aspartate subclass of glutamate receptor channels in rat hippocampal neurons. *J Pharmacol Exp Ther* 288 (2):544–9.

Yildiz, F., Erden, B. F., Ulak G., Utkan, T., and Gacar, N. 2000. Antidepressant-like effect of 7-nitroindazole in the forced swimming test in rats. *Psychopharmacology (Berl)* 149 (1):41–4.

Yildiz Akar, F., Ulak, G., Tanyeri, P., Erden, F., Utkan, T., and Gacar, N. 2007. 7-Nitroindazole, a neuronal nitric oxide synthase inhibitor, impairs passive-avoidance and elevated plus-maze memory performance in rats. *Pharmacol Biochem Behav* 87 (4):434–43.

Yildiz Akar, F., Celikyurt, I. K., Ulak, G., and Mutlu, O. 2009. Effects of L-arginine on 7-nitroindazole-induced reference and working memory performance of rats. *Pharmacology* 84 (4):211–18.

Yu, Y. W., Chen, T. J., Wang, Y. C., Liou, Y. J., Hong, C. J., and Tsai, S. J. 2003. Association analysis for neuronal nitric oxide synthase gene polymorphism with major depression and fluoxetine response. *Neuropsychobiology* 47 (3):137–40.

Yun, H. Y., Gonzalez-Zulueta, M., Dawson, V. L., and Dawson, T. M. 1998. Nitric oxide mediates *N*-methyl-D-aspartate receptor-induced activation of p21ras. *Proc Natl Acad Sci U S A* 95 (10):5773–8.

Zarate, C. A. Jr., Singh, J. B., Carlson, P. J., et al. 2006. A randomized trial of an *N*-methyl-D-aspartate antagonist in treatment-resistant major depression. *Arch Gen Psychiatry* 63 (8):856–64.

Zhang, J., Huang, X. Y., Ye, M. L., et al. 2010. Neuronal nitric oxide synthase alteration accounts for the role of 5-HT1A receptor in modulating anxiety-related behaviors. *J Neurosci* 30 (7):2433–41.

Zhang, R., Zhang, L., Zhang, Z., et al. 2001. A nitric oxide donor induces neurogenesis and reduces functional deficits after stroke in rats. *Ann Neurol* 50 (5):602–11.

Zhou, Q. G., Hu, Y., Hua, Y., et al. 2007. Neuronal nitric oxide synthase contributes to chronic stress-induced depression by suppressing hippocampal neurogenesis. *J Neurochem* 103 (5):1843–54.

Zhu, X. J., Hua, Y., Jiang, J., et al. 2006. Neuronal nitric oxide synthase-derived nitric oxide inhibits neurogenesis in the adult dentate gyrus by down-regulating cyclic amp response element binding protein phosphorylation. *Neuroscience* 141 (2):827–36.

Zomkowski, A. D., Hammes, L., Lin, J., Calixto, J. B., Santos, A. R., and Rodrigues, A. L. 2002. Agmatine produces antidepressant-like effects in two models of depression in mice. *Neuroreport* 13 (4):387–91.

Zou, L. B., Yamada, K., Tanaka, T., Kameyama, T., and Nabeshima, T. 1998. Nitric oxide synthase inhibitors impair reference memory formation in a radial arm maze task in rats. *Neuropharmacology* 37 (3):323–30.

18 Beta-Arrestins in Depression: A Molecular Switch from Signal Desensitization to Alternative Intracellular Adaptor Functions

Sofia Avissar and Gabriel Schreiber

CONTENTS

18.1 INTRODUCTION: ARRESTINS

The arrestins constitute a family of proteins that are capable of interacting with G-protein-coupled receptors (GPCRs) after their activation by agonists and subsequent phosphorylation by GPCR kinase (GRKs) (Stephen and Lefkowitz 2002). Arrestins recognize both GRK phosphorylation sites on the receptor and the active conformation of the receptor after agonist binding, which together drive robust arrestin association with the receptors (Luttrell and Lefkowitz 2002). Arrestin binding leads to uncoupling of the receptor from its cognate G protein in such a way that despite the continued activation of the receptor by the agonist, it cannot exchange the guanosine triphosphate group on the G-protein α subunit for guanosine diphosphate (Gainetdinov et al. 2004), eventually causing desensitization of G-protein signaling via downstream second messenger molecules (Stephen and Lefkowitz 2002; Gainetdinov et al. 2004; Lohse et al. 1990; Lefkowitz 2004) (Figure 18.1). Although most of the research regarding the desensitization process has been carried out using the β2-adrenoceptor as a model, it is now clear that this process regulates the function of many GPCRs,

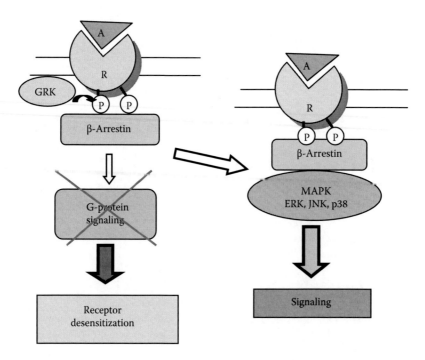

FIGURE 18.1 Classical GPCR desensitization pathway through receptor phosphorylation by GRK, followed by receptor–G-protein uncoupling by β-arrestin, and alternative pathway of β-arrestin-mediated intracellular signaling through MAPK cascade. A = Agonist, R = receptor, P = phosphorylated receptor.

including α- and β-adrenoceptors and muscarinic cholinergic, serotonergic, and dopaminergic receptors (Luttrell and Lefkowitz 2002; Kristen and Lefkowitz 2001). In addition to their role as regulators of GPCR desensitization, β-arrestins also interact with proteins of the endocytic machinery, such as clathrin and the adaptor protein AP-2, and thus promote internalization of receptors via clathrin-coated vesicles (Goodman et al. 1996; Laporte et al. 1999) (Figure 18.1). β-Arrestins are also involved in both receptor downregulation (Gagnon et al. 1998) and resensitization (Oakley et al. 1999; Zhang et al. 1997).

To date, four members of the arrestin gene family have been cloned (Freedman and Lefkowitz 1996). Arrestin-1 and arrestin-4, respectively known also as visual and cone arrestin, are expressed almost exclusively in the retina (Murakami et al. 1993), where they regulate photoreceptor function: rhodopsin and color opsins. By contrast, β-arrestin-1 and β-arrestin-2, known also as arrestin-2 and arrestin-3, respectively, are ubiquitously expressed proteins, regulating GPCRs. High β-arrestin-1 protein and mRNA levels were found in peripheral blood leukocytes. The abundant expression of β-arrestin-1 in peripheral blood leukocytes supports the suggestion of a major role for the GRK/β-arrestin system in regulating receptor-mediated immune functions (Parruti et al. 1993). Over the past decade, there has been a new appreciation regarding the capacity of β-arrestins to act not only as regulators of GPCR desensitization but also as multifunctional adaptor proteins that have the ability to signal through multiple mediators such as mitogen-activated protein kinases (MAPKs), Src, nuclear factor-κB (NF-κB), and phosphoinositide 3-kinase (DeWire et al. 2007; Rajagopal et al. 2010).

18.1.1   β-Arrestins: Mood Disorders and Antidepressants

A substantial body of evidence has accumulated indicating that β-arrestins play a major role in the pathophysiology of mood disorders as well as in the mechanism of action of antidepressants (for review, see Avissar and Schreiber 2006; Schreiber and Avissar 2007; Golan et al. 2009; Schreiber et al. 2009). β-Arrestin-1 levels were significantly elevated by selective serotonin reuptake inhibitors (SSRIs), selective norepinephrine reuptake inhibitors, and nonselective reuptake inhibitor antidepressants in the cortex and hippocampus of rats. This process became significant within 10 days and took 2–3 weeks to reach a maximal increase (Avissar et al. 2004). β-Arrestin-1 protein and mRNA levels in mononuclear leukocytes of untreated patients with major depression were significantly lower than those of healthy subjects, and the reduced levels of β-arrestin-1 protein and mRNA were significantly correlated with the severity of depressive symptomatology (Avissar et al. 2004; Matuzany-Ruban et al. 2005). The low β-arrestin-1 protein and mRNA levels were alleviated by antidepressant treatment, whereby normalization of β-arrestin-1 preceded and thus predicted clinical improvement (Matuzany-Ruban et al. 2005).

In a mouse model of anxiety/depression induced by chronic corticosterone, fluoxetine-reversible decreased expression of β-arrestin-1 and -2 was detected in the hypothalamus. Mice deficient in the gene encoding β-arrestin-2 displayed a reduced response to fluoxetine, suggesting that β-arrestin signaling is necessary for the antidepressant effects of fluoxetine (David et al. 2009). These findings support the implication of β-arrestin-1 in the pathophysiology of major depression and in the mechanism underlying antidepressant-induced receptor downregulation and therapeutic effects. β-Arrestin-1 measurements in patients with depression may potentially serve for biochemical diagnostic purposes and for monitoring and predicting response to antidepressants (Avissar and Schreiber 2006; Schreiber and Avissar 2007; Golan et al. 2009; Schreiber et al. 2009).

18.1.2 β-Arrestin Ubiquitination: Depressive Disorder and Antidepressants

Modification with chains of ubiquitin (Ub) constitutes a primary mechanism by which proteins are targeted for proteasomal degradation (Hershko and Ciechanover 1998). Ubiquitination involves the sequential action of Ub-activating enzyme (E1), Ub-conjugating enzymes (E2s), and Ub protein ligases (E3s) (Hershko and Ciechanover 1998; Pickart 2001).

The predominant role of ubiquitination is to target substrates for rapid degradation within a large protease complex, the 26S proteasome (Hershko and Ciechanover 1998). In order for this to occur, usually the substrate is modified with a polyubiquitin chain consisting of a minimum of four Ub molecules linked through lysine 48 (K48) on Ub itself, which can then be efficiently recognized by specific Ub receptors on the proteasome (Thrower et al. 2000). New discoveries indicate that ubiquitination does not always lead to degradation of the substrate and it can, in some cases, play a much broader role in regulating protein function by proteasome-independent processes (Sun and Chen 2004; Schnell and Hicke 2003; Wang and Dohlman 2006).

Previous studies have demonstrated that ubiquitination of specific proteins is selectively increased by antidepressants. Many RING finger domains simultaneously bind ubiquitination enzymes and their substrates and hence function as ligases. Ubiquitination in turn targets the substrate protein for degradation (Lorick et al. 1999; Joazeiro and Weissman 2000). *kf-1* is a gene that encodes a zinc-binding membrane protein with a RING-H2 finger domain at the C terminus (Yasojima et al. 1997). *kf-1* mediates E2-dependent ubiquitination and thus may play a role in modulating cellular protein levels as an E3 Ub ligase (Lorick et al. 1999). Expression of the *kf-1* gene is elevated in rat brain after chronic antidepressant treatments (Yamada et al. 2000), electroconvulsive therapy (Nishioka et al. 2003), and repetitive transcranial magnetic stimulation (Kudo et al. 2005), and mice lacking the *kf-1* gene were suggested as a suitable animal model for elucidating molecular mechanisms of psychiatric diseases such as anxiety/depression, and for screening novel anxiolytic/antidepressant compounds (Tsujimura et al. 2008). Hypericin was found to enhance ubiquitinylation of heat shock protein 90, but not of heat shock protein 70, constituting a mechanism for generating mitotic cell death in cancer cells (Blank et al. 2003).

Agonist-stimulated ubiquitinylation occurring on β-arrestins governs the stability of receptor–β-arrestin interactions (Shenoy and Lefkowitz 2003b; Perroy et al. 2004). β-Arrestin ubiquitinylation is crucial for both its endocytic and signaling functions (Shenoy and Lefkowitz 2003b; Shenoy et al. 2007). Ubiquitinylation of β-arrestin by Mdm2 promotes stabilization of receptor–arrestin complexes and augmentation of β-arrestin-dependent extracellular signal–regulated kinase (ERK) activation (Shenoy et al. 2007). Mdm2 and the deubiquitinating enzyme Ub-specific protease 33 function reciprocally and favor respectively the stability or lability of the receptor–β-arrestin complex, thus regulating the longevity and subcellular localization of receptor signalosomes (Shenoy et al. 2009). Ub-specific protease 33 plays a regulatory role by dissolving the receptor–arrestin signalosome by promoting deubiquitination and dissociation of β-arrestin-2, and by inhibiting β-arrestin-dependent ERK activation and regulating the extent of downstream signaling.

Immunoprecipitation experiments showed that antidepressants were able to increase coimmuno-precipitation of Ub with β-arrestin 2. Antidepressant-induced increase in β-arrestin-2 ubiquitinylation led to its degradation by the proteosomal pathway, as the proteasome inhibitor MG-132 prevented antidepressant-induced decreases in β-arrestin-2 protein levels (Golan et al. 2010). Antidepressant-induced ubiquitinylation of β-arrestin may promote stabilization of receptor–β-arrestin complexes and augmentation of β-arrestin-dependent ERK activation. Thus, these findings may have extended implications concerning a possible switch in postreceptor signaling related to tyrosine kinases, Src, and MAPKs and a variety of other signaling molecules that might play a role in major depressive disorder (MDD) or in antidepressant therapeutic effects.

18.1.3 Protein Kinase B, Glycogen Synthase Kinase-3, β-Arrestin-2, Lithium, and Antidepressants

Protein kinase B (PKB), also termed C-AKT or Akt, is a serine/threonine kinase regulated by phosphatidylinositol kinase–mediated signaling. Activation of PKB involves phosphorylation of a regulatory threonine residue (threonine-308) by 3-phosphoinositide-dependent protein kinase 1 and additional phosphorylation of Ser-473 by the target of rapamycin complex 2 kinase (Alessi and Cohen 1998; Jacinto et al. 2006). Glycogen synthase kinase (GSK)-3 α and GSK-3 β are two closely related serine/threonine kinases originally associated with the regulation of glycogen synthesis in response to insulin (Embi et al. 1980; Frame and Cohen 2001). These kinases are constitutively active and can be inactivated through the phosphorylation of single serine residues, serine-21 (GSK-3 α) and serine-9 (GSK-3 β), of their respective regulatory N-terminal domains (Cross et al. 1995). PKB has been shown to inhibit GSK-3 α and GSK-3 β in response to multiple hormones and growth factors, including brain-derived neurotrophic factor (BDNF), insulin-like growth factor, and insulin (Frame and Cohen 2001; Cross et al. 1995; Chen and Russo-Neustadt 2005). The PKB/GSK-3 signaling cascade can be regulated by a signaling complex scaffolded by β-arrestin-2 (Beaulieu et al. 2005, 2007). In the past 5 years, multiple lines of evidence have shown that lithium inhibits GSK-3 β, allowing accumulation of β-catenin and β-catenin-dependent gene transcriptional events (Klein and Melton 1996; Stambolic et al. 1996; Wada 2009; Jope 2003). Lithium is a competitive inhibitor of Mg^{2+}, directly inhibiting Mg^{2+}-ATP-dependent catalytic activity of GSK-3 (Ryves and Harwood 2001). In addition, lithium increases serine-21/serine-9 phosphorylation of GSK-3 α/β (Zhang et al. 2003). Lithium was recently found to regulate PKB/GSK-3 signaling and related behaviors in mice by disrupting a signaling complex composed of PKB, β-arrestin-2, and serine/threonineprotein phosphatase 2A. When administered to β-arrestin-2 knockout mice, lithium fails to affect PKB/GSK-3 signaling and induces behavioral changes associated with GSK-3 inhibition, as it does in normal animals.

These results point to a pharmacological approach to modulating GPCR function, which affects the formation of β-arrestin-mediated signaling complexes (Beaulieu et al. 2008; Beaulieu and Caron 2008). GSK-3 inhibitors, which allow stabilization of β-catenin, have been reported to exert antidepressant-like effects in animals (Kaidanovich-Beilin et al. 2004; Gould et al. 2006). Increased cell proliferation in the subgranular zone was achieved after chronic treatment with a high dose of the antidepressant venlafaxine. Significant increases in the nuclear level of hippocampal β-catenin in the subgranular zone were already detected after administration of a lower dose of the drug

(Mostany et al. 2008). No information about hippocampal β-catenin expression in tissues from depressed patients is available, but no changes have been found in cortical samples (Beasley et al. 2002).

18.2 MITOGEN-ACTIVATED PROTEIN KINASES

MAPKs constitute a family of serine/threonine kinases. These kinases are important mediators of signal transduction and play a key role in the regulation of many cellular processes, such as cell growth and proliferation, differentiation, and apoptosis (Kyosseva 2004). MAPKs are the terminal elements of highly conserved kinase cascades consisting of MAPK kinase kinases (MAPKKKs) (such as Raf-1), MAPK kinases (MAPKKs) (such as MEK1), and the MAPKs themselves. In mammalian cells, three major groups of MAPKs have been identified: (1) ERKs, stimulated by growth-related signals; (2) c-Jun N-terminal kinases (JNKs), activated by various stress stimuli; and (3) p38 MAPKs, activated by various stress stimuli (Figure 18.2). Because each kinase is activated through phosphorylation by the preceding kinase in the cascade, there can be a dozen or more enzymes at each level (Morrison and Davis 2003). Classically, activated MAPKs translocate to the nucleus, where they phosphorylate and activate transcription factors (Whitmarsh 2007), which regulate programs of transcription that lead to proliferation, differentiation, and other cellular processes. A growing list of cytosolic substrates for MAPKs has been identified in the past few years. Their phosphorylation leads to distinct but less well-characterized consequences such as changes in cell shape and motility.

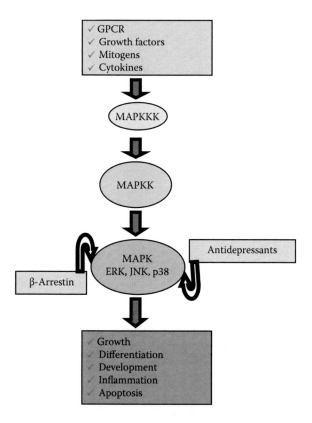

FIGURE 18.2 Elements of MAPK cascade: scaffolding roles for β-arrestins and targets for antidepressant drugs.

18.2.1   ERK: Scaffolding Roles of β-Arrestins

ERK belongs to the family of MAPKs that integrate signals received by membrane receptors and transfer them to the nucleus. The phosphorylation state and activity of ERK isoforms 1 and 2 (ERK1/2) are specifically modified by MAPK/ERK kinases (MEK1/2) (Nakielny et al. 1992). The ERK pathway is activated by neurotrophins and neuroactive chemicals to produce their effects on neuronal differentiation, survival, and on structural and functional plasticity (Schafe et al. 2000; Sweatt 2001; Zhu et al. 2002).

The MAPKs, particularly ERK1/2, have been the most extensively used model system to demonstrate the scaffolding roles of β-arrestins in signaling involving the GPCR superfamily. In the case of the ERK cascade, it has been demonstrated that upon activation of the type 1 angiotensin II receptor (AT1) (Luttrell et al. 2001), the tachykinin receptor NK1 (DeFea et al. 2000a), and protease-activated receptors (DeFea et al. 2000b), β-arrestin scaffolds the components of the ERK cascade, Raf-1, MEK1, and ERK1/2, into the receptor complex, leading to activation of ERK1/2. ERK1/2 and Raf-1 bind β-arrestin-2 directly, whereas the MAPKK MEK1 binds indirectly (DeWire et al. 2007). More recently, β-arrestin-dependent, G-protein-independent ERK1/2 activation by the β2-adrenoceptor has been described (Shenoy et al. 2006).

One effect of ERK activation is the phosphorylation and activation of Elk-1, an ETS domain–containing transcription factor involved in the transcription of genes that promote cell cycle progression. However, Elk-1-dependent transcription, which typically results from ERK activation, was decreased when β-arrestins were expressed (Tohgo et al. 2002). Confocal microscopy revealed that this effect is a result of the cytosolic retention of ERK by β-arrestins on endocytic vesicles (Luttrell et al. 2001; Tohgo et al. 2003), termed β-*arrestin signalosomes* (Shenoy and Lefkowitz 2003a).

Although these studies clarify that β-arrestin-dependent ERK does not have typical nuclear ERK functions, the precise downstream targets activated by β-arrestin-dependent ERK remain unknown. A wide variety of extracellular signals transduced via numerous cell surface receptors or integrins activate MAPKs, including ERK1/2, which in turn play a major role in the integration of multiple biological responses such as cell proliferation, differentiation, and survival. Thus, exquisite regulation of MAPK activation is crucial for generating the proper physiological outcomes from a particular stimulus (Ahn et al. 2004).

18.2.1.1   ERK: Depression and Antidepressants

The role for the ERK pathway in the molecular mechanism of depression is increasingly becoming the focus of a great deal of research, and a growing body of evidence indicates that the ERK pathway may participate in the neuronal modulation of depression. ERK1/2 phosphorylation has been hypothesized as an intracellular signaling mechanism mediating the efficacy of antidepressants in depressed humans and animal models of depression. Supporting evidence comes from studies in rodents or in vitro models and from postmortem studies of depressed suicide victims (Dwivedi et al. 2001, 2006).

Chronic forced-swim stress induced depressive-like behaviors and decreased the levels of cyclic adenosine monophosphate (cAMP)–responsive element–binding protein (CREB), phosphorylated ERK2 (P-ERK2), phosphorylated CREB (P-CREB), and ERK1/2 in the hippocampus and prefrontal cortex. Fluoxetine alleviated the depressive-like behaviors and reversed the disruptions of P-ERK2 and P-CREB in stressed rats. Moreover, it also exerted a mood-elevating effect and increased the levels of P-ERK2 and P-CREB in naive rats (Qi et al. 2008, 2009b). Blockade of MAPK signaling by acute administration of a MEK inhibitor (PD-184161) produced depressive-like behavior and blocked the behavioral actions of the antidepressants desipramine and sertraline. Heterozygous deletion of BDNF alone did not influence behavior but resulted in a depressive phenotype when combined with a low-dose MEK inhibitor (Duman et al. 2007). Acute elevated platform stress downregulates a putative BDNF/MEK/MAPK signaling cascade in the frontal cortex, whereas the antidepressants tianeptine and imipramine reversed the stress-induced downregulation of phosphorylated MEK (Ser-217/221) by increasing MEK and MAPK phosphorylation directly, thus offsetting the effects of stress (Qi et al. 2009a). The convergence of signaling pathways between monoamines and neuronal growth factors has attracted considerable interest. Monoamines have traditionally been associated with short-term signaling pathways, such as the modulation of cAMP and Ca^{2+} levels. In contrast, neuronal growth factors have traditionally been associated with signaling pathways, such as those for activation of ERK, which are known to induce long-term protective changes. It was therefore unclear how antidepressants that increase monoamines induce such changes as hippocampal neuroprotection and neurogenesis (Zhu et al. 2002; Cowen 2007). Recently, adult neurogenesis induced by antidepressants was demonstrated to be critical to antidepressant effects (Santarelli et al. 2003). Adult neurogenesis is regulated by several trophic and growth factors, such as BDNF and glial cell line–derived neurotrophic factor (GDNF) (Newton and Duman 2004). Changes in gene expression and signal transduction related to neuronal and glial plasticity and adaptations after chronic antidepressant treatment are important for the therapeutic effect of antidepressants (Duman 2004). Several different classes of antidepressants, including the tricyclic antidepressants amitriptyline, clomipramine, nortriptyline, and desipramine; the tetracyclic antidepressant mianserin; and the selective serotonin (5-hydroxytryptamine [5-HT]) reuptake inhibitors fluvoxamine and fluoxetine, increased GDNF production through acute activation of protein tyrosine kinase and ERK in rat C6 glioma cells and normal human astrocytes. Amitriptyline-induced phosphorylation of CREB was completely blocked by U-0126 (a MAPKK 1 inhibitor). Amitriptyline treatment also increased the expression of the luciferase reporter gene, regulated by cAMP-responsive elements. Amitriptyline-induced luciferase activity was completely inhibited by U-0126 in the same manner as phosphorylation of CREB. These results suggest that antidepressants acutely increase CREB activity in an ERK-dependent manner, which might contribute to gene expression, including that of GDNF in glial cells (Hisaoka et al. 2001, 2007, 2008). A recent report (Schmid et al. 2008) revealed a role for β-arrestin-2-mediated regulation of ERK in 5-HT_2 receptor signaling. Considering the association between 5-HT and the action of antidepressants, this study suggests that the 5-HT_{2A}–β-arrestin interaction may be particularly important for receptor function in response to endogenous serotonin levels, a fact that could have major implications in the development of drugs to treat neuropsychiatric disorders such as depression.

18.2.2   JNK: Scaffolding Roles of β-Arrestins

The JNK MAPKs respond to cell stress, growth factors, cytokines, and hormones and are involved in a variety of responses, including apoptosis, embryonic development, cell growth, and the immune response (Davis 2000; Kyriakis and Avruch 2001). JNK, similar to other MAPKs, is activated by a signaling module that includes MAPKKs and MAPKKKs. Components of JNK signaling modules include MAPKK 4 and MAPKK 7 and multiple MAPKKKs, including members of the MEK kinase and mixed lineage kinase families, apoptosis signal–regulating kinase 1, transforming growth factor-β–activated kinase 1, and tumor progression locus 2 (Davis 2000; Kyriakis and Avruch 2001). Once activated, depending on the cellular environment, JNKs can phosphorylate transcription factors, including c-Jun and ATF-2 (Ip and Davis 1998). JNKs can also phosphorylate nonnuclear substrates, including the Bcl-2 family member BAD (Yu et al. 2004) and the 14-3-3 protein (Tsuruta et al. 2004). β-Arrestin-2 specifically binds to JNK3 but not to JNK1 or JNK2 (McDonald et al. 2000) because of the unique extended N-terminal region of JNK3, which is absent in the other JNK isoforms (Guo and Whitmarsh 2008). It also binds to MAPKKK apoptosis signal–regulating kinase 1 and recruits MAPKK 4, thereby nucleating a JNK signaling module (McDonald et al. 2000). Phosphorylated, active JNK3 can only be detected in association with β-arrestin-2 and is excluded from the nucleus. Under certain conditions, β-arrestin-2 may also regulate JNK3 inactivation via the recruitment of MAPK phosphatases (Willoughby and Collins 2005).

18.2.2.1   JNK: Depression and Antidepressants

The JNK pathway is one of the early signaling cascades shown to mediate apoptotic death in response to a variety of stressful stimuli (Kyriakis et al. 1994; Derijard et al. 1995). In PC3 cells, desipramine caused apoptosis by inducing JNK-associated caspase-3 activation. Desipramine activated the phosphorylation of JNK. SP-600125 (a selective JNK inhibitor) partially protected cells from apoptosis caused by desipramine (Chang et al. 2008). SSRIs were found to activate the JNK pathway in C6 and neuroblastoma cell lines (Levkovitz et al. 2007b). SSRIs activated a number of insulin receptor substrate kinases, JNK included, leading to increased serine/threonine phosphorylation of insulin receptor substrate 1, with a concomitant reduction in its ability to undergo insulin-induced tyrosine phosphorylation. These findings suggest that SSRIs are potential inducers of insulin resistance by directly inhibiting the insulin signaling cascade (Levkovitz et al. 2007a). EGR-1 is a transcription factor involved in neuronal differentiation and astrocyte cell proliferation. Amitriptyline induces EGR-1 expression in rat C6 glioma cells. The amitriptyline-induced EGR-1 expression was mediated through serum response elements in the *EGR1* promoter. Serum response elements were activated by the ETS domain–containing transcription factor Elk-1 through the ERK and JNK–MAPK pathways. The inhibition of the ERK and JNK–MAPK signals attenuated amitriptyline-induced transactivation of GAL4–Elk-1 and the *EGR1* promoter (Chung et al. 2007). It is well known that the hypothalamic–pituitary–adrenal axis plays a significant role in the etiology of depression and in the mechanism of action of antidepressants. Various types of antidepressants have recently been found to attenuate the transcriptional activity of the glucocorticoid receptor stimulated by corticosterone. The JNK inhibitor SP-600125 reversed this effect (Otczyk et al. 2008).

18.2.3 p38: Scaffolding Roles of β-Arrestins

MAPK p38 is primarily involved in eliciting a transcriptional response to inflammatory cytokines, growth factors, and cellular stress. Although researchers have not directly shown that β-arrestins scaffold either p38 itself or upstream kinases nor that they lead to p38 phosphorylation, three reports have implicated β-arrestins in p38 activation (Sun et al. 2002; Miller et al. 2003; Bruchas et al. 2006). The chemokine receptor CXCR4-induced chemotactic response, which is β-arrestin dependent, is also sensitive to p38 inhibitors (Sun et al. 2002). Activated p38 was found to colocalize with β-arrestin-2 in features reminiscent of the signalosomes found with β-arrestins and ERK (Bruchas et al. 2006). Suppression of β-arrestin-2 blocks GPCR CXCR4-mediated activation of the MAPKs ERK and p38 (Chung et al. 2007), whereas β-arrestin-1 both induces activation of p38 and binds phosphorylated p38 (McLaughlin et al. 2006). It has recently been proposed that β-arrestin binds to phosphorylated p38 and negatively regulates Golgi transport via attenuation of p38. β-Arrestin negatively regulates Shiga toxin transport via an interaction with activated p38 and attenuation of its signaling (Skånland et al. 2009).

18.2.3.1 p38: Depression and Antidepressants

Apoptosis is developmentally and physiologically important and has been recognized as a key mode of cell death in response to cytotoxic treatment (Danial and Korsmeyer 2004). Cells undergo apoptosis through two major pathways: the extrinsic, death receptor pathway and the intrinsic, mitochondrial pathway. Certain cells have unique sensors, termed *death receptors*, on their surface. Death receptors detect the presence of extracellular death signals and, in response, rapidly fire up the cell's intrinsic apoptosis machinery (Adam-Klages et al. 2005). Caspases play a pivotal role in the regulation and conduction of apoptosis (Launay et al. 2005). The caspases implicated in apoptosis can be further divided into two functional subgroups based on their known or hypothetical roles in the process as initiator caspases (caspase-2, -8, -9, and -10) or effector caspases (caspase-3, -6, and -7) (Wang et al. 2005). Active caspases specifically process various substrates implicated in apoptosis and inflammation (Martinon and Tschopp 2004). Recent studies have shown that some antidepressants could affect cell viability. The SSRIs 6-nitroquipazine, zimelidine hydrochloride, and fluoxetine hydrochloride inhibited the proliferation of prostate cancer cells in a concentration-dependent manner (Abdul et al. 1995). Paroxetine and fluoxetine induced apoptosis in rat glioblastoma and human neuroblastoma cells (Levkovitz et al. 2007b; Spanova et al. 1997). The tricyclic antidepressants imipramine, clomipramine, and citalopram hydrobromide could evoke apoptosis in human peripheral lymphocytes (Xia et al. 1996, 1997) and acute myeloid leukemia HL60 cells via a caspase-3-dependent pathway (Xia et al. 1999). Both SSRIs and tricyclic antidepressants reduced the survival of mouse hippocampal cells (HT22) (Post et al. 2000). In biopsy-like Burkitt lymphoma cells, the SSRIs induced apoptosis, which was preceded by caspase activation (Serafeim et al. 2003). Although these studies have shown the antitumor and cytotoxic activities of some antidepressants, the underlying mechanisms were unknown. MG-63 human osteosarcoma cells treated with desipramine showed typical apoptotic features, including an increase in subdiploid nuclei and activation of caspase-3, indicating that these cells underwent apoptosis. The tricyclic desipramine and the SSRI paroxetine activated ERK, JNK, and p38. Although pretreatment of cells with PD-098059 (an ERK inhibitor) or SP-600125 (an inhibitor of JNK) did not inhibit cell death, the addition of SB-203580 (a p38 MAPK inhibitor) partially rescued cells from apoptosis. Desipramine- and paroxetine-induced caspase-3 activation required p38 MAPK activation (Lu et al. 2009; Chou et al. 2007). Human serotonin transporters inactivate 5-HT after release and are prominent targets for therapeutic intervention in mood, anxiety, and obsessive-compulsive disorders. Multiple human serotonin transporter–coding variants have been identified. When transporter modulation was examined by posttranslational regulatory pathways, cellular phenotypes associated with naturally occurring human serotonin transporter–coding variants were revealed, which were characterized by altered transporter regulation by means of a p38 MAPK–linked pathway (Prasad et al. 2005). This striking pattern of regulatory disruption by p38 may influence the risk of disorders attributed to compromised 5-HT signaling (Prasad et al. 2005).

18.3 β-ARRESTIN REGULATION OF TRANSCRIPTION AND ANTIDEPRESSANTS

18.3.1 INHIBITOR OF NUCLEAR FACTOR κBα

NF-κB is a ubiquitously expressed and highly regulated dimeric transcription factor that regulates genes involved in immunity, stress, and apoptosis. In the inactive state, NF-κB dimers are bound to the inhibitor of NF-κBα (IκBα) protein and thus retained in the cytoplasm. Extracellular signals stimulate phosphorylation of IκBα, which leads to its degradation and translocation of NF-κB to the nucleus, where it binds to promoters of target genes and activates transcription. β-Arrestins mediate cross-talk between GPCR and NF-κB signaling pathways. β-Arrestins bind directly to IκBα (Whitherow et al. 2004; Gao et al. 2004). The interaction with β-arrestin-2 prevents phosphorylation and degradation of IκBα and thus attenuates activation of NF-κB and transcription of NF-κB target genes (Whitherow et al. 2004; Gao et al. 2004). β-Arrestins not only bind to and stabilize IκBα but also interact with the IκB kinase complex that regulates its activation and thereby negatively regulates the NF-κB pathway (Whitherow et al. 2004). Recent results identify novel functions of β-arrestin-1 in binding to the β1γ2 subunits of heterotrimeric G proteins and promoting Gβγ-mediated Akt signaling for NF-κB activation (Yang et al. 2009).

Fluoxetine was found to induce production of nitric oxide in BV2 microglial cells through increased DNA binding activity of NF-κB (Ha et al. 2006). The tricyclic antidepressants amitriptyline and desipramine as well as the SSRI fluoxetine were found to induce cell death in hippocampal neurons and PC12 pheochromocytoma cells by a selective activation of ERK, the biosynthesis of the transcription factor Egr-1, and an increase in the transcriptional activity of NF-κB (Bartholomä et al. 2002). In contrast, clomipramine and imipramine were found to inhibit IκB degradation, nuclear translocation of the p65 subunit of NF-κB, and phosphorylation of p38 MAPK in the lipopolysaccharide-stimulated microglia cells. Moreover, clomipramine and imipramine were neuroprotective as the drugs reduced microglia-mediated neuroblastoma cell death in a microglia/neuron coculture. Therefore, these results imply that clomipramine and imipramine have anti-inflammatory and neuroprotective effects in the central nervous system by modulating glial activation (Hwang et al. 2008). Similarly, hypericin, the presumed active moiety within Saint John's wort treatment, inhibited NF-κB, causing accumulation of phosphorylated IκBα in U373 cells (Pajonk et al. 2005).

18.3.2 TRANSCRIPTION FACTORS: CREB, C-FOS, AND P27

Since their initial characterization, arrestins have been thought of as cytoplasmic proteins that could be recruited to the plasma membrane and endocytic compartments after receptor activation. Kang et al. (2005) revealed a novel unexpected function of β-arrestin-1 as a cytoplasm-nucleus messenger in GPCR signaling, showing that stimulation of a member of the GPCR family, δ opioid receptor, induced the nuclear translocation of β-arrestin-1. Selectively enriched at specific promoters such as those of the genes encoding cyclin-dependent kinase inhibitor p27 and proto-oncogene protein c-*fos*, β-arrestin-1 facilitated the recruitment of histone acetyltransferase p300, resulting in enhanced local histone H4 acetylation and in transcription of these genes. The authors suggest that β-arrestin-1 acts as a nuclear scaffold that recruits p300 to the transcription factor, CREB, which leads to increased acetylation of histone H4 and the reorganization of chromatin, thereby increasing gene expression. CREB orchestrates diverse neurobiological processes, including cell differentiation, survival, and plasticity. Alterations in CREB-mediated transcription have been linked with numerous central nervous system disorders, including depression, anxiety, addiction, and cognitive decline (Tanis et al. 2008). CREB has been implicated in signaling pathways relevant for pathogenesis and the therapy for depression. CREB is upregulated and activated in the hippocampus by chronic antidepressant treatment, similarly to neurogenesis (Gass et al. 2007). The findings described above were specific to β-arrestin-1 rather than to β-arrestin-2. As already noted, both β-arrestin-1 and β-arrestin-2 are able to shuttle between the cytoplasm and the nucleus. However, in contrast to β-arrestin-1, β-arrestin-2 possesses a strong nuclear export signal in its C terminus, which hinders its retention by the nucleus (Scott et al. 2002; Wang et al. 2003). Moreover, agonist stimulation failed to induce nuclear accumulation of a β-arrestin-1 mutant with a β-arrestin-2 nuclear export signal at its C terminus. It can thus be concluded that β-arrestin-1 may play a more important role in GPCR-mediated nuclear signaling. β-Arrestin-dependent GPCR signaling lasts longer than conventional G-protein-dependent signaling (Beaulieu et al. 2005; Ahn et al. 2004). Thus, the β-arrestin-dependent mechanism for transcriptional control as proposed by Kang et al. (2005) may be used under certain physiological situations when sustained signaling is needed (Beaulieu and Caron 2005). Neurotrophic hypotheses of depression and antidepressant efficacy argue that stress and antidepressants chronically regulate mood via opposing actions on intracellular and transcription factors associated with cellular growth, motility, and survival, primarily in the hippocampus (Duman et al. 1997). Consistent with this hypothesis, hippocampal CREB overexpression produces an antidepressant-like behavioral phenotype (Chen et al. 2001), whereas stress decreases BDNF expression and suppresses local CREB phosphorylation (Laifenfeld et al. 2005). Enhancing cortical and hippocampal P-CREB has been proposed as a common antidepressant mechanism (Blendy 2006), based largely on studies of antidepressant action in naive rodents (Thome et al. 2000). The fact that various types of antidepressants regulate c-*fos* immunoreactivity and the expression of c-*fos* mRNA in specific areas of the rat brain is a well-described phenomenon (Dahmen et al. 1997; Morinobu et al. 1997; Morelli et al. 1999; Miyata et al. 2005; Slattery et al. 2005; Kuipers et al. 2006). In addition, the antidepressant rolipram, a type IV phosphodiesterase inhibitor, was found to induce expression of p27 in malignant glioma cells (Chen et al. 2002). It should be noted that phosphodiesterase 4–selective inhibition by rolipram was recently found to facilitate the isoproterenol-induced membrane translocation of GRK2, phosphorylation of the β2-adrenoceptor by GRK2, membrane translocation of β-arrestin, and internalization of β2-adrenoceptors. In the absence of isoproterenol, rolipram-induced inhibition of phosphodiesterase 4 activity acted to stimulate PKA phosphorylation of GRK2, with consequential effects on GRK2 membrane recruitment and GRK2-mediated phosphorylation of the β2-adrenoceptor (Li et al. 2006). The above findings concerning the regulation of CREB and c-*fos* by antidepressants and of p27 by rolipram could possibly be mechanistically explained by the data on antidepressant-induced elevation of β-arrestin-1 protein and mRNA levels together with the data on selective enrichment of β-arrestin-1 at specific promoters such as those of the genes encoding p27 and c-*fos* (Kang et al. 2005).

18.4   GRKs, β-ARRESTINS, BIASED AGONISM: DEPRESSIVE DISORDER AND ANTIDEPRESSANTS

Concepts regarding the mechanisms by which drugs activate receptors to produce a physiological response have progressed beyond considering the receptor as a simple on–off switch. Current evidence suggests that the idea that agonists produce only varying degrees of receptor activation is obsolete and must be reconciled with available data to show that agonist efficacy has texture as well as magnitude. Thus, agonists can block system constitutive responses (inverse agonists), behave as positive and inverse agonists on the same receptor (protean agonists), and differ in the stimulus pattern they produce in physiological systems (ligand-selective agonists) (Kenakin 2001). Seven-transmembrane receptors (7TMRs), the most abundant drug targets, are critically regulated by β-arrestins, which both inhibit classic G-protein signaling and initiate distinct β-arrestin signaling. The interplay of G-protein and β-arrestin signals largely determines the cellular consequences of 7TMR-targeted drugs. Until recently, a drug's efficacy for β-arrestin recruitment was believed to be proportional to its efficacy for G-protein activities. This paradigm restricts 7TMR drug effects to a linear spectrum of responses, ranging from inhibition of all responses to stimulation of all responses. However, it is now clear that certain ligands can selectively activate G-protein or β-arrestin functions. Such ligands are referred to as biased ligands and the phenomenon as *biased agonism* or *ligand directed signaling* (Violin and Lefkowitz 2007). β-Arrestin-biased ligands are of particular interest because they offer the possibility of designing an entirely novel class of therapeutic agents (Violin and Lefkowitz 2007). Such ligands, in common with conventional antagonists, prevent agonist-activated G-protein signaling. However, in contrast with conventional antagonists, they simultaneously stimulate potentially beneficial effects of β-arrestin-mediated signaling (Shukla et al. 2008). There are still considerable gaps in our understanding of bias and in the different receptor conformations that are responsible for signaling to G proteins and β-arrestins. Some hypotheses explore regulated bias, in which G-protein versus β-arrestin signaling can be controlled by GRKs or other cofactors (Rajagopal et al. 2010). Given the complexity of β-arrestin function, the traditional concept of GRKs as redundant kinases is likely overly simplistic. GRK2, 3, 5, and 6 are ubiquitously expressed. The extent to which they provide redundancy versus functional specialization is largely unknown, but evidence for the latter has recently emerged in some receptor systems. For instance, GRK5/6 appear to be uniquely required for β-arrestin-mediated MAPK signaling (Zidar et al. 2009). Thus, angiotensin II–stimulated receptor phosphorylation, β-arrestin recruitment, and receptor endocytosis were all found to be mediated primarily by GRK2/3. In contrast, inhibiting GRK5/6 expression abolished β-arrestin-mediated ERK activation, whereas lowering GRK2 or 3 led to an increase in this signaling. Consistent with these findings, β-arrestin-mediated ERK activation was enhanced by overexpression of GRK5 and 6, and reciprocally diminished by GRK2/3 (Kim et al. 2005). Also, GRK2 and 3 were found responsible for most of the agonist-dependent V2 vasopressin receptor phosphorylation, desensitization, and recruitment of β-arrestins. In contrast, GRK5/6 mediated much less receptor phosphorylation and β-arrestin recruitment but appeared exclusively to support β-arrestin-2-mediated ERK activation. GRK2 suppression actually increased β-arrestin-stimulated ERK activation (Ren et al. 2005).

Growing evidence suggests that GPCR–G-protein coupling and its regulation by GRKs may be involved in the pathophysiology, diagnosis, and treatment of mood disorders (for review, see Avissar and Schreiber 2006; Schreiber and Avissar 2007; Golan et al. 2009; Metaye et al. 2005; Premont and Gainetdinov 2007). Using a convergent functional genomics approach, a series of candidate genes involved in the pathogenesis of mood disorders was identified, including GRK3, which was also found to be decreased in lymphoblastoid cell lines from a subset of bipolar patients (Niculescu et al. 2000). A single-nucleotide polymorphism in the promoter region of GRK3 was found to be associated with bipolar disorder (Barrett et al. 2003). The findings concerning the *GRK3* gene are in accord with evidence from genome-wide linkage surveys suggesting that the chromosome 22q12 region contains a susceptibility locus for bipolar disorder. Alterations of GRK2/3 levels were described in specimens of the prefrontal cortex collected from depressed subjects (Garcia-Sevilla et al. 1999; Grange-Midroit et al. 2003). Acute, but not prolonged, treatment with desipramine altered

GRK2/3 in rat brain (Miralles et al. 2002). Reduced platelet GRK2 levels detected in patients with major depression were normalized by mirtazapine treatment (Garcia-Sevilla et al. 2004). GRK2 protein and mRNA levels in mononuclear leukocytes of untreated patients with major depression were significantly lower than the measures in healthy subjects. The extent of reduction in GRK2 protein and mRNA levels in untreated patients with depression was found to be correlated with the severity of depressive symptoms assessed by Hamilton Depression Rating Scale score. The decreased GRK2 protein and mRNA levels were alleviated by antidepressant treatment. Normalization of GRK2 measures preceded and, thus, could predict clinical improvement by 1–2 weeks, segregating responders from nonresponders (Matuzany-Ruban et al. 2010). Similarly, platelet GRK2 and p-Ser670 GRK2 were reduced (36%–41%) in unmedicated MDD subjects, and GRK2 content correlated inversely with the severity of depression. Effective antidepressant treatments normalized platelet GRK2, and, notably, GRK2 upregulation discriminated between responder and nonresponder patients. Other findings revealed a modest reduction of platelet GRK3 and no alteration of platelet GRK5 content in untreated subjects with MDD (García-Sevilla et al. 2010). The fact that GRK2 is reduced in peripheral blood elements of yet untreated subjects with depression whereas GRK6 is unaltered suggests that, at least in these peripheral blood elements during the depressive state, signaling might favor β-arrestin-stimulated ERK activation and β-arrestin-mediated MAPK signaling. In such a system of a high GRK6/GRK2 ratio, one would expect a regulated bias toward desensitized G-protein-coupled second-messenger signaling and increased β-arrestin-mediated ERK signaling. Indeed, subjects with major depression had lower basal, cesium fluoride–stimulated, forskolin-stimulated, and Gpp(NH)p-stimulated platelet adenylate cyclase activity levels, and this relationship was dramatically attenuated by various types of drug use (Hines and Tabakoff 2005). Basal and forskolin-stimulated levels of cAMP were also found to be significantly reduced in lymphocytes of patients with depression (Mizrahi et al. 2004). Similarly, CREB protein activity and expression were significantly decreased in the neutrophils of drug-free MDD patients (Ren et al. 2010). GRK6/GRK2 ratio in brain tissue of patients with depression may differ from the ratio described in peripheral blood elements. Thus, a different pattern of biased agonism might be expected in the brain. The fact that antidepressants normalized the reduced GRK2 measures in peripheral blood elements of patients with depression (Matuzany-Ruban et al. 2010; García-Sevilla et al. 2010) suggests that antidepressants may alter GRK2/GRK6 ratio and thus affect a different pattern of biased agonism through their intricate effects both on GRKs and on β-arrestins.

ACKNOWLEDGMENTS

We appreciate the permission granted by Bentham Science Publishers and Prous Science, S.A.U., for the reproduction with modifications of some sections of this chapter, from our papers Golan, M., et al., *Curr Pharm Des* 15:1699–708, 2009 (Copyright © 2009, Bentham Science Publishers, all rights reserved), and Schreiber, G., et al., *Drug News Perspect*, 22, 467–80, 2009 (Copyright © 2009, Prous Science, S.A.U. or its licensors, all rights reserved).

REFERENCES

Abdul, M., Logothetis, C. J., and Hoosein, N. M. 1995. Growth-inhibitory effects of serotonin uptake inhibitors on human prostate carcinoma cell lines. *J Urol* 154: 247–50.

Adam-Klages, S., Adam, D., Janssen, O., and Kabelitz, D. 2005. Death receptors and caspases: Role in lymphocyte proliferation, cell death, and autoimmunity. *Immunol Res* 33: 149–66.

Ahn, S., Shenoy, S. K., Wei, H., and Lefkowitz, R. J. 2004. Differential kinetic and spatial patterns of beta-arrestin and G protein-mediated ERK activation by the angiotensin II receptor. *J Biol Chem* 279: 35518–25.

Alessi, D. R., and Cohen, P. 1998. Mechanism of activation and function of protein kinase B. *Curr Opin Genet Dev* 8: 55–62.

Avissar, S., and Schreiber, G. 2006. The involvement of G proteins and regulators of receptor-G protein coupling in the pathophysiology, diagnosis and treatment of mood disorders. *Clin Chim Acta* 366: 37–47.

Avissar, S., Matuzany-Ruban, A., Tzukert, K., and Schreiber, G. 2004. Beta-arrestin-1 levels: Reduced in leukocytes of patients with depression and elevated by antidepressants in rat brain. *Am J Psychiatry* 161: 2066–72.

Barrett, T. B., Hauger, R. L., Kennedy, J. L., et al. 2003. Evidence that a single nucleotide polymorphism in the promoter of the G protein receptor kinase 3 gene is associated with bipolar disorder. *Mol Psychiatry* 8: 546–7.

Bartholomä, P., Erlandsson, N., Kaufmann, K., et al. 2002. Neuronal cell death induced by antidepressants: Lack of correlation with Egr-1, NF-kappa B and extracellular signal-regulated protein kinase activation. *Biochem Pharmacol* 63: 1507–16.

Beasley, C., Cotter, D., and Everall, I. 2002. An investigation of the Wnt-signalling pathway in the prefrontal cortex in schizophrenia, bipolar disorder and major depressive disorder. *Schizophr Res* 58: 63–7.

Beaulieu, J. M., and Caron, M. G. 2005. Beta-arrestin goes nuclear. *Cell* 123: 755–7.

Beaulieu, J. M., and Caron, M. G. 2008. Looking at lithium: Molecular moods and complex behaviour. *Mol Interv* 8: 230–41.

Beaulieu, J. M., Sotnikova, T. D., Marion, S., Lefkowitz, R. J., Gainetdinov, R. R., and Caron, M. G. 2005. An Akt/beta-arrestin 2/PP2A signaling complex mediates dopaminergic neurotransmission and behavior. *Cell* 122: 261–73.

Beaulieu, J. M., Gainetdinov, R. R., and Caron, M. G. 2007. The Akt-GSK-3 signaling cascade in the actions of dopamine. *Trends Pharmacol Sci* 28: 166–72.

Beaulieu, J. M., Marion, S., Rodriguez, R. M., et al. 2008. A beta-arrestin 2 signaling complex mediates lithium action on behavior. *Cell* 132: 125–36.

Blank, M., Mandel, M., Keisari, Y., Meruelo, D., and Lavie, G. 2003. Enhanced ubiquitinylation of heat shock protein 90 as a potential mechanism for mitotic cell death in cancer cells induced with hypericin. *Cancer Res* 63: 8241–7.

Blendy, J. 2006. The role of CREB in depression and antidepressant treatment. *Biol Psychiatry* 59: 1144–50.

Bruchas, M. R., Macey, T. A., Lowe, J. D., and Chavkin, C. 2006. Kappa opioid receptor activation of p38 MAPK is GRK3- and arrestin-dependent in neurons and astrocytes. *J Biol Chem* 281: 18081–9.

Chang, H. C., Huang, C. C., Huang, C. J., et al. 2008. Desipramine-induced apoptosis in human PC3 prostate cancer cells: Activation of JNK kinase and caspase-3 pathways and a protective role of [Ca^{2+}]I elevation. *Toxicology* 250: 9–14.

Chen, A. C. H., Shirayama, Y., Shin, K. H., Neve, R. L., and Duman, R. S. 2001. Expression of cAMP response element binding protein (CREB) in hippocampus produces an antidepressant effect. *Biol Psychiatry* 49: 753–62.

Chen, M. J., and Russo-Neustadt, A. A. 2005. Exercise activates the phosphatidylinositol 3-kinase pathway. *Brain Res. Mol Brain Res* 135: 181–93.

Chen, T. C., Wadsten, P., Su, S., et al. 2002. The type IV phosphodiesterase inhibitor rolipram induces expression of the cell cycle inhibitors p21(Cip1) and p27(Kip1), resulting in growth inhibition, increased differentiation, and subsequent apoptosis of malignant A-172 glioma cells. *Cancer Biol Ther* 1: 268–76.

Chou, C. T., He, S., and Jan, C. R. 2007. Paroxetine induced apoptosis in human osteosarcoma cells: Activation of p38 MAP kinase and caspase-3 pathways without involvement of [Ca2+]i elevation. *Toxicol Appl Pharmacol* 218: 265–73.

Chung, E. Y., Shin, S. Y., and Lee, Y. H. 2007. Amitriptyline induces early growth response-1 gene expression via ERK and JNK mitogen-activated protein kinase pathways in rat C6 glial cells. *Neurosci Lett* 422: 43–8.

Cowen, D. S. 2007. Serotonin and neuronal growth factors — A convergence of signaling pathways. *J Neurochem* 101: 1161–71.

Cross, D. A., Alessi, D. R., Cohen, P., Andjelkovich, M., and Hemmings, B. A. 1995. Inhibition of glycogen synthase kinase-3 by insulin mediated by protein kinase B. *Nature* 378: 785–9.

Dahmen, N., Fehr, C., Reuss, S., and Hiemke, C. 1997. Stimulation of immediate early gene expression by desipramine in rat brain. *Biol Psychiatry* 42: 317–23.

Danial, N. N., and Korsmeyer, S. J. 2004. Cell death: Critical control points. *Cell* 116: 205–19.

David, D. J., Samuels, B. A., Rainer, Q. et al. 2009. Neurogenesis-dependent and -independent effects of fluoxetine in an animal model of anxiety/depression. *Neuron* 62: 479–93.

Davis, R. J. 2000. Signal transduction by the JNK group of MAP kinases. *Cell* 103: 239–52.

DeFea, K. A., Vaughn, Z. D., O'Bryan, E. M., Nishijima, D., Dery, O., and Bunnett, N. W. 2000a. The proliferative and antiapoptotic effects of substance P are facilitated by formation of a beta arrestin-dependent scaffolding complex. *Proc Natl Acad Sci U S A* 97: 11086–91.

DeFea, K. A., Zalevsky, J., Thoma, M. S., Dery, O., Mullins, R. D., and Bunnett, N. W. 2000b. Beta-arrestin-dependent endocytosis of proteinase-activated receptor 2 is required for intracellular targeting of activated ERK1/2. *J Cell Biol* 148: 1267–81.

Derijard, B., Raingeaud, J., Barrett, T., et al. 1995. Independent human MAP-kinase signal transduction pathways defined by MEK and MKK isoforms. *Science* 267: 682–7.

DeWire, S. M., Ahn, S., Lefkowitz, R. J., and Shenoy, S. K. 2007. Beta-arrestins and cell signaling. *Annu Rev
Physiol* 69: 483–510.

Duman, C. H., Schlesinger, L., Kodama, M., Russell, D. S., and Duman, R. S. 2007. A role for MAP kinase
signaling in behavioral models of depression and antidepressant treatment. *Biol Psychiatry* 61: 661–70.

Duman, R. S. 2004. Role of neurotrophic factors in the etiology and treatment of mood disorders. *Neuromolecular Med* 5: 11–25.

Duman, R. S., Heninger, G. R., and Nestler, E. J. 1997. A molecular and cellular theory of depression. *Arch
Gen Psychiatry* 54: 597–606.

Dwivedi, Y., Rizavi, H. S., Roberts, R. C., Conley, R. C., Tamminga, C. A., and Pandey, G. N. 2001. Reduced
activation and expression of ERK1/2 MAP kinase in the post-mortem brain of depressed suicide subjects.
J Neurochem 77: 916–28.

Dwivedi, Y., Rizavi, H. S., Conley, R. R., and Pandey, G. N. 2006. ERK MAP kinase signaling in postmortem brain of suicide subjects: Differential regulation of upstream Raf kinases Raf-1 and B-Raf. *Mol
Psychiatry* 11: 86–98.

Embi, N., Rylatt, D. B., and Cohen, P. 1980. Glycogen synthase kinase-3 from rabbit skeletal muscle. Separation
from cyclic-AMP-dependent protein kinase and phosphorylase kinase. *Eur J Biochem* 107: 519–27.

Frame, S., and Cohen, P. 2001. GSK3 takes centre stage more than 20 years after its discovery. *Biochem J* 359:
1–16.

Freedman, H. J., and Lefkowitz, R. J. 1996. Desensitization of G protein-coupled receptors. *Rec Prog Horm
Res* 51: 319–51.

Gagnon, A. W., Kallal, L., and Benovic, J. L. 1998. Role of clathrin-mediated endocytosis in agonist-induced
down-regulation of the beta2-adrenergic receptor. *J Biol Chem* 273: 6976–81.

Gainetdinov, R. R., Premont, R. T., Bohn, L. M., Lefkowitz, R. J., and Caron, M. G. 2004. Desensitization of G
protein-coupled receptors and neuronal functions. *Annu Rev Neurosci* 227: 107–44.

Gao, H., Sun, Y., Wu, Y., et al. 2004. Identification of β-arrestin2 as a G protein-coupled receptor-stimulated
regulator of NF-kB pathways. *Mol Cell* 14: 303–17.

Garcia-Sevilla, J. A., Escriba, P. V., Ozaita, A., et al. 1999. Up-regulation of immunolabeled alpha2A-adre-
noceptors, Gi coupling proteins, and regulatory receptor kinases in the prefrontal cortex of depressed
suicides. *J Neurochem* 72: 282–91.

Garcia-Sevilla, J. A., Ventayol, P., Perez, V., et al. 2004. Regulation of platelet alpha 2A-adrenoceptors, Gi proteins
and receptor kinases in major depression: Effects of mirtazapine treatment. *Neuropsychopharmacology*
29: 580–8.

García-Sevilla, J. A., Alvaro-Bartolomé, M., Díez-Alarcia, R., et al. 2010. Reduced platelet G protein-coupled
receptor kinase 2 in major depressive disorder: Antidepressant treatment-induced upregulation of GRK2
protein discriminates between responder and non-responder patients. *Eur Neuropsychopharmacol* 20:
721–30.

Gass, P., and Riva, M. A. 2007. CREB, neurogenesis and depression. *Bioessays* 29: 957–61.

Golan, M., Schreiber G., and Avissar S. 2009. Antidepressants, β-arrestins and GRKs: From regulation of signal desensitization to intracellular multifunctional adaptor functions. *Curr Pharm Des* 15: 1699–708.

Golan, M., Schreiber, G., Avissar S. 2010. Antidepressants increase β-arrestin2 ubiquitinylation and degradation by the proteasomal pathway in C6 rat glioma cells. *J Pharmacol Exp Ther* 332: 970–6.

Goodman, O. B. Jr., Krupnick, J. G., Santini, F., et al. 1996. Beta-arrestin acts as a clathrin adaptor in endocytosis of the beta2-adrenergic receptor. *Nature* 383: 447–50.

Gould, T. D., Picchini, A. M., Einat, H., and Manji, H. K. 2006. Targeting glycogen synthase kinase-3 in the
CNS: Implications for the development of new treatments for mood disorders. *Curr Drug Targets* 7:
1399–409.

Grange-Midroit, M., Garcia-Sevilla, J. A., Ferrer-Alcon, M., et al. 2003. Regulation of GRK 2 and 6, beta
arrestin-2 and associated proteins in the prefrontal cortex of drug-free and antidepressant drug-treated
subjects with major depression. *Brain Res Mol Brain Res* 111: 31–41.

Guo, C., and Whitmarsh, A. J. 2008. The beta arrestin-2 scaffold protein promotes c-Jun N-terminal kinase-3
activation by binding to its non-conserved N-terminus. *J Biol Chem* 283: 15903–11.

Ha, E., Jung, K. H., Choe, B. K., et al. 2006. Fluoxetine increases the nitric oxide production via nuclear factor
kappa B-mediated pathway in BV2 murine microglial cells. *Neurosci Lett* 397: 185–9.

Hershko, A., and Ciechanover, A. 1998. The ubiquitin system. *Annu Rev Biochem* 67: 425–79.

Hines, L. M., and Tabakoff, B. 2005. Platelet adenylyl cyclase activity: A biological marker for major depression and recent drug use. *Biol Psychiatry* 58: 955–62.

Hisaoka, K., Nishida, A., Koda, T., et al. 2001. Antidepressant drug treatments induce glial cell line-derived neurotrophic factor (GDNF) synthesis and release in rat C6 glioblastoma cells. *J Neurochem* 79: 25–34.

Hisaoka, K., Takebayashi, M., Tsuchioka, M., Maeda, N., Nakata, Y., and Yamawaki, S. 2007. Antidepressants increase glial cell line-derived neurotrophic factor production through mono-amine-independent activation of protein tyrosine kinase and extracellular signal regulated kinase in glial cells. *J Pharmacol Exp Ther* 321: 148–57.

Hisaoka, K., Maeda, N., Tsuchioka, M., and Takebayashi, M. 2008. Antidepressants induce acute CREB phosphorylation and CRE-mediated gene expression in glial cells: A possible contribution to GDNF production. *Brain Res* 1196: 53–8.

Hwang, J., Zheng, L. T., Ock, J., et al. 2008. Inhibition of glial inflammatory activation and neurotoxicity by tricyclic antidepressants. *Neuropharmacology* 55: 826–34.

Ip, Y. T., and Davis, R. J. 1998. Signal transduction by the c-Jun N-terminal kinase (JNK) — From inflammation to development. *Curr Opin Cell Biol* 10: 205–19.

Jacinto, E., Facchinetti, V., Liu, D., et al. 2006. SIN1/MIP1 maintains rictor–mTOR complex integrity and regulates Akt phosphorylation and substrate specificity. *Cell* 127: 125–37.

Joazeiro, C. A., and Weissman, A. M. 2000. RING finger proteins: Mediators of ubiquitin ligase activity. *Cell* 102: 549–52.

Jope, R. S. 2003. Lithium and GSK-3: One inhibitor, two inhibitory actions, multiple outcomes. *Trends Pharmacol Sci* 24: 441–3.

Kaidanovich-Beilin, O., Milman, A., Weizman, A., Pick, C. G., and Eldar-Finkelman, H. 2004. Rapid antidepressive-like activity of specific glycogen synthase kinase-3 inhibitor and its effect on beta-catenin in mouse hippocampus. *Biol Psychiatry* 55: 781–4.

Kang, J., Shi, Y., Xiang, B., et al. 2005. A nuclear function of beta-arrestin1 in GPCR signaling: Regulation of histone acetylation and gene transcription. *Cell* 123: 833–47.

Kenakin, T. 2001. Inverse, protein, and ligand-selective agonism: Matters of receptor conformation. *FASEB J* 15: 598–611.

Kim, J., Ahn, S., Ren, X. R., et al. 2005. Functional antagonism of different G protein-coupled receptor kinases for beta-arrestin-mediated angiotensin II receptor signaling. *Proc Natl Acad Sci U S A* 102: 1442–7.

Klein, P. S., and Melton, D. A. 1996. A molecular mechanism for the effect of lithium on development. *Proc Natl Acad Sci U S A* 93: 8455–9.

Kristen, L. P., and Lefkowitz, R. J. 2001. Classical and new roles of beta-arrestin in the regulation of G-protein-coupled receptors. *Nature Rev* 2: 727–30.

Kudo, K., Yamada, M., Takahashi, K., et al. 2005. Repetitive transcranial magnetic stimulation induces *kf-1* expression in the rat brain. *Life Sci* 76: 2421–9.

Kuipers, S. D., Trentani, A., Westenbroek, C., et al. 2006. Unique patterns of FOS, phospho-CREB and BrdU immunoreactivity in the female rat brain following chronic stress and citalopram treatment. *Neuropharmacology* 50: 428–40.

Kyosseva, S. V. 2004. Mitogen-activated protein kinase signaling. *Int Rev Neurobiol* 59: 201–20.

Kyriakis, J. M., and Avruch, J. 2001. Mammalian mitogen-activated protein kinase signal transduction pathways activated by stress and inflammation. *Physiol Rev* 81: 807–69.

Kyriakis, J. M., Banerjee, P., Nikolakaki, E., et al. 1994. The stress-activated protein kinase subfamily of c-Jun kinases. *Nature* 369: 156–60.

Laifenfeld, D., Kerry, R., Grauer, E., Klein, E., and Ben-Shachar, D. 2005. Antidepressant and prolonged stress in rats modulate CAM-L1, laminin, and pCREB, implicated in neuronal plasticity. *Neurobiol Dis* 20: 432–41.

Laporte, S. A., Oakley, R. H., Zhang, J., et al. 1999. The beta2-adrenergic receptor/beta-arrestin complex recruits the clathrin adaptor AP-2 during endocytosis. *Proc Natl Acad Sci U S A* 96: 3712–7.

Launay, S., Hermine, O., Fontenay, M., Kroemer, G., Solary, E., and Garrido, C. 2005. Vital functions for lethal caspases. *Oncogene* 24: 5137–48.

Lefkowitz, R. J. 2004. Historical review: A brief history and personal retrospective of seven-transmembrane receptors. *Trends Pharmacol Sci* 225: 413–22.

Levkovitz, Y., Ben-Shushan G., Hershkovitz, A., et al. 2007a. Antidepressants induce cellular insulin resistance by activation of IRS-1 kinases. *Mol Cell Neurosci* 36: 305–12.

Levkovitz, Y., Gil-Ad, I., Zeldich, E., Dayag, M., and Weizman, A. 2007b. Differential induction of apoptosis by antidepressants in glioma and neuroblastoma cell lines: Evidence for p-c-Jun, cytochrome c, and caspase-3 involvement. *J Mol Neurosci* 27: 29–42.

Li, X., Huston, E., Lynch, M. J., Houslay, M. D., and Baillie, G. S. 2006. Phosphodiesterase-4 influences the PKA phosphorylation status and membrane translocation of G-protein receptor kinase 2 (GRK2) in HEK-293beta2 cells and cardiac myocytes. *Biochem J* 394: 427–35.

Lohse, M. J., Benovic, J. L., Codina, J., Caron, M. G., and Lefkowitz, R. J. 1990. Beta-arrestin: A protein that regulates beta-adrenergic receptor function. *Science* 248: 1547–50.

Lorick, K. L., Jensen, J. P., Fang, S., Ong, A. M., Hatakeyama, S., and Weissman, A. M. 1999. RING fingers mediate ubiquitin-conjugating enzyme (E2)-dependent ubiquitination. *Proc Natl Acad Sci U S A* 96: 11364–9.

Lu, T., Huang, C. C., Lu, Y. C., et al. 2009. Desipramine-induced Ca-independent apoptosis in Mg63 human osteosarcoma cells: Dependence on P38 mitogen-activated protein kinase-regulated activation of caspase 3. *Clin Exp Pharmacol Physiol* 36: 297–303.

Luttrell, L. M., and Lefkowitz, R. J. 2002. The role of beta-arrestins in the termination and transduction of G-protein coupled receptor signals. *J Cell Sci* 115: 455–65.

Luttrell, L. M., Roudabush, F. L., Choy, E. W., Miller, W. M. E., Pierce, K. L., and Lefkowitz, R. J. 2001. Activation and targeting of extracellular signal regulated kinases by beta-arrestin scaffolds. *Proc Natl Acad Sci U S A* 98: 2449–54.

Martinon, F., and Tschopp, J. 2004. Inflammatory caspases: Linking an intracellular innate immune system to autoinflammatory diseases. *Cell* 117: 561–74.

Matuzany-Ruban, A., Avissar, S., and Schreiber, G. 2005. Dynamics of beta-arrestin1 protein and mRNA levels elevation by antidepressants in mononuclear leukocytes of patients with depression. *J Affect Disord* 88: 307–12.

Matuzany-Ruban, A., Golan, M., Miroshnik, N., Schreiber, G., and Avissar, S. 2010. Normalization of GRK2 protein and mRNA measures in patients with depression predict response to antidepressants. *Int J Neuropsychopharmacol* 13: 83–91.

McDonald, P. H., Chow, C. W., Miller, W. E., et al. 2000. Beta-arrestin 2: A receptor-regulated MAPK scaffold for the activation of JNK3. *Science* 290: 1574–7.

McLaughlin, N. J., Banerjee, A., Kelher, M. R., et al. 2006. Platelet-activating factor-induced clathrin-mediated endocytosis requires beta arrestin-1 recruitment and activation of the p38 MAPK signalosome at the plasma membrane for actin bundle formation. *J Immunol* 176: 7039–50.

Metaye, T., Gibelin, H., Perdrisot, R., and Kraimps, J. L. 2005. Pathophysiological roles of G-protein-coupled receptor kinases. *Cell Signal* 17: 917–28.

Miller, W. E., Houtz, D. A., Nelson, C. D., Kolattukudy, P. E., and Lefkowitz, R. J. 2003. G-protein-coupled receptor (GPCR) kinase phosphorylation and beta-arrestin recruitment regulate the constitutive signaling activity of the human cytomegalovirus US28 GPCR. *J Biol Chem* 278: 21663–71.

Miralles, A., Asensio, V. J., and Garcia-Sevilla, J. A. 2002. Acute treatment with the cyclic antidepressant desipramine, but not fluoxetine, increases membrane-associated G protein-coupled receptor kinases 2/3 in rat brain. *Neuropharmacology* 43: 1249–57.

Miyata, S., Hamamura, T., Lee, Y., et al. 2005. Contrasting Fos expression induced by acute reboxetine and fluoxetine in the rat forebrain: Neuroanatomical substrates for the antidepressant effect. *Psychopharmacology (Berl)* 177: 289–95.

Mizrahi, C., Stojanovic, A., Urbina, M., Carreira, I., and Lima, L. 2004. Differential cAMP levels and serotonin effects in blood peripheral mononuclear cells and lymphocytes from major depression patients. *Int Immunopharmacol* 4: 1125–33.

Morelli, M., Pinna, A., Ruiu, S., and Del Zompo, M. 1999. Induction of Fos-like-immunoreactivity in the central extended amygdala by antidepressant drugs. *Synapse* 31: 1–4.

Morinobu, S., Strausbaugh, H., Terwilliger, R., and Duman, R. S. 1997. Regulation of c-Fos and NGF1-A by antidepressant treatments. *Synapse* 25: 313–20.

Morrison, D. K., and Davis, R. J. 2003. Regulation of MAP kinase signaling modules by scaffold proteins in mammals. *Annu Rev Cell Dev Biol* 19: 91–118.

Mostany, R., Valdizán, E. M., and Pazos, A. 2008. A role for nuclear beta-catenin in SNRI antidepressant-induced hippocampal cell proliferation. *Neuropharmacology* 55: 18–26.

Murakami, A., Yajima, T., Sakuma, H., McClaren, M. J., and Inana, G. 1993. X-arrestin: A new retinal arrestin mapping to the X chromosome. *FEBS Lett* 334: 203–9.

Nakielny, S., Cohen, P., Wu, J., and Sturgill, T. 1992. MAP kinase activator from insulin-stimulated skeletal muscle is a protein threonine/tyrosine kinase. *EMBO J* 11: 2123–9.

Newton, S. S., and Duman, R. S. 2004. Regulation of neurogenesis and angiogenesis in depression. *Curr Neurovasc Res* 1: 261–7.

Niculescu, A. B., Segal, D. S., and Kuczenski, R. 2000. Identifying a serious of candidate genes for mania and psychosis: A convergent functional genomics approach. *Physiol Genomics* 9: 83–91.

Nishioka, G., Yamada, M., Kudo, K., et al. 2003. Induction of kf-1 after repeated electroconvulsive treatment and chronic antidepressant treatment in rat frontal cortex and hippocampus. *J Neural Transm* 110: 277–85.

Oakley, R. H., Laporte, S. A., Holt, J. A., Barak, L. S., and Caron, M. G. 1999. Association of beta-arrestin with G protein-coupled receptors during clathrin-mediated endocytosis dictates the profile of receptor resensitization. *J Biol Chem* 274: 32248–57.

Otczyk, M., Mulik, K., Budziszewska, B., et al. 2008. Effect of some antidepressants on the low corticosterone concentration-induced gene transcription in LMCAT fibroblast cells. *J Physiol Pharmacol* 59: 153–62.

Pajonk, F., Scholber, J., and Fiebich, B. 2005. Hypericin—An inhibitor of proteasome function. *Cancer Chemother Pharmacol* 55: 439–46.

Parruti, G., Peracchia, F., Sallese, M., et al. 1993. Molecular analysis of human beta arrestin-1: Cloning, tissue distribution, and regulation of expression. *J Biol Chem* 268: 9753–61.

Perroy, J., Pontier, S., Charest, P. G., Aubry, M., and Bouvier, M. 2004. Real-time monitoring of ubiquitination in living cells by BRET. *Nat Methods* 1: 203–8.

Pickart, C. M. 2001. Mechanisms underlying ubiquitination. *Annu Rev Biochem* 70: 503–33.

Post, A., Crochemore, C., Uhr, M., Holsboer, F., and Behl, C. 2000. Differential induction of NF-kappaB activity and neural cell death by antidepressants in vitro. *Eur J Neurosci* 12: 4331–7.

Prasad, H. C., Zhu, C. B., McCauley, J. L., et al. 2005. Human serotonin transporter variants display altered sensitivity to protein kinase G and p38 mitogen-activated protein kinase. *Proc Natl Acad Sci U S A* 102: 11545–50.

Premont, R. T., and Gainetdinov, R. R. 2007. Physiological roles of G protein-coupled receptor kinases and arrestins. *Ann Rev Physiol* 69: 511–34.

Qi, H., Mailliet, F., Spedding, M., et al. 2009a. Antidepressants reverse the attenuation of the neurotrophic MEK/MAPK cascade in frontal cortex by elevated platform stress: Reversal of effects on LTP is associated with GluA1 phosphorylation. *Neuropharmacology* 56: 37–46.

Qi, X. L., Lin, W. J., Li, J. F., et al. 2008. Fluoxetine increases the activity of the ERK-CREB signal system and alleviates the depressive-like behavior in rats exposed to chronic forced swim stress. *Neurobiol Dis* 31: 278–85.

Qi, X., Lin, W., Wang, D., Pan, Y., Wang, W., and Sun, M. 2009b. A role for the extracellular signal–regulated kinase signal pathway in depressive-like behavior. *Behav Brain Res* 199: 203–9.

Rajagopal, S. Rajagopal, K., and Lefkowitz RJ. 2010. Teaching old receptors new tricks: Biasing seven-transmembrane receptors. *Nat Rev Drug Discov* 9: 373–86.

Ren, X., Dwivedi, Y., Mondal, A. C., and Pandey, G. N. 2010. Cyclic-AMP response element binding protein (CREB) in the neutrophils of depressed patients. *Psychiatry Res* doi: 10.1016/j.psychres.2010.04.013.

Ren, X. R., Reiter, E., Ahn, S., Kim, J., Chen, W., and Lefkowitz RJ. 2005. Different G protein-coupled receptor kinases govern G protein and beta-arrestin-mediated signaling of V2 vasopressin receptor. *Proc Natl Acad Sci U S A* 102: 1448–53.

Ryves, W. J., and Harwood, A. J. 2001. Lithium inhibits glycogen synthase kinase-3 by competition with magnesium. *Biochem Biophys Res Commun* 280: 720–5.

Santarelli, L., Saxe, M., Gross, C., et al. 2003. Requirement of hippocampal neurogenesis for the behavioral effects of antidepressants. *Science* 301: 805–9.

Schafe, G. E., Atkins, C. M., Swank, M. W., Bauer, E. P., Sweatt, J. D., and LeDoux, J. E. 2000. Activation of ERK/MAP kinase in the amygdale is required for memory consolidation of Pavlovian fear conditioning. *J Neurosci* 20: 8177–87.

Schmid, C. L., Raehal, K. M., and Bohn, L. M. 2008. Agonist-directed signaling of the serotonin 2A receptor depends on beta-arrestin-2 interactions in vivo. *Proc Natl Acad Sci U S A* 105: 1079–84.

Schnell, J. D., and Hicke, L. 2003. Non-traditional functions of ubiquitin and ubiquitin-binding proteins. *J Biol Chem* 278: 35857–60.

Schreiber, G., and Avissar, S. 2007. Regulators of G-protein-coupled receptor-G-protein coupling: Antidepressants mechanism of action. *Expert Rev Neurother* 7: 75–84.

Schreiber, G., Golan, M., and Avissar S. 2009. β-Arrestin signaling complex as a target for antidepressants and a depression marker. *Drug News Perspect* 22: 467–80.

Scott, M. G., Le Rouzic, E., Perianin, A., et al. 2002. Differential nucleocytoplasmic shuttling of beta-arrestins: Characterization of a leucine-rich nuclear export signal in beta-arrestin2. *J Biol Chem* 277: 37693–701.

Serafeim, A., Holder, M. J., Grafton, G., et al. 2003. Selective serotonin reuptake inhibitors directly signal for apoptosis in biopsy-like Burkitt lymphoma cells. *Blood* 101: 3212–19.

Shenoy, S. K., and Lefkowitz, R. J. 2003a. Multifaceted roles of beta-arrestins in the regulation of seven-membrane-spanning receptor trafficking and signaling. *Biochem J* 375: 503–15.

Shenoy, S. K., and Lefkowitz, R. J. 2003b. Trafficking patterns of beta-arrestin and G protein-coupled receptors determined by the kinetics of beta-arrestin deubiquitination. *J Biol Chem* 278: 14498–506.

Shenoy, S. K., Drake, M. T., Nelson, C. D., et al. 2006. Beta-arrestin-dependent, G protein-independent ERK1/2 activation by the beta2 adrenergic receptor. *J Biol Chem* 281: 1261–73.

Shenoy, S. K., Barak, L. S., Xiao, K., et al. 2007. Ubiquitination of beta-arrestin links seven-transmembrane receptor endocytosis and ERK activation. *J Biol Chem* 282: 29549–62.

Shenoy, S. K., Modi, A. S., Shukla, A. K., et al. 2009. Beta-arrestin-dependent signaling and trafficking of 7-transmembrane receptors is reciprocally regulated by the deubiquitinase USP33 and the E3 ligase Mdm2. *Proc Natl Acad Sci U S A* 106: 6650–5.

Shukla, A. K., Violin, J. D., Whalen, E. J., Gesty-Palmer, D., Shenoy, S. K., and Lefkowitz, R. J. 2008. Distinct conformational changes in beta-arrestin report biased agonism at seven-transmembrane receptors. *Proc Natl Acad Sci U S A* 105: 9988–93.

Skånland, S. S., Walchli, S., and Sandvig, K. 2009. Beta-arrestins attenuate p38-mediated endosome to Golgi transport. *Cell Microbiol* 11: 796–807.

Slattery, D. A., Morrow, J. A., Hudson, A. L., Hill, D. R., Nutt, D. J., and Henry, B. 2005. Comparison of alterations in c-*fos* and Egr-1 (zif268) expression throughout the rat brain following acute administration of different classes of antidepressant compounds. *Neuropsychopharmacology* 30: 1278–87.

Spanova, A., Kovaru, H., Lisa, V., Lukasova, E., and Rittich, B. 1997. Estimation of apoptosis in C6 glioma cells treated with antidepressants. *Physiol Res* 46: 161–4.

Stambolic, V., Ruel, L., and Woodgett, J. R. 1996. Lithium inhibits GSK-3 activity and mimics signaling in intact cells. *Curr Biol* 6: 1664–8.

Stephen, J. P., and Lefkowitz, R. J. 2002. Arresting developments in heptahelical receptor signaling and regulation. *Trends Cell Biol* 12: 130–8.

Sun, L., and Chen, Z. J. 2004. The novel functions of ubiquitination in signaling. *Curr Opin Cell Biol* 16: 119–26.

Sun, Y., Cheng, Z., Ma, L., and Pei, G. 2002. Beta arrestin2 is critically involved in CXCR4-mediated chemotaxis, and this is mediated by its enhancement of p38 MAPK activation. *J Biol Chem* 277: 49212–9.

Sweatt, J. D. 2001. The neuronal MAP kinase cascade: A biochemical signal integration system subserving synaptic plasticity and memory. *J Neurochem* 76: 1–10.

Tanis, K. Q., Duman, R. S., and Newton, S. S. 2008. CREB binding and activity in brain: Regional specificity and induction by electroconvulsive seizure. *Biol Psychiatry* 63: 710–20.

Thome, J., Sakai, N., Shin, K., et al. 2000. cAMP response element-mediated gene transcription is upregulated by chronic antidepressant treatment. *J Neurosci* 20: 4030–6.

Thrower, J. S., Hoffman, L., Rechsteiner, M., and Pickart, C. M. 2000. Recognition of the polyubiquitin proteolytic signal. *EMBO J* 19: 94–102.

Tohgo, A., Pierce, K. L., Choy, E. W., Lefkowitz, R. J., and Luttrell, L. M. 2002. Beta-arrestin scaffolding of the ERK cascade enhances cytosolic ERK activity but inhibits ERK-mediated transcription following angiotensin AT1a receptor stimulation. *J Biol Chem* 277: 9429–36.

Tohgo, A., Choy, E. W., Gesty-Palmer, D., et al. 2003. The stability of the G protein-coupled receptor beta-arrestin interaction determines the mechanism and functional consequence of ERK activation. *J Biol Chem* 278: 6258–67.

Tsujimura, A., Matsuki, M., Takao, K., Yamanishi, K., Miyakawa, T., and Hashimoto-Gotoh, T. 2008. Mice lacking the *kf-1* gene exhibit increased anxiety- but not despair-like behavior. *Front Behav Neurosci* 2: 4–14.

Tsuruta, F., Sunayama, J., Mori Y., et al. 2004. JNK promotes Bax translocation to mitochondria through phosphorylation of 14-3-3 proteins. *EMBO J* 23: 1889–99.

Violin, J. D., and Lefkowitz, R. J. 2007. Beta-arrestin-biased ligands at seven-transmembrane receptors. *Trends Pharmacol Sci* 28: 416–22.

Wada, A. 2009. Lithium and neuropsychiatric therapeutics: Neuroplasticity via glycogen synthase kinase-3beta, beta-catenin, and neurotrophin cascades. *J Pharmacol Sci* 110: 14–28.

Wang, P., Wu, Y., Ge, X., Ma, L., and Pei, G. 2003. Subcellular localization of beta-arrestins is determined by their intact N domain and the nuclear export signal at the C terminus. *J Biol Chem* 278: 11648–53.

Wang, Y., and Dohlman, H. G. 2006. Regulation of G protein and mitogen-activated protein kinase signaling by ubiquitination: Insights from model organisms. *Circ Res* 99: 1305–14.

Wang, Z. B., Liu, Y. Q., and Cui, Y. F. 2005. Pathways to caspase activation. *Cell Biol Int* 29: 489–96.

Whitherow, D. S., Garrison, T. R., Miller, W. E., and Lefkowitz, R. J. 2004. β-arrestin inhibits NF-κB activity by means of its interaction with the NF-κB inhibitor IκBα. *Proc Natl Acad Sci U S A* 101: 8603–7.

Whitmarsh, A. J. 2007. Regulation of gene transcription by mitogen-activated protein kinase signaling pathways. *Biochem Biophys Acta* 1773: 1285–98.

Willoughby, E. A., and Collins, M. K. 2005. Dynamic interaction between the dual specificity phosphatase MKP7 and the JNK3 scaffold protein beta-arrestin 2. *J Biol Chem* 280: 25651–8.

Xia, Z., DePierre, J. W., and Nassberger, L. 1996. The tricyclic antidepressants clomipramine and citalopram induce apoptosis in cultured human lymphocytes. *J Pharm Pharmacol* 48: 115–16.

Xia, Z., Karlsson, H., DePierre, J. W., and Nassberger, L. 1997. Tricyclic antidepressants induce apoptosis in human T lymphocytes. *Int J Immunopharmacol* 19: 645–54.

Xia, Z., Bergstrand, A., DePierre, J. W., and Nassberger, L. 1999. The antidepressants imipramine, clomipramine, and citalopram induce apoptosis in human acute myeloid leukemia HL-60 cells via caspase-3 activation. *J Biochem Mol Toxicol* 13: 338–47.

Yamada, M., Yamada, M., Yamazaki, S., et al. 2000. Identification of a novel gene with RING-H2 finger motif induced after chronic antidepressant treatment in rat brain. *Biochem Biophys Res Commun* 278: 150–7.

Yang, M., He, R. L., Benovic, J. L., Ye, R. D. 2009. β-Arrestin1 interacts with the G protein subunits β1γ2 and promotes β1γ2-dependent Akt signaling for NF-kB activation. *Biochem J* 417: 287–96.

Yasojima, K., Tsujimura, A., Mizuno, T., et al. 1997. Cloning of human and mouse cDNAs encoding novel zinc finger proteins expressed in cerebellum and hippocampus. *Biochem Biophys Res Commun* 231: 481–7.

Yu, C., Minemoto, Y., Zhan, J. et al. 2004. JNK suppresses apoptosis via phosphorylation of the proapoptotic Bcl-2 family protein BAD. *Mol Cell* 13: 329–40.

Zhang, F., Phiel, C. J., Spece, L., Gurvich, N., and Klein, P. S. 2003. Inhibitory phosphorylation of GSK-3 in response to lithium. Evidence for autoregulation of GSK-3. *J Biol Chem* 278: 33067–77.

Zhang, J., Barak, L. S., Winkler, K. E., Caron, M. G., and Ferguson, S. S. G. 1997. A central role for beta-arrestins and clathrin-coated vesicle mediated endocytosis in beta2-adrenergic receptor resensitization. *J Biol Chem* 272: 27005–14.

Zhu, J. J., Qin, Y., Zhao, M., Van Aelst, L., and Malinow, R. 2002. Ras and Rap control AMPA receptor trafficking during synaptic plasticity. *Cell* 110: 443–55.

Zidar, D. A., Violin, J. D., Whalen, E. J., and Lefkowitz, R. J. 2009. Selective engagement of G protein coupled receptor kinases (GRKs) encodes distinct functions of biased ligands. *Proc Natl Acad Sci U S A* 106: 9649–54.

19 Brain-Derived Neurotrophic Factor and Major Depression

Kazuko Sakata

CONTENTS

19.1 INTRODUCTION

Over the past two decades, increasing evidence has indicated that the pathophysiology of major depressive disorder (MDD) and the action of antidepressants both involve brain-derived neurotrophic factor (BDNF). BDNF, a major neuronal growth factor in the brain, promotes neuronal survival and maturation, as well as plasticity during development and adulthood (Barde 1990; Thoenen 1995; Figurov et al. 1996). Reductions in BDNF levels may cause atrophy and cell loss in the hippocampus (HIP) and prefrontal cortex (PFC), which are morphological changes observed in depressed subjects. On the other hand, neuroprotective actions of BDNF that repair atrophy and

391

synaptic plasticity may be one of the mechanisms of antidepressant actions. BDNF has therefore been of interest for both clinical (human) and preclinical (mainly rodent) studies. As major depression is a complex disorder that does not result from either genetic or environmental influences alone but rather from both (Sullivan et al. 2000), studies have been aimed at determining the associations between *BDNF* gene-encoding variants (mutations) or BDNF levels and depressive behavior under environmental conditions such as stress and antidepressant treatments. This chapter reviews the current knowledge of the roles of BDNF in depression, focusing especially on *BDNF* gene regulation and the underlying molecular and cellular mechanisms.

19.2 BDNF HYPOTHESIS OF DEPRESSION

19.2.1 History

In the 1960s, the monoamine hypothesis of depression was developed based on the fact that many antidepressants increase monoamine levels in the brain (Schildkraut 1965; see Chapter 6). The hypothesis stated that depression was related to a reduction in monoamine levels and that restoration of monoamine levels was the underlying mechanism of antidepressant effects. However, this hypothesis is incomplete because there is a discrepancy in the time course of antidepressant effects—an increase in monoamine levels in the brain occurs within minutes and hours after ingestion of antidepressants, whereas the therapeutic effects take weeks. Therefore, the therapeutic effects of antidepressants have been hypothesized to involve long-term adaptive changes, including gene upregulation of neuroprotective factors such as BDNF.

Since 1995, accumulating evidence has pointed to the *BDNF* gene as a candidate for control of the action of antidepressant treatments and the pathophysiology of depression (for reviews, see Duman et al. 1997; Duman and Monteggia 2006; Castren and Rantamaki 2010). First, stress and glucocorticoids (stress hormones) reduce expression of BDNF mRNA in the HIP (Smith et al. 1995). Second, BDNF expression is decreased in the serum (Karege et al. 2002; Shimizu et al. 2003), HIP, and PFC of patients with MDD (Dwivedi et al. 2003), while this decrease is not seen in patients treated with antidepressants (Chen et al. 2001; Gonul et al. 2005; Karege et al. 2005a). Third, chronic, but not acute, administration of several classes of antidepressants increases BDNF mRNA in the HIP and frontal cortex, and the induction follows a time course observed for the therapeutic effects of antidepressant treatments (Nibuya et al. 1995). Fourth, direct infusion of BDNF protein into the midbrain (Siuciak et al. 1997), HIP (Shirayama et al. 2002), and intracerebroventricular region (Hoshaw et al. 2005) exerts antidepressant effects in rodents. These correlations between stress or antidepressant treatment and down- or upregulation of BDNF contributed to the development of the BDNF hypothesis of depression, that is, reduced BDNF levels lead to depression, and its restoration is a mechanism underlying antidepressant effects. Since BDNF is a major neuronal growth factor, plausible mechanisms are that reduced BDNF levels in response to stress can lead to atrophy and cell loss in the HIP and PFC, whereas increased BDNF expression induced by antidepressants may repair neuronal atrophy and increase neurogenesis and synaptic plasticity (for reviews, see Duman 2002; Nestler et al. 2002).

19.2.2 Causation Studies: Does BDNF Dysfunction Lead to Depression?

19.2.2.1 Previous Controversial Results

Although the BDNF hypothesis of depression is based largely on correlations between stress or antidepressant treatment and down- or upregulation, respectively, of BDNF, the key question of whether reduced BDNF levels can cause depression, rather than being a resulting phenomenon of depression caused by reduced neuronal activity, was not clear until recently. Over the past decade, a number of investigators sought direct evidence that reduced BDNF levels can cause depression using mice genetically manipulated in the *BDNF* gene and rodents injected with exogenous BDNF blockers. Because complete deletion of the *BDNF* gene results in severe developmental defects and mortality within postnatal 3 weeks, most studies used heterozygous or conditional BDNF knockout (KO) mice. Mice with compromised BDNF signaling would be expected to exhibit a depression-like phenotype if BDNF has a role in the pathogenesis of depression. However, most studies showed that mutant mice with reduced BDNF levels failed to exhibit depression-like behavior. For example, both heterozygous and conditional BDNF-KO mice were generally hyperactive and did not show depression-like behavior as determined by the forced swim test, while they exhibited blunted response to antidepressant treatments (Lyons et al. 1999; Kernie et al. 2000; Saarelainen et al. 2003; Chourbaji et al. 2004; Monteggia et al. 2004). In rats, region-selective knockdown of BDNF using virus systems in either the CA1 or the dentate gyrus of the HIP did not alter depression-related behaviors (Adachi et al. 2008), whereas BDNF knockdown in the ventral tegmental area (VTA) and nucleus accumbens (NAc) resulted in increased activity (Krishnan et al. 2007).

Only a limited number of studies reported depression-like behavior of BDNF mutant mice, although the conclusions about the role of BDNF in depression were unclear because of confounding findings. For example, heterozygous BDNF-KO mice displayed depression-like behavior, as shown by their decreased avoidance to foot shocks in the learned helplessness paradigm, although this effect could also be attributed to decreased pain sensitivity (MacQueen et al. 2001). Female (but not male) glial fibrillary acidic protein (GFAP)–dependent conditional BDNF-KO mice and Ca^{2+}/calmodulin-dependent protein kinase (CaMK) type II–dependent BDNF-KO mice showed depression-related behaviors, as measured by the sucrose preference test and the tail suspension test (Monteggia et al. 2007), and female (but not male) enolase-promoter-dependent rTA-inducible BDNF-KO mice exhibited decreased locomotor activity only when stressed (Autry et al. 2009). Although these findings may replicate epidemiological facts of sexual dimorphism in depression vulnerability, the lack of a depressive phenotype in male mice adds to the inconsistencies regarding the role of BDNF in mood regulation. Thus, the major consensus based on these reports has been that rodents with reduced BDNF levels are generally hyperactive and do not show depression-like behavior but rather show a blunted behavioral response to antidepressants (for review, see Duman et al. 2006; Castren and Rantamaki 2010). In addition to the lack of critical evidence that reduced BDNF levels lead to depression-like behavior, other reports suggested the opposite effect of BDNF—increasing depression—depending on the brain region. Eisch et al. (2003) reported that 1 week of BDNF infusions into the VTA produced depression-like behavior, whereas viral-mediated truncated tyrosine kinase receptor B (TrkB) expression, which functions as a blockade of BDNF signaling, in the NAc resulted in increased depression-like behavior, as determined by the forced swim test. This controversial report was also supported by findings that mice exposed to chronic social defeat stress, which displayed increased depression-like behavior, was coupled with an increase in BDNF protein levels in the NAc (Berton et al. 2006) and that a transgenic line of mice with increased BDNF levels displayed anxiety-like behavior (Govindarajan et al. 2006). These contradictory findings have suggested the need for reassessment of the BDNF hypothesis of depression (Groves 2007), as they imply bidirectional roles of BDNF depending on different brain regions—an increase in BDNF levels in the HIP and PFC has an antidepressant effect, while increases in the VTA and NAc exert a depressive effect.

19.2.2.2 Critical Components to Be Considered—Endogenous Gene Regulation of *BDNF*

The question remains: Is the hypothesis that the reduced BDNF levels lead to depression or depression-like behavior true? Before drawing a conclusion, the endogenous gene regulation of *BDNF* should be considered. During extensive studies over the past decade, one critical issue has not been fully considered: are the manipulations being used physiologically relevant? The manipulations (i.e., removing the BDNF protein-coding region under regulation of other gene promoters or exogenous BDNF knockdown/induction by viruses) may have reduced or induced BDNF expression levels to physiologically irrelevant levels (too little or too much compared to the physiological levels). These manipulations also may not have impacted endogenous BDNF levels in the relevant cells and brain regions that would be affected in depressed patients or stressed animals. For example, molecule X (e.g., GFAP) promoter–driven conditional BDNF-KO mice lack BDNF expression in molecule X–expressing cells. Although the protein X-promoter–driven BDNF-KO mice show unchanged behavioral phenotypes, this does not exclude the possibility that reduced endogenous BDNF levels in other cells and brain regions would lead to depression and depression-like behavior. These manipulations can be useful tools for understanding the region- and cell-specific role of BDNF; however, their outcome may not reflect the conditions in human patients. So, what are the possible conditions in patients with depression in terms of *BDNF* gene regulation? Which brain regions and which cells show reduced levels of BDNF expression in patients with depression? To better relate the conditions between human patients and animal models, understanding endogenous *BDNF* gene regulation is critical (see Section 19.3 regarding endogenous *BDNF* gene regulation).

19.2.2.3 Role of Activity-Dependent Promoter IV–Driven Expression of BDNF in Depression

Manipulation of the *BDNF* gene under endogenous BDNF promoters may determine whether reduced BDNF levels lead to depression. At least nine BDNF promoters regulate *BDNF* gene expression (see Section 19.3.1 for the *BDNF* gene structure). Interestingly, recent studies have implicated reduced functions of promoter IV (previously classified as promoter III) in stress conditions and in depressed patients. In rodents, social defeat stress and immobilization stress decrease the function of promoter IV through epigenetic regulation processes (Tsankova et al. 2006; Fuchikami et al. 2009). In the postmortem brain of suicidal subjects, increased methylation (which reduces transcription of the *BDNF* gene) at promoter IV and exon IV has been observed (Keller et al. 2010;

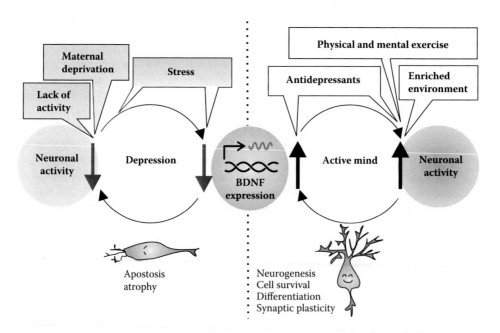

FIGURE 19.1 BDNF hypothesis in depression: involvement of activity-dependent of BDNF expression. Increased neuronal activity induced by physical and mental exercise, enriched environment, and medication upregulates BDNF expression, which in turn increases neuronal activity; and this positive feedback loop can keep active minds. In contrast, any disruption to BDNF expression caused by mutations at promoter regions, epigenetic regulation processes, stress, and reduced neuronal activity would lead to a decrease in neuronal activity and function, which in turn reduces activity-dependent BDNF expression. This vicious cycle of reduced BDNF expression and decreased neuronal activity may cause depression.

see Sections 19.4.5 and 19.5.1 for details). We recently found that mice lacking promoter IV–driven BDNF expression (Sakata et al. 2009) also displayed depression-like behavior, including reduced exploration, anhedonia, and behavioral despair (Sakata et al. 2010). Of the known promoters, promoter IV is the most responsive to neuronal activity and induces activity-dependent expression of BDNF (Timmusk et al. 1993; Shieh et al. 1998; Tao et al. 1998; see Section 19.3). This finding further postulates an intriguing hypothesis that includes the role of neuronal activity-dependent expression of BDNF and a potentially important feedback mechanism for sustaining neuronal activity; increased neuronal activity induces activity-dependent BDNF expression, which would induce neuronal activity to maintain active brain functions. Any disruption to activity-dependent BDNF expression would lead to a decrease in neuronal activity and function, which could lead to depression (Figure 19.1). Because decreased function of BDNF promoter IV may occur in real life via factors such as reduced neuronal stimuli, mutations in the promoter region, and epigenetic processes through stress, further research into the molecular and cellular mechanisms that determine how reduced activity-dependent promoter IV-driven expression of BDNF leads to depression is critical for understanding the pathogenesis of depression.

19.3 *BDNF* GENE STRUCTURE, EXPRESSION, AND FUNCTIONS

Because *BDNF* gene regulation is involved in stress and antidepressant responses, understanding the *BDNF* gene structure, its promoter functions, and epigenetic regulation of its gene transcription would provide critical insights into the pathophysiology of depression. Moreover, since new single nucleotide polymorphisms (SNPs) related to major depression have been recently found in the *BDNF* gene (Licinio et al. 2009; see Section 19.5.3 for details), understanding how these SNPs are involved in the processing of BDNF expression, such as in BDNF transcription, translation, trafficking, and secretion, would reveal possible mechanisms that underlie individual differences in depression susceptibility and responses to antidepressants.

19.3.1 STRUCTURE, PROTEIN PROCESSING, AND FUNCTIONS OF THE *BDNF* GENE

Since cloning of the *BDNF* gene in 1989 (Leibrock et al. 1989; Ernfors et al. 1990), the complex structure of the *BDNF* gene and its regulation have been revealed (see Figure 19.2a). Timmusk et al. (1993) originally reported four promoters that regulate *BDNF* gene expression. Recent studies have found at least nine functional promoters in both humans (Liu et al. 2005; Pruunsild et al. 2007) and rodents (Liu et al. 2006; Aid et al. 2007). Each promoter drives transcription of a short noncoding exon, which is spliced to the common protein-coding exon IX. The transcripts are polyadenylated at either of two alternative sites, leading to transcripts with either a short or a long 3′-untranslated region (UTR), which regulates trafficking (An et al. 2008) and activity-dependent stabilization (Fukuchi et al. 2010) of *BDNF* transcripts.

All *BDNF* transcripts are translated to the identical precursor protein (preproBDNF) in the endoplasmic reticulum. Following cleavage of the signal peptide, proBDNF is transported to the Golgi for sorting into either constitutive or regulated secretory vesicles (Mowla et al. 1999), converted into mature BDNF (mBDNF) by furin or prohormone convertases (Seidah et al. 1996), and secreted (Balkowiec et al. 2000). Secreted BDNF acts on its receptor, TrkB, to produce its neurotrophic effects (Soppet et al. 1991; Thoenen 1995). BDNF binding to TrkB receptor rapidly activates the PI3 kinase-Akt, Ras-mitogen-activated protein kinase (MAPK), and PLCγ pathways, thereby influencing transcriptional events that have multiple effects on cell cycle, neurite outgrowth, neuronal survival, and synaptic plasticity (for review, see Patapoutian and Reichardt 2001).

It has long been thought that only secreted mBDNF is biologically active. However, recent observations of proBDNF secretion (Nagappan et al. 2009) and its conversion to mature BDNF *in vitro* by extracellular proteases, such as plasmin and matrix mettalloproteinase-7, suggest that proBDNF may also be biologically active (Pang and Lu 2004), although the efficiency of intracellular/

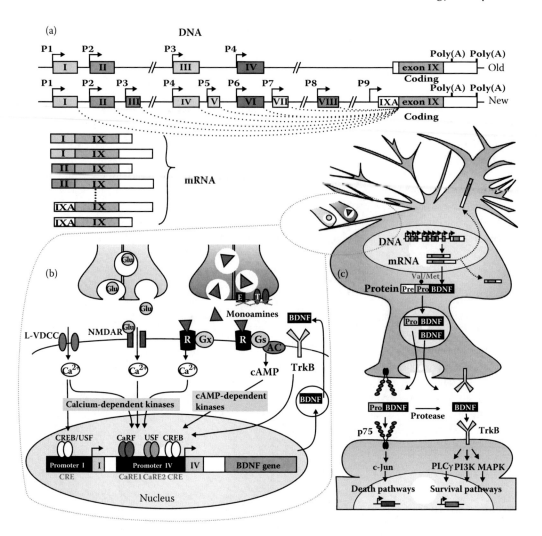

FIGURE 19.2 (See color insert.) (a) Structure of the *BDNF* gene and its promoters. Initial study revealed four promoters in the *BDNF* gene, and recent studies have found at least nine promoters (P1–P9). Accordingly, the nomenclature has been updated. Each promoter drives transcription starting from a small protein-non-coding exon (I-IXA) located immediately downstream of promoter. The transcripts are spliced to the common last exon (IX) which encodes BDNF protein. Each transcription unit uses one of two alternative polyadenylation signals in last exon. Thus, *BDNF* gene produces alternate forms of BDNF mRNA. (b) Three ways to induce of BDNF transcription: neuronal activity, antidepressants, and BDNF itself. Of the known promoters, promoters I and IV, located upstream of exons I and IV, respectively, are responsive to neuronal activity and Ca^{2+} influx. These promoters have binding sites of transcriptional factors, such as CREB, CaRF, and USF. Ca^{2+} influx activates these transcriptional factors via calcium-dependent kinases. Antidepressant treatments increase amount of neurotransmitters available at synaptic clefts by blocking transporters, which activate their receptors (R) and regulate cAMP or Ca^{2+} signaling cascades. This activates transcriptional factors and upregulates *BDNF* gene transcription. BDNF itself can also induce BDNF transcription via TrkB receptor signaling cascades. (c) BDNF protein processing and signaling pathways. BDNF transcripts are translated to a precursor protein (preproBDNF) in endoplasmic reticulum. PreproBDNF is processed to become proBDNF in the Golgi. ProBDNF is cleaved by proteases either intracellularly or extracellularly and becomes mature BDNF. When released, mature BDNF produces neuroprotective functions via TrkB receptor signaling pathways, whereas proBDNF produces apoptotic effects via p75 receptor signaling pathways. CaRE, calcium-responsive element; CRE, cAMP/calcium response element; T, transporter; E, degrading enzyme; AC, adenylate cyclase; CREB, CRE-binding protein; USF, upstream transcription factor; CaRF, calcium-response transcription factor; L-VDCC, L-type voltage-dependent calcium channel; Glu, glutamate.

extracellular cleavage is controversial (Matsumoto et al. 2008). Interestingly, proBDNF has been reported to induce neuronal apoptosis via activation of the p75^{NTR} receptor (Teng et al. 2005), whose activation can trigger increases in c-Jun N-terminal kinase and nuclear factor κB (Roux and Barker 2002). Thus, it is hypothesized that proBDNF and mature BDNF, both products of the *BDNF* gene, can elicit opposite effects of apoptosis or cell survival via the p75^{NTR} or TrkB receptors, respectively (Figure 19.2c) (for review, see Lu et al. 2005 and Greenberg et al. 2009). These receptors may be differentially involved in the pathophysiology of depression (for review, see Martinowich et al. 2007).

The physiological significance of activity-dependent secretion of BDNF has been highlighted because an SNP in the pro region of BDNF (Val66Met) results in marked impairments in dendritic trafficking and regulated secretion of BDNF (Egan et al. 2003; Chen et al. 2004; Chiaruttini et al. 2009), and the Val66Met SNP is thought to be involved in susceptibility to neuropsychiatric and neurodegenerative disorders. Human subjects carrying the Val66Met polymorphism in the *BDNF* gene exhibit impairments in hippocampal function and hippocampal-specific short-term memory (Egan et al. 2003). An increasing number of studies have investigated an association between the Val66Met SNP and susceptibility to major depression; however, results are still inconclusive (see Section 19.5.3).

19.3.2 Differential Expression of *BDNF* Exons at Baseline Conditions

BDNF gene expression is tightly regulated by its promoters with or without stimuli (see Section 19.4 regarding stimuli-dependent *BDNF* gene regulation). Promoter-selective BDNF expression can be detected by measuring the expression of its respective exons. The expression levels and pattern of the original four exons (I, II, IV, and VI in the new nomenclature; see Figure 19.2a) have been studied, whereas the newly found exons (III, V, VII, VIII, and XA in the new nomenclature) are less known, perhaps because of relatively low expression levels of the newly found exons compared with those of the original four exons (Pruunsild et al. 2007; Sakata K., unpublished data). Although all original four exons are expressed predominantly in the HIP and cortex, differential expression patterns of each exon have been observed when examined by *in situ* hybridization (Malkovska et al. 2006). For example, exon IV is expressed at relatively high levels in areas related to high synaptic plasticity, such as the cortex and HIP, whereas it is expressed strikingly at low levels in all regions of the thalamus and hypothalamus (see Figure 19.3a). On the other hand, exons I and II are expressed in a relatively small subset of cortical neurons in the cerebral cortex, whereas exons II and VI are expressed in the central medial thalamic nucleus. In the HIP, expression of exons IV and VI accounts for most of the BDNF expression in the CA1 region, whereas expression of exons I and II is highest in the CA3 region and lowest in the CA1 region. When the absolute amount of BDNF exon mRNA was determined by an RNase protection assay, exon IV was the most abundantly expressed in the cerebral cortex among all exons, whereas exon II was abundantly expressed in the brain stem and cerebellum (Timmusk et al. 1994; Figure 19.3b). Promoter-specific *BDNF* expression undergoes temporal regulation during development and diurnal regulation in different brain regions (Timmusk et al. 1994; Lauterborn et al. 1996). Furthermore, different distributions even within different parts of a cell have been found; exon IV expression is detected only in cell bodies, whereas exon VI expression is present in dendritic processes as well as in cell bodies of the visual cortex neurons (Pattabiraman et al. 2005). Recent studies have found that the different distributions of BDNF transcripts are attributable to different trafficking in the cells depending on the RNA structure. The short 3′-UTR mRNAs are restricted to somata, whereas the long 3′-UTR mRNAs are also localized in dendrites (An et al. 2008). Based on the different expression patterns regulated by each BDNF promoter, alterations in a certain BDNF promoter in specific neuronal subpopulations may contribute to specific behavioral phenotypes in depression, such as anxiety, anhedonia, and reduced attention. Rodent models with promoter-specific BDNF mutations may establish this possibility.

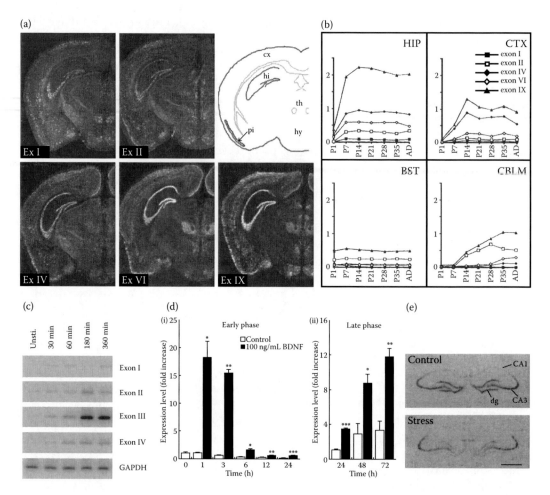

FIGURE 19.3 *BDNF* gene expression and its regulation by different promoters. (a) Differential expression of BDNF exons at baseline conditions. *In situ* hybridization showing mRNA expression of exons I, II, IV, VI, and IX. CA1 region, parafascicular thalamic nucleus, and cortex are striking examples of differential expression. Expression patterns of exons IV and VI are similar, except for low expression of exon IV in thalamus and hypothalamus. Exon VI shows a pronounced expression in parafascicular thalamic nucleus. Exons I and II are expressed in a relatively small subset of cortical neurons in cerebral cortex. Exons II and VI are expressed in central medial thalamic nucleus. In HIP, expression of exons IV and VI accounts for most of BDNF expression in CA1 region, in addition to being high in CA3 region. Exons I and II are highest in CA3 region and lowest in CA1 region. (Modified and used with permission from Malkovska, I., et al., *J Neurosci Res*, 83, 211–21, 2006. Copyright by John Wiley and Sons.) (b) Expression of exon-specific BDNF mRNAs in rat brain regions during postnatal development detected by RNase protection assay. Absolute amount of BDNF exon mRNA in picograms per microgram of total RNA is presented. (Modified and used with permission from Timmusk, T., et al., *Neuroscience*, 60, 287–91, 1994. Copyright by Elsevier.) (c) Neuronal activity-dependent expression of BDNF exons. Cultured cortical neurons were stimulated by 50 mM KCl, and expression levels of exons were detected by RT-PCR. Note the robust induction of exon IV transcripts. (Modified and used with permission from Tao, X., et al., *Neuron*, 20, 709–26, 1998. Copyright by Elsevier.) (d) Biphasic stimulation of BDNF exon IV–IX expression induced by BDNF in cultured cortical neurons. Three days after the change to serum-free medium, BDNF (100 ng/ml) or vehicle was added. Incubation was carried out for 1, 3, 6, 12, or 24 h (i; early phase) and for 24, 48, or 72 h (ii; late phase) in presence or absence of BDNF. Expression levels of BDNF exons IV to IX were measured by real-time RT-PCR. Level of expression (fold increase) was normalized to control at zero time for early phase and to control at 24 h for late phase. Data represent mean ± SE ($n = 3$); $*p < .05$, $**p < .01$, $***p < .001$ versus control at same time point.

19.4 GENE REGULATION OF *BDNF*

"Healthy factors" such as physical exercise (Neeper et al. 1995), learning training (Hall et al. 2000), being reared in an enriched environment (Falkenberg et al. 1992), and administration of antidepressants (Nibuya et al. 1995) all induce expression of BDNF, especially in the HIP and cortex. Because upregulation of endogenous BDNF expression in the brain could serve as a potential form of therapy for MDD (and many other mental disorders), understanding the mechanisms of how activity-dependent expression of BDNF occurs and how it contributes to neuronal adaptation in response to different stimuli is essential and has attracted a significant amount of interest. On the other hand, negative factors such as stress reduce BDNF expression in the HIP and PFC, and understanding the underlying mechanisms could provide significant insights into how depression occurs.

19.4.1 *BDNF* Gene Upregulation by Neuronal Activity

BDNF expression is regulated by neuronal activity. Zafra et al. (1990) initially found that kainic acid, a glutamate receptor agonist and seizure inducer, significantly increased BDNF mRNA levels (more than ~10 times) both *in vitro* and *in vivo*. Since then, *BDNF* gene upregulation by different neuronal stimuli has been extensively examined. Neuronal depolarization induced by high potassium (Zafra et al. 1990), glutamate (Zafra et al. 1991; Lindefors et al. 1992), and a GABA antagonist (bicuculine) (Zafra et al. 1991) has been shown to acutely induce BDNF mRNA expression (3 h after administering stimuli) in cultured neurons. *In vivo* conditions that increase neuronal activity in the brain, such as physical exercise (Neeper et al. 1995), learning tasks (Kesslak et al. 1998; Hall et al. 2000), and enriched environment (Falkenberg et al. 1992), as well as seizures inducing stimuli (Zafra et al. 1990; Isackson et al. 1991), have also been shown to induce BDNF mRNA expression in the HIP and cortical regions—the brain regions important for synaptic plasticity. Stimuli that induce long-term potentiation, a form of synaptic plasticity, have been shown to induce BDNF mRNA expression in the HIP, suggesting an important role of activity-dependent expression of BDNF in learning and memory (Patterson et al. 1992, 1996).

19.4.2 *BDNF* Gene Upregulation by Antidepressant Treatments

Different classes of antidepressants, as well as electroconvulsive shock treatment (ECT) also increase *BDNF* gene expression in the HIP and cortex, although longer time is required for the increase and the amount of increase by itself is not robust (about a 1.4-fold to 3-fold increase) compared with that observed with neuronal depolarization. Chronic (21-day), but not acute (1-day), administration of several different classes of antidepressants, including selective serotonin and norepinephrine selective reuptake inhibitors (SSRIs and NRIs), tricyclic antidepressants (TCA), tetracyclic antidepressants, and monoamine oxidase inhibitors (MAOI), increases BDNF mRNA expression in the HIP (Nibuya et al. 1995, 1996). Although ECT therapy is most often used for patients with severe major depression who have not responded to other antidepressant treatments, both acute (2-h) and chronic (10-day) ECT has been shown to increase BDNF mRNA approximately 2- to 3-fold in the HIP (Nibuya et al. 1995).

FIGURE 19.3 (Continued) (Modified and used with permission from Yasuda, M., et al., *J Neurochem*, 103, 626–36, 2007. Copyright by John Wiley and Sons.) (e) BDNF hybridization in HIP in rats stressed for 7 consecutive days and control rats. Note the robust decrease in BDNF mRNA expression in dentate gyrus (dg) in stressed rat. (Modified and used with permission from Smith, M.A., et al., *J Neurosci*, 15, 1768–77, 1995. Copyright by Society for Neurosciece.) CTX, cerebral cortex; BST, brain stem; CBLM, cerebellum; P, postnatal day; AD, adult; GAPDH, glyceraldehyde 3-phosphate dehydrogenase; Unsti., unstimulated; Ex, exon.

Whereas many antidepressants are supposed to increase monoamine levels in the brain, direct application of monoamines (serotonin, norepinephrine, dopamine) to hippocampal neurons had no effect in inducing BDNF mRNA expression in an acute phase (3 h) (Zafra et al. 1990). One explanation may be that monoamines may exert gradual neuromodulatory effects, acting through G-protein-coupled receptors and ligand-gated ion channels. This would contrast with the known effects of glutamate- and potassium-induced neuronal depolarization that rapidly increases intra-cellular calcium (see Section 19.4.3). In addition, monoamines affect cells and brain regions other than hippocampal neurons. For example, norepinephrine causes a robust increase in BDNF mRNA levels in cultured astrocytes, in which basal BDNF levels without stimuli are so low as to be almost undetectable (Zafra et al. 1992). Dopaminergic effects on *BDNF* gene upregulation are observed in the striatum, where dopamine receptors are abundantly expressed; dopamine and dopamine D_1 receptor agonists upregulate BDNF expression in striatal neuronal and astroglial cells (Kuppers and Beyer 2001), and *in vivo* dopaminergic stimulation by levodopa upregulates BDNF expression in the striatum (Okazawa et al. 1992). Serotonin (5-HT) acts through a large family of receptors, and its effects depend on its receptor subtypes. An agonist for G_s-protein-linked 5-HT$_6$ receptors that stimulate adenylate cyclase upregulates BDNF mRNA levels (1.5-fold increase) relatively acutely (after 18 h) (de Foubert et al. 2007), whereas a 5-HT$_{2A/2C}$ agonist upregulates BDNF mRNA expression (2-fold increase) in the parietal cortex but decreases it in the dentate gyrus of the HIP 3 h after intraperitoneal injection (Vaidya et al. 1997).

19.4.3 SIGNALING PATHWAYS INVOLVED IN AN INDUCTION OF *BDNF* TRANSCRIPTION

The difference in the induction efficacy of BDNF transcription—the robust and fast induction by neuronal activity (usually more than a 6-fold increase within 3 h after administering stimuli) and the slow and moderate induction by antidepressant treatment (approximately a 1.4- to 3-fold increase over several weeks)—can be explained by the operation of different intracellular signaling path-ways (Figure 19.2b). The induction of *BDNF* transcription involves at least two second-messenger cascades: the calcium-signaling cascade and the cyclic adenosine monophosphate (cAMP)–signal-ing cascade. Neuronal activity-dependent induction of *BDNF* transcription depends on intracellular levels of calcium, as a calcium channel blocker (i.e., nifedipine) dose-dependently abolishes the potassium depolarization–mediated and kainic acid–mediated induction of BDNF mRNA, whereas increasing intracellular calcium levels by calcium ionophores (e.g., Bay-K8644 that opens L-type calcium channels) robustly increases BDNF mRNA levels (more than 5-fold) in hippocampal neu-rons (Zafra et al. 1992). While calcium influx also occurs through NMDA receptors (glutamate receptors), NMDA receptors mediate maintenance of the normal levels of BDNF mRNA, whereas non-NMDA receptors mediate an increase in BDNF mRNA above normal levels (Zafra et al. 1991). The increase in intracellular calcium levels induces BDNF mRNA expression by activating calm-odulin and CAMK type II and IV (CaMKII and CaMKIV), but not protein kinase C (PKC), where the induction is blocked by calmodulin antagonists or a CaMKII/IV inhibitor but not by PKC antago-nists, although a PKC activator, phorbol ester, can slightly induce BDNF mRNA transcription (Zafra et al. 1992).

On the other hand, induction of BDNF mRNA levels by treatments with antidepressants and ECT may involve a cAMP cascade, as chronic treatments increase cAMP levels through stimula-tory GCPRs (Menkes et al. 1983). This suggestion is supported by the fact that an increase in cAMP levels by an activator (i.e., forskolin) of adenylate cyclase, an enzyme that produces cAMP, increases BDNF mRNA expression slightly in hippocampal neurons (~1.5-fold) and robustly in astrocytes that otherwise express very low amounts of BDNF mRNA (Zafra et al. 1992). In addition, chronic administration of inhibitors (e.g., rolipram) of phosphodiesterase (PDE), a catalyzing enzyme of cAMP, also slightly increases BDNF mRNA in the HIP (~1.5-fold) (Nibuya et al. 1996).

Both calcium and cAMP cascades can activate a transcription factor, cAMP response element–binding protein (CREB), by calcium-dependent protein kinases or cAMP-dependent protein kinase

(PKA) (Mayr and Montminy 2001; West et al. 2001), whereas increased calcium levels can activate another transcription factor, calcium-response factor (CaRF) (Tao et al. 1998). Activation of these transcriptional factors leads to the induction of BDNF mRNA expression (Figure 19.2b). It is interesting to note that CaRF is required for neuronal activity-driven increased BDNF mRNA levels (Tao et al. 2002), while CREB is required for antidepressant-induced increases in BDNF mRNA levels (Conti et al. 2002).

Interestingly, these two second-messenger pathways—calcium and cAMP cascades—can interact with each other and augment the induction of *BDNF* transcription. For example, an increase in cAMP by forskolin and PDE inhibitors alone induces only slight (1.5-fold) expression of BDNF mRNA; however, it markedly augments an induction of *BDNF* transcription by NMDA receptor–independent neuronal depolarization (6- to 12-fold) in hippocampal neurons (Zafra et al. 1992) and shortens the time required for the induction of BDNF mRNA expression (Nibuya et al. 1996). On the other hand, stimulation of CREB-dependent *BDNF* transcription by cAMP requires activation of NR2B-containing NMDARs and an increase in intracellular calcium levels (Almeida et al. 2009). This suggests that a combination of neuronal activity and antidepressant treatments can augment each other in inducing BDNF mRNA expression by acting on different signaling pathways. Indeed, a combination of physical exercise and antidepressant treatment has an additive, potentiating effect on BDNF mRNA expression in the HIP (Russo-Neustadt et al. 1999). Accumulating evidence has shown that treatments that use other signaling pathways also induce BDNF mRNA. For example, chronic administration of atypical antipsychotic agents (clozapine and olanzapine), which are used for augmentation of antidepressants, induces BDNF mRNA levels in the HIP (Bai et al. 2003). In addition, administering a group II metabotropic glutamate receptor (mGluR2/3) agonist acutely (3 h) induces BDNF mRNA levels in the cortex and HIP (Di Liberto et al. 2010). Furthermore, hormones such as estrogen and thyroid hormone also upregulate BDNF mRNA levels (Sohrabji et al. 1995; Koibuchi et al. 1999). Interestingly, stimulation of TrkB with BDNF itself induces promoter IV–driven BDNF mRNA expression in a biphasic manner (1–3 and 24–72 h after BDNF application; Figure 19.3d) in cortical neurons; secondary induction no longer requires exogenous BDNF application, indicating an autoactivation mechanism of BDNF mRNA induction (Yasuda et al. 2007). The primary activation is controlled by the extracellular signal-regulated kinase/mitogen-activated protein kinase (ERK/MAPK) pathway, whereas the secondary induction is controlled by the calcium signaling pathway. Therefore, understanding the signaling pathways involved in the induction of *BDNF* transcription and applying the knowledge to antidepressant intervention may greatly aid the development of effective treatments of major depression in the future.

19.4.4 PROMOTER-SPECIFIC GENE REGULATION BY NEURONAL ACTIVITY AND ANTIDEPRESSANTS

Of the known BDNF promoters, promoter IV is the most responsive to neuronal activity by depolarization and induces activity-dependent expression of BDNF *in vitro* and *in vivo*. Promoter IV contains calcium-responsive elements and cAMP/calcium response elements (CREs) to which CaRF and CREB bind, respectively (Timmusk et al. 1993; Shieh et al. 1998; Tao et al. 1998, 2002) (Figure 19.2b). Although CREB can be activated by phosphorylation via several kinases including CaMKs, PKA, and MAPK upon neuronal depolarization (Impey et al. 1998), induction of exon IV via activation of CREB by calcium influx requires CaMK cascades, specifically CaMKIV but not CaMKII (Shieh et al. 1998). On the other hand, CaRF can induce calcium- and neuron-selective exon IV transcriptional activity via phosphorylation by CaMKII and MAPK (Tao et al. 2002). Alternatively, calcium influx by neuronal depolarization phosphorylates methyl-cytosine (CpG) binding protein 2 (MeCP2), a transcriptional repressor, and leads to the release of its repressor complex (MeCP2–histone deacetylase [HDAC1]–mSin3A) from BDNF promoter IV (Chen et al. 2003; Martinowich et al. 2003). Interestingly, activity-dependent induction of promoter IV–driven BDNF expression correlates with a decrease in CpG methylation within the regulatory region of BDNF promoter IV (Martinowich et al. 2003). This epigenetic regulation demands special attention: the gene silencing effect by DNA methylation at the promoter IV region may be involved in the pathophysiology of major depression (see Section 19.4.5), while mutations in the *MeCP2* gene cause a majority of neurodevelopmental disorders, including Rett syndrome (Amir et al. 1999).

In addition to promoter IV, promoter I is responsive to neuronal activity to induce activity-dependent expression of BDNF transcripts. Promoter I contains an upstream transcription factor (USF)–binding/CRE to which USF1/2 and CREB bind (Tabuchi et al. 2002). Whereas promoter IV is activated by the calcium influx through either NMDA receptors or L-type voltage-dependent calcium channels (L-VDCC), promoter I is predominantly activated by the calcium influx by L-VDCC (Tabuchi et al. 2000) (Figure 19.2b). Strong seizure-inducing stimuli such as by kainic acid induce BDNF mRNA expression through almost all BDNF promoters except promoter III and VI in the HIP (Metsis et al. 1993; Aid et al. 2007).

Interventions that can be used for antidepressant treatments activate different promoters in different brain regions (Table 19.1). For example, chronic administration of desipramine (TCA) induces exons I and IV (promoter I– and promoter IV–driven BDNF expression) in the frontal cortex and HIP (Dwivedi et al. 2006), whereas duloxetine (serotonin–norepinephrine reuptake inhibitor) enhances levels of exons I and IV in the frontal cortex but not in the HIP (Calabrese et al. 2007). Phenelzine (MAOI, for both MAO_A and MAO_B) increases exon I transcripts in both frontal cortex and HIP, whereas exon IV is upregulated only in the HIP (Dwivedi et al. 2006). Tranylcypromine (MAOI, with a slight preference for MAO_B) is reported to enhance exon II expression (Dias et al. 2003) or exon I in the HIP (1.5-fold) (Khundakar and Zetterstrom 2006). Fluoxetine (SSRI) induces only promoter II–driven BDNF expression very slightly (1.2-fold) selectively in the HIP (Dwivedi et al. 2006), although other studies have reported no significant effect on any promoter-specific BDNF expression (Dias et al. 2003; Calabrese et al. 2007). The effects of both acute and chronic ECT are most prominent on exons I and II in hippocampal and cortical subfields (1.5- to 4-fold) (Dias et al. 2003). In addition, in the HIP, long-term (28–56 days) physical exercise increases exon I (2.5-fold), exon II (1.5-fold), and exon III (2.3-fold) (Adlard et al. 2004; Zajac et al. 2009); and 8 weeks of enriched environment rearing induces BDNF transcripts through exons IV and VI in male animals, and exon III in female animals (1.5-fold) (Zajac et al. 2009). The mechanisms of how different conditions of antidepressant interventions lead to promoter-specific induction of BDNF mRNA expression remain to be addressed.

19.4.5 PROMOTER-SPECIFIC DOWNREGULATION OF BDNF EXPRESSION

Negative environmental effects such as acute and chronic physical stress (Smith et al. 1995), psychological stress (Rasmusson et al. 2002), and chronic alcohol intake (MacLennan et al. 1995) reduce the expression of BDNF. Glucocorticoid (a stress hormone) prevents activity-dependent increases of BDNF mRNA in hippocampal neurons (Cosi et al. 1993), whereas glucocorticoid deficiency by adrenalectomy markedly increases exon I and II expression in the HIP (Lauterborn et al. 1998). It is interesting to note that both promoters I and II have AP1 sites, whereas glucocorticoid receptors, prototypical nuclear receptors, prevent their transcriptional activity by interacting with transcription factors such as AP-1 (Karin 1998). Although glucocorticoid is one factor of stress-induced repression of *BDNF* transcription, additional components of the stress response must also be involved in stress-induced reduction of BDNF mRNA levels because stress can still decrease BDNF mRNA levels in the dentate gyrus in adrenalectomized rats (Smith et al. 1995). Further studies have demonstrated that the reduction in BDNF mRNA involves epigenetic regulation processes. For instance, a single immobilization stress decreases levels of acetylated histon (H) 3, but not H4, at promoters I, IV, and VI and reduces exon I and IV mRNA levels (Fuchikami et al. 2009), while social defeat stress, which can lead to depression-like behavior in mice, produces long-lasting suppression of BDNF transcripts IV and VI in the HIP, accompanied with a robust increase in histone 3-lysine 27 dimethylation, a repressive histone modification state, on the chromatin of the respective promoters (Tsankova et al. 2006). Intriguingly, chronic imipramine reverses this downregulation

TABLE 19.1
Promoter-Specific *BDNF* Gene Regulation by Antidepressants and Stress Conditions

Treatment	Drug	Type	Brain Region	Exon					Reference
				I	II	IV	VI	IX	
Antidepressants (chronic)	Fluoxetine	SSRI	FC	—	—	—	—	—	Dwivedi et al. (2006)
			HIP	—	↑	—	—	—	Dwivedi et al. (2006) Dias et al. (2003)
			FC, HIP	—	—	—	—	—	Calabrese et al. (2007)
			FC, HIP	—	—	—	—	—	Conti et al. (2002)
			HIP (DG)	↑	—	—	—	↑	Khundakar and Zetterstrom (2006)
	Duloxetine	SNRI	FC	↑	↑2c	↑	↓	↑	Calabrese et al. (2007)
			HIP	—	—	—	—	↑	Molteni et al. (2009)
	Desipramine	TCA	FC, HIP	↑	—	↑	—	↑	Dwivedi et al. (2006)
			Ctx, Amg	—	—	↑	—	nd	Dias et al. (2003)
			FC, HIP	nd	nd	nd	nd	↑	Conti et al. (2002)
			HIP (DG)	—	nd	—	—	↑	Khundakar and Zetterstrom (2006)
	Imipramine	TCA	HIP	—	—	↑	↑	↑	Tsankova et al. (2006)
	Tranylcypromine	MAOI	HIP, Pir Ctx	nd	↑	—	—	nd	Dias et al. (2003)
			HIP	↑	—	nd	nd	nd	Russo-Neustadt et al. (2000)
			HIP (DG)	↑	nd	—	—	↑	Khundakar and Zetterstrom (2006)
	Phenelzine	MAOI	FC	↑	—	—	—	↑	Dwivedi et al. (2006)
			HIP	↑	—	—	↑	↑	Dwivedi et al. (2006)
ECT			HIP	↑	↑	↑	↑	nd	Dias et al. (2003)
			Ctx	↑	↑	↑	—	nd	Dias et al. (2003)
			Amg	↑	↑	—	—	nd	Dias et al. (2003)
			HIP (DG)	↑	↑	↑	↑	nd	Dias et al. (2003)
			Ctx, Amg	↑	↑	—	—	nd	Dias et al. (2003)
Stress	Social defeat		HIP	—	—	↓	↓	↓	Tsankova et al. (2006)
	Immobilization		HIP	↓	—	↓	—	↓	Fuchikami et al. (2009)
	Immobilization (acute)		HIP (DG)	↓	↓	↓	↓	↓	Nair et al. (2007)
	Immobilization (chronic)		HIP (DG)	↑	↑	↓	↓	↓	Nair et al. (2007)
	Maternal separation (acute)		HIP	↓	—	—	—	—	Nair et al. (2007)
	Maternal separation (chronic, p14)		HIP	—	↑	—	—	—	Nair et al. (2007)
	Maternal separation (chronic, p21)		HIP	—	—	—	↑	↑	Nair et al. (2007)
	Corticosterone (21 days)			—	↓	—	↓	↓	Dias et al. (2003)
	Prenatal dexamethasone		PVN	nd	nd	↓	nd	nd	Hossain et al. (2008)
	Swim stress (acute)		HIP	—	—	↑	—	—	Molteni et al. (2009)

Note: —, no change; ↑, increase; ↓, decrease; nd, no data; Fc, frontal cortex; HIP, hippocampus; Ctx, cortex; Amg, amygdala; Pir, piriform cortex; PVN, paraventricular nucleus; DG, dentate gyrus; SSRI, selective serotonin reuptake inhibitor; SNRI, serotonin-norepinephrine reuptake inhibitor; TCA tricyclic antidepressant; MAOI, monoamine oxidase inhibitor; ECT, electroconvulsive shock therapy.

by increasing histone acetylation (a more euchromatic state) at these promoters via downregulation of histone deacetylase 5 (Tsankova et al. 2006). These results highlight chromatin remodeling and epigenetic regulation in the *BDNF* gene, which are important mechanisms controlling long-term adaptive brain changes associated with major depression and its treatment.

19.4.6 EFFECTS OF EARLY-LIFE STRESS ON BDNF EXPRESSION

In addition to the effect of stress in adulthood, long-lasting repression of *BDNF* transcription during development resulting from early-life stress has recently been noted. Roceri et al. (2004) reported that repeated maternal deprivation from postnatal day (PND) 2 to PND 14 produces short-term upregulation of BDNF expression in the HIP and PFC, as measured on PND 17, whereas at adulthood, maternal deprivation results in a selective reduction of BDNF expression in the PFC. Importantly, dynamic methylation of exon IV may likely be a mechanism mediating *BDNF* gene expression during development and thus is susceptible to environmental insults (Dennis and Levitt 2005). Recently, Roth et al. reported that infant maltreatment (the first postnatal week in rats) results in methylation of the *BDNF* DNA at both exon IV and IX DNA regions throughout the life span to adulthood that dovetails reduced *BDNF* gene expression in the adult PFC but not in the HIP (for reviews, see Roth et al. 2009; Roth and Sweatt 2011). Furthermore, offspring derived from maltreated female rats exhibited the same abusive behaviors as their mothers did and showed greater DNA methylation in the PFC and HIP, indicating a generation-to-generation transmission of the perpetuation of maternal behavior and previously acquired DNA methylation pattern (Roth et al. 2009). Interestingly, further cross-fostering experiments showed no significant change in DNA methylation patterns, in the direction from either maltreated offspring to normally caring mothers or from normally treated offspring to maltreating mothers. Thus, the changes in DNA methylation acquired through maltreating experiences are not easy to reverse by switching to a well-caring environment. In addition, conditions during pregnancy also affect the BDNF levels in offspring. For example, treatment with dexamethasone (a synthetic glucocorticoid) to mothers during pregnancy abolished the acoustic stress-induced upregulation of the BDNF exon IV and TrkB mRNA expression without changing the basal expression levels in the hypothalamic paraventricular nucleus, the brain region that releases corticotropin-releasing hormone leading to stimulation of glucocorticoid secretion in the hypothalamic–pituitary–adrenal (HPA) axis (see Chapter 9 for details about the HPA axis; Hossain et al. 2008). Tozuka et al. (2010) recently reported that offspring from obese mothers treated with a high-fat diet exhibited reduced levels of BDNF mRNA and dendrite arborization in the HIP between PND 21 and PND 35. On the other hand, enriched environments during early development may have opposite effects on *BDNF* gene regulation, as infant mice raised in communal nests where they experienced higher levels of maternal care had an increased propensity for social interaction in adulthood that is correlated with increased hippocampal BDNF protein levels (Branchi et al. 2006). Further studies that aim to identify epigenetic mechanisms of BDNF involved between environment and gene regulation in detailed spatial and temporal aspects will provide new leads toward understanding the pathophysiology and treatment of depression.

19.5 BDNF AND DEPRESSION IN HUMAN STUDIES

19.5.1 STUDIES WITH POSTMORTEM HUMAN BRAIN SAMPLES

Postmortem analysis has detected decreased BDNF and TrkB expression in the HIP and PFC of depressed patients who committed suicide and increased levels in patients medicated with antidepressants before death (Chen et al. 2001; Dwivedi et al. 2003; Karege et al. 2005a). Although untreated depression is the number one cause for suicide (WHO 2009), more recent epigenetic research has revealed increased methylation (that reduces expression of BDNF mRNA) at the DNA regions of BDNF promoter IV and exon IV in the Wernicke area of the postmortem brains from suicide subjects compared with nonsuicide control subjects without changes in genome-wide methylation levels, suggesting a novel link between epigenetic alteration in the brain and suicidal behavior (Keller et al. 2010). Interestingly, proBDNF expression is reduced in the right, but not the left, HIP of patients with MDD, suggesting that the lateralization of changes may be involved in the pathophysiology of depression (Dunham et al. 2009), although further detailed studies on the role of proBDNF in major depression are needed. Because the neuroprotective role of BDNF has been specifically suggested as an antidepressant effect, reduced BDNF expression in patients with MDD may explain neuronal atrophy and loss of neuronal and glial cells observed in their brain (Rajkowska 2000).

19.5.2 SERUM BDNF LEVELS IN LIVING PATIENTS WITH DEPRESSION

Given the difficulty of obtaining brain samples of living patients, measuring BDNF levels in human peripheral blood garners special attention because it may serve as a possible clinical biomarker for depression. BDNF expression is detectable in serum blood, although its expression levels are much lower compared with that in brain tissues (Yamamoto and Gurney 1990; Radka et al. 1996). Accumulating evidence shows that serum BDNF in living depressed patients is reduced with a negative correlation to the severity of depression but can be restored after pharmacological antidepressant treatment (Karege et al. 2002; Shimizu et al. 2003; Aydemir et al. 2005; Gervasoni et al. 2005; Karege et al. 2005b; for meta-analyses, see Brunoni et al. 2008; Sen et al. 2008). These results suggest a possibility that serum BDNF levels may serve as a biomarker for depression levels and its recovery. Whether serum BDNF levels reflect brain BDNF levels remains unclear without direct evidence. However, a blood–brain association for BDNF is implicated by several studies; BDNF levels in the brain and serum undergo similar changes during aging processes in rats (Karege et al. 2002) and the reduction in human serum BDNF levels is associated with age-related hippocampal volume loss and declines in spatial memory performance (Erickson et al. 2010). BDNF protein has been reported to cross the blood–brain barrier (Poduslo and Curran 1996; Pan et al. 1998; Pardridge et al. 1998), although this remains controversial. The major source of serum BDNF is suggested to be from platelets (Rosenfeld et al. 1995), although production of BDNF in leukocytes has been observed and may also contribute to BDNF levels in the serum. Recent studies have shown that BDNF mRNA levels in peripheral leukocytes are also reduced in depressed patients (Cattaneo et al. 2010; Pandey et al. 2010). It will be interesting to use leukocytes as a source for measuring gene regulation of BDNF in each individual, although BDNF promoter activities involved in leukocytes may not reflect those in the brain (Kruse et al. 2007). A very recent study has reported significantly reduced levels of tissue-type plasminogen activator, an enzyme implicated in controlling processing of proBDNF to a mature form of BDNF (Pang et al. 2004), in the serum of patients with late-onset geriatric depression (Shi et al. 2010). This suggests that impaired processing of BDNF protein may be causing reduced BDNF levels in the serum under certain conditions such as aging. Furthermore, Schmidt and Duman (2010) reported that peripheral BDNF administration by subcutaneous injection produced antidepressant-like effects in rats, accompanied with elevated BDNF levels in the HIP, and increased levels of phosphorylated CREB and ERK, the downstream targets of BDNF-TrkB signaling. These results implicate the antidepressant-like effect of peripheral BDNF, although precise mechanisms underlying its effects remain to be determined.

19.5.3 GENETIC RESEARCH OF *BDNF* IN DEPRESSION

Although no specific depression-associated genes have yet been identified, it is estimated that 40% to 50% of depression vulnerability has a genetic component (Sullivan et al. 2000). The *BDNF* gene has been investigated as a candidate gene for susceptibility to mood disorders, especially an SNP in the gene that leads to an amino acid substitution of valine (Val) to methionine (Met) at codon 66 (Val66Met) of the proBDNF protein. This SNP has deserved attention because preclinical studies have shown that the BDNF Val66Met SNP influences the function of BDNF by reducing trafficking and activity-dependent secretion of BDNF protein in the brain (Egan et al. 2003; Chen et al. 2004, 2006; Chiaruttini et al. 2009), and a mouse knock-in model carrying the Val66Met variant exhibits increased anxiety-related behaviors (Chen et al. 2006). Whether these animals also show depressive behavior remains to be investigated. Regarding the effects of Val66Met SNP in human subjects with MDD, studies have produced inconsistent results, likely depending on the size and the ethnicity of the studied populations. For example, in American and German populations, the Val allele has been reported to be associated with bipolar disorder (Sklar et al. 2002), neuroticism (Sen et al. 2003), and MDD (Schumacher et al. 2005; Sarchiapone et al. 2008). In contrast, in Asian populations, no significant difference in the genotype and/or allele frequencies of the Val66Met polymorphism has been observed between major depression and control groups (Tsai et al. 2003; Iga et al. 2007). Further studies carried out using much larger samples (in 70,000–90,000 nonclinical English sub- jects) have failed to confirm the association between this polymorphism and mood status (Surtees et al. 2007; Willis-Owen et al. 2005). Other studies have suggested that this polymorphism is not related to the development of major depression but to clinical features of major depression; the Met allele has been related to psychotic features of patients with MDD in a Japanese population (Iga et al. 2007), to anxious temperament (harm avoidance) in American populations (Jiang et al. 2005), to geriatric depression in Chinese and American populations (Hwang et al. 2006; Taylor et al. 2007), and to Alzheimer's disease–related depression in an Italian population (Sarchiapone et al. 2008). Interestingly, a recent study by Petryshen et al. (2009) has revealed substantial BDNF Val66Met allele and haplotype diversity among global populations, and this diversity may explain the varying results between association studies using different ethnic populations. Further studies that consider population diversity in the BDNF genomic region are required to elucidate whether this SNP has a role in major mental illness.

Although serum BDNF can be measured as a possible marker for MDD, studies have shown contradictory results that Met-carriage show reduced (Ozan et al. 2010) or increased (Lang et al. 2009) serum BDNF levels. *In vivo* brain imaging studies have demonstrated reduced hippocampal volumes in patients with MDD (Sheline et al. 1996). Interestingly, significantly smaller hippocam- pal volumes have been observed for patients with MDD and for controls carrying the BDNF-66Met allele compared with subjects homozygous for the BDNF-66Val allele. This suggests that BDNF- 66Met allele carriers with smaller hippocampal volumes might be susceptible to MDD (Bueller et al. 2006; Frodl et al. 2007).

In addition to Val66Met susceptibility to MDD, an association between the Val66Met polymor- phism and treatment response has been further studied. Meta-analyses found that patients with the Met allele respond better to treatment for major depression (Kato and Serretti 2008; Zou et al. 2010). However, another study reported no association between genetic variation in BDNF (Val66Met and other 12 SNPs) and antidepressant treatment response or remission in a German cohort and the Sequenced Treatment Alternatives to Relieve Depression samples (Domschke et al. 2010).

Recent studies have identified new SNPs in the *BDNF* gene. Six SNPs and two haplotypes (one including Val66, another near exon VIIIh) are associated with MDD and eight SNPs are associ- ated with the antidepressant response (Licinio et al. 2009). Interestingly, a 5′-UTR intronic variant located near exon VIIh (SNP s12273539) and a 5′-UTR intronic variant located near exon V (SNP rs61888800) show the most significant effects in depression and antidepressant response, respec- tively. In addition, other studies have found novel SNPs at BDNF promoter I, –281A and G-712A, and have suggested that –281A protects against anxiety and that G-712A is associated with sub- stance abuse (Jiang et al. 2005; Zhang et al. 2006). Very recently, a combination of several inde- pendent risk alleles within the TrkB gene has been associated with suicide attempts among patients with MDD, suggesting that the BDNF–TrkB pathway plays a role in the pathophysiology of suicide (Kohli et al. 2010).

Whereas a single SNP may not account for a depression phenotype, an interaction among SNPs (haplotypes) in the *BDNF* gene, as well as between *BDNF* and other genes, may lead to suscep- tibility to MDD. Since genetic and environmental factors contribute to the etiology of depression

(Kendler 1998), investigating the interaction between those SNPs and environmental factors may provide further insight in the etiology of MDD. For example, variation in the promoter region of the serotonin transporter gene (5-HTTLPR, see Chapter 7) is also recognized as a risk factor for MDD, and a significant three-way interaction among BDNF Val66Met genotype, 5-HTTLPR, and maltreatment history in predicting depression has been reported; children with the Met allele of the *BDNF* gene and two short alleles of 5-HTTLPR have the highest depression scores, but the vulnerability associated with these two genotypes is evident only in the maltreated children (Kaufman et al. 2006). Further studies on gene–gene and gene–environment interaction using larger sample sizes stratified by ethnicity may reveal new insights into the effects of genetic variations of BDNF in depression.

19.6 INFLUENCE OF BDNF IN DEPRESSION

BDNF activates the TrkB signaling pathway to provide neurotrophic effects both in structural increases, such as neurogenesis, neurite arborization, and synaptogenesis, and in neuronal plasticity (e.g., long-term potentiation) and learning and memory (Thoenen 1995; Castren and Rantamaki 2010), which would be affected by changes in BDNF levels.

19.6.1 BDNF Effects on Neurogenesis

One of the hypothesized causes of depression is structural changes and neuronal damage in the HIP (Duman et al. 1997). MRI studies have shown that hippocampal volume is reduced in patients with MDD, whereas results for other brain regions are inconsistent (Videbech et al. 2004). Reduced hippocampal volume in depressed patients may also result from decreased neurogenesis, a process that includes proliferation, differentiation, and survival of neural progenitor cells. Chronic stress (e.g., chronic unpredictable stress, restraint stress, foot shock stress, and social defeat stress) has been shown to reduce neurogenesis in the dentate gyrus and damage neurons especially in the dentate gyrus, and CA3 and CA1 pyramidal cell layers in the HIP (Smith et al. 1995; Gould and Tanapat 1999). Many of these same stressors also lead to robust and rapid downregulation of BDNF mRNA and protein expression in the HIP (Smith et al. 1995; Ueyama et al. 1997; Nibuya et al. 1999; Rasmusson et al. 2002; Pizarro et al. 2004; Gronli et al. 2006). Thus, downregulation of BDNF may contribute to stress-induced neuronal damage and reduced neurogenesis in the HIP. However, the effect of reduced BDNF on neurogenesis has not been made clear until recently because studies using BDNF heterozygous mice reported conflicting results; the heterozygotes exhibited no change in neuronal survival (Rossi et al. 2006) or decreased survival of newborn neurons (Lee et al. 2002; Sairanen et al. 2005) with either reduced (Lee et al. 2002) or increased (Sairanen et al. 2005) cell proliferation. Recently, direct evidence has been reported that BDNF knockdown in the dorsal dentate gyrus causes a significant reduction in differentiation of newborn progenitors into immature neurons, without affecting proliferation and cell survival measured at 24 h and 7 days after BrDU (a marker for proliferating cells) injections (Taliaz et al. 2010). Remarkably, the BDNF dentate gyrus knockdown was sufficient to produce depression-like behavior in rats. Whereas this manipulation used an exogenous system, we have also obtained similar results using an endogenous system in mutant mice. The knock-in mice that lack endogenous promoter IV–driven BDNF expression (BDNF-KIV mice; Sakata et al. 2009) show decreased differentiation and reduced survival of newborn neurons (21 days) without changes in cell proliferation (Jha and Sakata, unpublished data, 2010), and this is accompanied by depression-like behavior (Sakata et al. 2010). These results suggest that reduced BDNF expression may cause impaired neurogenesis, which would be one of the pathophysiologies of depression. Adult neurogenesis was shown to be functional and is suggested to be critical for learning and memory (Gage 2002). Reduction of neurogenesis and any structural change in the HIP would cause impaired synaptic plasticity. Supporting this hypothesis, we have observed that protein synthesis–dependent long-term synaptic plasticity in the HIP is disrupted in our BDNF promoter IV mutant mice (Sakata et al. 2005; Sakata and Lu 2007). Disrupted long-term synaptic plasticity would result in lack of reward learning, which could contribute to MDD (Duman 2002).

In contrast to the effects of stress, chronic administration of different classes of antidepressants (e.g., SSRIs, NRIs, TCAs, MAOIs) and ECT increases neurogenesis in the adult HIP (Malberg et al. 2000), following the time course for the therapeutic action of these medications as well as for the upregulation of *BDNF* gene expression in the HIP (Duman et al. 2006). Hippocampal neurogenesis is required for behavioral effects of antidepressants (Santarelli et al. 2003). Interestingly, chronic administration of BDNF directly into the HIP increases neurogenesis *in vivo* (Scharfman et al. 2005; Schmidt and Duman 2007). *In vitro*, an addition of BDNF to central nervous system–derived neuronal precursors influences differentiation but not proliferation (Ahmed et al. 1995). We recently observed that reduced differentiation and survival of neurons in the HIP of BDNF-KIV mice can be reversed by 3 weeks of enriched environmental interventions, and that this change was accompanied by a marked increase in cell proliferation and behavioral recovery of depression-like phenotypes (Jha and Sakata, unpublished data, 2010). As BDNF-KIV mice retain other BDNF promoters than promoter IV, it is possible that increased BDNF expression through other promoters by the treatment may be involved in this reversal effects (this remains to be addressed), and antidepressant effects may be through increased BDNF levels and neurogenesis in the HIP.

19.6.2 ANTIDEPRESSANTS' EFFECT ON BDNF FUNCTIONS WITHOUT *BDNF* GENE INDUCTION

Without *BDNF* gene induction, antidepressants can induce a variety of cellular processes in different systems through cAMP (Menkes et al. 1983), including incorporation of TrkB into the neuronal plasma membrane (Meyer-Franke et al. 1998) and secretion of neurotrophins and other peptides (Goodman et al. 1996) that act synergistically with BDNF. Interestingly, acute, as well as chronic, antidepressant treatments rapidly increase TrkB phosphorylation within 30 to 60 min of drug administration in the PFC and HIP (Saarelainen et al. 2003; Rantamaki et al. 2007). This acute antidepressant-induced activation of TrkB unlikely involves an increase in BDNF transcription, as this takes at least 2 to 3 h (Zafra et al. 1990), but would rather involve local protein synthesis and release of BDNF. A very recent study has shown that BDNF long 3'-UTR, which is mainly localized in dendrites, is involved in rapid activity-dependent translation of BDNF protein (Lau et al. 2010). Whether antidepressant treatment actually induces translation of *BDNF* transcripts with long 3'-UTR remains to be investigated. This antidepressant-induced TrkB phosphorylation activates phospholipase-Cγ signaling and leads to the phosphorylation of CREB (Castren et al. 2007; Rantamaki et al. 2007), which can induce transcription of genes involved in cell proliferation and neuronal growth, such as other growth factors as well as BDNF itself (Yasuda et al. 2007). These gene transcriptions can produce long-lasting adaptation in the brain that may contribute to antidepressant effects. In addition, antidepressant-induced cAMP levels can contribute to the enhanced potentiation of secretion of other transmitters in response to *BDNF*, which would serve as a "gating" control for structural and functional plasticity of the nervous system (Boulanger and Poo 1999).

19.7 FUTURE DIRECTIONS FOR BDNF HYPOTHESIS OF DEPRESSION

Despite the remarkable progress in understanding gene regulation and function of BDNF during the past two decades, knowledge of the precise mechanisms of how reduced or increased BDNF levels and functions lead to depression and antidepressant effects remains incomplete. Many questions remain. For example, where in the brain and when is *BDNF* gene regulation critical for causing depression and antidepressant effects? Because BDNF expression is regulated in different brain regions with different time courses by different conditions, such as neuronal activity, stress, and antidepressants (Metsis et al. 1993; Timmusk et al. 1994; Malkovska et al. 2006; see Section 19.3 for details), understanding promoter-selective BDNF expression and focusing on when and where it works may provide further insights into the roles of BDNF in depression and actions of antidepressants. Whereas lack of promoter IV–driven BDNF expression leads to depression-like behavior and perhaps is involved in suicide, the roles of other BDNF promoters in depression have not been determined. How different conditions regulate the epigenetic states of the *BDNF* gene also remains largely unknown. While increased BDNF levels seem to exert opposite depression or antidepressant effects depending on the brain region (antidepressant effects in the HIP and PFC and depression-like effects in the VTA and NAc), these opposite effects may be brought about by activating different BDNF promoters. In addition, based on the recent "yin and yang" hypothesis of BDNF (Lu et al. 2005) that postulates that mature-BDNF-TrkB signaling provides neurotrophic effects whereas proBDNF-p75 signaling leads to apoptotic effects, the questions of whether BDNF has both depression and antidepressant effects by being proBDNF or mature BDNF remains to be investigated. Whether short- or long-form BDNF transcripts that localize in the cell body or dendrites are involved in different antidepressant efficacies also remains to be investigated.

The new SNPs and haplotypes in the *BDNF* gene related to depression (Licinio et al. 2009) may account for the different efficacy of BDNF functions in individuals. It would be interesting to investigate how these SNPs and haplotypes in individuals affect susceptibility to depression and responses to antidepressants, in the context of different environmental and treatment conditions. These BDNF variants can affect BDNF processing, such as transcription, trafficking, and secretion, and its functions; and rodent models with specific *BDNF* gene manipulations (e.g., knock-in of SNPs) may help to address these questions. Although animal models are useful, to avoid any artifact effects, it is critical to consider whether experimental conditions can be related to physiological conditions observed in depressed human patients. Selecting tests for a depression-like behavior should also be carefully justified as there is a discrepancy between antidepressant efficacy between animal models and human patients. Animals exhibit behavioral responses shortly after administration of antidepressants in the most current behavioral models (e.g., forced swim test, tail suspension test, learned helplessness test), whereas therapeutic effects of antidepressants in human patients require chronic treatments. Because no single behavioral paradigm using animals will sufficiently mimic all aspects of the complex disease in humans, using multiple behavioral tests may be required to verify a model of depression.

In conclusion, the evidence reviewed here supports the BDNF hypothesis of depression, suggesting a new insight that divergent mechanisms of *BDNF* gene regulation and processing are involved in the pathogenesis of depression and antidepressant actions. It is still unclear why some people are more susceptible to MDD, whereas others better tolerate the same stress, and why there are so many nonresponding patients in whom antidepressant treatments fail to exert sufficient positive effect. Individual genetic and epigenetic variations in the *BDNF* gene may explain many of the differences between individuals, and might be useful as a predictor of the antidepressant treatment efficacy. Moreover, since BDNF is not a "solo" player in the pathogenesis of depression and antidepressant actions, interactions and mutual effects between BDNF and other molecules must be investigated further. Understanding the underlying mechanisms provides the advantage of being able to optimize potential targets for novel drugs or other antidepressant interventions (e.g., exercise, psychotherapy), possibly by combining particular (epi-)genetic and environmental factors that act complementarily. Applying knowledge between preclinical and clinical studies is critical to further advance our understanding of the role of BDNF in the pathophysiology and treatment of major depression.

REFERENCES

Adachi, M., M. Barrot, A. E. Autry, et al. 2008. Selective loss of brain-derived neurotrophic factor in the dentate gyrus attenuates antidepressant efficacy. *Biol Psychiatry* 63:642–9.

Adlard, P. A., V. M. Perreau, C. Engesser-Cesar, et al. 2004. The timecourse of induction of brain-derived neurotrophic factor mRNA and protein in the rat hippocampus following voluntary exercise. *Neurosci Lett* 363:43–8.

Ahmed, S., B. A. Reynolds, and S. Weiss. 1995. BDNF enhances the differentiation but not the survival of CNS stem cell–derived neuronal precursors. *J Neurosci* 15:5765–78.

Aid, T., A. Kazantseva, M. Piirsoo, et al. 2007. Mouse and rat BDNF gene structure and expression revisited. *J Neurosci Res* 85:525–35.

Almeida, L. E., P. D. Murray, H. R. Zielke, et al. 2009. Autocrine activation of neuronal NMDA receptors by aspartate mediates dopamine- and cAMP-induced CREB-dependent gene transcription. *J Neurosci* 29:12702–10.

Amir, R. E., I. B. Van den Veyver, M. Wan, et al. 1999. Rett Syndrome is caused by mutations in X-linked MECP2, encoding methyl-CpG-binding protein 2. *Nat Genet* 23:185–8.

An, J. J., K. Gharami, G. Y. Liao, et al. 2008. Distinct role of long 3′ UTR BDNF mRNA in spine morphology and synaptic plasticity in hippocampal neurons. *Cell* 134:175–87.

Autry, A. E., M. Adachi, P. Cheng, et al. 2009. Gender-specific impact of brain-derived neurotrophic factor signaling on stress-induced depression-like behavior. *Biol Psychiatry* 66:84–90.

Aydemir, O., A. Deveci, and F. Taneli. 2005. The effect of chronic antidepressant treatment on serum brain-derived neurotrophic factor levels in depressed patients: A preliminary study. *Prog Neuropsychopharmacol Biol Psychiatry* 29:261–5.

Bai, O., J. Chlan-Fourney, R. Bowen, et al. 2003. Expression of brain-derived neurotrophic factor mRNA in rat hippocampus after treatment with antipsychotic drugs. *J Neurosci Res* 71:127–31.

Balkowiec, A., and D. M. Katz. 2000. Activity-dependent release of endogenous brain-derived neurotrophic factor from primary sensory neurons detected by ELISA in situ. *J Neurosci* 20:7417–23.

Barde, Y. A. 1990. The nerve growth factor family. *Prog Growth Factor Res* 2:237–48.

Berton, O., C. A. McClung, R. J. Dileone, et al. 2006. Essential role of BDNF in the mesolimbic dopamine pathway in social defeat stress. *Science* 311:864–8.

Boulanger, L., and M. M. Poo. 1999. Gating of BDNF-induced synaptic potentiation by cAMP. *Science* 284:1982–4.

Branchi, I., I. D'Andrea, M. Fiore, et al. 2006. Early social enrichment shapes social behavior and nerve growth factor and brain-derived neurotrophic factor levels in the adult mouse brain. *Biol Psychiatry* 60:690–6.

Brunoni, A. R., M. Lopes, and F. Fregni. 2008. A systematic review and meta-analysis of clinical studies on major depression and BDNF levels: Implications for the role of neuroplasticity in depression. *Int J Neuropsychopharmacol* 11:1169–80.

Bueller, J. A., M. Aftab, S. Sen, et al. 2006. BDNF Val66Met allele is associated with reduced hippocampal volume in healthy subjects. *Biol Psychiatry* 59:812–15.

Calabrese, F., R. Molteni, P. F. Maj, et al. 2007. Chronic duloxetine treatment induces specific changes in the expression of BDNF transcripts and in the subcellular localization of the neurotrophin protein. *Neuropsychopharmacology* 32:2351–9.

Castren, E., and T. Rantamaki. 2010. The role of BDNF and its receptors in depression and antidepressant drug action: Reactivation of developmental plasticity. *Dev Neurobiol* 70:289–97.

Castren, E., V. Voikar, and T. Rantamaki. 2007. Role of neurotrophic factors in depression. *Curr Opin Pharmacol* 7:18–21.

Cattaneo, A., L. Bocchio-Chiavetto, R. Zanardini, et al. 2010. Reduced peripheral brain-derived neurotrophic factor mRNA levels are normalized by antidepressant treatment. *Int J Neuropsychopharmacol* 13:103–8.

Chen, B., D. Dowlatshahi, G. M. MacQueen, et al. 2001. Increased hippocampal BDNF immunoreactivity in subjects treated with antidepressant medication. *Biol Psychiatry* 50:260–5.

Chen, W. G., Q. Chang, Y. Lin, et al. 2003. Derepression of BDNF transcription involves calcium-dependent phosphorylation of MeCP2. *Science* 302:885–9.

Chen, Z. Y., P. D. Patel, G. Sant, et al. 2004. Variant brain-derived neurotrophic factor (BDNF) (Met66) alters the intracellular trafficking and activity-dependent secretion of wild-type BDNF in neurosecretory cells and cortical neurons. *J Neurosci* 24:4401–11.

Chen, Z. Y., D. Jing, K. G. Bath, et al. 2006. Genetic variant BDNF (Val66Met) polymorphism alters anxiety-related behavior. *Science* 314:140–3.

Chiaruttini, C., A. Vicario, Z. Li, et al. 2009. Dendritic trafficking of BDNF mRNA is mediated by translin and blocked by the G196a (Val66Met) mutation. *Proc Natl Acad Sci U S A* 106:16481–6.

Chourbaji, S., R. Hellweg, D. Brandis, et al. 2004. Mice with reduced brain-derived neurotrophic factor expression show decreased choline acetyltransferase activity, but regular brain monoamine levels and unaltered emotional behavior. *Brain Res Mol Brain Res* 121:28–36.

Conti, A. C., J. F. Cryan, A. Dalvi, et al. 2002. cAMP response element–binding protein is essential for the upregulation of brain-derived neurotrophic factor transcription, but not the behavioral or endocrine responses to antidepressant drugs. *J Neurosci* 22:3262–8.

Cosi, C., P. E. Spoerri, M. C. Comelli, et al. 1993. Glucocorticoids depress activity-dependent expression of BDNF mRNA in hippocampal neurones. *Neuroreport* 4:527–30.

de Foubert, G., M. J. O'Neill, and T. S. Zetterstrom. 2007. Acute onset by 5-HT (6)-receptor activation on rat brain brain-derived neurotrophic factor and activity-regulated cytoskeletal-associated protein mRNA expression. *Neuroscience* 147:778–85.

Dennis, K. E., and P. Levitt. 2005. Regional expression of brain derived neurotrophic factor (BDNF) is correlated with dynamic patterns of promoter methylation in the developing mouse forebrain. *Brain Res Mol Brain Res* 140:1–9.

Di Liberto, V., A. Bonomo, M. Frinchi, et al. 2010. Group II metabotropic glutamate receptor activation by agonist Ly379268 treatment increases the expression of brain derived neurotrophic factor in the mouse brain. *Neuroscience* 165:863–73.

Dias, B. G., S. B. Banerjee, R. S. Duman, et al. 2003. Differential regulation of brain derived neurotrophic factor transcripts by antidepressant treatments in the adult rat brain. *Neuropharmacology* 45:553–63.

Domschke, K., B. Lawford, G. Laje, et al. 2010. Brain-derived neurotrophic factor (BDNF) gene: No major impact on antidepressant treatment response. *Int J Neuropsychopharmacol* 13:93–101.

Duman, R. S. 2002. Pathophysiology of depression: The concept of synaptic plasticity. *Eur Psychiatry* 17 (Suppl 3):306–10.

Duman, R. S., G. R. Heninger, and E. J. Nestler. 1997. A molecular and cellular theory of depression. *Arch Gen Psychiatry* 54:597–606.

Duman, R. S., and L. M. Monteggia. 2006. A neurotrophic model for stress-related mood disorders. *Biol Psychiatry* 59:1116–27.

Dunham, J. S., J. F. Deakin, F. Miyajima, et al. 2009. Expression of hippocampal brain-derived neurotrophic factor and its receptors in Stanley Consortium brains. *J Psychiatr Res* 43:1175–84.

Dwivedi, Y., H. S. Rizavi, R. R. Conley, et al. 2003. Altered gene expression of brain-derived neurotrophic factor and receptor tyrosine kinase B in postmortem brain of suicide subjects. *Arch Gen Psychiatry* 60:804–15.

Dwivedi, Y., H. S. Rizavi, and G. N. Pandey. 2006. Antidepressants reverse corticosterone-mediated decrease in brain-derived neurotrophic factor expression: Differential regulation of specific exons by antidepressants and corticosterone. *Neuroscience* 139:1017–29.

Egan, M. F., M. Kojima, J. H. Callicott, et al. 2003. The BDNF Val66Met polymorphism affects activity-dependent secretion of BDNF and human memory and hippocampal function. *Cell* 112:257–69.

Eisch, A. J., C. A. Bolanos, J. de Wit, et al. 2003. Brain-derived neurotrophic factor in the ventral midbrain–nucleus accumbens pathway: A role in depression. *Biol Psychiatry* 54:994–1005.

Erickson, K. I., R. S. Prakash, M. W. Voss, et al. 2010. Brain-derived neurotrophic factor is associated with age-related decline in hippocampal volume. *J Neurosci* 30:5368–75.

Ernfors, P., C. F. Ibanez, T. Ebendal, et al. 1990. Molecular cloning and neurotrophic activities of a protein with structural similarities to nerve growth factor: Developmental and topographical expression in the brain. *Proc Natl Acad Sci U S A* 87:5454–8.

Falkenberg, T., A. K. Mohammed, B. Henriksson, et al. 1992. Increased expression of brain-derived neurotrophic factor mRNA in rat hippocampus is associated with improved spatial memory and enriched environment. *Neurosci Lett* 138:153–6.

Figurov, A., L. D. Pozzo-Miller, P. Olafsson, et al. 1996. Regulation of synaptic responses to high-frequency stimulation and LTP by neurotrophins in the hippocampus. *Nature* 381:706–9.

Frodl, T., C. Schule, G. Schmitt, et al. 2007. Association of the brain-derived neurotrophic factor Val66Met polymorphism with reduced hippocampal volumes in major depression. *Arch Gen Psychiatry* 64:410–6.

Fuchikami, M., S. Morinobu, A. Kurata, et al. 2009. Single immobilization stress differentially alters the expression profile of transcripts of the brain-derived neurotrophic factor (BDNF) gene and histone acetylation at its promoters in the rat hippocampus. *Int J Neuropsychopharmacol* 12:73–82.

Fukuchi, M., and M. Tsuda. 2010. Involvement of the 3′-untranslated region of the brain-derived neurotrophic factor gene in activity-dependent mRNA stabilization. *J Neurochem* 155:1222–33.

Gage, F. H. 2002. Neurogenesis in the adult brain. *J Neurosci* 22:612–3.

Gervasoni, N., J. M. Aubry, G. Bondolfi, et al. 2005. Partial normalization of serum brain-derived neurotrophic factor in remitted patients after a major depressive episode. *Neuropsychobiology* 51:234–8.

Gonul, A. S., F. Akdeniz, F. Taneli, et al. 2005. Effect of treatment on serum brain-derived neurotrophic factor levels in depressed patients. *Eur Arch Psychiatry Clin Neurosci* 255:381–6.

Goodman, L. J., J. Valverde, F. Lim, et al. 1996. Regulated release and polarized localization of brain-derived neurotrophic factor in hippocampal neurons. *Mol Cell Neurosci* 7:222–38.

Gould, E., and P. Tanapat. 1999. Stress and hippocampal neurogenesis. *Biol Psychiatry* 46:1472–9.

Govindarajan, A., B. S. Rao, D. Nair, et al. 2006. Transgenic brain-derived neurotrophic factor expression causes both anxiogenic and antidepressant effects. *Proc Natl Acad Sci U S A* 103:13208–13.

Greenberg, M. E., B. Xu, B. Lu, et al. 2009. New insights in the biology of BDNF synthesis and release: Implications in CNS function. *J Neurosci* 29:12764–7.

Gronli, J., C. Bramham, R. Murison, et al. 2006. Chronic mild stress inhibits BDNF protein expression and CREB activation in the dentate gyrus but not in the hippocampus proper. *Pharmacol Biochem Behav* 85:842–9.

Groves, J. O. 2007. Is it time to reassess the BDNF hypothesis of depression? *Mol Psychiatry* 12:1079–88.

Hall, J., K. L. Thomas, and B. J. Everitt. 2000. Rapid and selective induction of BDNF expression in the hippocampus during contextual learning. *Nat Neurosci* 3:533–5.

Hoshaw, B. A., J. E. Malberg, and I. Lucki. 2005. Central administration of Igf-I and BDNF leads to long-lasting antidepressant-like effects. *Brain Res* 1037:204–8.

Hossain, A., K. Hajman, K. Charitidi, et al. 2008. Prenatal dexamethasone impairs behavior and the activation of the BDNF exon IV promoter in the paraventricular nucleus in adult offspring. *Endocrinology* 149:6356–65.

Hwang, J. P., S. J. Tsai, C. J. Hong, et al. 2006. The Val66Met polymorphism of the brain-derived neurotrophic-factor gene is associated with geriatric depression. *Neurobiol Aging* 27:1834–7.

Iga, J., S. Ueno, K. Yamauchi, et al. 2007. The Val66Met polymorphism of the brain-derived neurotrophic factor gene is associated with psychotic feature and suicidal behavior in Japanese major depressive patients. *Am J Med Genet B Neuropsychiatr Genet* 144B:1003–6.

Impey, S., K. Obrietan, S. T. Wong, et al. 1998. Cross talk between ERK and PKA is required for Ca^{2+} stimulation of CREB-dependent transcription and ERK nuclear translocation. *Neuron* 21:869–83.

Isackson, P. J., M. M. Huntsman, K. D. Murray, et al. 1991. BDNF mRNA expression is increased in adult rat forebrain after limbic seizures: Temporal patterns of induction distinct from NGF. *Neuron* 6:937–48.

Jiang, X., K. Xu, J. Hoberman, et al. 2005. BDNF variation and mood disorders: A novel functional promoter polymorphism and Val66Met are associated with anxiety but have opposing effects. *Neuropsychopharmacology* 30:1353–61.

Karege, F., G. Perret, G. Bondolfi, et al. 2002. Decreased serum brain-derived neurotrophic factor levels in major depressed patients. *Psychiatry Res* 109:143–8.

Karege, F., G. Bondolfi, N. Gervasoni, et al. 2005a. Low brain-derived neurotrophic factor (BDNF) levels in serum of depressed patients probably results from lowered platelet BDNF release unrelated to platelet reactivity. *Biol Psychiatry* 57:1068–72.

Karege, F., G. Vaudan, M. Schwald, et al. 2005b. Neurotrophin levels in postmortem brains of suicide victims and the effects of antemortem diagnosis and psychotropic drugs. *Brain Res Mol Brain Res* 136:29–37.

Karin, M. 1998. New twists in gene regulation by glucocorticoid receptor: Is DNA binding dispensable? *Cell* 93:487–90.

Kato, M., and A. Serretti. 2008. Review and meta-analysis of antidepressant pharmacogenetic findings in major depressive disorder. *Mol Psychiatry* 15:473–500.

Kaufman, J., B. Z. Yang, H. Douglas-Palumberi, et al. 2006. Brain-derived neurotrophic factor-5-HTTLPR gene interactions and environmental modifiers of depression in children. *Biol Psychiatry* 59:673–80.

Keller, S., M. Sarchiapone, F. Zarrilli, et al. 2010. Increased BDNF promoter methylation in the Wernicke area of suicide subjects. *Arch Gen Psychiatry* 67:258–67.

Kendler, K. S. 1998. Anna-Monika-Prize paper. Major depression and the environment: A psychiatric genetic perspective. *Pharmacopsychiatry* 31:5–9.

Kernie, S. G., D. J. Liebl, and L. F. Parada. 2000. BDNF regulates eating behavior and locomotor activity in mice. *EMBO J* 19:1290–300.

Kesslak, J. P., V. So, J. Choi, et al. 1998. Learning upregulates brain-derived neurotrophic factor messenger ribonucleic acid: A mechanism to facilitate encoding and circuit maintenance? *Behav Neurosci* 112:1012–19.

Khundakar, A. A., and T. S. Zetterstrom. 2006. Biphasic change in BDNF gene expression following antidepressant drug treatment explained by differential transcript regulation. *Brain Res* 1106:12–20.

Kohli, M. A., D. Salyakina, A. Pfennig, et al. 2010. Association of genetic variants in the neurotrophic receptor–encoding gene NTRK2 and a lifetime history of suicide attempts in depressed patients. *Arch Gen Psychiatry* 67:348–59.

Koibuchi, N., H. Fukuda, and W. W. Chin. 1999. Promoter-specific regulation of the brain derived neurotropic factor gene by thyroid hormone in the developing rat cerebellum. *Endocrinology* 140:3955–61.

Krishnan, V., M. H. Han, D. L. Graham, et al. 2007. Molecular adaptations underlying susceptibility and resistance to social defeat in brain reward regions. *Cell* 131:391–404.

Kruse, N., S. Cetin, A. Chan, et al. 2007. Differential expression of BDNF mRNA splice variants in mouse brain and immune cells. *J Neuroimmunol* 182:13–21.

Kuppers, E., and C. Beyer. 2001. Dopamine regulates brain-derived neurotrophic factor (BDNF) expression in cultured embryonic mouse striatal cells. *Neuroreport* 12:1175–9.

Lang, U. E., R. Hellweg, T. Sander, et al. 2009. The Met allele of the BDNF Val66Met polymorphism is associated with increased BDNF serum concentrations. *Mol Psychiatry* 14:120–2.

Lau, A. G., H. A. Irier, J. Gu, et al. 2010. Distinct 3′UTRs differentially regulate activity-dependent translation of brain-derived neurotrophic factor (BDNF). *Proc Natl Acad Sci U S A* 107:15945–50.

Lauterborn, J. C., S. Rivera, C. T. Stinis, et al. 1996. Differential effects of protein synthesis inhibition on the activity-dependent expression of BDNF transcripts: Evidence for immediate-early gene responses from specific promoters. *J Neurosci* 16:7428–36.

Lauterborn, J. C., F. R. Poulsen, C. T. Stinis, et al. 1998. Transcript-specific effects of adrenalectomy on seizure-induced BDNF expression in rat hippocampus. *Brain Res Mol Brain Res* 55:81–91.

Lee, T. H., H. Kato, S. T. Chen, et al. 2002. Expression disparity of brain-derived neurotrophic factor immunoreactivity and mRNA in ischemic hippocampal neurons. *Neuroreport* 13:2271–5.

Leibrock, J., F. Lottspeich, A. Hohn, et al. 1989. Molecular cloning and expression of brain-derived neurotrophic factor. *Nature* 341:149–52.

Licinio, J., C. Dong, and M. L. Wong. 2009. Novel sequence variations in the brain-derived neurotrophic factor gene and association with major depression and antidepressant treatment response. *Arch Gen Psychiatry* 66:488–97.

Lindefors, N., M. Ballarin, P. Ernfors, et al. 1992. Stimulation of glutamate receptors increases expression of brain-derived neurotrophic factor mRNA in rat hippocampus. *Ann N Y Acad Sci* 648:296–9.

Liu, Q. R., D. Walther, T. Drgon, et al. 2005. Human brain derived neurotrophic factor (BDNF) genes, splicing patterns, and assessments of associations with substance abuse and Parkinson's disease. *Am J Med Genet B Neuropsychiatr Genet* 134B:93–103.

Liu, Q. R., L. Lu, X. G. Zhu, et al. 2006. Rodent BDNF genes, novel promoters, novel splice variants, and regulation by cocaine. *Brain Res* 1067:1–12.

Lu, B., P. T. Pang, and N. H. Woo. 2005. The yin and yang of neurotrophin action. *Nat Rev Neurosci* 6:603–14.

Lyons, W. E., L. A. Mamounas, G. A. Ricaurte, et al. 1999. Brain-derived neurotrophic factor-deficient mice develop aggressiveness and hyperphagia in conjunction with brain serotonergic abnormalities. *Proc Natl Acad Sci U S A* 96:15239–44.

MacLennan, A. J., N. Lee, and D. W. Walker. 1995. Chronic ethanol administration decreases brain-derived neurotrophic factor gene expression in the rat hippocampus. *Neurosci Lett* 197:105–8.

MacQueen, G. M., K. Ramakrishnan, S. D. Croll, et al. 2001. Performance of heterozygous brain-derived neurotrophic factor knockout mice on behavioral analogues of anxiety, nociception, and depression. *Behav Neurosci* 115:1145–53.

Malberg, J. E., A. J. Eisch, E. J. Nestler, et al. 2000. Chronic antidepressant treatment increases neurogenesis in adult rat hippocampus. *J Neurosci* 20:9104–10.

Malkovska, I., S. G. Kernie, and L. F. Parada. 2006. Differential expression of the four untranslated BDNF exons in the adult mouse brain. *J Neurosci Res* 83:211–21.

Martinowich, K., D. Hattori, H. Wu, et al. 2003. DNA methylation-related chromatin remodeling in activity-dependent BDNF gene regulation. *Science* 302:890–3.

Martinowich, K., H. Manji, and B. Lu. 2007. New insights into BDNF function in depression and anxiety. *Nat Neurosci* 10:1089–93.

Matsumoto, T., S. Rauskolb, M. Polack, et al. 2008. Biosynthesis and processing of endogenous BDNF: CNS neurons store and secrete BDNF, not pro-BDNF. *Nat Neurosci* 11:131–3.

Mayr, B., and M. Montminy. 2001. Transcriptional regulation by the phosphorylation-dependent factor CREB. *Nat Rev Mol Cell Biol* 2:599–609.

Menkes, D. B., M. M. Rasenick, M. A. Wheeler, et al. 1983. Guanosine triphosphate activation of brain adenylate cyclase: Enhancement by long-term antidepressant treatment. *Science* 219:65–7.

Metsis, M., T. Timmusk, E. Arenas, et al. 1993. Differential usage of multiple brain-derived neurotrophic factor promoters in the rat brain following neuronal activation. *Proc Natl Acad Sci U S A* 90:8802–6.

Meyer-Franke, A., G. A. Wilkinson, A. Kruttgen, et al. 1998. Depolarization and cAMP elevation rapidly recruit TrkB to the plasma membrane of CNS neurons. *Neuron* 21:681–93.

Molteni, R., F. Calabrese, A. Cattaneo, et al. 2009. Acute stress responsiveness of the neurotrophin BDNF in the rat hippocampus is modulated by chronic treatment with the antidepressant duloxetine. *Neuropsychopharmacology* 34:1523–32.

Monteggia, L. M., M. Barrot, C. M. Powell, et al. 2004. Essential role of brain-derived neurotrophic factor in adult hippocampal function. *Proc Natl Acad Sci U S A* 101:10827–32.

Monteggia, L. M., B. Luikart, M. Barrot, et al. 2007. Brain-derived neurotrophic factor conditional knockouts show gender differences in depression-related behaviors. *Biol Psychiatry* 61:187–97.

Mowla, S. J., S. Pareek, H. F. Farhadi, et al. 1999. Differential sorting of nerve growth factor and brain-derived neurotrophic factor in hippocampal neurons. *J Neurosci* 19:2069–80.

Nagappan, G., E. Zaitsev, V. V. Senatorov Jr., et al. 2009. Control of extracellular cleavage of proBDNF by high frequency neuronal activity. *Proc Natl Acad Sci U S A* 106:1267–72.

Nair, A., K. C. Vadodaria, S. B. Banerjee, et al. 2007. Stressor-specific regulation of distinct brain-derived neurotrophic factor transcripts and cyclic AMP response element-binding protein expression in the postnatal and adult rat hippocampus. *Neuropsychopharmacology* 32:1504–19.

Neeper, S. A., F. Gomez-Pinilla, J. Choi, et al. 1995. Exercise and brain neurotrophins. *Nature* 373:109.

Nestler, E. J., M. Barrot, R. J. DiLeone, et al. 2002. Neurobiology of depression. *Neuron* 34:13–25.

Nibuya, M., S. Morinobu, and R. S. Duman. 1995. Regulation of BDNF and TrkB mRNA in rat brain by chronic electroconvulsive seizure and antidepressant drug treatments. *J Neurosci* 15:7539–47.

Nibuya, M., E. J. Nestler, and R. S. Duman. 1996. Chronic antidepressant administration increases the expression of cAMP response element binding protein (CREB) in rat hippocampus. *J Neurosci* 16:2365–72.

Nibuya, M., M. Takahashi, D. S. Russell, et al. 1999. Repeated stress increases catalytic TrkB mRNA in rat hippocampus. *Neurosci Lett* 267:81–4.

Okazawa, H., M. Murata, M. Watanabe, et al. 1992. Dopaminergic stimulation up-regulates the in vivo expression of brain-derived neurotrophic factor (BDNF) in the striatum. *FEBS Lett* 313:138–42.

Ozan, E., H. Okur, C. Eker, et al. 2010. The effect of depression, BDNF gene Val66Met polymorphism and gender on serum BDNF levels. *Brain Res Bull* 81:61–5.

Pan, W., W. A. Banks, M. B. Fasold, et al. 1998. Transport of brain-derived neurotrophic factor across the blood–brain barrier. *Neuropharmacology* 37:1553–61.

Pandey, G. N., Y. Dwivedi, H. S. Rizavi, et al. 2010. Brain-derived neurotrophic factor gene and protein expression in pediatric and adult depressed subjects. *Prog Neuropsychopharmacol Biol Psychiatry* 34:645–51.

Pang, P. T., and B. Lu. 2004. Regulation of late-phase LTP and long-term memory in normal and aging hippocampus: Role of secreted proteins TPA and BDNF. *Ageing Res Rev* 3:407–30.

Pardridge, W. M., D. Wu, and T. Sakane. 1998. Combined use of carboxyl-directed protein pegylation and vector-mediated blood–brain barrier drug delivery system optimizes brain uptake of brain-derived neurotrophic factor following intravenous administration. *Pharm Res* 15:576–82.

Patapoutian, A., and L. F. Reichardt. 2001. Trk receptors: Mediators of neurotrophin action. *Curr Opin Neurobiol* 11:272–80.

Pattabiraman, P. P., D. Tropea, C. Chiaruttini, et al. 2005. Neuronal activity regulates the developmental expression and subcellular localization of cortical BDNF mRNA isoforms in vivo. *Mol Cell Neurosci* 28:556–70.

Patterson, S. L., L. M. Grover, P. A. Schwartzkroin, et al. 1992. Neurotrophin expression in rat hippocampal slices: A stimulus paradigm inducing LTP in CA1 evokes increases in BDNF and NT-3 mRNAs. *Neuron* 9:1081–8.

Patterson, S. L., T. Abel, T. A. Deuel, et al. 1996. Recombinant BDNF rescues deficits in basal synaptic transmission and hippocampal LTP in BDNF knockout mice. *Neuron* 16:1137–45.

Petryshen, T. L., P. C. Sabeti, K. A. Aldinger, et al. 2009. Population genetic study of the brain-derived neurotrophic factor (BDNF) gene. *Mol Psychiatry* 15:810–5.

Pizarro, J. M., L. A. Lumley, W. Medina, et al. 2004. Acute social defeat reduces neurotrophin expression in brain cortical and subcortical areas in mice. *Brain Res* 1025:10–20.

Poduslo, J. F., and G. L. Curran. 1996. Permeability at the blood–brain and blood–nerve barriers of the neurotrophic factors: NGF, CNTF, NT-3, BDNF. *Brain Res Mol Brain Res* 36:280–6.

Pruunsild, P., A. Kazantseva, T. Aid, et al. 2007. Dissecting the human BDNF locus: Bidirectional transcription, complex splicing, and multiple promoters. *Genomics* 90:397–406.

Radka, S. F., P. A. Holst, M. Fritsche, et al. 1996. Presence of brain-derived neurotrophic factor in brain and human and rat but not mouse serum detected by a sensitive and specific immunoassay. *Brain Res* 709:122–301.

Rajkowska, G. 2000. Postmortem studies in mood disorders indicate altered numbers of neurons and glial cells. *Biol Psychiatry* 48:766–77.

Rantamaki, T., P. Hendolin, A. Kankaanpaa, et al. 2007. Pharmacologically diverse antidepressants rapidly activate brain-derived neurotrophic factor receptor TrkB and induce phospholipase-Cgamma signaling pathways in mouse brain. *Neuropsychopharmacology* 32:2152–62.

Rasmusson, A. M., L. Shi, and R. Duman. 2002. Downregulation of BDNF mRNA in the hippocampal dentate gyrus after re-exposure to cues previously associated with footshock. *Neuropsychopharmacology* 27:133–42.

Roceri, M., F. Cirulli, C. Pessina, et al. 2004. Postnatal repeated maternal deprivation produces age-dependent changes of brain-derived neurotrophic factor expression in selected rat brain regions. *Biol Psychiatry* 55:708–14.

Rosenfeld, R. D., L. Zeni, M. Haniu, et al. 1995. Purification and identification of brain-derived neurotrophic factor from human serum. *Protein Expr Purif* 6:465–71.

Rossi, C., A. Angelucci, L. Costantin, et al. 2006. Brain-derived neurotrophic factor (BDNF) is required for the enhancement of hippocampal neurogenesis following environmental enrichment. *Eur J Neurosci* 24:1850–6.

Roth, T. L., and J. D. Sweatt. 2011. Epigenetic marking of the BDNF gene by early-life adverse experiences. *Horm Behav* 59:315–2.

Roth, T. L., F. D. Lubin, A. J. Funk, et al. 2009. Lasting epigenetic influence of early-life adversity on the BDNF Gene. *Biol Psychiatry* 65:760–9.

Roux, P. P., and P. A. Barker. 2002. Neurotrophin signaling through the P75 neurotrophin receptor. *Prog Neurobiol* 67:203–33.

Russo-Neustadt, A., R. C. Beard, and C. W. Cotman. 1999. Exercise, antidepressant medications, and enhanced brain derived neurotrophic factor expression. *Neuropsychopharmacology* 21:679–82.

Saarelainen, T., P. Hendolin, G. Lucas, et al. 2003. Activation of the TrkB neurotrophin receptor is induced by antidepressant drugs and is required for antidepressant-induced behavioral effects. *J Neurosci* 23:349–57.

Sairanen, M., G. Lucas, P. Ernfors, et al. 2005. Brain-derived neurotrophic factor and antidepressant drugs have different but coordinated effects on neuronal turnover, proliferation, and survival in the adult dentate gyrus. *J Neurosci* 25:1089–94.

Sakata, K., and B. Lu. 2007. Activity-dependent transcription of BDNF through promoter 3 contributes to function in prefrontal cortex. *Society for Neuroscience Meeting (Abstract)* 638.4/EEE18.

Sakata, K., N. Woo, J. Wu, et al. 2005. Activity-dependent transcription of BDNF through promoter 3 contributes selectively to long-term hippocampal plasticity and long-term memory. *Society for Neuroscience Meeting* (Abstract) 611.4/F14.

Sakata, K., N. H. Woo, K. Martinowich, et al. 2009. Critical role of promoter IV–driven BDNF transcription in GABAergic transmission and synaptic plasticity in the prefrontal cortex. *Proc Natl Acad Sci U S A* 106:5942–7.

Sakata, K., L. Jin, and S. Jha. 2010. Lack of promoter IV–driven BDNF transcription results in depression-like behavior. *Genes Brain Behav* 9:712–21.

Santarelli, L., M. Saxe, C. Gross, et al. 2003. Requirement of hippocampal neurogenesis for the behavioral effects of antidepressants. *Science* 301:805–9.

Sarchiapone, M., V. Carli, A. Roy, et al. 2008. Association of polymorphism (Val66Met) of brain-derived neurotrophic factor with suicide attempts in depressed patients. *Neuropsychobiology* 57:139–45.

Scharfman, H., J. Goodman, A. Macleod, et al. 2005. Increased neurogenesis and the ectopic granule cells after intrahippocampal BDNF infusion in adult rats. *Exp Neurol* 192:348–56.

Schildkraut, J. J. 1965. The catecholamine hypothesis of affective disorders: A review of supporting evidence. *Am J Psychiatry* 122:509–22.

Schmidt, H. D. and R. S. Duman. 2007. The role of neurotrophic factors in adult hippocampal neurogenesis, antidepressant treatments and animal models of depressive-like behavior. *Behav Pharmacol* 18:391–418.

Schmidt, H. D., and R. S. Duman. 2010. Peripheral BDNF produces antidepressant-like effects in cellular and behavioral models. *Neuropsychopharmacology* 35:2378–91.

Schumacher, J., R. A. Jamra, T. Becker, et al. 2005. Evidence for a relationship between genetic variants at the brain-derived neurotrophic factor (BDNF) locus and major depression. *Biol Psychiatry* 58:307–14.

Seidah, N. G., S. Benjannet, S. Pareek, et al. 1996. Cellular processing of the neurotrophin precursors of Nt3 and BDNF by the mammalian proprotein convertases. *FEBS Lett* 379:247–50.

Sen, S., R. M. Nesse, S. F. Stoltenberg, et al. 2003. A BDNF coding variant is associated with the neo personality inventory domain neuroticism, a risk factor for depression. *Neuropsychopharmacology* 28:397–401.

Sen, S., R. Duman, and G. Sanacora. 2008. Serum brain-derived neurotrophic factor, depression, and antidepressant medications: Meta-analyses and implications. *Biol Psychiatry* 64:527–32.

Sheline, Y. I., P. W. Wang, M. H. Gado, et al. 1996. Hippocampal atrophy in recurrent major depression. *Proc Natl Acad Sci U S A* 93:3908–13.

Shi, Y., J. You, Y. Yuan, et al. 2010. Plasma BDNF and TPA are associated with late-onset geriatric depression. *Psychiatry Clin Neurosci* 64:249–54.

Shieh, P. B., S. C. Hu, K. Bobb, et al. 1998. Identification of a signaling pathway involved in calcium regulation of BDNF expression. *Neuron* 20:727–40.

Shimizu, E., K. Hashimoto, N. Okamura, et al. 2003. Alterations of serum levels of brain-derived neurotrophic factor (BDNF) in depressed patients with or without antidepressants. *Biol Psychiatry* 54:70–5.

Shirayama, Y., A. C. Chen, S. Nakagawa, et al. 2002. Brain-derived neurotrophic factor produces antidepressant effects in behavioral models of depression. *J Neurosci* 22:3251–61.

Siuciak, J. A., D. R. Lewis, S. J. Wiegand, et al. 1997. Antidepressant-like effect of brain-derived neurotrophic factor (BDNF). *Pharmacol Biochem Behav* 56:131–7.

Sklar, P., S. B. Gabriel, M. G. McInnis, et al. 2002. Family-based association study of 76 candidate genes in bipolar disorder: BDNF is a potential risk locus. Brain-derived neutrophic factor. *Mol Psychiatry* 7:579–93.

Smith, M. A., S. Makino, R. Kvetnansky, et al. 1995. Stress and glucocorticoids affect the expression of brain-derived neurotrophic factor and neurotrophin-3 mRNAs in the hippocampus. *J Neurosci* 15:1768–77.

Sohrabji, F., R. C. Miranda, and C. D. Toran-Allerand. 1995. Identification of a putative estrogen response element in the gene encoding brain-derived neurotrophic factor. *Proc Natl Acad Sci U S A* 92:11110–14.

Soppet, D., E. Escandon, J. Maragos, et al. 1991. The neurotrophic factors brain-derived neurotrophic factor and neurotrophin-3 are ligands for the TrkB tyrosine kinase receptor. *Cell* 65:895–903.

Sullivan, P. F., M. C. Neale, and K. S. Kendler. 2000. Genetic epidemiology of major depression: Review and meta-analysis. *Am J Psychiatry* 157:1552–62.

Surtees, P. G., N. W. Wainwright, S. A. Willis-Owen, et al. 2007. No association between the BDNF Val66Met polymorphism and mood status in a non-clinical community sample of 7389 older adults. *J Psychiatr Res* 41:404–9.

Tabuchi, A., R. Nakaoka, K. Amano, et al. 2000. Differential activation of brain-derived neurotrophic factor gene promoters I and III by Ca- signals evoked via L-type voltage-dependent and N-methyl-D-aspartate receptor Ca^{2+} channels. *J Biol Chem* 275:17269–75.

Tabuchi, A., H. Sakaya, T. Kisukeda, et al. 2002. Involvement of an upstream stimulatory factor as well as cAMP-responsive element–binding protein in the activation of brain-derived neurotrophic factor gene promoter I. *J Biol Chem* 277:35920–31.

Taliaz, D., N. Stall, D. E. Dar, et al. 2010. Knockdown of brain-derived neurotrophic factor in specific brain sites precipitates behaviors associated with depression and reduces neurogenesis. *Mol Psychiatry* 15:80–92.

Tao, X., S. Finkbeiner, D. B. Arnold, et al. 1998. Ca^{2+} influx regulates BDNF transcription by a CREB family transcription factor-dependent mechanism. *Neuron* 20:709–26.

Tao, X., A. E. West, W. G. Chen, et al. 2002. A calcium-responsive transcription factor, CaRF, that regulates neuronal activity-dependent expression of BDNF. *Neuron* 33:383–95.

Taylor, W. D., S. Zuchner, D. R. McQuoid, et al. 2007. Allelic differences in the brain-derived neurotrophic factor Val66Met polymorphism in late-life depression. *Am J Geriatr Psychiatry* 15:850–7.

Teng, H. K., K. K. Teng, R. Lee, et al. 2005. ProBDNF induces neuronal apoptosis via activation of a receptor complex of P75ntr and sortilin. *J Neurosci* 25:5455–63.

Thoenen, H. 1995. Neurotrophins and Neuronal Plasticity. *Science* 270:593–8.

Timmusk, T., K. Palm, M. Metsis, et al. 1993. Multiple promoters direct tissue-specific expression of the rat BDNF gene. *Neuron* 10:475–89.

Timmusk, T., N. Belluardo, H. Persson, et al. 1994. Developmental regulation of brain-derived neurotrophic factor messenger RNAs transcribed from different promoters in the rat brain. *Neuroscience* 60:287–91.

Tozuka, Y., M. Kumon, E. Wada, et al. 2010. Maternal obesity impairs hippocampal BDNF production and spatial learning performance in young mouse offspring. *Neurochem Int* 57:235–47.

Tsai, S. J., C. Y. Cheng, Y. W. Yu, et al. 2003. Association study of a brain-derived neurotrophic-factor genetic polymorphism and major depressive disorders, symptomatology, and antidepressant response. *Am J Med Genet B Neuropsychiatr Genet* 123B:19–22.

Tsankova, N. M., O. Berton, W. Renthal, et al. 2006. Sustained hippocampal chromatin regulation in a mouse model of depression and antidepressant action. *Nat Neurosci* 9:519–25.

Ueyama, T., Y. Kawai, K. Nemoto, et al. 1997. Immobilization stress reduced the expression of neurotrophins and their receptors in the rat brain. *Neurosci Res* 28:103–10.

Vaidya, V. A., G. J. Marek, G. K. Aghajanian, et al. 1997. 5-HT2A receptor-mediated regulation of brain-derived neurotrophic factor mRNA in the hippocampus and the neocortex. *J Neurosci* 17:2785–95.

Videbech, P., B. Ravnkilde, L. Gammelgaard, et al. 2004. The Danish pet/depression project: Performance on Stroop's test linked to white matter lesions in the brain. *Psychiatry Res* 130:117–30.

West, A. E., W. G. Chen, M. B. Dalva, et al. 2001. Calcium regulation of neuronal gene expression. *Proc Natl Acad Sci U S A* 98:11024–31.

WHO. 2009. http://www.who.int/mental_health/prevention/suicide/suicideprevent/en/index.html.

Willis-Owen, S. A., J. Fullerton, P. G. Surtees, et al. 2005. The Val66Met coding variant of the brain-derived neurotrophic factor (BDNF) gene does not contribute toward variation in the personality trait neuroticism. *Biol Psychiatry* 58:738–42.

Yamamoto, H., and M. E. Gurney. 1990. Human platelets contain brain-derived neurotrophic factor. *J Neurosci* 10:3469–78.

Yasuda, M., M. Fukuchi, A. Tabuchi, et al. 2007. Robust stimulation of TrkB induces delayed increases in BDNF and arc mRNA expressions in cultured rat cortical neurons via distinct mechanisms. *J Neurochem* 103:626–36.

Zafra, F., B. Hengerer, J. Leibrock, et al. 1990. Activity dependent regulation of BDNF and NGF mRNAs in the rat hippocampus is mediated by non-NMDA glutamate receptors. *EMBO J* 9:3545–50.

Zafra, F., E. Castren, H. Thoenen, et al. 1991. Interplay between glutamate and gamma-aminobutyric acid transmitter systems in the physiological regulation of brain-derived neurotrophic factor and nerve growth factor synthesis in hippocampal neurons. *Proc Natl Acad Sci U S A* 88:10037–41.

Zafra, F., D. Lindholm, E. Castren, et al. 1992. Regulation of brain-derived neurotrophic factor and nerve growth factor mRNA in primary cultures of hippocampal neurons and astrocytes. *J Neurosci* 12:4793–9.

Zajac, M. S., T. Y. Pang, N. Wong, et al. 2009. Wheel running and environmental enrichment differentially modify exon-specific BDNF expression in the hippocampus of wild-type and pre-motor symptomatic male and female Huntington's disease mice. *Hippocampus* 20:621–36.

Zhang, H., F. Ozbay, J. Lappalainen, et al. 2006. Brain derived neurotrophic factor (BDNF) gene variants and Alzheimer's disease, affective disorders, posttraumatic stress disorder, schizophrenia, and substance dependence. *Am J Med Genet B Neuropsychiatr Genet* 141B:387–93.

Zou, Y. F., D. Q. Ye, X. L. Feng, et al. 2010. Meta-analysis of BDNF Val66Met polymorphism association with treatment response in patients with major depressive disorder. *Eur Neuropsychopharmacol* 20:535–44.

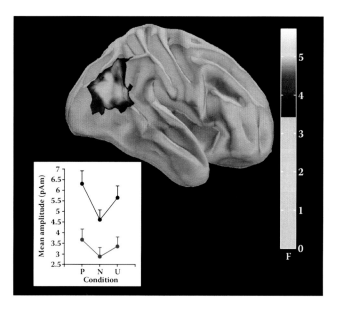

FIGURE 2.1 See figure in text for full caption.

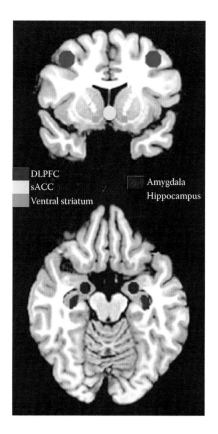

FIGURE 3.1 See figure in text for full caption.

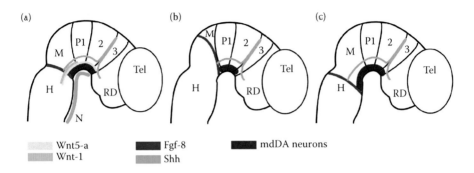

FIGURE 5.2 See figure in text for full caption.

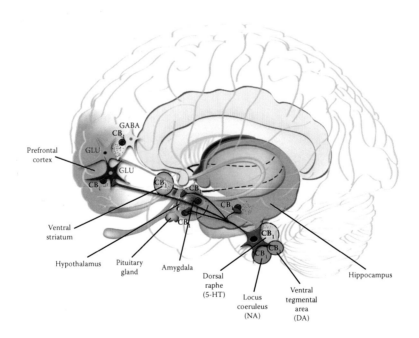

FIGURE 10.1 See figure in text for full caption.

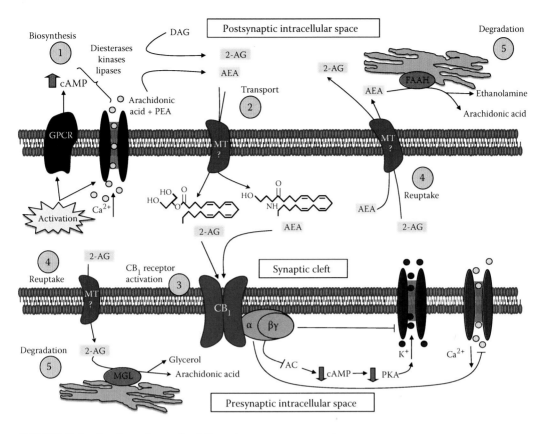

FIGURE 10.2 See figure in text for full caption.

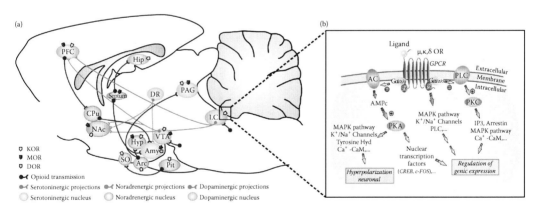

FIGURE 11.1 See figure in text for full caption.

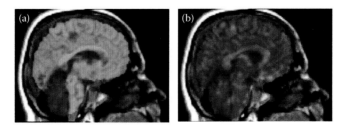

FIGURE 15.1 See figure in text for full caption.

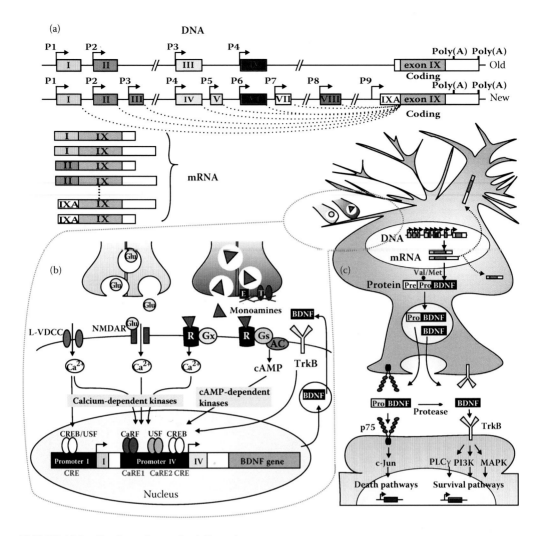

FIGURE 19.2 See figure in text for full caption.

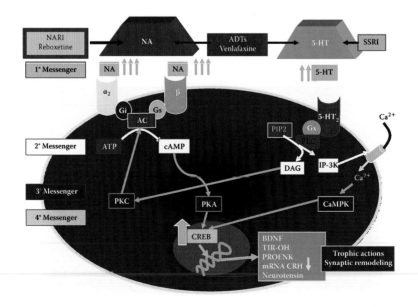

FIGURE 22.1 See figure in text for full caption.

20 Role of Cyclic Nucleotide Phosphodiesterases in Depression and Antidepressant Activity

Ying Xu, Han-Ting Zhang, and James M. O'Donnell

CONTENTS

20.1 INTRODUCTION

Phosphodiesterases (PDEs) are a superfamily of enzymes that catalyze the hydrolysis of cyclic nucleotides, including cyclic adenosine monophosphate (cAMP) and cyclic guanosine monophosphate (cGMP). PDEs are classified into 11 families (PDE1–11) based on their substrate specificities, kinetics, allosteric regulators, and amino acid sequences (Beavo 1995; Conti and Beavo 2007; O'Donnell and Zhang 2004; Soderling and Beavo 2000). Each family is encoded by one to four genes, leading to a total of 21 PDE isoforms (Table 20.1). Based on their substrates, PDE families can be classified as cAMP-specific (e.g., PDE4, PDE7, and PDE8), cGMP-specific (e.g., PDE5, PDE6, and PDE9), or dual substrate (e.g., cAMP and cGMP) PDEs (e.g., PDE1, PDE2, PDE3, PDE10, and PDE11) (Beavo and Brunton 2002; Conti 2000; Francis et al. 2009; Mehats et al. 2002; Zhang 2009). PDEs contain

TABLE 20.1
PDE Superfamily and Respective Substrates

Family	Subtype	Substrate	References
PDE1	1A, 1B, 1C	cAMP/cGMP	Nagel et al. (2006), Bender and Beavo (2006), Miller et al. (2009)
PDE2	2A	cAMP/cGMP	Wu et al. (2004), Hajjhussein et al. (2007)
PDE3	3A, 3B	cAMP/cGMP	Bender and Beavo (2006), Pyne et al. (1986)
PDE4	4A, 4B, 4C, 4D	cAMP	Conti et al. (2003), Houslay et al. (2007), Houslay (2010), O'Donnell and Zhang (2004), Zhang (2009)
PDE5	5A	cGMP	Lin (2004), Heikaus et al. (2009), Rapoport et al. (2000)
PDE6	6A, 6C, 6D, 6G	cGMP	Lugnier (2006)
PDE7	7A, 7B	cAMP	Han et al. (1997), Torras-Llort and Azorin (2003)
PDE8	8A, 8B	cAMP	Wang et al. (2008)
PDE9	9A	cGMP	Van Staveren et al. (2003)
PDE10	10A	cAMP/cGMP	Fujishige et al. (1999), Loughney et al. (1999), Soderling et al. (1999)
PDE11	11A	cAMP/cGMP	Francis (2005), Yuasa et al. (2001), Weeks et al. (2007)

three functional domains: the regulatory N terminus, the conserved catalytic domain, and the C terminus (Bolger et al. 1994; Keravis and Lugnier 2010; Thompson 1991). The N-terminal domains of these enzymes vary considerably and have regions that exert an inhibitory regulation of the catalytic core and/or control of PDE subcellular localization (Houslay and Adams 2003; Sonnenburg et al. 1995). The N termini may contain different sequences, such as a calmodulin-binding site in PDE1, cGMP-binding sites in PDE2, and a transducin-binding domain in PDE6, or similar structural and functional features such as phosphorylation sites for protein kinases in PDE1, PDE2, PDE3, PDE4, and PDE5, which are important for regulating the activity of the enzymes.

Almost all mammalian cells express PDEs at different levels, typically multiple PDE families and isoforms from the same family. Virtually all PDEs are expressed in the brain, some at high levels, suggesting their importance in overall central nervous system (CNS) function. The distributions of PDE isoforms in the brain are usually brain region specific, indicating that they may have different roles in CNS functions. Specifically, PDE2, PDE4, and PDE9 are widely and highly expressed in the brain, particularly the limbic system, including the olfactory cortex, hippocampus, and amygdala (Cherry and Davis 1999; Juilfs et al. 1997; Perez-Torres et al. 2000; Repaske et al. 1993; van Staveren et al. 2003; van Staveren and Markerink-van Ittersum 2005). This distribution pattern is consistent with the roles of these PDEs in depression (Li et al. 2009; Zhang 2009), anxiety (Masood et al. 2008, 2009; Zhang et al. 2008), and memory (Barad et al. 1998; Boess et al. 2004; Burgin et al. 2010; Rutten et al. 2009; Zhang 2010; Zhang et al. 2000, 2004). In contrast, PDE1B, PDE7A, PDE10A, and PDE11A are highly expressed in a specific brain region, that is, in the striatum for all but PDE11A, which is highly expressed in the dorsal root ganglia and ventral hippocampus (Kelly et al. 2010; Lakics et al. 2010; Seeger et al. 2003; Soderling et al. 1998; van Staveren et al. 2003), whereas the expression of PDE3, PDE5, and PDE8 is relatively low in the brain, particularly PDE3, which is primarily expressed in peripheral tissues.

The involvement of PDEs in antidepressant actions was first reported in the early 1970s. Studies then showed that tricyclic antidepressants altered PDE activity (Berndt and Schwabe 1973; Janiec et al. 1974; Rysànek et al. 1973). However, the direct demonstration for the role of PDE in antidepressant activity was not reported until 10 years later, when Wachtel and another group showed that PDE4 inhibitors produced antidepressant-like effects on behavior in mice (Przegalinski and Bigajska 1983; Wachtel 1983). Studies since then have not only provided strong evidence for antidepressant activity of PDE4 inhibitors but also initiated extensive work on inhibitors of PDE4, as well as other PDE families, as novel, putative antidepressants, given that they are completely different from classic antidepressant drugs in terms of structures and mechanisms. The role of PDE4 in antidepressant activity and depression has been well established (Fujimaki et al. 2000; Li et al. 2009; O'Donnell and Zhang 2004; Zhang 2009; Zhang et al. 2002). The contribution of other PDEs in mediating antidepressant effect remains less clear.

In recent years, the nucleus accumbens (NAc), a brain structure critical for the reward response, has been considered an important modulator in the pathophysiology of depression, particularly for anhedonia (loss of interest or pleasure) observed in depression. Neurochemical changes in the NAc can produce depressive- or antidepressant-like effects on behavior (Nestler et al. 2002; Shirayama and Chaki 2006). Therefore, PDEs with high expression in the striatum, such as PDE1, PDE7, and PDE10, may also be involved in depression or in the mediation of antidepressant activity. In this chapter, we review the advances in PDE studies and analyze the possibilities of PDEs, including PDE2, PDE4, PDE5, PDE9, PDE10, and PDE11, as targets for treatment of depressive disorders.

20.2 PHOSPHODIESTERASE 2

PDE2, a dual-substrate PDE family, was identified by Beavo et al. (1971) and shown to hydrolyze both cGMP and cAMP (Beavo et al. 1971; Erneux et al. 1981). Mammalian PDE2 is a 105-kDa homodimer that exists in particulate and soluble forms. It is coded for by a single gene and has three variants: PDE2A1, PDE2A2, and PDE2A3 (Rosman et al. 1997; Sonnenburg et al. 1991; Yang et al. 1994). PDE2A has two GAF domains in the N terminus, which regulate dimerization and catalytic activity in response to cGMP, and one C-terminal catalytic domain (Martinez et al. 2002). The rate of cAMP hydrolysis by PDE2 is increased 6-fold in the presence of cGMP (Martins et al. 1982; Lugnier 2006), suggesting that PDE2 provides a point of cross-talk between cAMP and cGMP signaling.

20.2.1 Distribution in the Brain and Its Relationship to Depression

PDE2 is highly expressed in the limbic system, including the amygdala, hippocampus, and hypothalamus (Blokland et al. 2006; Repaske et al. 1993), which are important for emotional and cognitive function. This distribution pattern suggests that PDE2 may be involved in the mediation of emotion and memory. PDE2 is also found in the olfactory epithelia, olfactory sensory neurons, and olfactory bulb and tubercle (Juilfs et al. 1997). The distribution of PDE2 in the olfactory system could be another indication of the involvement of PDE2 in mood regulation. Because neuronal projections from olfactory bulbs to the limbic system exert a major influence on emotional behavior, PDE2 may play a role in anxiety and depression. However, this has not been investigated systematically.

Given the high expression of PDE2 in neurons of the olfactory system (Juilfs et al. 1997), bilateral olfactory bulbectomy, a model of depression in rodents, may be an ideal approach to the identification of the role of PDE2 in depression. Animals with olfactory bulbectomy show well-characterized changes in behavior and physiology, including sleep disruption, cognitive impairment, hyperactivity in response to mild stress or "agitation-like" behavior, and increases in corticosteroid levels, which resemble clinical syndromes and symptoms of depression (Masini et al. 2004; Primeaux and Holmes 1999; Possidente et al. 2000; Yamamoto et al. 1997). In addition, olfactory bulbectomy produces changes in the monoaminergic systems throughout the brain that are relevant to neurochemical

TABLE 20.2

Overview of Antidepressant- and Anxiolytic-Like Effects of PDE Inhibitors

PDE Inhibitors	Drug	Model	Task	Result	Reference
PDE2	BAY 60-7550	L-Buthionine-(S,R)-sulfoximine-induced oxidative stress in mice	Elevated plus maze; open field test; holeboard test	Reversal of oxidative stress–induced anxiety-like behavior	Masood et al. (2008)
	ND7001	Same as above	Elevated plus maze; open field test	Reversal of oxidative stress–induced anxiety-like behavior	Masood et al. (2008)
PDE4	Rolipram, ICI 63,197, and Ro 20-1724	ICR mice	Antagonism of reserpine-induced hypothermia or hypokinesia; potentiation of yohimbine lethality	Antidepressant-like effects; rolipram was as potent as ICI 63,197 but 5- to 15-fold more potent than Ro 20-1724	Wachtel (1983)
	Rolipram, piclamilast, and CDP840	Male SD rats	DRL 72-s schedule	All but CDP840 produced antidepressant-like effects, i.e., reduced response rate and increased reinforcement rate	O'Donnell (1993); O'Donnell and Frith (1999); Zhang et al. (2005b, 2006)
	Rolipram	Male NIH-Swiss mice	FST; reserpine-induced hypothermia	Antidepressant-like effects, i.e., decreased immobility in the FST and reversal of reserpine-induced hypothermia	Li et al. (2003); Saccomano et al. (1991)
		Male ICR mice	FST; TST; elevated plus maze; light–dark transition; holeboard	Chronic administration of rolipram produced antidepressant- and anxiolytic-like effects	Li et al. (2009)
		Male SD rats	Learned helplessness	Antidepressant-like effect, i.e., rolipram blocked escape failures in the presence of imipramine	Itoh et al. (2003, 2004); Ye et al. (2000)
		Patients with MDD	A multicenter double-blind study of rolipram	Antidepressant effect; daily dosage of 3 × 0.5 mg was more potent than 3 × 0.25 or 3 × 1 mg; good tolerance.	Fleischhacker et al. (1992); Hebenstreit et al. (1989); Zeller et al. (1984)
		APP/PS1 transgenic mouse	Contextual and cued fear conditioning	Rolipram improved emotion-related memory and basal synaptic transmission	Gong et al. (2004)

PDE	Agent/Model	Subjects	Test	Result	Reference
PDE4	PDE4B knockout mice		Holeboard test; light–dark transition test; open field test	Anxiety-like behavior	Zhang et al. (2008)
			FST; TST; locomotor activity; passive avoidance; elevated plus maze	Decreased immobility in the FST but in not TST; decreases in prepulse inhibition and baseline motor activity and increases in locomotor response to amphetamine; no effect on behaviors of passive avoidance or elevated plus maze	Siuciak et al. (2008); Zhang et al. (2008)
	PDE4D knockout mice		FST; TST; elevated plus maze; MCC test	Decreased immobility in both FST and TST; no change in elevated plus maze or MCC	Zhang et al. (2002)
PDE5	Sildenafil (Viagra)	M-AChR antagonist–treated rats;	FST; TST	Sildenafil produced antidepressant-like effect when M-AChR was blocked but prevented antidepressant-like effects of memantine	Kaster et al. (2005b); Almeida et al. (2006)
		Flinders sensitive line rats			Brink et al. (2008); Liebenberg et al. (2010); Ribeiro and Sebastiao (2010)
		Male ICR mice	Elevated plus maze	Anxiety-like effect: decreased the percentage of time spent in open arms	Kurt et al. (2004)
		Gonadally intact and castrated male Wistar rats	Open field test	Anxiolytic-like effect: increased exploration in the central area of the chamber	Solis et al. (2008)
	Zaprinast	Male SD rats	DRL 72-s schedule	Antidepressant-like effect: decreased response rate but unaltered reinforcement rate	O'Donnell and Frith (1999)
PDE9		Depressed Mexican American human subjects	Double-blind, pharmacogenetic study of antidepressant response	Polymorphisms in PDE9A are associated with the diagnosis of MDD	Wong et al. (2006)
	Bay 73-6691	Young adult Wistar rats	Passive avoidance task	Reversal of aversive affection, possibly anxiety	van der Staay et al. (2008)
PDE10	Papaverine	Wild-type mice	Light–dark test	Chronic PDE10A inhibition produced a variety of behavioral and central neurochemical deficits, which were exacerbated by stress	Hebb et al. (2008)
PDE11		Depressed Mexican American human subjects	Double-blind, pharmacogenetic study of antidepressant response	Polymorphisms in PDE9A are associated with the diagnosis of MDD	Wong et al. (2006)

Note: DRL, differential-reinforcement-of-low-rate; M-AChR, muscarinic acetylcholinergic receptor; MCC, multicompartment chamber; MDD, major depressive disorder; TST, tail-suspension test.

changes presumed to be associated with depression (Grecksch et al. 1997; Mudunkotuwa and Horton 1996). All these behavioral, physiological, and neurochemical changes are independent of anosmia (Sieck and Baumbach 1974) and can be reversed by chronic antidepressant treatment (Mudunkotuwa and Horton 1996).

20.2.2 ROLE IN STRESS AND ANXIETY

PDE2 is also expressed in the rat adrenal cortex and pituitary gland (Menniti et al. 2006; Nikolaev et al. 2005). It may function as a regulator of release of stress hormones, such as aldosterone and glucocorticoids (Morley-Fletcher et al. 2004). Glucocorticoids are adrenal steroid hormones released in response to physical and psychological stressors. Such secretion is the final step in the neuroendocrine cascade triggering from stimulation of the hypothalamus and pituitary. Elevated corticosterone in rodents (cortisol in primates) after stress is the hallmark of feedback inhibition of the limbic–hypothalamic–pituitary–adrenal axis, which is also observed in clinically depressed patients (Morley-Fletcher et al. 2004). Excessive exposure to either glucocorticoids or stress accelerates hippocampal volume decreases and atrophy, as evidenced by loss of neurons, dysfunction of neuroplasticity, worsening of hippocampal-dependent cognitive impairment, and depression (Xu et al. 2009). The adrenal cortex, pituitary, and limbic system contain relevant concentrations of PDE2, indicating that reduction of PDE2 activity through pharmacological inhibition of this enzyme may alter stress-induced disorders, such as depression and anxiety, via increases in intracellular cGMP concentrations.

The first, relatively selective PDE2 inhibitor identified was erythro-9-(2-hydroxy-3-nonyl)adenine (EHNA), which inhibits the enzyme with a K_i value of 1 μM. EHNA is also a potent inhibitor of adenosine deaminase, which limits its use in vivo as a PDE2 inhibitor (Rivet-Bastide et al. 1997). Inhibition of PDE2A by EHNA in cerebral cortical neurons increases NMDA receptor-mediated cGMP (Hajjhussein et al. 2007; Masood et al. 2008; Suvarna and O'Donnell 2002). Similarly, the novel PDE2 inhibitor BAY 60-7550, an EHNA analogue with greater potency and higher selectivity for PDE2 relative to EHNA, increases basal cGMP concentrations in cultured neurons and hippocampal slices (Boess et al. 2004; Hepp et al. 2007). It also enhances the induction of LTP in the hippocampus (Boess et al. 2004) and facilitates memory (Boess et al. 2004; Domek-Lopacinska and Strosznajder 2008; Rutten et al. 2009). BAY 60-7550 is able to fully reverse the working memory deficit induced by the NMDA receptor inhibitor MK-801 (Boess et al. 2004). Most recently, it has been found that BAY 60-7550 and another PDE2 inhibitor, ND7001, exert anxiolytic-like effects in the elevated plus-maze, holeboard, and open field tests in mice (Table 20.2) (Masood et al. 2009). In addition, these two PDE2 inhibitors also prevent the anxiogenic-like effects induced by oxidative stress in these three behavioral tests (Masood et al. 2008). The anxiolytic-like effect of PDE2 inhibitors appears to be associated with nitric oxide (NO)–cGMP signaling mediated by NMDA receptors. The results so far indicate that inhibition of PDE2 appears to be a promising pharmacotherapy for memory deficits, anxiety, and other mood disorders including depression.

20.3 PHOSPHODIESTERASE 4

PDE4 has been studied as a potential target for treatment of depressive disorders for nearly three decades. The antidepressant activity of PDE4 inhibitors such as rolipram has been extensively studied in various preclinical experiments (Li et al. 2009; Mizokawa et al. 1988; Przegalinski and Bigajska 1983; Wachtel 1983) and clinical trials (Table 20.2) (Fleischhacker 1992; Horowski and Sastre-y-Hernandez 1985). The PDE4D subtype appears to be of particular importance in mediating antidepressant activity (Zhang et al. 2002), although roles for PDE4A and PDE4B cannot be ruled out. However, side effects, particularly emesis, have greatly limited the clinical utility of PDE4 inhibitors for depressive disorders. Recent findings suggest that it may be possible to develop allosteric inhibitors of PDE4D that may not have significant emetic activity (Burgin et al. 2010).

20.3.1 STRUCTURE AND ACTIVITY REGULATION

PDE4 specifically hydrolyzes cAMP with a low K_m and plays a critical role in the control of intracellular cAMP concentrations (Bolger et al. 1996; Nemoz et al. 1997). Mammalian PDE4 is encoded by four genes (*PDE4A*, *PDE4B*, *PDE4C*, and *PDE4D*); each gene results in 3 to 11 variants that are generated by differential splicing for the N terminus (Chandrasekaran et al. 2008; Cheung et al. 2007; Houslay et al. 2007; Lynex et al. 2008; Mackenzie et al. 2008; O'Donnell and Zhang 2004; Wallace et al. 2005). To date, 25 variants have been identified, which are classified into four categories: long form, short form, super-short form, and dead-short form. All PDE4 variants have a highly conserved, catalytic domain in the subtype-specific C terminus, except for dead-short-form PDE4s, whose catalytic domain is truncated in both N and C termini. Long-form PDE4s are characterized by upstream conserved regions 1 and 2 (UCR1 and UCR2), two unique conserved domains in the variant-specific N-terminal region (Figure 20.1); short-form PDE4s (e.g., PDE4B2, PDE4D1, PDE4D2) lack an intact UCR2 and super-short-form PDE4s (PDE4A1, PDE4B5, PDE4D6, and PDE10) only have a portion of UCR2 (i.e., an N-terminally truncated UCR2); dead-short-form PDE4 (i.e., PDE4A7) lacks both UCR1 and UCR2 and is catalytically inactive (Houslay et al. 2007;

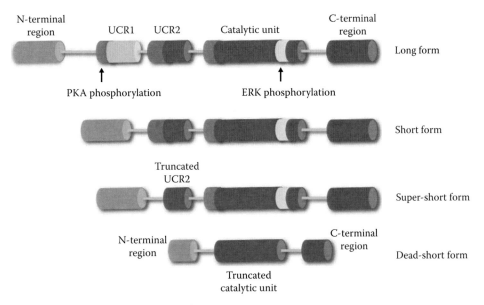

PDE4A: 4A1 (s-short), 4A5, 4A7, (d-short), 4A8, 4A10, 4A11

PDE4B: 4B1, 4B2 (short), 4B3, 4B4, 4B5 (s-short)

PDE4C: 4C1, 4C2, 4C3

PDE4D: 4D1 (short), 4D2 (short), 4D3, 4D4, 4D5, 4D6 (s-short), 4D7, 4D8, 4D9, 4D10 (s-short), 4D11

FIGURE 20.1 PDE4 splice variants. PDE4 is encoded by four distinct genes, comprising four subtypes (PDE4A, PDE4B, PDE4C, and PDE4D). Each PDE4 subtype can be alternatively truncated into multiple forms in N terminus, leading to at least 25 different splice variants. Based on structure of unique N-terminal region, PDE4 variants are divided into four categories: long-form, short-form, super-short (s-short)-form, and dead-short (d-short)-form PDE4s. Long-form PDE4s have upstream conserved regions 1 and 2 (UCR1 and UCR2) and a PKA phosphorylation site; short-form PDE4s simply have intact UCR2; super-short-form PDE4s only have an N-terminally truncated UCR2; dead-short-form PDE4s have an inactive catalytic unit truncated at both N and C termini, with no UCR domains. All but dead-short-form PDE4s have a catalytic unit in C terminus containing an ERK phosphorylation site with the exception of PDE4A. (Modified from Zhang, H. T., *Curr Pharm Des*, 15, 1687–98, 2009. With permission.)

Johnston et al. 2004; Zhang 2009). The other characteristic of long-form PDEs is the conserved phosphorylation sites: one is for cAMP-dependent protein kinase A (PKA) in the UCR1 domain (MacKenzie et al. 2002; Sette and Conti 1996) and the other is for extracellular signal-regulated kinases (ERK) in the catalytic domain, with the exception of PDE4A variants that do not have the ERK phosphorylation site (Baillie et al. 2000). The PKA phosphorylation site provides an important mechanism of cellular desensitization for cAMP signaling via increasing cAMP degradation, given that this site plays a positive role in the regulation of hydrolytic activity (Oki et al. 2000). The ERK phosphorylation site provides another mechanism by which cellular cAMP signaling is regulated via the alteration of PDE4 activity. This site regulates PDE4 hydrolytic activity in a variant-specific fashion; ERK phosphorylation inhibits activity of long-form PDE4s, weakly inhibits or does not alter activity of super-short-form PDE4s, and increases activity of short-form PDE4s (Baillie et al. 2000; Hoffmann et al. 1999; Houslay et al. 2007; MacKenzie et al. 2000). Since ERK signaling is increased by antidepressant treatment (Gourley et al. 2008; Qi et al. 2008), variant-dependent changes in PDE4 activity in response to ERK phosphorylation may contribute to antidepressant efficacy. The different sensitivities of PDE4 variants to ERK activation and their structural features indicate that PDE4 subtypes and their splice variants may differentially mediate CNS functions, including antidepressant activity.

20.3.2 DISTRIBUTION IN BRAIN

PDE4 is widely distributed in the brain in rats (Iwahashi et al. 1996; Kaulen et al. 1989; Perez-Torres et al. 2000), mice (Cherry and Davis 1999), nonhuman primates (Lamontagne et al. 2001), and humans (DaSilva et al. 2002; Perez-Torres et al. 2000; Schneider et al. 1986). However, the distribution of PDE4 varies across brain regions. For instance, PDE4 exhibits a high distribution in the hippocampus, frontal cortex, and olfactory bulb (Dlaboga et al. 2006; Kaulen et al. 1989; Zhao et al. 2003), suggesting a role of PDE4 in the mediation of memory and antidepressant activity (Song and Leonard 2005; Zhang et al. 2001). In addition, PDE4 is distributed differentially in the brain compared to adenylyl cyclase (AC), another important enzyme in the regulation of cellular cAMP levels (Gehlert et al. 1985; Poat et al. 1988). Specifically, in contrast to the high levels of PDE4 in the limbic system, AC is expressed at the highest levels in dopamine-rich brain regions, including the striatum, nucleus accumbens, and substantia nigra. This differential distribution may contribute to cAMP signal compartmentalization (Houslay 2010) and account at least partially for the different effects of the PDE4 inhibitor rolipram and the AC activator forskolin on animal behaviors. Although both drugs increase cellular cAMP levels in the brain (Schneider 1984; van Staveren et al. 2001; Wachtel et al. 1987), forskolin does not produce an antidepressant-like effect (O'Donnell 1993) or reverse scopolamine-induced memory deficit (Zhang and O'Donnell 2000).

The distributions of PDE4 subtypes are brain region specific, indicating differential roles of PDE4 subtypes in CNS functions. PDE4A, PDE4B, and PDE4D are highly but differentially distributed in the brain, whereas PDE4C is primarily located in peripheral tissues (Cherry and Davis 1999; Perez-Torres et al. 2000). More specifically, PDE4A and PDE4D share a similar distribution pattern in the brain: both are highly expressed in the cerebral cortex, olfactory bulb, hippocampal formation, and brainstem (Cherry and Davis 1999; Engels et al. 1995; Perez-Torres et al. 2000), suggesting that these two subtypes may be important in the mediation of antidepressant activity and memory. This is supported by studies from Zhang and colleagues showing that PDE4D is important in the mediation of antidepressant activity and memory enhancement (Li et al. 2011; Zhang et al. 2002). In contrast, PDE4B is the predominant PDE4 subtype in the striatum, amygdala, hypothalamus, and thalamus. It also exhibits a distribution pattern similar to AC (Perez-Torres et al. 2000), which is high in dopamine-rich brain regions (Gehlert et al. 1985; Poat et al. 1988). The distribution feature of PDE4B appears to account for the modest antidepressant-like behavior and little memory effects in mice deficient in PDE4B (Siuciak et al. 2008; Zhang et al. 2008) and, on the other hand, for the contributions of PDE4B to anxiety (Zhang et al. 2008), schizophrenia (Millar et al. 2005), and the antipsychotic effect of rolipram (Siuciak et al. 2007).

It has been noted that PDE4D is highly expressed in the nucleus of solitary tract and area postrema (Cherry and Davis 1999; Lamontagne et al. 2001; Perez-Torres et al. 2000), which are critical for emetic responses (Ariumi et al. 2000), and PDE4B is exclusively expressed in the nucleus accumbens (Perez-Torres et al. 2000), a structure commonly believed to mediate the effects of drugs of abuse (Koob 1992). These results indicate that PDE4D may be responsible for emetic responses usually observed after administration of PDE4 inhibitors and that PDE4B may be important for mediating aspects of drug dependence and abuse. The contribution of PDE4D to emesis has been demonstrated using mice deficient in PDE4D (Li et al. 2011; Robichaud et al. 2002).

20.3.3 Role in Antidepressant Activity

The antidepressant-like effects of PDE4 inhibitors started to be highlighted in the early 1980s. In 1983, Helmut Wachtel reported for the first time that PDE4 inhibitors such as rolipram and Ro 20-1724 produced antidepressant-like effects, including blockade of reserpine-induced hypothermia/hypokinesia and potentiation of yohimbine-induced lethality in mice (Wachtel 1983). Two months later, a Polish group reported similar results (Przegaliński and Bigajska 1983), which strongly supported Wachtel's findings. Follow-up studies have demonstrated that PDE4 inhibitors are active in a variety of preclinical tests sensitive to antidepressants, including antimuricidal activity in olfactory bulbectomized rats (Mizokawa et al. 1988), decreased immobility in the forced-swim test (Li et al. 2003; Saccomano et al. 1991; Zhang et al. 2002, 2006) and tail-suspension test (Li et al. 2009; Zhang et al. 2002), increased reinforcement rate, and decreased response rate in the differential-reinforcement-of-low-rate schedule (O'Donnell 1993; O'Donnell and Frith 1999; O'Donnell et al. 2005; Zhang et al. 2005a, 2006). The profile of rolipram effects is consistent with its efficacy in the treatment of major depression in clinical trials (Hebenstreit et al. 1989; Horowski and Sastre-y-Hernandez 1985; Zeller et al. 1984), although it has not advanced to clinical use because of emesis and gastrointestinal disturbance (Scott et al. 1991).

Antidepressant activity of rolipram exhibits several characteristics. First, rolipram produces antidepressant effects in various species, including rodents (Overstreet et al. 2005; Zhang et al. 2002, 2005b, 2006) and humans (Bobon et al. 1988; Fleischhacker et al. 1992; Zeller et al. 1984), leading to translation of experimental results from preclinical tests to clinical trials. Second, the effect of rolipram appears to be stereo-specific (Wachtel 1983), which is in agreement with stereo-specificity of rolipram binding (Kaulen et al. 1989; Schneider et al. 1986) and PDE4 inhibition (Ohsawa et al. 1998). The (−)-isomer of rolipram is approximately 10 to 15 times more potent than the (+)-isomer in terms of antidepressant-like actions. This is consistent with the results from the muricidal behavioral test, in which the (−)-isomer of rolipram is much more potent than the (+)-isomer (Mizokawa et al. 1988), and memory tests, in which (−)-rolipram is 200 times more potent than (+)-rolipram (Egawa et al. 1997). Third, antidepressant efficacy is greater when rolipram is administered repeatedly than it is given acutely (Li et al. 2009; Zhang et al. 2002), at least in the mouse forced-swim test (Figure 20.2). Finally, the antidepressant-like effect of rolipram is enhanced when coadministered with classic antidepressants (Itoh et al. 2004), selective serotonin reuptake inhibitors, or β-adrenergic receptor agonists (Zhang et al. 2005b). Therefore, development of dual inhibitors that inhibit PDE4 and transport of serotonin (SERT) (Cashman et al. 2009) may represent a new direction of antidepressant development, given that dual inhibitors may overcome the compensatory increase in PDE4 expression seen with repeated treatment with antidepressant drugs (Ye et al. 2000; Dlaboga et al. 2006).

Although the mechanisms whereby PDE4 inhibitors produce antidepressant activity remain to be fully elucidated, studies to date suggest that PDE4 inhibition-induced activation of cAMP signaling is their major mechanism of antidepressant actions. PDE4 plays a critical role in the control of intracellular cAMP levels in the brain (Ye and O'Donnell 1996; Zhang et al. 2002). Inhibition of PDE4 increases concentrations of intracellular cAMP, which triggers the downstream signal cascade, including activation of PKA and subsequent phosphorylation and activation of cAMP-

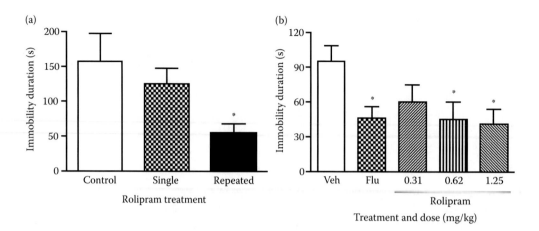

FIGURE 20.2 Antidepressant-like effects of rolipram in forced-swim and tail-suspension tests in mice. (a) Repeated, but not single, treatment with rolipram decreased immobility duration in forced-swim test. Rolipram (0.5 mg/kg) or its vehicle (saline containing 10% DMSO as control) was injected 30 min before the test for acute treatment or once daily for 8 consecutive days for repeated administration (testing was carried out 30 min after the last rolipram injection). (Redrawn from Zhang, H. T., et al., *Neuropsychopharmacology*, 27, 587–95, 2002. With permission.) (b) Chronic treatment with rolipram or fluoxetine (Flu) decreased immobility in tail-suspension test. Rolipram (0.31–1.25 mg/kg), fluoxetine (10 mg/kg), or vehicle (Veh) was administered intraperitoneally once a day for 22 days before the test, which was performed 1 h after the last injection. (Redrawn from Li, Y. F., et al., *Neuropsychopharmacology*, 34, 2404–19, 2009. With permission.) Values shown are means ± SEM. *$p < .05$ vs. control (Veh); $n = 5$–7 (a) or 8–9 (b).

responsive element–binding protein (CREB) (Conti and Beavo 2007; Houslay et al. 2005, 2007; Li et al. 2009). This signaling pathway appears to be the common mechanism undertaking antidepressant actions of different classes of antidepressants (Conti and Blendy 2004; D'Sa and Duman 2002; Duman 1998). Chronic treatment with rolipram or other antidepressants upregulates the cAMP signaling cascade, including increased components of cAMP-dependent protein kinase (Nestler et al. 1989), phosphorylation of CREB (Nibuya et al. 1996), and its downstream target brain-derived neurotrophic factor (BDNF) (Conti et al. 2002; Nibuya et al. 1996). These effects appear to contribute to rolipram-induced increases in hippocampal neurogenesis, which appear to be required for antidepressant-like effects on behavior (Santarelli et al. 2003), and the antidepressant- and anxiolytic-like effects of rolipram (Li et al. 2009; Zhang et al. 2002).

It has been noted that chronic administration of antidepressants increases PDE4 activity and levels of PDE4 mRNA and protein in the brain (Andersen et al. 1983; Dlaboga et al. 2006; D'Sa et al. 2005; Li et al. 2011; Suda et al. 1998; Takahashi et al. 1999; Ye et al. 1997; Zhao et al. 2003). This is considered as a compensatory response to the increase in cAMP signaling and suggests that PDE4 is a component of signaling pathways active by antidepressants. On the other hand, PDE4 activity is increased in the brain in learned-helplessness rats, an animal model of depression (Itoh et al. 2003, 2004); this is reversed by chronic treatment with antidepressants including rolipram. Therefore, targeting PDE4 causes alteration of cAMP signaling and modulation of antidepressant activity.

20.3.4 Role of Subtypes in Antidepressant Activity

Because of the lack of highly selective inhibitors of individual PDE4 subtype, the progress in identification of the roles of PDE4 subtypes in CNS functions has been slow. This situation started to be changed when Marco Conti's team generated mice deficient in PDE4D (Jin et al. 1999) and later, mice deficient in PDE4B (Jin and Conti 2002). Most recently, they also generated mice deficient in PDE4A, although the phenotype of this line has not been identified yet. It has been shown using

PDE4-subtype knockout mice that PDE4B and PDE4D are differentially involved in certain CNS functions. PDE4B is important in mediating antipsychotic effects (Siuciak et al. 2007, 2008) and anxiety-like behavior (Zhang et al. 2008), whereas PDE4D primarily mediates antidepressant activity (Zhang et al. 2002) and memory (Li et al. 2011; Rutten et al. 2008). Specifically, mice deficient in PDE4D display decreases in immobility in the forced-swim and tail-suspension tests compared with the wild-type littermates (Figure 20.3) (Zhang et al. 2002). They also show enhancement of hippocampal-dependent memory (Li et al. 2011), although they even exhibit impairment of fear conditioning (Rutten et al. 2008), which is associated with amygdala-dependent memory. Mice deficient in PDE4B, in contrast, display a moderate antidepressant-like effect in the forced-swim test, but not in the tail-suspension test (Siuciak et al. 2008; Zhang et al. 2008); they do not show any changes in memory tests. The different roles of PDE4B and PDE4D appear to be consistent with their distribution patterns in the brain. PDE4D is highly expressed in the hippocampus and olfactory system, whereas expression of PDE4B in these brain regions is more limited (Perez-Torres et al. 2000). The opposite distribution pattern of PDE4B and PDE4D is observed in the striatum and amygdala, in which PDE4B is highly expressed (Cherry and Davis 1999; Perez-Torres et al. 2000). In addition, the major role of PDE4D in antidepressant activity may also be attributed to its critical contributions to cAMP hydrolysis and CREB phosphorylation in the brain (Li et al. 2011; Zhang et al. 2002).

Although the phenotype of PDE4A knockout mice remains unknown, the role of PDE4A in antidepressant activity cannot be ruled out, given that PDE4A is also highly expressed in depression-associated regions such as the hippocampus, olfactory bulb, and frontal cortex (Cherry and Davis 1999; Perez-Torres et al. 2000). In addition, pharmacological and neurochemical studies also have shown that PDE4A is involved in antidepressant activity (Miro et al. 2002; Suda et al. 1998; Takahashi et al. 1999; Ye et al. 1997, 2000). Repeated treatment with electroconvulsive shock or different pharmacological classes of antidepressants increases expression of PDE4A in both the

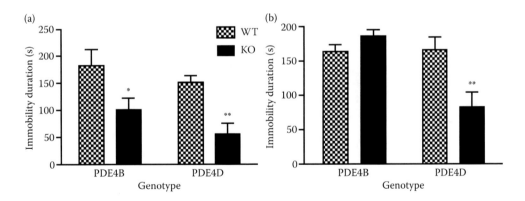

FIGURE 20.3 Antidepressant-like behavior of mice deficient in PDE4B or PDE4D (KO) in forced-swim test (FST; a) and tail-suspension test (TST; b). Compared with wild-type (WT) controls, PDE4D KO mice displayed decreased immobility duration in both FST and TST. In contrast, PDE4B KO mice displayed a similar effect only in FST. In FST, mice were individually placed in a cylinder (45 cm high × 20 cm diameter) containing water (22–23°C, 28 cm in depth) and allowed to swim for 6 min. Duration of immobility, which was defined as floating in an upright position without additional movement other than that necessary for the animal to keep its head above the water, was recorded. In TST, each mouse was suspended for 6 min using adhesive tape placed 1 cm from tip of its tail. Duration of immobility, which was defined as passive hanging without movement, was recorded. Bars shown represent means ± SEM. *$p < .05$; **$p < .01$ vs. corresponding WT control (two-tailed Student's t test); $n = 7$–15. [(a) Redrawn from Zhang, H. T., et al., *Neuropsychopharmacology*, 27, 587–95, 2002, and Zhang, H. T., et al., *Neuropsychopharmacology*, 33, 1611–23, 2008. With permission. (b) Modified from Zhang, H. T., *Curr Pharm Des*, 15, 1687–98, 2009. With permission.]

hippocampus and the frontal cortex (Takahashi et al. 1999; Ye et al. 2000). Therefore, it is reasonable to believe that PDE4A also may be involved in antidepressant activity. Further studies using PDE4A knockout mice are required to address this issue.

20.3.5 Role of Splice Variants in Antidepressant Activity

Assuming PDE4A and PDE4D are important PDE4 subtypes for mediating antidepressant activity, there is still a long way to go before we are able to identify the role of specific variants of these two subtypes, which include 6 PDE4As and 11 PDE4Ds. Fortunately, RNA interference, a powerful technique for gene silencing (Cullen 2005; Zeringue and Constantine-Paton 2004), has provided an effective approach to identifying the role of PDE4 splice variants. This has been achieved in vitro using small interfering RNAs that target a specific variant of PDE4B or PDE4D (Li et al. 2006, Lynch et al. 2005). Most recently, Han-Ting Zhang and colleagues applied lentiviral vectors harboring miRNAs that target long-form PDE4D variants in vivo in mice (Li et al. 2011). They found that the PDE4D miRNAs downregulated PDE4D4 and PDE4D5 in the hippocampus, increased hippocampal neurogenesis and pCREB, and enhanced memory. Since adult neurogenesis in the hippocampus is required for some antidepressant effects (Santarelli et al. 2003) and rolipram-induced increases in neurogenesis and cAMP/CREB signaling in the hippocampus contribute to antidepressant activity (Li et al. 2009; Nakagawa et al. 2002), knockdown of long-form PDE4D variants in depression-associated brain regions would most likely result in antidepressant activity. Although deficiency of PDE4D causes emetic-like responses in mice (Robichaud et al. 2002), knockdown of PDE4D in the hippocampus does not produce a similar effect (Li et al. 2011). Recent work suggests that selective, allosteric inhibitors of PDE4D may have psychopharmacological effects without significant emetic activity (Burgin et al. 2010). Because of the concern of the role of PDE4D in emetic responses, PDE4A variants, which are not expressed in the emesis-related brain structures, the area postrema and nucleus of solitary tract (Cherry and Davis 1999; Perez-Torres et al. 2000), may have promise as the targets for antidepressants. Overall, the findings to date are able to provide a possibility for development of PDE4 inhibitors that retain therapeutic profiles but do not cause major side effects such as emesis.

20.4 PHOSPHODIESTERASE 5

20.4.1 Structure and Brain Distribution

PDE5 was first identified and characterized from platelets by Hamet and colleagues (Coquil et al. 1980; Hamet and Coquil 1978) and its activity was originally described by Francis et al. (1980). However, PDE5 had not received significant attention until it was found to be a regulator of vascular smooth muscle contraction and, more importantly, as the target for treatment of erectile dysfunction (Bender and Beavo 2006).

PDE5 has GAF-A and GAF-B domains in the N terminus, which are responsible for substrate specificity and enzyme activity (Martinez et al. 2002). The GAF-A domain has low nanomolar affinities for cGMP ($K_d \leq 10$ nM); it increases catalytic activity of PDE5 when it selectively binds to cGMP (Martins et al. 1982). The allosteric cGMP binding enhances phosphorylation of PDE5 through cGMP-dependent protein kinase, which in turn increases PDE5 activity and cGMP binding affinity of GAF-A (Francis et al. 2002). There are three variants of PDE5, PDE5A1, PDE5A2, and PDE5A3, which are distinguished based on their N-terminal domains. PDE5A1 and PDE5A2 are ubiquitous, whereas PDE5A3 is specifically expressed in smooth muscles (Lin 2004). PDE5A1 and PDE5A2 have distinct N-terminal sequences in humans, canines, and rats (Kotera et al. 1998); they exhibit similar kinetic properties, including K_m, V_{max}, and inhibition by the PDE5 inhibitor zaprinast. The PDE5A2 transcript is an inducible variant of PDE5A in rats; it is increased by cAMP in cultured rat vascular smooth muscle cells (Kotera et al. 1999).

PDE5 transcripts are primarily expressed in peripheral tissues, particularly those enriched with blood vessels, including heart, kidney, lung, and intestine (Giordano et al. 2001; Kotera et al. 2000). They are also expressed at relatively high levels in the brain, including the cerebellum, olfactory bulb, cerebral cortex, and hippocampus (Garthwaite and Boulton 1995; Kleppisch 2009), indicating that PDE5 may be involved in CNS functions, including memory and antidepressant activity.

20.4.2 ROLE IN NO-cGMP SIGNALING AND ITS CONNECTION TO CNS FUNCTION

It has been shown that PDE5 plays a role in mood-related memory consolidation of object information through alteration of the NO-cGMP signaling pathway (Hartell 1996; Prickaerts et al. 2004). PDE5 is involved in alteration of monoaminergic neurotransmission, particularly serotonin via NO-dependent mechanisms (Puerta et al. 2009). NO is an important messenger in the nervous system; it targets the soluble isoform of guanylyl cyclase (sGC) (Jesse et al. 2008; Murad 1994), which catalyzes the formation of cGMP, whose breakdown is catalyzed by PDEs, such as PDE2 and PDE5. cGMP binding activates PKG, which phosphorylates serines and threonines on many cellular proteins, including CREB, frequently leading to changes in their activity or function, subcellular localization, or regulatory features (Francis et al. 2010). Therefore, regulation of NO-cGMP signaling appears to be the major mechanism whereby PDE5 mediates CNS functions. PDE5 may be a novel therapeutic strategy for depression and/or mood-related memory impairment (Wong et al. 2006).

Studies have also examined other mechanisms underlying the interactions between PDE5 and related CNS diseases. Inhibition of PDE5 and subsequent activation of PKG prevent dysfunction of the serotonergic system and increases in free radical formation (Abdul and Butterfield 2007; Aguirre et al. 1999; Green et al. 2003; Lee et al. 2003). Puerta et al. (2009) have shown that sildenafil reverses serotonin depletion–induced behavioral changes, which are involved in activation of PKG and mitochondrial ATP-sensitive K^+ channels. These effects may also be associated with improved NOS coupling and reduction of reactive oxygen species signaling. PDE5 inhibitors such as sildenafil and tadalafil, which readily cross the blood–brain barrier and are relatively well tolerated, may be promising treatments for mood-related disorders.

20.4.3 ROLE IN ANTIDEPRESSANT ACTIVITY

Zaprinast has been used as a PDE5 inhibitor, although it is even more potent for inhibition of PDE6 (Zhang et al. 2005b). It has been shown that zaprinast produces a slight antidepressant-like effect on rat differential-reinforcement-of-low-rate behavior (O'Donnell and Frith 1999), that is, it modestly decreases response rate and increases reinforcement rate, in contrast to rolipram, which robustly decreases response rate and increases reinforcement rate (Zhang et al. 2005b, 2006). Consistent with the effect of zaprinast on differential-reinforcement-of-low-rate behavior, sildenafil and tadalafil, two selective PDE5 inhibitors that are structurally distinct, have been found to exert potent antidepressant-like effects in a genetic animal model of depression, namely, the Flinders sensitive line rats (Liebenberg et al. 2010) and in the rat forced-swim test (Brink et al. 2008), but the effect requires the blockade of central muscarinic cholinergic receptors. This may be attributed to the enhancement by sildenafil of cholinergic transmission in the brain (Devan et al. 2004; Patil et al. 2004). The results indicate that PDE5 inhibitors may produce antidepressant-like effects primarily via activation of NO-cGMP signaling when the muscarinic cholinergic neurotransmission is reduced in the brain.

Antidepressant activity of PDE5 inhibitors appears to be supported by clinical observations. Double-blind, placebo-controlled studies have shown that vardenafil, a potent, selective PDE5 inhibitor developed for treatment of erectile dysfunction, significantly reduces symptoms of depression (Hatzichristou et al. 2005). Sildenafil also has a similar therapeutic effect (Moncada et al. 2009). Nevertheless, the effects of PDE5 inhibitors on depression could be complex. Increased cGMP levels by PDE5 inhibitors such as sildenafil prevent the antidepressant-like effect of adenosine or the

NMDA receptor antagonist memantine in the forced-swim and tail-suspension tests (Almeida et al. 2006; Kaster et al. 2005b). These indicate that inhibition of NO synthesis and decreases in cGMP levels may exhibit antidepressant-like effect, as demonstrated by reduced duration of immobility in the mouse model of despair tests (Jesse et al. 2008). Since inhibition of K^+ channels and the consequent increase in neurotransmitter release produce an antidepressant-like effect observed in the forced-swim test (Kaster et al. 2005a), the blockade of antidepressant activity by PDE5 inhibitors may be associated with activation of the channels in response to activated of NO-cGMP signaling caused by PDE5 inhibition.

20.4.4 Role in Anxiety and Schizophrenia

In addition to depression, PDE5-mediated NO-cGMP signaling also plays a role in the regulation of anxiety. Sildenafil has been shown to decrease the percentage of time spent in open arms in the plus-maze test in mice (Kurt et al. 2004), indicating an anxiogenic-like effect of sildenafil. This appears to be mediated by NO-cGMP signaling pathway. On the other hand, Solís and co-workers (2008) used open field exploration, an animal test used in anxiety behavior studies, to examine the effect of chronic treatment with sildenafil on anxiety-like and risk-taking behaviors in gonadally intact and castrated male Wistar rats. They found that chronic sildenafil administration increased the number of entries and time spent in the center area in both gonadally intact and castrated animals; this is independent of endogenous androgens. The results suggest that repeated treatment with PDE5 inhibitors may exert anxiolytic effects. Although it is not known what accounts for the contrary findings, animal species and/or the duration of PDE5 inhibitor treatment may need to be taken into account.

PDE5 may also contribute to the pathophysiology of schizophrenia, which consists of positive symptoms, such as hallucination and paranoia, and negative symptoms, including social or occupational dysfunction and disorganized speech and thinking. Negative symptoms are considered to contribute more to poor functional outcomes and poor quality of life for patients with schizophrenia (Akhondzadeh 2001; Buckley and Stahl 2007; Erhart et al. 2006). NMDA receptor hypofunction results in excessive activation of dopaminergic neurotransmission in the mesolimbic system of patients with schizophrenia, indicating that enhancing NMDA function may serve as a therapeutic strategy for schizophrenia-related negative and positive disorders (Seeman 2009; Stahl 2007). Most recently, studies have shown that PDE5 inhibitors such as sildenafil are effective in treatment of negative symptoms of schizophrenia (Akhondzadeh et al. 2011). Combination of sildenafil and risperidone, an atypical antipsychotic drug, even has a significant superiority over risperidone alone in controlling the negative symptoms. Sildenafil is found to increase cGMP without directly affecting the activity of NMDA receptors (Reneerkens et al. 2009; Siuciak 2008; Zhang 2010). Nevertheless, it is possible that targeting PDE5 to increase cGMP may selectively prevent behavioral deficits due to NMDA receptor hypofunction in schizophrenia.

20.5 PHOSPHODIESTERASE 9

PDE9 was first identified in 1998 by the teams of Beavo and Cheng independently and simultaneously (Fisher et al. 1998; Soderling et al. 1998). PDE9A is the single isoform in this family. It is a high-affinity and cGMP-specific PDE enzyme. To date, five N-terminal splice variants (PDE9A1–5) have been identified in humans and mice (Guipponi et al. 1998). Structural studies of PDE9A have shown that its cDNAs of different splice variants share a high percentage of amino acid identity in the catalytic domain. However, despite its highest specificity for cGMP among all the PDEs (Soderling et al. 1998), PDE9A lacks the GAF domain, whose binding to cGMP usually activates the catalytic activity (Esposito et al. 2009).

PDE9A is highly distributed in rodents' kidney, human spleen, and, at lower levels, in the brain and other peripheral organs in both species. Studies using in situ hybridization have shown that PDE9A mRNA is widely expressed throughout the brain, which closely resembles the expression of sGC and NOS, indicating the cooperation of these enzymes to regulate the cGMP levels through NO/sGC–cGMP signaling (Andreeva et al. 2001).

Although there are a few studies on the role of PDE9 in cognition, little is known about the role of PDE9 in depression. Wong et al. (2006) have revealed that PDE9A is possibly involved in major depressive disorder (MDD) using SNP genotyping in a study investigating the association between PDEs and their susceptibility to MDD. However, this finding was not confirmed by another group (Cabanero et al. 2009). Studies using animal tests are required for verification of the relationship between PDE9A and depression.

It has also been reported that the PDE9A inhibitor Bay 73-6691 is able to attenuate the scopolamine-induced memory deficit in a passive avoidance task (van der Staay et al. 2008) and reverse cognition deficits in schizophrenia (Zhang 2010). Given the changes in Ca^{2+} levels in accordance with calmodulin-dependent NOS activity (Jesko et al. 2003), PDE9 may be involved in the regulation of brain function via the common pathway of Ca^{2+}/NO/cGMP.

Besides the known PDE9 inhibitors BAY 73-6691 and SCH 51866, several other compounds have been found to penetrate the blood–brain barrier and increase cGMP levels in the brain through inhibition of PDE9 (Verhoest et al. 2009). However, pharmacological profiles of these compounds remain to be further evaluated, and as yet little is known about the side effects and isoform specificity of most PDE9 inhibitors.

20.6 PHOSPHODIESTERASE 10

PDE10 was first isolated and characterized by Beavo's team (Soderling et al. 1999). It is a dual-substrate PDE enzyme that hydrolyzes both cAMP and cGMP. PDE10 has four major isoforms (PDE10A1–4), which differ in N and C termini but share the identical GAF domains. PDE10A is highly expressed in the rodent striatum, hippocampus, and olfactory tubercle, particularly in the striatum (Hebb et al. 2004; Hu et al. 2004; O'Conner et al. 2004; Siuciak et al. 2006; Xie et al. 2006). More specifically, PDE10 is expressed at particularly high levels in γ-aminobutyric acid–containing medium spiny projection neurons of the mammalian striatum, which is the core site for information integration in the brain and regulates motoric, appetitive, and cognitive processes (Menniti et al. 2006; Seeger et al. 2003). Dysfunction of this region might involve multiple CNS disorders, such as Parkinson's disease, Huntington's disease, and related psychiatric disorders. Recently, PDE10A has emerged as a promising target for CNS disorders.

It has been shown in a study using PDE10A knockout mice that PDE10A plays a role in schizophrenia-related behavior but appears not to be involved in depressive- or anxiety-like behavior (Table 20.2) (Siuciak et al. 2008). However, since PDE10A is high in NAc, which has been demonstrated to mediate some antidepressant-like effects on behavior (Nestler et al. 2002; Shirayama and Chaki 2006), PDE10A may contribute to a certain extent to depression and antidepressant activity. Although more work is needed to verify this, the contribution of PDE10 to cognition has been explicitly demonstrated in recent years.

It appears that PDE10A is an important regulator of cyclic nucleotides in the striatum and may be involved in the physiological regulation of motor and cognitive function (Lehericy and Gerardin 2002). PDE10A may regulate striatal output by reducing the sensitivity of medium spiny neurons to glutamatergic excitation (Siuciak et al. 2006). Moreover, inhibition of PDE10 may result in changes in cyclic nucleotide signaling that is involved in schizophrenia (Siuciak 2008). This finding is also supported by studies showing that the PDE10 inhibitor papaverine is effective in improving executive function deficits associated with schizophrenia (Rodefer et al. 2005; Siuciak et al. 2006). Chronic PDE10A inhibition can induce a series of behavioral and central neurochemical deficits in the animal models that are exacerbated under stress (Hebb et al. 2004). All these data indicate the implication of PDE10 in CNS disorders.

Recent studies have shown that the effects of PDE10A inhibitors on psychiatric disorders might involve PKA/PKG-dependent signal pathways. Menniti and co-workers (2006) found that PDE10A inhibition effectively increases phosphorylation of CREB and expression of BDNF in primary cultures of cortical and hippocampal neurons. However, further studies on the functional significance of these effects as well as the on downstream signaling mechanisms are still needed.

20.7   PHOSPHODIESTERASE 11

PDE11, the newest member of the mammalian PDE superfamily, was first characterized by a British group in 2000 (Fawcett et al. 2000). About half a year later, the team of Beavo and a Japanese group identified four splice variants (PDE11A1–A4), which contain the same catalytic domain but have different lengths and/or compositions of their amino termini (Hetman et al. 2000; Yuasa et al. 2001). PDE11A1 (56 kDa) and PDE11A2 (66 kDa) are amino-truncated versions of PDE11A3 (78 kDa) and PDE11A4 (105 kDa), which differ at their N termini. PDE11 has a single gene (*PDE11A*); it is expressed at high levels primarily in peripheral tissues, including the skeletal muscle, prostate, kidney, liver, pituitary, salivary glands, and testis (Lakics et al. 2010), and also at relatively high levels in the hippocampus (Kelly et al. 2010). The PDE11A splice variants exhibit different expression patterns in mammalian tissues. For instance, PDE11A3 is specifically expressed in testis, whereas PDE11A4 is particularly abundant in the prostate (Yuasa et al. 2001). Because of the lack of selective PDE11 inhibitors, the function of PDE11A remains largely unknown. The enrichment of PDE11A in the hippocampus, particularly in the ventral hippocampus, indicates that this PDE family may be involved in cognition and depression. Most recently, Kelly and co-workers (2010) have shown that mice deficient in PDE11A display subtle psychiatric disease-related deficits, including hyperactivity, increased sensitivity to NMDA receptor antagonists, and social behavior deficits such as impairment of social odor recognition memory and social avoidance. In addition, PDE11A also has been shown important in cAMP/CREB signaling in adrenocortical tumors and associated with Cushing's syndrome (Horvath et al. 2006; Stratakis and Boikos 2007). Comparably, little is known about the potential contribution of PDE11A to depression. The only study to date that has touched this point is that by Wong and co-workers, who have revealed by using genotyped SNPs of PDEs that the *PDE11A* gene is associated with MDDs and the antidepressant response (Wong et al. 2006). Future studies are required to verify the role of PDE11A in antidepressant activity using mice deficient in PDE11A (Kelly et al. 2010).

20.8   CONCLUDING REMARKS

PDEs have been studied for more than six decades since the first report on nonspecific PDE in rats and rabbits (Zittle and Reading 1946). Although progress has been dramatic, especially during the past 30 years, the roles of PDEs in CNS functions are not fully understood. This is at least partially attributable to the lack of highly selective PDE inhibitors and the complexity of this superfamily, which has 11 families and 21 isoforms with more than 100 splice variants. Based on the studies to date, we summarized in Table 20.2 the antidepressant-like effects of inhibitors of PDE4 and other PDEs, together with anxiolytic profiles of some PDE inhibitors. Overall, although several PDEs, including PDE1, PDE2, PDE4, PDE5, PDE9, PDE10, and PDE11, have been shown to be involved in cognition and/or anxiety to different extents, only PDE4 has been demonstrated to be a target for treatment of depression. The major side effect emesis caused by PDE4 inhibitors has directed PDE studies to identifying the roles of PDE4 subtypes and splice variants in the mediation of antidepressant activity as well as other therapeutic profiles. It has also opened a window in the research on PDEs as potential targets for drugs with therapeutic benefits for CNS diseases, including depressive disorders.

ACKNOWLEDGMENTS

Work performed in the authors' laboratories was supported by research grants from NARSAD Young Investigator Awards (2006 and 2008 to H. T. Zhang), NIA (AG031687 to H. T. Zhang), and the NIMH (MH051175, MH040694 to J. M. O'Donnell and MH088480 to J. M. O'Donnell and C. G. Zhan).

REFERENCES

Abdul, H. M., and Butterfield, D. A. 2007. Involvement of PI3K/PKG/ERK1/2 signaling pathways in cortical neurons to trigger protection by cotreatment of acetyl-L-carnitine and alpha-lipoic acid against HNE-mediated oxidative stress and neurotoxicity: Implications for Alzheimer's disease. *Free Radic Biol Med* 42: 371–84.

Aguirre, N., Barrionuevo, M., Ramirez, M. J., Del Rio, J., and Lasheras, B. 1999. Alpha-lipoic acid prevents 3,4-methylenedioxy-methamphetamine (MDMA)–induced neurotoxicity. *Neuroreport* 10: 3675–80.

Akhondzadeh, S. 2001. The 5-HT hypothesis of schizophrenia. *IDrugs* 4: 295–300.

Akhondzadeh, S., Ghayyoumi, R., Rezaei, F., et al. 2011. Sildenafil adjunction therapy to risperidone in the treatment of the negative symptoms of schizophrenia: A double-blind randomized placebo-controlled trial. *Psychopharmacology (Berl)* 213(4): 809–15.

Almeida, R. C., Felisbino, C. S., Lopez, M. G., Rodrigues, A. L., and Gabilan, N. H. 2006. Evidence for the involvement of L-arginine-nitric oxide-cyclic guanosine monophosphate pathway in the antidepressant-like effect of memantine in mice. *Behav Brain Res* 168: 318–22.

Andersen, P. H., Klysner, R., and Geisler, A. 1983. Cyclic AMP phosphodiesterase activity in rat brain following chronic treatment with lithium, imipramine, reserpine, and combinations of lithium with imipramine or reserpine. *Acta Pharmacol Toxicol (Copenh)* 53: 337–43.

Andreeva, S. G., Dikkes, P., Epstein, P. M., and Rosenberg, P. A. 2001. Expression of cGMP-specific phosphodiesterase 9A mRNA in the rat brain. *J Neurosci* 21: 9068–76.

Ariumi, H., Saito, R., Nago, S., Hyakusoku, M., Takano, Y., and Kamiya, H. 2000. The role of tachykinin NK-1 receptors in the area postrema of ferrets in emesis. *Neurosci Lett* 286: 123–6.

Baillie, G. S., MacKenzie, S. J., McPhee, I., and Houslay, M. D. 2000. Sub-family selective actions in the ability of Erk2 MAP kinase to phosphorylate and regulate the activity of PDE4 cyclic AMP-specific phosphodiesterases. *Br J Pharmacol* 131: 811–19.

Barad, M., Bourtchouladze, R., Winder, D. G., Golan, H., and Kandel, E. 1998. Rolipram, a type IV–specific phosphodiesterase inhibitor, facilitates the establishment of long-lasting long-term potentiation and improves memory. *Proc Natl Acad Sci U S A* 95: 15020–5.

Beavo, J. A. 1995. Cyclic nucleotide phosphodiesterases: Functional implications of multiple isoforms. *Physiol Rev* 75: 725–48.

Beavo, J. A., and Brunton, L. L. 2002. Cyclic nucleotide research — still expanding after half a century. *Nat Rev Mol Cell Biol* 3: 710–18.

Beavo, J. A., Hardman, J. G., and Sutherland, E. W. 1971. Stimulation of adenosine 3V, 5V-monophosphate hydrolysis by guanosine 3V,5V-monophosphate. *J Biol Chem* 246: 3841–6.

Bender, A. T., and Beavo, J. A. 2006. Cyclic nucleotide phosphodiesterase: Molecular regulation to clinical use. *Pharmacol Rev* 58: 488–520.

Berndt, S., and Schwabe, U. 1973. Effect of psychotropic drugs on phosphodiesterase and cyclic AMP level in rat brain in vivo. *Brain Res* 63: 303–12.

Blokland, A., Schreiber, R., and Prickaerts, J. 2006. Improving memory: A role for phosphodiesterases. *Curr Pharm Des* 12: 2511–23.

Bobon, D., Breulet, M., Gerard-Vandenhove, M. A., et al. 1988. Is phosphodiesterase inhibition a new mechanism of antidepressant action? A double blind double-dummy study between rolipram and desipramine in hospitalized major and/or endogenous depressives. *Eur Arch Psychiatry Neurol Sci* 238: 2–6.

Boess, F. G., Hendrix, M., van der Staay, F. J., et al. 2004. Inhibition of phosphodiesterase 2 increases neuronal cGMP, synaptic plasticity and memory performance. *Neuropharmacology* 47: 1081–92.

Bolger, G. B., Rodgers, L., and Riggs, M. 1994. Differential CNS expression of alternative mRNA isoforms of the mammalian genes encoding cAMP-specific phosphodiesterases. *Gene* 149: 237–44.

Bolger, G. B., McPhee, I., and Houslay, M. D. 1996. Alternative splicing of cAMP-specific phosphodiesterase mRNA transcripts. Characterization of a novel tissue-specific isoform, RNPDE4A8. *J Biol Chem* 271: 1065–71.

Brink, C. B., Clapton, J. D., Eagar, B. E., and Harvey, B. H. 2008. Appearance of antidepressant-like effect by sildenafil in rats after central muscarinic receptor blockade: Evidence from behavioural and neuroreceptor studies. *J Neural Transm* 115: 117–25.

Buckley, P. F., and Stahl, S. M. 2007. Pharmacological treatment of negative symptoms of schizophrenia: Therapeutic opportunity or cul-de-sac? *Acta Psychiatr Scand* 115: 93–100.

Burgin, A. B., Magnusson, O. T., Singh, J., et al. 2010. Design of phosphodiesterase 4D (PDE4D) allosteric modulators for enhancing cognition with improved safety. *Nat Biotechnol* 28: 63–70.

Cabanero, M., Laje, G., Detera-Wadleigh, S., and McMahonm F. J. 2009. Association study of phosphodiesterase genes in the Sequenced Treatment Alternatives to Relieve Depression sample. *Pharmacogenet Genomics* 19: 235–8.

Cashman, J. R., Voelker, T., Johnson, R., and Janowsky, A. 2009. Stereoselective inhibition of serotonin reuptake and phosphodiesterase by dual inhibitors as potential agents for depression. *Bioorg Med Chem* 17: 337–43.

Chandrasekaran, A., Toh, K. Y., Low, S. H., Tay, S. K., Brenner, S., and Goh, D. L. 2008. Identification and characterization of novel mouse PDE4D isoforms: Molecular cloning, subcellular distribution and detection of isoform-specific intracellular localization signals. *Cell Signal* 20: 139–53.

Cherry, J. A., and Davis, R. L. 1999. Cyclic AMP phosphodiesterases are localized in regions of the mouse brain associated with reinforcement, movement, and affect. *J Comp Neurol* 407: 287–301.

Cheung, Y. F., Kan, Z., Garrett-Engele, P., et al. 2007. PDE4B5, a novel, super-short, brain-specific cAMP phosphodiesterase-4 variant whose isoform-specifying N-terminal region is identical to that of cAMP phosphodiesterase-4D6 (PDE4D6). *J Pharmacol Exp Ther* 322: 600–9.

Conti, M. 2000. Phosphodiesterases and cyclic nucleotide signaling in endocrine cells. *Mol Endocrinol* 14: 1317–27.

Conti, M., and Beavo, J. 2007. Biochemistry and physiology of cyclic nucleotide phosphodiesterases: Essential components in cyclic nucleotide signaling. *Annu Rev Biochem* 76: 481–511.

Conti, A. C., and Blendy, J. A. 2004. Regulation of antidepressant activity by cAMP response element binding proteins. *Mol Neurobiol* 30:143–55.

Conti, A. C., Cryan, J. F., Dalvi, A., Lucki, I., and Blendy, J. A. 2002. cAMP response element–binding protein is essential for the upregulation of brain-derived neurotrophic factor transcription, but not the behavioral or endocrine responses to antidepressant drugs. *J Neurosci* 22: 3262–8.

Conti, M., Richter, W., Mehats, C., Livera, G., Park, J. Y., and Jin C. 2003. Cyclic AMP-specific PDE4 phosphodiesterases as critical components of cyclic AMP signaling. *J Biol Chem* 278: 5493–6.

Coquil, J. F., Franks, D. J., Wells, J. N., Dupuis, M., and Hamet, P. 1980. Characteristics of a new binding protein distinct from the kinase for guanosine 3,5-monophosphate in rat platelets. *Biochim Biophys Acta* 631: 148–65.

Cullen, B. R. 2005. RNAi the natural way. *Nat Genet* 37: 1163–5.

Domek-Lopacinska, K., Strosznajder, J. B. 2008. The effect of selective inhibition of cyclic GMP hydrolyzing phosphodiesterases 2 and 5 on learning and memory processes and nitric oxide synthase activity in brain during aging. *Brain Res* 1216: 68–77.

D'Sa, C., and Duman, R. S. 2002. Antidepressants and neuroplasticity. *Bipolar Disord* 4: 183–94.

D'Sa, C., Eisch, A. J., Bolger, G. B., and Duman, R. S. 2005. Differential expression and regulation of the cAMP-selective phosphodiesterase type 4A splice variants in rat brain by chronic antidepressant administration. *Eur J Neurosci* 22: 1463–75.

DaSilva, J. N., Lourenco, C. M., Meyer, J. H., Hussey, D., Potter, W. Z., and Houle, S. 2002. Imaging cAMP-specific phosphodiesterase-4 in human brain with R-[^{11}C]rolipram and positron emission tomography. *Eur J Nucl Med Mol Imaging* 29: 1680–3.

Devan, B. D., Sierra-Mercado, D. Jr, Jimenez, M., Bowker, J. L., Duffy, K. B., Spangler, E. L., and Ingram, D. K. 2004. Phosphodiesterase inhibition by sildenafil citrate attenuates the learning impairment induced by blockade of cholinergic muscarinic receptors in rats. *Pharmacol Biochem Behav* 79(4): 691–9.

Dlaboga, D., Hajjhussein, H., and O'Donnell, J. M. 2006. Regulation of phosphodiesterase-4 (PDE4) expression in mouse brain by repeated antidepressant treatment: Comparison with rolipram. *Brain Res* 1096: 104–12.

Duman, R. S. 1998. Novel therapeutic approaches beyond the serotonin receptor. *Biol Psychiatry* 44: 324–35.

Egawa, T., Mishima, K., Matsumoto, Y., Iwasaki, K., Iwasaki, K., and Fujiwara, M. 1997. Rolipram and its optical isomers, phosphodiesterase 4 inhibitors, attenuated the scopolamine-induced impairments of learning and memory in rats. *Jpn J Pharmacol* 75: 275–81.

Engels, P., Abdel'Al, S., Hulley, P., and Lübbert, H. 1995. Brain distribution of four rat homologues of the *Drosophila* dunce cAMP phosphodiesterase. *J Neurosci Res* 41: 169–78.

Erhart, S. M., Marder, S. R., and Carpenter, W. T. 2006. Treatment of schizophrenia negative symptoms: Future prospects. *Schizophr Bull* 32: 234–7.

Erneux, C., Couchie, D., Dumont, J. E., et al. 1981. Specificity of cyclic GMP activation of a multisubstrate cyclic nucleotide phosphodiesterase from rat liver. *Eur J Biochem* 115: 503–10.

Esposito, K., Reierson, G. W., Luo, H. R., Wu, G. S., Licinio, J., and Wong, M. L. 2009. Phosphodiesterase genes and antidepressant treatment response: A review. *Ann Med* 41: 177–85.

Fawcett, L., Baxendale, R., Stacey, P., et al. 2000. Molecular cloning and characterization of a distinct human phosphodiesterase gene family: PDE11A. *Proc Natl Acad Sci U S A* 97: 3702–7.

Fisher, D. A., Smith, J. F., Pillar, J. S., St Denis, S. H., and Cheng, J. B. 1998. Isolation and characterization of PDE9A, a novel human cGMP-specific phosphodiesterase. *J Biol Chem* 273: 15559–64.

Fleischhacker, W. W., Hinterguber, H., Bauer, H., et al. 1992. A multicenter double-blind study of three different doses of the new cAMP-phosphodiesterase inhibitor rolipram in patients with major depressive disorder. *Neuropsychobiology* 26: 59–64.

Francis, S. H. 2005. Phosphodiesterase 11 (PDE11): Is it a player in human testicular function? *Int J Impot Res* 17: 467–8.

Francis, S. H., Lincoln, T. M., and Corbin, J. D. 1980. Characterization of a novel cGMP binding protein from rat lung. *J Biol Chem* 255: 620–6.

Francis, S. H., Bessay, E. P., Kotera, J., et al. 2002. Phosphorylation of isolated human phosphodiesterase-5 regulatory domain induces an apparent conformational change and increases cGMP binding affinity. *J Biol Chem* 277: 47581–7.

Francis, S. H., Corbin, J. D., and Bischoff, E. 2009. Cyclic GMP-hydrolyzing phosphodiesterases. *Handb Exp Pharmacol* 191: 367–408.

Francis, S. H., Busch, J. L., and Corbin, J. D. 2010. cGMP-dependent protein kinases and cGMP phosphodiesterases in nitric oxide and cGMP action. *Pharmacol Rev* 62: 525–63.

Fujimaki, K., Morinobu, S., and Duman, R. S. 2000. Administration of a cAMP phosphodiesterase 4 inhibitor enhances antidepressant-induction of BDNF mRNA in rat hippocampus. *Neuropsychopharmacology* 22: 42–51.

Fujishige, K., Kotera, J., and Omori, K. 1999. Striatum- and testis-specific phosphodiesterase PDE10A isolation and characterization of a rat PDE 10A. *Eur J Biochem* 266: 1118–27.

Garthwaite, J., and Boulton, C. L. 1995. Nitric oxide signaling in the central nervous system. *Annu Rev Physiol* 57: 683–706.

Gehlert, D. R., Dawson, T. M., Yamamura, H. I., and Wamsley, J. K. 1985. Quantitative autoradiography of [³H] forskolin binding sites in the rat brain. *Brain Res* 361: 351–60.

Giordano, D., De Stefano, M. E., Citro, G., Modica, A., and Giorgi, M. 2001. Expression of cGMP-binding cGMP-specific phosphodiesterase (PDE5) in mouse tissues and cell lines using an antibody against the enzyme amino-terminal domain. *Biochim Biophys Acta* 1539: 16–27.

Gong, B., Vitolo, O. V., Trinchese, F., Liu, S., Shelanski, M., and Arancio, O. 2004. Persistent improvement in synaptic and cognitive functions in an Alzheimer mouse model after rolipram treatment. *J Clin Invest* 114: 1624–34.

Gourley, S. L., Wu, F. J., Kiraly, D. D., et al. 2008. Regionally specific regulation of ERK MAP kinase in a model of antidepressant-sensitive chronic depression. *Biol Psychiatry* 63: 353–9.

Grecksch, G., Zhou, D., Franke, C., et al. 1997. Influence of olfactory bulbectomy and subsequent imipramine treatment on 5-hydroxytryptaminergic presynapses in the rat frontal cortex: Behavioral correlates. *Br J Pharmacol* 122: 1725–31.

Green, A. R., Mechan, A. O., Elliott, J. M., O'Shea, E., and Colado, M. I. 2003. The pharmacology and clinical pharmacology of 3,4-methylenedioxymethamphetamine (MDMA, "ecstasy"). *Pharmacol Rev* 55: 463–508.

Guipponi, M., Scott, H. S., Kudoh, J., et al. 1998. Identification and characterization of a novel cyclic nucleotide phosphodiesterase gene (PDE9A) that maps to 21q22.3: Alternative splicing of mRNA transcripts, genomic structure and sequence. *Hum Genet* 103: 386–92.

Hajjhussein, H., Suvarna, N. U., Gremillion, C., Chandler, L. J., and O'Donnell, J. M. 2007. Changes in NMDA receptor-induced cyclic nucleotide synthesis regulated the age-dependent increase in PDE4A expression in primary cortical cultures. *Brain Res* 1149: 58–68.

Hamet, P., and Coquil, J. F. 1978. Cyclic GMP binding and cyclic GMP phosphodiesterase in rat platelets. *J Cyclic Nucleotide Res* 4: 281–90.

Han, P., Zhu, X. Y., and Michaeli, T. 1997. Alternative splicing of the high affinity cAMP-specific phosphodiesterase (PDE7A) mRNA in human skeletal muscle and heart. *J Biol Chem* 272: 16152–7.

Hartell, N. A. 1996. Inhibition of cGMP breakdown promotes the induction of cerebellar long-term depression. *J Neurosci* 16: 2881–90.

Hatzichristou, D., Cuzin, B., Martin-Morales, A., et al. 2005. Vardenafil improves satisfaction rates, depressive symptomatology, and self-confidence in a broad population of men with erectile dysfunction. *J Sex Med* 2: 109–16.

Hebb, A. L. O., Robertson, H. A., and Denovan-Wright, E. M. 2004. Striatal phosphodiesterase mRNA and protein levels are reduced in Huntington's disease transgenic mice prior to the onset of motor symptoms. *Neuroscience* 123: 967–81.

Hebb, L. O., Robertson, H. A., and Denovan-Wright, E. M. 2008. Phosphodiesterase 10A inhibition is associated with locomotor and cognitive deficits and increased anxiety in mice. *Eur Neuropsychopharmacol* 18: 229–63.

Hebenstreit, G. F., Fellerer, K., Fichte, K., et al. 1989. Rolipram in major depressive disorder: Results of a double-blind comparative study with imipramine. *Pharmacopsychiatry* 22: 156–60.

Heikaus, C. C., Pandit, J., and Klevit R. E. 2009. Cyclic nucleotide binding GAF domains from phosphodiesterases: Structural and mechanistic insights. *Structure* 17: 1551–7.

Hepp, R., Tricoire, L., Hu, E., Gervasi, N., Paupardin-Tritsch, D., Lambolez, B., and Vincent, P. 2007. Phosphodiesterase type 2 and the homeostasis of cyclic GMP in living thalamic neurons. *J Neurochem* 102: 1875–86.

Hetman, J. M., Soderling, S. H., Glavas, N. A., and Beavo, J. A. 2000. Cloning and characterization of PDE7B, a cAMP-specific phosphodiesterase. *Proc Natl Acad Sci U S A* 97: 472–6.

Hoffmann, R., Baillie, G. S., MacKenzie, S. J., Yarwood, S. J., and Houslay, M. D. 1999. The MAP kinase ERK2 inhibits the cyclic AMP-specific phosphodiesterase HSPDE4D3 by phosphorylating it at Ser579. *EMBO J* 18: 893–903.

Horowski, R., and Sastre-y-Hernandez, M. 1985. Clinical effects of the neurotropic selective cAMP phosphodiesterase inhibitor rolipram in depressed patients: Global evaluation of the preliminary reports. *Curr Ther Res Clin Exp* 38: 23–39.

Horvath, A., Giatzakis, C., Robinson-White, A., et al. 2006. Adrenal hyperplasia and adenomas are associated with inhibition of phosphodiesterase 11A in carriers of PDE11A sequence variants that are frequent in the population. *Cancer Res* 66: 11571–5.

Houslay, M. D. 2010. Underpinning compartmentalised cAMP signalling through targeted cAMP breakdown. *Trends Biochem Sci* 35: 91–100.

Houslay, M. D., and Adams, D. R. 2003. PDE4 cAMP phosphodiesterases: Modular enzymes that orchestrate signalling cross-talk, desensitization and compartmentalization. *Biochem J* 370: 1–18.

Houslay, M. D., Schafer, P., and Zhang, K. Y. 2005. Keynote review: Phosphodiesterase-4 as a therapeutic target. *Drug Discov Today* 10: 1503–19.

Houslay, M. D., Baillie, G. S., and Maurice, D. H. 2007. cAMP-specific phosphodiesterase-4 enzymes in the cardiovascular system: A molecular toolbox for generating compartmentalized cAMP signaling. *Circ Res* 100: 950–66.

Hu, H., McCaw, E. A., Hebb, A. L., Gomez, G. T., and Denovan-Wright, E. M. 2004. Mutant huntingtin affects the rate of transcription of striatum-specific isoforms of phosphodiesterase 10A. *Eur J Neurosci* 20: 3351–63.

Itoh, T., Abe, K., Tokumura, M., Horiuchi, M., Inoue, O., and Ibii, N. 2003. Different regulation of adenylyl cyclase and rolipram-sensitive phosphodiesterase activity on the frontal cortex and hippocampus in learned helplessness rats. *Brain Res* 991: 142–9.

Itoh, T., Tokumura, M., and Abe, K. 2004. Effects of rolipram, a phosphodiesterase 4 inhibitor, in combination with imipramine on depressive behavior, CRE-binding activity and BDNF level in learned helplessness rats. *Eur J Pharmacol* 498: 135–42.

Iwahashi, Y., Furuyama, T., Tano, Y., Ishimoto, I., Shimomura, Y., and Inagaki, S. 1996. Differential distribution of mRNA encoding cAMP-specific phosphodiesterase isoforms in the rat brain. *Brain Res Mol Brain Res* 38: 14–24.

Janiec, W., Kroczak-Dziuba, K., and Herman, Z. S. 1974. Effect of phenothiazine neuroleptic drugs and tricyclic antidepressants on phosphodiesterase activity in rat cerebral cortex. *Psychopharmacologia* 37: 351–8.

Jesko, H., Chalimoniuk, M., and Strosznajder, J. B. 2003. Activation of constitutive nitric oxide synthase(s) and absence of inducible isoform in aged rat brain. *Neurochem Int* 42: 315–22.

Jesse, C. R., Bortolatto, C. F., Savegnago, L., Rocha, J. B., and Nogueira, C. W. 2008. Involvement of L-arginine-nitric oxide-cyclic guanosine monophosphate pathway in the antidepressant-like effect of tramadol in the rat forced swimming test. *Prog Neuropsychopharmacol Biol Psychiatry* 32: 1838–43.

Jin, S. L., and Conti, M. 2002. Induction of the cyclic nucleotide phosphodiesterase PDE4B is essential for LPS-activated TNF-alpha responses. *Proc Natl Acad Sci U S A* 99: 7628–33.

Jin, S. L., Richard, F. J., Kuo, W. P., D'Ercole, A. J., and Conti, M. 1999. Impaired growth and fertility of cAMP-specific phosphodiesterase PDE4D-deficient mice. *Proc Natl Acad Sci U S A* 96: 11998–2003.

Johnston, L. A., Erdogan, S., Cheung, Y. F., et al. 2004. Expression, intracellular distribution and basis for lack of catalytic activity of the PDE4A7 isoform encoded by the human PDE4A cAMP-specific phosphodiesterase gene. *Biochem J* 380: 371–84.

Juilfs, D. M., Fulle, H. J., Zhao, A. Z., Houslay, M. D., Garbers, D. L., and Beavo, J. A. 1997. A subset of olfactory neurons that selectively express cGMP-stimulated phosphodiesterase (PDE2) and guanylyl cyclase-D define a unique olfactory signal transduction pathway. *Proc Natl Acad Sci U S A* 94: 3388–95.

Kaster, M. P., Ferreira, P. K., Santos, A. R., and Rodrigues, A. L. 2005a. Effects of potassium channel inhibitors in the forced swimming test: Possible involvement of L-arginine-nitric oxide-soluble guanylate cyclase pathway. *Behav Brain Res* 165: 204–9.

Kaster, M. P., Rosa, A. O., Santos, A. R., and Rodrigues, A. L. 2005b. Involvement of nitric oxide-cGMP pathway in the antidepressant-like effects of adenosine in the forced swimming test. *Int J Neuropsychopharmacol* 8: 601–6.

Kaulen, P., Bruning, G., Schneider, H. H., Sarter, M., and Baumgarten, H. G. 1989. Autoradiographic mapping of a selective cyclic adenosine monophosphate phosphodiesterase in rat brain with the antidepressant [³H]rolipram. *Brain Res* 503: 229–45.

Kelly, M. P., Logue, S. F., Brennan, J., et al. 2010. Phosphodiesterase 11A in brain is enriched in ventral hippocampus and deletion causes psychiatric disease-related phenotypes. *Proc Natl Acad Sci U S A* 107: 8457–62.

Keravis, T., and Lugnier, C. 2010. Cyclic nucleotide phosphodiesterases (PDE) and peptide motifs. *Curr Pharm Des* 16: 1114–25.

Kleppisch, T. 2009. Phosphodiesterases in the central nervous system. *Handb Exp Pharmacol* 191: 71–92.

Koob, G. F. 1992. Drugs of abuse: Anatomy, pharmacology and function of reward pathways. *Trends Pharmacol Sci* 13: 177–84.

Kotera, J., Fujishige, K., Akatsuka, H., Imai, Y., Yanaka, N., and Omori, K. 1998. Novel alternative splice variants of cGMP-binding cGMP-specific phosphodiesterase. *J Biol Chem* 273: 26982–90.

Kotera, J., Fujishige, K., Imai, Y., et al. 1999. Genomic origin and transcriptional regulation of two variants of cGMP-binding cGMP-specific phosphodiesterases. *Eur J Biochem* 262: 866–73.

Kotera, J., Fujishige, K., Michibata, H., et al. 2000. Characterization and effects of methyl-2- (4-aminophenyl)-1, 2-dihydro-1-oxo-7- (2-pyridinylmethoxy)-4-(3,4,5-trimethoxyphenyl)-3-isoquinoline carboxylate sulfate (T-1032), a novel potent inhibitor of cGMP-binding cGMP-specific phosphodiesterase (PDE5). *Biochem Pharmacol* 60: 1333–41.

Kurt, M., Bilge, S. S., Aksoz, E., Kukula, O., Celik, S., and Kesim, Y. 2004. Effect of sildenafil on anxiety in the plus-maze test in mice. *Pol J Pharmacol* 56: 353–7.

Lakics, V., Karran, E. H., and Boess, F. G. 2010. Quantitative comparison of phosphodiesterase mRNA distribution in human brain and peripheral tissues. *Neuropharmacology* 59: 367–74.

Lamontagne, S., Meadows, E., Luk, P., et al. 2001. Localization of phosphodiesterase-4 isoforms in the medulla and nodose ganglion of the squirrel monkey. *Brain Res* 920: 84–96.

Lee, S. Y., Andoh, T., Murphy, D. L., and Chiueh, C. C. 2003. 17beta-Estradiol activates ICI 182,780-sensitive estrogen receptors and cyclic GMP-dependent thioredoxin expression for neuroprotection. *FASEB J* 17: 947–8.

Lehericy, S., and Gerardin, E. 2002. Normal functional imaging of the basal ganglia. *Epileptic Disord* 4: S23–30.

Li, X., Witkin, J. M., Need, A. B., and Skolnick, P. 2003. Enhancement of antidepressant potency by a potentiator of AMPA receptors. *Cell Mol Neurobiol* 23: 419–30.

Li, X., Huston, E., Lynch, M. J., Houslay, M. D., and Baillie, G. S. 2006. Phosphodiesterase-4 influences the PKA phosphorylation status and membrane translocation of G-protein receptor kinase 2 (GRK2) in HEK-293beta2 cells and cardiac myocytes. *Biochem J* 394: 427–35.

Li, Y. F., Huang, Y., Amsdell, S. L., Xiao, L., O'Donnell, J. M., and Zhang, H. T. 2009. Antidepressant- and anxiolytic-like effects of the phosphodiesterase-4 inhibitor rolipram on behavior depend on cyclic AMP response element binding protein-mediated neurogenesis in the hippocampus. *Neuropsychopharmacology* 34: 2404–19.

Li, Y. F., Cheng, Y. F., Huang, Y., et al. 2011. Phosphodiesterase-4D knockout and RNAi-mediated knockdown enhance memory and increase hippocampal neurogenesis via increased cAMP signaling. *J Neurosci* 31: 172–83.

Liebenberg, N., Harvey, B. H., Brand, L., and Brink, C. B. 2010. Antidepressant-like properties of phosphodiesterase type 5 inhibitors and cholinergic dependency in a genetic rat model of depression. *Behav Pharmacol* 21: 540–7.

Lin, C. S. 2004. Tissue expression, distribution, and regulation of PDE5. *Int J Impot Res* 16: 8–10.

Loughney, K., Snyder, P. B., Uher, L., Rosman, G. J., Ferguson, K., and Florio, V. A. 1999. Isolation and characterization of PDE10A, a novel human 3′,5′-cyclic nucleotide phosphodiesterase. *Gene* 234: 109–17.

Lugnier, C. 2006. Cyclic nucleotide phosphodiesterase (PDE) superfamily: A new target for the development of specific therapeutic agents. *Pharmacol Ther* 109: 366–98.

Lynch, M. J., Baillie, G. S., Mohamed, A., et al. 2005. RNA silencing identifies PDE4D5 as the functionally relevant cAMP phosphodiesterase interacting with beta arrestin to control the protein kinase A/AKAP79-mediated switching of the beta2-adrenergic receptor to activation of ERK in HEK293B2 cells. *J Biol Chem* 280: 33178–89.

Lynex, C. N., Li, Z., Chen, M. L., et al. 2008. Identification and molecular characterization of a novel PDE4D11 cAMP-specific phosphodiesterase isoform. *Cell Signal* 20: 2247–55.

MacKenzie, S. J., Baillie, G. S., McPhee, I., Bolger, G. B., and Houslay, M. D. 2000. ERK2 mitogen-activated protein kinase binding, phosphorylation, and regulation of the PDE4D cAMP-specific phosphodiesterases. The involvement of COOH-terminal docking sites and NH$_2$-terminal UCR regions. *J Biol Chem* 275: 16609–17.

MacKenzie, S. J., Baillie, G. S., McPhee, I., MacKenzie, C., Seamons, R., McSorley, T., Millen, J., Beard, M. B., van Heeke, G., and Houslay, M. D. 2002. Long PDE4 cAMP specific phosphodiesterases are activate by protein kinase A-mediated phosphorylation of a single serine residue in Upstream Conserved Region 1 (UCR1). *Br J Pharmacol* 136(3): 421–33.

Mackenzie, K. F., Topping, E. C., Bugaj-Gaweda, B., et al. 2008. Human PDE4A8, a novel brain-expressed PDE4 cAMP-specific phosphodiesterase that has undergone rapid evolutionary change. *Biochem J* 411: 361 9.

Martinez, S. E., Wu, A. Y., Glavas, N. A. et al. 2002. The two GAF domains in phosphodiesterase 2A have distinct roles in dimerization and in cGMP binding. *Proc Natl Acad Sci U S A* 99: 13260–5.

Martins, T. J., Mumby, M. C., and Beavo, J. A. 1982. Purification and characterization of a cyclic GMP-stimulated cyclic nucleotide phosphodiesterase from bovine tissues. *J Biol Chem* 257: 1973–9.

Masini, C. V., Holmes, P. V., Freeman, K. G., Maki, A. C., and Edwards, G. L. 2004. Dopamine overflow is increased in olfactory bulbectomized rats: An in vivo microdialysis study. *Physiol Behav* 81: 111–19.

Masood, A., Nadeem, A., Mustafa, S. J., and O'Donnell, J. M. 2008. Reversal of oxidative stress–induced anxiety by inhibition of phosphodiesterase-2 in mice. *J Pharmacol Exp Ther* 326: 369–79.

Masood, A., Huang, Y., Hajjhussein, H., et al. 2009. Anxiolytic effects of phosphodiesterase-2 inhibitors associated with increased cGMP signaling. *J Pharmacol Exp Ther* 331: 690–9.

Mehats, C., Andersen, C. B., Filopanti, M., Jin, S. L., and Conti, M. 2002. Cyclic nucleotide phosphodiesterases and their role in endocrine cell signaling. *Trends Endocrinol Metab* 13: 29–35.

Menniti, F. S., Faraci, W. S., and Schmidt, C. J. 2006. Phosphodiesterases in the CNS: Targets for drug development. *Nat Rev* 5: 660–70.

Millar, J. K., Pickard, B. S., Mackie, S., et al. 2005. DISC1 and PDE4B are interacting genetic factors in schizophrenia that regulate cAMP signaling. *Science* 310: 1187–91.

Miller, C. L., Oikawa, M., Cai, Y., et al. 2009. Role of Ca^{2+}/calmodulin-stimulated cyclic nucleotide phosphodiesterase 1 in mediating cardiomyocyte hypertrophy. *Circ Res* 105: 956–64.

Miro, X., Perez-Torres, S., Artigas, F., Puigdomenech, P., Palacios, J. M., and Mengod, G. 2002. Regulation of cAMP phosphodiesterase mRNAs expression in rat brain by acute and chronic fluoxetine treatment. An in situ hybridization study. *Neuropharmacology* 43: 1148–57.

Mizokawa, T., Kimura, K., Ikoma, Y., et al. 1988. The effect of a selective phosphodiesterase inhibitor, rolipram, on muricide in olfactory bulbectomized rats. *Jpn J Pharmacol* 48: 357–64.

Moncada, I., Martínez-Jabaloyas, J. M., Rodriguez-Vela, L., et al. 2009. Emotional changes in men treated with sildenafil citrate for erectile dysfunction: A double-blind, placebo-controlled clinical trial. *J Sex Med* 6: 3469–77.

Morley-Fletcher, S., Darnaudery, M., Mocaer, E., et al. 2004. Chronic treatment with imipramine reverses immobility behaviour, hippocampal corticosteroid receptors and cortical 5-HT1A receptor mRNA in prenatally stressed rats. *Neuropharmacology* 47: 841–7.

Mudunkotuwa, N. T., and Horton, R. W. 1996. Desipramine administration in the olfactory bulbectomized rat: Changes in brain beta-adrenoceptor and 5-HT2A binding sites and their relationship to behavior. *Br J Pharmacol* 117: 1481–6.

Murad, F. 1994. Regulation of cytosolic guanylyl cyclase by nitric oxide: The NO-cyclic GMP signal transduction system. *Adv Pharmacol* 26: 19–33.

Nagel, D. J., Aizawa, T., Jeon, K. I., Liu, W., Mohan, A., and Wei, H. 2006. Role of nuclear Ca^{2+}/calmodulin-stimulated phosphodiesterase 1A in vascular smooth muscle cell growth and survival. *Circ Res* 98: 777–84.

Nakagawa, S., Kim, J. E., Lee, R., et al. 2002. Regulation of neurogenesis in adult mouse hippocampus by cAMP and the cAMP response element–binding protein. *J Neurosci* 22: 3673–82.

Nemoz, G., Sette, C., and Conti, M. 1997. Selective activation of rolipram-sensitive, cAMP-specific phosphodiesterase isoforms by phosphatidic acid. *Mol Pharmacol* 51: 242–9.

Nestler, E. J., Terwilliger, R. Z., and Duman, R. S. 1989. Chronic antidepressant administration alters the subcellular distribution of cyclic AMP-dependent protein kinase in rat frontal cortex. *J Neurochem* 53: 1644–7.

Nestler, E. J., Barrot, M., DiLeone, R. J., Eisch, A. J., Gold, S. J., and Monteggia, L. M. 2002. Neurobiology of depression. *Neuron* 34: 13–25.

Nibuya, M., Nestler, E. J., and Duman, R. S. 1996. Chronic antidepressant administration increases the expression of cAMP response element binding protein (CREB) in rat hippocampus. *J Neurosci* 16: 2365–72.

Nikolaev, V. O., Gambaryan, S., Engelhardt, S., Walter, U., and Lohse, M. J. 2005. Real-time monitoring of the PDE2 activity of live cells: Hormone-stimulated cAMP hydrolysis is faster than hormone-stimulated cAMP synthesis. *J Biol Chem* 280: 1716–19.

O'Conner, M. S., Steiner, J. M., Roussel, A. J., et al. 2004. Evaluation of urine sucrose concentration for detection of gastric ulcers in horses. *Am J Vet Res* 65: 31–9.

O'Donnell, J. M. 1993. Antidepressant-like effects of rolipram and other inhibitors of cyclic adenosine monophosphate phosphodiesterase on behavior maintained by differential reinforcement of low response rate. *J Pharmacol Exp Ther* 264: 1168–78.

O'Donnell, J. M., and Frith, S. 1999. Behavioral effects of family-selective inhibitors of cyclic nucleotide phosphodiesterases. *Pharmacol Biochem Behav* 63: 185–92.

O'Donnell, J. M., and Zhang, H. T. 2004. Antidepressant effects of inhibitors of cAMP phosphodiesterase (PDE4). *Trends Pharmacol Sci* 25: 158–63.

O'Donnell, J. M., Marek, G. J., and Seiden, L. S. 2005. Antidepressant effects assessed using behavior maintained under a differential-reinforcement-of-low-rate (DRL) operant schedule. *Neurosci Biobehav Rev* 29: 785–98.

Ohsawa, F., Yamauchi, M., Nagaso, H., Murakami, S., Baba, J., and Sawa, A. 1998. Inhibitory effects of rolipram on partially purified phosphodiesterase 4 from rat brains. *Jpn J Pharmacol* 77: 147–54.

Oki, N., Takahashi, S. I., Hidaka, H., and Conti, M. 2000. Short term feedback regulation of cAMP in FRTL-5 thyroid cells. Role of PDE4D3 phosphodiesterase activation. *J Biol Chem* 275: 10831–7.

Overstreet, D. H., Friedman, E., Mathé, A. A., and Yadid, G. 2005. The Flinders Sensitive Line rat: A selectively bred putative animal model of depression. *Neurosci Biobehav Rev* 29: 739–59.

Patil, C. S., Jain, N. K., Singh, V. P., and Kulkarni, S. K. 2004. Cholinergic-NO-cGMP mediation of sildenafil-induced antinociception. *Indian J Exp Biol* 42: 361–7.

Perez-Torres, S., Miro, X., Palacios, J. M., Cortes, R., Puigdomenech, P., and Mengod, G. 2000. Phosphodiesterase type 4 isozymes expression in human brain examined by in situ hybridization histochemistry and [³H]rolipram binding autoradiography. Comparison with monkey and rat brain. *J Chem Neuroanat* 20: 349–74.

Poat, J. A., Cripps, H. E., and Iversen, L. L. 1988. Differences between high-affinity forskolin binding sites in dopamine-rich and other regions of rat brain. *Proc Natl Acad Sci U S A* 85: 3216–20.

Possidente, B., Lumia, A. R., McGinnis, M. Y., Pratt, C., and Page, C. 2000. Chronological dimensions of an olfactory bulbectomized rodent animal model for agitated depression. *Biol Rhythm Res* 31: 416–34.

Prickaerts, J., Steinbusch, H. W., Smits, J. F., and de Vente, J. 1997. Possible role of nitric oxide-cyclic GMP pathway in object recognition memory: Effects of 7-nitroindazole and zaprinast. *Eur J Pharmacol* 337: 125–36.

Prickaerts, J., Sik, A., van Staveren, W. C., et al. 2004. Phosphodiesterase type 5 inhibition improves early memory consolidation of object information. *Neurochem Int* 45: 915–28.

Primeaux, S. D., and Holmes, P. V. 1999. Role of aversively-motivated behavior in the olfactory bulbectomy syndrome. *Physiol Behav* 67: 41–7.

Przegalinski, E., and Bigajska, K. 1983. Antidepressant properties of some phosphodiesterase inhibitors. *Pol J Pharmacol Pharm* 35: 233–40.

Puerta, E., Hervias, I., Goni-Allo, B., Lasheras, B., Jordan, J., and Aguirre, N. 2009. Phosphodiesterase 5 inhibitors prevent 3,4-methylenedioxy-methamphetamine-induced 5-HT deficits in the rat. *J Neurochem* 108: 755–66.

Pyne, N. J., Cooper, M. E., and Houslay, M. D. 1986. Identification and characterization of both the cytosolic and particulate forms of cyclic GMP-stimulated cyclic AMP phosphodiesterase from rat liver. *Biochem J* 234: 325–34.

Qi, X., Lin, W., Li, J., et al. 2008. Fluoxetine increases the activity of the ERK-CREB signal system and alleviates the depressive-like behavior in rats exposed to chronic forced swim stress. *Neurobiol Dis* 31: 278–85.

Rapoport, M., van Reekum, R., and Mayberg, H. 2000. The role of the cerebellum in cognition and behavior: A selective review. *J Neuropsychiatry Clin Neurosci* 12: 193–8.

Reneerkens, O. A., Rutten, K., Steinbusch, H. W., Blokland, A., and Prickaerts, J. 2009. Selective phosphodiesterase inhibitors: A promising target for cognition enhancement. *Psychopharmacology* 202: 419–43.

Repaske, D. R., Corbin, J. G., Conti, M., and Goy, M. F. 1993. A cyclic GMP-stimulated cyclic nucleotide phosphodiesterase gene is highly expressed in the limbic system of the rat brain. *Neuroscience* 56: 673–86.

Ribeiro, J. A., and Sebastiao, A. M. 2010. Caffeine and adenosine. *J Alzheimers Dis* 20: S3–15.

Rivet-Bastide, M., Vandecasteele, G., Hatem, S., et al. 1997. cGMP-stimulated cyclic nucleotide phosphodi-esterase regulates the basal calcium current in human atrial myocytes. *J Clin Invest* 99: 2710–18.

Robichaud, A., Stamatiou, P. B., Jin, S. L., et al. 2002. Deletion of phosphodiesterase 4D in mice short-ens alpha(2)-adrenoceptor-mediated anesthesia, a behavioral correlate of emesis. *J Clin Invest* 110: 1045–52.

Rodefer, J. S., Murphy, E. R., and Baxter, M. G. 2005. PDE10A inhibition reverses subchronic PCP-induced deficits in attentional set-shifting in rats. *Eur J Neurosci* 21: 1070–6.

Rosman, G. J., Martins, T. J., Sonnenburg, W. K., Beavo, J. A., Ferguson, K., and Loughney, K. 1997. Isolation and characterization of human cDNAs encoding a cGMP-stimulated 3V, 5V-cyclic nucleotide phospho-diesterase. *Gene* 191: 89–95.

Rutten, K., Misner, D. L., Works, M., et al. 2008. Enhanced long-term potentiation and impaired learning in phosphodiesterase 4D-knockout (PDE4D) mice. *Eur J Neurosci* 28: 625–32.

Rutten, K., Van Donkelaar, E. L., Ferrington, L., et al. 2009. Phosphodiesterase inhibitors enhance object memory independent of cerebral blood flow and glucose utilization in rats. *Neuropsychopharmacology* 34: 1914–25.

Rysànek, K., König, J., Spànkovà, H., and Mlejnkovà, M. 1973. Effect of tricyclic antidepressants on phos-phodiesterase. Correlation between aggregability and thrombocyte metabolism. *Act Nerv Super (Praha)* 15: 126–27.

Saccomano, N. A., Vinick, F. J., Koe, B. K., et al. 1991. Calcium-independent phosphodiesterase inhibitors as puta-tive antidepressants: [3-(bicycloalkyloxy)-4-methoxyphenyl]-2-imidazolidinones. *J Med Chem* 34: 291–8.

Santarelli, L., Saxe, M., Gross, C., et al. 2003. Requirement of hippocampal neurogenesis for the behavioral effects of antidepressants. *Science* 301: 805–9.

Schneider, H. H. 1984. Brain cAMP response to phosphodiesterase inhibitors in rats killed by microwave irra-diation or decapitation. *Biochem Pharmacol* 33: 1690-3.

Schneider, H. H., Schmiechen, R., Brezinski, M., and Seidler, J. 1986. Stereospecific binding of the antidepres-sant rolipram to brain protein structures. *Eur J Pharmacol* 127: 105–15.

Scott, A. I., Perini, A. F., Shering, P. A., and Whalley, L. J. 1991. In-patient major depression: Is rolipram as effective as amitriptyline. *Eur J Clin Pharmacol* 40: 127–9.

Seeger, T. F., Bartlett, B., Coskran, T. M., et al. 2003. Immunohistochemical localization of PDE10A in the rat brain. *Brain Res* 985: 113–26.

Seeman, P. 2009. Glutamate and dopamine components in schizophrenia. *J Psychiatry Neurosci* 34: 143–9.

Sette, C., and Conti, M. 1996. Phosphorylation and activation of a cAMP specific phosphodiesterase by the cAMP-dependent protein kinase. Involvement of serine 54 in the enzyme activation. *J Biol Chem* 271: 16526–34.

Shirayama, Y., and Chaki, S. 2006. Neurochemistry of the nucleus accumbens and its relevance to depression and antidepressant action in rodents. *Curr Neuropharmacol* 4: 277–91.

Sieck, M. H., and Baumbach, H. D. 1974. Differential effects of peripheral and central anosmia producing techniques on spontaneous behavior patterns. *Physiol Behav* 13: 407–25.

Siuciak, J. A. 2008. The role of phosphodiesterases in schizophrenia: Therapeutic implications. *CNS Drugs* 22: 983–93.

Siuciak, J. A., McCarthy, S. A., Chapin, D. S., et al. 2006. Genetic deletion of the striatum-enriched phosphodi-esterase PDE10A: Evidence for altered striatal function. *Neuropharmacology* 51: 374–85.

Siuciak, J. A., Chapin, D. S., McCarthy, S. A., and Martin, A. N. 2007. Antipsychotic profile of rolipram: Efficacy in rats and reduced sensitivity in mice deficient in the phosphodiesterase-4B (PDE4B) enzyme. *Psychopharmacology (Berl)* 192: 415–24.

Siuciak, J. A., McCarthy, S. A., Chapin, D. S., and Martin, A. N. 2008. Behavioral and neurochemical charac-terization of mice deficient in the phosphodiesterase-4B (PDE4B) enzyme. *Psychopharmacology (Berl)* 197: 115–26.

Soderling, S. H., Bayuga, S. J., and Beavo, J. A. 1998. Identification and characterization of a novel family of cyclic nucleotide phosphodiesterases. *J Biol Chem* 273: 15553–8.

Soderling, S. H., Bayuga, S. J., and Beavo, J. A. 1999. Isolation and characterization of a dual-substrated phos-phodiesterase gene family: PDE10A. *Proc Natl Acad Sci U S A* 96: 7071–6.

Soderling, S. H., and Beavo, J. A. 2000. Regulation of cAMP and cGMP signaling: New phosphodiesterases and new functions. *Curr Opin Cell Biol* 12(2): 174–9.

Solis, A. A., Bethancourt, J. A., and Britton, G. B. 2008. Chronic sildenafil (Viagra) administration reduces anxiety in intact and castrated male rats. *Psicothema* 20: 812–17.

Song, C., and Leonard, B. E. 2005. The olfactory bulbectomised rat as a model of depression. *Neurosci Biobehav Rev* 29: 627–47.

Sonnenburg, W. K., Mullaney, P. J., and Beavo, J. A. 1991. Molecular cloning of a cyclic GMP-stimulated cyclic nucleotide phosphodiesterase cDNA. Identification and distribution of isozyme variants. *J Biol Chem* 266: 17655–61.

Sonnenburg, W. K., Seger, D., Kwak, K. S., Huang, J., Charbonneau, H., and Beavo, J. A. 1995. Identification of inhibitory and calmodulin-binding domains of the PDE1A1 and PDE1A2 calmodulin-stimulated cyclic nucleotide phosphodiesterases. *J Biol Chem* 270: 30989–1000.

Stahl, S. M. 2007. Novel therapeutics for schizophrenia: Targeting glycine modulation of NMDA glutamate receptors. *CNS Spectr* 12: 423–7.

Stratakis, C. A., and Boikos, S. A. 2007. Genetics of adrenal tumors associated with Cushing's syndrome: A new classification for bilateral adrenocortical hyperplasias. *Nat Clin Pract Endocrinol Metab* 3: 748–57.

Suda, S., Nibuya, M., Ishiguro, T., Suda, H. 1998. Transcriptional and translational regulation of phosphodiesterase type IV isozymes in rat brain by electroconvulsive seizure and antidepressant drug treatment. *J Neurochem* 71: 1554–63.

Suvarna, N. U., and O'Donnell, J. M. 2002. Hydrolysis of N-methyl-D-aspartate receptor-stimulated cAMP and cGMP by PDE4 and PDE2 phosphodiesterases in primary neuronal cultures of rat cerebral cortex and hippocampus. *J Pharmacol Exp Ther* 302(1): 249–56.

Takahashi, M., Terwilliger, R., Lane, C., Mezes, P. S., Conti, M., and Duman, R. S. 1999. Chronic antidepressant administration increases the expression of cAMP-specific phosphodiesterase 4A and 4B isoforms. *J Neurosci* 19: 610–18.

Thompson, W. J. 1991. Cyclic nucleotide phosphodiesterases: Pharmacology, biochemistry and function. *Pharmacol Ther* 51: 13–33.

Torras-Llort, M., and Azorin, F. 2003. Functional characterization of the human phosphodiesterase 7A1 promoter. *Biochem J* 373: 835–43.

van der Staay, F. J., Rutten, K., Barfacker, L., et al. 2008. The novel selective PDE9 inhibitor BAY 73-6691 improves learning and memory in rodents. *Neuropharmacology* 55: 908–18.

van Staveren, W. C., Markerink-van Ittersum, M., Steinbusch, H. W., and de Vente, J. 2001. The effects of phosphodiesterase inhibition on cyclic GMP and cyclic AMP accumulation in the hippocampus of the rat. *Brain Res* 888: 275–86.

van Staveren, W. C., Steinbusch, H. W., Markerink-van Ittersum, M., et al. 2003. mRNA expression patterns of the cGMP-hydrolyzing phosphodiesterases types 2, 5, and 9 during development of the rat brain. *J Comp Neurol* 467: 566–80.

van Staveren, W. C., and Markerink-van Ittersum, M. 2005. Localization of cyclic guanosine 3′,5′-monophosphate-hydrolyzing phosphodiesterase type 9 in rat brain by nonradioactive in situ hybridization. *Methods Mol Biol* 307: 75–84.

Verhoest, P. R., Proulx-Lafrance, C., Corman, M., et al. 2009. Identification of a brain penetrant PDE9A inhibitor utilizing prospective design and chemical enablement as a rapid lead optimization strategy. *J Med Chem* 52: 7946–9.

Wachtel, H. 1983. Potential antidepressant activity of rolipram and other selective cyclic adenosine 3′,5′-monophosphate phosphodiesterase inhibitors. *Neuropharmacology* 22: 267–72.

Wachtel, H., Löschmann, P. A., Schneider, H. H., and Rettig, K. J. 1987. Effects of forskolin on spontaneous behavior, rectal temperature and brain cAMP levels of rats: Interaction with rolipram. *Neurosci Lett* 76: 191–6.

Wallace, D. A., Johnston, L. A., Huston, E., et al. 2005. Identification and characterization of PDE4A11, a novel, widely expressed long isoform encoded by the human PDE4A cAMP phosphodiesterase gene. *Mol Pharmacol* 67: 1920–34.

Wang, H., Yan, Z., Yang, S., Cai, J., Robinson, H., and Ke, H. 2008. Kinetic and structural studies of phosphodiesterase-8A and implication on the inhibitor selectivity. *Biochemistry* 47: 12760–8.

Weeks, J. L., Zoraghi, R., Francis, S. H., and Corbin, J. D. 2007. N-terminal domain of phosphodiesterase-11A4 (PDE11A4) decreases affinity of the catalytic site for substrates and tadalafil, and is involved in oligomerization. *Biochemistry* 46: 10353–64.

Wong, M. L., Whelan, F., Deloukas, P., et al. 2006. Phosphodiesterase genes are associated with susceptibility to major depression and antidepressant treatment response. *Proc Natl Acad Sci U S A* 103: 15124–9.

Wu, A. Y., Tang, X. B., Martinez, S. E., Ikeda, K., and Beavo, J. A. 2004. Molecular determinants for cyclic nucleotide binding to the regulatory domains of phosphodiesterase 2A. *J Biol Chem* 279: 37928–38.

Xie, Z., Adamowicz, W. O., Eldred, W. D., et al. 2006. Cellular and subcellular localization of PDE10A, a striatum-enriched phosphodiesterase. *Neuroscience* 139: 597–607.

Xu, Y., Lin, D., Li, S., et al. 2009. Curcumin reverses impaired cognition and neuronal plasticity induced by chronic stress. *Neuropharmacology* 57: 463–71.

Yamamoto, T., Jin, J., and Watanabe, S. 1997. Characteristics of memory dysfunction in olfactory bulbecto-mized rats and the effects of cholinergic drugs. *Behav Brain Res* 83: 57–62.

Yang, Q., Paskind, M., Bolger, G., et al. 1994. A novel cyclic GMP stimulated phosphodiesterase from rat brain. *Biochem Biophys Res Commun* 205: 1850–8.

Ye, Y., and O'Donnell, J. M. 1996. Diminished noradrenergic stimulation reduces the activity of rolipram-sensitive, high-affinity cyclic AMP phosphodiesterase in rat cerebral cortex. *J Neurochem* 66: 1894–902.

Ye, Y., Conti, M., Houslay, M. D., Farooqui, S. M., Chen, M., and O'Donnell, J. M. 1997. Noradrenergic activity differentially regulates the expression of rolipram-sensitive, high-affinity cyclic AMP phospho-diesterase (PDE4) in rat brain. *J Neurochem* 69: 2397–404.

Ye, Y., Jackson, K., and O'Donnell, J. M. 2000. Effects of repeated antidepressant treatment of type 4A phosphodiesterase (PDE4A) in rat brain. *J Neurochem* 74: 1257–62.

Yuasa, K., Kanoh, Y., Okumura, K., and Omori, K. 2001. Genomic organization of the human phosphodiesterase PDE11A gene. Evolutionary relatedness with other PDEs containing GAF domains. *Eur J Biochem* 268: 168–78. (Published online, Dec. 2000.)

Zeller, E., Stief, H. J., Pflug, B., and Sastre-y-Hernandez, M. 1984. Results of a phase II study of the antidepressant effect of rolipram. *Pharmacopsychiatry* 17: 188–90.

Zeringue, H. C., and Constantine-Paton, M. 2004. Post-transcriptional gene silencing in neurons. *Curr Opin Neurobiol* 14: 654–9.

Zhang, H. T. 2009. Cyclic AMP-specific phosphodiesterase-4 as a target for the development of antidepressant drugs. *Curr Pharm Des* 15: 1687–98.

Zhang, H. T. 2010. Phosphodiesterase Targets for Cognitive Dysfunction and Schizophrenia—a New York Academy of Sciences Meeting. *IDrugs* 13: 166–8.

Zhang, H. T., and O'Donnell, J. M. 2000. Effects of rolipram on scopolamine-induced impairment of working and reference memory in the radial-arm maze tests in rats. *Psychopharmacology (Berl)* 150: 311–16.

Zhang, H. T., Frith, S. A., Wilkins, J., and O'Donnell, J. M. 2001. Comparison of the effects of isoproterenol administered into the hippocampus, frontal cortex, or amygdala on behavior of rats maintained by differential reinforcement of low response rate. *Psychopharmacology (Berl)* 159: 89–97.

Zhang, H. T., Huang, Y., Jin, S. L. C., et al. 2002. Antidepressant-like profile and reduced sensitivity to rolipram in mice deficient in the PDE4D phosphodiesterase enzyme. *Neuropsychopharmacology* 27: 587–95.

Zhang, H. T., Zhao, Y., Huang, Y., Dorairaj, N. R., Chandler, L. J., and O'Donnell, J. M. 2004. Inhibition of the phosphodiesterase 4 (PDE4) enzyme reverses memory deficits produced by infusion of the MEK inhibitor U0126 into the CA1 subregion of the rat hippocampus. *Neuropsychopharmacology* 29: 1432–9.

Zhang, H. T., Huang, Y., Mishler, K., Roerig, S. C., and O'Donnell, J. M. 2005b. Interaction between the antidepressant-like behavioral effects of beta adrenergic agonists and the cyclic AMP PDE inhibitor rolipram in rats. *Psychopharmacology (Berl)* 182: 104–15.

Zhang, H. T., Zhao, Y., Huang, Y., et al. 2006. Antidepressant-like effects of PDE4 inhibitors mediated by the high-affinity rolipram binding state (HARBS) of the phosphodiesterase-4 enzyme (PDE4) in rats. *Psychopharmacology (Berl)* 186: 209–17.

Zhang, X., Feng, Q., and Cote, R. H. 2005a. Efficacy and selectivity of phosphodiesterase-targeted drugs in inhibiting photoreceptor phosphodiesterase (PDE6) in retinal photoreceptors. *Invest Ophthalmol Vis Sci* 46: 3060–6.

Zhang, H. T., Huang, Y., Masood, A., et al. 2008. Anxiogenic-like behavioral phenotype of mice deficient in phosphodiesterase 4B (PDE4B). *Neuropsychopharmacology* 33: 1611–23.

Zhao, Y., Zhang, H. T., and O'Donnell, J. M. 2003. Antidepressant-induced increase in high-affinity rolipram binding sites in rat brain: Dependence on noradrenergic and serotonergic function. *J Pharmacol Exp Ther* 307: 246–53.

Zittle, C. A., and Reading, E. H. 1946. Ribonucleinase and non-specific phosphodiesterase in rat and rabbit blood and tissues. *J Franklin Inst* 242: 424–8.

21 Vascular Depression: A Neuropsychological Perspective

Benjamin T. Mast, Brian P. Yochim,
Jeremy S. Carmasin, and Sarah V. Rowe

CONTENTS

21.1 INTRODUCTION

Depression is a major source of disability worldwide, particularly among older adults who also have significant medical comorbidity and greater rates of cognitive impairment. Older adults with depression typically experience greater rates of subjective memory problems (Minett et al. 2008) and objective cognitive impairment on neuropsychological testing (Burt et al. 1995). Moreover, depressive symptoms are linked with greater medical comorbidity (Lyness et al. 1996). This triad of vulnerability (depression, cognitive dysfunction, and medical comorbidity) places them at particularly high risk for disability and mortality. Within this context, the vascular depression hypothesis examines one specific mechanism that links physical health conditions with greater rates of depression, and accompanying cognitive impairment.

Although the proposed link between vascular risk and depression has a long history, it was most explicitly articulated by Alexopoulos et al. (1997), who hypothesized that cerebrovascular risk in the form of conditions such as hypertension, diabetes, hyperlipidemia, and heart disease plays a role in the onset, maintenance, and exacerbation of depression among older people. Some have proposed that these chronic conditions contribute to small-vessel disease in the frontal and subcortical regions of the brain, which play a significant role in the regulation of mood (Lyness et al. 1998). In this context, it is noteworthy that other models highlight the key role of the anterior cingulate cortex (ACC) and orbitofrontal cortex (OFC) in depression and have emphasized related executive dysfunction as a consequence of this dysfunction in depressed patients (Alexopoulos 2003; Mayberg 2003).

Although it is also well established that this same mechanism contributes to changes in cognition and can lead to dementia, there has been little explicit focus on the neuropsychological aspects of the vascular depression hypothesis, including the extent to which patients with vascular depression demonstrate greater cognitive impairment and risk for dementia. In this chapter, we review the

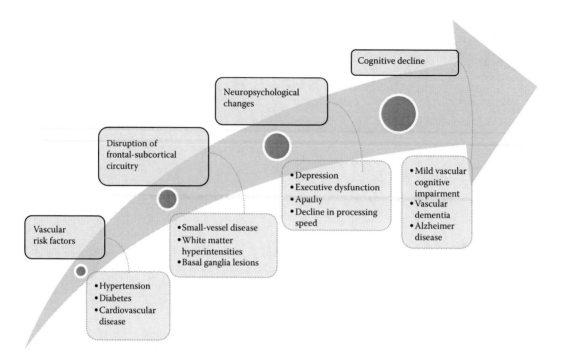

FIGURE 21.1   Impact of chronic vascular burden on depression and neuropsychological outcomes.

evidence for the vascular depression hypothesis with particular attention to the neuropsychological consequences and correlates (see Figure 21.1).

21.2   CLINICAL RISK FACTOR STUDIES

There has been considerable but nonuniversal support for the link between vascular burden and depression in later life. These studies have examined the relationship between vascular burden and depression in diverse samples, including well-functioning community-dwelling elders, psychiatric inpatients, primary care patients, and frail geriatric medical rehabilitation patients (Hickie et al. 2003; Lyness et al. 1998, 1999; Mast et al. 2004a, 2004b, 2008; Zimmerman et al. 2009). Most of these studies demonstrate a significant relationship between the severity of vascular burden and the severity of depressive symptoms both cross-sectionally and longitudinally. However, not all have demonstrated that this relationship is independent of general medical comorbidity and levels of disability (Lyness et al. 1998, 1999, 2000). Thus, whereas some studies show a unique connection between vascular burden and depression, others explain this relationship as a function of the person's overall level of functioning and health. This is an ongoing issue within vascular depression research, but the notion that vascular diseases specifically contribute to depression is further supported by a large body of neuroimaging research suggesting that cerebrovascular risk represents one of many depressogenic mechanisms. Thus, deepening our understanding of the relationship between cerebrovascular risk and depression has implications for treatment, as overall vascular burden may help identify which elders are most likely to experience depression.

21.3   NEUROIMAGING FOUNDATIONS

White and gray matter brain lesions (WMLs and GMLs) were linked to late-life depression more than 20 years ago, when Coffey et al. (1990) observed that elderly patients with depression had more severe WMLs and more frequent GMLs than controls on magnetic resonance imaging (MRI) (Coffey et al. 1990). Despite the use of other imaging techniques and investigations of potential bio-markers for vascular depression, WMLs, including white matter hyperintensities (WMHs), remain one of the most frequently investigated biomarkers in vascular depression neuroimaging. To date, much of the research has focused on WMLs in the frontal and subcortical regions (Campbell and Coffey 2001; Takahashi et al. 2008). Although a variety of brain regions and neural networks could be affected by cerebrovascular disease, the frontal and subcortical regions as well as their con-nections are likely to be most influential in the context of depression given the role of OFC and ACC in regulating mood (Mayberg 2003). With regard to subcortical regions, basal ganglia lesions have specifically been linked to cardiovascular changes and increased depression in older adults (Steffens et al. 1999). Older adults with depression have also evidenced higher levels of lesions in frontal medial orbital areas compared with older adults without depression (MacFall et al. 2001). Furthermore, Steffens et al. (2002) demonstrated that vascular disease, as evidenced by cortical and subcortical WMLs and basal ganglia lesions at baseline assessment, was related to depression in older adults over time. These frontal and subcortical lesions have been consistently linked with depression in late life.

Extensive WMLs would presumably lead to less functional connectivity within the brain, and multiple imaging methods have been used to draw inferences regarding connectivity. For example, functional MRI has demonstrated links between subcortical lesions in areas such as the caudate nucleus and reductions in cerebral blood flow (Hickie et al. 2007). Volumetric MRI studies have demonstrated links between volume reductions in key regions and the presence and severity of depression in late life. Caudate volume reduction, particularly the anterior caudate nucleus, was linked to more severe depression (Butters et al. 2009). Depression in older adults has also been associated with volume loss in frontal and temporal regions (Dotson et al. 2009).

Diffusion tensor imaging (DTI) is another method used to explore the extent to which functional connectivity is disrupted in late-life depression. Researchers using DTI in studying depression typi-cally calculate fractional anisotropy (FA), or the tendency of water to move in one direction (in these studies, it refers to the self-diffusion of water into brain tissue), measured on a scale of 0 to 1. An FA value of 0 indicates an isotrophic flow, where water is moving equally in all directions. An FA value of 1 indicates a completely anisotrophic flow, with water moving entirely in one direction without branching. Lower FA may reflect a brain losing its healthy white matter tissue structure and reduced connectivity (Alexopoulos and Kelly 2009). Researchers using DTI have found lower FA in depressed elders in both medial and lateral frontal regions, including the ACC, superior frontal gyri, and middle frontal gyrus (Bae et al. 2006). Lower FA in cortico-striato-limbic networks may predispose elders to develop depression and respond poorly to medication (Alexopoulos et al. 2008). Low FA in WMH was associated with high diastolic blood pressure, with the latter considered a vulnerability for late-life depression (Hoptman et al. 2009). These connections between depression and low FA seem to indirectly support the vascular depression hypothesis, although some excep-tions exist: higher frontal FA was linked to nonremittance to medication in depressed elders in one study (Taylor et al. 2008). Although the vascular depression approach suggests that vascular risk factors disrupt white matter tracts and connectivity, more research is clearly needed.

21.4 NEUROPATHOLOGICAL STUDIES AND GENOTYPING

Neuropathological and genetic studies have also sought to further understand the vascular depres-sion hypothesis by examining both a genetic vulnerability toward this syndrome and the vascular basis for the lesions commonly found in imaging studies. Neuropathological studies have suggested that the WMLs described in vascular depression research often have a vascular etiology (Firbank et al. 2005; O'Brien et al. 2003a; Thomas et al. 2001, 2002). In one neuropathology study, Thomas et al. (2002) found ischemic deep WMHs to be far more common in depressed elders than in controls, particularly in the dorsolateral prefrontal cortex. Regenold et al. (2007) linked deep WMLs to major depression via myelin staining, albeit in a sample of primarily middle-aged adults. Additionally, poststroke depression was linked to cerebrovascular risk factors in a postmortem study. The authors obtained tissue samples from 960 older adults with a clinical diagnosis of poststroke depression. They found that the chronic accumulation of lacunar infarcts in the thalamus, basal ganglia, and deep white matter may be a better predictor of poststroke depression than a single macroinfarct (Santos et al. 2009).

Genetic vulnerability toward depression has also been investigated in vascular depression imaging research. Subjects with HTTLPR heterozygotes (l/s) have more WMLs compared with subjects with other forms of the gene among depressed elders; this finding may be related to increases in hypertension in the HTTLPR heterozygous group (Steffens et al. 2008a). Additionally, subjects with either form of the gene (l/s or s/s) have more WMHs in frontolimbic and other regions and may be less likely to remit than noncarriers (Alexopoulos et al. 2009).

21.5 COGNITIVE FEATURES OF VASCULAR DEPRESSION

Before discussing cognition in vascular depression, it is important to note that cerebrovascular burden has been found to relate to cognitive functioning in several studies that did not explore the additional role of depression. In a neuroimaging study of patients with and without dementia (Tullberg et al. 2004), the frontal lobes were found to sustain the most damage from ischemic vascular disease and to have the most WMHs compared with other areas of the brain. The WMHs in this sample were related to poorer executive functioning, regardless of the location of the WMHs. Further evidence of a link between cerebrovascular risk factors and executive functioning was found by Brady et al. (2001), who found that a modified Framingham Stroke Risk Profile was associated with decline in verbal fluency over a 3-year interval. In 24 patients with vascular dementia, Cohen et al. (2002) found that subcortical hyperintensity volume was associated with executive and psychomotor performance but not related to performance in other cognitive domains. Whole-brain volume, on the other hand, was associated with performance in other cognitive domains but not with psychomotor or executive functioning. Lastly, patients with nonamnestic mild cognitive impairment (MCI) (i.e., MCI patients whose prominent deficit is not in memory) were more likely to have increased vascular risk compared to amnestic MCI patients (He et al. 2009). In summary, cerebrovascular disease seems to predominantly affect the frontal and subcortical areas of the brain and appears to be more closely linked to executive functioning compared to other cognitive domains.

The findings described in the previous section implicate disruption of frontal–subcortical circuits in late-life depression, particularly in patients with greater vascular disease. However, as noted above, the disruption of these same circuits can also be associated with other important clinical problems, including poorer executive functioning, slower processing speed, and poorer physical functioning, particularly in light of the well-recognized role that the ACC and other frontal–subcortical circuits play in regulation of mood, cognition, and behavior (Paus 2001; Allman et al. 2001; Cummings 1993). As a result, depressed patients with disruption of these circuits may also be at greater risk for cognitive decline and other behavioral dysregulation.

WMLs have been tied to cognitive decline and dementia in patients with late-life depression (Steffens et al. 2007). Depressed or remitted elders, found to have more WMLs than controls in the anterior-prefrontal region, were less likely to complete a complex task of planning or organization (Potter et al. 2007). Furthermore, WMLs in deep white matter and periventricular areas have been linked to increased cognitive deficits (memory, executive functioning, and processing speed) in depressed elders (Köhler et al. 2010). Another study found a similar relationship between late-life depression and cognitive deficits in elders with WMLs in various white matter tracts, including the superior longitudinal fasciculus, fronto-occipital fasciculus, uncinate fasciculus, and inferior longitudinal fasciculus (Sheline et al. 2008).

Observed cognitive and emotional changes that appear to be related to disruption of frontal–subcortical networks have led researchers to posit new syndromes. For example, subcortical ischemic depression has been defined as a combination of deep WMHs and subcortical gray matter hyperintensities and has been hypothesized as distinct from other late-life depressions because of vascular etiology (Krishnan et al. 2004). Similarly, Cook and colleagues have described subclinical structural brain disease (SSBD), which is characterized by cortical atrophy, deep WMHs, and periventricular hyperintensities (PVHs). In SSBD, each symptom slowly increases with normal aging and leads to greater disconnection. Although not specifically linked with vascular depression, cerebrovascular risk factors have been linked with SSBD, particularly increases in both deep WMHs and PVHs (Cook et al. 2004), which in turn have been associated with cognitive deficits in later life (Cook et al. 2002).

Frontal lobe regions have been linked to both geriatric depression and executive dysfunction, and executive dysfunction appears to be a key clinical feature of geriatric depression. Older adults with depression have demonstrated poorer executive functioning than controls across a variety of executive tasks, including fluency tasks (i.e., controlled oral word association), the Stroop Color–Word Interference task, and the Wisconsin Card Sorting task (Alexopoulos et al. 1997; Kindermann et al. 2000). The significant overlap between depression and executive dysfunction in late life has led Alexopoulos and colleagues to suggest a potentially distinct geriatric depressive syndrome: depression–executive dysfunction (DED) syndrome (Alexopoulos et al. 2002; Alexopoulos 2003). DED has implications for treatment, as depressed geriatric patients with poor executive functioning have higher levels of disability (instrumental activities of daily living) and poorer response to antidepressant medications (Alexopoulos et al. 2002; Kalayam and Alexopoulos 1999).

Findings of executive dysfunction in patients with vascular depression have been consistent with neuroanatomical findings related to executive dysfunction. Lacunes and WMLs have been correlated with metabolic rates in the dorsolateral prefrontal cortex, where metabolic activity has also been correlated with executive functioning (Reed et al. 2004). WMHs in subcortical regions have been linked with executive dysfunction even in older adults without dementia (Kramer et al. 2002). Stroke patients who experienced depressive symptoms and executive dysfunction were more likely to have infarcts in the basal ganglia and within frontal–subcortical tracts, than were stroke patients without depression and executive dysfunction (Vataja et al. 2005).

Processing speed may be another key cognitive domain affected by vascular depression. Sheline et al. (2006) investigated cognitive functioning in a sample of 155 patients with major depressive disorder. In this sample, depression symptoms were not correlated with vascular burden. Vascular burden predicted processing speed ($p = .06$) but not language, working memory, episodic memory, or executive functioning. Depression severity significantly predicted processing speed, language, working memory, and executive functioning, but not episodic memory. However, the relationship between depression severity and language, working memory, and executive functioning was found to be mediated by decreased processing speed. Deficient processing speed in vascular depression would be consistent with the location of lesions that are common in patients with vascular depression.

Findings regarding working memory and vascular depression have been mixed. In a sample of 198 patients with major depressive disorder, both the Framingham Stroke Risk Profile (D'Agostino et al. 1994) and intima medial thickness were found to be predictors of a composite measure of executive functioning (Smith et al. 2007). Framingham Stroke Risk Profile scores, but not intima medial thickness, significantly predicted working memory. Whereas this study demonstrated a significant relationship between vascular disease and both executive functioning and working memory in patients with major depressive disorder, another study indicated that vascular risk factors correlate more highly with the domains of processing speed and executive function than with working memory and language processing (Sheline et al. 2010). In fact, in the second study, working memory was unrelated to vascular risk (Sheline et al. 2010).

Thus, researchers have found a relationship between vascular depression and impaired processing speed, executive functioning, and possibly working memory. However, the nature of the relationship between vascular risk factors, depression, and cognitive changes needs further study. Vascular diseases may put individuals at risk for both depression and cognitive decline/dementia, and the most prominent cognitive changes in each are consistent with frontal–subcortical dysfunction, with particular emphasis on executive dysfunction and other frontal syndromes such as apathy and disinhibition (Cummings 1993). It is also important to recognize that many of the risk factors implicated in the vascular depression hypothesis have long been recognized as risk factors for vascular dementia and, now, vascular cognitive impairment (O'Brien 2006; O'Brien et al. 2003b). Moreover, vascular risk factors are now recognized as also increasing risk for other forms of dementia such as Alzheimer disease (Kalaria 2003). Researchers must attempt to ascertain whether cognitive impairment leads to depression or is a proxy representation of brain dysfunction that also produces depressive symptoms.

Along these lines, Mast et al. (2004c) found a significant interaction whereby medical patients with impaired executive functioning and vascular burden experienced significant symptoms of depression at baseline and 6 and 18 months after baseline. Executive dysfunction did not independently predict depressive symptoms but interacted with cerebrovascular risk factors such that patients with both executive dysfunction and high vascular burden showed more symptoms of depression than patients with only high vascular burden or executive dysfunction alone. Mast et al. (2004c) suggested that cerebrovascular risk factors may only predict depression symptoms when they are associated with frontostriatal dysfunction.

Vascular depression, vascular dementia, and vascular cognitive impairment are all likely to be characterized by a prominent frontal/executive dysfunction presentation. Yet few longitudinal studies have investigated the extent to which vascular depression specifically increases risk for cognitive decline and dementia over time. Although some studies have found a relationship between depressive symptoms and executive dysfunction in the context of vascular disease, it is unclear whether depressive symptoms precede a decline in executive functioning or executive dysfunction precedes depressive symptoms. It is also likely that vascular disease increases the risk of both depressive symptoms and executive dysfunction simultaneously (Alexopoulos 2003).

Some studies have found baseline depression to predict future development of cognitive impairment. In a sample of medical rehabilitation patients, greater symptoms of depression at baseline were found to predict lower verbal fluency 3 and 6 months later (Yochim et al. 2006). The number of cerebrovascular risk factors, although correlated with increased depressive symptoms, did not predict a measure of executive function (verbal fluency). These findings suggest that depressive symptoms may precede the development of executive dysfunction.

Fuhrer et al. (2003) assessed the relationship between baseline depressive symptoms, hypertension, and incident dementia over 8 years. They found that higher depressive symptomatology was found in those who developed vascular dementia than in those with no dementia or dementia due to Alzheimer's disease (AD) or other causes. They found that in men, hypertension was related to higher depressive symptoms. Depressed men with hypertension were at greater risk of developing dementia than depressed men with normal blood pressure. They hypothesized that depression in older men may be more likely to reflect a vascular etiology and that this vascular pathology might worsen the dementing process. It was proposed that a type of depression in late life may reflect underlying vascular damage in the brain that may "add to, amplify, or even accelerate the degeneration of the brain due to, or in conjunction with, the AD process" (p. 1062). Depression may be a marker of a dementia process rather than a causal risk factor (Jorm 2000, 2001).

Similar gender-specific findings occurred in a study by Cervilla et al. (2000). In a sample of older adults with hypertension, baseline depression predicted cognitive performance 9 to 12 years later among men but not among women. A weakness of this study was that the only cognitive measure used at Time 2 was the Mini Mental State Examination (MMSE). These findings add support to the idea that the relationship between vascular depression and cognitive decline may be more pronounced among men than women.

Another study utilizing a sample of 161 older adults (mean age = 69 years) with depression were followed for 2 years, and 20 of the participants developed dementia (Steffens et al. 2007). Change in WML volume over 2 years was a significant predictor of development of dementia, after controlling for age and baseline MMSE score. APOE genotype was unrelated to development of dementia, which was consistent with other findings that this genotype bears little relationship with cognitive impairment among older adults with depression (Butters et al. 2003). These findings suggest that, among patients with depression, the presence of WMLs significantly predict cognitive decline.

21.6 TREATMENT OUTCOMES

Vascular risk and executive dysfunction may impact treatment outcomes in those with depressive disorders. Depression in older adults with WMLs and poorer neuropsychological function were less likely to remit in a 12-week trial of sertraline (Sheline et al. 2010). Specifically, episodic memory, processing speed, executive function, and language predicted treatment response, whereas working memory did not. In fact, using selective serotonin reuptake inhibitors was linked to worsening WMLs over a 5-year study (Steffens et al. 2008b). On the other hand, despite attempts via structural MRI, another study was unable to link failed monotherapy response to WMLs; however, the authors cautioned that the vascular abnormalities observed in periventricular regions were significant and might have subtly impacted the treatment nonresponders (Baldwin et al. 2004).

21.7 FUTURE DIRECTIONS

There are several methodological changes that can be incorporated into future studies. Cognitive functioning could be examined on a continuum as opposed to placing patients into diagnostic categories. Studies often investigate patients with MCI or dementia without incorporating cognition as a continuous variable. In the same vein, depressive symptomatology can be explored as a continuous variable, in depressed as well as nondepressed samples. Moreover, people with cognitive impairment are typically excluded from depression studies (e.g., Sheline et al. 2006, 2010; Steffens et al. 2007), and people with significant depression are typically excluded from dementia and MCI studies (Steffens et al. 2006). Expanding the range of subjects in this manner will increase understanding regarding the relationship between depressive symptoms and cognitive functioning in patients with significant vascular disease.

REFERENCES

Alexopoulos, G. S. 2002. Frontostriatal and limbic dysfunction in late-life depression. *Am J Geriatr Psychiatry* 10 (6):687–95.

Alexopoulos, G. S. 2003. Vascular disease, depression, and dementia. *J Am Geriatr Soc* 51:1178–80.

Alexopoulos, G. S., and Kelly, R. E. 2009. Research advances in geriatric depression. *World Psychiatry* 8 (3):140–9.

Alexopoulos, G. S., Meyers, B. S., Young, R. C., Kakuma, T., Silbersweig, D., and Charlson, M. 1997. Clinically defined vascular depression. *Am J Psychiatry* 154:562–5.

Alexopoulos, G. S., Murphy, C. F., Gunning-Dixon, F. M., Latoussakis, V., Kanellopoulos, D., Klimstra, S., et al. 2008. Microstructural white matter abnormalities and remission of geriatric depression. *Am J Psychiatry* 165 (2):238–44.

Alexopoulos, G. S., Murphy, C. F., Gunning-Dixon, F. M., Glatt, C. E., Latoussakis, V., Kelly, R. E., et al. 2009. Serotonin transporter polymorphisms, microstructural white matter abnormalities and remission of geriatric depression. *J Affect Disord* 119 (1–3):132–41.

Allman, J. M., Hakeem, A., Erwin, J. M., Nimchinsky, E., and Hof, P. 2001. The anterior cingulate cortex. The evolution of an interface between emotion and cognition. *Ann N Y Acad Sci* 935:107–17.

Bae, J. N., MacFall, J. R., Krishnan, K. R., Payne, M. E., Steffens, D. C., and Taylor, W. D. 2006. Dorsolateral prefrontal cortex and anterior cingulate cortex white matter alterations in late-life depression. *Biol Psychiatry* 60 (12):1356–63.

Baldwin, R., Jeffries, S., Jackson, A., Sutcliffe, C., Thacker, N., Scott, M., et al. 2004. Treatment response in late-onset depression: Relationship to neuropsychological, neuroradiological and vascular risk factors. *Psychol Med* 34 (1):125–36.

Brady, C. B., Spiro, A., McGlinchey-Berroth, R., Milberg, W., and Gaziano, J. M. 2001. Stroke risk predicts verbal fluency decline in healthy older men: Evidence from the Normative Aging Study. *J Gerontol Psychol Sci* 56B:P340–P346.

Burt, D. B., Zembar, M. J., and Niederehe, G. 1995. Depression and memory impairment — A metaanalysis of the association, its pattern, and specificity. *Psychol Bull* 117:285–305.

Butters, M. A., Sweet, R. A., Mulsant, B. H., et al. 2003. APOE is associated with age-of-onset, but not cognitive functioning, in late-life depression. *Int J Geriatr Psychiatry* 18:1075–81.

Butters, M. A., Aizenstein, H. J., Hayashi, K. M., Meltzer, C. C., Seaman, J., Reynolds, C. F., et al. 2009. Three-dimensional surface mapping of the caudate nucleus in late-life depression. *Am J Geriatr Psychiatry* 17 (1):4–12.

Campbell, J. J., and Coffey, C. E. 2001. Neuropsychiatric significance of subcortical hyperintensity. *J Neuropsych Clin N* 13 (2):261–88.

Cervilla, J. A., Prince, M., Joels, S., and Mann, A. 2000. Does depression predict cognitive outcome 9 to 12 years later? Evidence from a prospective study of elderly hypertensives. *Psychol Med* 30:1017–23.

Coffey, C. E., Figiel, G. S., Djang, W. T., and Weiner, R. D. 1990. Subcortical hyperintensity on magnetic resonance imaging: A comparison of normal and depressed elderly subjects. *Am J Psychiatry* 147 (2):187–9.

Cohen, R. A., Paul, R. H., Ott, B. R., et al. 2002. The relationship of subcortical MRI hyperintensities and brain volume to cognitive function in vascular dementia. *J Int Neuropsychol Soc* 8:743–52.

Cook, I. A., Leuchter, A. F., Morgan, M. L., Conlee, E. W., David, S., Lufkin, R., et al. 2002. Cognitive and physiologic correlates of subclinical structural brain disease in elderly healthy control subjects. *Arch Neurol* 59 (10):1612–20.

Cook, I. A., Leuchter, A. F., Morgan, M. L., Dunkin, J. J., Witte, E., David, S., et al. 2004. Longitudinal progression of subclinical structural brain disease in normal aging. *Am J Geriatr Psychiatry* 12 (2):190–200.

Cummings, J. L. 1993. Frontal-subcortical circuits and human behavior. *Arch Neurol* 50:873–880.

D'Agostino, R. B., Wolf, P. A., Belanger, A. J., and Kannel, W. B. 1994. Stroke risk profile: Adjustment for antihypertensive medication: The Framingham study. *Stroke* 25:40–3.

Dotson, V. M., Davatzikos, C., Kraut, M. A., and Resnick, S. M. 2009. Depressive symptoms and brain volumes in older adults: A longitudinal magnetic resonance imaging study. *J Psychiatry Neurosci* 34 (5):367–75.

Firbank, M. J., O'Brien, J. T., Pakrasi, S., et al. 2005. White matter hyperintensities and depression—Preliminary results from the LADIS study. *Int J Geriatr Psychiatry* 20:674–9.

Fuhrer, R., Dufouil, C., and Dartigues, J. F. 2003. Exploring sex differences in the relationship between depressive symptoms and dementia incidence: Prospective results from the PAQUID study. *J Am Geriatr Soc* 51:1055–63.

He, J., Farias, S., Martinez, O., Reed, B., Mungas, D., and DeCarli, C. 2009. Differences in brain volume, hippocampal volume, cerebrovascular risk factors, and apolipoprotein E4 among mild cognitive impairment subtypes. *Arch Neurol* 66:1393–9.

Hickie, I., Simons, L., Naismith, S., Simons, J., McCallum, J., and Pearson, K. 2003. Vascular risk to late-life depression: Evidence from a longitudinal community study. *Aust N Z J Psychiatry* 37:62–5.

Hickie, I. B., Naismith, S. L., Ward, P. B., Little, C. L., Pearson, M., Scott, E. M., et al. 2007. Psychomotor slowing in older patients with major depression: Relationships with blood flow in the caudate nucleus and white matter lesions. *Psychiatry Res* 155 (3):211–20.

Hoptman, M. J., Gunning-Dixon, F. M., Murphy, C. F., Ardekani, B. A., Hrabe, J., Lim, K. O., et al. 2009. Blood pressure and white matter integrity in geriatric depression. *J Affect Disord* 115 (1–2):171–6.

Jorm, A. F. 2000. Is depression a risk factor for dementia or cognitive decline? A review. *Gerontology* 46:219–27.

Jorm, A. F. 2001. History of depression as a risk factor for dementia: An updated review. *Aust N Z J Psychiatry* 35:776–81.

Kalaria, R. N. 2003. Vascular factors in Alzheimer's disease. *Int Psychogeriatr* 15 (Suppl 1):47–52.

Kalayam, B., and Alexopoulos, G. 1999. Prefrontal dysfunction and treatment response in geriatric depression. *Arch Gen Psychiatry* 56:713–8.

Kindermann, S. S., Kalayam, B., Brown, G. G., Burdick, K. E., and Alexopoulos, G. S. 2000. Executive functions and P300 latency in elderly depressed. Patients and control subjects. *Am J Geriatr Psychiatry* 8:57–65.

Köhler, S., Thomas, A. J., Lloyd, A., Barber, R., Almeida, O. P., and O'Brien, J. T. 2010. White matter hyperintensities, cortisol levels, brain atrophy and continuing cognitive deficits in late-life depression. *Br J Psychiatry* 196 (2):143–9.

Kramer, J. H., Reed, B. R., Mungas, D., Weiner, M. W., and Chui, H. C. 2002. Executive dysfunction in subcortical ischaemic vascular disease. *J Neurol Neurosurg Psychiatry* 72:217–20.

Krishnan, K. R., Taylor, W. D., McQuoid, D. R., MacFall, J. R., Payne, M. E., Provenzale, J. M., et al. 2004. Clinical characteristics of magnetic resonance imaging-defined subcortical ischemic depression. *Biol Psychiatry* 55 (4):390–7.

Lyness, J. M., Bruce, M. L., Koenig, H. G., et al. 1996. Depression and medical illness in late life: Report of a symposium. *J Am Geriatr Soc* 44:198–203.

Lyness, J. M., Caine, E. D., Cox, C., King, D. A., Conwell, Y., and Olivares, T. 1998. Cerebrovascular risk factors and later-life major depression. Testing a small-vessel brain disease model. *Am J Geriatr Psychiatry* 6:5–13.

Lyness, J. M., Caine, E. D., King, D. A., Conwell, Y., Cox, C., and Duberstein, P. R. 1999. Cerebrovascular risk factors and depression in older primary care patients: Testing a vascular brain disease model of depression. *Am J Geriatr Psychiatry* 7:252–8.

Lyness, J. M., King, D. A., Conwell, Y., Cox, C., and Caine, E. D. 2000. Cerebrovascular risk factors and 1-year depression outcome in older primary care patients. *Am J Psychiatry* 157:1499–501.

MacFall, J. R., Payne, M. E., Provenzale, J. E., and Krishnan, K. R. R. 2001. Medial orbital frontal lesions in late-onset depression. *Biol Psychiatry* 49:803–6.

Mast, B. T., Neufeld, S., MacNeill, S. E., and Lichtenberg, P. A. 2004a. Longitudinal support for the relationship between vascular risk factors and late-life depressive symptoms. *Am J Geriatr Psychiatry* 12:93–101.

Mast, B. T., MacNeill, S. E., and Lichtenberg, P. A. 2004b. Post-stroke and clinically-defined vascular depression in geriatric rehabilitation patients. *Am J Geriatr Psychiatry* 12:84–92.

Mast, B. T., Yochim, B., MacNeill, S. E., and Lichtenberg, P. A. 2004c. Risk factors for geriatric depression: The importance of executive functioning within the vascular depression hypothesis. *J Gerontol Med Sci* 59A:1290–4.

Mast, B. T., Miles, T., Penninx, B. W., et al. 2008. Vascular disease and future risk of depressive symptomatology in older adults: findings from the Health, Aging, and Body Composition study. *Biol Psychiatry* 64:320–6.

Mayberg, H. S. 2003. Modulating dysfunctional limbic-cortical circuits in depression: Towards development of brain-based algorithms for diagnosis and optimised treatment. *Brain Med Bull* 65:193–207.

Minett, T. S. C., Da Silva, R. V., Ortiz, K. Z., and Bertolucci, P. H. F. 2008. Subjective memory complaints in an elderly sample: A cross-sectional study. *Int J Geriatr Psychiatry* 23:49–54.

O'Brien, J. T. 2006. Vascular cognitive impairment. *Am J Geriatr Psychiatry* 14:724–33.

O'Brien, J. T., Thomas, A., English, P., Perry, R., and Jaros, E. 2003a. A prospectively studied clinicopathological case of 'vascular depression'. *Int J Geriatr Psychiatry* 18:656–7.

O'Brien, J. T., Erkinjuntti, T., Reisberg, B., et al. 2003b. Vascular cognitive impairment. *Lancet Neurol* 2:89–98.

Paus, T. 2001. Primate anterior cingulate cortex: Where motor control, drive and cognition interface. *Nat Rev Neurosci* 2:417–24.

Potter, G. G., Blackwell, A. D., McQuoid, D. R., Payne, M. E., Steffens, D. C., Sahakian, B. J., et al. 2007. Prefrontal white matter lesions and prefrontal task impersistence in depressed and nondepressed elders. *Neuropsychopharmacology* 32 (10):2135–42.

Reed, B. R., Eberling, J. L., Mungas, D., Weiner, M., Kramer, J. H., and Jagust, W. J. 2004. Effects of white matter lesions and lacunes on cortical function. *Arch Neurol* 61:1545–50.

Regenold, W. T., Phatak, P., Marano, C. M., et al. 2007. Myelin staining of deep white matter in the dorsolateral prefrontal cortex in schizophrenia, bipolar disorder, and unipolar major depression. *Psychiatry Res* 151:179–88.

Santos, M., Kövari, E., Hof, P. R., et al. 2009. The impact of vascular burden on late-life depression. *Brain Res Rev* 62:19–32.

Sheline, Y. I., Barch, D. M., Garcia, K., et al. 2006. Cognitive function in late life depression: Relationships to depression severity, cerebrovascular risk factors and processing speed. *Biol Psychiatry* 60:58–65.

Sheline, Y. I., Price, J. L., Vaishnavi, S. N., Mintun, M. A., Barch, D. M., Epstein, A. A., et al. 2008. Regional white matter hyperintensity burden in automated segmentation distinguishes late-life depressed subjects from comparison subjects matched for vascular risk factors. *Am J of Geriatr Psychiatry* 165 (4): 524–32.

Sheline, Y. I., Pieper, C. F., Barch, D. M., et al. 2010. Support for the vascular depression hypothesis in late-life depression: Results of a 2-site, prospective, antidepressant treatment trial. *Arch Gen Psychiatry* 67:277–85.

Smith, P. J., Blumenthal, J. A., Babyak, M. A., et al. 2007. Cerebrovascular risk factors, vascular disease, and neuropsychological outcomes in adults with major depression. *Psychosom Med* 69:578–86.

Steffens, D. C., Helms, M. J., Krishnan, K R, and Burke, G. L. 1999. Cerebrovascular disease and depression symptoms in the cardiovascular health study. *Stroke* 30 (10):2159–66.

Steffens, D. C., Krishnan, K. R., Crump, C., and Burke, G. L. 2002. Cerebrovascular disease and evolution of depressive symptoms in the cardiovascular health study. *Stroke* 33 (6):1636–44.

Steffens, D. C., Otey, E., Alexopoulos, G. S., et al. 2006. Perspectives on depression, mild cognitive impairment, and cognitive decline. *Arch Gen Psychiatry* 63:130–8.

Steffens, D. C., Potter, G. G., McQuoid, D. R., et al. 2007. Longitudinal magnetic resonance imaging vascular changes, apolipoprotein E genotype, and development of dementia in the neurocognitive outcomes of depression in the elderly study. *Am J Geriatr Psychiatry* 15:839–49.

Steffens, D. C., Taylor, W. D., McQuoid, D. R., and Krishnan, K. R. 2008a. Short/long heterozygotes at 5HTTLPR and white matter lesions in geriatric depression. *Int J Geriatr Psychiatry* 23 (3):244–8.

Steffens, D. C., Chung, H., Krishnan, K. R., Longstreth, W. T., Carlson, M., and Burke, G. L. 2008. Antidepressant treatment and worsening white matter on serial cranial magnetic resonance imaging in the elderly: The Cardiovascular Health Study. *Stroke* 39 (3):857–62.

Takahashi, K., Oshima, A., Ida, I., Kumano, H., Yuuki, N., Fukuda, M., et al. 2008. Relationship between age at onset and magnetic resonance image-defined hyperintensities in mood disorders. *J Psychiatr Res* 42 (6):443–50.

Taylor, W. D., Kuchibhatla, M., Payne, M. E., MacFall, J. R., Sheline, Y. I., Krishnan, K. R., et al. 2008. Frontal white matter anisotropy and antidepressant remission in late-life depression. *PLoS One* 3 (9):e3267.

Thomas, A. J., Ferrier, I. N., Kalaria, R. N., Perry, R. H., Brown, A., and O'Brien, J. T. 2001. A neuropathological study of vascular factors in late-life depression. *J Neurol Neurosurg Psychiatry* 70:83–7.

Thomas, A. J., O'Brien, J. T., Davis, S., et al. 2002. Ischemic basis for deep white matter hyperintensities in major depression: A neuropathological study. *Arch Gen Psychiatry* 59:785–92.

Tullberg, M., Fletcher, E., DeCarli, C., et al. 2004. White matter lesions impair frontal lobe function regardless of their location. *Neurology* 63:246–53.

Vataja, R., Pohjasvaara, T., Mantyla, R., et al. 2005. Depression–executive dysfunction syndrome in stroke patients. *Am J Geriatr Psychiatry* 13:99–107.

Yochim, B. P., MacNeill, S. E., and Lichtenberg, P. A. 2006. "Vascular depression" predicts verbal fluency in older adults. *J Clin Exp Neuropsychol* 28:495–508.

Zimmerman, J. A., Mast, B. T., Miles, T., and Markides, K. S. 2009. Vascular risk and depression in the Hispanic Established Population for the Epidemiologic Study of the Elderly (EPESE). *Int J Geriatr Psychiatry* 24:409–16.

22 Neurobiological Basis for Development of New Antidepressant Agents

Cecilio Álamo and Francisco López-Muñoz

CONTENTS

22.1 INTRODUCTION

The pharmacotherapeutic approach to affective disorders, from the perspective of current scientific pharmacology, has its origin in the 1950s, the same decade in which the first antipsychotic and anxiolytic agents appeared, and what has been termed the "revolution of psychopharmacology" (López-Muñoz et al. 2000). The great historical breakthroughs in the treatment of mood disorders can focus mainly on the discovery and therapeutic utilization of tricyclic antidepressants (ADTs) (Fangmann et al. 2008) and monoamine-oxidase inhibitors (MAOIs) (López-Muñoz et al. 2007) as antidepressants in the 1950s, and in the definitive application of lithium salts in the treatment and prophylaxis of bipolar disorders during the 1960s (López-Muñoz et al. 2005). The introduction of the so-called atypical antidepressants, heterocyclic of the "second generation" (maprotiline, nomifensine, trazodone, and mianserine, among others) during the 1970s did not, contrary to what was thought at that time, assume transcendental advances from the therapeutic point of view nor from its safety profile, except in individual cases, with respect to classic antidepressants. However, from the end of the 1980s, the addition to the antidepressant arsenal of a series of new families of drugs has contributed a therapeutic advance in some cases and, in others, a disappointment in the treatment of depression. As positive events among these new antidepressants we can consider, above all, selective serotonin reuptake inhibitors (SSRI), as well as noradrenaline and serotonin reuptake inhibitors (NSRI), noradrenergic and specific serotonergic antidepressants (NaSSA), and noradrenaline reuptake inhibitors (NARI). In contrast, agents that combine the inhibition of the reuptake of serotonin with the blocking of the postsynaptic 5-HT$_2$ receptors (nefazodone) and that of the selective and reversible inhibitors of monoamine-oxidase (RIMA) have not been well accepted. Table 22.1 lists the various families of antidepressant drugs, with the drug prototype and with respect to their action mechanisms.

The introduction of ADTs inaugurated a new era in the treatment of depression, and they continue today to be agents of reference in clinical research, where imipramine is still the gold standard of antidepressant efficacy in multiple comparative clinical studies involving new agents, such as in antidepressant therapy, where the same rates of efficacy are maintained as with the rest of antidepressants that appeared later. Its main drawback is its high incidence of adverse effects, which is attributable to a pharmacodynamic nonspecificity, and its lack of "chemical cleanliness," because it blocks multiple types of receptors. While ADTs continue to be used clinically in a very important way, MAOIs have suffered a reduction in their wide usage, largely due to their problems with interactions with food rich in tyramine and other psychostimulant drugs, which can lead to a tragic hypertensive crisis. The clinical emergence of SSRIs made possible, as opposed to earlier drugs, considerable advantages, especially from the perspective of general tolerability, safety in overdose, and the profile of differential adverse effects, which allowed the opening of the field of antidepressive therapy to nonpsychiatric options and increased the spectrum of indication beyond the proper antidepressants (Álamo et al. 2000). In reality, these agents have facilitated the treatment of depression in primary care–based treatment from the standpoint of therapeutic efficacy in terms of major acceptance of medication on the part of the patients.

Despite the therapeutic importance of antidepressants, they are still a fundamentally empirical agent of therapy that, without a doubt, has paved the way for the knowledge of some etiopathogenic bases of depression (see Chapter 6). Still, the mechanisms that have been explored during the past 50 years center mainly on the field of monoamines (noradrenaline and serotonin) (López-Muñoz and Álamo 2009). Perhaps this is the fact, as has been mentioned earlier, that conditions a similar antidepressive efficacy, about 70% of patients, which determines—for all of them—a slow start action, generally superior to the 2 weeks. On the other hand, during these decades, it has been observed that therapeutic effects of antidepressants are relatively homogenous in terms of percentages between agents that act against different mechanisms, including opposites. The monoaminergic explanation of depression and of the antidepressive treatment is subject to multiple contradictions (lapse at the beginning of action, relative efficacy, inhibiting action of the reuptake of monoamines by the

TABLE 22.1

Classification of Antidepressant Drugs, According to Their Mechanism of Action

Family	Acronym	Drug Prototype
Reuptake inhibitors of 5-HT and NA with blocking action of different receptors	ADT	Imipramine
Irreversible MAO inhibitors	MAOI	Phenelzine
Reuptake inhibitors of NA with blocking action of different receptors		Maprotiline
Autoreceptor antagonists α$_2$		Mianserin
Selective reuptake inhibitors of DA	DSRI	Bupropion
Selective reuptake inhibitors of 5-HT	SSRI	Fluoxetine
Reversible inhibitors of MAO	RIMA	Moclobemide
Reuptake inhibitors of 5-HT receptor and 5-HT$_2$ blockers		Nefazodone
Antagonists of self- and heteroreceptors α$_2$ and 5-IIT$_2$ y 5HT$_3$ receptors	NaSSA	Mirtazapine
Reuptake inhibitors of NA and 5-HT	NSRI	Venlafaxine
Selective inhibitors of NA reuptake	NARI	Reboxetine
Agonists of melatonin receptors and 5-TH$_{2C}$ receptors		Agomelatine

TABLE 22.2
Antidepressant Drugs in Development

Principal Action	Product	Company	Development Phase	Other Actions
SSRI	Vilazodone	PGx Health	Phase III	5-HT$_{1A}$ agonist
	Lu-AA-21004	Lundbeck/Takeda	Phase III	5-HT$_{1A}$ partial agonist; 5-HT$_3$ antagonist
	F-2695	Pierre Fabre	Phase II	NARI
	Lu AA24530	Lundbeck	Phase II	–
	TGBA01AD	Fabre-Kramer Pharm.	Phase II	5-HT$_2$/5-HT$_{1A}$ agonist; 5-HT$_{1D}$ mod.
NARI	LY2216684	Eli Lilly	Phase III	–
	Levomilnacipram	Forest/Pierre Fabré	Phase III	–
DA reuptake inhibitor	DOV-21947	Dov Pharm./Merck	Phase II	NARI; SSRI
	SEP-227162	Sepracor	Phase III	NARI; SSRI
	NS-2359	NeuroSearch/GSK	Phase II	DA agonist; NARI; SSRI
	SEP-225289	Sepracor	Phase II	NARI; SSRI
	NSD-788	GSK/NeuroSearch	Phase I	NARI; SSRI
	RG7166	Roche	Phase I	NARI; SSRI
RIMA	CX157	CeNeRx BioPharma	Phase II	–
5-HT/DA modulator	RX-10100	Rexahn	Phase II	–
5-HT$_{1A}$ agonist	Adatanserin	Fabre-Kramer	Phase II	5-HT$_2$ antagonist
	PRX-00023	Epix	Phase II	Sigma antagonist
5-HT$_{1A}$ partial agonist	MN-305	MediciNova	Phase II	–
5-HT$_1$ antagonist	GSK-588045	GSK	Phase II	–
	GSK-163090	GSK	Phase I	–
	GSK-588045	GSK	Phase I	–
5-HT$_{2A}$ antagonist	Netamiftide	Innapharma	Phase II	–
β$_3$ agonist	Amibegron	Sanofi-Aventis	Phase III	–
D$_2$ agonist	Pardoprunox	Solvay	Phase II	5-HT$_{1A}$ agonist
Nicotinic α$_4$ agonist	CP-601927	Pfizer	Phase II	–
Nicotinic/ACh antagonist	TC-5214	Targacept/ AstraZeneca	Phase II	–
	TC-2216	Targacept	Phase I	–
Sigma agonist	OPC-14523	Otsuka/Vela	Phase II	5-HT$_{1A}$ agonist; SSRI
Sigma-1 agonist	SA-4503	M's Sci/Eisai	Phase II	–
NMDA modulator	GLYX-13	Naurex	Phase I	–
NMDA antagonist	AZD6765	AstraZeneca	Phase II	–
	EVT-101	Evotec	Phase II	–
	Traxoprodil	Pfizer	Phase II	–
	RO4917523	Roche	Phase II	–
AMPA modulator	ORG 26576	Schering Plough	Phase II	–
	Coluracetam	BrainCells	Phase II	Enhance choline uptake
AMPA antagonist	E2508	Eisai	Phase I	–
mGluR2 antagonist	RG1578	Roche	Phase I	–
mGluR5 Antagonist	RG7090	Roche	Phase II	–
CRF-1 antagonist	ONO-2333Ms	Ono Pharmaceuticals	Phase II	–
	Verucerfont	Neurocrine Biosci./ GSK	Phase II	–
	Emicerfont	GSK/Neurocrine Biosci.	Phase II	–

(continued)

TABLE 22.2 (Continued)
Antidepressant Drugs in Development

Principal Action	Product	Company	Development Phase	Other Actions
CRF-1 antagonist	CP 316,311	Pfizer	Phase II	–
	GW876008	GSK	Phase II	–
	BMS-562086	Bristol-Myers Squibb	Phase II	–
	JNJ-19567470	Janssen (J&J)	Phase I	–
	TS-041	Taisho	Phase I	–
	SSR 125543	Sanofi-Aventis	Phase I	–
	GSK-586529	GSK	Phase I	–
GR-II antagonist	Mifepristone Corcept	Corcept	Phase III	–
GR-II antagonist	ORG 34517	Organon	Phase II	Antagonist lipocortins synthesis
V-1B antagonist	SSR-149415[a]	Sanofi-Aventis	Phase II	–
NK-1 antagonist	Casopitant	GSK	Phase III	–
	Orvepitant	GSK	Phase II	–
	Vestipitant	GSK	Phase II	–
	SSR-240600	Sanofi-Aventis	Phase II	–
	GSK-424887	GSK	Phase I	SSRI
NK-2 antagonist	Saredutant[a]	Sanofi-Aventis	Phase III	–
NK-3 antagonist	SSR-146977	Sanofi-Aventis	Phase I	–
Enkephalinergic modulator	AZD2327	AstraZeneca	Phase II	–
	AZD7268	AstraZeneca	Phase II	–
TAAR1 partial agonist	RG7351	Roche	Phase I	–
PDE4 inhibitor	ND-1251	Evotec	Phase I	–
	GSK-356278	GSK	Phase I	–
P38-kinase inhibitor	Losmapimod	GSK	Phase II	–
MIF$_1$ pentapeptide analog	Nemifitide	Tetragenex Pharm.	Phase II	–
M$_1$/M$_2$ agonist	Tasimelteon	Vanda Pharmaceuticals	Phase I	–
FAAH inhibitor	SSR411298	Sanofi-Aventis	Phase II	–
	URB597	Schering Plough	Phase I	–
Omega-3-acid ethyl esters	Lovaza®	GSK	Phase II	–
Phenylalanine derivative	YKP-10A	SK Corp./J&J	Phase II	–
Unidentified	ODS-II	Genopia Biomed	Phase II	–
	ND-7001	Evotec	Phase I	–
	PSN0041	Psylin Neurosciences	Phase I	–
	BTG-1640	Abiogen Pharma	Preclinical	–

Sources: PharmaProjects, PJB Publications Ltd, 2010; R&D Insight, ADIS International, 2010; www.neurotransmitter.net/newdrugs.html; www.phrma.org/medicines_in_development; Mathew, S. J., et al., *Neuropsychopharmacology*, 33, 2080–92, 2008.

Note: 5-HT, serotonin; ACh, acetylcholine; AMPA, alpha-amino-3-hydroxy-5-methyl-4- isoxazole-propionic acid; CRF, corticotropin-releasing factor; DA, dopamine; FAAH, fatty acid amide hydrolase; GR, glucocorticoid receptor; M, melatonergic receptor; mGluR, metabotropic glutamate receptor; MIF, pentapeptide analog of melanocyte-inhibiting factor; NA, noradrenaline; NARI, selective noradrenaline reuptake inhibitor; NK, neurokinin; NMDA, N-methyl-D-aspartate; PDE4, phosphodiesterase-IV; RIMA, reversible inhibitor of monoamine oxidase; SSRI, selective serotonin reuptake inhibitor; TAAR1, trace amine-associated receptor 1; V-1B, vasopressin 1B.

[a] Recently discontinued.

majority of antidepressants against the boost observed in the case of tianeptine) and tries to explain that an antidepressive acts because it increases in the synaptic cleft the rate of monoamines (serotonin and/or noradrenaline), which seems as simple as trying to center the "action mechanism" of a switch on a computer. Recently, new theories have been elaborated that aim to explain not only how antidepressants work but also why the therapeutic measures are efficient in a particular percentage of patients, whereas they are not in the rest, as well as the causes of different types of depression.

In Table 22.2, we detail the products that are today—within the field of pharmacology of depression—being developed, in different phases of clinical research, specifying their main pharmacological action, as well as other action mechanisms that possibly play some important role in their pharmacological activity.

22.2 FROM SYNAPTIC PHARMACOLOGY TO INTRANEURONAL PHARMACOLOGY

As previously noted, during the 1960s and 1970s, the majority of studies aimed at demonstrating the effects of psychotropic drugs on the central nervous system were centered on the extracellular aspects of synaptic transmission (Figure 6.10), fundamentally implying the interaction of the neurotransmitter with its receptor. This interaction was the result of the action on the inhibition of the reuptake systems (ADT) and of metabolism (MAOIs), which increased the levels of monoamines in the synaptic cleft, facilitating its action on the receptor.

In the past 30 years, the monoaminergic hypothesis of depression has marked the development of new antidepressant agents (López-Muñoz and Álamo 2009). In short, this hypothesis is based on the capacity of reserpine, which as we know produces a depletion of monoamines to induce depression in some patients (Schildkraut 1965; Bunney and Davis 1965; Coppen 1967). The fact that the known antidepressants to this day provoke an increase in monoamines in the synaptic cleft supports the hypothesis of a decrease of monoaminergic neurotransmission in the pathogenesis of depression. Based on these facts, the mechanisms explored for the discovery of new antidepressants to date have been centered almost exclusively on provoking an increase in monoamines in the synaptic cleft. These biochemical changes occur rapidly and can be detected after the first dose of the antidepressant drug. However, the therapeutic effect is not produced until after a few weeks of treatment, which suggests that this effect occurs after a series of adaptations at a neuronal level, as a result of the chronic administration of these agents.

This fact brought about in the 1980s the theory of receptor adaptation. According to this theory, the persistent activation of receptors, as a result of the elevation of serotonin and noradrenaline in the synaptic cleft, brought the same (5-HT_2 and β-adrenergic receptors) to downregulation, a phenomenon that coincides in time with the onset of the therapeutic effect of an antidepressant (Sulser et al. 1978). However, the facts that this regulator phenomenon is not universal for all antidepressants and that the blockers of β-adrenergic receptors are devoid of antidepressant effects and may even induce depression in some subjects (Paykel et al. 1982) question the possibility that this adaptive receptor mechanism is the only thing responsible for the therapeutic effect.

As a result of this comes a greater appreciation of the complexity of synaptic transmission, with the acceptance that the regulation of the union of the neurotransmitter with the receptor is only a small part of the effects of the neurotransmitters on their target neurons, behaving as an interrupter of the biochemical mechanism responsible for the antidepressant effect. In reality, it is becoming increasingly clear that neurotransmitters regulate all the processes occurring inside the neurons. Additionally, during this decade, it was recognized that the effects of neurotransmitters on target neurons are not direct but are rather produced through biochemical cascades of intracellular messengers. Among these intracellular messengers are proteins that are linked to membranes, such as G proteins (GP), second messengers such as cyclic adenosine monophosphate (cAMP) or intracellular calcium, and phosphorylized proteins that, changing the allocation of phosphates of all types of neuronal proteins, alter their function, being responsible for the wide spectrum of biological responses that are produced in the neuron, including changes in gene expression (Figure 6.10).

In the 1990s, Blier and De Montigny (1994) involved the 5-HT$_{1A}$ receptor in the common mechanism responsible for the antidepressant activity. According to these authors, the different groups of antidepressants, including electroconvulsive therapy (ECT) and through different mechanisms, increased serotonergic transmission on a hippocampal level. Thus, it is known that the delay in the onset of the action of SSRIs is attributable to the time necessary to desensitize the 5-HT$_{1A}$ autoreceptors present in the cellular body of the serotonergic neurons. The desensitizing of these autoreceptors permits the enhancement of postsynaptic serotonergic neurotransmission. At present, it is thought that this mechanism may be necessary, but insufficient, to explain the antidepressant effect and that the intervention of additional factors must be taken into consideration (Duman et al. 1997).

Additionally, antidepressants are able to regulate other subtypes of monoamine receptors, which, rather than being related to the therapeutic effect, seems a secondary biochemical correlate to the increase of monoamines in the synaptic cleft and/or to the receptor block which, additionally, provokes. Among the observed adaptations in a more consistent manner, in the cerebral cortex of the rat and through chronic treatment with antidepressants, downregulation of the postsynaptic receptors (α_2, 5-HT$_2$, β) has been found. However, it is important to emphasize that, today, there is no convincing evidence of the regulation of adrenergic or serotonergic receptors per se being the only thing responsible for the therapeutic effect of antidepressant drugs. Therefore, rather than arrive at the conclusion that the regulation of the receptors is the mechanism responsible for the action of antidepressant agents, it seems more plausible that it be interpreted as a marker of a more complex adaptation mechanism.

In this way, in the 1990s, it was considered that the mechanisms of synaptic transmission were even more complex. Today, it is known that the regulation of the union of the neurotransmitter with the receptor and the mentioned processes of second messengers form only a small part of the mechanisms responsible for the neuronal response. Additionally, neurons have a large amount of tyrosine kinase proteins embedded in the cellular membranes, which serve as receptors for neurotrophins and other growth factors (Figure 6.10).

In the context of this knowledge, the most recent information suggests that the majority of antidepressant agents initially interact with extracellular proteins, called receptors, on a synaptic level, which, through intracellular messengers, are responsible for numerous actions of these drugs. Additionally, these intracellular messengers play a central role in the mediation of the long-term effects that these drugs have on brain function, thanks to phenotypic neuronal changes, such as the downregulation of receptors, protein synthesis, and release of neurotransmitters—a logical consequence of the modifications in the gene expression (Figure 22.1). These adaptative mechanisms provoked by antidepressants are fundamental to the discovery of new forms of action in this field. Currently, studies of cellular and molecular biology have given way to knowledge of modifications in the systems of intracellular transduction and the regulation of the expression of specific genes produced as a result of the sustained action of antidepressants. We have moved, therefore, from a superficial and synaptic psychopharmacology, which we can call an intracellular psychopharmacology, which explains some of the data found in previous decades.

At present, a priority goal in psychopharmacology is to discover with precision the nature and circuits responsible for the modifications of the neuronal function that causes depression, as well as the adaptation mechanisms that antidepressants initiate, to correct and normalize the behavioral, cognitive, affective, and neurovegetative disorders observed in these areas. To this end, research should not be limited to classic aminergic mechanisms (reuptake and metabolism) or more modern ones (receptor mechanisms) and should explore other bits of knowledge that biochemistry and molecular and genetic biology are constantly contributing. Some of the most recent information about these mechanisms is explored next.

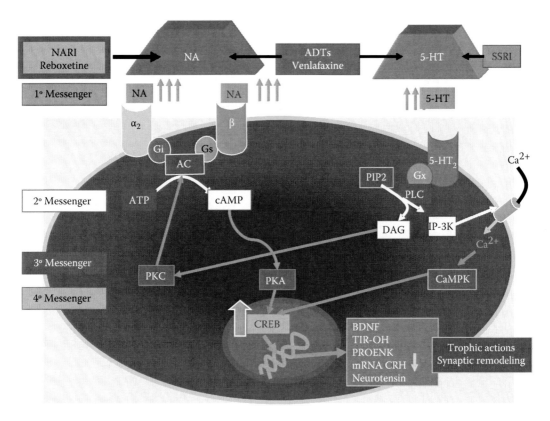

FIGURE 22.1 (See color insert.) Biochemical effects of antidepressant drugs. 5-HT, serotonin; AC, adenylcyclase; BDNF, brain-derived neurotrophic factor; CaMPK, calcium- and calmodulin-dependent protein kinase; CREB, cAMP response element–binding protein; CRH, corticotropin-releasing hormone; DAG, diacylglycerol; IP-3K, inositol triphosphate; NARI, noradrenaline reuptake inhibitors; PIP2, phosphatidylinositol 4,5-bisphosphate; NA, noradrenaline; PKA, protein kinase A; PKC, protein kinase C; PLC, phospholipase C; PROENK, proenkephaline; SSRI, selective serotonin reuptake inhibitors; TIR-OH, tyrosine hydroxylase.

22.3 POSTRECEPTOR MECHANISMS OF INTRACELLULAR TRANSDUCTION AS ACTION TARGETS OF ANTIDEPRESSANT DRUGS

As previously noted, the increase of monoamines in the synaptic cleft starts a series of intracellular neurochemical changes, which—according to what has recently been postulated—decisively influence the therapeutic effect of antidepressant drugs. These intracellular effects of antidepressants have led to the proposal of the hypothesis of these agents, through the increase in the action of monoamines (first messengers) on their corresponding receptors and, with independence from the transduction pathway of second messengers (cAMP, diacylglycerol, etc.), would have a common convergence point in protein kinases (PKs; PKA, PKC, CaMPK), considered third messengers, which would control the genetic expression, phosphorylization of transcription factors that could be considered fourth messengers (Figure 22.1). According to this hypothesis, SSRIs, ADTs, and NARIs would have a common intracellular mechanism, remote from their point of action in the synaptic cleft, which would modulate, through modifications of gene expression, the synthesis of particular substances such as proenkephaline, neurotensin, brain-derived neurotrophic factor (BDNF), or enzymes such as tyrosine hydroxylase, which ultimately will provoke changes in functional adaptation, trophic actions, or synaptic remodeling, likely to counter possible anomalies present in depression. All of these factors undoubtedly constitute a huge advance toward the knowledge of what would be the beginning of psychopharmacology of the present millennium.

22.3.1 Role of GP

The role of GP in the modulation of receptor signals has attracted considerable interest in recent years. This is because these proteins constitute the first postreceptor step in the most important intracellular transduction pathways (Figure 22.1). In fact, GP that mediate signaling between receptors and second-messenger systems have also been investigated in patients with depression, both in postmortem studies of the brain and in studies of peripheral-blood cells. Moreover, recent results confirm and extend previous findings on platelet G-protein-coupled receptor kinase 2 (GRK2) downregulation in patients with major depressive disorder (MDD), as basal reduction correlating with severity of depression, and its normalization after effective antidepressant treatment. Effective antidepressant treatments normalized platelet GRK2, and, notably, GRK2 upregulation discriminated between responder and nonresponder patients (García-Sevilla et al. 2010). Although these systems are clearly affected, no consistent picture has emerged because there are numerous forms of GP that vary in different areas of the brain (Belmaker and Agam 2008).

In this sense, it has been hypothesized that the basis of these alterations of the GP could be related, in some cases, to a genetic disorder or to modifications produced by distinct hormones or by situations, such as stress, which can alter the homeostatic mechanisms that involve some of these hormones. In fact, different experiments have shown that both the α subunit of the G_S protein and the G_I protein are under the control of glucocorticoids. Thus, it is known that lithium and some antidepressants change the expression of type II glucocorticoid receptors in the rat brain, and the activity of different subfractions of GP. These changes could have a relationship to the mood alterations detected in certain periods such as puberty, premenstrual dysphoric syndrome, menopause, and various stressful events present throughout life, in which hormonal influence is the norm (Bourin and Baker 1996).

Antidepressant drugs, in chronic administration, are capable of modifying the function of different fractions of these GP (Ozawa and Rasenik 1991; Lesch et al. 1991; Karege et al. 1998), which has made some people think that in depression, a disorder of the superfamily of GP coupled to the receptors could exist and that antidepressants would act to modify this dysfunction of GP, presumably toward "normalcy." Moreover, it has been shown that in mice lacking the G_Z protein (knockout mice), the experimental antidepressant effect of reboxetine and desipramine, both antidepressant inhibitors of the reuptake of noradrenaline, disappears in the forced-swim test (FST). This fact seems to indicate the need for this G_Z protein in order to manifest the antidepressive effect of these noradrenergic agents (Jing Yang et al. 2000).

Another convincing line of evidence came from the work of Donati and Rasenick (2003), which clearly demonstrated that chronic antidepressant treatments facilitate the activation of adenylcyclase (AC) by increasing the coupling of $G_S\alpha$ with this enzyme. They were able to show that this effect of antidepressants occurs even in the presence of downregulation of β-receptors. It was proposed that the increased coupling with AC is due to a membrane redistribution of G_S and to an altered interaction with tubulin-based cytoskeleton.

Therefore, it has been thought that a therapeutic objective in dealing with affective disorders could involve the action of pharmacological agents on GP. It is conceivable that the synthesis of these future medications could be obtained from a better understanding of the specific interactions of the bacterial and viral toxins in the GP system, but at the moment, there are no drugs that fit this profile and that would be sufficiently specific to obtain the desired effects, without provoking important adverse side effects (Bourin and Baker 1996). It is usually a good barometer of the importance of these findings that researchers in the pharmaceutical industry are developing agents capable of directly modulating these GP, either stimulating, as with the R-1204 (Roche Laboratories), or inhibiting, as in the case of GPCR (Roche Laboratories). However, these agents have been discontinued, and for the moment, these attempts have not yielded results from the therapeutic perspective.

22.3.2 ROLE OF PKS

In an intracellular step beyond GP, PKs are a very important intermediate link. A number of studies implicate intracellular signal transduction in the pathophysiology of depression. One key set of mechanisms involves phosphorylation enzymes, including PKA and PKC. The two PKs affect the cAMP response element–binding protein (CREB).

Several research groups have demonstrated deficiencies in PKA and PKC in a significant subset of patients with MDD. For example, certain depressed patients have deficient PKA and PKC protein levels. This has also been tested functionally by demonstrating lower binding of cAMP to PKA and decreased phosphorylation of target proteins such as CREB (Shelton et al. 2009). Moreover, it is known that prolonged, but not acute, administration of antidepressants can modify the cAMP phosphorylization dependent on concrete cerebral areas, thanks to the activation of PKs, especially PKA, which regulates transcription factors that influence the expression of additional specific proteins (Brunello and Racagni 1998) (Figure 22.1). Thus, it has been found that the chronic administration of a large variety of antidepressants decreases the production of cAMP stimulated by forskolin in the hippocampus (Newman et al. 1993), whereas other antidepressants such as imipramine or tranylcypromine increase the activity of nonsoluble PKA and decrease the cytosolic fraction of the frontal cortex (Nestler et al. 1989) and desmethylimipramine increases the regulatory subunits of PKA in the soluble fraction of the frontal cortex of the rat (Brunello et al. 1990). Similarly, chronic administration of fluoxetine or desmethylimipramine increases the union of cAMP to the regulatory subunit of PKA (Pérez et al. 1994).

On the other hand, fractionation of cellular compartments from the brain of rats treated with imipramine revealed that PKA activity was increased in the nuclei-enriched fraction (P1) and decreased in the cellular cytosol, suggesting that imipramine induces translocation of the catalytic PKA subunit from cytoplasm to the nucleus. However, recent experiments show that the nuclear PKA activity seemed to be somewhat increased only in the hippocampus and only by pronoradrenergic drugs. Pérez et al. (2000), in a series of different studies, demonstrated that chronic antidepressant treatments activate PKA in the microtubule compartment, increasing the phosphorylation of microtubule-associated protein (MAP$_2$), a major substrate of PKA. However, this change in PKA activation is restricted to the microtubule compartment and does not implicate changes in PKA in the neuronal nucleus (Tardito et al. 2006).

It is important to note that some of the steps of the different intracellular pathways can be modified in differential forms by different types of antidepressants. For example, reboxetine does not change the phosphorylization of MAP$_2$ cAMP dependent, as opposed to what happens with ADTs and SSRIs. In the same way, whereas ADTs and SSRIs are able to enhance the union of cAMP to the regulatory subunit of PK, the reboxetine does exactly the opposite. In contrast, the behavior of reboxetine is identical to that of the SSRIs in reference to the increase in the phosphorylizing action of calcium and calmodulin-dependent PK. Therefore, despite the fact that different types of antidepressants can share common central mechanisms, the way that the mechanism sites are affected can be different (Brunello and Racagni 1998), which can help explain the observed differences among antidepressants, both from a therapeutic point of view and from that of their adverse effects. Some of the steps from different intracellular pathways that we see can be modified in a differential form by different types of antidepressants (Table 22.3).

In summary, there is consistent and robust evidence for an upregulation of this signaling cascade by antidepressant treatments at the membrane level, with increased coupling of $G_S\alpha$ and AC, and at the level of microtubules, with increased binding of cAMP and activation of PKA. However, there is no consistent evidence for a robust activation of nuclear PKA that may mainly or exclusively account for the upregulation of cAMP response element (CRE)–dependent gene expression observed by independent studies (Tardito et al. 2006).

TABLE 22.3
Neurobiochemical Effects of Antidepressant Drugs

Substrate	NARIs	SSRIs	TCAs
NA-dependent cAMP	↓	↓↔	↓
β-Adrenergic receptors	↓	↓↔	↓
Link cAMP to RII-PK[a]	↓	↑	↑
MAP$_2$ phosphorylation	↔	↑	↑
CaM-kinase II activity	↑	↑	No data

Source: Brunello, N., and Racagni, G., *Hum Psychopharmacol,* 13, S13–9, 1998.

Note: CaM-kinase II, calcium- and calmodulin-dependent protein kinase; MAP$_2$, microtubule-associated protein cAMP dependent; NA, noradrenaline; NARI, noradrenaline reuptake inhibitors; SSRI, selective serotonin reuptake inhibitors; TCAs, tricyclic antidepressants.

[a] Regulation of protein kinases via binding of cAMP to the regulatory subunit of these proteins.

22.3.3 ROLE OF CREB

CREB is a transcription factor affected by cAMP in the cell. Multiple kinases such as PKA phosphorylate CREB at serine 133 which activates it. However, it is becoming increasingly clear that CREB can be phosphorylated by a vast array of PKs, which in turn are activated by a variety of signals (Figure 22.2) (Tardito et al. 2006). Phosphorylation enhances recruitment of proteins such as CREB-binding protein and formation of a transcriptional complex. This complex promotes histone acetylation through recruitment of histone acetyltransferases, which results in chromatin remodeling and RNA synthesis. Recruitment of different kinases, depending on the type of stimulus, as well as of different phosphorylation events on CREB and other transcriptional coactivators may lead to different patterns of gene expression (Deisseroth and Tsien 2002). CREB regulates the transcription of a number of diverse set of genes, including transcription factors (cFos), intracellular messengers (AC VIII), neurotransmitter enzymes (tyrosine hydroxylase), peptides (somatostatin), growth factors (BDNF), and apoptotic regulators (Bcl-2) (McKernan et al. 2009).

Levels of CREB and phospho-CREB were reduced in postmortem studies of the cortexes of patients who had MDD and had not taken antidepressants, as compared with controls (Belmaker and Agam 2008). Moreover, there is a reduction in mRNA and protein levels of CREB, CRE-DNA binding activity in the prefrontal cortex and hippocampus of suicide victims compared with normal control subjects, regardless of diagnosis, and these alterations were significantly important in the prefrontal cortex of teenage suicide victims compared with normal control subjects (Dwivedi et al. 2003; Tardito et al. 2006). On the other hand, CREB is also present in peripheral tissue, and there are several studies that show alterations of CREB mRNA in the peripheral cells, such as lymphocytes, leukocytes, and neutrophils of depressed patients. Recently, Ren et al. (2010) showed that depression may be associated with a decrease in protein expression of CREB in the nuclear fraction of neutrophils. There is a significant correlation between CREB protein levels and severity of depression (Hamilton Depression Rating Scale scores), which suggests that CREB may play an important role in the pathophysiology of depression, and one of the mechanisms for the therapeutic actions of antidepressant drugs may be related to their ability to increase CREB levels.

Regardless of the route involved and the type of antidepressant, it seems that some intracellular biochemical changes occur in the action of all antidepressants, such as the modifications of CREB (Meyer and Habener 1993; Ghosh and Greenberg 1995). Duman and his collaborators were pioneers in the investigation of the effects of antidepressants treatment on CREB expression and

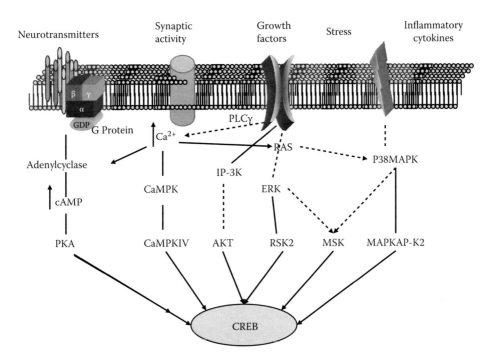

FIGURE 22.2 Signaling pathways regulating CREB phosphorylation. AKT, serine/threonine kinase; CaMPK, calcium- and calmodulin-dependent protein kinase; GDP, guanosine diphosphate; CREB, cAMP response element–binding protein; ERK, extracellular signal–regulated kinase; IP-3K, inositol triphosphate; MAPKAP-K2, MAP-kinase-activated protein kinase-2; MAPK, mitogen-activated protein kinase; MSK, mitogen- and stress-activated protein kinase; PKA, protein kinase A; PLC, phospholipase C; RAS, monomeric G protein; RSK2, ribosomal S6 kinase-2.

function (Nibuya et al. 1996). These authors found that chronic but not acute administration of several drugs, representative of different classes of antidepressants, as well as ECT, upregulated the expression of CREB mRNA in the hippocampus. They also showed that after repeated administration of fluoxetine, an SSRI agent, or ECT, the amount of CREB bound to CRE regulatory element was increased. This CREB system is subject to the action of antidepressants that enhances the serotonergic and/or noradrenergic transmission (Figure 22.1). It has also been revealed that desipramine (an NARI), fluoxetine (a serotonin reuptake inhibitor), or tranylcypromine (a MAOI) administered chronically to transgenic rats (CRE-lacZ) increased the activity of CREB on the genetic transcription in diverse limbic and hypothalamic areas (Thome et al. 2000). These data indicate that antidepressants, independently from the synaptic mechanism initially involved, provoke an increased regulation in CREB, possibly in a deficit state in depressive subjects, and that the time necessary for the production of the induction of CREB is chronologically related to the therapeutic effect of antidepressant treatment.

These results on the regulation of CREB by antidepressants were complemented by data from postmortem brain analysis, showing that the level of CREB protein was higher in the temporal cortex of depressed patients treated with antidepressants at the time of death than in the temporal cortex of untreated patients (Dowlatshahi et al. 1998).

Other experimental data confirm the importance of CREB in the etiopathogeny of depression. CREB regulates the *c-fos* gene, responsible for the c-fos protein, which is a component of the AP-1 transcription factor (activating protein 1). The importance of AP-1 comes from its capacity to regulate the gene expression of tyrosine hydroxylase, the limiting step in the synthesis of catecholamines, proenkephaline, and neurotensin. In the same way, the increase in CREB decreases the expression of mRNA of the corticotropin-releasing hormone (CRH), a hormone elevated in some types of depression (Dahmen et al. 1997). It has also been shown that the administration of CREB in the hippocampus, using the herpes virus as a vector, produces an experimental antidepressive effect (Chen et al. 2001a). All these data suggest that the antidepressant effect induced by agents that elevate the levels of CREB in the hippocampus (antidepressants, ECT) may be mediated by this substance (Tamura et al. 2002) and that CREB constitutes a molecular target for the study of new antidepressant agents.

There is some additional evidence that CREB plays an important role in the antidepressive action mechanism. Actually, it is possible to increase the functional pathway of cAMP through an inhibition of phosphodiesterase (PDE) (Fujimaki et al. 2000). Phosphodiesterase-4 (PDE4), one of 11 PDE enzyme families, specifically catalyzes hydrolysis of cAMP; it has four subtypes (PDE4A–D) with at least 25 splice variants. PDE4 plays a critical role in the control of intracellular cAMP concentrations. PDE4 inhibitors produce antidepressant actions in both animals and humans via enhancement of cAMP signaling in the brain. Rolipram, a PDE4 inhibitor, exhibits antidepressive properties in experimental models in rodents, as well as in some studies done in depressive patients (Zhang et al. 2002; Li et al. 2009). Activity of rolipram at the PDE4D subtype may represent a significant aspect of its action because several NSRIs with proven antidepressant efficacy share this characteristic. Thus, inhibition of PDE4 may represent a shared mechanism for efficacy among different classes of antidepressants (Halene and Siegel 2007). The adverse effects, especially nausea and emesis, have made it so that rolipram is not a therapeutic reality. However, some studies have been conducted to identify subtypes of PDE4 that are not expressed in areas related to vomiting (area postrema) and that localize in a selective form in limbic areas, which would constitute a good therapeutic target for new antidepressants, as well as for enhancing agents of the response of other antidepressants (Duman 2000). Although there is still a long way to go before PDE4 inhibitors with high therapeutic indices become available for treatment of depressive disorders, important advances have been made in the development of PDE4 inhibitors as antidepressants (Table 22.2). First, limited but significant studies point to PDE4D as the major PDE4 subtype responsible for antidepressant-like effects of PDE4 inhibitors, although the role of PDE4A cannot be excluded. On the other hand, PDE4D may contribute to emesis, the major side effect of PDE4 inhibitors. For this reason, identification of roles of PDE4D splice variants in mediating antidepressant activity is particularly important. Recent studies have demonstrated the feasibility of identifying cellular functions of individual PDE4 variants. Moreover, mixed inhibitors of PDE4 and PDE7 or PDE4 and serotonin reuptake have been developed and may be potential antidepressants with minimized side effects. Finally, relatively selective inhibitors of one or two PDE4 subtypes have been synthesized using a structure- and scaffold-based design (Zhang 2009). Various PDE inhibitors are still in the preclinical phase: trequinsin (PDE3 inhibitor), RO 20-1724 (PDE4 inhibitor), and zaprinast (PDE5 inhibitor) (Halene and Siegel 2007).

In any case, the provided data bring to light the role of CREB as a key element in the action mechanism of antidepressants.

22.4 NEW THERAPEUTIC TARGETS IN ANTIDEPRESSANT ACTION MECHANISM

As has been noted, the antidepressant action mechanisms explored during the past 40 years have focused primarily on the field of monoamines. Even nowadays, the development of new antidepressant agents acting by changes in monoamine levels continues (Table 22.2). Recently, new theories have been developed that try to explain not only how antidepressants act but also why the therapeutic means are efficient in a particular percentage of patients but not in others, as well as what the causes of the different types of depression are. The results of these works conclude that the possibility that there is not a single biochemical mechanism that can explain the antidepressive effect of all substances currently available is clearly evident. Hence, the research to elucidate the action mechanism of antidepressants continues in other directions, as well as to find a new locus of antidepressive action. In this way, the role of neurotrophic factors, CRH and glucocorticoids, excitatory amino acids, and NMDA receptors (*N*-methyl-ᴅ-aspartate) are being studied, as well as other possible targets (opioid system, interleukins, P substance, nitric oxide, etc.).

22.4.1 ROLE OF NEUROTROPHIC FACTORS IN ACTION MECHANISM OF ANTIDEPRESSANTS

In this line, different types of neurotrophins, as well as their corresponding receptors, can constitute two targets of importance in the development of knowledge of the basic mechanisms involved in depression and in its possible treatment. Thus, there is evidence indicating that stress decreases the expression of some neurotrophins in the hippocampus and in the frontal cortex, and that it can increase the vulnerability of the neurons to different forms of aggression. The family of neurotrophins include, among others, neurotrophin-3, neurotrophin-4, and also BDNF, the neurotrophic protein that seems to play an important role in the mechanisms surrounding the phenomenon of depression and the effect of antidepressants and can act, in adult animals, to protect the neurons from stress. BDNF is critical for axonal growth, neuronal survival, and synaptic plasticity, and its levels are affected by stress and cortisol. Antidepressant drugs and ECT upregulate BDNF and other neurotrophic and growth factors. A single bilateral infusion of BDNF into the dentate gyrus has antidepressant-like effects (Shirayama et al. 2002; Belmaker and Agam 2008).

On the other hand, it is known that BDNF displays antidepressive properties in different experimental models of depression (Siuciak et al. 1996). Furthermore, the chronic administration of antidepressants, among which are SSRIs, selective reuptake inhibitors of some of these amines, MAOIs, and atypical antidepressants, as well as ECT, increases the expression of BDNF and of the neurotrophin receptors (trkB, tropomyosin-related kinase B) in the hippocampus, with these effects coinciding temporarily with the therapeutic effect (Nibuya et al. 1996). BDNF is also able to increase the growth of noradrenergic and serotonergic neurons, and to protect these neurons from neurotoxic effect. Therefore, these facts help the role of BDNF in the actions of antidepressants (Duman et al. 1997; Álamo et al. 1998; Altar 1999; Skolnick 1999; Duman 2000).

A postmortem study of patients with depression who had committed suicide showed that BDNF was reduced in the hippocampus (Karege et al. 2005). Furthermore, in autopsies of subjects treated with antidepressants, an increase in BDNF has been found in different zones of the hippocampus compared with the levels in subjects not treated (Chen et al. 2001b). These data are consistent with the increases in CREB observed in autopsies of patients treated with antidepressants and show that the increase in BDNF is secondary to the increase in CREB induced by antidepressant treatment, not only in laboratory animals but also in humans (Dowlatshahi et al. 1998). In a study of 10 patients who were treated for 12 weeks with a dual reuptake inhibitor, improvement in depressive symptoms was correlated with increases in BDNF levels, and the BDNF levels of remitted patients had normalized to the same level observed in healthy controls (Aydemir et al. 2005). Response to various SSRI and NSRI treatments has been similarly associated with restoration of normative BDNF values (Gonul et al. 2005; Maletic et al. 2007). Recently, Molendijk et al. (2010) found serum BDNF levels to be low in antidepressant-free depressed patients relative to controls and to depressed patients who were treated with an antidepressant. BDNF levels of fully remitted persons were comparable with those of controls. The antidepressant-associated upregulation of serum BDNF in depressed patients was confined to SSRIs and St John's wort.

From a therapeutic perspective, the possibility of increasing the expression of BDNF by inhibiting the metabolism of cAMP has been evaluated, through the chronic administration of PDE inhibitors (rolipram, RO-20-1724) in a prolonged form. In these conditions, an increase in the expression of CREB and BDNF is observed in the hippocampus of the rat, a result not produced through acute administration (Nibuya et al. 1996; Fujimaki et al. 2000). In this sense, it is known that rolipram inhibits PDE4, and it is of great interest to learn if its manipulation is responsible for the antidepressant effect of other agents. Thus, a study sponsored by the NIMH (National Institute of Mental Health, United States), in which rolipram (C11) was used, marked to compare, through positron emission tomography (PET), the levels of PDE before and after the treatment with SSRI and in healthy controls. There is great interest to patent and there are several different agents being studied, both to inhibit PDE4 (RO 20-1724), PDE3 (trequinsin) and PDE5 (zaprinast) (Halene and Siegel 2007), in preclinical phases, as we have discussed previously.

Moreover, the data allow the search of analogues of BDNF that are able to act on trkB receptors and initiate the intracellular mechanisms, supposedly responsible for the effect of this neurotrophin. However, this strategy does not seem easy, since BDNF is a very small protein (14 kDa), which joins with its receptor through dimerization, and therefore, synthesizing molecules with these characteristics is not simple. Additionally, the behavior of BDNF is not always identical in all neuronal circuits, and whereas it exerts antidepressive effects on the level of the hippocampus, the increase at the ventral tegmental area (VTA) is antagonistic. Because of this, there could be more of a future in the use of trkB transduction proteins, which are predominantly expressed in the circuits related to depression (Berton and Nestler 2006).

Since BDNF peripheral levels appear to be a biomarker of depression and antidepressant response, Schmidt and Duman (2010) determined the influence of peripheral BDNF administration on experimental depression-like behavior. Furthermore, they examined adult hippocampal neurogenesis as well as hippocampal and striatal expression of BDNF, extracellular signal–regulated kinase (ERK), and CREB in order to determine whether peripherally administered BDNF produces antidepressant-like cellular responses in the brain. Peripheral BDNF administration increased mobility in the FST, attenuated the effects of chronic unpredictable stress on sucrose consumption, decreased latency in the novelty induced hypophagia test, and increased time spent in the open arms of an elevated plus maze test. Moreover, adult hippocampal neurogenesis was increased after chronic, peripheral BDNF administration. They also found that BDNF levels as well as expression of pCREB and pERK were elevated in the hippocampus of adult mice receiving peripheral BDNF. Taken together, these results indicate that peripheral/serum BDNF may not only represent a biomarker of MDD but also have functional consequences on molecular signaling substrates, neurogenesis, and behavior; this opens the possibility for its therapeutic use. However, clinical studies to support their peripheral administration are required.

Several studies suggest that BDNF is the link among stress, neurogenesis, and hippocampal atrophy in depression. However, BDNF may be related not only to depression but also to multiple psychiatric disorders (Belmaker and Agam 2008).

Recent studies demonstrate that neuropeptide VGF (nonacronymic), which is also a CREB-dependent gene, is upregulated by factors that increase synaptic activity, including neurotrophin, antidepressant drugs, hippocampal-dependent learning paradigms, and voluntary exercise. The decrease in immobility time in the FST induced by the dose of VGF was comparable to that reported for BDNF and neuropeptide Y (NPY), suggesting that VGF is as effective as those factors. Likewise, Thakker-Varia and Alder (2009) have demonstrated the effectiveness of a VGF-derived peptide (TLQP-62) as an antidepressant-like agent in the FST behavioral model of depression. Furthermore, VGF is reduced in models of depression. The expression of VGF in human psychiatric diseases is an area of focus. VGF has antidepressant-like properties in animal models that respond to acute treatment; however, the demonstration that VGF is effective in behavioral paradigms that respond to chronic antidepressant treatment, which better reflect the time course of antidepressant drug actions in humans, is still lacking. The identification of the receptor for VGF is critical for understanding its molecular and cellular actions. VGF may represent a common element for the monoamine, hypothalamic–pituitary–adrenal (HPA) axis, and neurotrophin hypotheses because they are regulated by antidepressants and stress as well as BDNF (Thakker-Varia and Alder 2009). In fact VGF, like BDNF, may enhance activity-dependent plasticity in emotional processing networks that are compromised in depression and is therefore a good candidate for future study.

All these data support the neurotrophin hypothesis of depression, which states that decreased levels of nerve growth factors mediate the neuroanatomical damage observed during stress and that increased levels mediate the antidepressant-induced reversal of damage.

22.4.2 HPA Axis: Its Role in Depression and Action Mechanism of Antidepressant Drugs

The discovery a couple decades ago of the role of CRH, also known as CRF (corticotropin-releasing factor), in the control of the reactions of the HPA axis, as well other zones of the central nervous system, under stress, has revolutionized our knowledge of the neurobiology of depression and has opened up interesting therapeutic perspectives. In depression, a hyperactivity of the HPA axis has been described (Figure 22.3), manifested in an increase in plasma levels and urinary and cerebrospinal fluid (CSF) cortisol, an increase in adrenocorticotropic hormone (ACTH), an increase in the levels of CRH in CSF and in brains of postmortem depressive patients (Mitchell 1998), and a decrease in the control of negative feedback of glucocorticoid. CRH seems to act not only as a liberating factor for ACTH but also as a neurotransmitter that mediates the endocrine actions through the HPA axis, emotional reactions through the amygdaloid nuclei, cognitive and behavioral responses through cortical neurons, and autonomic responses, thanks to the connection between the amygdaloid areas and brainstem nuclei, especially the locus coeruleus (Heim and Nemeroff 1999).

The relation between HPA axis and stress could also contribute to explain the etiology of affective disorders and the effects of antidepressant agents (Duman 2000) (Figure 22.4). One of the most contrasting facts provoked by stress is the ability to provoke atrophy of hippocampal neurons (Magarinos et al. 1996). This atrophy, possibly reversible (Manji et al. 2000), is observed in CA3 pyramidal neurons after being subjected to 2–3 weeks of immobilization stress or by prolonged social isolation. The atrophy of CA3 neurons is observed equally through the exposure of experimental animals to high levels of glucocorticoids (Magarinos and McEwens 1995). Additionally, the role of corticoids would not be limited to a destructive effect of neurons and their functional structure, but would also have an inhibiting effect of neurogenesis (Manji et al. 2000). According to this hypothesis, some types of neurons, such as CA3, that are more vulnerable to stress and other neuronal populations that are sensitive, additionally, to hypoxia, ischemia, or hypoglycemia cannot regenerate because the elevated levels of glucocorticoids would be slowing the restorative neurogenesis (Duman 2000). From a future therapeutic perspective, it is illustrative that cyanoketone, a blocker of the synthesis of glucocorticoids, antagonizes the atrophy induced by stress. This fact is indicative of the endogenous role that corticoids play in the dendritic atrophy caused by stress

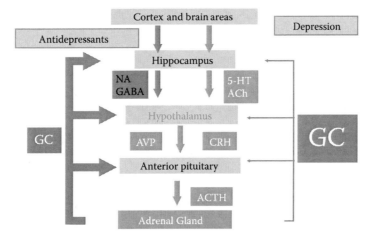

FIGURE 22.3 Schematic representation of HPA axis and control mechanisms in depressive subjects and after antidepressant treatment. 5-HT, serotonin; ACh, acetylcholine; ACTH, adrenocorticotropic hormone; AVP, arginine vasopressin; CRH, corticotropin-releasing hormone; GC, glucocorticoids; NA, noradrenaline.

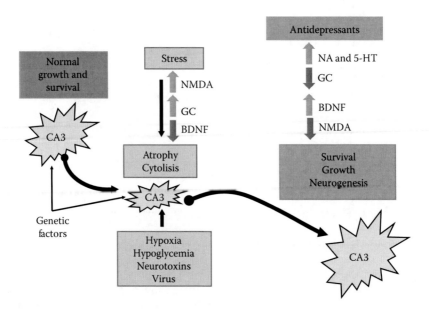

FIGURE 22.4 Hypothetical role of stress on neuronal growth and survival (CA3 hippocampal neurons) and action of antidepressant drugs. 5-HT, serotonin; BDNF, brain derived neurotrophic factor; GC, glucocorticoids; NA, noradrenaline; NMDA, *N*-methyl-D-aspartate.

(McEwen et al. 1995) and opens, as has been noted earlier, an important field of therapeutic research (Sapolsky 2000).

The accumulation of biological and clinical evidence, which today has considered the alteration of the HPA system as an integral part of the pathophysiology of depression, has prompted diverse researchers to become interested in the possible antidepressive role of agents that antagonize the effects of glucocorticoids (McDougle and Pelton 1996; Bosker et al. 2004). In this sense, it is known that long-term treatment with antidepressants increases the binding of glucocorticoids to their cerebral receptors; additionally, it increases the gene expression of the corticosteroid receptor and ultimately increases the number of corticoid receptors in these areas of the central nervous system, which leads to the normalization of cortisolemia (McQuade and Young 2000) (Figure 22.3). Similarly, it is known that ECT normalizes the levels of CRF in CSF, which is the same as what fluoxetine and certain ADTs do. These results are consistent with the fact that the hyperactivity of the HPA axis provokes a low regulation of glucocorticoid receptors, which is reversed by the antidepressive treatment. This suggests that antidepressants decrease the hyperactivity of the HPA axis as a result of the enhancement of negative feedback, which normalizes the elevated glucocorticoid levels (Figure 22.3) (Barden 1996; Owens and Nemeroff 1999; O'Brien et al. 2001).

The importance of the involvement of the HPA axis in depression and its modification by antidepressant treatment has stimulated the search for new approaches to depression (Bosker et al. 2004). Is seems logical then, that it has tried to act against depression, modifying the cortisol levels in a direct form, by inhibiting their synthesis. In this line, three inhibitor agents of the synthesis of corticosteroids, metyrapone, aminoglutethimide, and ketoconazole, as well as an antagonist of the glucocorticoid receptor, mifepristone, have been studied. Unfortunately, the data obtained, despite being promising in some subtypes of depression (Gallagher et al. 2008), were nonconclusive because they were not controlled studies with a sufficient number of patients; this, together with toxicity and scarce efficacy, made it so that, for the moment, we cannot count these agents in the antidepressant therapeutic arsenal (Price et al. 1996; Bosker et al. 2004).

Moreover, in recent years, there has been enormous interest in the development of CRF antagonists as potential antidepressants and anxiolytics. There are currently approximately a dozen of antagonists in the preclinical phase of development, and some of them have reached clinical phases (Table 22.2). The antagonists of the CRF1 receptor, such as R121919, have shown efficacy in depressed patients, but because of pharmacokinetic and liver problems, many of them had to be abandoned, something that has also happened with many drugs that act as neuropeptide receptors.

Recently, Valdez (2009) revised clinical trials with CRF1 antagonists. In an open label trial designed to assess the safety of NBI-30775 (also known as R121919), patients with MDD did not show significant alterations in liver enzymes or heart rate and showed reduced indications of anxiety and depression, which returned after discontinuation of treatment (Zobel et al. 2000). A subsequent trial has shown that 30 days of treatment with NBI-30775 did not affect various endocrine measures; the treatment was well tolerated (Kunzel et al. 2003) and normalized electroencephalogram sleep patterns in patients diagnosed with MDD (Held et al. 2004). Whereas this study showed NBI-30775 to be relatively safe, an unpublished study in the UK found elevated liver enzyme levels in two patients, leading to the termination of the development of NBI-30775 (Valdez 2009). Ising et al. (2007) have recently conducted a phase I clinical trial to assess the effects of another CRF1 receptor antagonist, NBI-34041, on neuroendocrine function. NBI-34041 reversed the increases in ACTH and cortisol because of intravenous CRF or psychosocial stress but did not impair normal HPA axis function. Although these initial results demonstrate that NBI-34041 is effective in reversing physiological responses to stress without affecting basal hormone levels, further work is clearly needed to assess the safety and efficacy of this drug in the treatment of depression.

On the other hand, many of these agents have been researched for decades and none has been commercialized, which is the biggest source of frustration in this field (Berton and Nestler 2006; Norman and Burrows 2007). In its last annual report, *Neurocrine* said it believed that the novelty and specificity of the CRF mechanism of action represented an opportunity to improve on the widely used SSRIs. At the same time, CRFs would offer better efficacy and fewer side effects when compared with the older benzodiazepines. However, the failure of these drugs so far to establish any real concrete proof of efficacy means that this hypothesis looks unlikely to be proven. According to pipeline data from *Evaluate Pharma*, only four CRF1 antagonists are in clinical development, whereas 18 preclinical or clinical candidates have been abandoned and none has successfully completed phase II trials. The latest setback means GlaxoSmithKline is likely to be seriously considering its commitment to this space, particularly given its decision earlier this year to exit neuroscience research (Table 22.2).

22.4.3 Implications of Glutamate in Antidepressive Action Mechanism

In the line of searching for agents that act through different mechanisms from the classic, there exist data that speak of the participation of glutamate in the action mechanism of antidepressant drugs. Glutamate is one of the key excitatory neurotransmitters in the mammalian brain and has been under study for several decades for a range of neuropsychiatric illnesses. Several studies in postmortem patients with mood disorders have reported intracellular protein changes associated with the postsynaptic density of the NMDA receptor complex. One study found that reduced expression of SAP102 corresponded with a reduction in expression of subunits NR1 and NR2A in the hippocampus, striatum, and thalamus of patients with depression. In a more recent study, NR2A and PSD95 protein levels in the lateral amygdala were significantly increased in depressed subjects compared with controls (Machado-Vieira et al. 2009).

In the 1950s it was observed that some anti-infective agents with properties antagonistic to the NMDA receptor, such as cycloserine, improved mood (Berton and Nestler 2006). Early reports described the action of antidepressants on glutamatergic receptors and the antidepressant-like effects of NMDA antagonists in animal models. But fundamentally, the importance of the NMDA receptor in the antidepressive spectrum comes from the capacity that various antidepressants have to modulate this receptor (Skolnick et al. 1996). In fact, various ADT agents (imipramine and desipramine), MAOI (clorgiline and tranylcypromine), SSRI (fluoxetine, sertraline, citalopran), and second-generation agents (mianserine), administered in the long term (14 days), antagonize the locus of the glycine of the NMDA receptor in cortical membranes (Paul et al. 1992; Skolnick 1999).

Moreover, various modulators of the NMDA complex receptor, antagonists (2-aminophospho-heptanoic acid, dizolcipine), and partial agonists of glycine (ACPC) mimic the effects of antidepressants in various preclinical models predictive of antidepressive action (Trullas and Skolnick 1990; Trullas et al. 1991; Skolnick 1999) and provoke a hyporegulation of cortical β-adrenergic receptors (Paul et al. 1992). Additionally, with ACPC, the partial agonist of glycine, at 2 weeks of treatment, effects achieved were similar to those obtained at 5 weeks of imipramine treatment (Skolnick 1999). It is important to remember that an advancement in the onset of antidepressive action is one of the most sought-after objectives in antidepressive research and therapy.

From a clinical point of view, a conceptual test has been conducted in depressed patients treated with an infusion of ketamine, a modulator in the channel of the NMDA receptoral complex. These patients, who were resistant to conventional antidepressive therapy, manifested an improvement in mood, maintained for 72 h after the administration of ketamine. This pilot study is consistent with the preclinical data observed with antagonists of the NMDA receptor (Berman et al. 2000). However, the competitive antagonists of the NMDA receptors, such as ketamine or phencyclidine, present the inconvenience of exhibiting psychotomimetic properties. There may be greater interest in agents of lower affinity, such as memantine, used in neurodegenerative processes or partial agonists of the glycine receptor, such as ACPC, or agents that act more than others as the locus of the NMDA receptor, such as eliprodil (Layer et al. 1995) and its analogue, infeprodil, which act on the polyamine site and have not shown psychotomimetic properties (Skolnick 1999).

Emerging data indicate that glutamate release appears to be important in the antidepressant-like effects of riluzole (2-amino-6-(trifluoromethoxy) benzothiazole), a drug indicated for the treatment of amyotrophic lateral sclerosis. Riluzole has diverse effects within different components of the glutamatergic system. However, its best-known mechanism of action involves the inhibition of glutamate release and its capacity to stimulate hippocampal BDNF synthesis and has been associated with increased numbers of recently created cells in the granule cell layer (Katoh-Semba et al. 2002). Preclinical studies support the antidepressant properties of riluzole, as evidenced by its antidepressant-like effects in the unpredictable stress paradigm (Banasr et al. 2006). Clinical use and clinical investigations have examined the antidepressant profile of riluzole. A 6-week open label monotherapy study in 19 patients with treatment-resistant MDD (mean daily dose of 169 mg) found that riluzole had significant antidepressant effects (Zarate et al. 2004). More recently, riluzole (mean dose of 95 mg/day) was added to an ongoing antidepressant therapy in 10 patients with treatment-resistant MDD; significant improvement in depressive symptoms was noted after 6 to 12 weeks of treatment (Sanacora et al. 2007). The most common side effects reported with riluzole were fatigue, nausea, and weight loss. These preliminary findings need to be confirmed in a controlled study (Machado-Vieira et al. 2009).

It is possible that a more selective subunit NMDA antagonist could be another target for antidepressants, with better tolerability than a nonselective one. In fact, significant reductions in the expression of NR2A and NR2B were observed in depressed subjects as compared with psychiatrically healthy controls. Consequently, modification of NMDA receptor signaling represents a novel approach for the development of effective antidepressant medication (Feyissa et al. 2009). Preclinically, NR2B-selective antagonists have also been reported to exhibit antidepressant activity. There are, however, no currently available subunit selective NR2B antagonists for clinical use, but a study by Maeng et al. (2008) found antidepressant-like effects with a subunit selective NR2B antagonist in rodents. Brain-penetrant NR2B antagonists currently being developed include indole-2-carboxamides, benzimidazole-2-carboxamides, and HON0001 (Machado-Vieira et al. 2009). Recently a randomized, double-blind study in patients with refractory MDD treated with CP-101,606 was reported (Preskorn et al. 2008). In this study, CP-101,606 was generally well tolerated and produced a greater antidepressant effect than placebo. A further clinical study in refractory depression has been initiated with the selective NR2B antagonist MK-0657 (Mony et al. 2009).

On the other hand, the stimulation of the glutamatergic AMPA (2-amino-3-(5-methyl-3-oxo-1,2- oxazol-4-yl)propanoic acid) receptor increases the expression of BDNF and neurogenesis at the hippocampal level. Additionally, both fluoxetine and reboxetine increase the expression of AMPA receptors on the neuronal surface. All this has stimulated the study of agents that act on this receptor in experimental models of depression. There are now several AMPA receptor–positive modulator or AMPA receptor potentiator (ARP) compounds under development (Table 22.2). These agents, unlike agonists, do not activate AMPA receptors themselves but slow down the rate of receptor desensitization and/or deactivation in the presence of an agonist. A series of modulators, which are generically called ampakines, with different chemical structures have been synthesized, including benzoylpiperidines (CX-516, CX-546), pyrrolidone (piracetam, aniracetam), benzothiazides (cyclothiazide), or biarylpropylsulfonamides (LY392098, LY404187). This compound showed an experimental activity similar to that of ADTs and SSRIs and with a faster onset (Du et al. 2006; Bleakman et al. 2007). To date, no placebo-controlled clinical trials with ARPs for the treatment of depression have been published (Zarate et al. 2010).

Although data on the effects of antidepressants on kainate (KA) receptors are sparse, chronic fluoxetine treatment was found to affect transcription GluR5 and GluR6l levels in the prefrontal cortex and hippocampus. To date, no KA modulator has been tested in the treatment of mood disorders. However, one recent study found that individuals with MDD who had a *GRIK4* gene polymorphism (rs1954787) were more likely to respond to treatment with the antidepressant citalopram than those who did not have this allele (Paddock et al. 2007).

Several studies have also suggested that mGluR1, mGluR2/3, and mGluR1/3 agonists have anxiolytic, antidepressant, and neuroprotective properties in rodents. LY341495 of the group II mGluRs agonists dose-dependently decreases immobility time in animal models of depression, and MGS-0039, an antagonist of this receptor, appears to be effective in the learned helplessness model of depression. However, the mode of antidepressant-like activity for drugs that target the mGluR1, mGluR2, or mGluR5 antagonists is uncertain (Zarate et al. 2010). The selective group III mGluR agonist ([1*S*,3*R*,4*S*]-1-aminocyclopentane-1,3,4-tricarboxylic acid) and the mGluR8 agonist ([*RS*]-4-phosphonophenylglycine) both induced significant antidepressant-like effects in rodents. Taken together, the results suggest that many of these compounds may be promising for the treatment of mood disorders. So far, however, there are no published clinical studies in mood disorders demonstrating their efficacy (Zarate et al. 2010).

Current data suggest that sigma receptors are involved in multiple processes effecting antidepressant-like actions in vivo and in vitro. Sigma receptors are widely distributed in the body and in the brain; they are found in significant concentrations in limbic and endocrine areas that have been implicated in the pathophysiology of depression, including the hippocampus, frontal cortex, hypothalamus, and olfactory bulb. In addition, the involvement of sigma receptors in the actions of antidepressant drugs was first suggested by observations that most marketed antidepressant drugs (ADTs, sertraline, bupropion, mianserine, MAOIs, iprindole, trazodone, nomifensine, and fluvoxamine) bind to these receptors. More recently, direct evidence in human volunteers for the in vivo binding of fluvoxamine to sigma receptors has been reported (Ishikawa et al. 2007). In the same way, agonists of this receptor, such as igmesine, SA4503, UMB23, and UMB82, show efficacy in experimental models of depression (Skuza and Rogóz 2007). Igmesine was tested in a 6-week, multicenter, double-blind, placebo-controlled study in 348 patients meeting the *Diagnostic and Statistical Manual of Mental Disorders (Fourth Edition)* criteria for MDD and was as effective and tolerated as fluoxetine (Volz and Stoll 2004). Although, sigma-1 receptor modulators may be a future therapeutic option (Table 22.2), either as individual agents or as adjuvants, in the treatment of mental disorders, the topic needs further preclinical and clinical exploration (Fishback et al. 2010).

However, other studied agents such as desmethylimipramine, paroxetine, fluvoxamine, and sertraline decrease the density of the sigma receptor in acute treatment, although in contrast, it increases in chronic treatment (Paul et al. 1994). According to these facts, the mechanism of the antidepressant action of sigma receptor agonists is not known, but a hypothesis has been developed. Sigma receptor antagonists have the ability to promote key neural adaptations that are characteristic of antidepressant drugs. In fact, several antidepressants could provoke adaptive changes in the NMDA receptors as a consequence of the modulation of the sigma receptors, which in turn would modulate the neuronal response to the stimulation of the locus of the glycine present in the NMDA receptor (Skolnick 1999). According to this, the adaptation of the NMDA receptors is necessary for the action of conventional antidepressants, and it seems logical to think that the involved mechanisms in the cited adaptation could be a good source for the study of new antidepressant molecules that act beyond the aminergic mechanisms. In this sense, it is noteworthy that there are important interrelations between the NMDA receptor and BDNF. It is known that this neurotrophin exerts a neuronal protective effect and that, on the contrary, the NMDA receptor, by facilitating the entry of calcium into the cell, exerts an aggressive effect on the neuron. Antidepressants increase, through CREB, the BDNF levels and have demonstrated that the exposure of granular cells to BDNF reduces the levels of protein and of the mRNA of the NMDA-2_A and NMDA-2_C receptors. The magnitude of this decrease of mRNA is within the range of that observed from the chronic administration of imipramine; additionally, the NMDA receptor changes induced by BDNF are accompanied by a decrease in the entry of calcium in the neuron, in a similar way to that of the antagonists of the NMDA receptor (Dildy and Leslie 1989; Brandoli et al. 1998; Fishback et al. 2010).

22.4.4 OTHER POSSIBLE ACTION MECHANISMS OF ANTIDEPRESSANTS, BEYOND MONOAMINERGIC HYPOTHESIS

Some authors have also implicated the opioid system in depression and in the action mechanism of antidepressants. The actions of endogenous opioids and opiates are mediated by three receptor subtypes (mu, delta, and kappa), which are coupled to different intracellular effector systems. Moreover, antidepressants that increase the availability of noradrenaline and serotonin through the inhibition of the reuptake of both monoamines lead to the enhancement of the opioid pathway. In this context, endogenous opioid peptides are coexpressed in brain areas known to play a major role in affective disorders and in the action of antidepressant drugs. Furthermore, it is also known that ECT increases the plasma levels of β-endorphins and that the administration of amitriptyline increases the concentration of leu-enkephaline in the hypothalamus and spinal cord. Additionally, an increase has been detected in the density of type μ opioid receptors, both in the frontal cortex and in the caudate nucleus of suicide victims, a density that decreases when antidepressants are chronically administered (Berrocoso et al. 2009). Indeed, it has been known for years that some opioids can have certain antidepressant properties. In fact, some authors call for the approval of buprenorphine in resistant depression (Bodkin et al. 1995; Callaway 1996). Finally, the fact that exogenous and endogenous opiates and the inhibitors of the catabolism of enkephalines are effective in predictive models of antidepressive activity (Gibert-Rahola et al. 1995; Broom et al. 2002; Berrocoso et al. 2009) supports the possibility that, in the future, antidepressant agents could be obtained that act through opioid mechanisms, without the addiction problems that they present. For the moment, we do not know of any opioid in development as an antidepressant.

In parallel with the rapid development of new immunological methods, the mechanisms of immunity are gaining more importance in the knowledge of psychiatric disorders (Kronfol and Remick 2000). A substantial body of evidence has suggested that depressive disorder is associated with the immune system. A cytokine imbalance hypothesis has been proposed for the pathophysiology of depressive disorder. This hypothesis is strongly supported by the observation that the treatment of patients with cytokines can produce symptoms of depression in the management of several forms of cancer and hepatitis C. On the other hand, numerous studies have reported that patients with endogenous depressive disorder exhibit higher-than-normal levels of various circulating inflammatory cytokines and their receptors, including interleukin-6, interleukin-1β, TNF-α, and interferon-β. Some studies have reported that the extent of increase in the cytokine levels is positively correlated with the severity of depressive symptoms and that the normalization of the cytokine levels is associated with the remission of symptoms (Nishida et al. 2009). In this sense, there are indicators that cascade of cytokines could be activated by neuronal processes that have a close theoretical bridge between stress and its influence on immunity, in depressive disorders (Capuron and Dantzer 2003). Interleukin-6 can stimulate neurons in vitro to release dopamine and probably other catecholamines. In fact, the peripheral implementation of interleukin-6 in laboratory animals empowers dopaminergic and serotonergic mechanisms in the hippocampus and in the frontal cortex. Additionally, elevated levels of interleukin-6 have been detected in the serum of patients affected by MDD, which correlated with the levels of plasmatic cortisol and which speaks in favor of this exercise having some stimulating effect on the HPA axis. On the other hand, the increase of interleukin-6 is associated with a decrease of tryptophan in depressed patients, which seems to emphasize its influence in serotonergic mechanisms. In this vein, it is important to note that, in depressed patients, a decrease of interleukin-6 has been observed from fluoxetine treatment, which opens the hypothesis of antidepressant action through immunological mechanisms, which could possibly be a starting point in the future of psychopharmacology (Müller and Ackenheil 1998).

Furthermore, it has been shown on an experimental level that long-term antidepressant treatment reduces the cerebral levels not only of interleukins but also of prostaglandin E_2 (PGE$_2$). The importance of this observation is that the presynaptic release of biogenic amines is reduced by PGE$_2$ and the plasmatic levels of this are elevated in depressed patients (Calabrese et al. 1986). Consequently, it is not illogical to postulate that antidepressants normalize the central aminergic neurotransmission, thanks to the reduction in the cerebral levels of PGE$_2$, through an inhibition of the cerebral cyclooxygenase activity. Some researchers have planned to use these anti-inflammatory drugs for treating psychiatric disorders for the neuroprotective effects of the drugs. In preclinical research, considerable evidence has accumulated indicating that anti-inflammatory drugs celecoxib may exert preclinical antidepressant effects (Müller and Schwarz 2008). In clinical trials, additional treatment with celecoxib, which inhibits the production of PGE$_2$ and proinflammatory cytokines, has shown significantly positive effects on the therapeutic action of reboxetine alone with regard to depressive symptomatology (Müller et al. 2006). These changes, together with those described in relation to the role of glucocorticoid receptors, could be responsible for the normalization of the defective central neurotransmission during antidepressant treatment.

Substance P, a neurokinin that acts on NK-1 receptors, has sparked an enormous interest in psychiatry because the antagonists of these receptors seem to exhibit anxiolytic and experimental antidepressive properties (Kramer 2000; Herpfer and Lieb 2005). MK-869 (aprepitant) was the first NK-1 antagonist studied from a clinical point of view, known for its antiemetic properties, although, curiously, despite the important role of substance P in pain, it lacks antinociceptive properties in humans. Its antidepressive properties have been shown in a controlled study to be superior to placebo and to have a similar effect to paroxetine, but with fewer side effects (Herpfer and Lieb 2005). However, in phase II and phase III studies, a high response to placebo was observed, and the efficacy of MK-869 was not confirmed despite near-maximum receptor occupancy (Stahl 1999; Herpfer and Lieb 2005). In randomized, double-blind phase II trials, three different NK-1 receptor antagonists, MK869, L759274, and CP122721, reduced symptoms of depression in clinically diagnosed outpatients significantly more than placebo. Moreover, the antidepressant effects of MK869 and CP122721 were similar to those of paroxetine and fluoxetine as comparators, respectively, but better in regard to adverse side effects (Ebner et al. 2009). On the other hand, two other studies failed to demonstrate the antidepressant actions of NK-1 receptor antagonists as well as of active comparators over placebo, making it difficult to interpret these negative findings. Consequently, Merck stopped its NK-1 receptor antagonist program, and several other pharmaceutical companies (e.g., Pfizer, Takeda-Abbott, Roche) followed Merck's decision despite initially promising results, whereas GlaxoSmithKline continued its program with other novel compounds (Table 22.2) (Berton and Nestler 2006; Ebner et al. 2009).

Recently, the NK-2 receptor antagonist SR48968 (saredutant) made it into clinical phase III trials for the treatment of major depression. The efficacy, safety, and tolerability of saredutant up to 52 weeks of treatment were evaluated in adult as well as elderly patients. So far, two of the multicenter, double-blind, placebo-controlled studies revealed positive results. However, in May 2009, Sanofi-Aventis published its quarterly results and announced the cessation of its saredutant development project for the treatment of major depression.

Arginine-vasopressin (AVP) provokes symptoms very similar to depression, which is also related to its plasmatic levels and in CSF. Several antidepressants are able to reduce the AVP levels. In this sense, it is known that an antagonist of the V-1 receptors of vasopressin, D(CH2)5Tir(CH3)AVP, and SSR-149415 exhibit experimental antidepressive activity, so results are expected from clinical trials with these agents (Norman and Burrows 2007). Sanofi-Aventis (July 2008 press release) has decided to discontinue the development of SSR-149415 in clinical phase II (ClinicalTrials.gov Identifier NCT0036149).

The neuropeptide Y (NPY) could also be a therapeutic target for the treatment of depression inasmuch as alterations in its activity, genetic expression, and function have been found in various experimental models. Furthermore, the chronic treatment with antidepressants modifies the function of this neuropeptide, and it has shown antidepressive activity in different models of depression. Clinical studies that put in evidence a possible antidepressive role in humans are necessary (Norman and Burrows 2007).

Finally, it must be noted that various work groups have focused their research on other parts of the depression-antidepressant puzzle, highlighting, among others, somatostatin (Kakigi et al. 1992), cholinergic and nicotinic mechanisms, estrogen, testosterone, leptin, melatonin, and nitric oxide (NO) (Table 22.2). Among those last mentioned, the one that stands out is agomelatine, which is a new antidepressant agent that explores the melatoninergic mechanisms, acting as an antagonist on the MT_1 and MT_2 receptors of melatonin. Additionally, it is able to block the $5-HT_{2C}$ receptors, which increases the release of dopamine and noradrenaline in the prefrontal cortex. These effects seem to be responsible for the antidepressive action and for improving the altered sleep profile of depressed patients (Stahl 2007).

However, without minimizing the importance of these contributions, the problem that concerns us probably does not have only one key and is the combination of all efforts with which, finally, in this century, we will be able to open the "Pandora's box" of affective disorders and their treatment. While this is happening, we must make progress in the employment and utilization of the existing therapeutic arsenal in order to minimize symptoms, prevent relapses, and help patients live with the suffering and discomfort of their diseases, as well as improve their quality of life and those of their family members (Álamo and Cuenca 1995).

22.5 CONCLUSIONS

The advances achieved in the field of psychopharmacology of depression in the past 50 years have been incredibly important, although not definitive. Treatment, which continues to be based on a fundamentally empirical therapy, as corresponds with a pathology whose physiopathology has not been fully revealed, is symptomatic and not curative. However—thanks to the antidepressant drugs that have made getting to know some basics of affective disorders possible—their treatment does not have one single key and actually requires the combined use of all efforts: pharmacological (which also allows for better use and acceptance of psychotherapeutic measures), psychotherapeutic, social, and familiar. In this sense, the actual progress made to this day in new antidepressant agents has been based—more than in an important differentiation from the point of therapeutic efficacy—on a greater acceptance of the medication on the part of patients, which facilitates the completion of therapy, prevents relapses, and, above all, improves their quality of life. However, we all know that there still exist problems in the treatment of these patients and that future perspectives in the field of antidepressants are aimed at the development of more effective drugs that improve clinical practice in terms of a greater percentage of patients, combat pathologies resistant to therapy, and perform faster, with greater specificity in their actions and free of side effects.

REFERENCES

Álamo, C., and E. Cuenca. 1995. ¿Avalancha de psicofármacos en la década del cerebro? *Arch Neurobiol* 58:337–9.

Álamo, C., F. López-Muñoz, and E. Cuenca. 1998. Contribución de los antidepresivos y los reguladores del humor al conocimiento de las bases neurobiológicas de los trastornos afectivos. *Psiquiatria.COM* (electronic journal) 2(3):Art. 7. Available at: http://www.psiquiatria.com/psiquiatria/vol2num3/art_7.htm.

Álamo, C., F. López-Muñoz, and E. Cuenca. 2000. Impacto de los SSRI en la terapéutica psiquiátrica: Patrones de empleo y prescripción. *Psicopatología* 20:63–101.

Altar, C. A. 1999. Neurotrophins and depression. *Trends Pharmacol Sci* 20:59–61.

Aydemir, O., A. Deveci, and F. Taneli. 2005. The effect of chronic antidepressant treatment on serum brain-derived neurotrophic factor levels in depressed patients: A preliminary study. *Prog Neuropsychopharmacol Biol Psychiatry* 29:261–5.

Banasr, M., G. Chowdhury, R. S. Duman, K. Behar, and G. Sanacora. 2006. Antidepressant-like effects in the unpredictable stress paradigm. *Neuropsychopharmacology* 31:S159–60.

Barden, N. 1996. Modulation of glucocorticoid receptor gene expression by antidepressant drugs. *Pharmacopsychiatry* 29:12–22.

Belmaker, R. H., and G. Agam. 2008. Mechanisms of disease. Major depressive disorder. *N Engl J Med* 58:55–68.

Berman, R. M., A. Cappiello, and A. Anand. 2000. Antidepressant effects of ketamine in depressed patients. *Biol Psychiatry* 47:351–4.

Berrocoso, E., P. Sánchez-Blázquez, J. Garzón, and J. A. Mico. 2009. Opiates as antidepressants. *Curr Pharm Des* 15:1612–22.

Berton, O., and E. J. Nestler. 2006. New approaches to antidepressant drug discovery: Beyond monoamines. *Nature Rev Neurosci* 7:137–51.

Bleakman, D., A. Alt, and J. M. Witkin. 2007. AMPA receptors in the therapeutic management of depression. *CNS Neurol Disord Drug Targets* 2:117–26.

Blier, P., and C. de Montigny. 1994. Current advances and trends in the treatment of depression. *Trends Pharmacol Sci* 15:220–6.

Bodkin, J. A., G. L. Zornberg, and S. E Lukas. 1995. Buprenorphine treatment of refractory depression. *J Clin Psychopharmacol* 15:549–67.

Bosker, F. J., B. H. C. Westerink, T. I. F. H. Cremers, et al. 2004. Future antidepressants. What is the pipeline and what is missing? *CNS Drugs* 18:705–32.

Bourin, M., and G. B. Baker. 1996. Do G proteins have a role in antidepressant actions? *Eur Neuropsychopharmacol* 6:49–53.

Brandoli, C., A. Sanna, M. A. De Bernardi, P. Follesa, G. Brooker, and I. Mochetti. 1998. BDNF and basic fibroblast growth factor down regulated NMDA receptor function in cerebellar granule cells. *J Neurosci* 18:7953–61.

Broom, D. C., E. M. Jutkiewicz, K. C. Rice, J. R. Traynor, and J. H. Woods. 2002. Behavioral effects of delta-opioid receptor agonists: Potential antidepressants? *Jpn J Pharmacol* 90:1–6.

Brunello, N., and G. Racagni. 1998. Rationale for the development of noradrenaline reuptake inhibitors. *Hum Psychopharmacol* 13:S13–19.

Brunello, N., J. Pérez, D. Tinelli, A. C. Rovescalli, and G. Racagni. 1990. Biochemical and molecular changes in rat cerebral cortex after chronic antidepressant treatment. *Pharmacol Toxicol* 66:112–20.

Bunney, W. E., and J. Davis. 1965. Norepinephrine in depressive reactions: A review. *Arch Gen Psychiatry* 13:483–94.

Calabrese, J. R., A. G. Skwerer, B. Barana, et al. 1986. Depression, immunocompetence, and prostaglandins of the E series. *Psychiatry Res* 17:41–7.

Callaway, E. 1996. Buprenorphine for depression: The un-adoptable orphan. *Biol Psychiatry* 39:989–90.

Capuron, L., and R. Dantzer. 2003. Cytokines and depression: The need for a new paradigm. *Brain Behav Immunol* 17 (Suppl. 1):S119–24.

Coppen, A. 1967. The biochemistry of affective disorders. *Br J Psychiatry* 113:1237–64.

Chen, A. C., Y. Shirayama, K. H. Shin, et al. 2001a. Expression of the cAMP response element binding protein (CREB) in hippocampus produces an antidepressant effect. *Biol Psychiatry* 49:753–62.

Chen, B., D. Dowlatshahi, G. M. MacQueen, et al. 2001b. Increased hippocampal BDNF immunoreactivity in subjects treated with antidepressant medication. *Biol Psychiatry* 50:260–5.

Dahmen, N., Ch. Fehr, S. Reuss, Ch. Hiemke. 1997. Stimulation of immediate early gene expression by desipramine in rat brain. *Biol Psychiatry* 42:317–23.

Deisseroth, K., and R. W. Tsien. 2002. Dynamic multiphosphorylation passwords for activity-dependent gene expression. *Neuron* 34:179–82.

Dildy, J. E., and S. W. Leslie. 1989. Ethanol inhibits NMDA induced increases in free intracellular Ca^{++} in dissociated brain cells. *Brain Res* 499:383–7.

Donati, R. J., and M. M. Rasenick. 2003. G protein signaling and the molecular basis of antidepressant action. *Life Sci* 73:1–17.

Dowlatshahi, D., G. M. MacQueen, J. F. Wang, and L. T. Young. 1998. Increase temporal cortex CREB concentrations and antidepressant treatment in major depression. *Lancet* 352 (9142):1754–5.

Du, J., R. Machado Vieira, S. Maeng, K. Martinowich, H. Manji, and C. Zarate. 2006. Enhancing AMPA to NMDA throughput as a convergent mechanism for antidepressant action. *Drug Discov Today Ther Strateg* 3:519–526.

Duman, R. S. 2000. Molecular and cellular determinants of stress an antidepressant treatment. In *Cerebral Signal Transduction: From First to Fourth Messengers*, ed. M. E. A. Reith, 223–47. Totowa, NJ: Humana Press Inc.

Duman, R. S., G. R. Heninger, and E. J. Nestler. 1997. A molecular and cellular theory of depression. *Arch Gen Psychiatry* 54:597–606.

Dwivedi, Y., J. S. Rao, H. S. Rizavi, et al. 2003. Abnormal expression and functional characteristics of cyclic adenosine monophosphate response element binding protein in postmortem brain of suicide subjects. *Arch Gen Psychiatry* 60:273–82.

Ebner, K., S. B. Sartori, and N. Singewald. 2009. Tachykinin receptors as therapeutic targets in stress-related disorders. *Curr Pharm Des* 15:1647–74.

Fangmann, P., H. J. Assion, G. Juckel, C. Álamo, and F. López-Muñoz. 2008. Half a century of antidepressant drugs. On the clinical introduction of monoamine oxidase inhibitors, tricyclics and tetracyclics: Part II. Tricyclics and tetracyclics. *J Clin Psychopharmacol* 28:1–4.

Feyissa, A. M., A. Zyga, C. A. Stockmeier, and B. Karolewicza. 2009. Reduced levels of NR2A and NR2B subunits of NMDA receptor and PSD-95 in the prefrontal cortex in major depression. *Prog Neuropsychopharmacol Biol Psychiatry* 33:70–5.

Fishback, J. A., M. J. Robson, Y. Xu, and R. Matsumoto. 2010. Sigma receptors: Potential targets for a new class of antidepressant drug. *Pharmacol Ther* 127:271–82.

Fujimaki, K., S. Morinobu, and R. S. Duman. 2000. Administration of a cAMP phosphodiesterase 4 inhibitor enhances antidepressant-induction of BDNF mRNA in rat hippocampus. *Neuropsychopharmacology* 22:42–51.

Gallagher, P., N. Malik, J. Newham, et al. 2008. Antiglucocorticoid treatments for mood disorders. *Cochrane Dat System Rev* 1, Art. CD005168. doi: 10.1002/14651858.CD005168.pub2.

García-Sevilla, J. A., M. Álvaro-Bartolomé, R. Díez-Alarcia, et al. 2010. Reduced platelet G protein-coupled receptor kinase 2 in major depressive disorder: Antidepressant treatment-induced upregulation of GRK2 protein discriminates between responder and non-responder patients. *Eur Neuropsychopharmacol* 20:721–30.

Ghosh, A., and M. E. Greenberg. 1995. Calcium signaling in neurons: Molecular mechanisms and cellular consequences. *Science* 268:239–47.

Gibert-Rahola, J., P. Tejedor Real, and J. A. Micó. 1995. Inhibidores del catabolismo de las encefalinas: Una nueva perspectiva en el tratamiento de los trastornos afectivos. *Psiquiatr Biol* 2:13–23.

Gonul, A. S., F. Akdeniz, F. Taneli, O. Donat, C. Eker, and S. Vahip. 2005. Effect of treatment on serum brain-derived neurotrophic factor levels in depressed patients. *Eur Arch Psychiatry Clin Neurosci* 255:381–6.

Halene, T. B., and S. J. Siegel. 2007. PDE inhibitors in psychiatry — future options for dementia, depression and schizophrenia? *Drug Discov Today* 12:870–8.

Heim, C., and C. B. Nemeroff. 1999. The impact of early adverse experiences on brain systems involved in the pathophysiology of anxiety and affective disorders. *Soc Biol Psychiatry* 46:1–15.

Held, K., H. Kunzel, M. Ising, et al. 2004. Treatment with the CRH1-receptor-antagonist R121919 improves sleep-EEG in patients with depression. *J Psychiatr Res* 38:129–36.

Herpfer, I., and K. Lieb. 2005. Substance P receptor antagonists in psychiatry. Rationale for development and therapeutic potential. *CNS Drugs* 19: 275–93.

Ishikawa, M., K. Ishiwata, K. Ishii, et al. 2007. High occupancy of sigma-1 receptors in the human brain after single oral administration of fluvoxamine: A positron emission tomography study using [11C] SA4503. *Biol Psychiatry* 62: 878–83.

Ising, M., U. S. Zimmermann, H. E. Kunzel, et al. 2007. High-affinity CRF1 receptor antagonist NBI-34041: Preclinical and clinical data suggest safety and efficacy in attenuating elevated stress response. *Neuropsychopharmacology* 32:1941–9.

Jing Yang, P. J., J. Wu, M. A. Kowalska, et al. 2000. Loss of signaling through the G protein GZ results in abnormal platelet activation and altered responses to psychoactive drugs. *PNAS* 97:9984–9.

Kakigi, T., K. Makeda, H. Kaneda, and K. Chichara. 1992. Repeated administration of antidepressant drugs reduces regional somatostatin concentrations in rat brain. *J Affect Disord* 25:215–20.

Karege, F., P. Bovier, R. Stepanian, and A. Malafosse. 1998. The effect of clinical outcome on platelet G proteins of major depressed patients. *Eur Neuropsychopharmacol* 8:89–94.

Karege, F., G. Vaudan, M. Schwald, N. Perroud, and R. La Harpe. 2005. Neurotrophin levels in postmortem brains of suicide victims and the effects of antemortem diagnosis and psychotropic drugs. *Brain Res Mol. Brain Res* 136:29–37.

Katoh-Semba, R., T. Asano, H. Ueda, et al. 2002. Riluzole enhances expression of brain-derived neurotrophic factor with consequent proliferation of granule precursor cells in the rat hippocampus. *FASEB J* 16: 1328–30.

Kramer, M. S. 2000. Update on substance P (NK-1 receptor) antagonists in clinical trials for depression. *Neuropeptides* 34:255.

Kronfol, Z., and D. G. Remick. 2000. Cytokines and the brain: Implication for clinical psychiatry. *Am J Psychiatry* 157:683–94.

Kunzel, H. E., A. W. Zobel, T. Nickel, et al. 2003. Treatment of depression with the CRH-1-receptor antagonist R121919: Endocrine changes and side effects. *J Psychiatr Res* 37:525–33.

Layer, R. T., P. Popik, T. Olds, and P. Skolnick. 1995. Antidepressant-like actions of the polyamine site NMDA antagonist, eliprodil (SL-82.0715). *Pharmacol Biochem Behav* 52:621–7.

Lesch, K. P., C. S. Aulakh, T. J. Tolliver, J. L. Hill, and D. L. Murphy. 1991. Regulation of G proteins by chronic antidepressant drugs in rat brain: Tricyclics but not chorgyline increase Goa subunits. *Eur J Pharmacol Mol Sect* 207:361–4.

Li, Y., Y. Huang, S. L. Amsdell, L. Xiao, J. M. O'Donnell, and H. T. Zhang. 2009. Antidepressant- and anxiolytic-like effects of the phosphodiesterase-4 inhibitor rolipram on behavior depend on cyclic AMP response element binding protein-mediated neurogenesis in the hippocampus. *Neuropsychopharmacology* 34:2404–19.

López-Muñoz, F., and C. Álamo. 2009. Monoaminergic neurotransmission: The history of the discovery of antidepressants from 1950s until today. *Curr Pharm Des* 15:1563–86.

López-Muñoz, F., C. Álamo, and E. Cuenca. 2000. La "Década de Oro" de la Psicofarmacología (1950–1960): Trascendencia histórica de la introducción clínica de los psicofármacos clásicos. *Psiquiatria. COM* (electronic journal) 4 (3). Available at: http://www.psiquiatria.com/psiquiatria/revista/47/1800/?++interactivo.

López-Muñoz, F., C. Álamo, and E. Cuenca. 2005. Historia de la Psicofarmacología. In: *Tratado de Psiquiatría*, vol. 2, ed. J. Vallejo and C. Leal. 1709–36. Barcelona: Ars Medica.

López-Muñoz, F., C. Álamo, G. Juckel, and H. J. Assion. 2007. Half a century of antidepressant drugs. On the clinical introduction of monoamine oxidase inhibitors, tricyclics and tetracyclics. Part I: Monoamine oxidase inhibitors. *J Clin Psychopharmacol* 27:555–9.

Machado-Vieira, R., G. Salvadore, L. B. Ibrahim, N. Diaz-Granados, and C. A. Zarate. 2009. Targeting glutamatergic signaling for the development of novel therapeutics for mood disorders. *Curr Pharm Des* 15:1595–611.

Maeng, S., C. A. Zarate CA, J. Du, et al. 2008. Cellular mechanisms underlying the antidepressant effects of ketamine: Role of alpha-amino-3-hydroxy-5-methylisoxazole-4-propionic acid receptors. *Biol Psychiatry* 63: 349–52.

Magarinos, A. M., and B. S. McEwens. 1995. Stress induced atrophy of apical dendrites of hippocampal CA3c neurons: Involvement of glucocorticoid secretion and excitatory amino acid receptors. *Neuroscience* 69: 89–98.

Magarinos, A. M., B. S. McEwens, G. Flugge, and E. Fuchs. 1996. Chronic psychosocial stress causes apical dendritic atrophy of hippocampal CA3 pyramidal neurons in subordinate tree shrews. *J Neurosci* 16: 3534–40.

Maletic, V., M. Robinson, T. Oakes, S. Iyengar, S. G. Ball, and J. Russell. 2007. Neurobiology of depression: An integrated view of key findings. *Int J Clin Pract* 61:2030–40.

Manji, H. K., G. J. Moore, G. Rajkowska, and G. Chen. 2000. Neuroplasticity and cellular resilience in mood disorders. *Mol Psychiatry* 5:578–93.

Mathew, S. J., H. K. Manji, and D. S. Charney. 2008. Novel drugs and therapeutic targets for severe mood disorders. *Neuropsychopharmacology* 33:2080–92.

McDougle, Ch. J., and G. H. Pelton. 1996. Antiglucocorticoids as treatments for depression. *CNS Drugs* 5:311–20.

McEwen, B. S., D. Albeck, H. Cameron, et al. 1995. Stress and the brain. A paradoxical role for adrenal ste-roids. In *Vitamins and Hormones*, 51, 371.402. New York: Academic Press.

McKernan, D. P., T. G. Dinan, and J. F. Cryan. 2009. "Killing the blues": A role for cellular suicide (apoptosis) in depression and the antidepressant response? *Progr Neurobiol* 88:246–63.

McQuade, R., and A. H. Young. 2000. Future therapeutic targets in mood disorders: The glucocorticoid recep-tor. *Br J Psychiatry* 177:390–5.

Meyer, T. E., and J. F. Habener. 1993. Cyclic adenosine 3'5'-monophosphate response element-binding protein (CREB) and related transcription-activating deoxyribonucleic acid-binding proteins. *Endocr Rev* 14:269–90.

Mitchell, A. J. 1998. The role of CRF in depressive illness: A critical review. *Neurosci Biobehav Rev* 22:635–52.

Molendijk M. L., B. A. Bus, P. Spinhoven, et al. 2010. Serum levels of brain-derived neurotrophic factor in major depressive disorder: State–trait issues, clinical features and pharmacological treatment. *Mol Psychiatry* Doi: 10.1038/mp.2010.98.

Mony, L., J. N. C. Kew, M. J. Gunthorpe, and P. Paoletti. 2009. Allosteric modulators of NR2B-containing NMDA receptors: Molecular mechanisms and therapeutic potential. *Br J Pharmacol* 157:1301–17.

Müller, N., and M. Ackenheil. 1998. Psychoneuroimmunology and the cytokine action in the CNS: Implications for psychiatric disorders. *Prog Neuropsychopharmacol Biol Psychiatry* 22:1–33.

Müller, N., and M. J. Schwarz. 2008. COX-2 inhibition in schizophrenia and major depression. *Curr Pharm Des* 14:1452–65.

Müller, N., M. J. Schwarz, S. Dehning, A. Douhe, et al. 2006. The cyclooxygenase-2 inhibitor celecoxib has therapeutic effects in major depression: Results of a double-blind, randomized, placebo controlled, add-on pilot study to reboxetine. *Mol Psychiatry* 11:680–4.

Nestler, E. J., R. Z. Terwilliger, and R. S. Duman. 1989. Chronic antidepressant administration alters the subcellular distribution of cyclic AMP dependent protein kinase in rat frontal cortex. *J Neurochem* 53:1644–7.

Newman, M. E., B. Lerer, and B. Shapira. 1993. 5-HT$_{1A}$ receptor-mediated effects of antidepressants. *Prog Neuro-Psychopharmacol Biol Psychiatry* 17:1–8.

Nibuya, M., E. J. Nestler, and R. S. Duman. 1996. Chronic antidepressant administration increases the expres-sion of cAMP response element-binding protein (CREB) in rat hippocampus. *J Neurosci* 16:2365–72.

Nishida, A., T. Miyaoka, T. Inagaki, and J. Horiguchi. 2009. New approaches to antidepressant drug design: Cytokine-regulated pathways. *Curr Pharm Des* 15:1683–7.

Norman, T. R., and G. D. Burrows. 2007. Emerging treatments for major depression. *Exp Rev Neurother* 7:203–13.

O'Brien, D., K. H. Skelton, M. J. Owens, and C. B. Nemeroff. 2001. Are CRF receptor antagonists potential antidepressants? *Hum Psychopharmacol Clin Exp* 16:81–7.

Owens, M. J., and C. B. Nemeroff. 1999. Corticotropin-releasing factor antagonists in affective disorders. *Expert Opin Invest Drugs* 8:1849–58.

Ozawa, H., and M. M. Rasenik. 1991. Chronic electroconvulsive treatment augment coupling of the GTP-binding protein Gs to the catalytic moiety of adenyl cyclase in a manner similar to that seen with chronic antidepressant drugs. *J Neurochem* 56:330–8.

Paddock, S., G. Laje, D. Charney, et al. 2007. Association of GRIK4 with outcome of antidepressant treatment in the STAR*D cohort. *Am J Psychiatry* 164:1181–8.

Paul, I. A., R. Trullas, P. Skolnick, and G. Nowak. 1992. Down-regulation of cortical β-adrenoceptors by chronic treatment with functional NMDA antagonists. *Psychopharmacology* 106:285–7.

Paul, I. A., G. Nowak, R. T. Laayer, P. Popik, and P. Skolnick. 1994. Adaptation of the *N*-methyl-D-aspartate receptor complex following chronic antidepressant treatments. *J Pharmacol Exper Ther* 269:95–102.

Paykel, E. S., R. Fleminger, and J. P. Watson. 1982. Psychiatric side effects of antihypertensive drugs other than reserpine. *J Clin Psychopharmacol* 2:14–39.

Pérez, J., D. Tinelli, E. Bianchi, N. Brunello, and G. Racagni. 1994. Cyclic AMP binding proteins in the rat cortical cortex after administration selective 5-HT and NE reuptake blockers with antidepressant activity. *Neuropsychopharmacology* 41:57–64.

Pérez, J., D. Tardito, S. Mori, G. Racagni, E. Smeraldi, and R. Zanardi. 2000. Abnormalities of cAMP signaling in affective disorders: Implication for pathophysiology and treatment. *Bipolar Disord* 2:27–36.

Preskorn, S. H., B. Baker, S. Kolluri, F. S. Menniti, M. Krams, and J. W. Landen. 2008. An innovative design to establish proof of concept of the antidepressant effects of the NR2B subunit selective *N*-methyl-D-aspartate antagonist, CP-101,606, in patients with treatment-refractory major depressive disorder. *J Clin Psychopharmacol* 28:631–7.

Price, L. H., R. T. Malison, C. J. McDougle, and G. H. Pelton. 1996. Antiglucocorticoids as treatments for depression. Rationale for use and therapeutic potential. *CNS Drugs* 5:311–20.

Ren, X., Y. Dwivedi, A. C. Mondal, and G. N. Pandey. 2010. Cyclic-AMP response element binding protein (CREB) in the neutrophils of depressed patients. *Psychiatry Res.* doi:10.1016/j.psychres.2010.04.013.

Sanacora, G., S. F. Kendell, Y. Levin, et al. 2007. Preliminary evidence of riluzole efficacy in antidepressant treated patients with residual depressive symptoms. *Biol Psychiatry* 61:822–5.

Sapolsky, R. M. 2000. Glucocorticoids and hippocampal atrophy in neuropsychiatric disorders. *Arch Gen Psychiatry* 57:925–35.

Schildkraut, J. J. 1965. The catecholamine hypothesis of affective disorders: A review of supporting evidence. *Am J Psychiatry* 122:509–22.

Schmidt, H. D., and R. S. Duman. 2010. Peripheral BDNF produces antidepressant-like effects in cellular and behavioral models. *Neuropsychopharmacology* 35:2378–91.

Shelton, R. C., E. Sanders-Bush, D. H. Manier, and D. A. Lewis. 2009. Elevated 5-HT2a receptors in post-mortem prefrontal cortex in major depression is associated with reduced activity of protein kinase A. *Neuroscience* 158:1406–15.

Shirayama, Y., A. C. Chen, S. Nakagawa, D. S. Russell, and R. S. Duman. 2002. Brain derived neurotrophic factor produces antidepressant effects in behavioral models of depression. *J Neurosci* 22:3251–61.

Siuciak, J. A., D. Lewis, S. J. Wiegand, and R. M. Lindsay. 1996. Antidepressant-like effect of brain-derived neurotrophic factor. *Pharmacol Biochem Behav* 56:131–7.

Skolnick, P. 1999. Antidepressants for the new millennium. *Eur J Pharmacol* 375:31–40.

Skolnick, P., R. T. Layer, P. Popik, G. Nowak, I. A. Paul, and R. Trullas. 1996. Adaptation of *N*-methyl-D-aspartate (NMDA) receptors following antidepressant treatment: Implications for the pharmacotherapy of depression. *Pharmacopsychiatry* 29:23–6.

Skuza, G., and Z. Rogóz. 2007. Antidepressant-like effect of combined treatment with selective - receptor agonists and a 5-HT1A receptor agonist in the forced swimming test in rats. *Pharmacol Report* 59:773–7.

Stahl, S. M. 1999. Peptides and psychiatry: Part 3. Substance P and serendipity: Novel psychotropics are a possibility. *J Clin Psychiatry* 60:140–1.

Stahl, S. M. 2007. Novel mechanism of antidepressant action: Norepinephrine and dopamine disinhibition (NDDI) plus melatonergic agonism. *Int J Neuropsychopharmacol* 10:575–8.

Sulser, F., J. Vetulani, and P. Mobley. 1978. Mode of action of antidepressant drugs. *Biochem Pharmacol* 27:257–61.

Tamura, T., S. Morinobu, Y. Okamoto, A. Kagaya, and S. Yamawaki. 2002. The effects of antidepressant drugs treatments on activator protein-1 binding activity in the rat brain. *Prog Neuropsychopharmacol Biol Psychiatry* 26:375–81.

Tardito, D., J. Perez, J. Tiraboschi, L. Musazzi, G. Racagni, and M. Popoli. 2006. Signaling pathways regulating gene expression, Neuroplasticity, and neurotrophic mechanisms in the action of antidepressants: A critical overview. *Pharmacol Rev* 58:115–34.

Thakker-Varia, S., and J. Alder. 2009. Neuropeptides in depression: Role of VGF. *Behav Brain Res* 197:262–78.

Thome, J., N. Sakai, K. Shin, et al. 2000. cAMP response element mediated gene transcription is upregulated by chronic antidepressant treatment. *J Neurosci* 20:4030–6.

Trullas, R., and P. Skolnick. 1990. Functional antagonist at the NMDA receptor complex exhibit antidepressant actions. *Eur J Pharmacol* 185:1–10.

Trullas, R., T. Folio, A. Young, R. Miller, K. Boje, and P. Skolnick. 1991. 1-Aminocyclopropanecarboxylates exhibit antidepressant and anxiolytic actions in animal models. *Eur J Pharmacol* 203:379–85.

Valdez, G. R. 2009. CRF Receptors as a potential target in the development of novel pharmacotherapies for depression. *Curr Pharm Des* 15:1587–94

Volz, H. P., and K. D. Stoll. 2004. Clinical trials with sigma ligands. *Pharmacopsychiatry* 37 (3): S214–20.

Zarate, C. A., J. L. Payne, J. Quiroz, et al. 2004. An open-label trial of riluzole in patients with treatment-resistant major depression. *Am J Psychiatry* 161:171–4.

Zarate, C. A., R. Machado-Vieira, I. Henter, L. Ibrahim, et al. 2010. Glutamatergic modulators: The future of treating mood disorders? *Harv Rev Psychiatry* 18:293–303.

Zhang, H. T. 2009. Cyclic AMP-specific phosphodiesterase-4 as a target for the development of antidepressant drugs. *Curr Pharm Des* 15:1688–98.

Zhang, H. T., Y. Huang, S. L. Jin, et al. 2002. Antidepressant-like profile and reduced sensitivity to rolipram in mice deficient in the PDE4D phosphodiesterase enzyme. *Neuropsychopharmacology* 27:587–95.

Zobel, A. W., T. Nickel, H. E. Kunzel, et al. 2000. Effects of the high-affinity corticotropin-releasing hormone receptor 1 antagonist R121919 in major depression: The first 20 patients treated. *J Psychiatr Res* 34:171–81.

Index

Page numbers followed by f and t indicate figures and tables, respectively.